Ecological Implications of Minilivestock

Potential of Insects, Rodents, Frogs and Snails

Editor
Maurizio G. Paoletti
Dipartimento de Biologia
Università di Padova
Padova, Italy

Science Publishers, Inc.
Enfield (NH), USA   Plymouth, UK

SCIENCE PUBLISHERS, INC.,
Post Office Box 699
Enfield, New Hampshire 03748
United States of America

Internet site: *http://www.scipub.net*

sales@scipub.net (marketing department)
editor@scipub.net (editorial department)
info@scipub.net (for all other enquiries)

Library of Congress Cataloging-in-Publication Data

Ecological implications of minilivestock; role of rodents, frogs, snails, and insects for sustainable development/editor, Maurizio G. Paoletti.
p. cm.
Includes bibliographical references and index.
ISBN 1-57808-339-7
1. Minilivestock. 2. Minilivestock--Ecology I, Paoletti, M.G.
SF140.M54E29 2004
639--dc22 2004056452

ISBN 1-57808-339-7

Published by Science Publishers, Inc. Enfield, NH, USA
Printed in India.

Foreword

Professor M.G. Paoletti and his colleagues present readers with a thoughtful outlook on MINILIVESTOCK ECOLOGY. The many chapters provide stimulating and timely suggestions about expanding the world food supply to include a variety of minilivestock and improved biodiversity management.

Indeed, finding new food sources to improve the food supply for the expanding world population is critical. The World Health Organization of the United Nations recently reported that more than 3 billion people in the world are malnourished as a result of a diet deficient in calories, protein, major vitamins, as well as the minerals of iron and iodine. Further, WHO reports that humans are dying from deficiencies in any one or a combination of nutrient shortages.

The world food supply, especially cereal grains, has been declining for about 20 years, according to data compiled by the Food and Agriculture Organization of the United Nations. FAO, has further noted that grains constitute about 80% of the world food supply.

Meanwhile, the world human population now stands at more than 6.3 billion and is rapidly increasing, at about a quarter million people each day. It has been projected that the world population will double in a mere 50 years, based on the current growth rate (1.3%) of the world population. Surely all future generations of humans deserve an adequate and nutritional food supply.

Although there are more than 15 million species of plants, animals, and microbes on earth, more than 90% of the world food supply comes from just 15 crop species and 8 livestock species.

In the United States and other developed nations, large quantities of meat, milk, and eggs are consumed. To provide this animal protein, farmers must maintain large numbers of livestock. In fact, the US livestock population outweighs the human population by more than 5 times. Of practical concern is that these livestock consume more than 250 million metric tons of cereal grains each year. This is sufficient food grain to feed more than 800 million people.

One way to augment the human food supply is to increase the diversity of plant and animal species used as food. As proposed by Professor Paoletti and his coauthors, the opportunity exists to utilize a wide variety of small animals as nutritious food. These animals include arthropods (especially insects and spiders), earthworms, snails, frogs and various rodents and reptiles. In fertile soil, the arthropods outweigh livestock by 10 times (1,000 kg/ha) and earthworms outweigh livestock 30 times (3,000 kg/ha). Granted there are problems in harvesting these small livestock, but the authors suggest technologies for harvesting them. Of course, a major advantage of the minilivestock is that they do not have to be fed grains.

Indeed, such minilivestock are already a part of many people's diet in various parts of the world today. The actual types of animals eaten depend on their availability in the environment and the culture of the people. These small animals contribute valuable protein and other essential nutrients, such as vitamin B_{12}. In cultures where these livestock are a part of the daily diet, the people have devised palatable ways of preparing these small animals.

As Professor Paoletti and his colleagues suggest, we need to diversify and augment the world's food resources, instead of relying by and large on just 8 livestock species for proteinic requirements, most of which rely on grain for feed. A movement to include the consumption of minilivestock would be a major step in the right direction.

David Pimentel
(Cornell University, Ithaca, N.Y., USA)

Preface

Food that is eaten locally in the countryside, in less developed rural regions and villages is often very different to the food eaten in urban areas. A knowledge of biodiversity and the rearing of minilivestock for both food and medicine is in many countries much more than simply a tradition or folkloristic feature for tourists. Though this attitude and awareness of a multitude of small animals is mainly a prerogative of hunter-gatherers, there are rural communities and periurban populations which still show an interest in these resources from uncultivated and marginal areas. Millions of people live on biodiversity from the wild and on bush food and this resource is a means of survival that is unknown or perceived as unpleasant to people living in large urbanized communities.

This traditional knowledge, which in most cases is neither written nor reported in the majority of agricultural or gastronomic texts, is crucial in maintaining the utilization of a diverse range of plants and animals of various sizes. There is also the danger of this knowledge being permanently dispersed and lost. Hunting and gathering is a sophisticated strategy of semidomestication in which the management of vegetation, key plants, fire, dissemination and fallow is appropriately used to facilitate these hidden animal resources. Women and children in the less affluent village communities utilize these resources extensively. Gathering of these resources does not require sophisticated tools; it involves only minimal risk and is in most cases easily monitored. Therefore, is this only famine food? Or emergency food? There is no certainty of this, but in many cases these terms are merely a simplistic way of labelling this gap in our knowledge. When local groups come into contact with civilized urban cultures, the traditional food and the use of biodiversity give way to new methods of adapting and using resources. When villagers attempt to sell their food at the local markets, the food is either not accepted or not given proper consideration and value. For instance, when a visitor asks villagers whether they eat insects, in most cases the villagers will respond that 'their neighbours do'. Generally, the utilization of such resources has a tendency to become secretive. In several cases, travellers and at times even anthropologists have overlooked such attitudes towards small animals. In today's world, insects and other small invertebrates are rarely regarded as a resource but more as a pest. This book attempts to improve the status of such resources and strives to improve and reinforce the appreciation of this reality.

Populations are in general well informed of the species in their areas and the potential problems of decline when these resources are under pressure. The destruction of biodiversity becomes imminent when communities have no knowledge of it. If hunting or collecting in the wild is the starting point in the use of these small creatures, an increase in demand for this can become crucial for local wild stocks. In order to understand domestication practices, it is essential to study the process of evolution of these local communities. Utilizing small creatures as food is a potential alternative in situations where food is a scarce. Such utilization is imperative in order to develop a more sustainable basis for resources in the future which requires less space, less energy and has fewer side effects on the environment. Concerns could be addressed by carrying out more research using an interdisciplinary approach and paying more attention to the local cultures—their values and knowledge. An informed approach towards domestication needs to be developed in which consumers and citizens are involved in increasing their knowledge about the different components of biodiversity.

Acknowledgements

Many people have been supportive of the idea of putting this book together, including the local communities' leaders and the local people who were sometimes surprised to find an outsider taking interest in an aspect of their lives which has hither to been a closed book. The chapters' authors and others who have helped in improving a previous version of each manuscript are:

Graziella Allegri, Matteo Bevilacqua, Geoff Bond, Keith S. Brown, Piergiovanni Bracchi, Sandra Bukkens, Barbara Burlingame, Piero Cardellini, Giuseppe M. Carpaneto, Hugo Cerda, Prudeucio Chacon, Jose Clavijo Michelangeli, Regina Pdo Constantino, Darna Dufour, Folco Giusti, Robert G. Glew, Guido Govoni, Sil-Soo Han, Jaques E.Hardouin, Jean Haxaire, Ferran Jori, Kenneth E. Lee, Danilo Lopez Hernandez, L.J. Elmslie, Françoise Malaisse, Paolo Mazzei, Benno Meyer-Rochow, Jun Mitshuashi, Ana Moreno, Tim Myles, Giovanni Onore, Emanuela and Giulia Paoletti, David Pimentel, Laura Pizzoferrato, Marco Prosdocimi, Carlo Renieri, Kenneth Ruddle, Mila Tommaseo Ponzetta, Fulvio Ursini, Arnold Van Huis, Roberto Villa, Alan L.Yen, Starford Zeut, S. Zhenjun. Alberto Collavo and Mirella Soratroi have helped in editing and translating the original documents; Claudio Friso and Renzo Mazzaro have improved some figures and images; Yelinda Araujo, Erika Buscardo, Alessandra D'Incà, Stefano Vanzin and Gabriele Volpato have provided information or collected specific databases.

Contents

List of Contributors

Araujo, Yelinda
INIA-Mérida, Edo
Mérida, Venezuela
E-mail: yelindaaraujo@yahoo.com

Bosse, Rebecca
NIOSH Laboratory
Cincinnati, OH, USA

Bukkens, Sandra G.F.
INRAN
Via Ardeatina 546
00178 Roma, Italy
E-mail: sgfbukkens@hotmail.com

Cerda, Hugo
Simon Rodriguez University
Apartado Postal 47.925
Caracas 1941-A, Venezuela
E-mail: hugocerda@hotmail.com

Chuang, Lu-Te
Ross Products Division
Abbott Laboratories
Columbus, OH, USA

Collavo, Alberto
Dept. of Biology
Padova University
Via U. Bassi 58/b
35121 Padova, Italy
E-mail: albertoc@bio.unipd.it

DeFoliart, Gene R.
Dept. of Entomology
University of Wisconsin
Madison, WI 53706, USA

Dreon, Angelo Leandro
Dept. of Biology
Padova University
Via U. Bassi 58/b
35121 Padova, Italy
E-mail: leandro.dreon@libero.it

Dufour, Darna L.
Dept. of Anthropology
University of Colorado
Boulder, CO 80309, USA
E-mail: Darna.Dufour@colorado.edu

Edderai, David
Projet DABAC
BP 2572 Yaoundé, Cameroun
E-mail: david-edderai@netcourrier.com

Elmslie, Leslie J.
Via Orti Gianicolensi 5
00152 Roma, Italy
E-mail: lelmslie@pelagus.it

Fielding, Denis
University of Edinburgh
Centre for Tropical Veterinary Medicine
Easter Bush, Rosling
Midlothian, EH25 9RG
E-mail: Denis.Fielding@ed.ac.uk

Giaccone, Valerio
Dept. of Public Health
Comparative Pathology and Veterinary Hygiene
Padova University
Agripolis, Legnaro Padova,
Italy
E-mail: valerio.giaccone@unipd.it

Glew, Robert H.
Dept. of Biochemistry and Molecular Biology
University of New Mexico School of Medicine
Albuquerque, NM, USA
E-mail: rglew@salud.unm.edu

Govoni, Guido
Dept. of Biology
Padova University
Via U. Bassi 58/b
35121 Padova, Italy
E-mail: bw.guido@libero.it

Hardouin, Jacques E.
BEDIM
Facultè Universitaire des Sciences Agronomiques
Passage des Déportés, 2
B-5030 Gembloux, Belgium
E-mail: goorickx.m@fsagz.ac.be

Houben, Patrick
DABAC
BP 9129 Libreville
Gabon, Africa
E-mail: dabac@assala.net

Huang, Yung-Sheng
Ross Products Division
Abbott Laboratories
Columbus, OH, USA

Jori, Ferran
Programme ECONAP
CIRAD EMVT, TA 39/E
Campus International de Baillarguet
Montpellier 34398
Cedex 5, France
E-mail: jori@cirad.fr

Lanfranchi, Givovanni B.
Dipartimento di Scienze dell'Antichità
Università di Padova
Piazza Capitaniato 7
35100 Padova, Italy
E-mail: gb.lanfranchi@pop.unipd.it

Malaisse, François
Laboratoire d'Ecologie
Faculté universitaire des Sciences Agronomiques de Gembloux
Passage des Déportés, 2
B1-5030 Gembloux, Belgium
E-mail: malaisse.f@fsagx.ac.be

Meyer-Rochow, Victor B.
School of Engineering & Science
International University Bremen
D-28725 Bremen
P.O. Box 750561, Germany
E-mail: b.meyer-rochow@iu-bremen.de

Mitsuhashi, Jun
Dept. of Bioscience
Tokyo University of Agriculture
1-1 Sakuragaoka 1-chome
Setagaya-ku, Tokyo 156-8502, Japan
E-mail: junmths@nodai.ac.jp

Negroni, Gianluigi
Alveo Co-operative Society
Via Guerrazzi 18
40100 Bologna, Italy
E-mail: sig1564@iperbole.bologna.it

Onore, Giovanni
Departmento de Biologia
Pontificia Universidad Católica del Ecuador
Apartado 17-01-2184
Quito, Ecuador
E-mail: GONORE@ puce.edu.ec

Paoletti, Maurizio Guido
Dept. of Biology
Padova University
Via U. Bassi 58/b
35121 Padova, Italy
E-mail: paoletti@civ.bio.unipd.it

Pemberton, Robert W.
Invasive Plant Research Laboratory
Agricultural Research Service, USDA
3205 College Ave.
Ft. Lauderdale, FL 33314, USA
E-mail: bobpem@saa.ars.usda.gov

Ramos-Elorduy, Julieta
Instituto de Biologia UNAM
Aptdo. postal 70-153
04510 México D.F. Mexico
E-mail: relorduy@ibunam.ibiologia,unam.mx

Tchibozo, Severin
Laboratoire d'Ecologie Appliquée
Faculté des Sciences Agronomiques
Université d'Abomey-Calavi 04
B.P. 0385
Cotonou, Bénin
Tel.: and Fax + 229 303084
E-mail: Tchisev@avu.org

Tommaseo-Ponzetta, Mila
Zoology Dept.
University of Bari
Via Orabona 4
70125 Bari, Italy
E-mail: m.tommaseo@biologia.uniba.it

Van Huis, Arnold
Laboratory of Entomology
Wageningen University
P.O. Box 8031
6700 EH Wageningen, The Netherlands
E-mail: Arnold.vanHuis@wur.nl

Wiwatpanich, Kanvee
Dept. of Community Medicine
Faculty of Medicine
Thammasat University
Pathumthani 12120, Thailand
E-mail: kanvee@alpha.tu.ac.th

Wenling, Cheng
College of Resources and Environmental Sciences
China Agricultural University
100 094 Beijing, P.R. China
E-mail: chengwenling88@sina.com

Xiwu, Gao
Dept. of Entomology
Beijing Agricultural University
100 094 Beijing, P.R. China

Yen, Alan L.
Arthur Rylah Institute for Environmental Research
Dept. of Sustainability and Environment
PO Box 137, Heidelberg
Victoria, Australia 3084
E-mail: ayen@satlink.com.au

Yhoung-Aree, Jintana
Institute of Nutrition
Mahidol University at Salaya
Puthamonthon 4 Road
Nakhon Pathom 73170, Thailand
E-mail: jyaree@maildozy.com

Yonghua, Zhao
Institute of Medicinal Plants
Chinese Academy of medical Sciences
100 094 Beijing, P.R. China

Zhenjun, Sun
College of Resources and Environmental Sciences
China Agricultural university
100 094 beijing, P.R. China
E-mail: sun108@cau.edu.cn

Zhi-Yi, Luo
Shanghai Institute of Entomology
Chinese Academy of Sciences
Chungkin Road (S) 225
Shanghai 200 025, P.R. China

Zimian, Ding
Institute of Medicinal Plants
Chinese Academy of Medical Sciences
100 094 Beijing, P.R. China

1

Minilivestock, Environment, Sustainability, and Local Knowledge Disappearance

MAURIZIO GUIDO PAOLETTI AND ANGELO LEANDRO DREON

Dept. of Biology, Padova University, Via U. Bassi 58/b, 35121, Padova, Italy.
E-mail: paoletti@civ.bio.unipd.it

Abstract

A multitude of small animal creatures have been adopted as food by local human cultures, especially in the tropics. If many species are potentially edible not all are eaten and sometimes individual species are avoided and considered poisoning (the two different stages of the same caterpillar in the cover page of the book). In areas with major biodiversity different ethnic groups utilize different species. If loss of species and biodiversity can be the effect of improper management, when target invertebrates or smaller animals are considered as food or regarded as semidomesticates, in most cases they are protected from major decline. At the same time, hunting and collecting pressure on wild species can be decreased if semidomestication takes place successfully. The higher profile of nutrients, vitamins, proteins and fats and especially the poliunsaturated fatty acids confer to such a myriad of small creatures far more interest than what has been paid yet. The local knowledge is a crucial milestone of these sensibilities and traditional unwritten cultures. To maintain and promote locally and extend to the local markets this sensibility and interest towards minilivestoks must become a priority. Major urban populations as well have to better acknowledge the importance of local knowledge and biodiversity resources. Educational programs have also to be carefully developed.

Key words: Edible invertebrates, Minilivestock, Local knowledge, Sustainable farming, Biodiversity, Semidomestication

Introduction

On the planet Earth, 1.8 million animal species are known, the bulk of which are represented by small creatures, the majority comprising invertebrates. However, estimations for insects alone range from 10 to 30 million species. The discrepancy between known and unknown exemplifies how few species are used in Western culture as food and the myriad species adopted by natives, especially in the tropics, and still in use and under potential domestication in developing countries. This book is devoted to minilivestock, covering most of the species, its essence and its potential. Most of this knowledge is based on local knowledge that needs to be maintained and reinforced, especially in the tropics. Most of this knowledge is likely to disappear as it is challenged by other issues. Accordingly, the connection that people have with many species is based on semidomestication rather than domestication. Furthermore, producing new domestication processes is just one important perspective and is one initiative being implemented in different parts of the world. Protection of the environment, in most cases, goes along with local cultures and their heritage. In most situations the contacts have created some misunderstandings regarding the local use of species, its marginalization, and its disappearance. Furthermore, local knowledge has to become everyone's patrimony in order to make people living in large towns aware of the fact that what they eat represents the real biodiversity of the planet instead of the few traditional 8 livestock and 15 plant species that provide most food. Most of the floral and faunal species on the planet are located in the tropics, especially in forested areas, and regions and areas covered by forest and savanna are the focus of this volume. Rural areas and developing countries are involved in such processes as revision and enlargement of the use of species. If some groups such as vegetarians or vegans reject totally animal food, in future, minilivestock will represent a compromise to maintain diversity in the environment and access to food for developing countries and new opportunities for the developed world in which environmental protection issues are of paramount importance.

Adopting Small Animals as Food: The Minilivestock

Diversity of local knowledge includes ecological reference to a multitude of organisms, comprising invertebrates, small mammals, amphibians, reptiles, and birds, that constitute the everyday source for hunter-gatherers and hunter gatherers-horticulturalists of our recent and remote past. In the tropics small animals including invertebrates together with a multitude of plants and fungi are collected in the wild and still provide food for millions of people (Prance, 1984; NAC, 1989, 1996; Scoones et al., 1992; Hardouin and Stiévenart, 1993; Hladic et al.,1996; Malaisse, 1997; Paoletti and Bukkens, 1997). For instance, in the Ivory Coast about 8 kg per person is the amount of wild meat consumed per year

(Caspary and Momo, 1998). It has been calculated that in West Africa 30 million people eat around one million tons of forest animals or about 100 g per day per person (Whitfield, 2003). In Central Africa insects provide over 50% of the animal protein consumed (Bahuchet, 1978). Invertebrates such as insects, spiders, crustaceans, snails or earthworms represent a consistant share of the diet in many human groups (Ramos-Elorduy, this volume). These wild and semidomesticated species depend on nonconventional management of forest (Posey, 1993; Etkin, 2000). For example, in Central America: "forest gardens probably originated with the ancient Maya and played a very important role in the domestication or semidomestication of many plant and animal species" (Gomez-Pompa and Bainbridge 1995). In some cases, these key food species under particular care (semidomesticates) were linked to human activities, e.g. slash and burn. Forest gardens and fallows have been documented as preferred pasture for key hunted birds such as Caracidae (Zent, 1988 in Melnyk, 1995) and game (Melnyk, 1995). Fruiting plants sometimes offer byproducts such as small invertebrates; palms in the tropical countries produce both edible fruits and palm worms (e.g., in Amazon *Mauritia flexuosa*). In Papua New Guinea palms (especially *Metroxylon* sp.) produce *sago* and weevil grubs (Tommaseo-Ponzetta and Paoletti, this volume). Several other trees on which fruits are harvested host edible insects, in particular caterpillars, namely *Inga* sp., *Bertholletia excelsa, Manihot esculenta, Pourouma guianensis, Caryodendron orinocense*, etc. (Paoletti and Dufour, this book; Onore, this book). More investigations are needed to realize the potential of small-scale rearing projects. Leguminous plants such as *Inga* appear more interesting. Insects, rodents and other small vertebrates and invertebrates are not just emergency foods in most cases, but provide a substantial base for everyday subsistence. Burning is a tool in many conditions for hunting some species (in the case of Guajibo and Yukpa tribes of Colombia and Venezuela) including small rodents, grasshoppers, and lizards in savanas (Ruddle, 1973; Paoletti and Dufour, this book). In rural communities this range of small animals represents a regular part of their diet (Hardouin, this volume). In villages, different people have access to different foods. In most cases women and children are involved in collecting small animals (Dufour, 1987, 2000).

Various authors have observed the different amounts and quality of foods utilized by different groups in villages. In many cases children and women rely on small animals that are easily collected with limited effort and very limited risk (Dufour, 2000; Tommaseo-Ponzetta, this volume). In addition, time allocation to collect the small invertebrates is efficient compared with fishing and hunting.

Table 1.1 compares the energy expended to collect invertebrates, to fish, and to hunt other larger animals in the Amazon. Efficiency in collecting small invertebrates is considerably high. In some cases, possibly insects are nutritious because of nutrients such as fatty acids, protein, and vitamins. Chimpanzees devote up to 39% of their foraging time to collecting insects even though they

Table 1.1: Mean foraging efficiencies for some invertebrates compared with fish and game (Paoletti et al., 2000)

Animal	Foraging efficiency[a] g/h	Foraging success rate	Data source
Earthworm *Motto*	388	100%	Ye'Kuana: Direct measurement in villages Toki and Guatamo (May 1998 and Jan. 1999)
Earthworm *Kuru*	446	100%	Ye'Kuana: Direct measurement in village Buena Vista (Jan. '99)
Syntermes spp.	300??	~100%	Tukanoans [Dufour (5) and Dufour (unpublished)]
Syntermes aculeosus	200	99%	Ye'Kuana: Direct measurement in village Toki (J. '99)
Atta spp.	200	99%	Tukanoans [Dufour (5) and Dufour (unpubl)]
Caterpillars	300	99%	Tukanoans [Dufour (unpubl)]
Fish	494	73%[c]	Tukanoans [Dufour (unpubl)]
Fish	927	?	Various Amerindians [Beckerman (28)][d]
Game	1,003	?	Various Amerindians [Beckerman (28)][e]

[a]Does not include time needed to go to the target place to collect invertebrates. Does include travel and search time for fish and game.
[b]Per cent of foraging trips successful in procuring organisms sought.
[c]Based on returns from 78 fishing trips observed at various times of the year.
[d]Mean of annual means of five Amerindian groups (Table 8.2 in Beckerman, 1994).
[e]Mean of annual means of 14 Amerindian groups (Table 8.3 in Beckerman, 1994).

represent less than 5% of their diet, consisting mainly of plant materials (Hladick, 1973). Insecta clearly are of significant value to chimpanzees.

In Vaupes, Colombia, Tukanoan women are more dependent on a diet based on insects than adult males (Dufour, 1987). Among the elderly women of Yanomamo Indians in Alto Orinoco, Venezuela, large spiders (*Theraphosa apophysis*), (Plate VII, 5; Paoletti and Dufour, this book) are a delicacy (pers. comm. G. Bortoli, 2002). The Ye'Kuana in Alto Orinoco recognize earthworms as gourmet (*motto* and *kuru*) food for women during the first month following parturition. These women consume primarily a diet composed of cassava and earthworms (Paoletti et al., 2003).

Posey (1984, 1993) has reported that stingless bees and several plants provide valuable food resources in Brazilian Amazon. Native people manage these resources that the newcomer would normally consider wild. For instance, the earthworm (*Andiorrhinus motto*) (an anellid belonging to Glossoscolecidae, only recently described scientifically) is normally farmed by the Ye'Kuana (in Alto Orinoco). They collect adults and cocoons during April-May (at the time of reproduction) from river banks and introduce them into stream banks where they are absent. One year later they return and collect the inseminated earthworms (Moreno and Paoletti, 2004).

Why have humans concentrated attention on the domestication of large animals rather than small creatures? Perhaps the ritualistics involved in hunting large game (as observed by many researchers) set the base for building the power structure in small human communities. Clearly collecting termites is not as dramatic as bringing home a large deer or wild pig to the people in a village. Hunters in the Western culture have long spent incredible amounts of money to bring home large exotic game exhibited as trophies. Paying attention to wild species and domestication is more than a sequence of biological-ecological rules.

From the practical point of view however, it has been suggested that animal domestication was developed based on six main features or conditions (Diamond, 1997, 2002):

—diet has to be efficient (which is why most carnivores are not normally domesticated to feed humans);
—growth rate: domesticates must grow quickly;
—problems of captive breeding: difficulties encountered in reproduction;
—nasty disposition: solitary animals do not accept captivity;
—tendency to panic: these animals create problems;
—social structure: have a well-developed dominance hierarchy.

These important features discriminate many animals as not good candidates for domestication but possibly another condition is at the base of the domestication procedure, namely efficiency in use of space and availability in the area of abundant wild game.

In many local conditions wild animals have been managed to become a regular source of food (Redford and Padoch, 1992; Hladic et al., 1996; Bodmer et al., 2003). Most wild animals are so closely associated with local human populations as to be very interesting and potential candidates for domestication, existing already in a state of semidomestication. For instance, regular hunting of game and procurement of small birds as a strategy for obtaining meat became a mainstream tradition not only in the tropics, but also in Europe (for instance in Italy, for migrating small birds; Sweden and Norway for large reindeer or Austria and Switzerland for roebuck and chamois) in which large wild animals or small game and wild fish became stable resources for villagers and various hunting strategies could support the high amount of meat consumed. The same holds true for wild or semidomesticated plant materials. Hunting and particular techniques of trapping have in many cases been regulated or banned in order to protect species against decline and environmental impoverishment. One case is the net-catching of small birds in Italy; it was severely restricted or banned in the last decades due to the pressure on migratory birds. The picture of a trapped robin (Plate I, 1) now induces great sorrow. In the past the everyday meat provision of many villagers in some rural and forested areas, for instance in Italy (such as Friuli, Veneto, Lombardia, Toscana), was trapped small birds. Therefore, in most countries, laws definitely ban or regulate these kinds of hunting.

In spite of regulations seeking to ban or regulate bushmeat hunting (in the countryside in most developed and undeveloped countries), people still appreciate small-size prey including small birds, mammals, fish, frogs and particularly in the tropics, invertebrates such as snails, crustaceans, sometimes spiders, and earthworms but especially insects. Many different species of insects have gained space in this volume (see chapters by De Foliart, Ramos-Elorduy, Malaisse, Yen, Meyer-Rochow, Paoletti and Dufour and Onore).

In some cases, agricultural development and landscape simplification and impoverishment of diversity due to pesticides, tillage, and large machinery have consistently reduced the real potential of these "nonconventional" food resources. A combination of landscape degradation due to the presence of extensive and intensive agricultural fields and dramatic decline of wild forest vegetation, natural grassland, and hedgerows may have been the key cause in Italy, as well as in other countries of small bird decline rather than hunting or poaching (Whitfield, 2003). These topics, landscape change and hunting pressure, have not been properly investigated or are snared in political debate.

Minilivestock can move toward a more sustainable method for using biodiversity as a food source. New candidates for domestication that have been the object of traditional foods offer a real potential. Many such candidates are tamed as pets in village houses (see in Govoni et al., this book, Plate III, 1 and 2). They are adaptable to small-scale operations (Malaisse, 1997; Jori et al., this volume). They have value for developing countries but could also be used in developed areas.

In Europe, few insects are traditionally exploited for food, except for the bee (*Apis mellifera*) and a cheese maggot (of the fly *Piophila casei* L.) that is appreciated in some seasoned local cheeses at least in Italy.

In Western Friuli near Tramonti di Sotto, people in summer traditionally ate raw the abdomen of several species of colorful butterflies (families Zygaenidae and Ctenuchidae, genera *Zygaena* and *Syntomis*) Plate I, 2. Apparently in the same area some people in the past have occasionally consumed *Tenebrio* sp. larvae together with pupae of *Bombyx mori*.

Several chapters in the present volume discuss over 2,000 species of insects utilized as food but possibly many more are consumed. Most of the species utilized in the wild could be domesticated in some way and become small-farm activities. In this way pressure on wild animals could be reduced (this volume: Hardouin; DeFoliart).

Local Knowledge: Important or Useless?

Local knowledge applied to the use of biodiversity (species) needs to be extensively recorded, studied, understood and monitored. There is a risk that this knowledge will become extinct before it can be properly recorded. We need unwritten knowledge of living creatures, plants, and animals and the way they support traditional people. It is important that this knowledge be available to future generations and all people (Posey and Dutfield, 1996; Laird, 2002). Loss of local languages and oral cultures could effect the disappearance of the base of knowledge of living creatures, plants, and animals and the way they support people.

Examples of collection of folk knowledge include the pioneering work of Posey (1984) and numerous recent communications, sometimes limited to local situations but very important (Malaisse, 1997; Reid et al., 1992; Yen et al., 1997; Blanco et al., 2002; Otero et al., 2002; Rodriguez et al., 2002; Meyer-Rochow, this volume). If recording this knowledge is important and welcome as well as its transfer to local schools, so, too, is the biological integrity of the land and living space left for the communities—aspects all too often controversial.

It is imperative that traditional knowledge be maintained for coming generations. This knowledge has possibilities for subsistence and potential for new resources to be properly handled, domesticated, and marketed. Subsistence from local resources provides economic bases for living (sale for local markets, ethnotourism, ecotourism, fair trade, insect farming for butterfly farms).

It is equally important to share this information with larger (urban) communities in order to promote small-scale businesses of the original resources and support the villagers. It is amazing the loss of knowledge among inhabitants of towns regarding the different species adopted as food in the countryside. For instance, many fruits and small animals eaten in villages throughout the state of Amazonas, Venezuela, are not known in the state capital Puerto Ayacucho. The rural people market only a few products, including yucca, plantain, banana, papaya, a few palm fruits, occasionally among the several others found in villages *tupirillo*, *yarayara*, and very rarely insects such as caterpillars known as *opomoschi* (*Erinnyis ello*) by the Yanomamo (or as *raciukiu* for the Ye'Kuana at Guatamo), *Atta* ants or *Syntermes* termites, grasshoppers, earthworms (such as *motto* or *kuru*—gourmet food for the Ye'Kuana Indians in Alto Orinoco) or iguanas (*Iguana iguana*; Plate I, 3) are exhibited. Caimans and crocodiles such as the Orinoco crocodile (*Crocodylus intermedius*; Plate I, 4), small rodents (see this volume: Govoni et al.; Jori et al.), snakes or tortoises such as *Peltocephalus dumerilianus* (Plate I, 5) and *Geochelone denticulata* (Plate I, 6) appreciated by the Guajibo and the Piaroa among the other American Indians in Amazon rarely appear in the town market. Frogs (Negroni, this volume) and toads are appreciated meat by different tropical groups. For instance *Bufo marinus* (see Plate III, V) is appreciated by the Yanomamo Indians in Alto Orinoco. In Australia, however, the same introduced toad is blamed as poisoning pest for wild animals and humans.

Local knowledge is the base for future progress in food resources in most rural areas in the tropics and elsewhere but more attention, appropriate development and research are absolute requirements.

Risks for Loss of Species

In densely populated areas, especially islands, pressure on some species has in some cases led to complete annihilation of a species (Wilson, 1988). Improper management of habitat, limited traditional knowledge, and change of land use are the major causes of most species extinctions. In other words, improper pressure on biodiversity can cause loss of species and in many situations species extinction has been the final result (especially in Islands). However, in most cases species loss is the result of loss of poorly managed landscape, together with intensive farming or fire. Vegetation decline seems the major cause of decline of biodiversity. Progressing toward domestication of small animals (minilivestock) could be the way to decrease pressure on wild stock of various wild animals and also maintaining and enriching local traditions. Furthermore, small animals better optimize the limited amount of land available in most countries to small-scale farmers (Beets, 1997). Table 1.2 assesses the feed needed for obtaining one unit of meat and animals such as crickets appear more efficient than larger livestock such as beef, pork, and poultry.

Table 1.2: Examples of food conversion for some minilivestock under farming conditions

Animals		Food conversion*	Reference
Snails	(*Helix adspersa*)	1.2-2	Elmslie, 2004
Crickets	(*Acheta domesticus*)	1.7	Collavo et al., 2002
Chicken		2	Collavo et al., 2002
Pork		3.8	Collavo et al., 2002
Beef		7	Collavo et al., 2002

*Dried food required for fresh available meat.

Knowledge and Coevolution with Natural Resources

Many dramatic cases are known in which optimal resource management stabilized human populations and vice versa, lack of resource management created loss of biodiversity and disappearance of resources. Emblematic of the latter is the mammal macrofauna overkill in North-South America executed by hunters colonizing the new continent of America (Martin, 1973; Diamond, 2002).

Martin demonstrated that overkill of 24 genera of North American mammal macrofauna was apparently effected by the first groups of hunters around 10,000 years B.C. (Martin, 1973; Simmons, 1996) invading the North American continent as a consequence of their inexperience in dealing with resources new to them (large mammals). In fact it has been suggested that the collapse of species was due to human migrations. In Africa similar collapses of species did not take place perhaps as a result of better and longer coevolution of humans and macrofauna (Simmons, 1996; Diamond, 1997) or perhaps because greater attention was paid to the array of different resources such as small animals, including fish and aquatic mollusks, for instance in the Rift Valley Lake District in Africa (Broadhurst et al., 1998). Also, the abundant diversity of invertebrates played a role, a sizable amount of which is consumed by chimpanzees and hominids alike even if evidence for the hominids is not easy to find (Hladic, 1973; Bond, 2000; Tommaseo-Ponzetta, this book).

Why Minilivestock?

The smallest animals (vertebrates and invertebrates) are more numerous and biodiverse, abundant and in many cases more easily collected or hunted. Some have been less readily domesticated and farmed. Only a few have been extensively domesticated, e.g. the guinea pig (*Cavia porcellus*) in the Andean region of Peru, rabbit (in southern Europe) silkworms (*Bombyx mori* L.) in China, and bees (*Apis mellifera*) especially in the Mediterranean regions and to some extent in other areas including the tropics (Bodenheimer, 1951).

Why have only large animals proven more successful and thus the base for frequent domestication? Diamond suggested that larger animals in particular fit all the requirements for the process of domestication (Diamond, 2002). In addition, most domesticated species provide various byproducts, such as milk in the case of cows, sheep, and goats; eggs from chickens, ducks, and geese; honey, propolis and wax from honeybees; and silk from silkworms (in addition to the edible stages: nymphs are traditionally eaten, especially in China).

If, for instance the number of current farms growing small vertebrate animals in Brazil totals 347, most are focused on just rodents such as capybara (*Hydrochaeris hydrochaeris*), nandu (*Rhea americana*), peccaries (*Tayassu angulatus, T. tajacu*), agouties and pacas (*Agouti* sp., *Dasyprocta* spp.) (Jori, 2002; Govoni et al., this volume).

Some birds are easily kept as pets inside villages in different tropical countries including the Amazon, for instance the Muscovy duck (Plate I, 7) under study at FUDECI in Puerto Ayacucho, the black curassow (Plate I, 8), and various guans (Plate I, 9) readily observed in Piaroa and Yanoamo villages in Alto Orinoco just to mention a few.

If many wild species have been under attention and could be considered the first steps in taming, only a few have been developed to assess their real potential and problems for use in local communities. Pacas and agouti are among the few that have been investigated (Govoni et al., this book). Currently, only a few local studies have been made with insects. One has been done in an Amerindian village with palm worms. In Alto Orinoco, Cerda et al. (2001) demonstrated that there is excellent potential for creating small-scale farms, specially designed to attract visitors, even if the production scale and use of resources need to be better investigated to avoid disproportionate pressure on palms.

Educational Programs

"Once, in a local primary school, at Thuong Lo, of the Katu people in the mountains near Hue, one of us (MGP) asked the pupils, using an interpreter, if they ate insects. Everybody looked at me as though I were a "stupid man". I persisted asking had they tasted the small creamy legless bug (possibly three genera of

Curculionidae live inside bamboo according to Dr. C.O'Brien: *Otidocephalus, Cyrtotrachelus*, and *Macrocheirus*) thriving inside the bamboo shoots along field margins and in the forest close to the village. Now everyone was waving his/ her arms enthusiastically, saying this bug was a very good raw snack. I asked many people living in the town of Hue, even school teachers, food scientists, and agronomists and not one was aware of this local custom" (Paoletti and Pimentel, 2003) (Plate I, 10).

Implementing maintenance and growth of local knowledge about biodiversity use and management is an ambitious task for developed and undeveloped countries (Laird, 2002).

Much effort has to be directed toward this isssue at different levels in order to maintain and improve local knowledge: 1) promote teaching with the village elders involved; 2) provide practical examples of use and adoption of biodiversity; 3) provide materials that range from lists of species, booklets, small facilities (gardens) or field trips, cooking sections, and rearing sections. All these activities have to include an active component of different age villagers. Extending traditional knowledge to successive generations is a very difficult but important task. It could be the route way for development of new opportunities starting just from tradition.

In general, shamans and medicine men prefer to extend their knowledge to someone in their family to continue their profession within the village territory. However, in each village such knowledge is not strictly linked to one or a few persons. For instance, fungi and insects are in some cases (Guajibo and Piaroa Indians) better known by women and children.

To build strong links between local traditional knowledge, modern knowledge, and resource evaluation is a very important issue.

Nutritional Arguments

Wild food has recently been upgraded due to the recognition that it has antioxidants, vitamins, nutrients, polyunsaturated fatty acids, carbohydrates and amino acids that are highly valued for healthy human nutrition (Brand-Miller and Holt, 1998; Bond, 2000; Zeghichi et al., 2003; Yen, this volume 1; Bukkens). Essential and trace nutrients appear to be present and easily assimilated (this volume: Bukkens; Van Huis).

Proteins, fatty acids, especially the PUFA (polyunsaturated fatty acids) seem to be well represented in small animals (earthworms, spiders, caterpillars, termites, palm worms, and others, (Bukkens, this book; Paoletti et al., 2003, 2003a).

Table 1.3 presents data on the PUFA content in some foods and invertebrates. Remarkably earthworms, snails, and insects contain consistent amounts of essential PUFAs, an observation that could favor their large consumption in areas (tropics) in which fatty acids in particular may be limited (Dufour, 2000). Anthropophagy and especially consumption of brains and medulla have been

Table 1.3: Data from current literature and our research concerning Polyunsaturated Fatty Acids (PUFA) content in different organisms. Note the higher content in aquatic organisms including aquatic insects.

Species and habitat	**Fat (g/100 g)**	**Linolenic acid 18:2ω6**	**α-Linolenic 18:3ω3**	**Arachidonic acid 20:4ω6**	**DHA 22:6ω3**	**Reference**
Lake Malawi African freshwater						
Mbelele (catfish)	10.3 d	n.a.	n.a.	4.3 d	8.6 d	Pauletto et al. (1996a)
Njenu (carp)	4.9 d	n.a.	n.a.	1.8 d	7.8 d	Pauletto et al. (1996a)
Mfui (local sp.)	1.1 d	n.a.	n.a.	8.0 d	19.1 d	Pauletto et al. (1996a)
Kambale (local sp.)	1.8 d	n.a.	n.a.	5.9 d	13.3 d	Pauletto et al. (1996a)
Australian tropical freshwater						
Bream meat	1.6	n.a.	n.a.	5.3	5.6	Sinclair (1992)
Bream fat	91	n.a.	n.a.	2.0	1.5	Sinclair (1992)
Fish from northern Australia coastal waters						
Sand whiting (*Sillago*)		1.1% w	n.a.	4.0% w	14.2% w	Sinclair et al. (1983)
Barramundi (*Lates*)		2.0% w	n.a.	1.8% w	18.1% w	Sinclair et al. (1983)
Spotted bat fish (*Drephane*)		1.1% w	n.a.	5.6% w	11.1% w	Sinclair et al. (1983)
Red snapper (*Lutjanus*)		1.9% w	n.a.	2.5% w	21.8% w	Sinclair et al. (1983)
Catfish (*Arius*)		1.8% w	n.a.	3.0% w	10.8% w	Sinclair et al. (1983)
Trevally (*Caranx*)		1.0% w	n.a.	2.9% w	10.8% w	Sinclair et al. (1983)
Queenfish (*Scomberoides*)		1.6% w	n.a.	1.5% w	27.5% w	Sinclair et al. (1983)
Mullet (*Liza*)		1.3% w	n.a.	1.3% w	11.6% w	Sinclair et al. (1983)
Skippy (*Carangoides*)		1.4% w	n.a.	1.8% w	6.4% w	Sinclair et al. (1983)
Threadfin salmon (*Poladactylus*)		0.5% w	n.a.	1.4% w	22.5% w	Sinclair et al. (1983)
Tropical marine						
Australian barramundi	0.3	n.a.	n.a.	14.5	16.2	Mann et al. (1995)
Indian halibut	1.7	n.a.	n.a.	6.3	10.4	Pauletto et al. (1996a)
Freshwater temperate						
Unspecified	(oil)	n.a.	n.a.	3.3	8.0	Innis et al. (1995)
Higher-latitude marine						
N. Atlantic mackerel	(oil)	n.a.	n.a.	0.4	7.7	Pauletto et al. (1996a)
Temperate aquatic invertebrates						
Mollusk (not specified)	>0.1	n.a.	n.a.	2.3 w	22.0 w	Cunnane et al. (1993)
Squid (not specified)	>0.1	n.a.	n.a.	5.8 w	21.3 w	Cunnane et al. (1993)
Tropical invertebrate (Anellida: Glossoscolicidae)						
Motto (*Andiorrhinus motto*) body	0.66 d	7.6 d	1.7 d	15.7 d	0.6 d	Paoletti et al. (2003)
Andiorrhinus motto	1.01% d	5.29% d	1.26% d	0.95% d	0.07% d	Nov. 2002 M.G. Paoletti
Kuru (*Andiorrhinus kuru*) body	1.05 d	9.2 d	0.8 d	23.0 d	0.2 d	Paoletti et al. (2003)
Gastropoda						
Lumaca (Helix spp.) Italia	0.68	14.65	1.19	12.51	2.05	Novelli and Bracchi (in press)

Table 1.3: (Contd.)

Table 1.3: (Contd.)

Species and habitat	**Fat (g/100 g)**	**Linoleic acid 18:2ω6**	**α-Linol-enic 18:3ω3**	**Arachi-donic Acid 20:4ω6**	**DHA 22:6ω3**	**Reference**
INSECTS						
House cricket (*Acheta domesticus*)	??	20.84% d	1.31% d	0.34% d	—	Collavo (2002)
Cicada (*Magicicada septendecim*)		26.57%	—	trace	n.a.	Hoback et al. (1999)
Seri (*Termites*) soldiers	1.28 d	24.85 d	6.9 d	0.75 d	n.a.	Paoletti et al. (2003)
Lepidoptera						
Caterpillar (*Imbrasia epimethea*)		7.0%	35.1%	traces	n.a.	Kokondi et al. (1987a)
Caterpillar (*Imbrasia ertli*)		20.0%	11.01%	—	n.a.	Santos Oliviera et al. (1976)
Caterpillar (*Imbrasia truncata*)		7.6%	36.8%	—	n.a.	Kokondi et al. (1987a)
Caterpillar (*Nudaurelia oyemensis*)		5.7%	35.6%	0.3%	n.a.	Kokondi et al. (1987a)
Caterpillar (*Usta terpsichore*)		27.2%	2.8%	—	n.a.	Santos Oliviera et al. (1976)
Silkworm (pupae)	30.1	4.2%	25.7%	n.a	n.a.	Rao (1994)
Macrotermis bellicosus	46.1	34.42%	3.85%			Ukhun and Osasona (1985)
Rhyncophorus palmarum (larva)	48.19% d	1.28% d	0.76% d	—	—	
Syntermes aculeosus	1.74% d	19.77% d	6.93% d	0.77% d	—	Nov. 2002 M.G. Paoletti
Termite 1	2.61% d	16.19% d	0.27% d	0.6% d	0.05% d	Nov. 2002 M.G. Paoletti
Termite 2	5.27% d	11.33% d	0.64% d	0.57% d	0.03% d	Nov. 2002 M.G. Paoletti
Termite 3	5.90% d	7.72% d	0.24% d	0.58% d	—	Nov. 2002 M.G. Paoletti
Atta cephalotes	2.01% d	29% d	1.45% d	0.15% d	0.11% d	Nov. 2002 M.G. Paoletti
Eciton burchelli	1.83% d	1.48% d	1.03% d	6.11% d	0.07% d	Nov. 2002 M.G. Paoletti
Theraphosa apophysis	7.31% d	15.98% d	0.63% d	6.76% d	—	Nov. 2002 M.G. Paoletti
Termitary 1	0.27% d	13% d	1.33% d	0.74% d	0.25% d	Nov. 2002 M.G. Paoletti
Termitary 2	0.09% d	13.25% d	3.65% d	0.77% d	0.77% d	Nov. 2002 M.G. Paoletti
Termitary 3	0.3% d	12.26% d	1.5% d	0.8% d	—	Nov. 2002 M.G. Paoletti
Grayling (*Oncorhynchus nerka*)						
Heart		5.7*	9.1*	12.5*	53.7*	Ballantyne (2003)
Dorsal		3.9*	7.0*	2.3*	10.3*	Ballantyne (2003)
Egg		5.7*	9.6*	8.4*	35.1*	Ballantyne (2003)
Intestine		44.4*	80.0*	13.1*	23.6*	Ballantyne (2003)
Liver		8.6*	13.8*	14.2*	37.8*	Ballantyne (2003)
Kidney		5.0*	6.7*	12.4*	35.5*	Ballantyne (2003)
Gut		19.2*	20.6*	6.5*	3.1*	Ballantyne (2003)
Ephemeroptera		49.7*	94.4*	17.1*	4.7*	Ballantyne (2003)
Trichoptera		62.0*	65.1*	13.4*	6.7*	Ballantyne (2003)
Plecoptera		58.2*	106.7*	14.8*	0.4*	Ballantyne (2003)

Table 1.3: (Contd.)

Table 1.3: (Contd.)

Species and habitat	Fat (g/100 g)	Linoleic acid 18:2ω6	α-Linolenic 18:3ω3	Arachidonic acid 20:4ω6	DHA 22:6ω3	Reference
Microalgae						
Scenedesmus abundans	1.482 d	0.158 d	0.307d	n.a.	—	Isik et al. (1999)
Monoraphidium minitum	5.044 d	0.465 d	1.538 d	n.a.	—	Isik et al. (1999)
Chlorella vulgaris	7.662 d	1.857 d	2.676 d	n.a.	0.016 d	Isik et al. (1999)
Rotifers and *Tilapia*						
Brachionus calyciflorus (fed *S. abundans*) A	3.605 d	1.292 d	1.135 d	n.a.	n.a.	Isik et al. (1999)
Brachionus calyciflorus (fed *M. minitum*) B	3.789 d	0.785 d	1.387 d	n.a.	n.a.	Isik et al. (1999)
Brachionus calyciflorus (fed *C. vulgaris*) C	3.983 d	1.129 d	1.309 d	n.a.	n.a.	Isik et al. (1999)
Tilapia zillii (fed A)	4.152 d	0.347 d	0.122 d	n.a.	1.072 d	Isik et al. (1999)
Tilapia zillii (fed B)	4.455 d	0.166 d	0.119 d	n.a.	1.438 d	Isik et al. (1999)
Tilapia zillii (fed C)	4.492 d	0.353 d	0.114 d	n.a.	1.295 d	Isik et al. (1999)
Vegetable oils						
Soybean		54%	7%	n.a.	n.a.	Innis (1996)
Safflower		76%	0.5%	n.a.	n.a.	Innis (1996)
Sunflower		68%	1%	n.a.	n.a.	Innis (1996)
Corn		54%	1%	n.a.	n.a.	Innis (1996)
Olive		10%	1%	n.a.	n.a.	Innis (1996)
Canola		22%	10%	n.a.	n.a.	Innis (1996)
Palm		10%	1%	n.a.	n.a.	Innis (1996)
Cottonseed		54%	1%	n.a.	n.a.	Innis (1996)
Peanut		32%	—	n.a.	n.a.	Innis (1996)
Linseed		16%	54%	n.a.	n.a.	Innis (1996)

Salmon with skin contains up to 19.9 g%
n.a.: not analyzed
%: percentage of total fatty acid
If not specified data refers to weight percentage (g/100 g)
—: not detected
d: dry weight
w: wet weight
*: mg/100 g ? dry weight?
[v]November 2002 Data Refer to Alto Ounoco Edible Invertebrates

remarked as a practice for obtaining fat and requirements of PUFA. It might be said that insects could represent a sort of "alternative" to the problematic attitude that has provoked *kuru* disease in Papua New Guinea (Creutzfeldt-Jakob disease) (http://www.cjdfoundation.org/CJDInfo.html).

For example, it was discovered that the Asmat in New Guinea consume a large amount of palm worms (*Rhynchophorus ferrugineus*) (see Plate IX, 4 and X, 4), which are very rich in fat content; furthermore, the legless larvae "mimic" human brain segments somewhat:

"Among the coastal Asmat, larvae are also represented as stylized decorative motifs on shields, and various artistic objects are devoted to them, e.g., special bowls to be used on ritual occasions. In earlier times, such bowls carried human brains, presently substituted by sago grubs" (Tommaseo-Ponzetta and Paoletti, this book).

This could be a ritual and also effective way in which essential fatty acids could be recovered directly from minilivestock, the palm worms, rather than from human brain segments.

Minilivestock as Healing Food?

We offer four chapters in this volume devoted to traditional use of small invertebrates as medicine in China (this book: Zimian et al., Zhenjun; Zhenjun and Wenling) and South Korea (Pemberton, this book). Many other invertebrates have been claimed to have medicinal curative properties. For instance, in China scorpions (*Buthus martensi*) (Plate I, 11) have long been considered an important medicinal prescription and apparently only recently, in the last few decades, become a specialty in specialized restaurants in large cities such as Peking (Han Chun Ru, pers comm. 1996). To be sure, invertebrates represent an important source for products and compounds including antibiotics of great significance for human health care.

For instance, leafcutter ants especially those of genus *Atta,* a much relished snack ant in the Amazon, bear strains of *Streptomyces* (Cameron et al., 1999; Schultz, 1999).

Risks and Limits

Not all plants nor all animals can be consumed or eaten by humans even though potentially good nutritional resources for the simple reason that they are poisonous or unsafe. In China, Africa, and South America each human group, region, and village has prohibited eating of certain animals. These rules reflect ecological, behavioral (Bejossi, 2004), and practical as well as toxicological reasons. Some species are indicated as famine food or food in case of emergency. Some reasons may also be political: good relationships with neighbors or protection of key species from high hunting pressure (Milton, 1997). Apparently for the Piaroa in Alto Orinoco, Venzuela, the tapir (danto) albeit rated as the most preferred meat is not allowed to be hunted (personal interviews at Caño Tigre, Alto Orinoco)!

Some animals are not eaten, for instance *mamocorisina* (Sphingidae, possibly a form of *Isognathus leachi*) because considered by the Yanomamo, who have named and eat over 25 different caterpillars, a very poisonous one (Plate I, 12 and I, 13 and cover). However, no data are available on this species to indicate whether the poison is inherent in the caterpillar or assimilated from its host plant. Some evidence could be that this species lives on an Apocynacean poisonous plant, which is rich in alcaloids and used by Ye'Kuana to prepare *barbasco* (a plant extract utilized in fishing).

So, we cannot recommend eating any and every insect available in a tropical forest or found in temperate regions.

There are other reasons for caution in small-scale or large-scale production of minilivestock (this volume: Giaccone; Collavo et al.). These reasons concern hygienic conditions of rearing and the possibility of contamination with pesticides or pathogens.

Development of a new method for collection and small-scale production has potential risks for the environment and humans. Therefore, conscientious monitoring must be ensured.

References

Bahuchet S. 1978. Introduction à l'ethnoécologie des pygmées Aka de la Lobaye Empire Centre Africain. Thèse Ecole Supérieure des Hautes Etudes, 348 pp.

Ballantyne A. 2003. Personal communication.

Beets W.C. 1997. The need for an increased use of small and minilivestock in integrated smallholder farming systems. Ecol. Food Nutr., 36: 237–245.

Blanco H.A., Garrido S. and Signi A. 2002. Guni Matsiadari Uwasio. El rio: sustento de Nuestra vida. Museo Etnologico de Amazonas, Puerto Ayacucho, Venezuela, 36 pp.

Bodenheimer, F.S. (1951). Insects as Human Food. Dr. W. Junk Publication, The Hague, The Netherlands.

Bodmer R.E., Pezo Lozano E., and Fang, T.G. 2003. Economic analysis of wildlife use in the Peruvian Amazon. *In*: Silvius K., Bodmer R., and Fragoso, J. (eds.). People in Nature: Wildlife Conservation in South and Central America. Columbia University Press, New York, NY.

Bond G. 2000. Natural Eating. Griffin Publ. Group, Torrance, CA, USA.

Brand-Miller J.C. and Holt H.A. 1998. Australian aboriginal plant foods: a consideration of their nutritional composition and health implication. Nutr. Res. Rev. 11: 5–23.

Broadhurst C.L., Cunnane S.C., and Crawford M.A. 1998. Rift Valley lake fish and shellfish provided brain-specific nutrition for early Homo. British J. Nutr., 79: 3–21.

Bukkens G.F. 2004. Insects in the human diet: nutritional aspects (this volume).

Cameron R.C., Scott J.A., Summerbell R.C., and Malloch D. 1999. Fungus-growing ants use antibiotic-producing bacteria to control garden parasites. Nature 398: 701–704.

Caspary H.U. and J.J. Momo 1998. La chasse villageoise en Cote d'Ivoire. Resultats dans le cadre de l'étude Filière et Viande de Brousse (Enquetes Chasseurs)—Rapport préliminaire, Version 1 DPN et BM/DAE/Abidjan, Cote d'Ivoire, 117 pp.

Cerda H., Martinez R., Briceno N., Pizzoferrato L., Manzi P., Tommaseo-Ponzetta M., Marin O., and Paoletti M.G. 2001. Palm worm (Insecta, Coleoptera, Curculionidae: *Rhynchophorus palmarum*), traditional food in Amazonas, Venezuela: nutritional composition, small scale production and tourist palatability. Ecol. Food Nutr. 40(1): 13–32.

Collavo A. 2002. Minilivestock: Utilizzo di *Acheta domesticus* come fonte alimentare non convenzionale. Tesi di Laurea,Università di Padova, Dip. Biologia, Padova, Italy.

Diamond J. 1997. Guns, Germs and Steel. Norton Co., New York, N.Y 480 pp.

Diamond J. 2002. Evolution, consequences and future of plant and animal domestication. Nature, 418: 700–707.

Dufour, D. L. 1987. Insects as food: a case study from the northwest Amazon. Amer. Anthropol. 89: 383–396.

Dufour D.L. 2000. Use of tropical rainforests by native Amazonians. *In*: Goodman A.H., Dufour D.L., and Pelto G.H. (eds.) Nutritional Anthropology. Mayfield Publ. Co., Mountain View, CA, USA, pp. 77–86.

Elmslie L.J. 2004. Snail collection and small-scale production in Africa and Europe (this volume).

Etkin N. 2000. Eating on the Wild Side. University of Arizona Press, Tucson, AZ, USA.

Gomez-Pompa A. and Bainbridge D.A. 1995. Tropical forestry as if people mattered. *In*: A. Lugo and C. Lowe (eds.) Tropical forests: management and ecology. Springer Verlag. 408-422.

Hardouin J. 2004. Minilivestock: environment, education, research, and economics (this volume).

Hardouin, J. and Stievenart, C. 1993. Invertebrate (Minilivestock) Farming; CTA (Center of Technical Cooperation): Wageningen, The Netherlands, 223 pp.

Hladic C.M. 1973. Alimentation et activite d'un groupe de chimpanzes reintroduits en foret Gabonaise. Terre Vie 27 : 343–413.

Hladic C.M., Hladic A., Pagezy H., Linares F.O., Koppert G.J.A., and Froment A. 1996. L'alimentation en la Forete Tropicale. Unisco, Paris, 2 vols.

Innis S.M.1996. Essential dietary lipids. *In:* Present Knowledge in Nutrition. ILSI Press, Washington DC, USA.

Isik O., Sarihan E., Kusvuran E., Gul O., and Erbatur O. 1999. Comparison of the fatty acid composition of the freshwater fish larvae *Tilapia ziili*, the rotifer *Brachionus calcyflorus*, and the microalgae *Scenedesmus abundans*, *Monoraphidium minitum* and *Chlorella vulgaris* in the algae-rotifer-fish larvae food chains. Aquaculture 174: 299–311.

Jori F. 2002. La produccion de animales silvestres en Brazil: una actividad en pleno desarrollo. BEDIM 11(2): 20–21.

Jori F., Edderai D., and Houben P. 2004. Potential of rodents for minilivestock in Africa (this volume).

Laird S. (ed.). 2002. Biodiversity and traditional knowledge. Equitable partnership in practice. Earthscan Publ., London, U.K. 504 pp.

Malaisse F. 1997. Se nourrir en forêt claire africane. Approche écologique et nutritionelle. Les Presses Agronomiques de Gembloux, 384 pp.

Martin P.S. 1966. Africa and Pleistocene overkill. Nature, 212: 339–342.

Martin P.S. 1973. The discovery of America. Science, 179: 969–74.

Melnyk M.A. 1995. The contributions of forest foods to the livelihoods of the Huottuja (Piaroa) people of southern Venezuela. PhD thesis, University of London, London, UK.

Menzel P. and D'Aluisio F. 1999. Man eating bugs. Ten Speed Press Berkeley.

Milton K. (1997) Real men don't eat deer, Discovery, June, 46–53.

Moreno A. and Paoletti M.G. (2004). *Andiorhinus kuru* n. sp., a giant earthworm (Oligochaeta: Glossoscolecidae) food resource for Makiritare Indians of the Alto Rio Padamo, Amazonas, Venezuela. Can. J. Zool.

NAC (National Academic Council) 1989. The Lost Crops of the Incas. National Academy Press, Washington DC, USA, 415 pp

NAC (National Academic Council) 1996. The Lost Crops of Africa. National Academy Press, Washington DC, USA, 383 pp.

Nor Aliza A.R., Rana R.L., Skoda S.R., Berkebile D.R., and Stanley D.W. 1999. Tissue polyunsaturated fatty acids and a digestive phospholipase A_2 in the primary screwworm, *Cochliomyia homnivorax*. Insect Biochem. Molec. Biol. 29: 1029–1038.

Nor Aliza A.R., Bedick J.C., Rana R.L., Tunaz H., Wyatt Hoback W., and Stanley D.W. 2001. Arachidonic and eicosapentaenoic acids in tissues of the firefly, *Photinus Pyralis* (Insecta, Coleoptera). *Comp. Biochem. Physiol.* Part A, 128: 251–257.

Novelli, E. and Bracchi P.G. 2002. Indagine sul valore dietetico-nutrizionale della lumaca. Confronto fra specie e fra soggetti raccolti in natura ed allevati. Da pubblicare.

Otero J.A., Jimenez J.A., Hernandez J., and Signi A. 2002. Dearuwä Wotjuja. Los Piaroa y la Selva. Museo Etnologico de Amazonas, Puerto Ayacucho, Venezuela 44 pp.

Paoletti, M.G. and Bukkens S. 1997. Minilivestock Ecology of Food and Nutrition, 36(2–4): 95–346.

Paoletti M.G., Buscardo E., VanderJagt D.J., Pastuszyn A., Pizzoferrato L., Huang Y-S., Chuang L-T., Millson M., Cerda H., Torres F., and Glew R.H. 2003. Nutrient content of earthworms consumed by Yekuana Amerindians of the Alto Orinoco of Venezuela. Proc. R. S. London B. 270: 249–257.

Paoletti M.G., Buscardo E., Vanderjagt D.J., Pastuszyn A., Pizzoferrato L., Huang Y-S., Chuang L-T., Millson M., Cerda H., and Glew R.H. 2003a. Nutrient content of termites (Syntermes soldiers) consumed by Makiritare Amerindians of Alto Orinoco, Venezuela. Ecol. Food Nutr., 42: 173–187.

Paoletti M.G., Dufour D.L., Cerda H., Torres F., Pizzoferrato L., and Pimentel D. 2000. The importance of leaf- and litter-feeding invertebrates as sources of animal protein for the Amazonian Amerindians. Proc. R. S. London B. 267: 2247–2252.

Paoletti M.G. and D. Pimentel. 2003. Natural resources, biodiversity, integrated management and regulation: General considerations and discussion. *MultiCiencia*, Campinas, Brazil.

Pauletto P., Puato M., Angeli M.T., Pessina A.C., Munhambo A., Bittolo-Bon G., and Galli C., 1996. Blood pressure serum lipids, and fatty acids in population on a lake-fish diet or a vegetarian diet in Tanzania. Lipids, 31: 309–312.

Pemberton R.W. 2004. Contemporary use of insects and other arthropods in traditional Korean medicine (hanbang) in South Korea and elsewhere (this volume).

Posey D.A. 1984. A preliminary report on diversified management of tropical forest by the Kayapo Indians of the Brazilian Amazon. Advances Econ. Bot. 11: 112–126.

Posey D.A. 1993. The importance of semi-domesticated species in post-contact Amazonia: Effects of the Kayapo Indians on the dispersal of flora and fauna. *In:* Hladic C.M., Hladick A., Linares O.F., Pegezy H., Semple A., and Hadley M. (eds.). Tropical Forests, People and Food. Parthenon Publ. Group, New York, pp. 63–71.

Posey D.A. and Dutfield G. 1996. Beyond Intellectual Property. Toward Traditional Resource Rights for Indigenous People and Local Communities. IDRC, Ottawa, Canada.

Prance G.T. 1984. The use of edible fungi by Amazonian Indians. Advances Econ. Bot., 1: 127–139.

Redford K.H. and Dorea J.G. 1984. The nutritional value of invertebrates with emphasis on ants and termites as food for mammals. J. Zool. Lond. 203: 385–395

Redford K.H. and Padoch C. 1992. Conservation of Neotropical Forests. Working from Traditional Resource Use. Columbia University Press, Columbia, NY.

Reid J., Baker L., Morton S.R., and the Mutitjulu Community. 1992. Traditional knowledge and ecological survey equals better land management. Search 23: 249–251.

Rodriguez W., Chipiaje A., Gaitan D., and Signi A. 2002. Wayapjio, pija liwaisi. La Sabana y sus Riquezas. Museo Etnologico de Amazonas, Puerto Ayacucho, Venezuela, 44 pp.

Ruddle, K. 1973. The human use of the insects: examples from the Yukpa, *Biotropica* 5(2): 94–101.

Schultz T.R. 1999. Ants, plants and antibiotics. Nature, 389: 747–748.

Scoones I., Melnyk M., and Pretty J.N. 1992. The hidden harvest: Wild foods and agricultural systems. A literature review and annotated bibliography. Swedish International Development Authority, The Worldwide Fund for Nature and the International Institute for Environment and Development, London.

Simmons I.G. 1996. Changing the Face of the Earth. Blackwell, Oxford, UK, 464 pp.

Sinclair, A.J., O'Dea K. , and Naughton J.M 1983. Elevated levels of arachidonic acid in fish from Northern Australian coastal waters. Lipids, 18: 877-881.

Tommaseo-Ponzetta M. 2004. Insects. Food for human evolution (this volume).

Udayasekara Rao, P. 1994. Chemical composition and nutritional evaluation of spent silk worm pupae. J. Agric. Food Chem., 42: 2201–2203.

Ukhun M.E., and Osasona M.A. 1985. Aspects of the nutritional chemistry of *Macrotermis bellicosus*. Nutrition Reports International, 32: 1121–1130.

Whitfield J. 2003. Forest mammals are food for people across Asia, Africa and South America. But with many species disappearing fast, can ecologists balance supply and demand? Nature, 26/09/03.

Wyatt Hoback W., Rana R.L., and Stanley D.W. 1999. Fatty acid composition of phospholipids and triacylglycerols of selected tissues, and fatty acid biosynthesis in adult periodic cicadas, *Magicicada septendecim*. Comp. Biochem. Physiol. Part A, 122: 355–362.

Yen A. L. 2004. Insect and other invertebrate foods of the Australian aborigines (this volume).

Yen A.L., Gillen J., Gillespie R., Vanderwal R. and the Mutitjulu Community. 1997. A preliminary assessment of Anangu knowledge of Central Australian invertebrates. Memoires Museum of Victoria, 56 (2): 631–634.

Zeghichi S., Kallithraka S., Simopoulos A.P., and Kypriotakis Z. 2003. Nutritional composition of selected wild plants in the diet of Crete. *In:* Simopoulos A.P. and Gopalan C. (eds.). Plants in Human Health and Nutrition Policy. World Rev. Nutr. Diet, Basel Karger 91: 22–40.

Zhenjun S. and Wenling C. 2004. Pharmaceutical value and use of earthworms (this volume).

Zimian D., Yonghua Z., and Xiwu G. 2004. Medicinal terrestrial arthropods in China (this volume).

2

The Minilivestock: Environment, Education, Research, and Economics

JACQUES E. HARDOUIN

Bureau for Exchange and Distribution of Information on Minilivestock (BEDIM), Faculté universitaire des Sciences agronomiques. Passage des Déportés, 2 ; B-5030 Gembloux, Belgium. E-mail: goorickx.m@fsagx.ac.b

Abstract

The field of minilivestock is permanently fluctuating, with some species (grasscutters, giant snails) nearly going out as their breeding is totally under control, and others coming in (snakes, bush fowl...). The development of minilivestock breeding is not only a tool within the tropical rural development process, but also a substitute to poaching, accepting the fact that the demand for a specific meat exists and is sufficiently high. Biological know-how is available and production systems can be established. Training, extension, education and research are required, as well as identification of and remedies for diseases and other ailments. A call to Faculties is formulated to suggest specific themes for student works, theses, etc.

At least one case is known of someone from West Africa having won a Golden Medal after patenting a technique to produce maggots as a meal replacement for raw protein in poultry feed.

Key Words: Biodiversity, constraints, diseases, productivity, profitability, minilivestock

Introduction

The concept of "minilivestock" is accepted today by most institutions as a normal tool of and component for rural development in the tropics. Relevant literature is rather abundant these days (Branckaert, 1995; Peters, 1986; Hardouin, 1997; Hardouin and Thys, 1997; Hardouin et al., 2002; Paoletti and Bukkens, 1997).

Minilivestock is never at a standstill, however, as it fluctuates in accordance with popular demand, progress in scientific and technical know-how, and people's ever rising expectation that more meat can be realized from indigenous animal species.

In this context, animal species never subjected to controlled breeding 20 years ago are commonly targeted nowadays for backyard or small-scale production. The time is now ripe for them to leave the sector of minilivestock and to be integrated into the normal breeding routine of indigenous species. The best example is probably the grasscutter *Thryonomys* sp.; it should appear now next to sheep and goat, or guinea fowl and rabbits in the normal course of lectures. The same applies to the guinea pig *Cavia porcellus*, far more abundant in Africa than usually thought and widely consumed though more or less ignored by universities and research institute staff, and even often by official veterinary and extension services. At the same time, species not mentioned in the first reference documents on minilivestock (Anonymous, 1991; Hardouin and Stiévenart, 1991; Branckaert, et al., 1992) are entering the sector of minilivestock due to the interest shown in them by people and attempts to breed them under control. Such has been the case these last years for peccaries, bush fowl, snakes, frogs, lizards, iguanas, maggots and others. In the present book, the main groups of minilivestock for which investigations have been undertaken in several countries and continents include rodents in Africa (*Thryonomys, Cricetomys, Atherurus*...) and in Latin America (*Hydrochoerus, Dasyprocta, Agouti, Myocastor*...) and guinea pigs (*Cavia*), peccaries (*Tayassu*), bush pig (*Potamochoerus*), bush fowl (*Francolinus*), frogs etc. The spectrum is probably broader for invertebrates, encompassing giant snails (*Achatinidae*), manure worms (*Eudrilus, Perionyx, Eisenia*...), and several insects. Among the latter, many are used as food or even delicacies such as palm grubs, termites, caterpillars, while others represent high-priced protein sources for pigs, poultry or fish, such as maggots, cockroaches, termites. A too often neglected usage of nice-looking insects prevails in arts and crafts (mounted in frames, included in plastic blocks, butterfly wings assembled for "paintings"...). Surprisingly, some hospitals of repute breed flies (Diptera, genus *Lucilia*) under special conditions to use the sterile maggots for cleaning septic wounds resistant to antiseptics and antibiotics, or removing dead tissues and augmenting the healing process.

Biodiversity and Environment

The demand for small wild animals is very high among people in tropical countries, even in towns. There is no need to describe the situation in forest villages as this food source has little impact on biodiversity as long as the captures meet the local demand, without external trade. However, town-dwellers also enjoy traditional dishes from their respective villages. This explains in part the pressure exerted by towns on wildlife leading in some places to the local disappearance of species (giant snails, bush rats...), which does not mean these species should be classified as threatened. On the other hand, one has to accept that the local fauna represents a normal source of animal protein for the population.

One of the objectives of minilivestock breeding is the replacement of poached meat by similar meat obtained under controlled techniques. Luckily, such a position is accepted nowadays by many funding and development agencies, which have understood that breeding snails, grasscutters, pacas, capybaras, frogs and so forth is not a utopia but a reality and a possibility if one expects a balance between biodiversity and the justified demand by people for meat obtained from their own animals. Among the vertebrates rodents are the most important, but other groups, such as wild pigs, birds, snakes and frogs are also present.

Personal field experience as well as many documented surveys showed indeed that any kind of bushmeat is higher priced than the meat of domestic animals (beef, mutton, pork, etc.), although the rank for any type of bushmeat (monkeys, antelopes, porcupines, snails, rodents) varies according to areas or ethnic groups.

Faunal biodiversity likewise includes invertebrates, among which at least three groups of animals are representated in the minilivestock sector. Giant land snails are well known, of which some species are already totally absent in some areas of their biotope around large towns. Excessive gathering without control meant rather quickly that collection for sales and consumption were realised on the parent stock and not just the progeny (Hardouin, 1995).

Manure worms are another group of animals playing an important role in biodiversity due to their role in the natural decomposition process of litter and, consequently, the production of humus and high-quality arable soils. Discussion of this phenomenon used by some organizations in the joint action of sanitation of urban garbage and compost with worm production has waned. In the wild, several species of annelids are consumed directly by many human populations, for which they represent an important dietary component. Furthermore, it was recently confirmed that such worms contain useful quantities of many nutrients critical to the health of the humans who consume them (Paoletti et al., 2003).

Books have been written on insects breeding and use. It suffices here to remind the reader of the vast material dealing with insects as food in many parts of the world that were published consequent to the International Symposium on Biodiversity in Agriculture held in Beijing, P.R. China in September 1995 (Paoletti and Bukkens, 1997). Very recently, entomophagy has drawn considerable interest from people in Europe, as shown at Gembloux, Belgium (Mignon, 2002).

Constraints to Further Development

One can consider that in 2004 minilivestock development has gained the consideration deserved, and that international authorities support it. Indeed the very last issue of the well-known CIRAD-GRET "Memento de l'Agronome" devotes for the first time a full subsection to nonconventional animals and minilivestock breeding (Brescia et al., 2002).

Many cases of minilivestock production exist today, from the earliest guinea pigs and grasscutters to the more recent cricetomas, giants snails, iguanas..., and emerging maggots, palm grubs, earth- and manure worms, frogs, a.o.

Basic knowledge is available and animal production specialists ought now to adapt zoological data to economic production schemes. In any case, farming approaches have to rely on what happens in the wild, trying initially to create a local environment not too different. Not only technical characteristics but also behavior, taming, quiet attitude of the people etc. are of utmost importance.

Quantitative development of minilivestock farms is closely associated with competence, which means in turn training, extension, education and research. Unfortunately, trainers in this field do not yet exist due to the fact that education in this area is lacking. Reluctance to embark on minilivestock education or research is quite understandable since the authorities involved are often themselves not well aware of the true situation.

In 1992, participants in the seminar "Invertebrate (Minilivestock) Farming" (Hardouin and Stiévenart, 1992) adopted recommendations for measures to make decision-makers aware of the potentialities of minilivestock and the need for research in this area.

Thanks to the European Commission, the FAO, several other institutions and the international association BEDIM, it has been possible to produce and distribute the 52-minute videotape "Minilivestock in the Tropical Forest Habitat". Since 1992, the same association has produced thanks to FAO, a bilingual semestral Bulletin of Information on Minilivestock, now widely distributed in tropical areas.

Jumping from gathering to controlled production means that related animals are no more to be considered wild fauna but bred animals. Hence they should progressively be scratched from the list of animals under the jurisdiction of Wildlife/Forestry Services and transferred to official services in charge of animal production and health. In other words, the status of the erstwhile wild fauna should be officially modified. However, since civil servants and officers of the Animal Production/Health Services are not themselves trained in this field, targeted postgraduate education courses will have to be developed.

Sad to say, the diseases and specific problems of rearing minilivestock are scarcely known. This situation is not surprising as the pioneers in minilivestock sought ways and means of controlling the entire life cycle of the animals. Luckily, in most cases troubles pertained to injuries and wounds, often associated with characteristics of remaining feral. As time passed and the density of animals kept in captivity increased, pathological problems also arose. Parasitism, infections, nutritional deficiencies etc. appeared but solutions were generally found.

Healthier minilivestock requires practical research on pertinent topics. This is a wonderful avenue for student work (dissertations, experiences, theses) and young scientists, as results are more quickly obtained in undiscovered fields (rodents, frogs, snakes, snails, earthworms, a.o.) than in classical sectors (cattle, pigs, poultry, a.o.). Hence this is indeed a call to the Faculties of Veterinary Medicine, Agricultural Sciences, Biological/Zoological Sciences, etc. as well as Research Foundations offering fellowships. Needless to say, practical research should be encouraged.

Similarly, costs of production of minilivestock are still poorly documented; Faculties and Research Institutions could supply short- and long-term sketches of profitability and markets for many different species. Comparisons could be made more significant were all the data expressed in at least one standard value applicable to all species, preferably the Relative Weight Productivity Index (RWPI) since it is expressed as a percentage and not in kilograms. The RWPI corresponds to the ratio (= %) between total weight (kg) produced in one year (365 days) and weight (kg) of a single adult reproductive female. Such an expression allows comparisons between grasscutters and cricetomas, or guinea pigs and pacas, or even earthworms and palm grubs.

Conclusions

Minilivestock breeding can no longer be kept on the back burner; it constitutes a ready tool for low investment production, using live renewable resources and local materials, becoming thereby a wonderful example for backyard or semiurban activities.

This rather new chapter in animal sciences has progressed tremendously in the last 15 years. The first protagonists were considered foolish for some years, even detrimental to the "conventional" animal production programs. This initial attitude has been fairly well overcome, and preparation made for the present and future. It was highly rewarding for me to learn from an enthusiastic and open-minded scientist of a the developing country that his minilivestock products presented at an Innovations and Inventions Show in West Africa won him the Golden Medal as best inventor in 2001 from the "Organisation Mondiale de la Propriété Industrielle" (Kone, pers. comm.). He had patented a reliable technique for producing maggots as a substitute for other raw protein sources in poultry feed.

The need for an up-to-date database concerned solely with minilivestock, already recommended in the 1992 Philippines Seminar (Hardouin and Stiévenart, 1992), today is a must if full use is to be made of the tremendous amount of literature and other documents stored by the international association BEDIM.

The international scientific community which showed confidence in minilivestock development was prescient. The time is now amenable to other specializations, of which two are already under study in the BEDIM world action program: i) Ecology and Human Nutrition, Department of Biology, University of Parma, Italy, and ii) Cultural Anthropology, Cultural Anthropology Centre, Free University Brussels, ULB Belgium.

Acknowledgment

The author is grateful to Maurizio G. Paoletti for requesting this paper and thanks Ms. Sandra Bukkens (INRAN, Rome, Italy) and the publisher for translational/ editorial improvements.

References

Anonymous 1991. Minilivestock: Little Known Small Animals with a Promising Economic Future. National Academy Press, Washington DC, USA: 450 pp.

Branckaert R.D. 1995. Rearing unconventional livestock species: a flourishing activity. World Anim. Rev. 83 (2): ii+iv.

Branckaert R.D., Hardouin J. and Stiévenart C. 1992. Utilisation du mini-élevage pour une production alimentaire durable. Roma, Italy: Staff working paper FAO/AGAP.

Brescia F., Chardonnet P., de Garine-Wichatisky M. and Jori F. 2002. Les élevages non-conventionnels. *In*: Mémento de l'Agronome. Paris : Ministère des Affaires étrangères (CIRAD-GRET), pp. 1617–1646.

De Visser J., Mensah G.A., Codjia J.T.C., and Bokonon-Ganta A.H. 2001. Grandes lignes sur l'élevage de l'aulacode en captivité étroite. In: Guide préliminaire de reconnaissance des rongeurs au Bénin, ReRE, 01 B.P. 526, Cotonou, Bénin, pp. 228–243.

Hardouin J. 1995. Minilivestock: from gathering to controlled production. Biodiv. Conserv. 4: 220–232.

Hardouin J. 1997. Developing minilivestock as source of human feed, animal food or revenue: a brief overview. Ecol. Food Nutr. 36: 95–107.

Hardouin J. and Stiévenart C. 1991. Le mini-élevage dans les pays tropicaux. CTA, Wageningen, The Netherlands, 34 pp.

Hardouin J. and Stiévenart C. 1992. Seminar Invertebrates (Minilivestock) Farming, the Philippines. November 1992, c/o BEDIM, Gembloux, Belgium.

Hardouin J. and Thys E. 1997. Le mini-élevage, son développement villageois et l'action de BEDIM. Biotechnol. Agron. Soc. Environ. 1 (2): 92–99.

Hardouin J., Thys E., Joiris V. and Fielding D. 2002. Minilivestock breeding with indigenous species in the tropics (unpubl.).

Paoletti M.G. and Bukkens S.G.F. (Guest eds.). 1997. Special Issue Minilivestock. Ecol. Food Nutr. 36 (2–4): 346 pp.

Paoletti M.G., Buscardo E., VanderJagt D.J., Pastuszyn A., Pizzoferrato L., Huang, Y-S, Chuang L-T, Millson M., Cerda H., Torres F., Glew, R. 2003. Nutrient content of Earthworms consumed by Yekuana Amerindians of the Alto Orinoco of Venezuela. *Proceedings Royal Society London B*. 270: 249–257.

Peters K.J. (1986). Constraints to unconventional livestock production. Proc. Intern. Workshop Constraints and Strategies Livestock Development and Improvement of Animal Health. G.T.Z. Jan. 1986, West Berlin, F.R. Germany, pp. 96–116.

3

Potential of Rodents for Minilivestock in Africa

Ferran Jori[1], David Edderai[2], and Patrick Houben[3]

[1]*Programme ECONAP, CIRAD EMVT, TA 30/E, Campus International de Baillarguet, Montpellier 34398, Cedex 5, France. E-mail: jori@cirad.fr*
[2]*Projet Développement d'Alternatives au Braconnage en Afrique Centrale (DABAC), BP 2572 Yaoundé, Cameroun. E-mail: david-edderai@netcourrier.com*
[3]*Projet Développement d'Alternatives au Braconnage en Afrique Centrale (DABAC), BP 9129 Libreville, Gabon. E-mail: dabac@assala.net*

Abstract

In many African countries several species of rodents are highly valued as a source of food and income for local people. Systems of rearing are fully developed, however, only in those species, such as the cane rat (*Thryonomys swinderianus*) whose biology is already well known. Semidomesticated breeds have been selected and periurban extension experiments have yielded satisfactory results in many countries. For other species, such as the brush-tailed porcupine (*Atherurus africanus*), its potential as minilivestock has yet to be fully assessed. It adapts well to captivity but shows little promise for rearing because female productivity is low. Giant rats or Cricetoma (*Cricetomys* spp.) are widely consumed and some countries have initiated research on them with promising results. Unfortunately, despite considerable improvements, no attempts have been made to develop extension programs. Even though rodent production is slowly developing in some parts of Africa, with obvious ecological and socioeconomic benefits, rodent farming projects are not the panacea and many problems still need to be solved to reach a large-scale production and to offer an alternative to the bushmeat trade.

Key Words: Rodents, Minilivestock, West Africa, Central Africa, Grasscutter, cane rat, brush-tailed porcupine, Giant rat, Eming rat

Introduction

There are more than 60 countries in the world where local populations obtain at least 20% of their animal protein from harvesting activities such as hunting and fishing (Robinson and Bennet, 2000). In sub-Saharan Africa, wildlife represents a susbtantial source of animal protein and income for a large part of the rural and urban populations (Martin, 1985; Peters, 1988; Wilkie and Carpenter, 1999; Robinson and Bennet, 2000). This feature is often overlooked in official statistics from these countries. It is particularly important, however, in areas where large extensions of tropical forest still remain, hunting traditions are still strong, and farming of domestic animals is limited (Feer, 1993; Chardonnet and Fritz, 1996; Robinson and Benett, 2000). In Africa, several recent studies have documented this situation and the importance of bushmeat trade, especially in the Congo Basin (Juste et al., 1995; Chardonnet et al., 1995; Wilkie and Carpenter, 1999;) where wildlife hunted for human consumption exceeds one million metric tons per annum (Robinson and Benett, 2000).

Bushmeat is sold in city markets where a large variety of wildlife species is offered (antelopes, primates, rodents and reptiles). This activity, also known as the bushmeat trade, is thought to have devastating effects on wildlife populations in the Congo Basin (Steel, 1994; Brugière, 1997; Wilkie and Carpenter, 1999; Auzel and Wilkie, 2000). Besides ecological implications, it is of public health concern since meat may remain several hours on market stalls without refrigeration before being sold and cooked. Furthermore, being an informal trade, there is no veterinary control of the meat. Rodents represent a high proportion of this bushmeat on sale in city markets (Steel, 1994; Juste et al., 1995; Caspary, 1999; Fa, 2000), particularly in areas of heavy human pressure (Chardonnet and Fritz, 1996; Wilkie and Carpenter, 1999).

Within this context, several authors have mentioned that wildlife meat production, together with other measures, could contribute to relieving pressure on wildlife populations (Hardouin and Thys, 1997; Caspary, 1999; Wilkie and Carpenter, 1999; Auzel and Wilkie, 2000).

Several authors have recognized in the last decade the importance in tropical countries of breeding local species for food, even if they are not conventional domestic animals (Asibey, 1974; Feer, 1993). Hardouin (1995) and Hardouin and Thys (1997) described as minilivestock the production of small species of wildlife, that can be reared on a small scale for animal or human consumption, representing a source of additional income for local populations living in periurban areas where wildlife is scarce and hence can fetch high prices (Chardonnet et al., 1995; Steel, 1994; Jori, 1997; Wilkie and Carpenter, 1999). Rodents are often cited as minilivestock species with great potential, due to their supposedly high rate of reproduction and widespread popularity in certain areas of Africa (Table 3.1). Therefore, those rodents particularly popular for meat consumption have been studied over the last three decades in various African countries, in order to assess their potential as production animals: these are basically the cane rat or

Table 3.1: Prices of domestic and wild meat in some African markets from Nigeria (Bendel state) and Gabon (Libreville) (*Source:* JORI et al., 1998)

Species	Nigeria, 1977	Nigeria, 1982	Gabon,1996
**Atherurus africanus*	8.80 $/kg	15.00 $/kg	5.27 $/kg
**Thryonomys swinderianus*	8.00 $/kg	11.33 $/kg	4.80 $/kg
Tragelaphus spekei	3.20 $/kg carcass	4.35 $/kg carcass	1.70 $/kg carcass
**Cephalophus* spp.	6.80 $/kg	6.63 $/kg	2.30 $/kg
**Cricetomys* spp.	1.38 $/kg	1.50 $/kg	2.72 $/kg
Potamochoerus porcus	8.00 $/kg carcass	1.20 $/kg carcass	3.40 $/kg carcass
Beef	3.20 $/kg carcass		3.40 $/kg carcass
Pork	3.20 $/kg carcass		2.72 $/kg carcass
Lamb	3.20 $/kg carcass		4.25 $/kg carcass

*Whole carcass.

grasscutter (*Thryonomys swinderianus*), the brush-tailed porcupine (*Atherurus africanus*), and the giant rats or Cricetoma (*Cricetomys gambianus* and *C. emini*).

Despite consumption of certain rodents being very common in certain past and present cultures—such as dormice and guinea pigs in some countries (Carpaneto and Cristaldi, 1995; Chauca de Zaldivar, 1995)—the idea of consuming rodents is not at all attractive in Western culture. On the contrary, many ethnic groups in the tropics, especially from Afrotropical regions, even use small rodents (Muridae) as food (Carpaneto and Germi, 1992). In general, for Western people rodents are generally associated with common rats (*Rattus* spp.) and mice (*Mus* spp.). However, order Rodentia encompasses more than 2,015 different species, representing 43% of the total 4,629 described mammal species, and colonising most of the earth's extant habitats (Wilson and Reeder, 1993). This chapter provides an update on 3 rodent species in Africa that show good economic prospects for meat production and rearing and have been the target of research during the last 30 years: the cane rat or grasscutter, the brush-tailed porcupine, and Cricetoma.

Cane Rat or Grasscutter

The grasscutter is without doubt one of the most popular game species in West Africa (Asibey, 1974; Baptist and Mensah, 1986; Caspary, 1999). After the African porcupine from genus *Hystrix*, it is the largest of all African rodents. It belongs to Suborder Hystricognathi, Family Thryonomidae containing only one genus, *Thryonomys* with two species: *T. swinderianus* or giant cane rat and *T. gregorianus* or small cane rat (see Table 3.2). The giant cane rat has a thickset body, measuring 40 to 60 cm, and a 20–25 cm tail. Its average weight fluctuates between 2 and 4 kg in females and 3 to 6 kg in males. Its fur is a mixture of brown, reddish, and gray hair that varies depending on habitat (Jori et al., 1995).

Table 3.2: Taxonomic classification of useful African rodents (Wilson and Reeder, 1993)

Order		Rodentia	
Suborder	Hystricognathi		Sciurognathi
Family	Hystricidae	Thryonomyidae	Muridae
Subfamily			Cricetomyinae
Genus	*Atherurus*	*Thryonomys*	*Cricetomys*
Species	*A. africanus*	*T. swinderianus*	*C. gambianus*
		T. gregorianus	*C. emini*

Table 3.3: Main feeding resources used in African rodent farming

Cane rat	• Wild graminees: *Pennisetum purpureum, Panicum maximum, Brachiaria ruziziensis, Setaria* spp., *Hyparrhenia* spp., *Andropogon* spp... • Dry leaves of *Leucaena leucocephala* • cereals: corn, wheat, rice, sorgho... • tubers: manihot, taro, sweet potatoes... • nuts: palm nuts, peanuts... • fruits: pawpaw, bananas... • pellets: rabbit pellets, wheat...
Atherurus	• tubers: manihot, taro, potatoes, yams (*Dioscorea* spp.)... • grains and nuts: palm nuts, peanuts... • rabbit pellets • fruits: pawpaw, *Prunus*... • *Amaranthus* leaves
Cricetomys	• tubers: manioc, taro, potatoes... • grains: palm nuts, peanuts, soybean • fruits: bananas, pawpaws, goyavas • industrial by-products: beer, wheat, palm nut... • dry bread

It is a monogastric herbivore, easy to feed and a good food transformer of fiber into protein. Offspring are born with 4 incisors that grow continuously during the life of the animal. Gestation lasts 5 months. It is able to reproduce year long (Asibey, 1974) and two litters of between two to six youngsters may be produced per annum (Baptist and Mensah, 1986; Adjanahoun, 1989, 1992). Diet in captivity consists of 80% *Pennisetum purpureum* and *Panicum maximum* combined with a mixture of corn, wheat, and mineral salts. The composition of the mixture can be varied and several items are accepted by this species including corn grains, corn cob, rabbit pellets, sugar cane, etc. (see Table 3.3).

Sex determination is done at birth on the basis of the ano-genital distance (Schrage and Yewadan, 1995). Weaning of young animals can be performed from age 35 days onward. As shown in Table 3.4, sexual maturity is achieved at 1,500 g live weight (six months for females and five months for males on average) and breeding in paddocks is undertaken in polygamous groups of one male per 5

Table 3.4: Reproductive parameters of exploited tropical rodents in Africa (*Source:* Jori, 1997)

	Cane rat	**Brush-tailed porcupine**	**Giant rat**	**Bovine**
Adult weight (kg)	3	2.8	1.3	350
Gestation length (days)	150	120	30	290
Weight at birth (g)	120	135	20	28000
Litter size	4	1	3.5	1
Litters/year	1.8	2.3	6	1
Culling weight (kg)	3.5	2.5	1.5	360
Carcass offtake (%)	65	67	52	45
Weight at sexual maturity (kg)	1.5	2	0.6	180
Average mother weight (kg)	2.5	3.5	1.5	350

females. Individual cages are used to house males at rest and pregnant females (Plate II, 1). Oestrus is induced by male courtship and multiple ovulation is induced by copulation (Adjanohoun, 1992). Gestation lasts for 152 ± 2 days and can be determined between 4 weeks and 8 weeks of pregnancy by a simple test developed by Adjanohoun (1989, 1992), consisting of the visual examination of vaginal mucus characteristics. If vaginal mucus is red, brown, or yellow the female is pregnant in 100% cases (Edderai and Houben, 2001). This method seems to have more practical applications than the vaginal plug formation after mating suggested by some authors (Adu and Yeboah, 2000), which requires close monitoring of females in breeding groups. Whenever possible, females are separated from the group at the end of gestation and reintroduced into the breeding group after a period of two weeks from weaning. Preventive medication is limited to worm control every 6 months (Jori and Cooper, 2001). Routine management consists of forage supply, livestock inspection, gestation tests of females every fortnight (Adjanohoun, 1989, 1992) and prophylactic hygienic measures.

Animals can be genetically selected according to zootechnical performances such as prolificacy of females, body weight or docile behavior.

Technical Feasibility

From the beginning of the 1970s, several attempts to domesticate the cane rat in different West African countries such as Nigeria, Ghana, Benin or the Ivory Coast failed because the basic biology and behavior of the species was ignored at the time, resulting in high mortality and low reproductivity (Asibey, 1974; Baptist and Mensah, 1986).

Although performances can still improve, the technical feasibility of grasscutter production initially demonstrated in Benin (Schrage and Yewadan, 1995) is nowadays confirmed and a selected breed of animals adapted to captivity (Senou et al., 1999) widespread in many countries (Jori et al., 1995; Jori and Chardonnet, 2002; DABAC, 2003). Reproduction is sufficiently understood to allow a reproductive success of 95% of mated females (Edderai and Houben, 2001) and the animals achieve a marketable weight of 3.5 kg between 8 and 12 months (Edderai and Ntsame Nguema, 2000).

Technical feasibility of cane rat farming in rural conditions (Fig. 3.1) appears unquestionable and the volume of publications on the subject coming from different countries has increased in the last years (Jori and Chardonnet, 2002). Cane rat meat is still in great demand even though sourced from animals bred in captivity that can be culled and consumed under controlled conditions.

Fig. 3.1: Small scale grasscutter-farmer in Benin (photo F. Jori).

Economics

Economic viability depends on the socioeconomic context of the farm. If the farm is situated near urban centers where bushmeat prices and demand are high, a modest-size cane rat farm can definitely be profitable. In Libreville, the capital of Gabon, wild cane rat meat is sold at 2.8 US$/kg (1 US$ = 695 FCFA) but farmed animals are sold at 5 US$/kg without difficulty. Although cane rat meat is rarely found in the city markets in Gabon, it is known that the species is widely consumed and production currently lags behind demand (Jori and Chardonnet, 2002).

A World Bank study showed that small-scale cane rat farming with a yearly stock of 260 animals (40 reproductive females) was the most profitable system of animal exploitation in Ghana, followed by poultry and rabbit farming (Tutu et al., 1996). In Gabon, a farm of this size could easily reach a profitability threshold of between 350 and 400 US$/y with the sale of 14 to 20 animals for meat at 5 US$/kg. Although grasscutter farming can complement the domestic economy with a little extra work, the current level of production in Gabon hardly suffices for earning a living since most farms have less than 30 reproductive females and have not yet reached productivity threshold levels. However, revenues can be substantially increased by improving management methods and by selling live cane rats as stock founders or technical know-how to other farmers. Several authors in various African countries (e.g. Fantodji and Mensah, 2000; Dabogrogo, 2000) seem to agree that a small-scale farm of 40 reproductive females is the most profitable scale of production for that species and that well-managed cane rat farms can substantially contribute to local economies and are sufficiently profitable to ensure a livelihood. This is a crucial point for the development of grasscutter farming in Africa that deserves further analysis or investigation in years to come. Generally speaking, cane rat farming profits are variable depending on the country and location of the farm. Prospects for economic success are higher in periurban areas where demand for bushmeat is greater, transport costs limited, and game sold at high prices. In rural areas, hunting management of wild cane rats certainly shows more promise than farming since these rodents are abundant, and their capture reduces predation on and damage to feeding crops. Nevertheless, prices in rural areas are at least twice lower than those paid in urban centers (Steel, 1994). Therefore, spending money to produce animals that are abundant in the wild seems unrealistic unless hunting is prohibited and the law enforced.

Creation of a Demonstration Pilot Farm

A pilot station could be constructed in a periurban area with the same capacity and type of equipment and buildings that would be recommended for private farmers. One might be tempted to establish the initial livestock for this pilot station from animals captured in the wild. However, initial mortality in previous attempts reached 50% of the stock, zootechnical performances were low, and hence it might take years to achieve the desired level of performances from animals born in captivity (Yewadan and Schrage, 1992). It is therefore definitely recommended to initiate breeding activities with a strain of cane rat suitable for production (Jori et al., 1995; Adu et al., 2000). Such a strain is obtainable from preexisting cane rat farms in West or Central Africa. The staff in charge of the farm should be trained by experienced technicians in several aspects of cane rat farming. Since the 1980s Benin has reached a high level of expertise in captive breeding stimulated by an annual demand higher than 200,000 animals, thereby enabling extension of this activity to rural and periurban areas from 1989. Today, more than 500 cane rat farmers and 16,000 captive animals exist in that country (Mensah, 2000) and new farmers feel encouraged by an interesting autopromotional approach at the local level: experienced farmers teach this activity to new inexperienced ones and provide them with a founding breeding group of one male and four females. Gabon is another country where cane rat farming developed during the 1990s (Jori et al., 1995) is now spreading to neighboring countries such as the Republic of Congo and Cameroon (DABAC, 2003).

The high demand coming from other African countries proves the ease with which cane rat farming know-how can be extended to other parts of the continent.

Besides Benin, experimental and private grasscutter farms exist today in several countries in West and Central Africa, namely Togo, Ghana, Nigeria, Cameroon, Gabon, Ivory Coast, Burkina Faso, Republic of Congo, and South Africa (Jori et al., 1995; Mensah, 2000; Van der Merwe and Van Zyl, 2001; DABAC, 2003). The significant improvement in key production parameters obtained experimentally such as prolificacy of females (5 kits/litter), growth rate of cane rats (8 and 10 g/day for females and males respectively during the first 6 months), and number of weaned kits per annum (10 kits/female/y) described in the literature (Edderai and Nguema, 2000; Edderai and Houben, 2001), is probably the result of a cautious husbandry, improved diet, and genetic selection applied to that species over the last 25 years (Jori and Chardonnet, 2002; Yewadan, 1992). Tameness of the animals can improve considerably if measures to reduce stress are undertaken (MacCoy et al., 1997; Jori and Cooper, 2001). Some animals with a lighter colored or white fur may appear among the stock (Plate II, 2). According to Hemmer (1992), the appearance of lighter colored coats in wild mammals during the process of domestication is a sign of adaptation to captivity. However, further genetic studies are necessary to confirm this correlation.

Recommendations for Initiating an Extension Program

Once a pilot demonstration farm is running, the new farmers come to gather information and knowledge at the demonstration site. This farm should be productive enough to provide the candidate farmers with tamed animals and advice. As well as receiving an overview of daily routines, they are advised about materials and the skills necessary to build their facilities. Once these are ready, the farmers receive a more intensive training course in farming techniques, lasting two to four weeks depending on their level of education. Once technically trained, it is a good idea to start with a small group of castrated males at a subsidized price. After a couple of months, if the males are in good condition, candidates may receive on loan a reproductive group of one male and six females (two of which may be pregnant). Modalities of partnership can differ in each case. However, one possible option is that candidates commit themselves to return to the project a total of seven animals from the first three litters. Initially close monitoring of the project entails two visits per month, reduced subsequently to once a month (after the first litter) and later to bimonthly. With the extension method suggested here, the evolution of farmers and animals in captivity becomes exponential: in a project in Gabon, production increased from 40 animals in a single farm to 52 farms and more than 1,200 grasscutters in captivity after 6 years (Jori and Chardonnet, 2002).

Aside from Gabon, published evaluations of cane rat farming extension programs in Africa are scarce and relate to West African countries such as Benin and Ghana (Adu et al., 2000; Mensah, 2000). Mortality during the first extension trials in Benin was high, reaching 56.4% of the stock (Adoun, 1992; Yewadan,

1995). In Gabon, results among farmers were generally lower than those obtained in the demonstration pilot farm (Jori and Chardonnet, 2002). Mortality was higher and reproductive management to optimize production was lacking. Nevertheless, in some cases litter sizes were surprisingly high (see Table 3.5), reaching 4.6 kits/female/litter. A survey of spontaneous cane rat farms in Ghana (Adu et al., 2000) also reported outstanding prolificacy performances in 58 farms. A possible explanation might be that the animals received good feeding ratios. Adu and Yeboah (2000) suggested a positive role of maternal nutrition in increasing litter size in the grasscutter. Unfortunately, preweaning mortality was high and yearly litters per female low during the first months of the extension activities, reducing numeric productivity of the females. Like other agronomic activities, production performances can be very diverse among farmers and a large margin of improvement often occurs if measures are taken to reduce kit mortality and intensify reproductive management of females, in order to obtain more than one litter/female/y.

Table 3.5: Comparative parameters recovered during different cane rat extension programs in Africa

	[1]Gabon, 1996	[2]Gabon, 1998	[3]Benin, 1993	[4]Ghana, 1992
Number of farms surveyed	10	19	16	31
Total number of litters reported	62	142	—	—
Total number of kits born	283	603	—	—
Global mortality (%)	18.8	26	56.43	—
Preweaning mortality (%)	18.9	40	—	—
Mean litter size	4.47	4.59	—	4.63
Number of weaned kits/litter	3.62	3.04	—	—
Average reported litters /female/year	0.62	0.78	—	—
Numeric productivity/female*	2.24	2.37	—	—

[1]Chardonnet, 1996; [2]Jori, 1998; [3]Adoun, 1992; [4]Adu et al., 2000
* Numeric productivity of females is the number of weaned kits produced/female/year.

Environmental Impact

One of the expected goals of promoting grasscutter farming in Africa is that cane rat meat from captive farms can be offered as a market alternative, which could reduce the demand or consumption of other species of bushmeat (Blom et al., 1993; Wilkie and Carpenter, 1999; Auzel and Wilkie, 2000). Given this assumption, promotion of grasscutter farming around protected areas was expected to noticeably reduce hunting by local populations. Let it be noted that the possible impact of cane rat farming on the demand for wild-caught bushmeat or on the conservation status of bushmeat targeted species has never been demonstrated. However, that does not mean it is lacking. Although it is unlikely that cane rat farming alone can compete with the enormous volumes of bushmeat prevalent in Central Africa, we may expect that an enhancement in volume of cane rat meat from animals bred in captivity could draw a certain proportion of bushmeat consumers. This hypothesis can only be achieved if:

a) many more farms are created and the level of production improved;

b) development of bushmeat substitutes is accompanied by other measures such as taxation and price manipulation (Wilkie and Carpenter, 1999);

c) illegal hunting of protected species is effectively controlled.

From an environmental point of view, cane rat rearing in surrounding protected areas has generally failed due to the lack of profitability mentioned previously for rural areas. To increase motivation among local people living in buffer zones to farm these species, it might be possible to link production with tourist restaurants in areas where ecotourism has developed, inducing tourists to pay high prices to taste an exotic local dish.

Conclusion

Despite the opinion of some authors (Fa, 2000), substantial progress has been made in the domestication and exploitation of the cane rat in West and Central Africa. This activity is technically viable; productivity under experimental conditions is good and can certainly be improved. Productivity and profitability under local small-scale conditions is still lower than in experimental conditions, but the margin of improvement is important. Previous experience indicates that difficulties in promoting cane rat farming are social rather than technical: the level of success and technical appropriation varies according to the individual and the farming tradition of the country. In that sense, candidates from countries with agricultural tradition seem to be easier to motivate, since animal husbandry is a more widespread and traditional activity than in Central Africa. Nevertheless, the potential of cane rat farming for protein production, economic development, and changing the mentality of African populations toward wildlife use cannot be underestimated. The enormous demand of bushmeat provides evidence of the great potential of this activity in periurban areas of West and Central Africa, which will hopefully continue to develop, improve, enhance income earning capability, and increase farm-produced protein intake of periurban populations.

Brush-tailed Porcupine (*Atherurus africanus*)

The brush-tailed porcupine (*Atherurus africanus* Gray, 1842) is a rodent that lives in equatorial Africa and is particularly popular as a game species in its homeland. For that reason this species has often been suggested as a candidate for ranching or farming in order to produce local meat in a sustainable manner (Anadu et al., 1988; Blom et al., 1993; Feer, 1993; Vietmeyer, 1991). However, information on the biology of this species is scarce and very few trials have been undertaken so far to investigate its potential (Rahm, 1962). Since 1994, i.e. a period of six years, the French NGO Vétérinaires sans Frontières, has operated a rodent farming project in Gabon to study the potential of local rodents, including the African brush-tailed porcupine, for rearing in captivity at the local level (Jori et al., 2002). This trial provides the basis for discussing its potential as a minilivestock species.

Distribution and Natural History

The African brush-tailed porcupine (*Atherurus africanus* Gray, 1842) is a rodent from suborder Hystricomorpha, family Hystricidae (Table 3.2). It has an elongated body measuring 40 to 50 cm, short legs, and short rounded ears. An adult in Gabon averages 3 kg in weight in both sexes (Jori et al., 1998).

The species is found mainly in rainforest areas of the African continent, in particular West and Central Africa but also in forests of Kenya, Uganda, and the Democratic Republic of Congo up to an altitude of 3,000 m. It is also present on the island of Bioko (Equatorial Guinea) and other small islands off the African Atlantic coast. The species can live in the deep forest, tolerating secondary vegetation, but is fond of feeding on cultivated crops such as manioc, banana, palm trees, and peanut plantations. Brush-tailed porcupines also feed on forest fruits, roots, carrion, insects, and earthworms (Nowak and Paradiso, 1983). They prefer living near water sources and are good swimmers although they rarely cross these natural barriers. They are nervous, fast-moving animals that can climb trees and jump up to one meter. These rodents are strictly nocturnal and hide during the daytime in natural burrows such as old termite nests, abandoned dens of other animals, empty fallen tree trunks or between tree roots. Adult animals live in small family groups of up to eight to ten individuals of different age and sex in a network of several dens and resting sites. The mean home range is 13.4 ha (Emmons, 1983).

Commercial Use of Porcupine Meat

This African porcupine has few natural predators. Leopards, large eagles, and snakes (*Bitis gabonica*) are among the most common (Kingdom, 1974), but man is certainly its major predator. In Central and West Africa, the meat of the brush-tailed porcupine is consumed in large quantities. In Nigeria, like cane rats (*Thryonomys swinderianus*) it is one of the most popular bushmeat species (see Table 3.1). In Bendel State in 1982 it accounted for 19% of the total species sold by the roadside (Martin, 1983). A questionnaire reported brush-tailed porcupine to be the preferred meat of 29% of the respondents and it is the most expensive species per unit of bushmeat sold in that state (Anadu et al., 1988). In urban centers such as Kisangani (D.R. Congo), rodent meat is also one of the most important in terms of consumption (37% of total records), with *A. africanus* representing 10% of this amount (Colyn et al., 1987). In Equatorial Guinea, other surveys have shown that the bushmeat trade relies heavily on the brush-tailed porcupine and the blue duiker (*Cephalophus monticola)* (Fa et al., 1995; Fa et al., 2002).

In Gabon, the brush-tailed porcupine is also the most popular and abundant game meat species sold in the markets along with the blue duiker (*Cephalophus monticola).* It is also the most expensive meat per kilogram in many African cities (Table 3.1), reaching 5.3 US$/kg in the Libreville market in 1996. According to a WWF evaluation, it accounted for 27% of the recorded game meat in Libreville (Steel, 1994). The same survey gives a consumption estimate of at least 452,099 kg *Atherurus* meat/y in urban areas of Gabon (Jori et al., 1998).

Subsistence Hunting

The African brush-tailed porcupine is a species of choice for villagers hunting at night in Central Africa. Hunting or trapping is unselective, and there is no tendency to avoid shooting immature animals or pregnant females. In rural areas of Northeastern Gabon, brush-tailed porcupine accounted for 11% of all game caught (Lahm, 1993). In the D.R. Congo, *A. africanus* together with *Cricetomys emini* accounts for more that 70% of the species captured and consumed in rural areas (Colyn et al., 1987). It is therefore a favorite meat in most of the rainforest habitats of West and Central Africa. Like the cane rat in West Africa, consumption of its meat does not seem to be associated with any taboo or prohibition (Jori et al., 1995). Almost everything is consumed in this species, except the quills. Initial studies on the dressing-out percentage of *A. africanus* carcasses gave a carcass yield of more than 65% of the total body weight, which is higher than in most domestic and wild species (Table 3.4). Therefore, an adult animal can produce an average of 2,000 g of meat (Jori et al., 1998).

Technical Aspects of Porcupine Breeding

Very little research has been undertaken on the biology and knowledge of this species in captivity (Jori et al., 1998). Investigations first occurred in the D.R. Congo (former Zaire) in the 1960s (Rahm, 1962). But most of the research was produced, during the project implemented by Vétérinaires sans Frontières in Gabon (Jori et al., 1998; 2002; Houben et al., 2000; Edderai and Houben, 2000). There, brush-tailed porcupines are bred with success in two different kinds of paddocks. One consists of two 1 m^2 chambers, made of concrete equipped with top wooden doors for catching the animals and cleaning the facilities, and communicating through an inside hole (Fig. 3.2). This system, also used with cane rats, allows one of the chambers to be cleaned while the animals shelter in the other (Schrage and Yewadan, 1995). A wooden structure such as an empty tree trunk in which the animals can hide and feel secure is recommended, especially

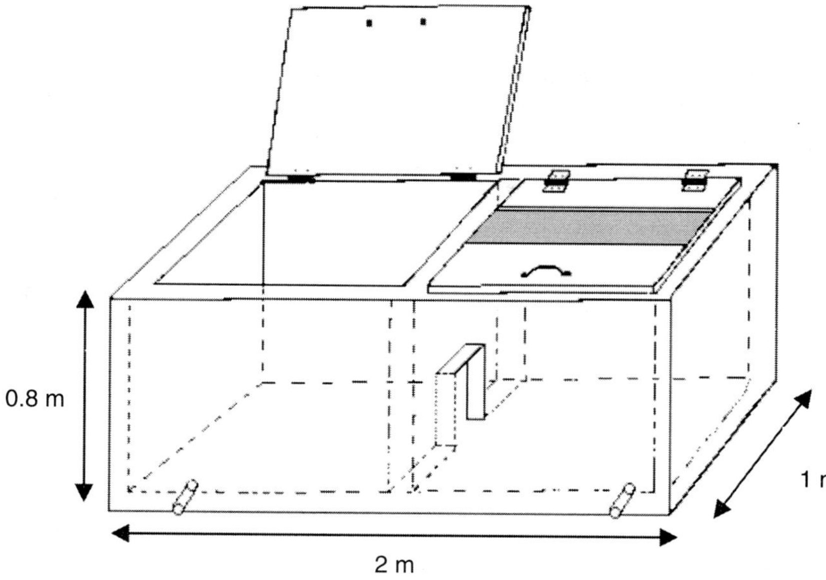

Fig. 3.2: Dimensions and schematic representation of a grasscutter or brush-tailed porcupine paddock made of concrete and wood (*Source:* Houben et al., 2000).

for breeding. Such a structure allows enough space for four or five adults and up to eight weaned young.

The second kind of paddock for breeding brush-tailed porcupine consists of a more natural facility, composed of an external 4 m^2 surface equipped with a 0.2 m^2 wooden hide in which the animals are readily confined (Fig. 3.3). The external walls of the paddock are made of concrete 65 cm in thickness and finished with a 1.15 m wire mesh. The floor can be made of soil since brush-tailed porcupines do not dig holes in the ground. However, some individuals might climb over the wire fence. It is therefore recommended to cover the paddock with wire fencing to prevent escapes.

Captive porcupines can be fed tubers (potatoes, manioc), fruits (pawpaw, palm nuts), dry bread, amaranth (*Amaranthus* spp.), and rabbit pellets (see Table 3.3). An adult individual can eat up to an average of 105 g dry matter per day. Deworming with 20 mg of Fenbendazole per kg twice a year prevents internal parasites. In fact, brush-tailed porcupines adapt readily and few fatalities occur. The fact of being naturally defended by its quills, confers on this animal a higher tolerance to stress compared with other rodents such as the cane rat. This behavior facilitates its management and manipulation (Plate II, 3 and 4) (Houben et al., 2000). Indeed, between 1997 and 1999 only 8.3% mortality was recorded among young animals before weaning (n = 48) and 7.4% subadults and adults (Edderai and Houben, 2000). These data are significantly lower than those registered during the first trials of cane rat domestication (Yewadan, 1992).

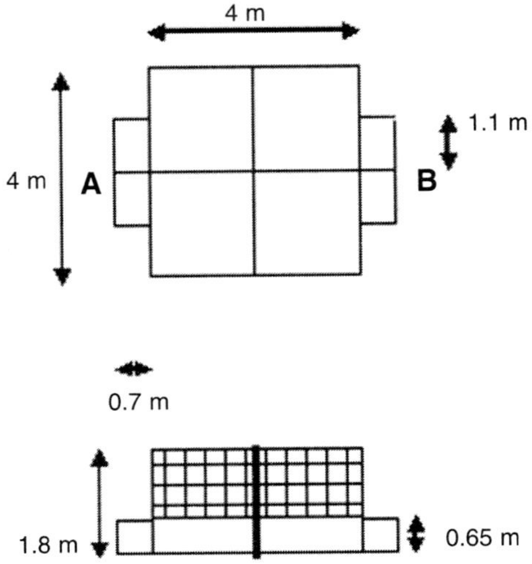

Fig. 3.3: Superior and sagittal view of a seminatural paddock for breeding *Atherurus africanus* (Source: Houben et al., 2000).

Reproduction

Sexual maturity is reached at 11 months by females and 12 months by males. Females are taken singly to the male paddock and precautions taken to protect each from male attacks. The female is isolated from the males after mating to prevent cannibalism of the offspring. Gestation continues for 100 to 110 days (Rahm, 1962). Polygamous groups have not given good results so far since males showed interest only in a single female. However, fertility in females was very high, reaching a success rate of over 80% (Edderai and Houben, 2000). Unfortunately, *Atherurus africanus* rarely gives more than one offspring per birth (Jori et al., 1998, 2002). This weak reproductive turnover is not abnormal in hytricognath rodents (Weir, 1974). However, what is exceptional in this species is that it is polyovular but most of the ova are probably lost before implantation (Jori et al., 2002). Such a low productivity is compensated by long life spans. In addition, the young are very active from the first hours of life. Weaning can be done at 45 days and the animals have usually reached commercial weight (2.5 kg) by ten months of age. Moreover, mortality during production is extremely low compared to the cane rat, rarely exceeding 10%.

Conclusion

Despite its very high popularity as game and its very good adaptability to captive conditions, the brush-tailed porcupine is not a prospective minilivestock species. Indeed, despite regular births in captivity, *Atherurus* females yield only one offspring and obtaining more than two pregnancies/y is difficult (Edderai and Houben, 2000). With such a low reproductive performance and the low success of polygamous groups, the cost of annual feeding and concrete facilities represent respectively 72.75% and 10.31% of the total yearly costs. The income generated by the farming of 9 females with a production turnover of 2 young/y hardly compensates for 24% production costs (Edderai and Houben, 2000).

These performances confirm the low capacity of *Atherurus africanus* as a meat-producing animal. However, some aspects of research on farming this animal still need to be trialed. In particular, the possibility of manipulating the natural behavior of the species by creating polygamous groups of females from an early age might possibly increase productivity of breeding groups in captivity. This method has proven successful in farming the paca (*Agouti paca*), a New World rodent with singleton litters (Smythe, 1996; Hosken, 1999; Govoni et al. this volume).

Cricetoma (*Cricetomys* spp.)

Giant rats or Cricetoma belong to the order Sciurognathi and are abundantly consumed in sub-Saharan Africa (Ajayi, 1974; Malekani and Paulus, 1989; Fa, 2000). Two species are known for the genus *Crycetomys* (see Table 3.2): the Gambian cricetoma (*Cricetomys gambianus*), with a wider distribution in areas of sudano-guinean savannas, and the Emin cricetoma (*Cricetomys emini*) more circumscribed in forested areas (Plate II, 5).

Cricetoma are eligible for minilivestock due to their small size, high reproductive turnover, and limited requirements of space. *C. gambianus* adapts much better to captive conditions than *C. emini*, which has given less satisfactory results to date and for which literature is scant. Average weight gains are also better for the Gambian rat (4.1 g/d) than for the Emin´s rat (2.9 g/d) (DGEG, 2000). Gambian rat studies have concentrated thus far on its utilization of local foodstuffs (Amubode, 1985), its behavior in captivity (Ajayi, 1975), and its reproductive performances (Tchoumboue et al., 2002). Several domestication trials have been initiated on that species in Nigeria, Cameroon, and D.R. Congo since the 1970s.

C. gambianus can be bred in wire cages, provided a dark chamber is maintained to recreate underground conditions (Ajayi, 1975). Cricetoma are particularly sensitive to heat and therefore facilities have to be cool, avoiding temperatures higher than 28°C (Malekani, 2001). In Nigeria, Ajayi (1974) and Anizoba (1982) showed that *C. gambianus* reaches sexual maturity at 5 months of age, with a weight from 1,000 to 1,500 g. After a gestation period of one month, females produced litters of 1 to 5 young, that could be weaned at 30 days of age. The University of Ibadan (Anizoba, 1982) showed that *C. gambianus* has almost the same capacity as rabbits for producing animal protein in tropical countries. In fact, the Gambian *Cricetomys* is able to reproduce year round, with a record of 5 litters observed in 9 months. However, recent reports from breeding trials in

Gabon show that these performances are difficult to reproduce and from 113 breeding trials, only 7 births were successful, resulting in 17 offspring (2.4 young/ female/ birth). There are certain areas in Africa where consumption of Cricetoma reaches important levels and where captive breeding programs could perhaps allow the provision of markets with captive bred meat. This is the case of Bioko Island in Equatorial Guinea where consumption of this species is reported to be 5,000 animals/y for a population of 50,000 inhabitants (Fa, 2000; Fa et al., 2002). It is no coincidence that in this country and on the West Coast of Nigeria and Cameroon, Cricetoma are popularly known as "ground beef". These rodents are also greatly prized in the D.R. Congo where 64% of 455 interviewees declared Cricetoma to be an excellent quality meat (Malekani and Paulus, 1989).

However, results obtained to date in captive breeding experiences are inconclusive and illustrate that more time and research are needed for fully adapting this species to captivity. Reproduction is not totally controlled, the fertility rate of females is low (DGEG, 2000), and neonatal mortality high due to frequent cannibalism (Ajayi, 1975; Tchoumboue et al., 2002). As mentioned, research must concentrate on creating a breed of Cricetoma adapted to captivity as was done for the grasscutter in Benin during the 1980s. This scarcity of data is probably the reason that precise information on the success of Cricetoma as minilivestock animals in extension programs in Africa is nonexistent. Other cultural barriers such as its rat resembling morphology or several African taboos related to their consumption (Ajayi, 1974), may also have deterred funding or technical feasibility, which would explain the lack of progress in farming this prolific rodent after 20 years of research.

General Considerations

The literature concerning some methods that have proven successful in domesticating some popular rodents in Africa has been reviewed. Despite many difficulties, several African populations remain extremely attracted to breeding these species. Sometimes, economic viability is not the main motive for rearing these animals, however: other factors such as social recognition or pride in producing African species domesticated and bred by Africans may also be considered. Another aspect, often overlooked is the social role that minilivestock can play by integrating women, children or old people in new domestic tasks that are a source of income for the household, on a level with guinea pig farming in Andean communities. It is also important to realize that farming might not be the only method nor the most suitable to all circumstances and situations. The introduction of these species of rodents in areas where they do not naturally exist is not recommended and should be avoided, since it can have several ecological implications that may be irreversible. Such is the case of the accidental introduction of the coypu (*Myocastor coypus*) in southern Europe, which escaped from breeding farms more than ten years ago and is still causing damage to the environment, having proved very difficult to eradicate from the ecosystem.

The species reported here and in particular the cane rat, are well adapted to small-scale farming with low economic risk and quick return. Therefore, in most cases, they are suitable for the implementation of extension programs in periurban areas or in rural areas with ready access to urban centers. However, in distant rural areas, all these species are important crop pests, usually attracted by agricultural crops and therefore existing in large densities near villages. In this situation, other systems of production such as organized trapping or hunting activities, can probably provide animal protein to be consumed or sold while protecting agricultural crops and may be more suitable and cost effective than farming. So far, no trials using this approach have been found in the literature, and applied research on this subject using local knowledge would certainly be very useful.

Another subject requiring further research for better development of rodent farming in Africa concerns the diseases and health of these species. Even in the case of the cane rat, whose technology is far better developed than that of other animals, few reports exist regarding the diseases and sanitary aspects of this species in captivity (Adu et al., 2000; Jori et al., 2001; Jori and Cooper, 2001). Some evidence of the role of wild rodents as vectors of zoonotic diseases has been described by several authors, e.g. leptospirosis (Baylet and Van Riel, 1971), hepatic capillariasis (Malekani et al., 1994), toxoplasmosis (Garin et al., 1971), and salmonellosis (Oboegbulem and Okoronkwo, 1990). Although these diseases do not seem to be a great threat to bushmeat consumers (African tradition to overcook the meat would prevent disease transmission), they represent a potential risk to persons manipulating carcasses and meat.

On the other hand, farming rodents implies that a large concentration of living animals are exposed to conditions of stress and concomitantly come into close contact with wild rodents or other wildlife carrying infectious agents. This situation can facilitate the emergence of diseases of sanitary importance for animal keepers and meat manipulators and can also have decimating effects on the captive breeding stock. Such is the case of outbreaks of pseudotuberculosis that have occurred in chinchilla farms (Wilkerson et al., 1997) or in guinea pig farms in Latin America (Gonzalez-Cardona et al., 1989). Similar suspected outbreaks of rat-transmitted diseases have occurred in cane rat farms in Gabon causing high mortality among captive stock. Research in veterinary aspects of captive African rodents is therefore urgently needed if rodent farming is to be developed in Africa with prospects of commercial exploitation. This and many other topics need to be explored in order to optimize them and other systems of rodent exploitation in Africa. However, the progress achieved during the last 30 years in this area is promising. Its continuity in the future should be maintained and certainly represents a significant challenge for agricultural research in the tropics in the 21st century.

References

Adjanohoun E. 1989. Contribution au développement de l'élévage de l'aulacode (*Thryonomys swinderianus*, Temminck 1827) et à l'étude de sa reproduction. Thèse doctorale, Maisons-Alfort, Paris, ACCT, 198 pp.

Adjanohoun E. 1992. Le cycle sexuel et la reproduction de l'aulacode (*Thryonomys swinderianus* Temminck, 1827). Mammalia, 56: 109–119.

Adoun C. 1992. Pre-extension of grasscutter production: Concepts and experience of the Benin/German Grasscutter Project. Proc. 1st Conf. Grasscutter Prod., Cotonou, Benin, pp. 203–209.

Adu E.K. and Yeboah S. 2000. The efficacy of the vaginal plug formation after mating for pregnancy diagnosis, and embryonic resorption in utero in the greater cane rat (*Thryonomys swinderianus* Temminck). Trop. Anim. Health and Prod. 32 (1): 1–10

Adu E.K., Alhassan W.S., and Nelson F.S. 2000. Smallholder farming of the greater cane rat, *Thryonomys swinderianus* Temminck, in southern Ghana: a baseline survey of management practices. Trop. Anim. Health and Prod., 31 (4): 223–32.

Ajayi S.S. 1974. Giant rats for meat and some taboos. Oryx, 12 (3): 379–380.

Ajayi S.S. 1975. Caging and breeding the African giant rat (*Cricetomys gambianus* Waterhouse). J. Inst. Anim. Technicians 25 (2): 75–81.

Amubode F.O. 1985. The relation of growth performance to dietary oil source in African Giant Rat (*Cricetomys gambianus* Waterhouse). Lab. Anim. 19 (2): 134–137.

Anadu P.A., Elamah P.O., and Oates J.F. 1988. The bushmeat trade in southwestern Nigeria: a case study. Human Ecol. 16: 199–208.

Anizoba M.A. 1982. Reproductive cycles of the African giant rat (*Cricetomys gambianus* Waterhouse) in the wild Rodentia. Rev. Zool. Africaine 96: 833–840.

Asibey E.O.A. 1974. The grasscutter, *Thryonomys swinderianus* Temminck 1827, in Ghana. Symp. Zool. Soc. London 34: 161–170.

Auzel P. and Wilkie D. 2000. Wildlife use in Northern Congo: Hunting in a Commercial Logging Concession. *In:* Robinson J.G. and Benett E.A. (eds.). Hunting for Sustainability in Tropical Forests, Columbia University Press, New York, pp. 413–426.

Baptist R. and Mensah G.A. 1986. The cane rat: farm animal of the future? World Anim. Rev., 60(4): 2–6.

Baylet R. and Van Riel. 1971. *Cricetomys gambianus*, nouvel hôte de conservation de leptospires. Médécine d'Afrique Noire 18 (11): 803–805.

Blom A., Alers M.P.T., Feistner A.T.C., Barnes R.F.W., and Barnes K.L., 1993. Primates in Gabon: Current status and distribution. Oryx 26: 223–225.

Brugiere D. 1997. Grande faune du Gabon. Situation actuelle et perspectives. Courrier nature 166: 34–38.

Carpaneto G.M. and Germi F. 1992. Diversity of mammals and traditional hunting in central African rain forests. Agric. Ecosys. Environ., 40: 335–354.

Carpaneto G.M. and Cristaldi M. 1995. Dormice and man: a review of past and present relations. Hystrix (n.s.) 6 (1–2): 303–330.

Caspary H.U. 1999. Wildlife utilization in Côte d'Ivoire and West Africa: Potentials and constraints for development cooperation, GTZ Publ., Eschborn, 183 pp.

Chardonnet P. and Fritz H. 1996. Chasse traditionnelle. *In:* Chardonnet P. (ed.).: Faune Sauvage Africaine: La ressource oubliée. IGF/CIRAD-EMVT, Luxembourg, vol. 1, pp. 88–126.

Chardonnet P., Fritz H., Zorzi N., and Féron E. 1995. Current importance of traditional hunting and major constrasts in wild meat consumption in sub-Saharan Africa. *In:* Bissonette J.A. and Kraussman P.R. (eds). Integrating People and Wildlife for a Sustainable Future. Wildlife Society, MD, Bethesda, USA, pp. 304–307.

Chauca de Zaldivar. 1995. La produccion de cuyes (*Cavia porcellus*) en los paises andinos. World Anim. Rev. 83 (2): 9–19.

Colyn M., Dudu A., and Mbaelele M., 1987. Data on small and medium scale game utilisation in the rainforest in Zaire. *In:* Proc. Int. Symp. Wildlife Utilisation in Africa. IGF/CIC, Zimbabwe, pp. 109–141.

DABAC. 2003. http://www.DABAC.cirad.fr

Dabogrogo. M. 2000. La Ferme pilote d'aulacodiculture de Ouagadougou. Actes du Séminaire International sur l'élévage intensif de gibier à but alimentaire en Afrique. Libreville, Gabon, pp. 167–170.

DGEG. 2000. Rapport d'activité annuel. Vétérinaires Sans Frontières, Lyon, 48 pp.

Edderai D. and Houben P. 2000. L'élevage d'athérures (*Atherurus africanus* Gray 1842) au Gabon. Actes du Séminaire International sur l'élévage intensif de gibier à but alimentaire en Afrique Libreville, Gabon, pp. 97–105.

Edderai D. and Ntsame Nguema M. 2000. Résultats zootechniques de la station d'aulacodiculture d'Owendo. Actes du Séminaire International sur l'Elevage Intensif de Gibier à But Alimentaire en Afrique, Libreville, Gabon, pp. 75–77.

Edderai D. and Houben P. 2001. La gestion de la reproduction en aulacodiculture. Synthèse des outils et méthodes existants. Prod. Anim. INRA, 14 (2): 97–103.

Emmons L.H. 1983. A field study of the African brush-tailed porcupine, *Atherurus africanus*, by radiotelemetry. Mammalia 47: 183–195.

Fa J.E. 2000. Hunting Animals in Bioko Island, West Africa: Sustainability and Future. In: Robinson J.G. and Benett E.A. (eds.). Hunting for Sustainability in Tropical Forests, Columbia University Press, New York, pp. 168–198.

Fa J.E., Juste J., Perez del Val J., and Castroviejo J. 1995. Impact of market hunting on mammal species in Equatorial Guinea. Cons. Biol. 9: 1107–1115.

Fa J.E., Juste J., Burn R.W., and Broad G. 2002. Bushmeat consumption of two ethnic groups in Bioko Isalnd, West Africa. Human Ecol. 30 (3): 397–416.

Fantodji A. and Mensah G.A. 2000. Rôle et impact économique de l'élevage de gibier en Côte d'Ivoire. Actes du Séminaire International sur l'élévage intensif de gibier à but alimentaire en Afrique. Libreville, Gabon, pp. 25–42.

Feer F. 1993. The potential for sustainable hunting and rearing of game in tropical forests. *In:* Hladik C.M., Hladik A., Linares O.F., Pagezy H., Semple A., and Hadley M. (eds.). Tropical Forests, People and Food. Man and Biosphere Series. UNESCO and Partenon Publ. Group, Paris, vol. 13, pp. 691–707.

Garin J.P., Baylet R., Despeignes J., Kien Tuong T., Rioche M., and Correa P. 1971. Recherches épidémiologiques sur la Toxoplasmose humaine et animale au Sénégal. Méd. Afrique Noire 18 (10): 751–754.

Gonzalez Cardona H.G., Neira-Rairan L.R., and Patino-Burbano R.E. 1989. Caracterizacion etiologica y clinicopatologica de la yersiniosis (Pseudotuberculosis) de los cuyes. Revista ICA, Colombia 24 (4): 438–450.

Hardouin J. 1995. Minilivestock: From gathering to controlled production. Biodiv. Cons., 4: 220–232.

Hardouin J. and Thys E. 1997. Le mini-élevage, son développement villageois et l'action de BEDIM. Biotech. Agron. Soc. et Environ. 1 (2): 92–99.

Hemmer H. 1992. Domestication—concept and consequences. Proc. 1st Int. Conf. Grasscutter Production. Benin, pp. 189–200.

Hosken F.M. 1999. Criação de pacas. Coleção Natureza e Negocios, 2. SEBRAE/MT, Cuiaba. 178 pp.

Houben P., Jori F., and Edderai D. 2000. L'élevage d'athérures (*Atherurus africanus* Gray 1842) au Gabon. Tropicultura G.A., 18(1): 40–43.

Jori F., 1997. Etude de faisabilité de l'élevage commercial d'espèces sauvages au Gabon. WWF/GEF/PNUD, pp. 76.

Jori F. and Cooper J.E., 2001. Spontaneous neoplasms in captive cane rats (*Thryonomys swinderianus* Temminck, 1827). Vet. Path., 38 (5): 102–104.

Jori F. and Chardonnet P. 2002. Cane rat farming in Gabon. Status and perspectives. *In:* Ebedes H., Reilly B., van Houven W., and Penzhorn B. (eds.). Sustainable Utilisation—Conservation in Practice. South Africa Game Rancher's Organisation, Pretoria South Africa, pp. 99–109.

Jori F., Mensah G.A., and Adjanohoun E. 1995. Grasscutter production. A model of rational exploitation of wildlife. Biodiv. Cons. 4 (3): 257–265.

Jori F., Lopez-Béjar M., and Houben P., 1998. The biology and use of the African brush-tailed porcupine (*Atherurus africanus* Gray, 1842) as a food animal. A review. Biodiv. Cons. 7 (11): 1417–1426.

Jori F., Cooper J.E., and Casal J., 2001. A survey of pathology in captive cane rats (*Thryonomys swinderianus*). Vet. Rec. 148: 624–628.

Jori F., Lopez-Béjar M., Mayor P., and Lopez C. 2002. Functional anatomy of the ovaries of wild brush tailed porcupines from Gabon. J. Zool. 256: 34–43.

Juste J., Fa J.E., Perez del Val J., and Castroviejo J., 1995. Market dynamics of bushmeat species in Equatorial Guinea. J. Appl. Ecol., 32: 454–467.

Kingdom, J. 1974. Old World Porcupines (Hystricidae). In Kingdom, J. Ed. Esat African Mammals: AN atalas of evolution in Africa, Vol. II, Part B, Academy Press, London, pp. 682–686.

Lahm S.A., 1993. Utilisation of forest resources and local variation of wildlife populations in Gabon. *In:* Hladik C.M., Hladik A., Linares O.F., Pagezy H., Semple A., and Hadley M. (eds.). Tropical Forests, People and Food. MAB Series, UNESCO, Paris, vol. 13, pp. 213–226.

MacCoy J., Jori F., and Stem C. 1997. Tranquillisation of captive cane rats (*Thryonomys swinderianus*) with a long acting neuroleptic (Pipothiazine palmithate). J. Vet. Pharm. Therap. 20: 233–239.

Malekani M. 2001. Guide technique d'élevage n°08. Les cricétomes. *In:* Hardouin J. (ed.) BEDIM, FUSAGx, Gembloux, Belgique, 8 pp.

Malekani M. and Paulus S.J. 1989. Quelques aspects sur la consommation du cricétome, *Cricetomys* (Rongeur), par des populations zaïroises, Tropicultura, 7 (4): 141–144.

Malekani M., Kumar V., and Pandey V.S. 1994. Hepatic capillariasis in edible *Cricetomys* spp. (Rodentia : Cricetidae) in Zaïre and its possible public health implications. Ann. Trop. Med. Parasitol. 88 (5): 569–572.

Martin G.H.G. 1983. Bushmeat in Nigeria as a natural resource with environmental implications. Environmental Conservation, 10: 125–131.

Martin G.H.G. 1985. Carcass composition and palatability of some wild animals commonly used as food. World Anim. Rev. 53: 40–53.

Mensah G.A. 2000. Présentation générale de l'élevage d'aulacodes; Historique et état de la diffusion en Afrique. Actes du Séminaire International sur l'élévage intensif de gibier à but alimentaire en Afrique. Libreville, Gabon, pp. 25–42.

Morosetti G. and Molé S. 1992. Notes on the role of wildlife in the epidemiology of zoonoses. Veterinary Public Health Reports, WHO/FAO/ISS/ IVZTe, Rome, Italy, 93 pp.

Noss A. 2000. Cable snares and nets in the Central African Republic. *In:* Robinson J.G. and Benett E.A. (eds). Hunting for Sustainability in Tropical Forests, Columbia University Press, New York, pp. 282–304.

Nowak R.M. and Paradiso J.L., 1983. Rodentia: Family Hystricidae. *In:* Nowak R.M. and Paradiso J.L. (eds.). Walker 's Mammals of the World. John Hopkins Univ. Press, Baltimore, MD, USA (4th ed.), pp. 794–796.

Oboegbulem S.I., and Okoronkwo I. 1990. Salmonellae in the African great cane rat (*Thryonomys swinderianus*). J. Wildlife Dis. 26: 119–121.

Peters K.J. 1988. Potential of unconventional livestock production. World Rev. of Anim. Prod. 24: 71–84.

Rahm U. 1962. L'élevage et la reproduction en captivité de *Atherurus africanus*. Mammalia 26: 1–9.

Robinson J.G. and Benett E.A. 2000. Hunting for Sustainability in Tropical Forests. Columbia Univ. Press, New York, NY, 571 pp.

Schrage R. and Yewadan L.T. 1995. Abrégé d'aulacodiculture. Deutsche Gesellschaft für Technische Zusammenarbeit (GTZ), Rossdorf, Germany, 103 pp.

Senou M., Yewadan L.T., and Behr K., 1999. Genetic parameters of some economic importance in the cane rat (*Thryonomys swinderianus*). J. Anim. Breed. and Genet. 116 (4): 305–310.

Smythe N. 1996. La domesticación y cría de la paca (*Agouti paca*). Guía de conservación, nº 26. FAO, Roma, 83 pp.

Steel E.L. 1994. Study of the value and volume of bushmeat commerce in Gabon. WWF Programme for Gabon, Libreville, Gabon.

Tchoumboue J., Niba A.T., Zango P., Dafem R., and Teguia A. 2002. Reproductive and growth performance of the Cricetoma (*Cricetomys gambianus*) under captivity. Tropicultura 20 (3): 130–134.

Tutu K.A., Ntiamoa-Baidu Y., and Asuming-Brempong S. 1996. Economics of living with wildlife in Ghana. *In:* Bojo, J. (ed.). The Economics of Wildlife. AFTES working paper n°19, The World Bank, no. 11, 34 pp.

Van der Merwe M. and Van Zyl A. 2001. Postnatal growth of the greater cane rat—*Thryonomys swinderianus*—in Gauteng, South Africa. Mammalia 65 (4): 495–507.

Vietmeyer N.D. 1991. Microlivestock: Little Known Small Animals with a Promising Economic Future. Natl. Res. Council, Natl. Acad. Press, Washington, DC.

Weir B.J. 1974. Reproductive characteristics of Hystricomorph rodents. *In:* Rowlands I.W. and Weir B.J. (eds.). The Biology of Hystricomorph Rodents. Zool. Soc. London, Acad. Press, London, UK.

Wilkerson M.J., Melendy A., and Stauber E. 1997. An outbreak of listeriosis in a breeding colony of chinchillas. J. Vet. Diag. Invest. 9: 320–323.

Wilkie D.S. and Carpenter J.F. 1999. Bushmeat hunting in the Congo Basin: an assessment of impacts and options for mitigation. Biodiv. Cons. 8: 927–955.

Wilson D.E. and Reeder D.M., 1993. Mammal Species of the World: A Taxonomic and Geographic Reference. Smithsonian Inst. Press, Washington DC (2nd ed.), 1207 pp.

Yewadan L.T. 1992. Systèmes d'élevage en aulacodiculture. Proc. 1st Conf. Grasscutter Prod., Cotonou, Benin, pp. 131–142.

Yewadan L.T. and Schrage R. 1992. Les performances zootechniques de l'aulacode (*Thryonomys swinderianus*) en captivité étroite. Proc. 1st Conf. Grasscutter Prod., Cotonou, Benin, pp. 165–175.

4

Rodent Farming in the Amazon: Experience with Amerindians in Venezuela

GUIDO GOVONI[1], DENIS FIELDING[2] AND MAURIZIO G. PAOLETTI[3]

[1,3] Dept. of Biology, Padova University, Via U. Bassi 58/b, 35121 Padova, Italy. E-mail: bw.guido@libero.it and paoletti@civ.bio.unipd.it
[2] University of Edinburgh, Centre for Tropical Veterinary Medicine, Easter Bush, Roslin, Midlothian, EH25 9RG. E-mail: Denis.Fielding@ed.ac.uk

Abstract

Minilivestock represents a promising sector of animal production as it provides the opportunity to cope with the increasing food demand in the developing world. In the state of Amazonas, Venezuela the Amerindian communities are facing conflicts with new settlers for the use of natural resources, which is leading to overexploitation of the existing biodiversity, a reduction in wildlife populations, and transformation of the rainforest into poor quality cultivable land. Reductions in the wild populations of pacas (*Agouti paca*) and agoutis (*Dasyprocta* spp.) induced the enactment of a statute to promote the protection of the paca, the most threatened species. Although subsistence hunting is allowed to indigenous people in the current situation, its original meaning has changed, resulting in progressive depletion of the wild populations. In this context, breeding under human control of pacas and agoutis could offer a chance to local people to increase their incomes and could also represent an opportunity to reduce hunting pressure.

This paper discusses the importance and role of pacas and agoutis in the life of some indigenous people of Venezuela, the realization of a rodent farming project in several Amerindian communities, and the results of a follow-up assessment carried out four years after the initial project's conclusion. Investigation of the local knowledge of the aforesaid rodents obtained by visiting indigenous communities and utilizing participatory techniques, represented an important source of information for drawing up key recommendations and guidelines for the project. Its realization was based on the traditional knowledge and interest shown by the people. Four years after the rodent farming project ended, activity is still underway in the communities formerly involved, as determined by a recent visit to the area. This continuing activity demonstrates the suitability of rodent breeding for the communities concerned. It is concluded that this new activity could represent an example of indigenous resource management which promotes positive interactions between local people and results in long-term sustainable utilization of the natural environment while concomitantly increasing food security.

Key Words: paca, agouti, minilivestock, rodent farming, Amazonas, Amerindians, indigenous knowledge, natural resources, biodiversity, sustainable utilization, food security

Introduction

The human population of the world continues to increase, resulting in ever greater demands on the finite resources available for food production. This situation is leading to the impoverishment of the natural environment and an accelerating reduction in biodiversity. Legislation by governments appears unable to reverse this trend. Food needs and low incomes act as the main factors which induce people, especially in developing countries, to extend their activities into new areas formerly free of human interference. If such areas are to be conserved, the challenge facing governments is to increase and stabilize food production and incomes for the poorest communities so that they have no need to widen their imprint on the ecosystems in which they live. This challenge has to be the main objective if we want to achieve long-lasting results not only in alleviating poverty, but in conservation of the natural environment and its biodiversity.

Animal husbandry represents one method of high-value food production. In traditional systems in the western world the range of animals produced is usually quite narrow. In developing countries this dependence on only a few species, generally domesticated elsewhere, may not be necessary or advisable. There are many wild species in developing countries that could, with appropriate interventions, be developed for food production. Sustainable utilization of these wild species began 15 years ago, albeit very slowly, and a new type of animal production is developing. This new form of animal production, involving numerous species, sometimes described as unconventional by industrialized countries, may represent a significant future means of increased food supply and income, especially for landless farmers and poor people in developing countries who cannot afford activities which require high inputs to initiate.

Properly managed production of wild species should lead to increased food security. Potentially it could also lead to the abandonment of uncontrolled hunting in favor of rational exploitation of the species concerned, thus contributing to conservation of wild populations facing the threat of extinction (Robinson and Redford, 1991).

Among mammals, order Rodentia has great potential as a greater source of human food. Rodents are found throughout the world with many wild species located in tropical forests and savannas. Indigenous people of many countries in Latin America, Africa, and Asia regularly consume wild rodents (Hardouin, 1997; Jori et al., 2004). Their utilization is due not only to a need for food, but is often part of the local culture. This should provide an excellent platform for greater use in future if suitable systems can be developed.

In Latin America several species are raised for human consumption. Among these, guinea pig (*Cavia porcellus* L.) represents the first rodent domesticated as attested to by the remains of hundred of animals discovered in the foundations of every house of a town dated as early as 2500 BC excavated in northern Peru (Kyle, 1987). Even today guinea pigs are bred throughout the Andean region by rural people for integrating into a diet based mainly on subsistence agriculture (Kyle, 1987). Many factors support this activity, such as reproduction rate, low investment, and relative disease resistance. Recent investigations have led to a remarkable improvement in terms of number of piglets born live and weaned as well as live weight (Cicogna et al., 1994). Further researches determined animal size and currently the live weight has moved from 0.75 kg to approximately 2 kg; interest in these rodents is evidenced by the development of some industrial-level farms raising more than a thousand reproductive females (Morales, 1994).

Capybara (*Hydrochaeris hydrochaeris*), the largest living rodent of the world with an average weight of 59 kg, represents probably the best known rodent raised in Latin America (Ojasti, 1991). Consumption of capybara meat is common throughout the range of its distribution and subsistence hunting is widely practiced. The main large-scale exploitation of this animal for city markets occurs in the Venezuelan and Colombian plains called *llanos* (Ojasti, 1991). Other countries such as Argentina, Uruguay, Brazil and Paraguay use capybaras mainly for their hides rather than meat (Ojasti, 1991). In Venezuela there is a long tradition of capybara consumption in the *llanos* where the rodents thrive in large cattle ranches without competition since the two species feed on different types of grass (Kyle, 1987).

Latin American rodents other than paca (*Agouti paca*) and agouti (*Dasyprocta* spp.) presented in this chapter, are locally consumed but are little known to animal production or health officers of western countries. The hutia (*Capromys* spp.), weighing up to 9 kg, merits mentioning. Several species are known, some of them characterized by arboreal behavior (Hardouin, 1995). The mara (*Dolichotis patagonum*), resembling a dwarfed antelope or a tall hare (8 kg average), is another rodent living in the desert plains of Argentina and in Patagonia (Hardouin, 1995).

Two endemic rodents of Latin America presented in this paper, the paca (*Agouti paca*) and agouti (*Dasyprocta* spp.) are well known for their tasty meat across the entire range of their distribution. In particular, they are intensively hunted by Amerindians in the state of Amazonas, Venezuela, traditionally for home consumption but more recently for sale to urban areas where the meat is highly prized.

In the last few years Amazonas has faced remarkable demographic growth due mainly to the immigration of mixed blood Venezuelans, called *criollos*. This immigration has caused conflicts with the local Amerindians concerning the use of natural resources. Excessive exploitation of the wild populations of the aforesaid rodents has occurred and the paca, in particular, is considered to be in danger of extinction in Venezuela, although technically hunting them is illegal. In 1996, a religious institution (Conferenza Episcopale Italiana) financed a project supporting preliminary trials of these two rodents in some villages in the area surrounding the capital. The objective was to encourage the breeding of pacas and agoutis in small farming units managed directly by Amerindians as a backyard activity, after they submitted a document attesting to their interest.

In addition the villages selected for the project have documented cases of semidomestication of paca and "pet" rodents had been kept in the village earlier (Plate III, 1).

This paper analyses the indigenous knowledge about pacas and agoutis, their roles in the life of Amerindian communities in Amazonas, and the implementation of a rodent farming project. It also discusses the results of a recent follow-up which took place four years after the original project's conclusion.

Animal Species Concerned: Paca (*Agouti paca*) and Agouti (*Dasyprocta* spp.)

Paca

Agouti paca (Plate III, 2; Fig. 4.1) is the second largest rodent in the world, the capybara (*Hydrochaeris hydrochaeris*) being first (Perez, 1983); in appearance it

Fig. 4.1: A paca (*Agouti paca*) at right and an agouti (*Dasyprocta* spp.) at left (photo G. Govoni).

looks like a giant guinea pig and adults weigh 6-14 kg (NRC, 1991). It is widely distributed in the Neotropics from central Mexico to northern Argentina and in Venezuela occurs throughout the country, except on Margarita Island (Perez, 1992).

Pacas are a mainly nocturnal, resting during daylight in underground burrows or in hollow trees (Govoni, et al., 2001), but sometimes active early morning and late afternoon (Smythe, 1970). Although they are terrestrial animals living in forests, pacas thrive in areas surrounded by rivers or streams, and water often represents the best means of escape from danger (NRC, 1991). Aggressiveness is a behavioral trait and constitutes the main obstacle to controlled breeding in captivity (Matamoros, 1982; NRC, 1991). Raising pacas in small enclosures and creating small social groups seems to overcome this problem. However, introduction of new animals which alter the preexisting social structure may induce aggressive behavior (Aguirre and Fey, 1981; Perez, 1983; Smythe, 1991).

Reproduction occurs throughout the year and pacas normally give birth once or twice per annum. Usually one offspring is recorded but twins are not unknown (Collett, 1981; Matamoros, 1982; Smythe, 1991).

Nutrition of pacas is mainly based on fruits according to seasonal and local availability and the diet determined more by seasonal abundance than specific taste. Investigations concerning fruits eaten by pacas have been carried out in various Latin American countries and the plant species they consume are listed in Tables 4.1 and 4.2 (Collett, 1981; Gallina, 1981). In captivity pacas feed on a

Table 4.1: Plants listed in Gallina's (1981) study in Mexico

Scientific name Family	**Species**	**Local common name (Mexico)**
Sapotaceae	*Pouteria sapota* Jacq.	Mamey
Myrtacea	*Eugenia uliginosa*	Canastilla
Cesalpinoideae	*Dialium guianense* (Aubl.) Sandwith.	Guapaque
Palmae	*Chamaedora tepejilote* Liebm.	Guatapil
Myristicaceae	*Virola guatemalensis* (Hemsl.) Warb.	—
Moraceae	*Brosimum alicastrum* Swartz	Ramon
Rubiaceae	*Coussarea* sp.	Rubin
Lauraceae	*Licaria capitata* Cham. et Schlecht.	—
Chrysobalanaceae	*Licania platypus* (Hemsl.) Fritsch	Cabeza de Mico
Guttiferae	*Calophyllum brasiliense* Cambess	Bari'
Hippocrataceae	*Salacia belizensis* Standl.	Wo'che
Sapotaceae	*Manilkara zapota* (L.) van Royen.	Chicozapote
Anacardiaceae	*Cymbopetalum penduliflorum* (Dunal) Baill	Orejuelo
Marantaceae	*Calathea lutea* (Aubl.) Schult.	Hoja Blanca
Meliaceae	*Spondias mombin* L.	Jobo
Annonaceae	*Guarea glabra* Vahl.	Cedrillo
Musaceae	*Musa paradisiaca* L.	Platano
Monimiaceae	*Mollinedia guatemalensis*	—

Table 4.2: Common plants utilized by the paca in Colombia

Scientific name Family	Local common name (Colombia)	Parts eaten
Anacardiaceae		
Spondias cf. *mombin* L.	hobo, jogo	Pc.*
Anonaceae		
Anona sp.	anon	F.*
Areaceae		
Attalea regia Mart.	cucurita	Pc.
Bactris spp.	cubarro	Pc.
Mauritia flexuosa L.	moriche	F.
Mauritiella sp.	moriche macho	F.
Oenocarpus mino	seje, pusuy	Pc.
O. cf. *polycarpa*	seje	Pc.
Socratea spp.	araco, chuapo	Pc.
Syagrus inajai (Spruce) Beccari	churrubay	Pc.
Burseraceae		
Bursera sp.	aguacatillo	F.
Protium sp.	nispero	F.
Chrysobalanaceae		
Licania sp.	guaray	F.
Unknown	pendare	F.
Cochlospermaceae		
Cochlospermum sp.	barbascal	I.*
Guttiferae		
Rheedia madrunno	madrono	F.
Lecythidaceae		
Eschweilera sp.	coco de mono	S.*
Gustavia sp.	tejo	F.
Leguminosae		
Dipteryx sp.	serrapia	S.
Unknown	tapir	S.
Meliaceae		
Guarea sp.	cedro	F.
Moraceae		
Ficus sp.	matapalo	F.
Ficus sp.		F.

*Pc.= pericarp, F.= whole fruit, S.= seeds, I.= inflorescence
Source: Collett, 1981.

wide range of foods, such as rice, beans, kitchen scraps, fish, concentrated food, etc. (Perez, 1983; NRC, 1991; Smythe, 1991). An investigation with captive pacas carried out at the Smithsonian Institute in Panama identified a great variety of

Table 4.3: Wild plants eaten by the paca in the Panama study

Scientific name	**Local common name (Panama)**	**Parts eaten**
Anacardium excelsum Bertero et Balb. ex Kunth	espave'	seeds
Astrocaryum standleyanum LH Bailey	chunga, chonta	fruit, seeds
Bactris major Jacq.	palma brava, uva	fruit, seeds
Byrsonima crassifolia L.	nance	fruit, leaves, bark
Cecropia spp.	guarumo	fruit, leaves
Chrysophyllum cainito L.	caimito	fruit
Enterolobium cyclocarpum (Jacq.) Griseb.	corotu'	fruit, leaves
Faramea sp.	madrono	fruit, leaves
Ficus spp.	higueron	fruit, leaves, bark
Guazuma ulmifolia Lam.	guacimo	fruit, leaves
Gustavia superba (Kunth) O. Berg	membrillo	fruit, seeds, leaves
Hymenaea courbaril L.	algarrobo	fruit, seeds, leaves (?)
Inga sp.	guava	fruit, seeds
Ipomea spp.	batatilla	fruit
Miconia bicolor	—	fruit, leaves
Oenocarpus sp.	—	fruit, seeds
Passiflora sp.	—	fruit
Prioria copaifera Griseb.	cativo	seeds, leaves
Scheelea zonensis LH Bailey	palma real	fruit, leaves
Socratea durissima (Oerst.) H. Wendl.	jira, chonta	fruit
Spondias mombin L.	jobo	fruit, leaves
Spondias purpurea L.	ciruela	fruit, leaves
Wulffia baccata (L.) Kuntze	—	leaves

Source: FAO, 1995.

wild and cultivated species of plants consumed (Tables 4.3 and 4.4) (FAO, 1995). This is an encouraging feature for raising pacas in captivity.

Agouti

This species (*Dasyprocta* spp.) (see Fig 4.1) shows many similarities with the paca in terms of distribution, behavior, and nutrition. Numerous characteristics of agoutis are similar to those of pacas. In shape and size it looks like a hare with short ears, and weighs up to 5 kg (Govoni et al., 2001).

The agouti is a diurnal animal with peak activity in the early morning and late afternoon (Emmons, 1990) although it may extend activity into darkness when intensively hunted (Eisenberg, 1989). Home range boundaries are scent-marked by secretions from anal glands (Smythe, 1983) which are smelly; this is a constraint to agouti farming if production near human habitation is intended (NRC, 1991).

Table 4.4: Cultivated plants eaten by the paca in the Panama study

Scientific name	Local common name (Panama)	Parts eaten
Abelmoschous esculentus (L.) Moench	naju'	fruit
Acacia mangium Willd.	—	leaves
Anacardium occidentale L.	maranon	fruit
Ananas comosus (L.) Merr.	piña	fruit, leaves
Anona muricata L.	guanabana	fruit
Artocarpus altilis (Parkinson) Fosberg.	fruta de pan	fruit, seeds
Bactris gasipaes Kunth.	piba', pejivae	fruit
Basella alba L.	espinaca	leaves
Brassica sp.	hoja de mostaza	leaves
Brassica oleracea L.	repollo	leaves
Cajanus cajan (L.) Millsp.	guandu'	fruit, leaves, bar
Capsicum spp.	pimenton	fruit
Carica papaya L.	papaya	fruit
Cassia fistula L.	caña fistula	fruit, seeds
Citrullus lanatus (Thunb.) Matsum. et Nakai	sandia	fruit, seeds
Citrus spp.	citricos	fruit, seeds
Cocos nucifera L.	coco	seeds
Cucumis melo L.	melon	fruit, seeds
Desmodius ovalifolium Wall.	pega-pega, oval	leaves
Dioscorea alata L.	ñame	roots, leaves (?)
Glycine max (L.) Merr.	soya	seeds, milling by-products
Gmelina minutiflora	melina	leaves
Grasses (various spp.)	pasto	seeds, leaves
Hibiscus sp.	papo	leaves, flowers
Ixora sp.	bouquet de novia	leaves, flowers
Leucaena leucocephala (Lam.) De Wit.	—	leaves (?)
Lycopersicon esculentum Mill.	tomate	fruit
Mangifera indica L.	mango	fruit, seeds, young leaves
Manihot esculenta Crantz	yuca	leaves, roots
Musa sapientum L.	gineo, platano	fruit, leaves (?)
Oryza sativa L.	arroz	seeds
Passiflora spp.	maracuya'	fruit
Persea americana Mill	aguacate	fruit, leaves, seeds
Phaseolus vulgaris L.	poroto	seeds
Pouteria sapota (Jacq.) H. Moore et Stearn	mamey	fruit
Psidium guajava L.	guayaba	fruit, leaves, bark
Solanum quitoense Lam.	naranjilla	fruit
Syzygium jambos (L.) Alston.	pomarosa	fruit
Syzygium malaccense (L.) Merrill et LM Perry	maranon curazao	fruit, seeds
Tamarindus indica L.	tamarindo	fruit, leaves, seeds
Vigna unguiculata (L.) Walp.	frjiol chiricano	leaves, seeds
Zea mays L.	maiz	seeds

Source: FAO, 1995.

Reproduction occurs throughout the year but analogous with the paca, birth clusters have been reported during the season of fruit abundance (Vergara, 1980; Henry, 1994). Parturition normally takes place twice each year and twins are a regular occurrence although single births and triplets are not unusual (Smythe, 1978; NRC, 1991; Henry, 1994).

Like the paca, the agouti is a frugivore but its diet also includes a wide variety of fruits, seeds, stalks, leaves, and roots. Field research carried out on Barro Colorado Island in Panama identified at least 36 species of plants eaten by agoutis (Smythe, 1983).

Role of Pacas and Agoutis in the Life of Amerindian Communities in Amazonas

Everyone in Amazonas, whether *criollo* or indigene is familiar with pacas and agoutis. This is advantageous for a project intervention in that knowledge of the rodents already exists within the communities. However traditional knowledge can be difficult to change and build upon. The aim of the original project investigation was to ascertain the generic level of knowledge as a basis for training, if needed, and to investigate the importance of the rodents in the economic life of Amerindians. A prior understanding of the social and ethnic background of the communities was thought to be extremely important before proposing any intervention. In fact the original idea for the project arose from the social problems encountered in the communities in the past.

Amazonas Situation: Demographic Growth and Social Implications

Amazonas state is considered the "last frontier" in Venezuela. According to the data of the Officina Central de Estadistica y Informatica (OCEI, 1992a, b), number of people living there has increased considerably in the last few years. This fast-increasing population is mostly due to the immigration of *criollos* because of improved road access to the area and secondly to the high birth rate. As a result, the population of Puerto Ayacucho, the capital of the state and the only urban center in the region, has increased dramatically. The potential employment opportunities in the urban areas have stimulated indigenous people to move to the periurban areas of the city and this trend is expected to continue.

Although the overall population density of Amazonas is the lowest in the country (0.5 inhabitants/km^2), the high concentration of people in Puerto Ayacucho has led to several problems:

- increased exploitation of surrounding resources;

• food shortages due to limited cultivable land with the consequent need for importation;

• progressive deforestation to obtain land for agriculture;

• uneven exploitation of natural resources between *criollos* and indigenous people;

• loss of cultural and ethnic identity of Amerindian groups.

This pattern of overexploitation is characterized by a destructive environmental impact and negative social consequences, both common to the colonization processes within the Amazonian region (Hecht, 1992; Lopez Hernandez et al., 1997; May, 1992).

Many indigenous groups have moved from the forests to settle in rural communities within the environs of Puerto Ayacucho. These communal relocations are funded by the government or religious institutions. Concrete houses have replaced the traditional *churhuata* built with palm leaves, wood and mud. The native agricultural system based on shifting cultivation and the "slash and burn" method to obtain a plot of cultivable land called *conuco*, is currently being forced toward a shorter cycle. The resting time allowed for *rastrojo*, the green fallow that allows the soil to recover from exploitation is now considerably reduced. Hence this traditional practice, once ecologically sustainable, is facing a dramatic drop in productivity, thereby promoting soil erosion and contributing to "savannization" of the rainforest (Melnyk, 1995; Pimentel et al., 1997; Netuzhilin et al., 1999).

Ethnic Groups in the Project Area

The ethnic groups living near the capital are mainly Piaroa, Guajibo, and Kuripako along with a small number of other tribes. The Guajibo are not autochthonous; they came from Colombia in the recent past and have lived traditionally in savanna areas mainly in the northern part of the state. Conversely the Piaroa, one of the most important ethnic groups of Venezuela, called *Uwotiuja* in their language, are natives still living in their original regions. The anthropologist Mansutti-Rodriguez (1990) identified the basins of the Marieta and Cuao rivers as the areas of origin and defined as Piaroa territories the basins of many rivers such as Sipapo, Cataniapo, Samariapo, Paria Grande, Paria Chiquito etc.—all tributaries of the Orinoco river in its middle reaches (Fig. 4.2). The region delimited by these rivers was the project area, mainly Piaroa territory, so they were the ethnic group involved in the project. The Piaroa, traditionally nomads, are now sedentary, but some individuals still tend to shift among communities for periods of variable length. They are typical forest people, as signified by the term *De'aruwa*, a word they coined to define themselves that means "owner of the forest" (Mansutti-Rodriguez, 1990).

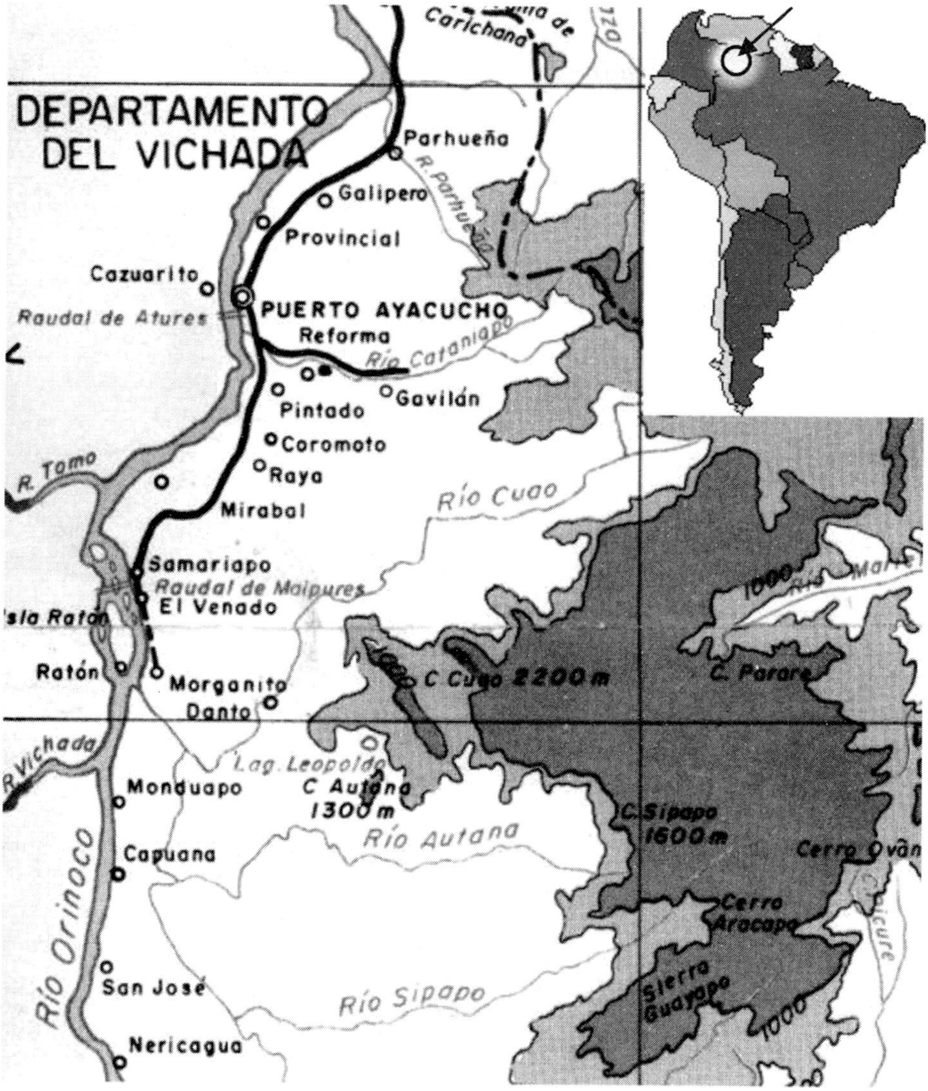

Fig. 4.2: Amazonas State and map of the project area.

Methods of Investigation

The information assembled during the project was collected with the support of the governmental institution *Fondo Nacional de Investigaciones Agropecuarias* (FONAIAP now INIA, *Instituto Nacional Investigaciones Agropecuarias*) of the Ministry of Agriculture and Livestock. The personal acquaintance of the village chiefs with the INIA staff appeared to be the only way to approach the indigenous people who, albeit living near the city remain extremely wary of foreigners, especially "westerners". Hence visits to communities beyond the usual range of INIA required one or two local guides to make the introductions.

Selection of the communities was opportunistic and influenced primarily by three factors:

- Limited time available.
- Willingness of the communities to cooperate with the project. This was determined by the INIA staff according to their experience with the communities in previous developmental activities.
- Availability of local guides willing to introduce two members of INIA and one of the authors of this article (G.G.) to villages located beyond the normal range of INIA activity.

The procedures employed to collect information followed insofar as possible the recommendations and philosophy of the PRA (Participatory Rural Appraisal) and RRA (Rapid Rural Appraisal) methodologies (Chambers et al., Pacey and Thrupp, 1989; IIED, 1994a, b). Because of the time constraint, however, individual semistructured interviews represented the main means of gathering information from the people. Group investigation methods employing PRA and RRA techniques such as transect walking, pie-charts, matrix ranking, and seasonal calendars were used only twice. Altogether 17 communities were visited, as listed in Table 4.5. However, the opportunity to realize four more one-month missions over a period of approximately two years allowed one of the authors (Govoni) to maintain continuity with the project. Once again the role of INIA

Table 4.5: Communities visited and the river basins in which they are located

River	Community	Ethnic group	No. Inhabitants
Cataniapo	Caño Tigre	Piaroa	60-70
	Babilla de Pintado	Piaroa	70-80
	Gavilan	Piaroa	not available
Orinoco	Coromoto del Toboga della Selva	Guajibo	not available
	Samariapo	Piaroa	not available
	Isla Raton	Piaroa	not available
Sipapo	Caño Gato	Piaroa	23
	Pendare	Piaroa	about 300
	Caño Veneno	Piaroa	about 20
	Piedra de Tonina	Piaroa	about 50
	Autana	Piaroa	not available
Cuao	Raudales del Danto	Piaroa	about 200
	Pauji	Piaroa	20
	Selva del Cuao	Piaroa	15
	Coromoto del Cuao	Piaroa	150
	St. Elena	Piaroa	40
None present	Pato Guaiabal	Guajibo	not available

Table 4.6: Timetable of missions and activities carried out

Missions	Main activities carried out
August–September 1996	• Evaluation of the first interventions carried out by FONAIAP from January 1996 • Collection of information about traditional knowledge on pacas and agoutis • Visit to several communities already involved with FONAIAP • First suggestions for pen building and animal purchase
August 1997	• Involvement of further Piaroa communities, considered the most reliable for raising animals • Organizations of participative investigations in some villages • Drawing up of guidelines for the minilivestock production project
June 1998	• Visits to the communities where rodent farming was underway • Sensitization of further villages beyond the usual range of activity of FONAIAP • Organization with FONAIAP of a simple "network" for exchanging information concerning rodent availability, technical advice and management support among villages • Veterinary research on hematic and intestinal parasites • Production of a videotape on the activities underway
November–December 1998	• Monitoring and evaluation of the farming units • Visits to communities which had taken up rodent farming outside the framework of activities financed by the project • Presentation of the videotape to some of the communities as feedback support for the people involved in the project
November 2002	• Information collection about the minilivestock production project • Visits to two communities where rodent farming was still being practiced

was fundamental in keeping contact with the communities involved. A timetable of visits and activities is shown in Table 4.6.

Rodent Minilivestock Production Project

In Venezuela the paca has been considered in danger of extinction since the 1970s and hence has been protected by law (Perez, 1983; Dickinson and Jorgenson, 1991) with alternating periods of temporary and total hunting prohibition. This conservation policy has achieved no consistent result due to lack of enforcement (Perez, 1983). A hunting ban on a species traditionally hunted and consumed in Venezuela merely prompted illegal commercial hunting to supply animals to urban areas where paca meat is highly prized (Govoni, 1998). The establishment of minilivestock production in the Puerto Ayacucho area, including the sale of animals, thus required obtaining special government authorization by the INIA for the project initiated in 1996.

Animal Health Considerations

Health status and sanitary aspects are fundamental issues when rearing animals. Two kinds of problems arise when working with wild species in captive conditions:

- Animal diseases: unknown pathologies may be encountered with consequent damage to the animal stock, which may be exacerbated by grouping animals in a restricted space. Such conditions could also augment existing diseases that are usually subclinical in the wild.
- Human health hazards may occur as a result of contact with, or consumption of, wild species that are potential reservoirs of pathogenic agents.

The former group includes a broad range of pathologies encompassing infectious and parasitic diseases, aberrant behavior, occasional outbursts (trauma, fights), and nutritional disorders. The latter concern mainly zoonotic diseases.

Animal diseases: The main problems threatening paca health in captivity have been identified as stress, intestinal parasites, and respiratory diseases (FAO, 1995).

Stress is considered a principal causal factor of mortality, especially at the beginning of the domestication process when even simple handling of animals can result in their death. Stress is also thought to be responsible for the mortality recorded in animals transferred from their usual place to a new one. Under these circumstances the presence of subclinical pathologies, such as endoparasitosis, can develop quickly and lead to mortality (FAO, 1995).

Human health hazards: Human health risks represent an important factor when rearing wild species, especially when a new development arises. As a general statement, rodents are often hosts of diseases such as leptospirosis, salmonellosis and yersiniosis (Hardouin, 1995), pathologies whose effects on humans are already known.

Neotropical echinococcosis represents another pathology closely associated with pacas and agoutis: *Echinococcus vogeli* and *Echinococcus oligarthrus* produce polycistic hydatids in several internal organs of rodents (D'Alessandro, 1984). Epidemiological investigations by the same author in Colombia on *E. vogeli* showed a high prevalence of infestation in pacas (25%) in comparison with other rodents. Therefore the paca is considered the elective host of the larval stage while the final host of the wild cycle is the bush dog (*Speothos venaticus*). Analogous results have been confirmed by Gardner and colleagues (1988) who also described the first record of *E. vogeli* in Bolivia. Human cases have likewise been recorded (Chigot et al., 1995; D'Alessandro et al., 1996) but probably the pathology is far more widespread than the diagnosed cases would suggest, especially among rural populations. The hazards for human health are represented by the habit of giving, the offal of hunted animals to village dogs which become hosts of the adult cestodes (D'Alessandro, 1984; Chigot et al., 1995), initiating establishment of an urban or semiurban cycle. Information obtained from interviews with hunters in Venezuela confirmed the common presence of internal lesions in pacas, mainly in the liver and referable to hydatid cysts. The hunters further confirmed the custom of giving offal and diseased organs to dogs.

Pacas can also harbor leishmaniasis and trypanosomiasis locally called "Chagas" disease (NRC, 1991); the occurrence of the latter is considered sporadic in rural areas of Venezuela, although this information has not been verified.

According to experiments carried out in Brazil (NRC, 1991) the agouti is highly susceptible to foot and mouth disease (FMD) and, if confirmed, this could be important in controlling this highly contagious disease in other livestock such as cattle.

Identification of Communities Involved in the Project

Many factors were considered in selecting the villages in which rodent farming units were to be set up:

1) interest shown by future farmers,
2) level of local knowledge of pacas and agoutis,
3) previous or current experience as paca keepers,
4) presence of suitable environmental conditions in the village environs,
5) village size,
6) compliance of the entire community,
7) accessibility.

Factors 1, 2 and 3 were the most important because the willingness and reliability of the people were considered crucial to the successful rearing of the animals. A natural environment, as mentioned in point 4, close to the conditions encountered in the wild (presence of water, fruit trees, abundant shade) was preferred to savanna areas with poor vegetation. Selection criteria 5 and 6, closely related, were aimed at avoiding conflicts or envy potentially arising between families. Therefore small communities were chosen, constituted mainly of relatives (Plate III, 3). These were similar to the traditional Piaroa settlements, often replaced nowadays by more populated villages of 100-350 people (Melnyk, 1995). Although local knowledge and experience were important technical supervision was also considered essential for further investigations concerning the minilivestock production to be carried out and hence some attention was given to criterion 7, accessibility.

Husbandry Methods

Almost all villagers preferred pacas rather than agoutis and therefore the following husbandry guidelines are mostly applicable to the former. However, as closely related species sharing the same environment, it is thought acceptable to utilize the same methods for agoutis with modifications where necessary.

Enclosures The farmers were dealing mainly with originally wild animals and therefore attention the surrounding environment was particularly emphasized. As Smythe (1991) asserted about the "founders" of a new colony, the main aim is to produce offspring which can be trained and domesticated rather than to modify the behavior of the parents. For this reason spacious pens (10 x 10 m) were designed to recreate the conditions, insofar as possible, encountered in the wild. A large water pond was also set up. Fruit trees within the pens played a duol role; protecting the animals from direct light and high temperatures and acting as a source of food (Plate III, 4).

Diet Pacas and agoutis, as mentioned before, feed on a great variety of fruits and plants. Indigenous people know their feeding behavior very well and therefore it appeared sensible to delegate animal feeding to the people, who fed the rodents products cultivated locally or gathered in the forest. A list of plants cultivated in the study and suitable for the rodents is given in Table 4.7 (Broder, 1984).

Table 4.7: Plants cultivated in the Piaroa *conuco* of the Caño Tigre community

Local common name	**Scientific name**	**Piaroa name**
Yuca dulce	*Manihot esculenta* Crantz	attahua ire
Yuca amarga	*Manihot esculenta* Crantz	te ire
Mapuey	*Dioscorea trifida* L.	ogh juare'
Ocumo	*Xanthosoma sagitifolium* (L.) Schott	muppa'
Batata	*Ipomoea batata* L.	muiriä
Name	*Dioscorea alata* L.	teasa juare'
Auyama	*Cucurbita moscata* Duch.	kähuiya
Platano	*Musa paradisiaca* L.	otoo paruru'
Topocho	*Musa* sp.	manu' paruru'
Cambur	*Musa* sp.	reo paruru'
Lechoza	*Carica papaya* L.	mäpäyä
Parchita	*Passiflora* sp.	chämure
Naranja	*Citrus* sp.	naranja
Cocura	*Pouruma cecropifolia*	nei'
Cana de azucar	*Sacharum oficinalis* L.	näjä
Tupiro	*Solanum tupiro*	nuä
Temare	*Pouteria caimito* (Ruiz et Pav.) Radlk.	jumare
Guama	*Inga splendens* Willd.	ruhuä
Merey	*Anacardium occidentale* L.	ärära'
Pijiguao	*Bactris gasipaes* Kunth.	päjare
Pina	*Ananas comusus* L.	käna'
Algodon	*Gossypium barbedense* L.	pujä
Maiz	*Zea mays* L.	yämin
Aji	*Capsicum frutescens* L.	ratte'
Cilantro	*Coriandum sativum* L.	cilantro
Aguacate	*Persea americana* Mill.	äppä
Onoto	*Bixa orellana* L.	mono'
Limon	*Citrus* sp.	limone'

Source: Broder, 1994.

Animal management: The daily activities of minilivestock production (cleaning, feeding, drinking) were assigned to one family, which was also responsible for the maintenance of the pen. The farmer families became the owners of the animals purchased by the project but they had to provide them food, look after them and be responsible for their good and conscientious management, and allow technical supervision by the project staff. As the food was already cultivated or gathered, the only input required was the labor in attending to the animals, while the outputs were represented by the sale of the offspring. The labor involved should not be underestimated. However, the families committed to the project spontaneously offered their cooperation and therefore used their spare time to look after the animals.

It was thought useful to constitute farming units of the same species and of no more than 5 animals, made up of 4 females and 1 male, to avoid overcrowded pens and to reduce intolerance problems. Furthermore, in the case of fighting between animals it was possible to separate one or more of them with a temporary wire net wall, reducing the enclosure size and leaving the animals meanwhile in sight of each other. This technique was also used at parturition. Particular attention was paid to the difficult task of forming settled polygamous groups. The management technique adopted was to keep the polygamous groups of animals together as far as possible unless specific problems of aggression arose.

Animal health: The sanitary aspect was an important part of the project in terms of animal disease and human risk control. Implementation of a program of disease control was difficult because the local people were not traditionally livestock keepers and therefore were not aware of the fundamental principles of animal health. As a result the basic sanitary requirements (deworming, ectoparasite control, prevention of respiratory diseases) was left to the technical staff of the project through periodic monitoring of the communities. Basic laboratory equipment was available at INIA as well as essential material to carry out simple investigations such as feces examinations and hemoparasite analysis . This was possible through the involvement of a biology student adequately trained to undertake routine investigations and monitoring of basic parameters of minilivestock production.

Results

The project was carried out from January 1996 to December 1998. Because of limited funding it was not possible to keep expatriate personnel in Venezuela. Only short sojourns were affordable. This constraint hampered activity coordination as well as routine monitoring.

The first mission took place August–September 1996 to analyse and identify the causes of the failure of the initial attempts to raise the rodents in captivity. When the project was initiated the animals were raised in very small enclosures near INIA buildings and not in the villages. The investigation revealed lack of involvement of indigenous people and the use of unsuitable husbandry methods. At that time visits to several Amerindian communities allowed the establishment of relationships with the people and collection of local knowledge about pacas and agoutis. This played a fundamental role in subsequent field research because acquaintance with the indigenous people enhanced their willingness to cooperate, thus considerably facilitating the investigations.

Gradually, the involvement of more communities allowed a wider exchange of information not only of indigenous knowledge, but also about rodents already kept by some villagers. The spontaneous creation of this unexpected "network" played an important role in accelerating the establishment of rodent units. Formation of the aforesaid polygamous groups became easier because animals could be exchanged according to needs and availability. However, obtaining animals was difficult because the rodents were hunted with tracking dogs and/or weapons. Live animals were rarely captured. Those caught uninjured were usually young ones (Plate III, 1). Young animals are easier to tame and they adapt well to captive conditions. But, taking into account that puberty occurs at approximately 9 months for females and 1 year for males (Collett, 1981), the interval between setting up a unit and producing offspring is quite long. At the end of the project rodent farming was established in 7 communities and 26 animals were being raised. Among these only 3 pacas were born in captivity. Results are detailed in Table 4.8.

Table 4.8: Communities involved and rodents raised

River basin	Community	Paca	Paca born in captivity	Agouti	Rodents/ village
Cataniapo	Caño Tigre	2			2
Sipapo	Caño Gato	3	1		4
	Pendare	3		1	4
	Caño Veneno	4	1	2	7
	Piedra de Tonina	2	1	1	4
	Autana	2		1	3
Cuao	Selva del Cuao	2			2
Total rodents		18	**3**	**5**	**26**

Only 3 of the 7 communities built pens as recommended by the project. The others set up enclosures utilizing local materials, especially wood. The recommended dimensions were not respected although basic requirements for space and water provision were met.

The basic principles of animal husbandry such as hygiene, feeding, watering, fundamental animal care, especially after parturition and at the time of introduction of new individuals, were sufficiently followed to ensure good animal management.

No serious problems of aggression were recorded during the project realization and the people looking after the animals were able to raise them with neither severe fights between pacas nor killing of offspring as reported in the literature (Matamoros, 1982). Simple and temporary interventions such as separation of animals as described above were successful in solving aggression problems. Indigenous knowledge about rodent behavior combined with good husbandry conditions resulted in successful breeding units.

Follow-up Four Years Later: Rodent Farming Still Underway

In November 2002 a follow-up expedition to the Minilivestock Production Project was carried out. The official conclusion of the project was December 1998, 4 years earlier. In-between no significant communication was maintained nor further funding sent. Attempts to obtain information about the rodent farming units proved futile. The 2002 mission, concerned with developing different studies, presented a good opportunity for visiting some of the rodent production units. Let it be noted that the INIA, due to modifications in the national development strategy, had considerably reduced its intervention and support range, especially for foreign-funded projects. Thus the rodent units received no support from the end of the project in December 1998.

Given this scenario it was wholly unexpected to find some minilivestock production units still well established and working. It was only possible to visit two of the communities concerned. The choice was opportunistic according to accessibility and the availability of transport. In both communities the pens, built following the project guidelines, held several pacas. Unfortunately no records had been kept to indicate total production over the period. However, the villagers reported that animals had been sold each year since the project ended, but getting quantitative data was definitely not possible.

Further information collated both from the INIA personnel and from indigenous people of communities formerly involved in the project confirmed that rodent farming is still underway and that pacas are regularly bred in other villages. All the interviewees asserted that only pacas are raised except in one village where agoutis are raised as well. It was not possible to verify the information concerning minilivestock production but the responses of the interviewees cross-checked and showed good uniformity.

Discussion

Pacas and agoutis are traditionally known throughout their Neotropical range and Robinson and Redford (1991) included both species among the most important mammals for subsistence hunters in this area. This study confirmed their important role in the economy and culture of the Amerindians but their functions differ because pacas are considered a source of cash while agoutis represent food. The sale of pacas can provide money to buy supplies, food or "city goods" such as stereo equipment and refrigerators, given that electric power is now largely available in some villages. Conversely agoutis, more common and their meat less prized, are usually eaten and their sale sporadic. As a result, people are more interested in rearing pacas than agoutis.

Setting up animal production units in Amerindian communities involved social and cultural aspects that couldn't be overlooked. A basic issue, for instance, is this: the villagers are traditionally agriculturists, hunters, and gatherers and not animal growers. Animal breeding is unknown in their culture; they normally keep some wild animals as pets but not for business. According to Piaroa tradition, animals do not belong to humans but have their home in certain mountains called *urou* where the animal spirits live in the form of wind. Hence, Piaroa people may eat animals but not become their owners (Mansutti-Rodriguez, 1997). Accordingly, any effort to raise animals would appear useless, but it is important to emphasize that the project was run among cultured communities, aware of the "western world", in which some traditions have been modified. As reported by Mansutti-Rodriguez (1997), the tapir is a sacred animal in Piaroa tradition, but all the people contacted in the present investigation considered it a most challenging hunting target. The pacas kept as pets were not eaten by their owners but were usually sold. In such a context paca farming fitted perfectly into the role attributed by the people to this rodent.

Several factors hindered the realization of a broader research effort. The irregular assistance provided to farmers represented the main constraint and the lack of adequate project funding and shortage of means (boat, cars, etc.) at INIA hampered a wider sensitization and deeper involvement of the communities concerned. Investigations such as health studies, productivity parameters, reproduction improvements, record keeping and cost-benefit analyses were not done. Moreover farmer education in building pens and in basic husbandry techniques were being put into practice in just a few communities; in the others most of the initiatives were left to the farmers. Hence the presence in some villages of unimproved pens (Fig. 4.3), suitable for keeping rodents true, but certainly not ideal for successful breeding of wild animals.

In situations such as those encountered in Amazonas and probably common to many developing countries time management is difficult because local people have only a general concept of time.

Methodologically, as far as possible, transfer of the technology approach, or "top/down" philosophy which has often led to the failure of development projects offering technology and "know-how" to people who do not recognize in those solutions the answers to their problems, has been avoided. Cooperation with the Amerindians rather than their simple assistance considered an important aspect of the project. It also afforded an opportunity to increase our understanding of local knowledge. From the interviews, for instance, new information emerged, namely the existence of two varieties of paca identified by different names in the Piaroa language—*Bua-tuwa* the larger and *Tua-tuwa* the smaller—

Fig. 4.3: A traditional unimporoved pen in a Piaroa village (photo G. Govoni).

although the validity of this assertion has no support in the literature. The rapid changes occurring even in remote regions tends to produce homogeneous environments and leads quickly to the disappearance of indigenous knowledge threatening the whole culture and traditions of rural populations. People of rural areas worldwide possess considerable knowledge of their natural environment and make use of a vast variety of plants and animals (Paoletti and Bukkens, 1997). With reference to Venezuela, for instance, Finkers (1986) listed 238 species of birds and 62 mammals eaten by the Yanomami, and 131 wild plants and 135 wild animals recognized by the Piaroa (Melnyk, 1995a).

In spite of the failure to achieve many of the "technical" objectives relevant to rodent farming, the biggest worry of the whole intervention, namely, the reliability of Amerindians in attending to their animals proved to be no problem at all, as demonstrated by the recent visit. The findings of the follow-up are not satisfactory as scientific data but they do confirm the interest of the indigenous people in raising rodents, despite the fact that neither economic support nor technical supervision were provided for over four years.

Wildlife Management: A Step Toward Resource Security?

In the current project wildlife conservation was thought most feasible by encouraging its sustainable utilization by communities rather than by banning its exploitation. A rational use of natural resources represents an important opportunity to improve the livelihood of indigenous people, alleviating the poverty often encountered in rural areas. In this context the fundamental challenge is to conserve the biodiversity, extremely rich in tropical forests, and to increase meanwhile food production, ensuring reasonable distribution and accessibility to a balanced diet for everyone (Branckaert, 1995). Based on this project it is too ambitious to assert that wildlife farming is the best option for using wildlife resources. It represents only one possible method to exploit rationally and in a sustainable manner the wild populations. But the conservation of the paca in the wild is not necessarily enhanced by the project.

As asserted by Hinojosa (1992), writing about wildlife management in the Amazonian lands of Ecuador, particular interest has to be given to the manner of associating the use of natural resources with the needs of the local population. So far, in the state of Amazonas, Venezuela, nothing has been done in this direction except to ban the hunting of endangered species. Managed hunting, even if partially effective, is not sufficient to avoid the massive exploitation that is occurring. The main factors hampering managed hunting are poor government support thus far for biodiversity conservation, both in terms of policies adopted and in terms of funding, lack of personnel to control hunting, and lack of infrastructures to establish a concerted network among the communities. Moreover, small animals living in the forest and hunted in high numbers represent a further obstacle to establishing in the short term a feasible plan. Nevertheless it would be desirable to associate both methods (farming and managed hunting) which could then act in a complementary way for a more effective impact on wildlife conservation as well as on the support of rural communities. As stated by Redford and Padoch (1992) "the fates of the forest and the rural poor are often inextricably linked" and therefore biological conservation must encompass the human population as well.

Because of funding constraints the small size of the project described here probably represents its main limitation. Nevertheless the interest shown by the involved communities and their constancy in carrying on minilivestock production represent a powerful stimulus for encouraging further work on a more quantitative basis. Although a cost-benefit study is not available, the continued presence of rodent farming units indicates the effectiveness of this backyard activity as a means of improving living conditions. The rodents act as a cash reserve for dealing with periods of food scarcity or unexpected problems. According to this principle, it is not unreasonable to assert that paca farming may represent a step toward food security for the Amerindians who traditionally depend on "slash and burn" agriculture, gathering, and hunting as the core of their existence.

Although previous trials with Amerindians in wildlife utilization have not been undertaken in Amazonas, an adequate management of the natural environment could be more profitable than cutting trees for short-term agriculture and cattle production, thus converting rainforest into a poorer savanna landscape managed by fire.

Conclusions

This project, undertaken in the state of Amazonas, Venezuela, provided new information on pacas and agoutis in terms of their importance in the life and customs of the local people and their traditional knowledge concerning these rodents. Unfortunately, more intensive research on some interesting productivity parameters was not possible. Nevertheless the project has shown the feasibility of such interventions in the Amazonian context. Taking into account the potential opportunities deriving from the wildlife management described, further interventions are justified to expand the findings of this small-scale trial. Such interventions are likely to contribute to improving the living conditions of the people involved and concomitantly contribute to a more sustainable utilization of wildlife and natural resources.

Acknowledgments

We gratefully acknowledge the financial support of Conferenza Episcopale Italiana (CEI), project 26/96, for giving us the opportunity to implement this project. We extend our thanks also to the staff of Instituto Nacional de Investigaciones Agropecuarias (INIA) of Puerto Ayacucho, Venezuela, for their help during the field work. Their support allowed us to visit the indigenous communities and the realization of participatory investigations which otherwise would not have been possible. In particular, special thanks to the agronomist, Frans Torres, and the technicians, Jesus Garcia and Ramiro Royero. Inigo Narbaria and the FUDECI offered invaluable inputs.

References

Aguirre G. and Fey E. 1981. Estudio preliminar del tepezcuintle (Agouti paca nelsoni) en la selva Lacandona, Chiapas. *In:* Reyes Castillo (ed.) Estudios Ecologicos en el Tropico Mexicano. Instituto de Ecologia, Mexico, pp. 45–54.

Branckaert R.D. 1995. Minilivestock: sustainable animal resource for food security. Biodiv. Cons. 4 (3): 336–338.

Broder J.C. 1984. El sistema de produccion del conuco Piaroa. Un estudio en la comunidad indigena de Caño Tigre. School for International Training. Venezuela: Natural and Cultural Ecology. Independent Study Project (unpubl.).

Chambers R., Pacey A., and Thrupp L.A. 1989. Farmer First. Farmer Innovation and Agricultural Research. Intermediate Technology Publ.; Southampton Row, London.

Chigot J.P., Menegaux F., Hammoud R., Nozais J. P., and Hoang C. 1995. Report of a case of *Echinococcus vogeli* infection successfully treated by surgery. Semaine des Hopitaux, 71 (13–14): 429–431.

Cicogna M., Castrovilli C., Rigoni M., and Crepaldi P. (1994). Effect of different breeding and mating systems on the productivity of the guinea pig, a minilivestock raised for meat production in the tropics. Rivista di Agric. Subtrop. Tropi. 88(1): 129–144.

Collett S.F. 1981. Population characteristics of *Agouti paca* (Rodentia) in Colombia. Publ. Museum Michigan State Univ. Biol. Series 5 (7): 485–602.

D'Alessandro A. (1984). Equinococcosis Neo-Tropical en el hombre y los animales. Medicina Tropical y Parasitologia, 1 (1) 7–16.

D'Alessandro A., Moraes M.A.P., and Raick A.N. 1996. Polycistic hydatid disease in Brazil: report of five new human cases and a short review of other published observations. Revi. Sociedade Brasileira Med. Trop., 29, (3): 219–228.

Dickinson J. and Jorgenson A. 1991. Demasiado de algo bueno: alternativas para la production animal sostenible. Ganaderia y recursos naturales in America Central, pp. 129–140.

Eisenberg J.F. (1989). Mammals of the Neotropics. Univ. Chicago Press, Chicago, IL vol. 1.

Emmons L.H. (1990). Neotropical Rainforest Mammals. A Field Guide. Univ. Chicago Press, Chicago, IL.

FAO, 1995. La domesticacion y cria de la paca (*Agouti paca*). Guia FAO Conservacion 26, Rome.

Finkers J. 1986. Los Yanomami y su sistema alimenticio (Yanomami Nii Pë). Vicariato Apostolico de Puerto Ayacucho Ed., 261 pp.

Gallina S. 1981. Contribucion al conocimiento de los habitos alimenticios del Tepezcuintle (Agouti paca Lin.) en Lancaja-Chansayab, Chiapas. Estudios Ecologicos en el Tropico Mexicano. In: Reyes Castillo P. (ed.). Instituto de Ecologia, Mexico, pp. 57–67.

Gardner S.L., Rausch R.L., and Camacho O.C.J. 1988. *Echinococcus vogeli* Rausch and Bernstein, 1972, from the paca, *Cuniculus paca* L. (Rodentia: Dasyproctidae), in the Departament de Santa Cruz, Bolivia. J. Parasitol. 74 (3): 399–402.

Govoni G. 1998. The Paca (*Agouti paca*): General characteristics and farming opportunities. Bull. BEDIM, 7(2): 16.

Govoni G., Fielding D., and Paoletti M.G. 2001. Paca (*Agouti paca*) and Agouti (*Dasyprocta* spp.)—minilivestock production in the Amazonas State of Venezuela: 1. Biology. Tropicultura, 19(2): 56–60.

Hardouin J. 1995. Minilivestock: from gathering to controlled production. *Biodiv. Cons.* 4(3): 220–232.

Hardouin J. 1997. Developing minilivestock as source of human food, animal feed or revenue: a brief overview. Ecol. Food Nutr. 36: 95–107.

Hecht S.B. 1992. Valuing land uses in Amazonia: Colonist agriculture, cattle and petty extraction in comparative perspective . *In:* Redford K.H. and Padoch C. (eds.), Conservation of Neotropical Forest. Working from Traditional Resource Use. Columbia Univ. Press, New York, NY, pp. 379–399.

Henry O. 1994. Saison de reproduction chez trois Rongeurs et un Artiodactyle en Guyane francaise, en fonction des facteurs du milieu et de l'alimentation. Mammalia. 58(2): 183–200.

Hinojosa F.C. 1992. The Cuyabeno Wildlife Reserve: Human needs and natural resource conservation in the Ecuadorian Amazon. *In:* Redford, K.H. and Padoch C. (eds.), Conservation of Neotropical Forest. Working from Traditional Resource Use. Columbia Univ. Press, New York, NY, pp. 245–258.

IIED. 1994a. RRA Notes. Special Issue on Training. Int. Inst. Environ. Develop., London, UK.

IIED. 1994b, RRA Notes. Special Issue on Livestock. Int. Inst. Environ. Develop., London, UK.

Jori F., Edderai D., and Houben P. 2004. Potential of Rodents For Minilivestock in Africa (this volume).

Kyle, R. 1987. A Feast in the Wild. Kudu Publ. Kidlington, Oxfordshire, UK, 203 pp.

Lopez Hernandez D., Garcia-Gaudilla M.P., Torres F., Chacon P., and Paoletti M.G. 1997. Identification, characterization and preliminary evaluation of Venezuelan Amazon production systems in Puerto Ayacucho Savanna-Forest Ecotone. Interciencia 22(6): 307–314.

Mansutti-Rodriguez A. 1990. Los Piaroa y su territorio. Documento de Trabajo, no. 8. Centro Venezolano de Investigaciones en Antropologia y Poblacion (CEVIAP). Haydee Seijas (ed.)

Mansutti-Rodriguez A. 1997. Los Gerentes de la Selva. La Iglesia en Amazonas Año XVIII, no. 77, pp. 44–48.

Matamoros Y. 1982. Investigaciones preliminares sobre la reproduccion, comportamiento, alimentacion y manejo del Tepezcuinte (*Cuniculus paca brisson*) en cautiverio. Actas VIII Congreso Latinoamericano de Zoologia, Merida,Venezuela, pp. 961–993.

May P.H. 1992. Common Property Resources in the Neotropics: Theory, Management Progress and an Action Agenda. *In:* Redford K.H. and Padoch C. (eds.), Conservation of Neotropical Forest. Working from Traditional Resource Use. Columbia Univ. Press, New York, NY, pp. 359–378.

Melnyk M. 1995. Productos forestales comestibles, una oportunidad para el desarrollo sustentable. Amazonas Modernidad en Tradicion. GTZ/CAIAH-SADA Amazonas, pp. 295–310.

Melnyk M. 1995a. Contribution of forest foods to the livelihoods of the Huottuya (PIAROA) people of Southern Venezuela. PhD Thesis, Imperial College, London, UK.

Morales E. 1994. The guinea-pig in the Andean economy: from household animal to market commodity. Latin Amer. Res. Rev. 29 (3): 129–142.

Netuzhilin I., Cerda H., Lòpez-Hernàdez D., Torres F., Chacon P., and Paoletti M.G. 1999. Biodiversity tools to evaluate sustainability in savanna-forest Ecotone in the Amazonas (Venezuela). *In:* Reddy M.V. (ed). Management of Tropical Agroecosystems and Beneficial Soil Biota. Science Publ. Inc., Enfield, New Hampshire, USA, pp. 291–352.

NRC (National Research Council) 1991. Microlivestock: little-known small animals with a promising economic future. NRC, National Acad. Press, Washington DC., pp. 193–205 and 263–269.

OCEI. 1992a. Tiempo de Resultados año 2, No. 2. Officina Central de Estadistica y Informatica, Caracas, Venezuela.

OCEI. 1992b. El Censo '90 en el Territorio Federal Amazonas. Officina Central de Estadistica y Informatica, Caracas, Venezuela.

Ojasti J. (1991). Human Exploitation of Capybara. *In:* Robinson J.G., and Redford K.H. (eds). Neotropical Wildlife Use and Conservation. Univ. Chicago Press, Chicago, IL, pp. 236–252.

Paoletti M.G. and Bukkens R.F. 1997. Minilivestock: Sustainable use of biodiversity for human food. Ecol. Food Nutr., 36(2–4): 95–396.

Perez E.M. 1983. La lapa: recurso natural aprovechable. Trabajo de Ascenso, Universidad Simon Rodriguez, Venezuela. 64 pp.

Perez, E.M. 1992. Agouti paca. Mammalia Species. Amer. Soc. Mammol., 404: 1–7.

Pimentel D., McNair M., Buck L., Pimentel M., and Kamil J. 1997. The value of the forests to world food security. Human Ecol., 25(1): 91–120.

Redford K.H. and Padoch C. 1992. Preface. Conservation of Neotropical Forest. Working from Traditional Resource Use. Columbia Univ. Press, New York, NY, 475 pp.

Robinson J.F. and Redford K.H. 1991. Subsistence and commercial uses of wildlife in Latin America. *In:* Robinson J.G. and Redford K.H. (eds). Neotropical Wildlife Use and Conservation. Univ. Chicago Press, Chicago, IL, pp. 6–23.

Smythe N. 1970. Ecology and behaviour of the agouti (*Dasyprocta punctata*) and related species on Barro Colorado Island, Panama. Unpubl. PhD diss., Univ. Maryland, College Park, MD, 201 pp.

Smythe N. 1978. The natural history of the Central American agouti (*Dasyprocta punctata*). Smithsonian Contribs. Zool. 257: 1–52.

Smythe N., 1983. *Dasyprocta punctata* and *Agouti paca* (Guatusa, Cherenga, Agouti, Tepezcuintle, Paca). Costa Rican Natu. Hist. Univ. Chicago Press, Chicago, IL, pp. 463–465.

Smythe N. 1991. Steps towards domesticating the paca (*Agouti=Cuniculus paca*) and prospects for the future. *In:* Robinson J.G. and Redford K.H. (eds). Neotropical Wildlife Use and Conservation. Univ. Chicago Press, Chicago, IL, pp. 202–216.

Vergara S. G. 1980. Cria Masiva y Explotacion del Aguti en Cautiverio. Revista Esso Agricola, pp. 1479–1496.

5

Frogs as Food

GIANLUIGI NEGRONI

Alveo Co-operative Society, Via Guerrazzi 18, 40100 Bologna, Italy
E-mail: sig1564@iperbole.bologna.it

Abstract

Frog populations are declining worldwide due to widespread environmental degradation and wild frog population capture. Frog leg imports are estimated to have been around 8,000 t y^{-1} in 2001 in the European Union (EU) member states and 30,000 throughout the world. Frog consumption is an ancient tradition in most wetland areas. France has the highest European consumption of frogs. The main frog ethnoculture traditions are described with possible commercial, small-scale, and traditional frog breeding techniques. There is no commercial frog production in Europe; consumption depends mostly on imported frogs. Production is concentrated in South America (farming) and Asia (especially for wild frogs).

Key Words: frog, frog farming, frog collecting, frog marketing, frogs and the environment

Introduction

Frog breeding and capture: a general view with particular reference to the European situation.

Why Breed Frogs? It is probable that man has fed on wild frogs, as on many other local edible animals (insects, snails...) for a very long time. In all marshland areas, frogs have been or are consumed. In some areas, the customs of the past remain the same today. Nowadays wild frogs are decreasing rapidly due to uncontrolled capture, increased pollution of the environment, and scarce protection by law (Anonymous, 1981; Hayes et al., 2002). The international frog market is constantly provided with wild frogs. Live and processed frogs are imported from several countries—India, Turkey, Bangladesh, Albania, China, Malaysia, Brazil, Indonesia and others—by consumer countries such as North America, Japan and the EU.

Generally, people eat frog thighs which are cooked in various ways, but in some non-EU countries people eat the whole frog, except for the entrails. Frogs, in addition, are also bred for scientific purposes, since they are excellent experimental animals.

Frog Biology

Frog biology is briefly summarized here for cultivated frogs (family Ranidae); the reader may kindly refer to specialized texts for an in-depth analysis. The *Ranidae* family, commonly called the true frog or typical frog, comprises more than 250 species (Negroni, 1991) distributed throughout the continents and includes the most well-known amphibians (genus *Rana*). Frog usually live close to water and when suddenly frightened take refuge by jumping into it. From these general characteristics we understand that frogs can live for long periods both in and out of water but must always have a humid environment.

Other genera are raised, e.g. *Xenopus* and *Leptodactylus*, but their performance on a commercial scale is generally not good. The giant Goliath frog (*Conraua goliath*) is the largest in this family, more than 34 cm (13.3 in) in length, and found in the African rainforest. It is very difficult to find; the author worked for one year in African rainforests and never saw one. Another strange frog is the burrowing frog (*Rana adspersa*) that lives a long part of its cycle underground waiting for the rainy season: it is common in the southern desert areas of Africa and is considered a local delicacy.

It may be noted that mature frogs have a large stomach, with strong muscles and a mouth that can swallow large prey species.

No frogs live in seawater (Negroni, 1997). Tables 5.1 to 5.5 list the most commonly eaten (according to one author's experience) and raised frogs divided according to continent (Negroni, 1991). The list needs to be continually updated as more data become available.

In many countries the *Rana catesbeiana* imported has been raised in frog farms; some have also escaped and adapted to the local climate. Importing frog species different from the local ones and escape of frogs from farms need to be strictly controlled and monitored to avoid biodiversity problems and exotic disease diffusion.

Data in Tables 5.1 to 5.5 were compiled from contributions made by many frog farmers, scientists, colleagues, and personal experience of the author during several field projects.

Table 5.1: South American frogs

Name	Nose-cloaca length cm (in)	Breeding	Note
Batrachophrynus microphthalmus Werner	20 (7.8)	Under study	Completely aquatic, cold resistant
Caudiverbera caudiverbera (L.)	20 (7.8)	No	CL
Ceratophrys cornuta (L.) (horned frog)	15 (5.9)	No	CL
Leptodactylus fallax Müller	15 (5.9)	No	CL
Leptodactylus labyrinthicus (Spix).	20 (7.8)	Under study	CL
Leptodactylus ocellatus L.	25 (9.8)	Intensive	CL, wetlands
Leptodactylus pentadactylus (Laurenti)	25 (9.8)	Under study	CL
Leptodactylus macrosternum Miranda-Ribeiro	20 (7.8)	No	CL
Telmatobius culeus (Garman) (Titicaca frog)	20 (7.8)	No	Completely aquatic, cold resistant
Bufo marinus (L.) (see Plate III, 5)	18	No	Alto Orinoco CL*

CL = Consumed locally.
*Considered poisoning species by Australians

Table 5.2: North American frogs

Name	Nose-cloaca length cm (in)	Breeding	Note
Rana aurora (Baird et Girard)	12 (4.7)	No	Not often CL
Rana catesbeiana Shaw	20 (7.8)	Intensive	Most diffused frog for farming, wetlands
Rana grylio (Stejneger)	16 (6.2)	No	CL
Rana pipiens Schreber	15 (5.9)	Intensive	Utilized for research, CL, wetlands
Rana heckscheri Wright	15 (5.9)	No	CL

CL = Consumed locally.

Capture of Wild Frogs and the World Market

Historically there have always been frogs and consequently frog consumers in the proximity of marshlands. Unfortunately, in developed countries, due to intensive capture and increased pollution, frog abundance is constantly decreasing. Vast land reclamation projects at the beginning of the last century took extensive natural areas away from frogs, leaving them only limited, wet, and often highly polluted areas.

Table 5.3: European frogs

Name	Nose-cloaca length cm (in)	Breeding	Note, all in humid and wetland areas
Rana ridibunda Pallas	10–15 (3.9–5.9)	extensive, small-scale	CL
Rana esculenta L.	10–15 (3.9–5.9)	"	"
Rana dalmatina Fitzinger	10–15 (3.9–5.9)	"	"
Rana lessonae Camerano	10–15 (3.9–5.9)	"	"
Rana temporaria L.	10–15 (3.9–5.9)	"	"
Rana graeca Boulenger	10–15 (3.9–5.9)	"	"
Rana latastei Boulenger	10–15 (3.9–5.9)	"	"

CL = Consumed locally.
Rana esculenta and the other species are often hybridized. The genus of the various *Rana* species in some EU areas is under debate by zoologists.

Table 5.4: Asian frogs

Name	Nose-cloaca length cm (in)	Breeding	Note
Glyphoglossus molossus Günther	10 (3.9)	No	CL
Rana acanthi Taylor	10 (3.9)	Extensive in rice fields	CL
Rana blythi Boulenger	10 (3.9)	No	CL
Rana hexadactyla Lesson	13 (5.1)	No	Exported
Rana magna Stejneger	13 (5.1)	No	CL (Philippines)
Rana moodiei Taylor	13 (5.1)	Extensive	CL (Philippines)
Rana tigrina Boulenger	25 (9.8)	Intensive and extensive	Most grown Asian frog, exported
Rana crassa Jerdon	20 (7.8)	Intensive	Exported
Rana limnocharis Boulenger	15 (5.9)	Intensive	Exported

CL = Consumed locally.

Table 5.5: African frogs

Name	Nose-cloaca length cm (in)	Breeding	Note
Conraua goliath (Boulenger)	> 30 (> 11.8)	No	The largest (giant) frog eaten, CL, very rare and present in the central African rainforest.
Conraua robusta Nieden	14 (5.5)	No	CL
Pyxicephalus adspersus Tschudi	22 (8.6)	No	Aggressive with painful bait, CL
Rana fuscigola Duméril and Bibron	> 8 (> 3.1)	No	CL
Rana vertebralis Hewitt	15 (5.9)	No	CL
Xenopus muelleri (Peters)	> 20 (> 7.8)	Intensive	CL, wetland area
Xenopus laevis (Daudin)	20 (7.8)	Intensive in lab.	CL

CL = Consumed locally.
Source of Tables 5.1 to 5.5 (Negroni, 1991).

As mentioned above, the market demand in the EU, North America, and Japan is greater than frog farm production, which is why thousands of tons are imported.

The present world market demand for frog meat appears to be around 30,000–35,000 t y^{-1} or about 200 million frogs (Aquaculture Magazine Buyer's Guide, 1997), but actually these figures underestimate the real market and frog consumption since worldwide local catches in swamps and wetlands consumed by local catchers, are not included in the official statistics.

Methods for frogs fishing vary. One of the commonest is flock cotton grafted on a normal hook that is baited. The frog attracted by this movement, bites the hook. It is a well-known fact that anurans prefer live food and mobile prey.

Another method consists of capturing of frogs at night with a small fish hand net, or any net, blinding them with a beam of light. However, this technique is easier in theory than in practice. As a matter of fact, the frog freezes in the beam of light, but as soon as it hears a sudden movement immediately disappears into the water where it is very difficult to catch. Nevertheless a little experience and skill improves capture results. Other, more or less complicated methods for capturing frogs exist but are not discussed here.

Wild frog capture is regulated by local laws that generally do not allow capture during the animal's reproduction period. In swamp areas or rice fields, frogs are usually captured by a single fisherman who can catch an average of 2-3 kg per day. It is easier to catch them while mating during the breeding season and also during the dry season, but both should be prohibited.

Wild Catch Management

Until now wild catches have produced the bulk of world consumption and are practiced mainly in Asia (notably Indonesia and China due to the low cost of manpower) but unfortunately without appropriate management of wild frog resources. The main causes of depletion of wild frog populations are the combination of overfishing, pollution, and habitat destruction.

These three factors have had a considerable influence on the decline of traditional frog populations in several areas. Frogs are also an important link in the chain of ecology of swamp areas and must be protected since they are predators of several species of insects, snails and any other prey they are able to swallow. At the same time they are prey for some birds, fishes and snakes. Proper management of frog resources must be organized in the frog-catching areas to protect them from pollution and overcollecting during the reproduction period, and by banning the destruction of wetlands.

Frogs and the Environment

Frogs are also used as environmental indicators since their presence guarantees clean water and limited pollution (Baker, 1998). Indeed, frogs have a very delicate skin that can absorb pollutants that kill them. The eggs are likewise very delicate and develop into tadpoles only in clean water. In developed nations chemicals used in agriculture create severe problems for frogs living in wetland areas close to or within the treated land area (Hayes et al., 2002).

Alternative to the Decline of Frog Populations in Nature

Increased consumption of frogs, their overfishing, decrease in wetland areas and pathogens (Piotrowski et al., 2004) are the cause of rarefaction in nature. For a long time trials to breed frogs produced significant practical results, mostly from the experimental point of view: from the commercial point of view the results in developed countries have been deceptive. In fact the frog is an anuran whose intensive breeding requires considerable labor, qualified and exceptional technical knowledge, and a climate with a hot season. In certain cases, above all in countries of the Northern Hemisphere, it is difficult but not impossible to access all the optimal conditions for a profitable frog farm: climate (long hot season), technical skills, low labor cost, frog feed availability, and appropriate marketing.

Generally tropical and subtropical countries have more suitable conditions for breeding frogs.

The Possibility of Breeding Frogs

For over fifty years many people have worked to breed frogs, especially in the United States and Japan, but some valid practical examples (generally in smaller and larger operations) can be seen in other countries such as Brazil, Taiwan, Thailand, and China (Negroni, 1996) and recently Malaysia and the Democratic Republic of Congo (on a much smaller scale).

The technology for medium to small breeding farms has been defined but is not without problems, e.g. finding quality tadpoles. Adult frogs, as is well known, prefer live food (Lester, 1988) but pelleted feed is also acceptable. Some 2000 commercial frog operations have been successful, with the most interesting examples found in Asia and South America.

Main Problems in Frog Breeding

The major problems in frog breeding can be summarized as follows: finding good-quality tadpoles, supplying an adequate quantity and quality of food, preventing theft, and avoiding a too high density of frogs (to limit stress and pathologies, particularly those of bacterial origin). We have to add to these technical considerations, all those aspects which derive from a good daily management of aquaculture farms, such as daily check of water quantity and quality, continuous availability of food and its adequate distribution, presence of grass in the raising area, health checks on frogs, etc. Overall a well-managed project is the basis for obtaining good technical and economical performances on a frog farm (Negroni, 2000).

Many breeders think that when tadpoles are placed in the water all the rest of the cycle proceeds without human intervention! This is incorrect as there are many cases in which the carelessness of the breeder has caused the end of a cycle and resulted in heavy economic losses. Sometimes intervention of a frog specialist is needed to solve specific technical problems.

The growing-out of frogs is considered the most difficult part of frog raising, requiring as in many aquaculture operations, the most care and attention.

Frog and Frog Product Marketing Possibilities

Food Consumption

Most frogs bred and captured throughout the world are consumed by humans. In European countries generally only small frog legs are utilized since this is the traditional European table size. But in France and North America, larger size legs, imported from Asian countries, are common. Frog thighs are sold and imported fresh or frozen and, depending on availability, labor cost and market demand; fresh and frozen thighs are more economical than the whole animal. Markets with the highest frog demand are located in traditional frog fishing areas, which are concentrated near marshlands. All the European wetland areas are traditionally consumers, from Sicily to the Danube delta and in particular in areas with a French cultural influence. Other important consumers and importers among EU member countries are Belgium, The Netherlands, Italy and Spain. Frog thighs are considered a "gourmet" food and always fetch a high price in restaurants. In non-European countries, the frog is usually eaten whole, after slaughtering, skinning, and degutting. In fact the quarter front-head-legs though slightly less muscular than the hind legs, is equally flavorful (see Appendix 5.1). In Europe frogs are mostly slaughtered at a weight of 60–80 g (see Appendix 5.2), but some non-European ethnic minorities prefer larger sizes slaughtered at 150 g body weight. In non-European countries, consumption of frogs weighing several kilos (captured in the wild) is not a rarity. European consumers believe that smaller frogs are superior in taste to larger ones (Texteria et al., 2001).

Another very interesting market for frogs is training and scientific research, with frogs sold not by the kilo but by numbers. In the United States several tens of thousands of frogs are sold annually for scientific and educational purposes at an average price (retail) of approximately $ 20 each, which can be considered an attractive price. Statistics about frog trade for research in Europe could not be found.

Some researchers need uniform groups, in certain cases clones. For example, some researchers require a particular group of *Rana pipens* that is susceptible to a specific disorder such Luke's adenocarcinoma.

Although there is a demand for frogs in scientific research in Europe, quality frog suppliers are not yet available; hence some research institutes produce their own frogs, mainly belonging to genus *Xenopus*. Since there is a lack of frog producers or frog distributors for scientific research in Europe, 13 • is the normal retail price for one adult frog in Italy for example, for research purposes.

Another interesting use of frogs is as environmental indicators due to the fact mentioned before that they are very sensitive anurans that cannot survive in polluted water. The skin, very delicate and permeable to pollutants, does not permit their survival in fouled water. As noted above, the eggs produced are also very delicate and even a low percentage of pollutants suffices to arrest development of the larval stages in them.

The Skin

Although it is not the first use to come to mind when we think of frog farming, utilization of the frog skin is a very promising business. In fact, appropriately treated frog skin is suitable for the fashion industry, even though not yet world renowned. Italy and other EU countries are the center of the world fashion industry, but Brazil and Thailand also use frog skins for fashion. Obviously an infrastructure system and a very high level of competence is required for skinning and tanning frog skins to obtain a product of consistently high quality.

I worked in a frog bracelet firm, facing such problems as maintenance of constant supply, high skin cost, low quality of raw skin, and poor processing; only one bracelet could be made from one large skin. Often the skins had holes and were unusable. The Italian market, one of the hardest to please for fashion leather, has accepted frog skin bracelets tanned in several colors. The gross price of one bracelet to the retailer was 5 • a few years back. Skin tanning and dyeing is a costly task especially if high-quality skin has to be obtained for realizing higher retail prices. Wallets, purses, belts, bikinis, jackets and other products can be manufactured from the amphibian processed skins; but it should be remembered that the fashion industry can change its materials and product supply almost overnight.

Possibility of Hind-Legs Quarter Utilization and Other Frog Products

Usually in Europe only the hind-legs quarter is utilized for human alimentation. In other countries, frog sausages and burgers are produced from the whole frog excluding the hind legs. The flesh has a pleasant flavor and is well accepted by consumers; specialties based on frog products can also be sold, e.g. "pâté de foie". Obviously, before starting production of this novelty, it is right and proper to carefully study the market and the consumer. South Americans have the most imaginative frog-derived products (as many as one hundred), proof of market demand.

Selling Tadpoles, Food for Frogs, and other Production Methods for Breeders

A good market exists for the supply of production resources to frog breeders who require tadpoles, frog parents, food for frogs, and equipment. Supplies must generally be supported by good technical assistance to avoid losses. A good pair of adult frogs may cost US $ 30 or more; tadpole prices are not known as there is no market for them at the moment but in Italy can be estimated at ca. 7 US cents.

Frogs as Pets

Every species of frog can be kept as a pet and the market is very promising; keeping a frog in a little aquarium or in a terrarium is a good hobby for adults and children alike. *Xenopus* albino species fetch nice prices in pet shops. Frogs with strange shapes, colors and sizes are much appreciated by people who keep pets. The pet market demand is expected to rise in future.

Frogs for Games

Frogs are also utilized in games: one of the oldest known games is "The Sacramento international frog jumping competition" held yearly in Sacramento (CA, USA). Local televisions broadcast the event. Competitors from all continents participate in the international jumping games entering several frog "champions" each; after three days of competition they select the ten finalists. The frogs are put on a plastic disc and set free, then the distance after three jumps is measured (sometimes the frog jumps backwards!). The frogs are then immediately captured by a person dressed as a "cowboy". The secret to having the most agile frog is manifold: generally big frogs are not needed, frogs of no more than 200 g perform well. In 2000 a boy of 17 won second prize against very well organized Japanese and Irish competitors (with more than 10 trained frogs each). He had captured the frog from a lake near his home. This is the spirit of the competition. In Europe no one has taken up the idea yet.

Frogs for Food Consumption in the EU and the USA

Frogs are mainly consumed as food in Europe and the USA. Gross prices in the USA are about 7 • kg^{-1} for imported live frogs f.o.b.; in Italy 4 • kg^{-1} for imported live frogs but 20 • kg^{-1} for clean fresh legs (supermarket price). Annual frog imports in the EU are variable, around 10,000 t y^{-1}. The total EU consumption purportedly includes the local frog catch (consumption by catchers) but no estimation is available of this. The level of local catch in importer countries is not measurable but without doubt is lower compared to imports. It is more a hobby because in the EU and the USA it is hardly economical to catch a few kilos of frogs per workday.

The general gross volume price in the EU for frozen legs is 7.50 • (± 2 •) and according to size (6–8, 41–50 • per kg).

Frog Import-Export in Europe

Frog legs import-export in value (•) and quantity (ton) from 1988 to 1995 from Eurostat. Main EU importer France (4,072 t) and Belgium-Luxemburg (3,547 t) followed by the Netherlands (455 t) and Italy (456 t). Among other countries only Spain merits mention for frog leg imports (257 t).

Eurostat statistics do not separate live frogs from frozen frog legs. We know that frogs imported from Asia (Indonesia, Taiwan, China, and Vietnam) are all packs of frozen legs, while those coming from Turkey, Albania, Egypt are generally whole live frogs. Frog exporters state that "live frogs" could be estimated at around 20% of the total imported.

Frog statistics Eurostat, 2001, reported that France (3,418 t), Belgium (2,633 t), and the Netherlands (1,222 t) were the main importers and Indonesia (3,971 t) and China (760 t) the main exporters. In 2001 the total imported frog legs was ca 8,000 t (mainly frozen) but there is no differentiation in the EU statistics.

The declared import of frogs (all types) in EU countries over a 10-year period ranged from almost 5,000 t y^{-1} to more than 10,000 t y^{-1} (Table 5.6).

In 2002, several traders requested a supply of frozen frog legs because of the scarce amount on the market due to the low supply from the main Asian producers. All Asian frogs are imported frozen.

Possibility of Increasing Frog Consumption in Europe

Without doubt a product labeled "frog" must have strong advertising support to stabilize and/or increase its consumption. In fact, some of the good qualities of frogs that give them an attractive marketing position are summarized below:

—flesh has a pleasant delicate taste, neither strong nor fishy, which appeals to the modern consumer;

—frog combines well with several types of fish and vegetables, for which hundreds of recipes are available;

—frog has a good biochemical and nutritive composition in line with modern diets with very low fat content (less than 1%);

—frog wild catch is declining in the world;

Table 5.6: EU imported frog products (according to Eurostat)

Quantity (t)	Year
10,024	1989
6,219	1990
8,144	1991
8,978	1992
4,870	1993
5,952	1994
10,237	1995
*	1996
6,968	1997
9,702	1998
*	1999
*	2000
8,014	2001

*Not available

—frogs are important in the swamp areas ecological cycle, so the wild catch must be regulated;

—a high-level commercial frog production technology exists to provide a supply for developing markets;

—frogs can be farmed in a sustainable manner to respect the needs of the modern consumer with appropriate certification;

—slaughtering methods are highly sophisticated so frogs do not suffer during the process.

All the foregoing support the frog's attractiveness as a product that has not been appropriately advertised. It is strange that frog consumption and farming are concentrated only in areas of traditional frog consumption, i.e., people living in wetland areas and a limited number of gourmets.

European frog consumption is definitely associated with frog meat availability to buyers and these are the areas needing consideration:

—a standard certified product with a known trademark for frog quality;

—elicitation of advertising from importers and producers;

—more attractive "frog products" for EU consumers, not just "frog legs";

—frog traceability.

The distribution system for frogs is not yet well organized in Europe, which adds to the final prices charged from consumers. Frogs are classified as "gourmet" in fish markets so quality, presentation and packaging are very important to attract more consumers. Frog is consumed more often in restaurants on special occasions than at home. Many consumers, especially women, are negatively influenced by the slimy appearance of frogs in the market but change their minds on seeing frog legs in a restaurant.

Many consumers are put off by the task of cleaning and cooking frogs, so processed frog meat would have a better consumer acceptance.

In my experience there are good opportunities for the frog trade. Consumers are currently searching for particular food delicacies and willing to pay more for them.

Frog Processing

Frogs are either imported frozen (legs) and prepacked or live, then processed and sold as fresh legs. The latter, like the frozen legs, are packaged in two forms given the attitude of some consumers (mainly in the EU), who prefer only the hind quarter.

EU Sanitary Legislation for Frog Imports

Interestingly the new EU food and hygiene legislation such as regulation 96/340/EC lays down specific public health conditions applicable to trade in and imports of frog legs intended for human consumption, while 91/493/EEC and 94/356/EC and others provide detailed rules for frog trade and processing (imported and in the EU) under the applied HACCP rules and regulations.

All the processing factories in the EU and abroad must adhere to the above legislation and regulations and have an EU number to enable periodic monitoring for hygienic compliance with requirements. Severe sanctions may be applied if legislative requirements are not observed.

Organoleptic and microbiological controls (aerobic plate count, coliforms, *E. coli*, *Staphylococcus aureus* analysis with ISO standards) must be done on representative samples for every batch processed. *Salmonella* could also be a problem in low-quality processing plants, originating from frog viscera in contact with the flesh.

Final Marketing Considerations

Frog imports already represent the major part of frog consumption in Europe; it is very difficult to evaluate the local European catch as it is consumed only locally or is merely a sport. The cost of labor in the developed world makes frog capture uneconomical. Non-EU countries export live and frozen frogs to Europe, Japan and North America. Frogs (generally all the consumed species) are also part of a "green" campaign in some European states (e.g. Germany) due to bad slaughtering conditions in the exporting countries; in others (France) they are well protected and breeding is controlled. Moreover frogs are considered a "gourmet" product and as such very difficult to clean and cook, so only specialized consumers are attracted to frog flesh. It would be very interesting were certification for farmed frogs (not caught in the wild) imported ones, and a traceability system imposed.

Frog Breeding: Technologies

As mentioned in other publications (Negroni, 1987), the breeding of frogs is divided into different systems: extensive, semiextensive, and intensive. As in other types of breeding, each of the aforesaid systems has its own special requirements, which depend both on the natural characteristics of the selected sites and on economic factors, e.g. the cost of raw materials and labor. The type of facilities could be divided into inundated (completely covered by water) or semidry (with only a canal or small area of water) (Flores Nava and Vera, 1999). In choosing technology for frog breeding, apart from the aforesaid factors, others could be mentioned among which the most important factor is the frog breeding entrepreneur. He hopes not to be confronted by natural, economic, and human obstacles and has to evaluate how best to avoid them: if his assessment is correct, his undertaking will be a technical and economic success. In Europe there are very few frog breeders and as the farm is generally a small-scale extensive system, the results of their activities are not known. South America and Southeast Asia have several examples of successful small owner-operators and commercial frog farms in inundated and semidry systems.

Extensive Frog Breeding

In developed countries, the catching and restocking of frogs in rice fields produce a remarkable quantity of these precious tadpoles. Obviously extensive systems of production as a result of biological and meteorological factors cannot guarantee a quantitative and qualitative production. Frog rice field restocking is important for ecology and in an integrated pest management system in rice fields frogs are very important in controlling rice pests (insects, snails...).

It must be remembered, however, that pond rice fields and marshes are subject to pollution or chemical treatment and thus not suitable in most cases for frog breeding.

Semiintensive Frog Breeding

This was the first step towards a less aleatory production dependent on weather and biological factors. This type of breeding began in Brazil in the 1980s and the technique has remained practically unchanged. It is worth consideration by small-scale frog farmers or as part of a larger scheme.

To defend frogs from external predators and to prevent their escape, barriers are installed. These animals prefer wide open spaces and as soon as an escape from a breeding area looks likely, follow their natural instinct. This kind of system provides a considerable increase in tadpoles in a central pond containing barriers, where after metamorphosis the animals remain until harvesting. This system is termed semidry.

Due to difference in size and feeding system, cannibalism is widespread. Feeding is based principally on insects attracted by the water and lights properly placed inside the enclosure. Larvae of fly, fish other insects and other animals that frogs like are provided. Waters are heavily fertilised so that phyto- and zooplankton can develop to feed tadpoles; they also like tender and appetising vegetable foliage (i.e., *Urtica* or large seaweed leaves), which they tear with their mouths and digest through their long intestine. Tadpole harvesting takes place in a little pond situated just below the larger one and is done with nets or by hand.

This last method has practical difficulties and well-trained staff are imperative to preclude damage to the tadpoles during the harvest.

Frog Intensive Closed Cycle Breeding

In the last twenty years, given the market demand, intensive frog breeding has increased significantly due to the decline in wild captives. With that type of breeding a satisfactory survival rate and in particular profitable productions were never reached except in small systems characterized by intensive manual labor and low cost of tadpoles. With new and better kinds of breeding, it is possible to reach elevated percentages of survival and standardized frog sizes for the internal market and for export. In fact, the greatest breeding is done in developing countries because of their favorable weather conditions and low production costs. Intensive systems provide a marked division into the various growing stages of the amphibians, so as to increase survival and productive performance, following the basic principles of animal production.

The biological growth cycle can be divided into these stages: reproduction, spawning, larval stage, premetamorphic stage, metamorphosis, adult frog. Each of these stages needs adequate containers, adapted to minimize the amphibian stress. For the last four stages, in particular the grow-out, several types of "patented" systems—inundated or semidry—have been designed to divide them into categories.

It must be remembered that metamorphosis, spawning, egg collections and grow-outs are all critical stages of intensive breeding. Each stage requires a particular type of feed (in paste or in pellets of different composition). Part of the feed may be composed of insect larvae or worms or live "moving" food, often employed in the final fattening stages of the adult tadpoles. In fact, the kinds of frogs bred nowadays come from the wild and have not been highly selected.

To summarize, frog breeding can be divided into an aquatic and a terrestrial stage, each of which presents its own problems that must be solved if good productive performance is to be realized. As in all zoo-technical intensive breeding, frog breeding needs net divisions between one growing stage and another; in fact amphibians should stay as long as possible with other frogs of the same size and age. This ensures a better productive performance, a lower rate of cannibalism, and good health.

The most common commercially raised frogs are, as far as the author knows *Rana catesbeiana* (American bullfrog) and *Rana tigrina* (Asian bullfrog), but many other frogs, acknowledged or not, are raised as pets and for scientific purposes.

Feeding

As mentioned in the previous section, one of the critical factors in frog breeding is feeding. In fact, frogs bred in nature after the metamorphosis nourish themselves by preying on live food. Two points should be observed: the first is that the food they prey on has a high protein content and the second, it moves. These two conditions must be reproduced artificially for bred frogs. The elevated protein content presents a feeding problem (we employ for elevated performances, integrated and expensive high protein feed). Different techniques have been elaborated. Experts in the Far East train the frogs to feed on non-live food, others employ a sort of physical or mechanical strategy to provide mobility to the feed, others give live food directly; fish, insects, larvae. This last system gives good results but entails higher cost as the production of live food is expensive. Local resources permit the breeder to use the food most readily available at that location and at better costing.

Integrated pellets are now produced with 40% pelleted crude protein content and fed to frogs in commercial farms during the grow-out period. Generally speaking, most of our knowledge is based on *Rana catesbeiana* farmed in South America, Mexico, and Southeast Asia (China and Taiwan).

Frog Pathologies

One of the biggest problems in frog breeding is that they readily fall ill, especially under crowded and stressful conditions. Many diseases, especially bacterial, can enter through their skin. Much frog breeding, mostly at the fattening stage, suffers some bacterial pathology with an elevated loss of animals (Poinar and Thomas, 1988; Giaccone, this volume). At the fattening stage it is very important to maintain excellent health conditions in the frog herd. Frog leg diseases are one of the most dangerous, arising from bacterial septicemia, and caused by stress-related diseases. Insects and nematodes are not a common cause of disease for frogs but insect larvae left in the tanks may eat tadpoles. Cannibalism is also a problem in overcrowded cages with a low environmental quality. Rectal prolapse (large intestine protrudes from the cloaca) is also a condition noted in several Thailand frog farms, attributed by some veterinarians to low digestibility of the feed. Rectal prolapse is more common in small-scale frog farms.

Buildings and Equipment

Different systems for frog breeding are installed depending on the system of breeding employed—intensive, semiintensive or extensive (Fig. 5.1). Galvanized iron or plastic enclosures, wooden containers and cages in cement with several compartments are used depending on the animals' developmental stage. For the larval stages, according to the number of tadpoles present, tanks of different shapes are utilised. Respecting their physiological needs and size, 20 tadpoles per liter with low water change can be obtained. For metamorphosis, tanks with an appropriate facility are used. This stage, prove to a high mortality rate, is not an economical one in breeding.

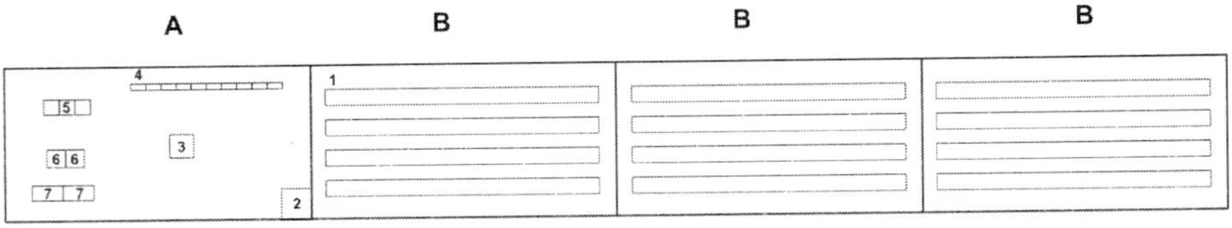

Fig. 5.1: Frog semiintensive semiwet system of breeding. a) Hatchery; (b) Growth divided into three phases, according to frog size; 1) water depth 0.2–1 m; 2) warehouses; 3) offices; 4) tadpole development, metamorphosis, max depth 1 m; 5) egg hatching area; 6) mating area; 7) separated parents (male/female).

In intensive systems (inundated or semidry), there should be 2 to 50 adult frogs per square meter (up to 100 and more can be included but with a high mortality risk). Preferably the water should be of good quality but both tadpoles and adult frogs are more resistant than carp to low-quality water. Inundated tanks for intensive and semi-intensive grow-outs are currently used in Taiwan and China for highly intensive frog production. Also multistoried tanks are used nowadays with a concentration of 50 and more frogs per square meter (on every layer). In the intensive system it is possible to distinguish between inundated and semidry according to the percentage of space inundated with water. The inundated system is preferred in Asia wherein the frog stays in the tank or pond with very little or no access to dry areas.

Genetic Factors

The frogs reared are generally randomly selected. Only for certain types of biomedical research are clones of a particular species of frog selected. However, the cost is very high except when large breeding areas can provide a consistent supply at budget cost for research and development .

To date the author knows of no commercial frog farm with sophisticated genetic selection but it would be interesting to investigate this possibility further. Currently, mass selection is the most used system.

Main Species Bred

The principal kinds of frogs bred in farms come from: North America; *Rana catesbeiana* frog (American bullfrog) and *Rana pipens*; South America, *Leptodactylus ocellatus*; and Asia; *Rana tigrina* (Asian bullfrog) which is more adapted to humid climates and *Rana hexadactyla.*

Possibilities of Frog Farming

The world frog import figure of 30,000 tons (Aquaculture Magazine Buyer's Guide, 1997) led us to consider the possibilities of frog raising: today several successful commercial frog farms are in operation. Some small semi-intensive farms with a low production in southern Europe and Southeast Asia seem to work well. Very intensive frog farms are found in Taiwan and mainland China, and in Brazil and Mexico (Teixteira et al., 2001). Frog farming technologies are now available that are suitable for every part of the world and applied research could collaborate to make these technologies more economically sound.

Frog Farming Options

In my opinion, frogs should preferably be raised in tropical or subtropical climate to take advantage of the temperature but raising frogs in any climate with a hot season should also be considered.

Smaller frog-raising units can make use of the local labor force to produce frogs for the local market in or close to the traditional areas of consumption. Schemes with a tadpole production unit can supply hundreds of small farms and provide technical advice, feed, tadpoles and eventually collect the adult frogs.

Large-scale production means commercial frog farms with specialized technicians and high capital input, besides well-organized marketing and processing systems.

Frog Frauds

Unfortunately, some societies do promise excessive technical and economical results from frog breeding. Alerting people to false claims is one objective here.

Certain frog species are also protected by some organizations; CITIES, FAO/OIE have issued international legislation on frogs. However, more controls are needed governing the capture of wild frogs, especially in Southeast Asia where the bulk of the catches take place. India has already prohibited frog export.

Conclusion

A brief and regrettably incomplete description of the various possibilities of breeding frogs in the world has been presented. The presentation is not exhaustive because considerable research and new technologies in frog farming are constantly emerging. In fact the frog industry is working hard on new technologies as amphibian breeding grows steadily with demand, and not just as a culinary item.

Because of the decrease in frog capture and the strong increase in the frog market, frog breeding has become a good investment, employing the latest technical innovations that permit breeding in intensive and semi-intensive systems with good technical and economic results. It is advisable to follow the advise of experts to ensure good economic results.

Frog breeding is not widespread in the world because it is relatively quite complex and until recently research was scant; with the disappearance of frogs in the wild and the growing market prices, the business has become more attractive. Europe is a very good market where the majority of frogs consumed are imported from non-European countries. Frog demand is strong and the market expected to expand.

References

Anonymous. 1981. The lucrative business of raising bull-frog. Philipp. Dev. 18–25.

Aquaculture Magazine Buyer's Guide, 1997. Asheville, USA.

Aubert C. 1987. Les grenouilles et l'éventualité de leur élevage en France. Ed. ITAVI, Paris, France, 39 pp.

Baker J. 1998. Frog culture and declining wild populations. World Aquaculture 29 (1): 14–17.

Barua G. 1988. Food of *Rana tigrina* tadpoles. Bangladesh J. Fish. 11: 61–69.

Donizete T.R. 1993. A technological revolution in frog culture in Brazil. Aquaculture Magazine, 19 (2): 42–48.

EU COMETEX Statistics Table 2001

Flores Nava A. and Gasca Leyva E. 1997. Use of artificial grazing substrates in bullfrog tadpole culture. Aquaculture 152 (1–4): 91–101.

Flores Nava A. and Munoz Vera P. 1999. Growth, metamorphosis and feeding behaviour of *Rana catesbeiana* Shaw, 1802 tadpoles at different regime densities. Aquaculture Res. 30 (5): 341–347.

Giaccone V. 2004. Hygiene and Health Features of "Minilivestock" (this volume).

Hardouin J. 1994. Commerce international de cuisses de granouilles dans la CEE de 1988 à 1992. Bull. Rech. Agron. Gembloux 29 (2): 217–245

Hardouin J. 1995. Un avenir pour la zootechnie de la grenouille en Afrique/Scope for frog production in Africa. Nature & Faune 11 (2): 2–7. FAO Regional Office for Africa, PO Box 1628, Accra, Ghana.

Hayes T., Haston K., Tsui M., Hoang A., Haeffele C., and Vonk A. 2002. Herbicides: Feminization of male frogs in the wild. Nature 419: 895 – 896.

Lester D. 1988. Raising bullfrogs on non-living food. Aquacult. Mag. 14 : 20–27.

Munyuli Bin Mushambanyi T. 2002. Elevage contrôlé des grenouilles au Kivu (République Démocratique du Congo). Cahiers Agric. 11: 269–274.

Negroni G. 1987. Frog culture. World Aquaculture, vol. 28 (1).

Negroni G. 1991. La mappa delle rane allevate nel mondo. Informatore zootecnico no. 24.

Negroni G. 1996. The basics of breeding frog. Infofish Int., no. 4.

Negroni G. 1997. Hopping into a new business. Resource (ASAE), April 1997.
Negroni G. 2000. Frog culture. Aquaculture Encyclopedia. J. Wiley and Sons, Inc., Chichester, UK.
Negroni G. and Farina L. 1993. L'èlevage de grenouilles, Cahiers d'Etudes-et-de-Recherches-Francophones-Agricultures. 2: 48–55.
Poinar G.O. Jr. and Thomas G.M. 1988. Infection of frog tadpoles by insect parasitic nematode. Experientia 44: 528–30.
Prier J.B. 1981. L'élevage des grenouilles. Ed. Dargaud (Paris, Bruxelles, Lausanne, Montréal), 80 pp.
Sukumaran N. 1986. Frog Farming. Seafood Export J. 18: 13–15.
Texteira R.D., dos Santos C.A.M.L. *et al.*, 2001. Globefish Research Programme. In: The World Market for Frog Legs, PAO, Roma.

Appendix 5.1

Chemical composition and energy value of *Rana esculenta* and other meat (G. Negroni and L. Farina, 1993)

	Protein %	Fat %	Glucid %	Calories 100 g (3.2 oz)	Calcium mg/ 100 g	Phosphate mg/ 100 g	Iron mg/ 100 g	Vit. A UI	Vit B_1 UI	Vit B_2 UI	Vit C UI
Frog	15.5	0.2	0	65	20	430	6	—	160	60	0
Horse	21.7	2.5	0.4	114	10	—	2.4	0	70	100	0
Beef	18.7	15	0	217	10	0	2.1	40	60	160	0
Chicken	16.7	14	1.3	198	12	200	1.5	400	100	100	0
Turkey	24.7	8.5	0	179	24	395	3.2	600	60	80	0
Rabbit	21.5	1.5	0	102	17	180	1.3	0	80	60	0
Pig	17.2	22	0.5	274	7	156	1.4	0	420	120	0
Liver (veal)	20.7	5	0.6	133	6	343	10.6	22.50	21	3.12	36
Brain (veal)	10.6	9	0	125	16	330	3.6	0	230	260	18
Mackerel	13.9	2.6	0.8	84	25	194	0.7	0	50	80	0
Trout	13.9	3	0.04	84	31	218	1	50	60	150	0
Lobster	16.2	2	1	88	60	280	0.80	—	150	180	0

Appendix 5.2

Frog soup recipe from the ancient frog consumption area of Po valley, Italy

Ravenna is a city in Romagna (part of the Emilia-Romagna region) where consumption of frogs is very popular. In fact, this soup is one of its specialities. For 6 persons buy 1.5 kg filleted frog. Fry lightly in butter finely chopped carrots, onion and celery, until tender. Add the frog and fry for one minute. Then add the necessary quantity of water to obtain a broth, add salt to taste and leave to cook for about 1 hour. Pass the broth through a strainer. You can use it to prepare any kind of soup, adding, if you like, pieces of the frog meat.

6

Snail Collection and Small-scale Production in Africa and Europe

Leslie J. Elmslie

Via Orti Gianicolensi 5, 00152 Roma, Italy. E-mail: lelmslie@pelagus.it

Abstract

Snails have been collected from wild populations, traded, and eaten since time immemorial, providing a significant source of food and income for poorer sections of the human population, and continue to do so today. Traditional collection of edible snails from wild populations does not threaten their existence. The value of wild snails to the collectors and to the country can be increased by appropriate regulations and reduced by inappropriate ones.

Snail farming is not a way of producing cheap food for the masses but can produce a premium product when appropriate systems are used. Edible snail consumption could be significantly increased if quality supplies were regularly available; however, the market for farmed snails is limited by consumer prejudice, as well as by price. Farmed snails probably account for less than 5% of world consumption at present.

Misguided attempts at snail farming have caused enormous economic damage not only by wasting resources but also by spreading agricultural pests.

Key Words: Snail exploitation, Snail farming, *Helix*, *Achatina*, *Archachatina*

Introduction

Snails have been collected from wild populations, traded, and eaten in many parts of the world since time immemorial. In fact snail shells found at ancient sites may be of considerable interest to archaeologists.

Today the two main centers of snail consumption are Europe (especially France, the center of a considerable international trade) and West Africa, but smaller quantities are eaten in a great many parts of the world, often quite locally.

The tradition of collecting and eating wild snails has declined over time in many areas—people prefer to spend their time in other ways—and in places changing land-use practices have reduced the number of snails in the wild. Concomitantly, interest in new types of gourmet food has grown and with it a potentially larger market for high-quality snails ready to eat. This combination of factors has created great interest in snail farming in recent years, though to date misguided projects have done vastly more damage by spreading pests and wasting resources than successful projects have produced benefits. This chapter constitutes a general guide for those who wish to know more about edible snails with a view to management of traditional collection and understanding the state of, and prospects for, snail farming.

Basic Biology

Few nonspecialists know much about the biology of snails in general or of the more commonly eaten species of land snails in particular. Furthermore, such sources as are easily accessible very often omit information critical to the successful management of snail populations for whatever purpose: exploitation of wild populations, snail farming, or pest control.

Each species of snail naturally has its own distinctive biology. There are also important differences within each species, corresponding roughly to the differences between the various strains of the same species of mammal: in size, growth rate, aspects of behavior, and so on. Not all species, let alone strains, have been studied in detail, and for practical purposes one often has to make the assumption that the species or strain one is dealing with is similar to a related one already studied in some detail.

Egg-laying and Hatching

With that premise, let us start from the egg. In the most widely eaten families, Helicidae and Achatinidae, the egg-laying process begins with the "mother" digging a hole in which to deposit the eggs. It does so in damp soil in damp weather. When the hole has been completed, the mother remains with its front end deep in the hole until the whole clutch has been laid. In *Helix*, the egg formation process begins after the hole has been dug. Oocytes are released from the ovary one by one at intervals, typically of 15–30 minutes, and are joined by sperm that the mother has stored from matings that may have taken place quite some time earlier and by eggwhite material that has accumulated in the albumen gland. As each egg passes along the oviduct, it receives a vitelline membrane and a proportion of calcium carbonate as a covering, followed by a sticky coating. As each egg is laid, it is added to the mass already present in the hole, held together by the sticky coating. The entire process is rather long; a snail that lays, say, 120 eggs at 20 minute intervals will take 40 hours to do so. A thorough description of what is known of the egg-laying process can be found in Tompa (1984).

The number of eggs in a clutch and their size vary greatly within, as well as between species. I have found fertile clutches of 3 to 189 eggs while working with *Helix adspersa*. The amount of material stored in the albumen gland must obviously play a key role in determining the maximum size of the clutch. The need to reconstitute the store of albumen must also influence the time between the laying of one clutch and the next. Egg size is largely constant for a given strain of a given species. The parent snail's investment in egg production is large; clutch weight can reach as much as 40% of the "mother's" weight after ovipositing.

In contrast to helicids, achatinids lay eggs very rapidly (Hodasi, 1979).

Hatching time is extremely variable (in *H. adspersa* from 7 days to 4 weeks). Temperature plays a role, but so do many other factors, such as the nutritional status of the eggs, and the surrounding environment. In general, in clutches that hatch early the individuals hatch almost simultaneously. In clutches that hatch later, some individuals emerge much earlier than others; early hatchlings will then eat their unhatched siblings, reaping considerable nutritional benefit (Baur, 1986; Elmslie, 1988). If the soil is excessively dry, eggs may shrivel, while contrarily, if excessively wet, the eggs may swell and burst.

The hatchlings may/may not emerge promptly from the hole in the soil where the egg were laid. Young hatched from eggs laid underground in autumn may not surface until spring. That they remain underground does not necessarily mean that they are inactive; they can and do eat soil and other materials and thereby grow.

Growth

The growth rate of young snails in the wild is extremely variable, to a degree almost incredible to people more familiar with birds or mammals. A snail 12 months old may be no larger than another of the same strain only 3 weeks old. Climatic factors play a role, but the most important roles are played by nutrition, followed by population density. Snails are notoriously "limited in their movements" and thus constrained to eat whatever they find in their proximity. When it finds a "good" food, the snail eats varaciously; if it finds only "bad" food, it eats just enough to remain alive. If only "bad" food is available, the young snail will be dwarfed, perhaps tiny when it reaches the biological age at which it is too old to grow further, and will not mature. It seems that biological age is a function of the time during which the snail is active, that is, not hibernating, estivating or temporarily inactivated by cold, heat or dryness. Chemicals of three distinct categories play a key role in convincing the snail to eat the food found: attractants, phagostimulants and antifeedants (Airey et al., 1989).

Population density limits snail growth in two ways: first, if the snails are abundant the "good" food is quickly eaten and only the "bad" remains. Secondly, when overcrowded snails depress each other's growth by some well-demonstrated (e.g. Oosterhoff, 1977; Heller and Ittiel, 1990) but not understood mechanism involving their mucus. Population density can produce permanent dwarfing in the same manner as poor nutrition.

Climatic effects seem to be less important for growth in the sense that the time the snail spends in inactivity because of cold, hot or dry ambient conditions seems to add little to its biological age; its capacity to grow and reach maturity later is therefore undiminished.

The various factors limiting growth result in a very wide range of size in a given cohort (snails from eggs laid at the same time). In suboptimal conditions the distribution is skewed forward, with a few individuals well grown and the majority much smaller (Elmslie, 1989).

In nature, snails are most active at night, when dew makes it easy for them to crawl, and after rain. Their activity seems to comprise elements of trail following, random "searching" and homing. Feeding activity tends to peak in the early part of the night and again just before dawn. Snails often have a home area in which they generally spend the night. *H. pomatia* has been found to migrate seasonally and rather variably over modest distances (Lind, 1989).

Growth of the young in many species and strains is constrained by a circannual cycle that governs the activity of both growing and adult snails. When snails are in their natural environment, the circannual cycle is in tune with climatic conditions that favour snail activity. For example, wild young *H. adspersa* in North Africa will stop growing and go into estivation in spring, when the weather becomes too hot and dry to support their activity. If the same snails are moved previously to a cooler and damper climate, they will nevertheless go into estivation at that time, and no amount of cool moist weather will lure them back to full activity until estivation has been completed (Elmslie, 1992). A strain's circannual rhythm can be modified or removed in captivity but only over several generations.

When a *Helix* snail has reached adult size, the edge of the shell turns outward to form a lip as the shell starts to harden. Thereafter the shell will increase in thickness and weight only.

The seasonal pattern of growth is similar in helicids of Europe and achatinids of West Africa. The snails emerge from hibernation or estivation in spring or after the first rains. All those that are immature then grow very rapidly until about the end of June, at which time growth ceases, even if ambient conditions are optimal. The main egg-laying season follows and thereafter the newly hatched snails likewise grow rapidly, while those hatched in the previous year show hardly any growth even if they have not matured.

Within the constraints imposed by that seasonal pattern, growth depends greatly on temperature. It seems that growth at a constant temperature or at a mean fluctuating around that temperature barely differs (Oosterhoff, 1977). The optimum mean temperature for the growth of *H. adspersa* and other helicids seems to be about 20°C. Achatinids grow faster at a higher temperature.

Reproduction

The most important species of edible snail in families Helicidae and Achatinidae are hermaphroditic. If self-fertilization takes place at all, it is of little consequence. In most matings both partners act as both male and female. On mating, the sperm go into indefinite storage along with those from previous matings. No specific time need elapse between a mating and subsequent oviposition. Chevallier (1985) presents a diagram showing dates of mating and laying for 60 individual *H. adspersa* from April to August. However, egg-laying does not seem to occur without prior mating.

In optimal conditions an *H. adspersa, H. lucorum* or *H. pomatia* snail may be ready to lay 4–8 weeks after shell lip formation, but will not do so until soil and atmospheric conditions are right. The soil must be moist and soft and the atmosphere persistently humid. *H. adspersa* may lay from zero to three clutches in a season depending largely on nutrition and environment.

Diapause and Circannual Rhythm

Hibernation and estivation can be given the common name of diapause. Physiologically, diapause is quite different from lethargy, which is the reduction or cessation of activity when ambient conditions are adverse (too cold, too hot or too dry). The snail makes a specific physiological preparation for diapause by accumulating reserves of glycogen. (Note that galactogen is stored in the albumen gland to be used as an energy source for the embryo, whereas glycogen is stored as the energy source for the parent snail itself.) The length of diapause is inherited (Cobbinah, 1995) but the internal circannual rhythm cycle can also apparently undergo modification in individual snails (Elmslie, 1992).

Species Consumed

Although a great number of species of snails are eaten locally in modest quantities in various parts of the world, by far the most significant quantities of snail meat consumed are taken from species of genus *Helix* or species of the achatinid family.

Achatinids

This family of giant snails is essentially African, though one species, *Achatina fulica*, has spread over much of southern Asia in the last hundred years, doing much damage (e.g. Canestri Trotti and Toffoletto, 1979). In the main consumption area, the coastal belt of West Africa, the most important species are *Achatina achatina* (L.) and *Archachatina marginata* (Swainson) (Plate IV, 1, 2, 3 and 4). Their distribution has been illustrated by Hodasi (1984). *A. achatina* is the most important species eaten in Ghana and *A. marginata* in Nigeria. Stievenart (1996) provides in-depth information on both species.

A. achatina (Plate IV, 1, 2) was studied by Hodasi (1979), who reared it in boxes in a laboratory corridor, so that its activity would follow the natural seasonal cycle, and fed it on leaves and fruit in addition to the soil in the box. He found that snails hatched from eggs laid in June began to lay eggs in May, 23 months later. That year, 32 snails laid an average of 112 eggs each, with individual clutch size ranging from 91 to 209 eggs. The following year the same 32 snails laid an average of 183 eggs each with clutch size ranging from 37 to 305. No eggs were laid thereafter. The egg-laying season extended from the beginning of May to July end. Eggs were laid one by one at quite a high speed; in one case 72 in 2.5 hours. Egg length ranged from 5.4 to 9.4 mm and width 4.2 to 9.4 mm. Eggs took from 10 to 31 days to hatch in moist cotton wool, in which environment the hatch rate was 77%.

Growth of the young snails was rapid for the first twelve months (except during the short diapause) and much slower thereafter. The figures in Hodasi's (1979) article lead one to think that a farmer would want to sell his young snails at about one year of age (or about 100–110 g), even though older and larger snails would fetch a higher individual price in the market.

Archachatina marginata differs markedly from *Achatina* species in eggs that are much bigger and fewer in number. Its shell is dumpier than that of *A. achatina* and it lacks the latter's dark wavy stripes (picture and table in Hodasi, 1984).

One of the most useful descriptions of the species remains that of Plummer (1975) who studied it in a laboratory in London. She reported egg lengths between 10.5 and 26 mm and widths from 9.5 to 19 mm. The average clutch size was 9 eggs, with a range of 3 to 10. Egg-laying is quick: one snail was observed to lay 6 eggs in 30 minutes. Growth rate seems to be very similar to that of *A. achatina*, with the snails reaching marketable size 12 months after hatching, though maximum value per snail would not be reached for another year. Plummer indicated a weight of 375–500 g at 3–4 years of age, of which the shell accounted for 65–170 g. Awesu (1988) studied the same species reared in cages in Nigeria and found that mature snails laid an average of 4 clutches of seven eggs per year. *A. marginata* (Plage IV, 3 and 4) thus seems to be the only species of potentially farmable snail in which the cost of the hatching egg would be a significant proportion of total costs.

Achatina fulica is best known as an agricultural pest, having spread from East Africa into India, Sri Lanka, and all of Southeast Asia, at times accidentally, at times as a result of misguided attempts at snail farming. It was introduced into the United States (accidentally or for the same reason) and subsequently eliminated at great expense. As late as the 1990s it was introduced into Brazil through misguided attempts at snail farming and also appeared in West Africa, almost certainly again introduced by would-be snail farmers and then allowed to escape.

Nevertheless, the species is an important edible snail. It is farmed in the UK, kept in cages in heated buildings to produce *Helix*-size snails for the British market. The wild (pest) populations of Southeast Asia are also exploited, the meat being exported as a cheaper alternative to *Helix* snail meat.

Monney (1994b) provided some data on the growth and reproduction of *A. fulica* farmed in the UK. Young snails reach a size suitable for the *Helix*-equivalent market at about 12 weeks of age and the first eggs are laid at 6–7 months of age at a body weight of about 40 g. Bodyweight reaches a plateau of about 60 g by 44 weeks of age. The average clutch size is 84 eggs (range 26–212). Life span was estimated as about a year and a half, contrasting with previous reports of a life span of six years in the wild. It is clear that many generations of intensive rearing have greatly changed the characteristics of the UK strain of *A. fulica*, reducing both final body size and life span. Such changes during domestication are to be expected but seem to have been particularly dramatic in this case.

Helix species

Helix adspersa, an easily recognizable species (Plate IV, 5) of apparently Mediterranean origin, has spread widely throughout the world following European colonization, to southern South America, California in North America where it is a major pest in orange groves, and Australia where it seems to be largely a garden snail. It is not found in areas with a dry or wet tropical climate nor those with cold winters, such as central Europe. French snail terminology distinguishes the "petit gris" or European *H. adspersa* and the "gros-gris" from North Africa, considered to be a subspecies. However, there are intermediate types and a great deal of variation within each group. A genuine farmed strain has been developed from the gros-gris (Chevallier, 1985, 1990) and has been in successful commercial use for many years. Inevitably, the farmed strain grows to a smaller maximum size than the largest wild specimens; however it is vastly more manageable and seems to tolerate somewhat higher stocking densities than unselected strains.

The body color of *H. adspersa* strains is of some commercial significance; white or at least pale gray is preferred to darker colors. The wild "gros-gris" is almost black but becomes much paler when farmed and fed compound meal. Other strains of the species have a much paler gray body in the wild and a "white" strain can be found in Illyria.

The adult weight of *H. adspersa* ranges from 4 to 30 grams according to the strain and to a lesser extent rearing conditions. Egg numbers vary correspondingly. Some strains need to estivate, some to hibernate and others have no fixed requirement for diapause (Elmslie, 1992). Bonnet et al. (1990: p. 72) indicated an average weight of 10.8 g (appropriate for the market) for 2,952 French *H. adspersa* reared in an outside pen from 6 weeks of age to lip formation.

H. adspersa differs from many other *Helix* species in preferring to spend its time above ground, spending its daily rest period and even its period of diapause attached to solid surfaces and/or in crevices. Nevertheless it will go underground for diapause if no crevices to its liking are found.

Though quite large quantities of *H. adspersa* are eaten, international trade in the species is limited, mostly from North Africa to southern Europe.

Helix pomatia is the most famous of the world's edible snails, thanks to its place in French cuisine as the white-fleshed Escargot de Bourgogne, the Burgundy snail. Its natural habitat is central Europe, from the middle of France eastward through Poland into the Ukraine and Russia and south to the Po valley and northwest Bulgaria. It seems to require both cold winter and hot summer weather for success and is not found in the milder coastal areas of France where it is replaced by *H. adspersa*. Compared with the latter species, *H. pomatia* lays fewer, though larger, eggs (typically 40–60) and is correspondingly less of an agricultural pest. These characteristics mean that the species lends itself more than others to open pen snail farming. Its distribution is often quite patchy, with no signs of rapid spread. In spite of being so prized as food, it has spread far less with European colonization than *H. adspersa*, but is found in parts of North America on both sides of the USA-Canadian border.

The adult size of *H. pomatia* seems to correlate with the number of hibernations that interrupt its growth. The shell growth marks on large specimens found in the Alps indicate that they have passed 4–5 winters before reaching adulthood, whereas at lower elevations the adults are smaller and reach adulthood after only 1, 2 or 3 winters.

A special feature of the species is the formation of a hard calcareous epiphragm as a protection during hibernation. The snail digs into the soil, quite deeply if the soil is friable, aligns itself with the shell mouth upward, and constructs the epiphragm.

Adults entering commerce in spring typically weigh 22–25 g. However, adults weighing over 30 g are not rare, while farmed adults rarely reach 20 g. The maximum weight is, of course, reached just before oviposition when the albumen gland content is maximal.

H. pomatia eggs are somewhat larger than those of *H. adspersa*, with more calcified shells, and take longer to hatch (rarely less than 3 weeks, sometimes up to 4). The peak egg-laying season in the wild is from late June to July end. Young from eggs laid in August or September may not emerge on the soil surface until the following spring, which has misled some observers into thinking that a very early spring breeding season occurs.

As with all animals, there are different strains with different agricultural characteristics. I have compared two such strains (Elmslie, 2001). Some enter diapause only as the weather turns cold, others do so in late summer and resume activity only in the following spring. This can lead to a certain confusion between estivation and hibernation.

In many ways, as the most highly prized species, *H. pomatia* would seem a natural candidate for economically successful snail farming, but the trend has been to replace it in traditional recipes with lower priced and more abundant species, in particular *H. lucorum* or *A. fulica* collected from the wild. Potential farmers have also been discouraged by the time it takes to adapt wild strains to their requirements and by the slightly longer time it takes to rear them to adult size compared with *H. adspersa* (the growth rate is much the same, but *H. pomatia* is bigger).

Helix lucorum forms part of a complex with *H. pomatia* that also includes many other related forms. Large quantities, several thousand tons a year, are collected from the wild in southeast Europe, Turkey etc., processed in the country of collection or in Greece and exported, chiefly to France to replace *H. pomatia* in traditional recipes. It is an easy snail to rear and has been farmed in Italy in areas where it is traditionally eaten, but farmers find it difficult to maintain a high enough price for their product and tend to abandon farming for trading in cheaper imported snails.

The flesh of *H. lucorum* is sometimes said to be tough, but this is probably because the snail is long lived and many of those collected from the wild may be quite old. Another disadvantage is the color of the flesh, which is never white and sometimes a rather off-putting shade of brown. In Bulgaria, the species tends to replace *H. pomatia* in areas that are drier in summer.

Wild populations of this species are abundant, but there is need to ensure that collection policies maximize their value, to collectors and exporting countries alike.

A very thorough study of a wild population of the species in northern Greece was published by Staikou et al. (1988). Combe and Abbes (1980) produced a note on the species in southern France. In field pens in Italy I found that a group of wild *H. lucorum* produced an average of 60 young to the 3-gram stage each year for four consecutive years and that the young began to lay eggs 13 months after hatching. In other words, they adapted to field pen farming immediately, without the selection/adaptation over a number of generations required by *H. pomatia* and most other species.

Many other species of snail are eaten and to a limited extent traded, especially in the Mediterranean area. Mention may be made of *Helix aperta*, highly valued in Sicily, Sardinia and southern Italy when estivating, *Helix cincta*, eaten in Greece, *Otala* spp., popular in Spain, and the closely related *Eobania vermiculata* in Italy.

Snails in Human Nutrition

Bonnet et al. (1990: p. 94) analyzed farmed snails (*H. adspersa*) and stated that they contain 81.6% moisture, 16.3% protein, 0.8% fat, and 1.3% mineral matter. Other analyses of the same and other species are similar (e.g. Gomot, 1998) and suggest snail meat should be a good source of animal protein for poor gatherers who cannot afford mammal or bird meat, at least if they eat significant quantities. Unfortunately the protein is less digestible and somewhat inferior in quality compared to vertebrate meat protein. Published analyses of the amino acids in *Helix* snails are not fully convincing. The figures presented by Cantoni et al. (1980) for example, show *H. pomatia* with 80% more methionine than *H. adspersa* and 60% more lysine than *H. adspersa* or *H. lucorum*.

Novelli and Bracchi (pers. comm.) analyzed the fatty acid composition of the small amount of fat and the mineral elements in the edible part (head and foot) of *Helix* species; their results are shown in Tables 6.1 and 6.2.

The cholesterol content of snail flesh seems to be similar to that of marine mollusks, about 80–150 mg/100 g, which is somewhat higher than that that of vertebrate meat, presumably because the edible part of the snail includes the central nervous system, in which cholesterol is most highly concentrated, whereas vertebrate meat does not.

All in all, it can be said that land snails are a nutritious food, in general comparable to marine mollusks—high in protein and low in saturated fats.

Still, a number of health risks are associated with eating snails. Wild snails habitually eat and thrive on plants and fungi that are poisonous, even extremely so, for humans. They must therefore be "purged" before eating, i.e., fed something safe such as flour, then retained until their intestine has been wholly

Table 6.1: Percentage of various fatty acids in the edible part (head + foot) of *Helix*

Fatty acid (% of total)	**Mean**	**Lowest**	**Highest**
C14:0 (myristic)	0.32	0.18	0.48
C14:1 (myristoleic)	0.12	0.08	0.16
C16:0 (palmitic)	7.43	5.99	9.11
C16:1 (palmitoleic)	0.26	0.12	0.62
C18:0 (stearic)	10.26	8.50	12.66
C18:1 (oleic)	12.14	9.65	16.05
C18:2 (linolenic)	14.65	11.64	24.15
C18:3 (linolenic)	1.19	0.64	2.55
C20:2 (eicosadienoic)	10.08	8.83	12.01
C20:3 (eicosatrienoic)	0.88	0.67	1.42
C20:4 (arachidonic)	12.51	10.55	15.33
C20:5 (eicosapentaenoic)	2.05	1.49	2.75
% saturated	26.10		
% monounsaturated	17.87		
% polyunsaturated	56.04		

Table 6.2: Concentration of mineral elements in the edible part (head + foot) of *Helix*

Metal	(concentration)	Mean	Lowest	Highest
Calcium	(mg/100 g)	499	297	957
Phosphorus	(mg/100 g)	186	156	216
Magnesium	(mg/100 g)	62	42	95
Potassium	(mg/100 g)	164	101	201
Sodium	(mg/100 g)	109	47	201
Copper	(mg/kg)	48	5	173
Iron	(mg/kg)	35	18	55
Manganese	(mg/kg)	3	2	4
Zinc	(mg/kg)	15	12	22

emptied. Being in contact with the ground, snails may be contaminated externally with bacteria from the soil or may ingest them. The most serious health risk presented by any of the commonly eaten snails is that from the parasitic worm *Angiostrongylus cantonensis* for which *A. fulica* is one of its intermediate hosts and the rat its definitive host. Man is an accidental host of the parasite, but there have been not a few deaths recorded from meningocephalitis from *Angiostrongylus cantonensis* infection, all in Southeast Asia and most probably due to eating *A. fulica* raw or not well cooked (Canestri Trotti and Toffoletto, 1979). There is no risk of infection from eating *A. fulica* reared in cages in the UK.

Snails have a system of protecting themselves against the toxic effects of heavy metals by isolating and storing them. This may bring the level of lead and other metals in the hepatopancreas of snails living in contaminated environments to above regulatory limits, but the resultant risk to human health appears to be minimal.

Snail Market

Snail have been collected from wild populations, traded and eaten from time immemorial, including in places where they no longer form part of the human diet.

Today, as mentioned earlier, the two main centers of consumption in Europe (especially France) and West Africa, but smaller quantities are eaten, often very locally in many parts of the world. Bonnet et al. (1990: p. 81) indicated that current annual consumption in France may be about 30,000 tons of live snail equivalent.

Traditionally collection and consumption of wild snails have generally been limited to specific seasons, i.e., those times of the year when snails are at their best—just before the egg-laying season and to a lesser extent at the start of hibernation or estivation.

Collection from the wild population is a low value activity—it takes a long time to collect a large quantity, especially of European species—and is generally undertaken by those who have no more profitable use for their time, the elderly, children, the unemployed, etc. It thus performs a valuable social function in providing food or income for the poorer sections of the population. Obviously there are no records of the quantities of snails eaten by those who collect them or traded informally, though these may well account for the largest part of world snail consumption.

In eastern Europe, several thousand tons of *Helix* snails are collected every spring for export. Some go live to consuming countries, others to processors, in recent years chiefly in northern Greece, who convert them to canned or frozen snail meat. There are regulations governing the time of year and the size of the snails that may be collected. Stepczak (1992) stated that in Poland, which exports *H. pomatia*, only snails that fail to pass through a 30 mm diameter hole can be exported and in the period up to the end of May. In Bulgaria the hole size is smaller, 28 mm.

Exporting and importing countries keep records of the snails exported, though these may be inaccurate as a result of misclassification, and some wholesale markets keep records of the live snails traded.

A distinction needs to be made between the market for snails in their own shell and that for snail meat. Snail meat, frozen or canned, can be used by restaurateurs for a variety of recipes, but probably most of it is used for the traditional Burgundy recipe in which a large shell contains a cooked snail (theoretically *H. pomatia)* and a rather larger quantity of garlic butter. There is quite a significant international market for empty shells (especially of *H. lucorum)* for use in this recipe, while the snail meat inside may be of any species, even *A. fulica* from Southeast Asia. The preference is for a tender snail meat, and the preferred species, in order of preference and price, are *H. pomatia, H. lucorum,* other *Helix* species, and definitely last, Southeast Asian *A. fulica.*

The International Trade Commission produced a very thorough report on the international snail market in 1979. In spite of the time that has elapsed since publication, it is still valuable. The main change that has taken place in the market since then is that processing—another labor-intensive activity—has tended to move out of France to lower wage countries.

Mention should be made of the market for "operculated" (hibernating) *H. pomatia* in northern Italy and parts of France where this snail is found and eaten. Traditionally, wild snails are dug up from known hibernation sites. Alternatively, active snails are collected in September and put into special enclosures to hibernate, which they do in late October; they are then dug up in early December, some to be eaten and some to be sold at a high price (several times that for "active" snails). There are two traditional fairs in northern Italy where such snails are sold. In recent years the market has been damaged by poor quality lots containing dead (and therefore rotten) snails. Since the snails are cooked with the epiphragm still in place, the rotten snails are only discovered at the dining table. A similar market for "operculated" (estivating) *H. aperta* exists in southern Italy, Sicily and Sardinia in the late summer period and snails are imported from Tunisia to supply it. It is possible to determine whether the snail in diapause is alive, by detecting its (slow) heartbeat, but this is hardly practicable on a commercial scale.

Also warranting a mention are "snail festivals", town festivals at which snails are a traditional dish. There are many of these in Italy for example, and although the quantities eaten are individually not huge, rarely more than a few hundred kilos, they are of considerable potential interest to snail farmers as a way of publicizing their output and attracting new consumers.

Apparently data have not been compiled on the West African snail markets, though it is known that the quantities eaten are large and that price and quantities available vary greatly from one time of the year to another. Live snails or smoked snail meat are commonly sold at the roadside where a major road passes through a forest area. Snail meat is a popular bar food and it would seem that this market could absorb a substantial additional supply. West Africans mostly prefer a chewable (tough) meat, which means the snails should be old. There is no information available on the possibility of selling substantial quantities of younger snails, say 12 months of age, that might be made available by snail farming.

As already said, snail farming produces high cost and, hopefully, high-quality snails. The natural market for farmed snails is therefore direct to restaurants, most of which prefer the snails to be at least part processed. There seems to be considerable room for expansion of this market in many countries but only if producers can maintain high quality. Distribution costs tend to be high since restaurants take only limited quantities. Producers aiming at this market need to have processing and storage facilities.

A final note of caution. Some snail-farming promoters produce misleading data on the snail market, usually indicating that it is easier to break into than it really is. Salghetti (1996) provided a remarkable example of fabricated data; a promoter provided him with a complete table of the number and size of Italian snail farms and their output by species and province, a work of pure fiction that could not have been honestly intended.

Pests and Diseases

Snail deaths are more commonly associated with parasitism than with bacterial or (identified) viral disease, and parasitism is often associated with a poor initial condition of the snails, attributable in turn to overcrowding or poor nutrition, which per se can cause heavy mortality. I have visited open pen snail farms in Italy where over a thousand young snails per square metre had hatched, and practically all had died because no adequate preparations had been made for either feeding or transferring them.

A common parasite of both wild and farmed snails is the mite *Riccardoella limacum*, which can be seen running in and out of the pneumostome (breathing hole) of infected snails. It feeds on the blood of the host while in the lung and heavy feeding by the mites results in profound pathological changes to the lung tissue, a 40% reduction in growth rate, and retardation of reproductive development in *H. adspersa* (Graham et al., 1996). *Archachatina marginata* is similarly affected by *Riccardoella* mites (Imevbore, 1992).

Nematode parasites of various species are frequently encountered. Imevbore (1992) discussed *Rhabditis* spp. Bonnet et al. (1990) have given good illustrations of a number of nematode species encountered in *H. adspersa*, some of which are very damaging.

There is no effective treatment for either mite or nematode parasites. General good hygiene and good husbandry are obviously the best prevention.

Dipteran parasites are very common in wild helicids in Europe. I have observed very high mortality from sarcophaga and sciomyzid parasitism in *H. adspersa* farmed in field pens in Italy. Adult snails that have recently laid eggs seem to be preferentially attacked. This may be one reason why wild snails are traditionally eaten just before the egg-laying season starts. I have also found phorid larvae to be common parasites of overwintering *Helix* of all ages; the flies emerge from the dead snails in spring. The fly *Alluadihella flavicornis* has been described by Cobbinah as a major parasite of *A. achatina* in Ghana. Eggs are laid on the tip of the shell, larvae feed on the snail and pupate attached to the inner walls of the shell.

The presence of even a very small percentage of live or dead snails containing larvae is extremely damaging to the marketability of a batch.

Numerous insects eat snails, including glowworm larvae *Lampyris noctiluca*, staphylinids, and various species of *Silpha*. I have observed plagues of *Silpha* spp. and of *Ocipus olens* in open pen snail farms in Italy (Baronio, 1974). Spraying the whole area with carbaryl solution solved the problem without measurable damage to the snails.

Few outbreaks of bacterial diseases in snails have been recorded. In fact, snail mucus contains an antibiotic substance or substances (Kubota et al., 1985). Nevertheless, large-scale mortality associated with infection by *Aeromonas* spp. has been reported both in wild *A. fulica* and in snail farms of *H. adspersa*.

Poisoning by heavy metals dissolved in the water supply has been reported in snail farms in the UK (Bradley and Runham, 1996).

The "pink egg syndrome" in *H. adspersa* is said to be associated with fungus infection by *Fusarium* spp. Such eggs rarely hatch and if they do produce unhealthy hatchlings. The syndrome appears to be associated with the parent rather than the soil in which the eggs are laid. However, Hodasi (1979) found that in his conditions eggs hatched much better in moist cotton wool than in soil, from which one can deduce that some soils may have adverse effects on the eggs laid in them.

Characteristics and Utilization of Wild Populations

Edible snail species lay many eggs. In a wild population, however, all those eggs do not augment the number of adults, which remains the same. It follows that a wild population is composed numerically mainly of young snails whose numbers decrease with age and size, though the adults may account for a large part of the total body weight.

In all countries, it is only the adults or near adults that are in demand for food. This preference is readily understandable in the case of *Helix* species, as even the adults are quite small in size and the shell of the young snails is quite fragile and easily broken in the collection and cooking process. This means that most of the population is of no value to collectors.

Wild populations are most likely to be a traditional food source in areas where the snails are abundant and other forms of meat are scarce (France seems to be the exception), and the human population has time to engage in such a low productivity occupation as snail collection and preparation as food. An abundant wild snail population is likely wherever the soil is high in calcium (if it is not, slugs predominate), the climate suitable for snail activity for much of the year, and the natural vegetation (and/or farm weeds) can support a high snail population. These conditions formerly existed in many parts of central Italy but conditions have now changed. Herbicides have eliminated a major food source for the snails, thin soils are no longer cultivated, and people are more inclined to buy quickly prepared beef or chicken than spend the great amount of time needed to collect and prepare local snails. Many of the areas where snails were once abundant have been abandoned to natural pasture and scrub. The uncultivated soil surface is inhospitable to snails and periodic burning of scrub devastates the population. Areas that once yielded large quantities of snails now yield nothing.

No good research seems to have been done on the value of wild snail populations to collectors, either for money or for eating, nor does there seem to be any good research on the actual or potential contribution from wild or managed snail populations to the output from a given area of land. This lack of research is undoubtedly due to the relatively low value in each case, as well as to the difficulty of measurement.

Collection of wild *Helix* snails is a low-value activity because the individual picked up is small and the snails are scattered and may be hidden in the vegetation. Collection is quickest just after first light after a damp or dewy night, when the collector can see well and the snails have not yet retired to rest for the daylight hours. It is also quickest at the time of year when the adults can most easily be distinguished from soft shelled juveniles without wasting time picking up the latter.

In eastern Europe, snail collectors are mostly pensioners, school children, gypsies or occasionally the unemployed. The quantities collected over the buying period vary greatly according to the weather and the individual's knowledge of the local snail population, with a maximum of about 5 kg per person per day. I have estimated that a very expert pensioner in Bulgaria might add up to 50% to his pension during the month when collection is allowed.

In rural West Africa, individual achatinid snails are about ten times the weight of the European *Helix*; more weight can be collected in a given time and collection consequently involves a wider cross section of the population. In remote areas the snails are collected chiefly for home consumption (Monney, 1994) while in more accessible areas most may be sold to add to the otherwise slight cash income of a typical subsistence farmer. There seem to have been no attempts to quantify either the nutritional or cash values involved. Monney cited a claim that snail meat may be the main source of "animal" protein in some parts of Nigeria, but for numerous reasons it seems extremely unlikely that there are many, if indeed any, such places.

The nutritional contribution of snails to the diet in most places where they are collected and eaten, though not necessarily negligible, is undoubtedly quite modest. In the first place, the nutritional value of snail flesh is less than that of fish or meat from mammals. Again, most people eat snail meat only occasionally and then in modest quantities. Thirdly, not everyone eats it. Apart from purely personal preferences, there are numerous and very varied traditions, superstitions or pieces of incorrect information that discourage snail meat consumption in sundry communities and categories of the population, some of which are mentioned by Monney (1994a,b).

Conservationists have sometimes expressed concern about the effect of collection on wild snail populations.

Traditional collection from wild populations that has been practiced for decades or centuries is sustainable by definition. Nevertheless concern has arisen as a result of the decline in snail populations that were formerly abundant. The causes are almost always to be found in land-use practices rather than collection, not least because collection from areas where snails are scarce is not worth the effort! Prevention of collection from localities where snails are abundant will not lead to an increased population in places where the snail population has declined because of changes in the local environment.

The critical factor determining whether a snail population is expanding or declining seems to be the number of young that survive to maturity rather than the number of eggs laid. Heller and Ittiel's (1990) report is particularly interesting in this connection. They found that removal of the adult snails from a population (they were eaten by a wild boar) led to their rapid replacement by faster growth of juveniles.

Snail Farming

The idea of farming snails seems to have a remarkably wide appeal, even, or perhaps especially, to those who know nothing about snails, farming, or the market!

This appeal is not new; there are records of snail farming in classical Roman times. The patchwork distribution of *H. pomatia* in northern Europe is thought to be in part the result of attempts at farming the species in the Middle Ages. Introduction of *H. pomatia* and *H. adspersa* to many parts of the world colonized by Europeans is also thought to be due in part to attempts at snail farming. Spread of the pest species *A. fulica* in Southeast Asia in the mid-1900s, and more recently and even more disastrously *Pomacea canaliculata* (e.g. Adalla and Morallo-Rejesus, 1989; Pallis et al., 1996), which has invaded rice fields from Indonesia to Kyushu, were largely due to misguided attempts at snail farming. In spite of the well-known pest status of the species, attempts to introduce *A. fulica* illegally into the United States for farming purposes still continue, and it has recently been misguidedly introduced into Brazil and West Africa for the same reason. In western Europe, very large amounts of public and private money have been wasted on setting up *Helix* species snail farms that fail to produce. (Fig. 6.1)

That said, snail farming is possible and in some circumstances can be profitable. There are some critical constraints.

• Snails, especially *Helix* snails are small. Individual handling of such small pieces of meat makes the product very expensive in terms of labour costs at all stages, including in the kitchen.

• Growing snails do not tolerate life at a high stocking density. The maximum feasible density in terms of body weight per unit area for *Helix* species is a fraction of that possible for broiler chickens. This means that containment systems must be cheap and unsophisticated if the farm is to be economically viable, which in turn limits the performance that can be expected from the snails.

• Snails take much longer than broiler chickens to reach marketable size.

• Although snails can eat a variety of low-cost wastes, a tolerable growth rate and reasonably even growth of a cohort can only be obtained if the snails are fed on a finely ground compound meal. Against this, they offer a very good rate of feed conversion (always under 2:1, often as low as 1.2:1)

• Snails, especially *Helix* snails, are eaten only when fully grown (unlike farmed birds or mammals). As snails grow at very variable rates, the difficulty of separating new adults from juveniles before the former start to lay can easily lead to a chaotic situation.

• Wild snails are adapted to life in the wild. Trying to farm them is a little like trying to farm jungle fowl!

• Most people have some instinctive understanding of other mammals and birds. They have no such understanding of snails and can only acquire it through experience, which takes time.

The net result of the first three constraints in particular is that commercial snail farming, unlike collection from the wild, is not a way of producing cheap food. At least as developed to date for *Helix* species in Europe, it can provide only a superior product for the luxury market.

At the time of this writing, the only country producing a significant output of farmed snails is France, where two different varieties of the species *H. adspersa* are farmed. In both cases the system involves a modification of the natural life cycle. Adult snails are made to hibernate artificially in late summer and autumn (in rooms kept at 5°C) and brought into activity in heated buildings with artificial lighting about the end of December, so that they lay eggs in the early months

Fig. 6.1: An example to avoid. This large "snail farm" in Italy received very substantial public assistance when it was set up by entrepreneurs with no experience, using unproven technology. Inevitably, it produced almost nothing, and the surviving snails were allowed to escape.

of the year. The only soil available is in numerous small pots, which the snails therefore use to lay their eggs. Each pot is removed as soon as eggs have been laid in it and kept in a warm environment until the young snails have hatched. After hatching, the young snails may be reared briefly indoors and then transferred to open pens or plastic covered tunnels or greenhouses, where they are fed on dry compound feed (Figs. 6.2 and 6.3). The pens are watered by spray irrigation, especially during the night, and no other watering system is required. The meal must be protected from wetting by the spray irrigation. Feeders are very simple: boards or slabs of plastic on which the meal is placed, plus spray-protection lids that can be raised when the meal supply is replenished. A simple hand scraper is used to remove any droppings from the feeder before the new supply of meal is added, normally twice a week. The young snails are ready to harvest by July. Most must then be processed and stored for the main selling season at Christmas. Variations on this method have been described by Bonnet et al. (1990) and Chevallier (1985, 1986, 1990). Improvements to the basic method have been made by commercial farmers, especially by Combe and Moulin who use their own self-sustaining strain developed from an original Algerian wild population (Chevallier, 1990). Contrarily, the native French strains used by Bonnet et al. (1990) are not self-sustaining; the artificial modification to the breeding season leads to the strain dying out in a few generations, so that new snails have to be brought in annually as breeders from natural populations. Technically, this system seems to be reasonably reliable even though mortality in the growing

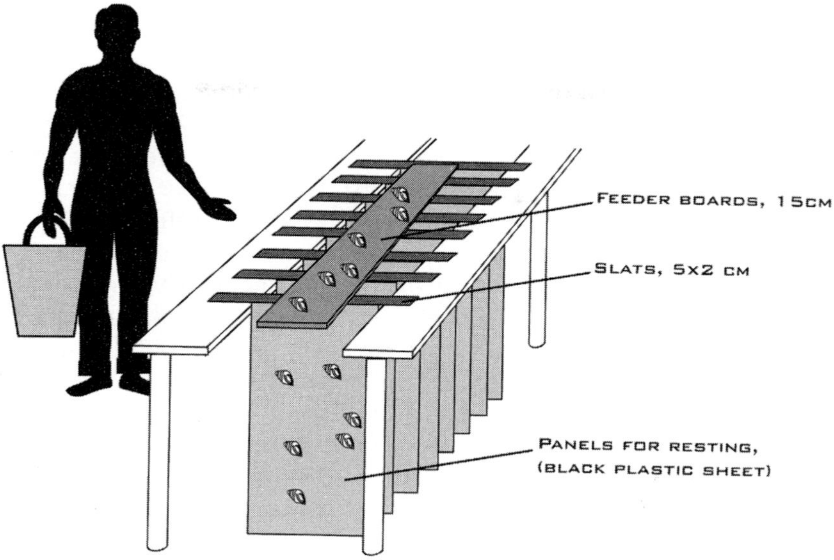

Fig. 6.2: This structure has been used successfully for the farmed strain of the so-called *Helix aspersa maxima*. During the egg-laying phase in a controlled environment the adult snails rest on the plastic sheets and climb to the feeders to eat. Another board, with holes to hold tubs of earth is placed parallel with the feeders on the slats. The snails lay their eggs in these tubs of earth, which are removed when there are signs that a snail has laid in them. The structure is housed in a building with temperature and humidity control and placed on a concrete floor, washed down as required to remove the fallen droppings.

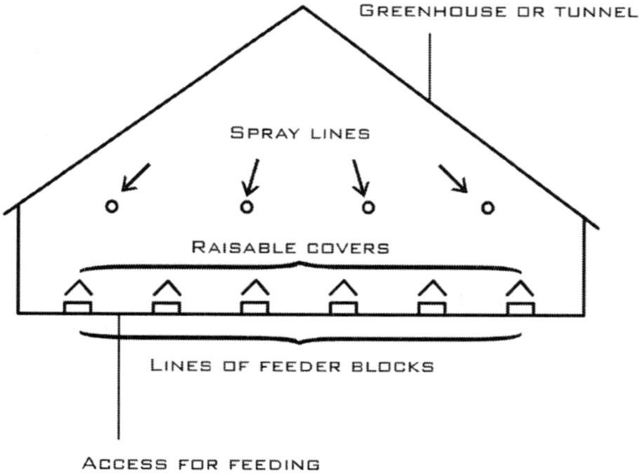

Fig. 6.3: From quite a young age, growing *Helix aspersa* can be reared in a greenhouse or tunnel fitted with spray lines for irrigation and water supply. Rectangular polystyrene blocks make suitable feeders; the gap between feeders should be no more than 1 m. The covers protect the feeder from the spray and are raised when the blocks are scraped and feed supply renewed twice a week. The covers can be raised individually or a whole line lifted by a pulley system. A water spray is the only water supplied.

pens seems to be quite variable from year to year. The biggest challenge growers face is economic.

An alternative containment system is the autocleaning cage developed by Gomot and De Grisse, a variant of which was subsequently manufactured and marketed in the UK by Hawkyard. In this system the snails are housed in cages arranged in batteries similar to those used for chickens. These are cleaned daily by water spray and food introduced into each cage by hand after cleaning. These cages have been used for farming *A. fulica* for some years. The young are reared to about the size of adult *Helix* and are of excellent quality. The farmed strain of the so-called *H. adspersa maxima* has also been successfully reared in cages. However, the initial technical success with snail battery cages in the UK succumbed to a variety of health and growth rate problems (Bradley and Runham, 1996; Graham et al., 1996) and many farms had to close.

I have published several descriptions of open pen snail farming in Italy in which *Helix* snails are raised in field pens in which food and shelter plants are grown. The pens are irrigated every evening to ensure that the snails are always active when the weather is warm enough, which in most of Italy means from March or April to October or November, with an interval mid-summer when temperatures are too high.

Normally the parent snails are introduced into clean, newly planted pens in late spring, as soon as the vegetation has grown sufficiently. They lay eggs in the soil and the young snails hatch. The parents and young must then be moved to fresh pens to ensure that the young are stocked at the right density per square meter and the parents maintained properly until sold. The snails hibernate naturally underground, awakening and resuming activity when the weather is sufficiently warm in spring. If the parents were of a suitable strain, the climate good, and the farm properly managed, most of the young hatched in the 1st year will reach adulthood (shell lip formation) in the 2nd year. They must then be promptly removed from the pens to preclude their laying and irremediably confusing the situation in the pens. However, if the original breeding snails are from wild populations of *H. pomatia*, only a modest proportion of the snails hatched in the 1st year will reach adulthood in the 2nd.

The labor requirement for this system is exceptionally low as long as the snails can nourish themselves adequately from the food plants in the pen, chiefly fodder rape. But when they can no longer do so, supplementary food (meal or forage) must be brought in from elsewhere and the labor requirement multiplies many times. The snail yield from plants growing in the pen is limited by their growth and by feed conversion ratios to less than 200 g/m^2, which is not enough to pay for the containment needed to keep the snails in the pen and predators out, so that additional food has necessarily to be supplied to the system. Novice farmers in particular find it very difficult to judge when and how much additional food is required, since snails will eat modest amounts of shelter plants, weeds, and the less palatable parts of food plants; results in early years of the farm are therefore usually poor and most farmers soon abandon their unrewarding enterprise.

Furthermore, the variability of egg production, and the farmer's lack of control over it in field pens, usually results in either too many or too few young being produced. Promoters invariably recommend too many breeding snails, so that *novice* farmers find themselves with far too many young, nowhere to put them, and no means of feeding them. At present only a few *established* farms use this system profitably. Those achieving some success are farming *H. pomatia*, more suited to this system than *H. adspersa* (N. Griglione), not least because the number of young snails produced per adult is less and reproduction is less likely to get out of hand.

I have used a variant of this system successfully for research purposes for many years, relying far more on compound feed than on plants grown in the pen. For reasons not yet clear to me, the yield has nonetheless never reliably exceeded 1 kg/m^2 year for any strain of the three *Helix* species reared (*H. adspersa, H. pomatia, H. lucorum*), definitely much less than that achieved in France with systems involving artificial hibernation.

Other, more reliable systems are being tried in Italy (F. Ballone) and seem to have given respectable yields per square meter, but the economic results are also inevitably dependent on producer ability to obtain a premium price for output.

Theoretically prospects for establishment of snail farming in West Africa thus look more promising than in western Europe. The snails eaten there are bigger, labor is cheaper, and the market gives the impression of being more elastic. What seems to be missing at present is a clear view of the economic role that snail farming might play and the sort of people who might farm snails with profit, in terms of either cash sales or a better diet.

One field for study that seems to have been neglected to date is that of improved methods for storing wild snails collected during the rainy season, when they are abundant and cheap, to the dry season when their value is several times greater. A commercial farming operation, if developed, would also presumably aim at selling as much as possible of its output at times of the year when prices are high.

Almost all those keeping snails in West Africa at present seem to be doing so either for study or as a hobby. They all use a natural life cycle and most rear the snails in cages made of local materials. In such cages the adult snails have access to pots filled with soil in which to lay eggs. The pots are removed when eggs have been laid in them and isolated until the young snails hatch, which are reared separated from the adults. Pens at ground level and pits have also been used, as illustrated in the short guide written by Cobbinah, a booklet very useful to anyone wanting to rear a few snails in their yard, or to acquire the experience that is an essential prerequisite to any commercial snail farming operation. However, the risk of predators of various kinds attacking young snails kept in ground level pens appears extremely high.

It is not possible to obtain accurate information on the total output from snail farming worldwide at present. A reasonable guess might be about one thousand tons of *H. adspersa* per year in France and about the same total amount of species in the rest of the world.

This compares with the estimate given by Bonnet et al. (1990: p. 81) of 30,000 tons live snail equivalent consumed every year in France alone.

Feeding Farmed Snails

As Runham (1975) remarked: "Digestion (by snails) at first sight appears inefficient, as the feces are often of the same color as the food and contain large pieces of apparently undigested material, yet utilization of ingested food appears to be very high." In fact, the food conversion ratio achieved with farmed *H. adspersa* is always less than 2 (units of meal weight to 1 unit of snail weight gain) and on occasion closer to 1.2. This is a much better conversion rate than achieved by any warm-blooded animal, and appears even more encouraging when one remembers that snail meal contains a very high percentage of the cheapest possible ingredient, calcium carbonate.

A number of proprietary snail meals are available on the market and give good results. For those who wish to prepare their own feed, nutritional information is scant, with none whatsoever available on requirements for amino acids and practically none for vitamins or trace minerals. One of the major difficulties in determining such requirements is that snails do not grow properly if they have no access to soil organic matter (Gomot, 1986; Elmslie, 1998). There is, however, information on the optimum content of calcium carbonate, which should be a little under 30%, with more available, not necessarily continuously. Snail meal should be finely ground. I have found that a hammer mill with a 1 mm screen produces an acceptable particle size and that a mixture of 17.5% wheat, 17.5% maize, 12.5% barley, 25% extracted soy meal, and 27.5% calcium carbonate, with a vitamin-trace mineral supplement to give 10,000 IU kg^{-1} vitamin A, yields tolerable results. Marks and Jess (1994) found that sow meal (composition not known) gave better growth than some commercial snail meals. They also found that plant protein sources (extracted soy meal) gave better growth than herring meal or milk powder, and that urea reduced growth.

Farmed snails should never have to travel more than 50 cm to reach food. If they do, some snails will not make the effort and variation in the cohort will increase.

Although snails can be reared on leaves, especially those of the cabbage family, carrots and fruit, the results are extremely uneven and commercially unsatisfactory. A glance at any cabbage or lettuce attacked by slugs or snails will show why. Snails can distinguish between the qualities of the different leaves in a way that humans cannot; they eagerly consume some leaves but of others on the same plant they eat only as much as they need to stay alive. Snail farmers can usefully offer waste fruit or vegetables to growing snails in the hope of compensating deficiencies in the basic diet of meal, but more than a very small proportion of these foods will reduce growth of juveniles and enormously increase variation within the cohort. On the other hand, fruits and vegetables could apparently constitute a substantial proportion of the diet of adult snails without much reduction in reproductive performance, but quantitative data are not available.

As might be expected from their general ecology, compensatory growth is a well-established phenomenon in edible snail species.

Selective Breeding in Snail Farming

All studies with *Helix* species indicate that inbreeding produces a rapid decline in performance, and sterility within a few generations. On the other hand, there is no evidence whatsoever for hybrid vigor. In fact, all studies to date indicate that hybridization or crossing, whether of inbred or wild strains, depresses performance (e.g. Gomot-De Vaufleury and Borgo, 2001). In snail farming practice, excessive inbreeding can easily be prevented by rearing the snails in groups and regrouping those selected for breeding in each generation.

Such information as is available indicates that adult body size is fairly strongly inherited but reproductive performance much less so, as in warm-blooded domesticated animals. In practical snail farming, however, conventional breeding objectives for warm-blooded farm animals become quite secondary. The primary aim of the snail farmer is to shorten the rearing period. This may require development of a farmed strain in which the snail's circannual rhythm is eliminated or brought into line with the farmer's requirements. This generally takes several generations of selection and results in a smaller adult size. This last size is not a disadvantage in the case of *H. adspersa maxima*, and in the case of the strain of *A. fulica* farmed in the UK constitutes a definite economic improvement (the smaller adults reach maturity earlier and eat less food). Parallel with the decrease in adult size, the life span shortens (e.g. Monney, 1994b).

On the other hand, a smaller adult size is a disadvantage in the case of *H. pomatia*, as the larger individuals fetch a higher price per kilo, and small adults may be rejected by the market. When the species is selected to complete its life cycle within 12 months (with only one hibernation) the adult size is reduced to below the market optimum.

The same effects (smaller size with fewer hibernations or estivations) would undoubtedly be found with any other species that might be farmed.

Yield

The yield achievable per unit area and the time taken to achieve it, are two of the most important constraints in snail farming.

There do not appear to be any reliable published data on yield per unit area from the collection of wild snails. Personal observations suggest that 20 g/m^2 year in the most favourable areas would be unusually high.

Jess and Marks (1995) studied various rearing densities for a cultivated strain of *Helix adspersa maxima* in small cages cleaned at different intervals. Under these conditions, densities equivalent to 400 snails per m^2 gave satisfactory results to lip formation. Higher densities had no adverse effect in the very early weeks of life.

Reliable data on output per unit area of pen from snail farms are scarce, indeed practically limited to one species, *H. adspersa*, as reared in France.

Bonnet et al. (1990: p. 72) quoted a yield of 3 kg/m^2 adult snails from "small pens" in 1988 and a remarkable 4.4 kg/m^2 in 1989. In the latter year, the yield from "large pens" was 2.1 kg/m^2. These figures seem to be the highest possible to achieve. The snails concerned were hatched from eggs laid indoors by artificially stimulated breeders and transferred to the outdoor small or large pen at about 6 weeks of age in April, to be harvested from July onward. They were, of course, fed entirely on meal. The 6-week old snails were initially stocked at the rate of 300/m^2 and both survival and lip formation rate were high in the years Bonnet cited. In commercial practice these two characteristics vary considerably from year to year. In general, it is considered satisfactory if 80% of the young snails placed into pens in April reach salable condition (i.e., shell lip formation which indicates adulthood).

Yields of 2 kg/m^2 have also been reported from snail farms in Italy using a system of platforms in plastic or netting tunnels (F. Ballone, pers. comm.), although the proportion of adult (lipped) snails in the total is not clear.

Nor is it certain whether the unnatural rearing cycle adopted by French growers of *H. adspersa* makes it possible to increase the yield per square meter. No direct comparisons seem to have been made. What is certain is that irrespective of conditions, overstocking reduces yields. If too many young are placed in the same pen, the result is always a drastic increase in the proportion that fail to grow properly, as well as a modest reduction in the adult size of those that do mature; beyond a certain point therefore the number of snails per square meter increases and the saleable yield falls. Snails stunted by overcrowding do not grow subsequently.

Working only with *Helix* snails hatched naturally in field pens in the summer or autumn of one year, hibernating naturally in the field, and maturing in the same or different field pens in the following year, I have been able to obtain hardly more than 1 kg/m^2 year of any *Helix* species (*H. pomatia, H. lucorum, H. adspersa*).

Yield from open pen snail farms that attempt to rear *Helix* species and rely on food plants growing in the pens, supplemented only by forage grown elsewhere (i.e., no compound meal) seem in practice only rarely to exceed 150 g/m^2 year.

Legislation

If there are laws and regulations pertaining to the exploitation of wild populations of snails, their aim should be to maximize the medium- to long-term value of these populations for the inhabitants of the district and the country as a whole, bearing in mind the fact that the economically weakest sections of society are most likely to collect wild snails and in some areas also to eat them. Laws and regulations on collection, if any, should also be based on genuine experimental work and not mere speculation or theory.

A surprising number of countries do in fact have laws and regulations, some good, some bad, some merely silly, concerning exploitation of wild snail populations.

Let us look at the good ones first. In parts of West Africa where snails are eaten, exploitation of the natural population is controlled by the traditional authorities, village councils or chiefs (Monney, 1994), who are likely to be well informed about local conditions and to know something about snails.

The people of Poland do not eat snails to any significant extent, but the country has well-designed and implemented regulations governing the collection of (*H. pomatia)* snails for export. The regulations are described by Stepczak (1992) and have resulted in Polish snails earning a good reputation and some of the highest prices. Collection is permitted only of snails too large to pass through a 30-mm diameter round hole and only from the time the animals emerge from hibernation in spring until May end. Buying stations are open only during this period, which is also that in which the snails are in their peak condition, i.e., before egg-laying. Extensive research by Stepczak and colleagues over many years has shown that collection as thus regulated does not reduce the snail population.

Bulgaria has an export trade similar to Poland and at the time of writing similar, but far less appropriate regulations. Collection is permitted only from mid-May through June. This means Bulgarian snails obtain only a very low price because by late May the quality of the snails has deteriorated and the market season is almost over. Bulgarian regulations are therefore against the national interest (income from the export of snails is perhaps half of what it could be) as well as against the interests of collectors (who are poorly rewarded for their work). It appears that the laws are based on a misunderstanding of the snail life cycle and the effects of collection.

France has both national and local regulations. The national ones prohibit the collection, transport, farming, and utilization of a number of minor helicid species and collection of *H. pomatia* in the period April 1st to June 30th. For the rest of the year collection of *H. pomatia* with a shell diameter of less than 30 mm is prohibited. Surprisingly, and in contrast to Poland, there seems to have been no research to verify the value of these regulations. French and Polish regulations prohibiting collection of adult *H. pomatia* that pass through the 30-mm diameter hole seem likely to favor a long term decline in the average adult size of the wild population.

Some authorities have prohibited traditional collection due to the mistaken notion that this will promote snail farming and help conservation. Given the difference in cost between collecting snails for oneself and buying farmed snails, such regulations merely encourage those who traditionally collected and ate snails to switch to other foods.

Some countries subsidize or otherwise encourage snail farming. In all countries except perhaps France, this has proved a serious mistake. In some cases, as already mentioned, the result has been a disastrous spread of pest species. In other cases, the result merely proved a failure. Italy has wasted considerable sums of taxpayer money on grants to persons with no experience to set up snail farms using unproven technology (Fig. 6.1). In contrast, the French regulations quoted by Bonnet et al. (1990) require applicants for grants to use a proven system and either to have real experience or to undergo training.

France, the largest consuming nation, seems to be the only country to have formulated laws pertaining to the snail market. These are described in the ITC report (1979) and in Bonnet el al. (1990). The regulations specify size classifications and require indication of the species. One might object that a statement of size and species is a rather inadequate description of quality. There is a vast difference in toughness, for example, between very young *A. fulica* and very old individuals of the same species. The regulations also take little note of the difficulty of distinguishing between *H. pomatia* and related species such as *H. lucorum* (Van Osselaer and Tursch, 2000).

Snail Farming in Practice: A Case Study

There are no large snail farms. All the successful ones are small and depend for commercial success on selling above the going market price on the basis of a local reputation for quality.

In considering whether to establish a snail farm, the first requisite is to determine whether a local demand exists for such a premium-priced product. If not, the snail farm will be uneconomical; at most it could be a hobby enterprise comparable, perhaps, to growing one's own vegetables.

If there is such a demand, the next step is to investigate ways in which it might be supplied. Accurate detailed information is in short supply; France is the only country with a government research station providing reliable data. Elsewhere, successful snail farmers exploiting niche markets are not likely to give away their own hard-won know-how to potential competitors, but the sector is infested with misguided or fraudulent promoters who "help" beginners to set up snail farms that fail. Furthermore, "snail farming" currently includes the farming of five or more species, so obviously there can be no single simple "method". The aspiring snail farmer may therefore still have to do a substantial amount of research and development, as did the pioneers.

One of the best known and most successful pioneer snail farms in Europe is that of Combe and Moulin near Digne in southern France. As mentioned earlier, for the last two decades or so they have been raising, processing, and marketing their own strain of species *H. adspersa* developed from wild North-African foundation stock. The yearly cycle starts with bringing of the breeding stock out of diapause to lay eggs in a heated building, furnished approximately as shown in Figure 6.2, in the early months of the year. When the eggs are hatched, the young snails are moved out to a greenhouse outfitted roughly as shown in Fig. 6.2. The firm has developed its own snail feed, which gives better results and costs less than that available commercially. By July the young snails are ready to harvest. A proportion are sold live, but the great majority are processed and frozen, many to be sold in the peak buying season, Christmas. Deliveries are made to restaurants and the like within a 100 km radius. Supermarkets are not potential customers as they prefer to buy cheaper imported (i.e. wild) snails.

In West Africa, as far as I know, there are no commercially successful snail farms at this time, though there are quite a number of people and organizations rearing snails as a hobby, or to see how it goes, and at least one firm in Lagos has combined an attempt at snail farming with an apparently more successful business in preparation of wild snails for the consumer. The future of snail farming in West Africa certainly depends on establishing rational objectives (i.e., profit rather than feeding the masses) and on the provision of economically focused training for would-be snail farmers.

Conclusions and Recommendations

1) Snails collected from the wild have provided a significant source of food and income for centuries and continue to do so, especially for poorer sections of society.

2) The value of wild snails to collectors and to the country can be increased by appropriate regulations and reduced by inappropriate ones. The aim should be to maximize the medium- to long-term value of such populations to collectors and the country.

3) Laws and regulations should be based on genuine experimental work, not mere theory.

4) Traditional collection of edible snails from wild populations does not threaten their existence.

5) Misguided attempts at snail farming have caused great economic damage by spreading agricultural pests.

6) Snail farming is not a means of producing cheap food for the masses. When appropriate systems are used, it can produce a premium product.

7) The market for edible snails is limited by consumer preference or prejudice, as well as price.

8) Edible snail consumption could be significantly increased if quality supplies were regularly available.

9) Would-be snail farmers should first assess their ability to market the product.

References

Adalla C.B. and Morallo-Rejesus B. 1989. The golden apple snail *Pomacea* sp., a serious pest of rice culture in the Philippines. Slugs and Snails in World Agriculture BCPC Mono 41: 417–422.

Airey W.J., Henderson I.F., Picket J.A., Scott G.C., Stephenson J.W., and Woodcock C.M. 1989. Novel chemical approaches to mollusc control. Slugs and Snails in World Agriculture BCPC Mono 41: 301–307.

Awesu M.O. 1988. Observations on some aspects of reproduction of *Archachatina marginata* (Swainson) in captivity in a culture pen. Snail Farm. Res. 2: 39–44.

Baronio P., 1974. Gli insetti nemici dei molluschigasteropodi. *Boll. Ist. Ent. Agr. Bologna* 32: 169–187.

Baur B. 1986. The influence of cannibalistic egg-eating on the growth of young *Arianta arbustorum* (L.). Snail Farm. Res. 1: 9–13.

Bonnet J-C., Aupinel P., and Vrillon J-L. 1990. L'escargot *Helix* aspersa biologie-élevage. INRA, Paris, 124 pp.

Bradley M.D. and Runham N.W. 1996. Heavy metal toxicity in the snail *Helix aspersa maxima* reared on commercial farms: cellular pathology. Slug and Snail Pests in Agriculture BCPC Symp. Proce. 66: 353–358.

Canestri Trotti G. and Toffoletto F. 1979. L'attuale distribuzione geografica di *Achatina fulica* e sua importanza come ospite di *Angiostrongylus cantonensis*. Quaderni di Elicicoltura. 8: 35–40. Comune di Borgo San Dalmazzo. Italy.

Cantoni C., Beretta G., Bianchi M.A., and Belloli A. 1980. Aminoacidi nei molluschi. Quaderni di Elicicoltura 9: 31–34. Comune di Borgo San Dalmazzo. Italy.

Chevallier H. 1985. L'élevage des escargots. Du Point Veterinaire (ed.). Maisons Alfort, 127 pp.

Chevallier H. 1986. Final fattening and outdoor breeding of the edible snail *Helix aspersa*. Snail Farm. Res. 1: 1–8.

Chevallier H. 1990. Breeding in outdoor pens of the edible snail *Helix aspersa maxima*. Snail Farm. Res. 3: 32–35.

Cobbinah J.R. 1995. Aestivation responses of three ecotypes of the Giant African Snail *Achatina achatina*. Abstracts, 12th Int. Malacological Congr., Vigo Spain, The abstracts for the International Malacological Congresses at Vigo and Siena were prepared by the Congress organizers and distributed to participants and other members of Unitas Malacologica. pp. 164–165.

Cobbinah J.R. Snail Farming in West Africa, a practical guide. Technical Centre for Agricultural and Rural Cooperation (CTA), ISBN 92 9081 1021

Combe J.C and Abbes G. 1980. Premières observations sur le comportement de *Helix lucorum* Linné en milieu naturel (Alpes de Haute Provence) et en élevage. Haliotis 10: 91–94.

Elmslie L.J. 1988. Studies on the feeding of newly hatched *Helix aspersa*. Snail Farm. Res. 2: 45–48.

Elmslie L.J. 1989. Snail farming in field pens in Italy. Slugs and Snails in World Agriculture BCPC Mono 41: 19–26 (density effect).

Elmslie L.J. 1992. Adaptation of snail species to farming systems, 2: Precocious aestivation of an introduced strain of *Helix aspersa* in central Italy. Snail Farm. Res. 4: 7–12.

Elmslie L.J. 1998. Humic acid: a growth factor for *Helix aspersa* Müller (Gastropoda: Pulmonata). J. Molluscan Studies 64: 400–401

Elmslie L.J. 2001. Strain differences in *H. pomatia* L. (Gastropoda: Pulmonata): diapause, digging, growth rate and shell fill. J. Molluscan Studies 67: 121–124.

Gomot A. 1998. Biochemical composition of *Helix* snails: influence of genetic and physiological factors. J. Molluscan Studies 64: 173–182.

Gomot-De Vaufleury A. and Borgo R. 2001 Experimental hybridisation between two sub-species of snails (*Helix aspersa aspersa* and *Helix aspersa maxima*): consequences for fertility, survival and growth. Invert. Reprod. Develop. 40: 2–3, 217–226.

Gomot A., Bruckert S., Gomot L., and Combe J.C. 1986. A contribution to the study of the beneficial effects of soil on the growth of *H. aspersa*. Snail Farm. Res. 1: 76–83.

Graham F., Runham N.W., and Ford J.B. 1996 Long term effects of *Riccardoella limacum* living in the lung of *Helix aspersa*. BPCP Symp. Proce. Slug and Snail Pests in Agriculture 66: 359–364.

Hawkyard Cage Systems. 72 High St., Ketton, Stamford, Lincolnshire, PE9 3TE England.

Heller J. and Ittiel H. 1990. Natural history and population dynamics of the land snail *Helix texta* in Israel. J. Molluscan Studies 56: 189–204.

Hodasi J.K.M. 1979. Life History studies of *Achatina achatina* (Linnè). J. Molluscan Studies 45: 328–339.

Hodasi J.K.M. 1984 . World Animal Review 52: 24–28.

Imevbore E.A. 1992. Investigation into the parasitic infestation and health problems of the African Giant Snail *Archachatina marginata* (Swainson). 11th Int. Malacological Cong., Siena, Italy.

ITC (International Trade Commission). 1979. The market for live snails and snail products with special reference to the French market. Geneva: (pamphlet)

Jess S. and Marks R.J. 1995. Population density effects on growth in culture of the edible snail *Helix aspersa* var. *maxima*. J. Molluscan Studies 61: 313–324.

Kubota Y., Watanabe Y., Otsuka H., Tamiya T., Tsuchiya T., and Matsumoto J.J. 1985. Purification and characterisation of an antibacterial factor from snail mucus. Comp. Biochem. Physiol. 82C: 345–348.

Lind H. 1989. Homing to hibernating sites in *Helix pomatia* involving detailed long term memory. Ethology 81: 221–234.

Marks R.J. and Jess S. 1994. Effects of dietary protein source and content on growth of *Helix aspersa* var. *maxima* snails. Snail Farm. Res. 5: 64–74.

Monney K.A. 1994a. Notable notes on giant African snails. Snail Farming Research 5: 1–13.

Monney K.A. 1994b. Effects of different dietary regimes on growth and reproductive function of farmed *Achatina fulica* Bowdich. Snail Farming Research 5: 14–22.

Oosterhoff L.M. 1977. Variation in growth rate in as an ecological factor in the land snail *Cepaea nemoralis* (L). Netherlands Journal of Zoology 27: 1–132.

Pallis F.V., Macatula R.F., and Browing L. 1996. Nicolosamide, an effective molluscicide for Golden Apple Snail control in Philippine rice production systems. Slug and Snail Pests in Agriculture BCPC Symposium Proceedings 66: 213–230.

Plummer J.M. 1975. Observations on the reproduction, growth and longevity of a laboratory colony of *Achachatina (Calachatina) marginata* (Swainson), subspecies ovum. Proceedings of the Malacological Society of London 41: 395–413.

Runham N.W. 1975. The alimentary canal. Pulmonates, Fretter and Peake, (eds.). Functional Anatomy and Physiology. 1: 53–104. Academic Press, London.

Salghetti A. 1996. L'elicicoltura in Italia: analysi strutturale ed economica. Parma Economica—March 1996: 69–108.

Staikou A., Lazaridou-Dimitriadou M., and Farmakis N. 1988. Aspects of the life cycle, population dynamics, growth and secondary production of the edible snail *Helix lucorum* Linnaeus 1758 (Gastropoda, Pulmonata) in Greece. J. Molluscan Studies 54: 139–156.

Stepczak K. 1992 Effects of exploitation of natural populations of *Helix pomatia* L. Snail Farming Research 4: 53–71.

Stievenart C. 1996. Morphologie coquillière, Croissance, Reproduction and estivation chez les escargots géants africains. PhD thesis, Prince Leopold Institute of Tropical Medicine, Department of Tropical Animal Production and Health. Antwerp, Belgium.

Tompa A.S. 1984. Land Snails (Stylommatophora). The Mollusca. Vol. 7. Reproduction. pages 48–127, Academic Press Inc.

Van Osselaer C. and Tursch B. 2000. Variability of the genital system of *Helix pomatia* L., 1758 and *H.lucorum* L. 1758 (Gastropoda: Stylommatophora). J. Molluscan Studies 66: 499–516.

7

Overview of Role of Edible Insects in Preserving Biodiversity

GENE R. DEFOLIART

Dept. of Entomology, University of Wisconsin, Madison, WI 53706, USA

Abstract

The principle adopted here is that factors pending to increase food and/or income for economically marginal rural families, while decreasing pressure for land clearing, pesticides, and intensive agriculture, will tend to favor preservation of biodiversity and sustainable future. The great diversity of habitats of insect species that serve as traditional foods presents an almost endless diversity of situations in which recognition and enlightened management of the food insect resource can result not only in better human nutrition, but simultaneously aid in maintaining diversity of habitats for other forms of life. Approaches include: 1) enhancing forest conservation and management by acting on the desire of local populations for protection of traditional insect foods (i.e., caterpillars in Zambia and Zaire); 2) reducing poaching in parks and wildlife preserves by allowing sustainable use of the food insect resources by the local people (i.e., caterpillars in Malawi); 3) reducing pesticide use by developing more efficient methods of harvesting pest species that are traditional foods (i.e., grasshoppers); 4) increasing environmental and economic efficiency by developing dual product systems (i.e., silks and silk moth larvae/pupae, honey and honeybee brood); 5) reducing organic pollution by recycling agricultural and forestry wastes into high-quality food or animal feedstuffs (i.e., fly larvae, palm weevils). Other relevant considerations are that some edible insect species enhance their local environment in various ways (i.e., leafcutter ants in S. America) or create additional diversity of species within the habitat (i.e., termites in Africa). Some, as shown in studies with crickets exhibit considerably higher food conversion efficiency than beef cattle when fed diets of similar quality. Finally, there is need for research on industrial scale mass production of edible insects, for increased recognition of the nutritional and environmental importance of insects by national governments, and for increased involvement of Western media and academia in dispelling unfounded cultural biases in the Western World toward insects as food.

Key Words: biodiversity, edible insects, entomophagy, food ecology, food insects

Introduction

Insects are among the traditional foods of many cultures around the world and have played an important role in the history of human nutrition. They are a good source of protein, fat (and thus calories and energy), unsaturated fatty acids, important minerals such as iron and zinc, and vitamins such as thiamin and riboflavin (DeFoliart, 1992). They are sold in village markets and selected favorites in urban restaurants; thus they are not only a source of nutrition, but a source of income for economically marginal rural populations. In approaching this subject, the precept is adopted that policies, principles, and practices that tend to increase food and/or income for economically marginal rural families, may decrease pressure for land clearing, intensive monoculture agriculture and/or pesticides, thereby favoring preservation of biodiversity. Economic incentives are inextricably linked to successful conservation effort and are included in the ensuing discussion.

Many, Taxonomically Diverse, Insect Species Used as Food

Worldwide the number of insect species consumed by humans probably totals well over 2,000 with 20 or 30 or more species constituting part of the cuisine in many individual countries. These insect species represent a great diversity of habitats, ranging from arid to rainforest to aquatic. The total number of edible insect species used in selected countries is shown in Table 7.1. These are very conservative estimates. Only those species were counted which could be assigned from the available published information, to a specific genus, family, and order. For example, the 32 species tabulated for Zimbabwe include only six species of Lepidoptera (caterpillars) although 12 to 15 vernacular names, mostly Shona names, have been recorded for caterpillars by four different investigators in that country. For China, the 46 species listed in Table 7.1 include one species each of the hymenopteran families Apidae, Vespidae, and Scoliidae, but Xiaoming (1990) stated that in Yunnan Province, larvae and pupae of five species of bees and wasps belonging to those three families are eaten by minority nationalities. The 27 species tabulated for Japan include only one species of Cicadidae, a fried cicada (*Graptopsaltria nigrofasciata*) known as semi, is found on menus of restaurants that specialize in foods of the Japanese Alps (Pemberton and Yamasaki,

Table 7.1: Number of edible insects reported from selected countries

Country	**Number of each taxon**			
	Orders	**Families**	**Genera**	**Species**
Asia				
Burma	7	14	17	17
China	10	30	36	46
India	7	17	22	24
Indonesia	8	15	20	25
Japan	11	19	22	27
Philippines	6	13	17	21
Thailand	10	31	69	80
Vietnam	8	18	20	24
Australian Region				
Australia (indigenous)	7	22	39	49
Papua New Guinea	11	22	31	34
Africa/Madagascar				
Congo	7	15	25	30
Madagascar	7	15	22	22
South Africa	7	16	32	36
Zaire	5	21	47	62
Zimbabwe	7	14	25	32
Western Hemisphere				
Brazil	7	14	19	23
Colombia	8	20	36	48
Mexico	10	42	99	136*
USA (formerly indigenous)	10	27	53	69

*Now known to total more than 200 species.

1995). On the other hand, Remington (1946) stated that among the very popular foods eaten in Japan are all species of Cicadidae. Some countries which obviously use as many edible insect species as those countries in Table 7.1 are not included because of inadequate taxonomic information for placement in the Table. For example, one observer in Cameroon stated that (in addition to other insects consumed) 21 species of caterpillars are eaten by the Pangwe of southern Cameroon; a worker in Zambia listed 18 vernacular names of caterpillars consumed in that country.

Number of Insect Species Used as Food Greatly Underreported for Most Countries

Despite the totals shown in Table 7.1, more field studies supported by expert taxonomic input are needed before we can fully appreciate the number and taxonomic diversity of insects used as food. The taxonomic identity of fewer than 20 species was known in Mexico prior to the early 1970s when Ramos-Elorduy (formerly de Conconi) launched her intensive studies. These studies, published in a long series of papers (see de Conconi et al., 1984 for example), reveal more than 200 edible species in various parts of Mexico. Similarly, the intensive study of edible caterpillars in southern Zaire by Malaisse and Parent (1980) more than doubled the number of edible species with known taxonomic identity in that country. Marais (1995) in Namibia noted that "different tribes and ethnic groups in the same environment do not necessarily make use of the same resources."

The findings of two studies conducted in Colombia and Venezuela are compared in Table 7.2, the first by Ruddle (1973) on insect foods of the Yukpa in the northeastern part of Venezuela, the second by Dufour (1987) on insect foods of the Tukanoans in southeastern Colombia. Samples from both studies were examined by taxonomists of the US Dept. of Agriculture's Insect Identification Laboratory, Beltsville, Maryland. The results suggested that at least 25 species are included in the diets of the Yukpa and at least 20 consumed by the Tukanoans, but with virtually no overlap in species recorded from the two areas. Wide differences are evident. Ruddle, for example, found seven or more species of grasshoppers used by the Yukpa while Dufour observed no grasshoppers in the Tukanoan cuisine. Conversely, Ruddle observed no termites, while Dufour recorded three species. The only overlap occurred in Formicidae for which the "*Atta* spp." reported by Ruddle undoubtedly include one or more of the three species of the leafcutter ants (*Atta*) reported by Dufour.

Desire of Local People to Protect Traditional Food Resources Favorable to Forest Management

In Zambia, late burning during the dry season (when it is very dry) can severely damage regeneration of the *miombo* woodland that dominates the country. Some trees are killed, regrowth is reduced, and erosion is increased. The best way to prevent this damage is by early burning. In part of this territory, is saturniid caterpillar (called *mumpa* in the local language) flourishes, which is highly relished and important both as a source of nutrition and source of income. *Mumpa* caterpillars feed on *Julbernardia paniculata* and several other common trees in the *miombo* woodland. One person can pick about 20 liters per day if the bush is rich in caterpillars. Seven days' picking will earn, if all are sold, the equivalent of a month's salary for a general laborer in Zambia. It is not surprising that people travel 200-300 km to pick caterpillars and traders come from Lusaka and the Copperbelt (900 km) to buy and sell them on their return at a much higher price.

In 1985, an agroforestry researcher, Stein Holden (1991), noticed that there were very few late bushfires in the areas where the *mumpa* caterpillars are found. In fact, on one occasion when Holden was travelling with Zambian companions in caterpillar territory and a late bushfire was seen, his companions said: "A stupid guy has set it on fire because he wants to destroy our caterpillars."

Table 7.2: Comparison of edible insect species used by the Yukpa in northeastern Colombia (Ruddle, 1973) and the Tukanoan (Tatuyo) in southeastern Colombia (Dufour, 1987)

Taxa	Yukpa (Ruddle)	Tukanoan (Dufour)
COLEOPTERA (beetles, weevils)		
Bruchidae	*Caryobruchus* spp.	
Buprestidae		*Euchroma gigantea*
Cerambycidae		*Acrocinus longimanus*
Curculionidae	*Anthonomus* spp.	*Rhynchophorus* spp.
Passalidae		Genus ?
Scarabaeidae	*Podischnus agenor*	*Megaceras crassum*
HYMENOPTERA (ants, bees, wasps)		
Fomicidae	*Atta* spp.	*Atta cephalotes*
		Atta laevigata
		Atta sexdens
Meliponidae	*Trigona clavipes*	
	Trigona trinidadensis	
Vespidae	*Mischocyttarus* spp.	*Apoica thoracica*
	Polistes canadensis	*Polybia rejecta*
	Polistes pacificus modestus	*Stelopolybia angulata*
	Polistes versicolor	
	Polybia ignobilis	
LEPIDOPTERA (caterpillars)		
Hesperiidae		Genus ?
Lacosomidae		Genus ?
Noctuidae	*Laphygma frugiperda*	
	Mocis repanda	
		Genus ?
	Genus ? (2 spp.)	
Notodontidae		Genus ?
Saturniidae		Genus ?
ORTHOPTERA (grasshoppers)		
Acrididae	*Aidemona azteca*	
	Orphulella spp.	
	Osmilia flavolineata	
	Osmilia spp.	
	Schistocerca spp.	
	Tropidacris latreillei	
Tettigoniidae	*Conocephalus angustifrons*	
DIPTERA (fly larvae)		
Stratiomyidae	*Chrysochlorina* spp.	
ISOPTERA (termites)		
Termitidae		*Syntermes parallelus*
		Syntermes snyderi
NEUROPTERA (dobsonflies, etc.)		
Corydalidae	*Corydalus* spp.	
TRICHOPTERA (caddisflies)		
Hydropsychidae	*Leptonema* spp.	

In areas where the *mumpa* caterpillars are found, the people are careful to burn early to protect them and concomitantly enhance woodland regeneration.

Several years earlier, in Zaire, a territorial administration in the Kwango District of southwestern Zaire commissioned a study (Leleup and Daems, 1969) to determine whether fluctuations and reduced tonnage of the most economically important caterpillars might be due to poorly timed burning. Optimum times for burning that would minimize caterpillar destruction were determined. One species involved, *Cirina forda,* is one of the most widely eaten in Africa. In addition to optimum times for burning, recommendations from the study included: 1) enforce the ban on felling trees in order to harvest the caterpillars; 2) prohibit the increasing practice of harvesting pupae; 3) encourage resowing attempts on a massive scale by collection of eggs prior to burning; and 4) create reserves of some small wooded savannas in which all harvesting for purposes of consumption would be forbidden.

Opening National Parks and Other Wildlife Preserves to Controlled Sustainable Use by Local People can Reduce Poaching

A study conducted in Malawi's Kasungu National Park (2,316 km^2) and in human settlements adjacent to its eastern boundary (Munthali and Mughogho, 1992) demonstrated the advantages of introducing economic incentives that integrate biological conservation with economic development for the rural people. Prior to the study, management practices for Kasungu and other protected areas stressed nonconsumptive utilization through ecotourism and law enforcement. For neighboring rural people, however, most of them families and their descendants who were resettled outside the Park when it was established in 1930, outdoor recreation is of low priority in their hierarchy of needs and anyway, the cost of entry to parks and reserves is more than they can afford. Furthermore, since the money from ecotourism goes into the central treasury, rural people view the management policies as favoring the most affluent rather than addressing their socioeconomic dependence on wildlife. They manifest their antagonism through illicit encroachment into protected areas.

In 1990, Malawi's Department of National Parks and Wildlife allowed 173 families (about 10% of all households around the Park) to harvest caterpillars in the Park, and simultaneously initiated modern beekeeping in the Park to diversify the rural communities' income base and to win their support for wildlife conservation programs. The caterpillars are two species of emperor moth (Saturniidae), *Gonimbrasia belina* and *Gynanisa maia,* which occur abundantly, with larvae available from about mid-October to December every year. Initially, 100% of the families practiced beekeeping and utilized saturnid caterpillars and other products of the forest such as game animals, small mammals, medicine, mushrooms, firewood and poles. Now, only 33% practice beekeeping outside the Park, due mainly they say to lack of bee forage. Caterpillars are nonexistent outside the Park because of the absence of forage tree species. According to the investigators, extensive agriculture (tobacco estates, and maize, beans and peanuts grown by small landholders for subsistence and cash) is the main cause of the rapid dwindling of Malawi's rich biodiversity, even though 22% of its total area is legally protected as national parks, wildlife and forest preserves.

It was found during the study that (similar to the situation in Zambia), significantly greater caterpillar yields were obtained from plots burned early every year followed by no burn, with the lowest yield after late burn as it is obviously destructive to the eggs and larvae as well as the foliage on which the caterpillars feed. Yields also varied significantly with forage tree height, with highest yield from height class 1–3 meters. The authors therefore recommend a rotation burning policy that promotes both good caterpillar yield and vegetation coppicing with more stems in the 1–3 m class. This height class has the added advantage that it puts the caterpillars within easy reach for harvesting. Relative to beekeeping productivity, both honey and wax yields were found to increase for 1 through 5 years, then decline, thus requiring modest investments in new hives and other equipment in year 1 and again in the fifth year.

Munthali and Mughogho (1992) used gross margin analysis (defined as output minus the variable associated costs expressed in money terms) as a measure of each enterprise's economic efficiency. Caterpillars and beekeeping had more than twice to several times the gross margin values of maize, beans, and peanuts. These wildlife-based enterprises not only produce earnings exceeding those from agriculture, they also do not compete directly for labor requisite for existing agricultural enterprises. Most families affirmed that they would have time to practice beekeeping and/or to harvest caterpillars even during the crop season. Of added importance, of the small landholder families in the study area, 50% run out of food stocks by November, which coincidentally, is when caterpillars and honey are in season.

Munthali and Mughogho (1992) concluded that the utilization of honey and caterpillars by the rural people in the Park is an important turning point in the history of wildlife management in Malawi. While taking full cognizance of the Park's primary purpose—to preserve the country's representative biotic communities, they indicated that: "The DNPW needs to take full advantage of the rural people's willingness to be allied with wildlife management programmes and consolidate it through the validation of sustainable land use practices."

Reducing Pesticide Use by Harvesting Food Pest Species that are Traditional Foods

According to Pemberton (1994), rice-field grasshoppers, primarily *Oxya velox*, locally called *metdugi*, were earlier a common food ingredient in Korea but consumption declined as insecticide use increased during the 1960s and 1970s.

Metdugi ceased to be found in Seoul markets whereas silkworm pupae (*Bombyx mori*) were almost always available. In Chahwang Myun (a district in Sanchung County) insecticide spraying began to decline in 1981, allowing *metdugi* populations to begin increasing, with some collected and sold once more in the local market at Sanchon in 1982.

Pemberton stated: "The decline in insecticide use and the desire of some Koreans to eat pesticide-free rice led to the development of organic rice farming in Chahwang Myun. This was economically viable because the yields of rice were the same in unsprayed fields as in sprayed fields, and organic rice sold (and still sells) for higher prices. In 1989, the Chahwang Agricultural Cooperative began buying dried *metdugi* from farmer collectors. In 1990, more than 600 families sold 1,744 liters of *metdugi* to the Cooperative at 5,000 Won per liter (US $6.98). The Cooperative sold them for 6,500 Won per liter (US $9.08); much of the 1990 sale went to a supermarket company in Pusan which divided the *metdugi* into 0.2 liter packages and sold these for 3,000 Won (US $4.19). By 1992 the Cooperative was paying US $9.91 per liter for *metdugi* and selling it at bulk rate for US $12.03 per liter. In addition to selling to the Cooperative, farmers sold *metdugi* at the local five-day markets (open one day every five days) and on the street.

Metdugi are most commonly collected by older women and usually from mid-October to early November. The collected *metdugi* are steamed or boiled, then dried in the sun for one day and in a room for two more days. Dried *metdugi* are sometimes eaten as such without seasoning, but usually they are pan-fried with or without oil after removing the legs and wings. Pemberton described another preparation: "During or after cooking, they are flavored with sesame oil and salt, or sesame oil and sugar, or soy sauce with or without sugar. I have also seen live ones fried whole. These turn red like shrimp as they cook. Many of these preparations produce a product with good snack food essence. They are bite-sized, crispy, crunchy, and salty and/or slightly sweet...". According to Pemberton, many Koreans consider *metdugi* a health food and, for older Koreans, it brings nostalgia—a taste of the past.

A one-liter package of *metdugi* purchased from the Cooperative was found to consist of three species, Oxya *velox* (84.5%), *Oxya sinuosa* (14.8%), and a single *Acrida lata*, a large species not expected to be found in *metdugi* although one of the species eaten in Korea.

According to Litton (1993), grasshoppers are a favorite food in many parts of the Philippines and therefore are not sprayed with chemical insecticides. Further, they are fed to chickens raised on pasture and no commercial feed: these chickens have a delicious taste and are sold at a much higher price than chickens fed on commercial feed. Insecticidal sprays were used recently, however, during a 1993–1994 outbreak of *Locusta migratoria*, with little success according to Philippine newspaper accounts (DeFoliart, 1995a). A movement began in some areas, apparently partly farmer-instigated and partly government-instigated, to harvest the insects for sale as food for humans and as animal feed supplements, primarily for cattle and fish (tilapia). Some farmers began using commercially available nets to catch locusts, and netting to provide protective cover for high value crops. In Nueva Ecija, the provincial board appropriated additional funds to pay for locusts caught by residents in affected towns. According to one report: "Locusts are selling [here] like hotcakes in San Antonio, [since] the insects are considered a gourmet's delight." In Zambales, the locust task force sponsored a locust cooking contest among housewives. In some areas, however, there was criticism that lack of community involvement by some local government officials hampered the campaign to harvest locusts for use as food and feed.

In Thailand: "Deep-fried, crispy grasshoppers are very well-liked by a lot of people" (Vara-asavapati et al., 1975; see chapter 20, this volume). According to Gorton (1988), villagers in northeast Thailand have been able to stem the insect damage to crops by "turning foe into food," a notable example being the grasshoppers or "flying shrimp" which feed on corn seedlings, sugarcane, rice and banana leaves. Government spraying programs involving "massive amounts of chemicals" became less and less successful while "being expensive and hazardous to health." Villagers sometimes ate insects dead from pesticides and as recently as 1986, stories circulated of villagers dying or falling seriously ill due to chemically treated insects. "Large cross-province shipments of grasshoppers killed by pesticides, then fried to a crisp in cooking oil were a lucrative business for some." Gorton (1988) reported that as a result of information from news reports and health officials, some districts and villages gave up pesticidal spraying in favor of grasshopper-catching competitions. Health and culinary experts gave demonstrations on the best way to clean and prepare the grasshoppers. Gorton added: "Those who can catch the insects in mass quantity are able to sell them on the village roadside or become involved with the lucrative 'export' trade to Bangkok".

Similarly, the November 13, 1983 edition of the *National Review,* published in Bangkok, described a campaign launched by local officials in which villagers in the Province of Prachinburi collected more than 10 tons of pest grasshoppers for use as food. The program was launched because chemical control efforts had been unsuccessful. The article stated: "Fried and crispy grasshoppers are, according to many people, delicious snacks and many food shops in Prachinburi and other provinces served them for their customers. For beer and whiskey drinkers, fried grasshoppers are marvelous. Grasshoppers have now become one of the exporting items of Prachinburi which has a long list of orders from traders who buy them at six baht a kilo.... Grasshoppers have become a favorite dish for many people...".

Anon. (1992) described the Thai grasshopper-collecting initiated in 1983 in more precise economic terms. The price of grasshoppers rose from US 12 cents per kg in 1983 to US $2.80 per kg in 1992. At local restaurants, once deep fried, they cost about US $6 kg^{-1}. A small farmer can earn up to US $120 per half-acre, twice as much as he can from corn. The trade in grasshoppers now averages about US $6 million per year. Because of the obvious benefits in containing the grasshopper population, the Thai government has publicized a number of grasshopper recipes.

Multiple-product Food Insect Systems can Increase Economic and Environmental Efficiency

The silkworm or mulberry silk moth, *Bombyx mori* (Lepidoptera: Bombycidae), furnishes an edible by-product, the pupa, that has long been considered a food delicacy in many parts of Asia. In some countries it has also been used as a high-protein animal feedstuff, especially in pond fish culture. In addition, there are more than a dozen species of "wild silk" producers in Asia and Africa (Lepidoptera: Lasiocampidae, Notodontidae, and Saturniidae) from which the pupae (or in one genus the larvae) likewise are available as a food by-product (DeFoliart, 1995b). For tribal people in northeastern India, the pupa of the eri silkworm, *Samia ricini,* is so highly regarded as a food delicacy that the cocoon is regarded more or less as a by-product (Chowdhury, 1982). Sericulture is carried on as a cottage industry involving about 40,000 families (Peigler, 1993). It is also a cottage industry in Nepal and, although the pupae are not eaten, there is interest in using them as feed for poultry and pond fish.

Neupane et al. (1990) investigated the rearing biology of *S. ricini* in Nepal and found that six generations are produced per year when the caterpillars are grown on castor leaves, although rearing is not recommended during the cold months, November to April, because the life cycle is then extended. This species is an excellent example of a multiple-product insect and of sustainable agricultural practice. The castor plant grows on poor soils, helping to prevent erosion; castor bean oil is sold for medicinal and industrial uses; excess leaves are fed to the caterpillars which produce silk and a pupa that is a high-protein food or animal feedstuff, and the caterpillar frass and other rearing residue can be used for pond fish culture.

Honeybees (Hymenoptera: Apidae) are another multiple-product insect group. In addition to their immense importance in pollination, modern products of the hive include honey, beeswax, pollen, propolis, royal jelly, venom (for treatment of severe sting allergies) and brood (larvae and pupae). The brood is not only relished as food in many indigenous cultures, but has proven nutritional value in the feeding of nonhuman animals, in particular songbirds. In the form of drone powder, brood has proven valuable in rearing certain insectivorous predators used in biological control programs (Schmidt and Bachmann, 1992).

Apis mellifera is by far the most widely distributed honeybee, but three species of wild *Apis—A. dorsata A. florea,* and *A. indica*—are important sources of honey, wax, and brood in Southeast Asia. Elsewhere, many species of stingless bees (subfamily Meliponinae) are important. In Mexico, the honey and brood of *A. mellifera* and at least eight species of Meliponinae belonging to genera *Melipona, Trigona, Partamona,* and *Lestrimelita* are used as food. The stingless bees are cultivated in small clay jars kept near the walls of houses (de Conconi, 1982). Nine species, *A. mellifera* and eight species of stingless bees are semi-domesticated or manipulated to some extent by the Kayapo in the state of Para, Brazil, who recognize 56 species of bees mainly on the basis of ecological niche and behavioral characteristics (Posey, 1983; Posey and Camargo, 1985). The brood as well as the honey of seven of the manipulated species (genera *Trigona, Oxytrigona, Scaptotrigona,* and *Tetragonisca)* are used as food. Bees, including stingless species, are also important in Africa. Although apiculture is not practiced in southern Shaba in Zaire, villagers are very fond of the brood and it is harvested with the honey. In addition to *Apis mellifera adansonii*, domesticated elsewhere in Zaire, Parent et al. (1978) call attention to *Meliponula bocandei* as a candidate for possible domestication and local apiculture. Five species of *Trigona* are also important as a source of both honey and brood in the clear forest of southern Shaba.

It has been suggested that *Apis mellifera* because of its good public image, might be a valuable tool in reshaping attitudes toward insects as food in the US (DeFoliart, 1989). Initially at least, honey producer associations could provide a ready-made marketing network. DeFoliart (1995b) has discussed the advantages of drone pupae production but research would be needed to develop efficient harvest procedures.

Reducing Organic Pollution by Recycling Agricultural and Forestry Wastes Directly into High-quality Food or Animal Feedstuffs

Throughout the tropics, palm weevil larvae (Curculionidae: *Rhynchophorus)* are widely regarded as a delicacy. There are three major species: *Rhynchophorus palmarum* in tropical America, *R. phoenicis* in Africa, and *R. ferrugineus* in Asia (see chapters 17 and 21). Indigenous people have long "farmed" the larvae, collecting them from palm logs and felled palms in the knowledge that they will develop to harvestable size within two to three months. Mercer (1994) has described in detail the present-day production of sago grubs *(Rhynchophorus ferrugineus papuanus)* in Papua New Guinea. These insects are serious pests of palms and in the Western Hemisphere, *Rhynchophorus* is the vector of the red-ring disease nematode, *Bursaphelenchus cocophilus.* Although insecticides have been used for control, emphasis is on cultural methods including elimination of breeding sites by reducing physical injury to palms, control of *Oryctes* beetles, destruction of infested, injured or decaying trees, and trapping of adult weevils (Hill, 1983).

In view of the above, DeFoliart (1990) discussed the use of trap logs for increasing larval production and proposed: "Palm worms would certainly seem worthy of wider publicity as traditional cuisine of gourmet quality, the kind of delicacy that could be promoted as tourist and urban fare by the best restaurants throughout the tropics and subtropics and eventually, maybe, even as an item of export. Could such wider promotion and use create more opportunities for employment and entrepreneurship in the rural countryside? Could, in fact, expanded markets provide a basis for attempting to combine increased palm-worm production with more efficient recycling of dead and diseased palms and as part of reduced pesticide integrated pest management (IPM) programs and disease control on coconut and other palm species?" Trials developed locally in Amazonas, Venezuela, proved promising (Cerda et al., 2001; Chapter 17, this volume).

Although not as universally admired as palm weevils, palm rhynoceros beetle larvae (Scarabaeidae: mainly genus *Oryctes*) are widely eaten by indigenous populations in Asia and Africa. Cultural methods similar to those for *Rhynchophorus,* biocontrol methods, and insecticides have been recommended for control. In reference to control, Bedford (1980) stated that although there is unanimity in advocating the destruction of breeding sites, "the methods are laborious, expensive, unpopular, and frequently ignored." In Western Samoa, dead coconut trunks were dumped into the sea, in Malaysia and the Ivory Coast they were cut into lengths and stacked for burning, while in Indonesia burying rather than burning was suggested. It should be mentioned that *Oryctes rhinoceros,* widely distributed in Asia and the Western Pacific, breeds not only in standing dead coconut palms, stumps and logs on the ground, but in other types of decaying wood, compost, dung heaps, rotting straw, rotting coconut husks, coffee and cacao pulp waste, and refuse from sugarcane factories, ricemills, sawmills, and other wastes from agricultural processing. Considering the similarity in cultural methods, it seems possible that *Oryctes* could also be incorporated into palm IPM programs, recycling an endless variety of tropical wastes into animal protein and fat.

In situations where poultry and other animal manures are not usable as fertilizer in the immediate vicinity where produced, their disposal poses problems of odor, fly production (which may require insecticide use) and N contamination of soil and water. Various fly species, especially the housefly, *Musca domestica* (Diptera: Muscidae), have received experimental attention for their ability to recycle these manures into useful feed products for poultry and livestock. To reduce the scope of the subject, further discussion here is limited mainly to poultry.

A chicken ranch with 25,000 caged layers produces 2,500 kg wet manure per day or 912.5 tons per year (El Boushy, 1991). Dry poultry manure is not, of itself, recyclable as a good feedstuff for poultry because of its low energy and high content of uric acid and nonproteinic nitrogen, neither of which can be utilized by monogastric animals. The true magnitude of the problem is indicated by the fact that in high production areas such as Georgia, USA, 60,000 hens per house (caged layers) is the preferred house size and most farms have several such houses (Sheppard, 1992). Housefly (*M. domestica*) pupal meal produced by biodegradation of poultry manure has been shown in numerous studies to be of high proteinic quality when fed to chicks (Teotia and Miller, 1973, 1974; and others). In addition, digestion of the manure by larvae reduces its water content and converts it into an odorless, loose, crumbly product that can be easily dried and used as a soil conditioner (Calvert et al., 1970; Miller et al., 1974; and others).

According to El Boushy (1986), poultry in the developing countries produce about 40.3 million metric tons of manure per year (about 26.2 Mt from layers and 14.1 Mt from broilers). With housefly pupae, based on a yield of 3.2%, a metric ton of manure can be converted to 32 kg of high-protein feedstuff. The 40.3 million metric tons per year could thus be converted to 1.3 Mt of fly pupae. El Boushy (1991) suggested increased consumption of poultry as poultry meat is most palatable and broilers are fast-growing, reaching a weight of 1.5 kg in 7 weeks with a food conversion of 1.8. The total volume of broiler meat produced in developing countries is estimated to be 7 million metric tons per year, providing an estimated consumption of 3 kg per capita per annum. El Boushy advocated development of local industries, as opposed to home slaughter, partly because of better utilization of by-products such as manure and offal, leading to the establishment of secondary rural industries. Since, unfortunately, no practical large-scale method of separating pupae from digested manure residue has yet been found, El Boushy (1991) proposed that the most practical procedure is to produce a mixture of pupae and manure residue, thus upgrading the latter to reasonable feedstuff quality.

Studies in Georgia, USA, using the soldier fly, *Hermetia illucens* (Diptera: Stratiomyidae), appear to have solved the problem of efficient harvest of pupae from manure under caged layers (Sheppard, 1992). This nonpest species has proven to be an excellent manure management agent capable of producing large quantities of high-quality animal feedstuff, almost completely precluding housefly development, and reducing manure residue volume by 50%. Prepupal soldier flies are self-collected as they crawl out of the manure basin seeking pupation sites. They crawl up a 40° slope on one wall of the basin, into a 1/2-inch slit in a 15-cm diameter PVC pipe at the top of the slope, then crawl to a container at the end of the pipe (in the experimental facility, they negotiated a 12-m pipe length). The authors estimated that the value of the dried larval feedstuff produced, and savings in the costs of insecticide, manure removal, and surface application would net a small 20,000 hen egg producer an extra US $7,360. They state that the system could be easily adapted to swine waste management and that soldier flies could be used to degrade many other organic wastes.

In El Salvador, Larde (1989, 1990) conducted a series of studies on the use of fly larvae to recycle coffee pulp, a noxious waste product with an offensive odor. Although up to 85% of the pulp is disposed of efficiently, the remainder induces flies and other insects and is thus a sanitation problem during the processing season. To avoid this, the pulp is covered with soil, lime, ash, or coffee shells (another by-product of processing), or sprayed with insecticides or buried in excavations. The two most promising species are *Ornidia obesa* (Syrphidae) and *Hermetia illucens,* but so far larval yields have been low, mainly due to formation of anaerobic zones in the pulp beds. Better ventilation of the substrate is necessary to preclude this problem. According to the author, banana wastes are among other residues that might successfully be used for larviculture of these two species. Relative to larviculture, the size of *O. obesa* and *H. illucens* compared to *M. domestica* is an advantage, i.e., 247 mg, 204 mg, and 11 mg respectively on a dry weight basis.

Additional Factors and Considerations

Edible insects not only occupy a great diversity of habitats, some create additional habitat diversity: Termites of many species are widely eaten in Africa. The tall termitaria in Shaba, southern Zaire, average 3–5 per ha and may cover 4.3 to 7.8% of the *miombo* woodland (Malaisse, 1974). Vegetation of termite mounds is unique and differs notably from that of the surrounding *miombo.* Termite hill flora number more than 200 species in both Zambia and southern Zaire but differ significantly from one region to another (Malaisse, 1978). Malaisse pointed out that three species of edible saturniid caterpillars—*Tagoropsis flavinata, Urota sinope*, and *Gonimbrasia zambesina*—feed only on Shaban termite hill plants. Three other species feeding on termite hill plants feed on other *miombo* plants as well.

Edible insect species enhance local environments in a variety of ways: There are many examples, but probably none on a more massive scale than the leafcutter ants (genus *Atta)* of the Western Hemisphere. Two species, *Atta cephalotes* and *A. sexdens*, are the most widely consumed, being relished across the northern half of South America. Colonies usually start in a forest clearing made, for example, by a fallen tree. The leafcutter nest is a slightly raised bare mound up to 15 m in diameter and 4 m deep (Hill, 1983), with many underground chambers in which the fungus gardens made of chewed leaf fragments and saliva are cultured. The fungus converts cellulose into carbohydrates which can be metabolized by the ants and, as noted by Hodgson (1955), achieve for *Atta* a preeminent position among rainforest fauna by allowing it to tap the virtually inexhaustible supply of cellulose in its environment. The ants forage for distances up to 150 m from the nest, using a number of semipermanent trails leading from it (Cherrett, 1968). In rainforests their actions remove only about 15% of leaf production but they turn over and aerate large quantities of soil (Moffett, 1995).

Leafcutters are serious pests of many cultivated trees and other crops but Cherrett (1968) proposed that within the forest, a "conservational grazing system" is practiced which evens out the grazing pressure around the nest and prevents overexploitation of the plant resources by providing periods of relief from intense grazing during which vegetation can recover. The clearing of tropical rainforest upsets this conservational foraging strategy. Citrus or cocoa, commonly planted in newly cleared forest areas, present a greatly reduced availability of forage, thus increasing the grazing pressure per plant. Repeated defoliation may kill the plant.

Some edible insects exhibit high food conversion efficiency: The food conversion efficiency of insects varies widely depending on species and type of food consumed (e.g., forbs, grasses, woody plant herbage, wood, organic wastes, etc.) (Slansky and Rodriguez, 1987). Nakagaki and DeFoliart (1991) estimated the food conversion efficiency of the cricket *Acheta domesticus,* when kept at a temperature of 30°F or higher and fed the high-quality diets used to bring conventional livestock species to market condition, to be about twice as high as those of broiler chicks and pigs, 4 times higher than sheep, and nearly 6 times higher than steers when losses due to dressing percentage and carcass trim were taken into account. High fecundity increases the advantage in favor of the insect. Female crickets lay an average of 1,200 to 1,500 eggs over a period of 3–4 weeks. By comparison, in beef production four animals exist in the breeding herd per market animal produced, thus giving crickets a true food conversion efficiency some 20 times better than for beef.

Need for Research on Mass Production of Edible Insects

The widespread use of insects as traditional foods, the diversity of substrates (some of them of low quality) which the insects use as food, and the high food conversion efficiency of some insects when they are fed high-quality diets, when wedded to the idea of industrial-scale mass-production (e.g. Kok, 1983; Kok et al., 1988) offers the possibility of tremendous impact by insects in meeting future world food needs without additional acreage for conventional intensive agriculture. Research is needed.

Need for Reeducation of the Western Public, Government Policy-Makers, and Agricultural Researchers about Insects as Food and Insects in General

Mercer (1994), working in Papua New Guinea, concluded that the predicted world protein shortage could be ameliorated by using insect protein, but that an educational program would be necessary "to overcome the taboos currently held in the West." He further stated (1995):

> The majority of my students are keen consumers of a whole range of insects when they return to their villages during vacation time. I have come to the conclusion that it is the West which is out of step in its aversion to insects as food.

Many other researchers and educators who have explored the actual and potential use of insects as food have also targeted the adverse Western attitude as the main problem. Reversing centuries old cultural taboos is a slow process but there are indications that the adverse attitude is beginning to undergo significant change and improvement. The BBC Natural History Unit, based in Bristol, England, recently completed two programs on insects as food, one on the mopane worm *(Gonimbrasia belina)* in Botswana and one on a variety of insects consumed in Thailand. Producer Rupert Barrington (1995) stated:

> The gratifying thing is that most people react with fascination rather than disgust when they see this footage. We are very much pushing the angle that this is no different than eating Crustaceans, so western attitudes toward eating insects are groundless.

As editor of *The Food Insects Newsletter* for the past eight years, I have been contacted by more than 300 mass-media organizations—newspapers, magazines, radio and television stations and networks, and film documentary makers, and some of them repeatedly regarding preparation of articles or films. The "tone" of these interviews is quite different now, much more positive than 20 years ago when I also had a period of concentrated contact with the mass-media on this subject. There are also many articles appearing now independent of me. The point is that the Western public is currently being bombarded with favorable exposure to the role of insects as food so that the idea is no longer so strange as it was formerly.

While the attitude of the mass-media is important, equally or more important is the fact that edible insects are now finding their way into the US educational system. I have been contacted by scores of teachers and professors, elementary to university level, who are now including discussion of insects as food in their biology courses, and by scores of students, elementary to university level, doing class papers (including science fair projects) on insects as food.

A number of zoos/insectaria, e.g. in Cincinnati, San Francisco, Montreal, and numerous universities and nature centers hold annual open houses at which insect snacks are offered to the public. It appears to have become fashionable for Western entomological societies to offer insect snacks or even whole dinners at their annual functions. Educational progress is underway but continued persistence by interested scientists and educators in providing information and new research is essential if the current educational momentum is to continue.

References

Anon. 1992. Sky prawns and other dishes. The food of the future. Expat. World 4: 1.

Barrington R. 1995. Letters. Food Insects Newsl. 8(2): 6.

Bedford G.O. 1980. Biology, ecology, and control of palm rhinoceros beetles. Ann. Rev. Entomol. 25: 309–339.

Calvert C.C., Morgan N.O., and Martin R.D., 1970. House fly larvae: biodegradation of hen excreta to useful products. Poult. Sci. 49: 588–589.

Cerda H., Martinez R., Briceno N., Pizzoferrato L., Manzi P., Tommaseo Ponzetta M., Marin O., and Paoletti M.G. 2001. Palm worm (Insecta, Coleoptera, Curculionidae: *Rhynchophorus palmarum*) traditional food in Amazonas, Venezuela: nutritional composition, small scale production and tourist palatability. Ecol. Food Nutr. 40: 13–32.

Cherrett J.M. 1968. The foraging behaviour of *Atta cephalotes* L. (Hymenoptera, Formicidae). I. Foraging pattern and plant species attacked in tropical rain forest. J. Anim. Ecol. 37: 387–403.

Chowdhury S.N. 1982. Seri silk industry. Directorate of Sericulture and Weaving, Govt. of India: Guwahati (Assam).

de Conconi J.R.E. 1982. Los Insectos Como fuente de proteinas en el futuro. Editorial Limusa: Mexico D.F.

de Conconi J.R.E., Pino J.M., Mayaudon C.M., Valdez F.R., Perez M.A., and Bourgas R.H. 1984. Protein content of some edible insects in Mexico. J. Ethnobiol. 4: 61–72.

DeFoliart G.R. 1989. The human use of insects as food and as animal feed. Bull. Entomol. Soc. Amer. 35: 22–35.

DeFoliart G.R. 1990. Hypothesizing about palm weevil and palm rhinoceros beetle larvae as traditional cuisine, tropical waste recycling and pest and disease control on coconut and other palms—can they be integrated? Food Insects Newsl. 3(2): 1, 3–4, 6 (reprinted in Principes 1993, 37: 42–47).

DeFoliart G.R. 1992. Insects as human food... nutritional and economic aspects. Crop Protect. 11: 395–399.

DeFoliart G.R. 1995a. In the Philippines, local press coverage of locust control efforts—spraying and/or eating. Food Insects Newsl. 8(l): 3–4.

DeFoliart G.R. 1995b. Edible insects as minilivestock. Biodiver. Conserv. 4: 306–321.

Dufour D.L. 1987. Insects as food: a case study from the northwest Amazon. Amer. Anthropologist 89: 383–397.

El Boushy A.R. 1986. Local processing industries offer food and beneficial byproducts to developing countries. Feedstuffs 20: 36–37.

El Boushy A.R. 1991. House-fly pupae as poultry manure converters for animal feed: a review. Bioresource Tech. 38: 45-49.

Gorton P. 1988. Villagers turn 'foe' into food. Agric. Inform. Develop. Bull. 198: 19–20.

Hill D.S. 1983. Agricultural Insect Pests of the Tropics and Their Control. Cambridge Univ. Press, Cambridge, London.

Hodgson E.F. 1955. An ecological study of the leaf-cutting ant, Atta cephalotes. Ecol. 36: 298–304.

Holden S. 1991. Edible caterpillars—a potential agroforestry resource? Food Insects Newsl. 4(2): 3–4.

Kok R. 1983. The production of insects for human food. J. Can. Inst. Food Sci. Tech. 16: 5–18.

Kok R., Lomaliza K., and Shivhare U.S. 1988. The design and performance of an insect farm-chemical reactor for human food production. Can. Agric. Eng. 30: 307–318.

Larde G. 1989. Investigation on some factors affecting larval growth in a coffee pulp, Biol. Wastes 30: 11–19.

Larde G. 1990. Recycling coffee pulp by *Hermetia illucens* (Diptera: Stratiomyidae) larvae. Biol. Wastes 33: 307–310.

Leleup N. and Daems H. 1969. Les chenilles alimentaires du Kwango. Causes de leur rarefaction et mesures preconisees pour y remedier. J. Agric. Trop. Bot Appl. 16: 1–21.

Litton E. 1993. Letters. Food Insects Newsl. 6(l): 3.

Malaisse F.P. 1974. Phenology of the zambezian woodland area, with emphasis on the miombo ecosystem. *In:* Leith H. (ed.). Phenology and Seasonality Modeling. Springer-Verlag, Ecol. Studies 8: 269–286.

Malaisse F. 1978. High termitaria. Monogr. Biologicae 31: 1281–1300.

Malaisse F. and Parent G. 1980. Les chenilles comestibles du Shaba meridional (Zaire). Nat. Beiges 61: 2–24.

Marais E. 1995. Letters. Food Insects Newsl. 8(2): 6–7.

Mercer C.W.L. 1994. Sago grub production in Labu swamp near Lac, Papua New Guinea. Klinkii (J. Forestry Papua New Guinea) 5(2): 30–34.

Mercer C.W.L. 1995. Letters. Food Insects Newsl. 8(2): 7.

Miller B.F., Teotia J.S., and Thatcher T.O. 1974. Digestion of poultry manure by *Musca domestica.* Brit. Poult. Sci. 15: 231–234.

Moffett M.W. 1995. Leafcutters: gardeners of the ant world. Nat. Geogr. 188(l): 99–111.

Munthali S.M. and Mughogho D.E.C. 1992. Economic incentives for conservation: Bee-keeping and Saturniidae caterpillar utilization by rural communities. Biodivers. Conserv. 1: 143–154.

Nakagaki B.J. and DeFoliart G.R. 1991. Comparison of diets for mass-rearing *Acheta domesticus* (Orthoptera: Gryllidae) as a novelty food, and comparison of food conversion efficiency with values reported for livestock. J. Econ. Entomol. 84: 891–896.

Neupane F.P., Thapa R.B., and Parajulee M.N. 1990. Life and seasonal histories of the eri silkworm, *Samia cynthia ricini* Hutt. (Lepidoptera: Saturniidae), in Chitwan, Nepal. J. Inst. Agric. Anita. Sci. 11: 113–120.

Parent G., Malaisse F., and Verstraeten C. 1978. Les miels dans la foret clair do Shaba meridional. Bull. Rech. Agron. Gembloux 13(2): 161–176.

Peigler R.S. 1993. Wild silks of the world. Amer. Entomol. 39: 151–161.

Pemberton R.W. 1994. The revival of rice-field grasshoppers as human food in South Korea. Pan-Pacific Entomol. 70: 323–327.

Pemberton R.W. and Yamasaki T. 1995. Insects: old food in new Japan. *Amer. Entomol.* 41: 227–229.

Posey D.A. 1983. Polk apiculture of the Kayapo Indians of Brazil. Biotropica 15: 154–158.

Posey D.A. and Camargo J.M.F. 1985. Additional notes on beekeeping of Meliponinae by the Kayapo Indians of Brazil. Ann. Carnegie Mus. 54: 247–274.

Remington C.L. 1946. Insects as food in Japan. Entomol. News 57: 119–121.

Ruddle K. 1973. The human use of insects: examples from the Yukpa, Biotropica 5: 94–101.

Schmidt J.O. and Bachmann S.L. 1992. Other products of the hive. In: Graham J.M. (ed.). The Hive and the Honeybee. Dadant and Sons, Hamilton, IL (USA), pp. 927–988.

Sheppard D.C. 1992. Large-scale feed production from animal manures with a non-pest native fly. Food Insects Newsl. 5(2): 3–6.

Slansky F., Jr., F. and Rodriguez J.G. 1987. Nutritional Ecology of Insects, Mites, Spiders, and Related Invertebrates. John Wiley & Sons, New York, NY.

Teotia J.S. and Miller B.F. 1973. Fly pupae as a dietary ingredient for starting chicks. Poult. Sci. 52: 1830–1835.

Teotia J.S. and Miller B.F. 1974. Nutritive content of house fly pupae and manure residue. Brit. Poult. Sci. 15: 177–182.

Vara-asavapati V., Visuttipart J. and Maneetorn C. 1975. Edible insects in north-east Thailand. Res. Note no. 7. University of Srinakarinvirote: Mahasarakam (in Thai).

Xiaoming C. 1990. Food Insects Newsl. 3(2): 6.

8

Insects: Food for Human Evolution

MILA TOMMASEO-PONZETTA

Zoology Dept., University of Bari, Via Orabona 4, 70125 Bari, Italy.
E-mail: m.tommaseo@biologia.uniba.it

Abstract

Scientific attention is currently focusing even more on the role of insects in the diet of traditional societies since insects are considered as a food rich in nutritional value. It is therefore reasonable to infer that insects played an essential role in the diet of our ancestors, of crucial importance in satisfying specific physiological requirements in the early stages of human evolution. Brain expansion in early *Homo* sp. required high-quality food and insects were a reliable alimentary source accessible to women and children as well as men.

The study of nonhuman primates in their natural environment and a cautious homology with the subsistence strategies of today's foragers, offer important behavioral models that integrate the data of entomological archaeology, a science still in its infancy. The importance of insects in prehistoric diets has been largely ignored by archaeological research, since their presence in the past has left such elusive traces as to be missed unless specifically sought.The available direct and indirect evidence is reviewed here.

Lastly, hypotheses are advanced for future research: DNA analysis and the application of the AMmtDB, the DNA databank of invertebrates, may be precious instruments for the identification and phylogenetic study of insect remains found in archaeological sites.

Key Words: entomophagy, human evolution, *Homo ergaster*, diet, primates, hunter-gatherers, archaeology, coprolites, tool wear, teeth, mtDNA.

Introduction

Hominid foraging and feeding behavior is of ever-increasing interest in the reconstruction of human evolution (Mann, 1981; Milton, 1987; Sept, 1992; Leonard and Robertson, 1994, 1997; Antón et al., 2002) and important new models have been proposed, which take into account various factors of morphology, energetics, metabolic needs, life history (especially that of females), and inferred social structure related to diet (Leonard and Robertson, 1994; Steudel, 1994; Aiello and Wheeler, 1995; O'Connell et al., 1999, 2002; Wrangham et al., 1999; Aiello and Wells, 2002; Leonard, 2003). Concomitantly, diet-related diseases of modern affluent societies have prompted nutritionists to look into human evolutionary history, in order to define what optimal nutritional models humans originally followed (Eaton and Konner, 1985; Eaton, 1992; Eaton et al., 1997, 2002). Paleolithic diet has been considered from varying perspectives (Ungar and Teaford, 2002) and so many data, hypotheses, and "recommendations" have been proposed that a complex and sometimes contradictory picture results (Milton, 2002).

When human evolution is considered, there is general agreement on the close relationship between a diet of high quality and the elevated metabolic requirements of an expanding brain, whose growth becomes especially noticeable with the emergence of *Homo ergaster* around 1.8 Mya (Wood and Collard, 1999). Human diet quality values are substantially higher than those of any nonhuman primate of similar size (Leonard and Robertson, 1994) and much of the present discussion focuses on the ways in which a larger brain evolved. More meat in the diet, which would have provided more animal proteins and fats—also indicated by a contemporary improvement in hominid material culture and behavior (Potts, 1996)—is the first plausible explanation (Leakey, 1971; Milton, 1999, 2002). However, it is not the only one: inclusion in the diet of carbohydrates, provided by plants with underground storage organs, is also convincing (Aiello and Wheeler, 1995). Tubers would have represented a particularly convenient food when hominid energetics, metabolism and occasional food stress in an increasingly dry tropical habitat, are taken into account (Aiello and Wells, 2002).

However, when the rapid expansion of the human brain is considered, the dietary role played by lipids, cholesterol (Mann, 1998) and especially long-chain polyunsaturated ω3 fatty acids is of particular importance. Since the human brain contains 600 g lipid/kg, with a profile consisting of equal proportions of arachidonic and docosahexaenoic acids (Broadhurst et al., 1998), an abundance of these essential fatty acids in the diet would have represented a precious resource for brain growth and development (Eaton, 1992; Leonard and Robertson, 1994; Chamberlain, 1996; Eaton et al., 1997, 2002; Johns, 1999). Since the most critical period for human brain growth takes place during fetal development and the first postnatal period, maternal nutrition and metabolic output through pregnancy and lactation would be crucial for brain size and evolution (Martin, 1989; Leonard and Robertson, 1994). Therefore, as suggested by Martin, this could only take place at a time when humans—women in particular—had access to reliable and abundant food resources. If a high-quality diet is not the only factor explaining the expansion of the human brain, it does appear to have made it possible. The behavioral strategies to ensure such a diet are also certainly of critical importance (Milton, 2002).

One convincing hypothesis has been put forward by Broadhurst et al. (1998) and Cunnane (1999), who identify the preferential reservoir for an early "shore-based diet" in the East African lacustrine and riverine environment from which most fossil hominids have been recovered (Lake Turkana, Olduvai Beds I and II, Hadar, and the Omo River). Aquatic resources are particularly rich in long-chain polyunsaturated fatty acids, and would thus have been fundamental for human brain growth and evolution. Although the consumption of aquatic resources at such an early age has left limited archaeological evidence (Leakey, 1971; Stewart, 1994), this hypothesis is plausible. Similarly, we are led to believe that harvesting minor food resources, such as invertebrates, provided a regular, reliable basis for hominid nutrition: in this early menu, insects must have represented not only an occasional snack but also a source of abundant, high-quality food, easily collected by women and children.

Today, in Western societies the idea of eating insects induces aversion, for a number of conscious or subconscious reasons, since insects are often associated with dirt, disease, parasitism, and poisonous stings or bites. Westerners would never think of eating insects other than out of curiosity or eccentricity, except for a few daring researchers interested in assessing their comestibility. This is one of the reasons why, until recently, the role of insects in the nutrition of humans and their evolutionary importance has been neglected (DeFoliart, this volume).

Only recently has the interest of scientists and the media in insects as food among traditional populations (e.g., the magazine, *Food Insects Newsletter*) led to an examination of their alimentary functions in prehistoric times (Sutton, 1990, 1995; Klein, 1999; McGrew, 2001; Tommaseo-Ponzetta, 2003). Since the human species underwent most of its evolution in tropical or subtropical areas where insects are calculated to represent around 80% of all species (Wheeler, 1990), it is reasonable to infer that these invertebrates played a major role in our early ancestors' diet. An overview and update of the present state of research, the main outlines of which were first put forward by Sutton (1990, 1995) and McGrew (2001), are presented here.

The evidence of insects as food in human evolution is both "hard" and "speculative": although the data deriving from archaeological entomology are still scant, they can be efficaciously integrated with those offered by primatology and ethnography.

The nutritional strategies of chimpanzees in their natural environment represent the nearest analogue of a common ancestor's feeding patterns. This approach is usually supplemented by homology with the ways by which present-day foragers and other traditional societies obtain their provisions. However, it must be borne in mind that millions of years separate present-day chimpanzees from a hypothetical "common ancestor", and that the culture of the last modern tropical foragers is in turn the result of a complex evolutionary history. In addition, the territories presently occupied by these populations are often mere sanctuary areas compared with the original ones.

Insects as High-quality Food

Insects represent a high-quality food, complementary to most vegetarian diets (DeFoliart, 1992, 1997; Ramos-Elorduy, 1996; Bukkens, 1997 and this volume) and usually not very hard to find. In common with other primates, man has shown his ability to overcome the physical or chemical defenses of social insects, such as termites or bees, and one may conclude with McGrew (1979) that the "insect hunt" requires more skill than strength. From the nutritional point of view, insects offer the advantage of often being organized in colonies (bees, ants, termites) or temporarily assembled (e.g. caterpillars and occasionally orthopterans), which makes their capture easier and more plentiful. Needless to say, these gregarious (social) insects are also the most sought after as food.

In addition, human and nonhuman primates have an opportunistic nutritional relationship with the products of insects or their harvests: honey is the most coveted of all. But the residual secretions of homopteran larvae, or the soil of termite nests, absorbed as a nutritional integrator and a natural remedy (Bodenheimer, 1951; Mahaney et al., 1996; Aufreiter et al., 2001), as well as the stores of grain harvested by some ants, are equally sought by traditional populations (Sweeny, 1947).

Data from Primatology and Ethnography

Insects in the Diet of Chimpanzees

The study of chimpanzees in their natural environment has provided essential data, such as their use of tools for termite or ant fishing and their occasional hunting (Goodall, 1963, 1986; Wrangham et al., 1999; Boesch and Boesch-Aschermann, 2000). However, since it is known that different groups of chimpanzees have different nutritional habits, we are warned against facile generalizations (McGrew, 1992; Whiten et al., 1999; Boesch and Boesch-Aschermann, 2000; Whiten, 2000). Termites (*Macrotermes bellicosus*) and the savanna ant (*Dorylus* [*Anomma*] *nigricans*) are the favorite insects of chimps in Gombe (McGrew, 1979), whereas the Gabon chimps studied by Hladik (Hladik and Viroben, 1974; Hladik, 1977) prefer ants (*Macromiscoides aculeatus, Oecophylla longinoda, Polyhachis militaris, Paltothyreus tarsatus, Camponotus* sp.). Other insects, orders Diptera,

Lepidoptera or Hemiptera, are occasional prey, and even scorpions (*Opisthacanthus lecomtei*) are appreciated.

In Gabon the chimps spend one-third of their time hunting for insects, capturing ants and termites. Although the product in terms of weight is trifling—a mere 4% of the total diet—these insects are rich in proteins, lipids, and essential amino acids such as histidine, leucine, lysine, and threonine, and are thus a complement to leaves which, in turn, are rich in cystine. In fact, eating insects is often combined with that of plant fibers: bark in Gabon, leaves in Gombe. Among ants, the Gabon chimps prefer those rich in lipids (reproducers and soldiers) to ones rich in proteins (workers) (Hladik, 1977). Hayden (1981) and Hladik and Simmen (1996) have suggested that this is due to taste sensitivity to lipids, which may have been essential in the evolution of the first hominids.

A certain number of behavioral characteristics appear to have remote roots: according to McGrew (1979), the amount of time spent by females and males in gathering insects differs significantly and is far greater in females. This difference is also significant when insects and parts of insects found in feces are considered (45% females, 26% males). Females have been described as having a protracted, systematic, and repetitive approach to the food (insects), which allows them to avoid competition with the dominant males for animal proteins. Thus, chimpanzee females may be considered prototypes of human females devoted to gathering (McGrew, 1981).

Insects in the Diet of Traditional Populations

Insects are part of the dietary traditions of tropical-environment populations that consider them "appetizing" and tasty. They also provide a supplement which may become vital in overcoming intermittent periods of scarce game, or even famine-circumstances which may well have occurred in the Plio-Pleistocene in an environment such as East Africa, where the climate was becoming increasingly seasonal (Speth, 1987). Even today, insects constitute nutritional compensation for those who, within a social group, i.e. women and children, are disadvantaged in terms of animal proteins, due to limited access to the larger game of a hunt. Like our cousins the chimpanzees, in most traditional societies, from Africa to New Guinea, the gathering of insects is relegated to women and children (Turnbull, 1965; Lee, 1968; Tommaseo-Ponzetta and Paoletti, 1997 and this volume).

Insects are also a valuable food during pregnancy: a pregnant woman needs extra nutritional energy when she is in a vulnerable condition. Insects represent rich food which can be caught and collected without too much effort. Also, food suitable for weaning is sometimes scarce or difficult to obtain: in the past insects—especially grubs, which are particularly fatty and also easy to catch—may have been a privileged source of nutritional elements satisfying specific human physiological needs during the early stages of growth (Crawford and Marsh, 1995).

Honey, like larvae, is also a rich food, easy to eat and digest, even for small children. The nutritional value of some of the most common edible insects in Africa can be found in Table 8.1 (see also Bukkens, this volume).

Table 8.1: Nutritional value of some food insects collected in Africa and South America.

Insects	**Moisture (%)**	**Energy (kcal)**	**Protein (g/100 g)**	**Fat (g/100 g)**
Mopanie worms (dried) *Gonimbrasia belina* [1]	6.1	444	56.8	16.4
Palm weevil larvae *Rhynchophorus phoenicis* [2]	10.75	562	20.34	41.73
Locust, *Locustana* spp. [1]	57.1 (raw)	N.A.	18.2	21.5
	48.7 (fried)	N.A.	30.0	10.0
	7.1 (flour)	436	47.5	22.9
Grasshoppers, *Zonocerus* sp.[1]	62.7 (raw)	170	26.8	3.8
	7.0 (grilled, grounded)	420	62.2	10.4
Flying ants (raw), females *Carebara* sp.[1]	60	N.A.	3.0	9.5
Flying ants (raw), males *Carebara* sp.[1]	60	N.A.	10.1	1.3
Termites, mature alates *Macrotermes subhyalinus* [2]	0.94	612	38.42	46.1
Termites, soldiers *Syntermes* spp. [3]	10.3	467	58.9	4.9

Composition per 100 g edible portion .
[1] Wu Leung et al., 1968
[2] Santos Oliveira et al., 1976
[3] Dufour and Sander, 2000

Edible insects are important dietary integrators, rich in calcium (e.g. *Rhynchophorus* sp.), iron (larvae of Mopanie, *Gonimbrasia belina*: 76.9 mg/100 g) and generally in vitamins, e.g. group B (Ramos-Elorduy et al., 1996; Paoletti et al., 2000). The danger of a lack of vitamin B_{12} or other vitamins not found in plants must have been a real threat to hominids, surviving as they did mostly on a vegetarian diet: we must recall that anemia as a result of parasitosis or dietary deficiencies is still one of the most frequent pathologies in tropical environments (WHO, 1997).

Insects and Human Evolution

Recent African discoveries of sparse prehominid fossils are indicative of a very ancient Miocenic root for the hominine clade, calculated between 6 and 7 Mya. The environmental scenario of these earliest possible ancestors—*Sahelanthropus tchadensis* (Brunet et al., 1995, 2002), *Orrorin tugenesis* (Senut et al., 2001), *Ardipithecus ramidus* (White et al., 1994; Haile-Selassie, 2001), and *Australopithecus anamensis* (Leakey et al., 1998) (Fig. 8.1) besides being indicative of larger spatial diffusion than previously imagined—suggests their persistent dependence on a forest habitat for food, shade, and protection. It is believed that their nutritional

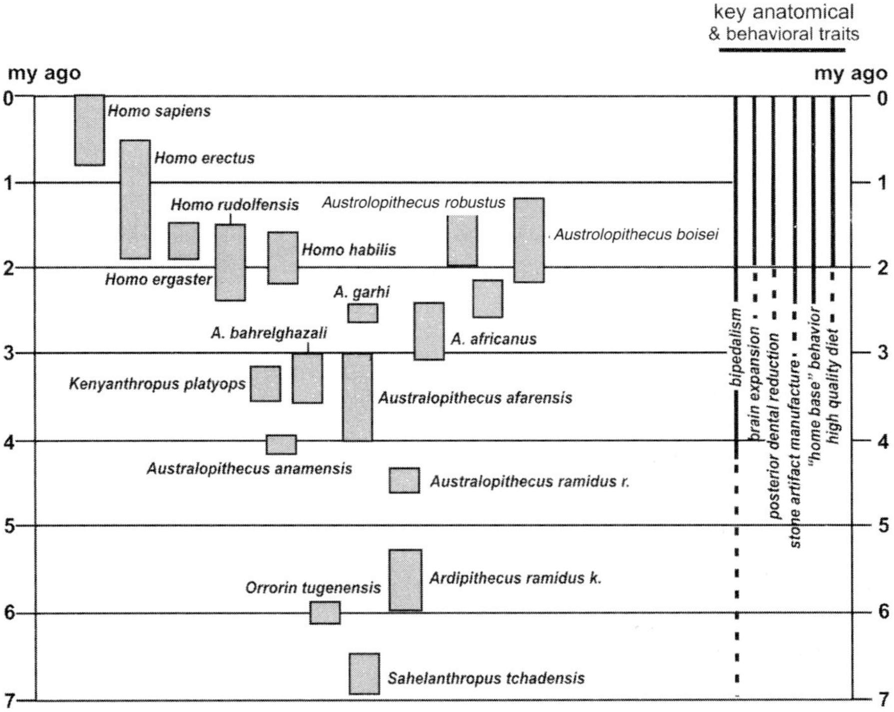

Fig. 8.1: Time extension of most commonly recognized hominid species, with indication of some key behavioral and anatomical traits (modified from Klein, 1999; Chiarelli, 2003).

strategies, due to their flexibility, must have been comparable to those of extant chimpanzees, and that insects and other invertebrates must have formed part of their diet.

The dietary use of these various types of animal food may have preceded and then accompanied the consumption of meat: the brain is a very spendthrift organ, which developed in hominids long before the creation of stone tools produced and used especially for obtaining meat and extracting bone marrow, but also employed for digging into termite mounds (see below). From *Australopithecus* (*A. afarensis, A. africanus*) to *Paranthropus* (*A. aethiopicus, A. robustus, A. boisei*) to *Homo habilis* (*H. rudolfensis* and *H. habilis* s.s.), the evolutionary trend which may be inferred from dental and mandibular features, when tested against external and internal characteristics of foods, shows that these hominids were not pre-adapted to meat-eating, but rather subsisted on soft fruits, while gradually including a growing number of various terrestrial items in their diet (Teaford and Ungar, 2000).

However, the African habitat underwent progressive cooling from 2.5 to 1.8 Mya and gradually developed into a more open environment (Speth, 1987; Vrba, 1988). It was at this time that hominids, implementing new foraging strategies, began to benefit by a richer diet. *Homo ergaster* is the first true representative of our genus. A brain expanding in size during evolution doubtless demanded a great quantity of energy, sought in high-quality food such as insects. The linoleic acid content of African termites (*Macrotermes bellicosus* and *M. subhyalinus*) is very high (34% and 43% respectively), superior to that of conventional meat (e.g. pork, 18.1%). The caterpillars *Imbrasia ertli* and *Gonimbrasia belina* have very high alfa-linoleic acid contents (20% and 38% respectively), much higher than those of pork (0.81%) and veal (1.86%) (Paoletti et al., 2003). Furthermore, the human brain is a consumer of sugar: extra glucose was certainly provided by honey.

Data from Entomological Archaeology

Archaeological Findings

Until now, archaeological research concerning the Plio-Pleistocene has largely neglected the importance of insects in the diet of prehistoric hominids since their presence in the past has left such elusive traces as to be overlooked unless specifically sought. In addition, most insects were procured—with some exceptions considered later—by using occasional and perishable tools. When insects were not directly eaten along the way, but collected and stored to be consumed/ shared at a later time, the containers used to hold them were also made of perishable materials such as skins, bark, wood, fruit shells, leaves or vegetal fibers—which leave no durable residues. Thus, although the chance of finding in archaeological digs traces of edible insects that can be linked with the remains of early hominids is highly improbable, more for behavioral than taphonomical reasons.

Only in more recent prehistory, when the use of fire and cooking were introduced, is it possible to find traces of insects roasted in large quantities in ashes or hot sand (Fowler and Walter, 1985).

Coprolites

Coprolites provide unquestionable evidence of the consumption of insects, since chitin is expelled without digestion, except by some prosimians (Cornelius et al., 1976; Kay and Sheine, 1979). Fragments of exoskeletons from the area of the mouth or the wings can be recovered and identified (Reinhardt and Bryant, 1992; Sutton, 1995).

In archaeological sites in South and North America, the presence of insects in the diet has been directly and frequently demonstrated. Insects have been recovered from human coprolites in several localities: Peru (Weir and Bonavia, 1985), Mexico (MacNeish, 1958; Callen, 1967), and states of Texas (Bryant, 1974) and Kentucky, USA (Webb and Baby, 1957; Yarnell, 1969), to mention just a few (see Reinhardt and Bryant, 1992). Studying coprolites from the Texas rock-shelter, Bryant (1974) found grasshoppers to be the most frequently eaten animal in the archaic Indian diet.

However, the study of very ancient coprolites, such as those found at Olduvai Gorge in Kenya (Leakey, 1971), in the caves of South Africa, or at Terra Amata (DeLumley, 1969; Trevor-Deutsch and Bryant, 1978) and Lazaret in France (Callen, 1969), presents exceptional technical difficulties. These fossilized remains, besides being rare, are often fragmentary, difficult to distinguish due to their highly variable shape or size from those of other mammals and, when analyzed, do not react to the usual reagents (Bryant and Williams-Dean, 1975; Kliks, 1978; Reinhardt and Bryant, 1992). Also, it is difficult if not impossible to identify insects eaten in their larval state. Furthermore, insects or their larvae found in feces may be coprophagous and should thus be distinguished from edible insects existing in a region at a certain period. We should also attempt to distinguish insects which formed part of the diet from those consumed for medicinal purposes.

Some inferences may be made from the presence of parasites in coprolites. Moore et al. (1969) noted that the eggs of certain worm parasites of insects (Acantocephala) in coprolites of the Great Basin (western US) might in itself indicate the consumption of insects.

In naturally mummified specimens of humans or animals, preserved thanks to particular environmental conditions, paleofeces may be recovered and their contents subjected to DNA analysis. No insect remains have been identified in the intestinal contents of the mummies studied by Brothwell (1996), Poinar et al. (1998, 2001), and Rollo et al. (2002) (nor were they looked for, given the cold habitat in which the subjects lived), but these studies have opened up a promising pathway for collecting direct information on the presence of insects in ancient diets.

Traces of Wear and Organic Remains on Tools

The study of traces of wear and tear on tools is a largely experimental discipline (Keeley, 1980; Keeley and Toth, 1981; Toth, 1985; Shipman, 2001). Recently, Backwell and d'Errico (2001) gave evidence that the striations found on Lower Paleolithic bone tools from Swartkrans and Sterkfontein in South Africa (2-1 Mya) were produced by digging into the hard crust of termite mounds to induce swarming of the insects, and that this may also apply to bone tool pieces found at Drimolen, an early hominid site nearby. Fossil evidence at these sites indicates that both *Homo* and *Australopithecus robustus* were the possible users of this bone technology, in accord with recent isotopic analyses, which revealed a consistent proportion of dietary C_4 carbon on these hominid remains, suggesting that they ate grass-eating animals (termites, maybe locusts or caterpillars) rather than the plants themselves. It is therefore proven that termites as food have been an important dietary resource since the earliest times.

The custom of grinding dried insects for flour is widespread throughout the world and possibly very old: it is probable that traces of wear typical of this activity can be found on grinding stones or molds, constituting a further source of data.

The study of organic residues found on the surface of prehistoric stone tools also opens up a promising research field. The remains of blood, hair, bone and cartilage can be detected on tools by means of detailed microscopic observations (Fullagar et al., 1996), and DNA extraction and analysis may follow, allowing specific characterization of these organic residues.

Recovery of DNA from mammalian residues found on the surface of stone tools from the site at Quina in France, going back to the Middle Paleolithic (35,000–65,000 BP), was carried out by Hardy et al. (1997) using the PCR technique (polymerase chain reaction). This method enables identifying not only human, but also boar/pig (*Sus scrofa*), artiodactyl and lagomorph sequences. However, protein preservation and DNA retrieval from ancient fossils remains controversial (Poinar and Stankiewicz, 1999), especially in hot climates where DNA probably cannot be preserved for more than 700–800 years, as studies of the DNA decay rate in *Papyrus* and human remains from Egyptian archaeological sites have recently indicated (Marota et al., 2002; Rollo, 2002).

Data from Human Fossils

Bone Biochemistry

Unfortunately, all the other evidence of insects in the diet is indirect, such as biochemical analysis of bone: it enables us to hypothesize that they were eaten, like any other type of fauna, but does not prove it, nor can such analysis clearly distinguish between the different types of animals eaten. For instance, the strontium/calcium (Sr/Ca) ratio in bone varies according to type of food absorbed. This ratio is higher for plants and lower for meat, since mammals discriminate against Sr compared with Ca: yet this indication needs to be further examined, since there is great variability between organisms, and the ratio decreases along the food chain (Sillen, 1992, 1994). Only studies of prey-predator pairs and the search for elements characteristic of insects, e.g., comparative studies between species of insect- and leaf-eating primates, will yield more accurate data.

Paleodietetics can also benefit by the study of isotopic signatures in the bone and tooth collagen of mineralized tissues (Drucker et al., 1999). Analysis of stable carbon isotopes (C_{12}/C_{13}) differentiates the relative proportions of plants which follow either the type of photosynthesis known as "in C_3", such as trees, bushes, shrubs and tubers in tropical savanna, from those following the type called "in C_4" (grasses/graminaceae) (Smith and Epstein, 1971). The isotopic carbon signature of carbonate hydroxylapatite, which may be conserved for millions of years in tooth enamel, reveals the amount of food coming from food chains based on plants whose photosynthesis is in C_3 from those based on plants whose photosynthesis is in C4 (Bocherens, 1999).

Isotopic carbon analysis of *Australopithecus robustus* from Swartkrans (Lee-Thorp and Van der Merwe, 1993; Lee-Thorp et al., 1994) and *Australopithecus africanus* from Makapansgat, South Africa (Sponheimer and Lee-Thorp, 1999), compared with other species of mammals whose diet is known, do show a predominance of food coming from plants in C_3, although there is also a significant proportion of plants in C_4. Since the dentition of these hominids does not fit the consumption of graminaceae (plants in C_4), the supply of C_4 food may have come from grass-eating animals, such as small invertebrates and insects, e.g., the termite *Trinervitermes trinervoides*. This hypothesis is consistent with the study of wear traces on bone tools by Backwell and d'Errico (2001) (Fig. 8.2).

Fig. 8.2: Wear traces on bones, perhaps produced by digging into termite mounds (from Backwell and D'Errico, 2001).

Skeletal Pathologies

Found at Koobi Fora (eastern Lake Turkana, Kenya), an adult female skeleton of *Homo erectus* about 1.6 Mya old (KNM ER 1808) had the long bones of the legs partly encrusted with new ossification (Leakey and Walker, 1985). This pathology was attributed to an excess of vitamin A, the source of which was assumed to be the liver of carnivores, consumed to a toxic level by these hominids (Walker et al., 1982). Skinner (1991) suggested that another reason for this pathology was bee brood-comb, a food extremely rich in vitamin A, the protracted ingestion of which could theoretically produce this same hypervitaminosis. This is not unlikely, since many traditional societies appreciate brood-comb even more than honey (Bodenheimer, 1951; Hocking and Matsumura, 1960).

Caries

The skull from Broken Hill (Kabwe, Zambia—"Rhodesia Man") presents a combination of archaic features and more modern characters. It is attributed to a period ranging from 500,000 to 200,000 years BP. For the first time the presence in human fossil remains of several dental caries was observed—an exceptional fact in *Homo erectus*—perhaps compatible with the habit of eating cariogenic carbohydrates, not only from plants but even more from honey (Newbrun, 1982).

Traces of Wear and Organic Remains on Teeth

As most edible insects are eaten in the form of larvae or, as adults, without chitinous appendages or wings, it is difficult to distinguish traces of wear on the occlusal surface of teeth which can definitely be related to the consumption of insects. Not much is known of the dental microwear of the earliest hominids, although several studies have been carried out on tooth wear in later australopithecines (Puech and Albertini, 1984; Ryan and Johanson, 1989; Ungar and Grine, 1991), showing how they were beginning to exploit savanna resources and were able to process a great variety of foods. Comparative studies of microwear patterns between insectivorous and noninsectivorous primates would give some useful indications and perhaps show whether chitin leaves recognizable traces on tooth enamel (McGrew, 2001).

Opal phytoliths were detected on the teeth of a Mio-Pleistocenic hominoid, *Gigantopithecus blacki* (around 400,000 BP), found in a North Vietnam cave in association with *H. erectus*, by Ciochon et al. in 1990. This study opens new perspectives on paleodietary research and implies the future possibility of identifying residues of food—and possibly of food insects—in hominid dental remains.

Upper Paleolithic in Europe

Products of Bees

In European regions, the most important evidence of insects being exploited in the Paleolithic relates to harvesting the products of bees (*A. mellifica* L.), which perhaps were far more plentiful than they are today.

<u>Honey:</u> As far as we know at present, the oldest wall paintings depicting the harvesting of honey come from the central Sahara (Zimbabwe), South Africa and Zambia, and go back about 20,000 years (Pager, 1973; Isack and Reyer, 1989).

There is evidence that honey was also exploited in Europe at the end of the last glaciation, between 20,000 and 13,000 BP. At that time, thanks to its mild climate, the Iberian Peninsula was certainly the most favorable habitat for *Apis mellifera mellifera* (Ruttner, 1952) and the Altamira Cave (14,000 years), in northern Spain, shows paintings of rope-ladders and ovoid shapes that may be interpreted as honeycombs (Pager, 1976).

Again in Spain, near Bicorp, in Araña Cave, between 7,000 and 4,000 years old, a honey-harvesting scene is unmistakably represented (Hérnandez-Pacheco, 1924) (Fig. 8.3). A similar harvesting technique is still in use among the Veddas of Ceylon (Spittel, 1924, cited in Bodenheimer, 1951).

Fig. 8.3: Wall painting representing honey collection in Araña Cave, Spain, 7000–4000 B.P. (from Hérnandez-Pacheco, 1924).

Propolis: Harvesting bee products is not limited just to honey. From a Late Paleolithic grave (12,000 years BP) discovered in 1988 in the Dolomites (northeastern Italy), evidence was found of the use of propolis mixed with plant resins (Mondini and Villabruna, 1988; Broglio, 1990; Broglio and Villabruna, 1991). In addition, buried at an altitude of 500 m, the skeleton of a hunter was found together with a small block of resin, in which pollen analysis revealed traces of propolis (Cattani, 1993, 1994). A similar series of funerary objects had been found the year before in the grave of a Mesolithic hunter (7,300 years) at Mondeval de Sora (Dolomites) at an altitude of 2,150 m (Guerreschi, 1990). The use of propolis, until then assumed to go back to ancient Egypt, is at least 12,000 years old. However, exactly how it was used is not known, owing to the multifunctional character of the substance.

Other Food Insects

Crickets: A bison bone in which a grasshopper (*Troglophilus*) had been carved was found in the Grotte des Trois Frères (France). Although the dietary use of this insect is not demonstrated, its accurate depiction proves that humans were very familiar with this species sometimes abundant in caves (Begouen, 1929, cited in Vandel, 1964).

Silkworms: References in Sanskrit literature date the silk industry in India to 1,000 or perhaps even 4,000 years BC (Cloudsley-Thompson, 1976). The imprint of a cocoon and part of the pupa (*Pachypasa otus*) it contained, were found in the tephrite overlying the Bronze Age village of Akrotiri, on the island of Santorini, Greece (Panagiotakopulu-Buckland et al., 1997), dated toward the middle of the second millenium BC. Identification of this silkworm indicates the presence and perhaps use of natural silk on the island as early as this remote time. However, we cannot advance an hypothesis regarding its use as food.

Future Possibilities of Entomological Archaeology

Entomological archaeology is a constantly evolving discipline (Buckland et al., 1996; Dobney et al., 1998) given to improved sifting techniques and chemical methods accompanied by detailed microscopic analysis, using an approach similar to that of paleopalinology.

Anaerobic soil conditions (usually very dry or very wet) allow practically unlimited conservation of organic remains (Scudeller-Baccelle and Nardi, 1991; Cresser et al., 1993), as shown by the different types of insects unearthed in Bronze Age or European Neolithic deposits (Robinson, 1991). So archaeologists should pay increasing attention to the possible remains of edible insects in digs.

An extensive bibliography on Quaternary insects compiled by P.C. Buckland, G.R. Coope, and J.P. Sadler (web version by Buckland) is available at the following address:

http://www.umu.se/envarchlab/BUGS/QBIB/QBIBFRAM.HTM

Databanks of Invertebrate DNA

Today, DNA databanks are indispensable stores of information and research tools. Their purpose is to collect and make available to the scientific community all the data on genetic sequences obtained for different organisms (plants and animals, including humans).

The Institute of Biomedical Technology of the Italian Research Council (CNR-ITB) of Milan and the Bioinformatic and Genomic Section of Bari (Italy), have recently set up the AMmtDB (Aligned Metazoan mitochondrial Data Bank), to collect all complete and aligned sequences of hitherto published mitochondrial genes of invertebrates (Lanave et al., 2002). Consultation of the AMmtDB, by allowing comparison of sequences, facilitates identification of various species of insects which may be found in archaeological digs and also the study of genetic variability and phylogenetic reconstruction of species. All data and sequences are available at the following address:

http://bighost.area.ba.cnr.it/mitochondriome

Conclusions

An increasingly detailed reconstruction of the paleo-environment and of seasonal differences and precipitation in different periods in the Southern Hemisphere would allow us to formulate correlative hypotheses on the frequency of the different species of edible insects over time. Recent studies have shown that the geographic and temporal availability of various types of food can determine the harvesting choices of primates and significantly influence the subsistence strategies of traditional societies. Further study of how nonhuman primates and traditional societies obtain provisions in their present environments would also enable us to estimate the opportunities of edible insect gathering by Pleistocene prehominids and hominids in at least partly comparable environments (Sept, 1993).

Data from entomological archaeology can then be more fully integrated into the bioecological models proposed by primatology and ethnography, for an ever more plausible reconstruction of the nutritive role of insects (and other small invertebrates) in human evolution.

References

Aiello L.C. and Wheeler P. 1995. The expensive tissue hypothesis: the brain and digestive system in human and primate evolution. Curr. Anthropol. 36: 199–221.

Aiello L.C. and Wells J.C.K. 2002. Energetics and the evolution of the genus *Homo*. Ann. Rev. Anthropol. 31: 323–38.

Antón S.C., Leonard W.R., and Robertson M.L. 2002. An ecomorphological model of the initial hominid dispersal from Africa. J. Human Evol. 43: 773–785.

Aufreiter S., Mahaney W.C., Milner M.W., Huffman M.A., Hancock R.G.V., Wink M., and Reich M. 2001. Mineralogical and chemical interactions of soil eaten by chimpanzees of the Mahale Mountains and Gombe Streams National Parks, Tanzania. J. Chem. Ecol. 27 (2): 285–311.

Backwell L.R. and d'Errico F. 2001. Evidence of termite foraging by Swartkrans early hominids. Proc. Natl. Acad. Sci., USA 98 (4): 1358–1363.

Bocherens H. 1999. Isotopes stables du carbon et reconstitution du régime alimentaire des hominidés fossiles: une revue. Bull. Mém. Soc. Anthropol. Paris, n.s.t., 11 (3–4): 261–287.

Bodenheimer F.S. 1951. Insects as Human Food. W. Junk, The Hague
Boesch C. and Boesch-Aschermann H. 2000. Chimpanzees of the Taï Forest: Behavioural Ecology and Evolution. Oxford Univ. Press, Oxford, UK.
Broadhurst C.L., Cunnane S.C., and Crawford M.A. 1998. Rift Valley lake fish and shellfish provided brain-specific nutrition for early Homo. Brit. J. Nutr. 79: 3–21.
Broglio A. 1990. La preistoria delle Dolomiti. *In:* Le Dolomiti, un patrimonio da tutelare e amministrare. Agordo (Bl), Comunità Montana Agordina.
Broglio A. and Villabruna A. 1991. Vita e morte di un cacciatore di 12.000 anni fa. Risultati preliminari degli scavi nei Ripari Villabruna (Valle del Cismon-Val Rosna, Sovramonte, Belluno). Odeo Olimpico. Vicenza, Academia Olimpica.
Brothwell D. 1996. European bog bodies: current state of research and preservation. *In:* Spindler K. et al. (eds.). Human Mummies: A Global Survey of Their Status and the Techniques of Conservation. Springer-Verlag: New-York, NY, pp. 161–172.
Brunet M., Beauvilain A., Coppens Y., Heintz E., et al. 1995. The first australopithecine 2,500 kilometres west of the Rift Valley (Chad). Nature 378: 273–275.
Brunet M., Guy F., Pilbeam D., Mackaye H.T., et al. 2002. A new hominid from the Upper Miocene of Chad, Central Africa. Nature 418: 145–151.
Bryant V.M. 1974. Prehistoric diet in southwest Texas: The coprolite evidence. Amer. Antiq. 39: 407–420.
Bryant V.M. Jr. and Williams-Dean G. 1975. The coprolites of man. Sci. Amer. 232: 100–109.
Buckland P.C., Amorosi T., Barlow L.K., Dugmore A.J., et al. 1996. Bioarchaeological and climatological evidence for the fate of Norse farmers in medieval Greenland. Antiquity 70: 88–96.
Bukkens S.G.F. 1997. The nutritional value of edible insects. *In:* Paoletti M.G. and Bukkens S.G.F. (eds), Minilivestock. Ecology of Food and Nutrition (s.i.) 36 (2–4): 287–320.
Bukkens S.G.F. 2004. Insects in the human diet: Nutritional aspects (this volume).
Callen E.O. 1967. Analysis of the Tehuacan coprolites. *In:* Byers D. (ed.), The Prehistory of the Tehuacan Valley, vol. 1: Environment and Subsistence. Austin: University of Texas Press, Austin, TX, pp. 261–289.
Callen E.O. 1969. Les coprolithes de la cabane acheuléenne du Lazaret: Analyse et diagnostic. Mém. de la Soc. Préhist. Française 7: 123–124.
Cattani L. 1993. Contenuto pollinico di materiali resinosi come elemento di corredo funebre. Antropologia Contemp. 16: 55–60.
Cattani L. 1994.Estudio polínico sobre resinas fósiles de edad paleolítica. X Simposio de palinología básica y aplicada (A.P.L.E.). Universitat de València, València, Spain, pp. 175–187.
Chamberlain J.G. 1996. The possible role of long-chain, omega-3 fatty acids in human brain phylogeny. Persp. Biol. Med. 39 (3): 436–445.
Chiarelli B. 2003. Dalla Natura alla Cultura. Principi di Antropologia Biologica e Culturale, vol. 1. Piccin Editore, Padova, Italy.
Ciochon R.L., Piperno D.R. and Thompson R.G., 1990. Opan Phytoliths found on the teeth of the extinct ape, *Gigantopithecus blacki:* implications for paleodietary studies. Proc. Natl. Acad. Sci. USA 87: 8120–24.
Cloudsley-Thompson J.L. 1976. Insects and History. Weidenfeld and Nicholson, London, UK.
Cornelius C., Dandrifosse G., and Jeniaux C. 1976. Chitinolytic enzymes of the gastric mucosa of *Periodicticus potto* (Primate: Prosimian): Purification and enzyme specificity. Int. J. Biochem. 7: 445–448.
Crawford M.A. and Marsh D. 1995. Nutrition and Evolution. New Canaan, CT, Keats Publishing.
Cresser M., Killham K., and Edwards T. 1993. Soil Chemistry and Its Applications. Cambridge Univ. Press, Cambridge, MA.
Cunnane S.C. 1999. A shore-based diet rich in energy and brain-specific nutrients made human brain evolution possible. Proc. Ghent Conf. Water and Human Evolution. Ghent, Belgium.
DeFoliart G.R. 1992. Insects as human food. Crop Protec., 11: 395–399.

DeFoliart G.R. 1997. An overview of the role of edible insects in preserving biodiversity. *In:* Paoletti M.G. and Bukkens S.G.F. (eds), Minilivestock: Ecology of Food and Nutrition (s.i.) 36 (2–4): 109–132.

DeFoliart G.R. 2004. Overview of role of edible insects in preserving biodiversity (this volume).

DeLumley H. 1969. A Paleolithic camp at Nice. Sci. Amer. 220: 42–50.

Dobney K.H., Kenward P., Ottaway P., and Donel L. 1998. Down, but not out: biological evidence for complex economic organisation in Lincoln in the late fourth century. Antiquity 72: 417–424.

Drucker D., Bocherens H., Mariotti A., Lévêque F., et al. 1999. Conservation des signatures isotopiques du collagène d'os et de dents du Pléistocène supérieur (Saint-Césaire, France): implications pour les reconstitutions des régimes alimentaires des néandertaliens. Bull. Mémo. Soc. Anthropol. Paris, n.s.t. 11 (3–4): 289–305.

Dufour D.L. and Sander J.B. 2000. Insects. In: Kiple K.F. and Ornelas K.C. (eds), The Cambridge World History of Food, vol. I, Cambridge: Cambridge Univ. Press, pp. 546–554.

Eaton S.B. 1992. Humans, lipids and evolution. Lipids 27: 814–820.

Eaton S.B. and Konner M. 1985. Paleolithic nutrition: A consideration of its nature and current implications. New England J. Med. 312 (19): 1253–1254.

Eaton S.B., Eaton S.B. III, and Konner M.J. 1997. Paleolithic nutrition revisited: A twelve-year retrospective on its nature and implications. European J. Clin. Nutr. 51: 207–216.

Eaton S.B., Eaton S.B. III, and Cordain L. 2002. Evolution, Diet and Health. *In:* Ungar P.S. and Teaford M.F. (eds). Human Diet, Its Origin and Evolution. Bergin and Garvey, Westport, CT, pp. 7–18.

Fowler C.S. and Walter N.P. 1985. Harvesting Pandora moth larvae with the Owens Valley Paiute. J. California and Great Basin Anthropol. 7: 155–165.

Fullagar R., Furby J., and Hardy B. 1996. Residues on stone artifacts: state of a scientific art. Antiquity 70: 740–745.

Goodall J. 1963. Feeding behavior of wild chimpanzees. A preliminary report. Symp. Zool. Soc. London 10: 39–47.

Goodall J. 1986. The Chimpanzees of Gombe: Patterns of Behaviour. Harvard Univ. Press, Cambridge, MA.

Guerreschi A. 1990. La scoperta di Mondeval de Sora ed alcune considerazioni sul Mesolitico di alta quota nelle Dolomiti. *In:* Le Dolomiti, un patrimonio da tutelare e amministrare. Agordo (Bl), Comunità Montana Agordina.

Haile-Selassie Y. 2001. A Late Miocene hominid from the Middle Awash, Ethiopia. Nature, 412: 178–181.

Hamilton W.J.III. 1987. Omnivorous primate diets and human overconsumption of meat. *In:* Harris M. and Ross E.B. (eds). Food and Evolution: Towards a Theory of Human Food Habits. Philadelphia, Temple Univ. Press, Philadelphia, PA, pp. 117–132.

Hardy B.L., Raff R.A., and Raman V. 1997. Recovery of Mammalian DNA from Middle Paleolithic stone tools. J. Archaeol. Sci. 24: 601–611.

Hayden B. 1981. Subsistence and ecological adaptations of modern hunter/gatherers. *In:* Harding R.S.O. and Teleki G. (eds). Omnivorous Primates. Gathering and Hunting in Human Evolution. Columbia Univ. Press, New York, NY, pp. 344–421.

Hérnandez-Pacheco E. 1924. Las peinturas prehistóricas de las Cuevas de Araña (Valencia). Museo Nac. Cienc. Nat., Madrid, Spain.

Hladik C.M. 1977. Chimpanzees of Gabon and chimpanzees of Gombe: some comparative data on the diet. *In:* Clutton Brock T.H. (ed.). Primate Ecology. Academic Press, London, UK.

Hladik C.M. and Viroben G. 1974. L'alimentation protéique du Chimpanzé dans son environnement forestier naturel. Comptes Rendus Acad. Sci. Paris 279: 1475–1478.

Hladik C.M. and Simmen B. 1996. Taste perception and feeding behavior in nonhuman primates and human populations. Evol. Anthropol. 5(2): 58–71.

Hocking B. and Matsumura F. 1960. Bee brood as food. Bee World 41: 113–128.

Isack H.A. and Reyer H.-U. 1989. Honey guides and honey gatherers: interspecific communication in a symbiotic relationship. Science 243: 1343–1346.

Johns T. 1999. The chemical ecology of human ingestive behaviors. Ann. Rev. Anthropol. 28: 27–50.

Kay R.F. and Sheine S.W. 1979. On the relationship between chitin particle size and digestibility in the primate *Galago senegalensis*. Amer. J. Phys. Anthropol. 55: 301–308.

Keeley L.H. 1980. Experimental Determination of Stone Tool Use. A Microwear Analysis. Univ. Chicago Press, Chicago, IL.

Keeley L.H. and Toth N. 1981. Microwear polishes on early stone tools from Kooby Fora, Kenya. Nature 293: 464–465.

Klein R.J. 1999. The Human Career. Human Biological and Cultural Origins. Univ. Chicago Press: Chicago, IL (2nd ed.).

Kliks M. 1978. Paleodietetics: a review of the role of dietary fiber in preagricultural human diets. *In:* Spiller G.A. and Amen R.J. (eds.). Topics in Dietary Fiber Research. Plenum Press, New York, NY.

Lanave C., Liuni, S., Licciulli F., and Attimonelli, M. 2002. Update of AMmtDB:a database of multi-aligned Metazoa mitochondrial DNA sequences. Nucleic Acids Res., 30 (1): 174–175.

Leakey L.S.B. 1971. Olduvai Gorge. Vol. III: Excavations in Beds I and II, 1960–1963. Cambridge Univ. Press, Cambridge, UK.

Leakey M.G., Feibel C.S., McDougall I., and Walker A. 1995. New four-million-year-old hominid species from Kanapoi and Allia Bay, Kenya. Nature 376: 565–571.

Leakey M.G., Feibel C.S., MacDougall I., Ward C.V., and Walker A. 1998. New specimens and confirmation of an early age for *Australopithecus anamensis*. Nature 393: 62–66.

Leakey R.E.F. and Walker A. 1985. Further hominids from the Plio-Pleistocene of Kooby Fora, Kenya. Amer. J. Phys. Anthropol. 67: 135–163.

Lee R.B. 1968. What hunters do for a living or how to make out on scarce resources. *In:* Lee R.B. and DeVore I. (eds.). Man the Hunter. Aldine, Chicago, IL, pp. 30–48.

Lee-Thorp J.A. and van der Merwe N.J. 1993. Stable carbon isotopes studies of Swartkrans fossils. *In:* Brain C.K. (ed.). Swartkrans. A cave's chronicle of early man. Transvaal Museum, Pretoria, Transvaal.

Lee-Thorp J.A., van der Merwe N.J., and Brain C.K. 1994. Diet of *Australopithecus robustus* at Swartkrans from stable carbon isotopic analysis. J. Human Evol. 27: 361–372.

Leonard W.R. 2003. Cibo per pensare. Le Scienze 413: 67–74.

Leonard W.R. and Robertson M.R. 1994. Evolutionary perspectives on human nutrition: The influence of brain and body size on diet and metabolism. Amer. J. Human Biol. 6: 77–88.

Leonard W.R. and Robertson M.R. 1997. Comparative primate energetics and hominid evolution. Amer. J. Phys. Anthropol. 102: 265–281.

MacNeish R.S. 1958. Preliminary archaeological investigations in the Sierra de Tamaulips, Mexico. Amer. Phil. Soc. Trans. 44: 5.

Mahaney W.C., Hancock R.G.V., Aufreiter S., and Huffman M.A. 1996. Geochemistry and clay mineralogy of termite mound soil and role of geophagy in chimpanzee of the Mahale Mountains, Tanzania. Primates 37 (2): 121–134.

Malaisse F. 1997. Se nourrir en forêt claire africaine. Approche écologique et nutritionelle. Les presses Agronomiques, Gembloux, Belgium.

Mann A.E. 1981. Diet and human evolution. *In:* Harding R.S.O. and Teleki G. (eds.). Omnivorous Primates. Columbia Univ. Press, NY, New York, pp. 10–36.

Mann F.D. 1998. Animal fat and cholesterol may have helped primitive man evolve a large brain. Persp. Biol. Med. 41 (3): 417–425.

Marota I., Basile C., Ubaldi M., and Rollo F. 2002. DNA decay rate in Papyri and human remains from Egyptian archaeological sites. Amer. J. Phys. Anthropol. 117: 310–318.

Martin R.D. 1989. Evolution of the brain in early hominids. Ossa 14: 49–62.

McGrew W.C. 1979. Evolutionary implications of sex differences in chimpanzee predation and tool use. *In:* Hamburg D.A. and McCown E.R. (eds.). The Great Apes. Benjamin/Staples, Menlo Park, CA, pp. 440–463.

McGrew W.C. 1981. The female chimpanzee as an evolutionary prototype. *In:* Dahlberg F. (ed.). Woman, the Gatherer. Yale Univ. Press, New Haven, CT, pp. 35–73.

McGrew W.C. 1992. Chimpanzee Material Culture. Cambridge Univ. Press, Cambridge, UK.

McGrew W.C. 2001. The other faunivory: Primate insectivory and early human diet. *In:* Stanford C.B. and Bunn H.T. (eds.). Meat-eating and Human Evolution. Oxford Univ. Press, Oxford, UK, pp. 160–178.

Milton K. 1987. Primate diets and gut morphology. Implications for hominid evolution. *In:* Harris M. and Ross E.B. (eds.). Food and Evolution: Toward a Theory of Human Food Habits. Temple Univ. Press, Philadelphia, PA, pp. 93–115.

Milton K. 1999. A hypothesis to explain the role of meat-eating in human evolution. Evol. Anthropol. 8 (1): 11–21.

Milton K. 2002. Hunter-gatherer diets: Wild foods signal relief from disease of affluence. *In:* Ungar P.S. and Teaford M.F. (eds.). Human Diet, Its Origin and Evolution. Bergin and Garvey: Westport, CT, pp. 111–122.

Mondini C. and Villabruna A. 1988. Notizie preliminari sulla sepoltura epigravettiana di Val Rosna. Archivio storico di Belluno, Feltre e Cadore, LIX: 117–137.

Moore J.G., Fry G.F., and Englert E. Jr. 1969. Thorny headed worm infection in North American prehistoric man. Science. 163: 1324–1325.

Newbrun E. 1982. Sugar and dental caries: a review of human studies. Science 217: 418–423.

O'Connell J., Hawkes K., and Blurton Jones N. 1999. Grandmothering and the Evolution of *Homo erectus*. J. Human Evol. 36: 461–485.

O'Connell J., Hawkes K., and Blurton Jones N. 2002. Meat-eating, grandmothering, and the evolution of early human diets. *In:* Ungar P.S. and Teaford M.F. (eds.). Human Diet, Its Origin and Evolution. Bergin and Garvey, Westport, CT, pp. 49–60.

Pager H. 1973. Rock paintings in Southern Africa showing bees and honey hunting. Bee World, 54: 61–68.

Pager H. 1976. Cave paintings suggest honey hunting activities in Ice Age times. Bee World 57: 9–14.

Panagiotakopulu-Buckland P.C., Day P.M., Doumas C., et al. 1997. A lepidopterous cocoon from Thera and evidence for silk in the Aegean Bronze Age. Antiquity 71: 420–429.

Paoletti M.G., Dufour D.L., Cerda H., Torres F., Pizzoferrato L., and Pimentel D. 2000. The importance of leaf- and litter-feeding invertebrates as sources of animal protein for the Amazonian Amerindians. Proc. Roy. Soc. B. 267: 2247–2252.

Paoletti M.G., Buscardo E., Vanderjagt D.J., Pastuszyn A., et al. 2003. Nutrient content of Termites (*Synthermes soldiers*) consumed by Makiritare Amerindians of the Alto Orinoco of Venezuela. Ecol. Food Nutr. 42: 173–187.

Poinar H.N. and Stankiewicz B.A. 1999. Protein preservation and DNA retrieval from ancient tissues. Proc. Natl. Acad. Sci. USA 96: 8426–8431.

Poinar H.N., Hofreiter M., Spaulding W.G., Martin P.S., et al. 1998. Molecular coproscopy: dung and diet of the extinct ground sloth *Nothrotheriops shastensis*. Science 281: 402–406.

Poinar H.N., Kuch M., Sobolik K.D., Barnes I., et al. 2001. A molecular analysis of dietary diversity for three archaic Native Americans. Proc. Natl. Acad. Sci. USA 98 (8): 4317–4322.

Potts R. 1996. Humanity's Descent: The Consequences of Ecological Instability. William Morro, New York. NY.

Puech P.F. and Albertini H. 1984. Dental microwear and mechanisms in early hominids from Laetoli and Hadar. Amer. J. Phys. Anthropol. 65: 87–91.

Ramos-Elorduy J. 1996. Rôle des insectes dans l'alimentation en forêt tropicale. *In:* Hladik C.M. et al. (eds.). Alimentation en forêt tropicale. UNESCO, Paris, pp. 371–381.

Reinhardt K.J. and Bryant V.M.Jr. 1992. Coprolite analysis. A biological perspective on archaeology. Adv. Archaeol. Method Theory 4: 245–288.

Robinson M. 1991. The Neolithic and Late Bronze Age insect assemblages. *In:* Needham S. (ed.). Excavation and Salvage at Runnymede Bridge, 1978: the Late Bronze Age waterfront site. British Museum/English Heritage, London, UK.

Rollo F. 2002. Il DNA dei Faraoni: realtà o leggenda? Scienze 412: 78–83.

Rollo F., Ubaldi M., Ermini L., and Marota I. 2002. Ötzi's last meals: DNA analysis of the intestinal content of the Neolithic glacier mummy from the Alps. Proce. Natl. Acad. Sci. USA 99 (20): 12594–12599.

Ruttner F., 1952. Alter und Herkunft der Bienenrassen Europas. Öst. Imker 2: 8–12.

Ryan A.S. and Johanson D.C. 1989. Anterior dental microwear in *Australopithecus afarensis*. J. Human Evol. 18: 235–268.

Santos Oliveira J.F., Passos de Carvalho J., Bruno de Sousa R.F.X., and Madalena Simão M., 1976. The nutritional value of four species of insects consumed in Angola. Ecol. Food Nutr. 5: 91–97.

Scudeller-Baccelle L. and Nardi S. 1991. Interaction between calcium carbonate and organic matter: An example from the Rosso Ammonitico Veronese (Veneto, north Italy). Chem. Geol. 93: 303–311.

Senut B., Pickford M., Gommery D., Mein P., Cheboi K., and Coppens Y. 2001. First hominid from the Miocene (Lukeino Formation, Kenya). C.R. Acad. Sci. Paris 332: 137–144.

Sept J. 1992. Archaeological evidence and ecological perspectives for reconstructing early hominid subsistence behavior. Adv. Archaeol. Method Theory 4: 1–56.

Shipman P. 2001. What can you do with a bone fragment? Proc. Natl. Acad. Sci. USA 98 (4): 1335–1337.

Sillen A. 1992. Strontium-calcium ratios (Sr/Ca) of *Australopithecus robustus* and associated fauna from Swartkrans. J. Human Evol. 23: 495–516.

Sillen A. 1994. L'alimentation des hommes préhistoriques. La Recherche, 25, 264: 384–390.

Skinner M. 1991. Bee brood consumption: An alternative explanation for hypervitaminosis A in KNM-ER 1808 (*Homo erectus*) from Kooby Fora, Kenya. J. Human Evol. 20: 493–503.

Smith B.N. and Epstein S. 1971. Two categories of 13C/12C ratios for higher plants. Plant Physiol. 47: 380–384.

Speth J.D. 1987. Early hominid subsistence strategies in seasonal habitats. J. Archaeol. Sci. 14: 13–29.

Spittel R.L. 1924. Wild Ceylon. Describing in particular the lives of the present-day Veddas. Ceylon (cited in Bondenheimer, 1951).

Sponheimer M. and Lee-Thorp J.A. 1999. Isotopic evidence for the diet of an early hominid, *Australopithecus africanus*. Science 283: 368–370.

Steudel K.L. 1994. Locomotor energetics and hominid evolution. Evol. Anthropol. 3: 421–428.

Stewart K.M. 1994. Early hominid utilisation of fish resources and implications for seasonality and behavior. J. Human Evol. 27: 229–245.

Sutton M.Q. 1990. Insect resources and Plio-Pleistocene hominid evolution. In: Posey D.A. et al. (eds), Ethnobiology: Implications and Applications. Proc. 1st Int. Cong. Ethnobiology (Belém, 1988), Museu Paraense Emilio Goeldi, Belém, Brazil, 1990.

Sutton M.Q. 1995. Archaeological aspects of insects use. J. Archaeol. Method Theory 1: 253–298.

Sweeney G. 1947. Food supplies of a desert tribe. Oceania 17: 289–299.

Teaford M.F. and Ungar P.S. 2000. Diet and the evolution of the earliest human ancestors. Proc. Natl. Acad. Sci. USA 97(25): 13506–13511.

Tommaseo-Ponzetta M. 2003. Rôle alimentaire des insectes dans l'evolution humaine. *In:* Motte-Florac E. and Thomas J.M.C. (eds.). Les insectes dans la tradition orale. Peeters-Selaf (Ethnosciences), Paris, pp. 269–286.

Tommaseo-Ponzetta M. and Paoletti M.G. 1997. Insects as food of the Irian Jaya populations. Ecol. Food Nutr. (s.i.) 36 (2–4): 321–346.

Tommaseo-Ponzetta M. and Paoletti M.G. 2004. Lessons from traditional foraging patterns in western Papua (Indonesia) (this volume).

Toth N. 1985. The Olduwayan reassessed: a close look at early stone artifacts. J. Archaeol. Sci. 12: 101–120.

Trevor-Deutsch B. and Bryant V.M. Jr. 1978. Analysis of suspected human coprolites from Terra Amata, Nice, France. J. Archaeol. Sci. 5: 387–390.

Turnbull C.M. 1965. The Mbuti Pygmies: An ethnographic Survey. Anthropological Papers American Museum Natural History 50: 139–282.
Ungar P.S. and Grine S.E. 1991. Incisor size and wear in *Australopithecus africanus* and *Paranthropus robustus*. J. Human Evol., 20: 313–340.
Ungar P.S. and Teaford M.F. 2001. The dietary split between apes and the earliest human ancestors. *In:* Tobias P.V. et al. (eds.). Humanity from African Naissance to Coming Millennia. Firenze University Press, Firenze, pp. 337–352.
Ungar P.S. and Teaford M.F. (eds), 2002. Human Diet. Its Origin and Evolution. Bergin and Garvey, Westport, CT, USA.
Vandel A. 1964. Biospéleologie. La Biologie des Animaux Cavernicoles. Gautier-Villars, Paris.
Vrba E.S. 1988. Late Pliocene climatic events and hominid evolution. *In:* F.E.Grine (ed.), Evolutionary history of the "Robust" australopithecines. Aldine de Gruyter, New York, NY, pp. 405–426.
Walker A. 1982. Diet and teeth. Dietary hypotheses and human evolution. Phil. Trans. Roy Soc. London B292: 57–64.
Walker A., Zimmerman M.R., and Leakey R.E.F. 1982. A possible case of hypervitaminosis A in *Homo erectus*. Nature 296: 248–250.
Webb W.S. and Baby R.S. 1957. The Adena People. Ohio State University Press, Columbus, OH.
Weir G.W. and Bonavia D. 1985. Coprolitos y dieta del preceràmico tardìo de la Costa Peruana. Bol. Inst. Francés Estudios Andinos 16: 85–140.
Wheeler Q. 1990. Insect diversity and cladistic constraints. Ann. Entomol. Soc. Amer. 83: 1031–1047.
White T.D., Suwa G., and Asfaw B. 1994. *Australopithecus ramidus*, a new species of early hominid from Aramis, Ethiopia. Nature 371: 306–312.
Whiten A. 2000. Primate culture and social learning. Cognitive Sci., s.i., 24: 477–508.
Whiten A., Goodall J., McGrew W.C., Nishida T., et al. 1999. Cultures in Chimpanzees. Nature 399: 682–685.
WHO (World Health Organisation). 1997. Tropical Disease Research. Progress 1995–1996. WHO, Geneva.
Wood B. and Collard M. 1999. The human genus. Science 284: 65–71
Wrangham R.W., Jones J.H., Laden G., Pilbeam D., and Conklin-Brittain N.L. 1999. The raw and the stolen: Cooking and the ecology of human origins. Curr. Anthropol. 40: 567–594.
Wu Leung W.T., Busson F., and Jardin C., 1968. Food Composition Table for Use in Africa. FAO, Rome/U.S. Department of Health, Education and Welfare, Bethesda, MA.
Yarnell R.A. 1969. Contents of human paleofeces. *In:* Watson P. (ed.). The Prehistory of Salts Cave, Kentucky. Illinois State Museum, Springfield, IL, pp. 1–56.

9

Minilivestock Consumption in the Ancient Near East: the Case of Locusts

GIOVANNI B. LANFRANCHI

Dipartimento di Scienze dell'Antichità, Università di Padova, Piazza Capitaniato 7, 35100 Padova, Italy. E-mail: gb.lanfranchi@unipd.it

Abstract

Reconstruction of minilivestock consumption in the Ancient Near East is extremely difficult due to the scarcity of precise archaeological data. However, it is attested to in a select group of texts stemming from ancient Israel and pre-Classical Syria and Mesopotamia. The dietary prescriptions prevailing in the biblical Hebrew culture generally forbade consumption of any kind of minilivestock, especially of insects and other small animals; contrarily, locust consumption was expressly allowed. The capture of locusts for consumption is attested to, as well as their killing to preclude severe crop damage, in some letters to Syrian and Assyrian kings (IInd and Ist millennium BC respectively). The high rank of the writers and addressees clearly underscores that locust consumption was an elitist custom. Some hints for preparation of unique locust recipes are found in scattered texts of various kinds, with a special soup made from locusts and pomegranates being perhaps the most renowned dish. In an 8th century BC Neo-Assyrian relief on a palace slab there is a vivid representation of the introduction of sets of locusts fixed on spits to a royal banquet. Their careful representation as completely preserved animals suggests their preparation at the table, perhaps by sauteing or some other procedures. Still, one cannot exclude that their ostentatious presentation was a prelude to the making of the elitist locust-and-pomegranate soup.

Key Words: locust, minilivestock consumption, pre-Classical Syria, pre-Classical Mesopotamia, Hebrew dietary prescriptions, Bible

Introduction

Mesopotamia, and especially southern Mesopotamia, is generally considered to have been the kernel for the development of agricultural customs which constitute the base of modern Western food habits. It is generally assumed that the cultivation of wheat, barley and flax, and the domestication of many mammals such as sheep, goats, pigs, and oxen began in the Xth millennium BC; from the VIIth millennium BC onwards, lentils, chickpeas, beans, and vetches were also introduced. Olive tree cultivation spread from the Palestinian coastal regions, and wine preparation seems to have originated in the Syrian and Iranian foothills. In general, the agricultural cultivation progressions were so impressive that already in the early IVth millennium BC urban development reached an important peak, aptly represented by the astonishing extension of the southern Mesopotamian Sumerian city of Uruk which was not inferior to that of Imperial Rome.

Human diet in the ancient Near East can be studied with some degree of confidence through a large set of data stemming from archaeological relics, from artistic imagery, and from written texts (cuneiform tablets from Mesopotamia and its neighboring regions, and the Biblical text).

As for archaeological relics, a large set of precious information on diet may be drawn from various elements, such as vessels and cooking pots (in terms of their physical composition, shape, and composition of the food remains preserved inside them), other instruments related to cooking (grinding stones, mortars, etc.), and buildings (destination, plan, materials, status of preservation, etc.). As for animal consumption, the problem is more complex because the physical preservation of some remains is practically impossible. Precious data may stem from bone remains, found—often in large numbers—in or around town or village buildings. However, these are generally limited to the skeletons of large mammalians or even large birds, which are apt to survive natural decay and consumption in underground locations. Relics of smaller mammalians and other small animals, subject to easier and faster decay, are extremely rare, and hence rarely reported in scientific publications. Further, such relics cannot be confidently attributed to human consumption (Hesse and Wapnish, in Collins, 2002: 462). In principle, they can be explained as remnants of consumption by animals of prey (such as owls), or even as the result of death in haunt due to accidental natural phenomena such as earthquakes or to sudden human destructive activity such as fire or collapse following the destruction of buildings. Due to this basic ambiguity, the rare findings of complete skeletons of lizards and snakes in archaeological excavations are not to be taken seriously as clear indicators for the reconstruction of human diet (Hesse and Wapnish, *ibid.*).

Unfortunately, artistic imagery too is of little help in this kind of research. Monumental iconography, preserved mainly in royal palaces or in important private buildings, is very rarely devoted to small animals in association with human consumption. When small animals are portrayed or reproduced, this is due almost regularly either to decorative purposes, or to religious and magic symbolisms (animals representing the image of gods or animals as amulets), as is the case with the well-known Egyptian scarabs and bees.

Written texts obviously offer a much larger set of information with regard to human diet in the ancient Near East. A great number of Mesopotamian, Syrian, and Hittite texts deal with animal husbandry, and with animal food preparation, consumption, and storage from the administrative point of view. However, here too small animals are rarely mentioned: the focus is mainly on large mammals, such as bovids, goats and sheep, and on fish and various kinds of birds.

Animal food consumption is also treated in religious texts. In the majority of these texts, the focus is mainly sacrificial animals. However, the choice of animals fit for sacrifice cannot be considered directly relevant to the human diet, even though it might be admitted that gods were thought to preferably eat what men currently ate (Scurlock, 2002a, in Collins: 392). In any case, here too small animals have very rarely been mentioned. More interesting for small animal consumption is a restricted group of religious texts which list dietary prescriptions, to be observed either always throughout the year or only during some specific periods or days.

The most revealing texts of this kind are the Biblical books of Leviticus and Deuteronomy. Here the consumption of almost all small animals, either terrestrial, aquatic or aerial, is drastically prohibited. In Leviticus, insectivores, rodents, saurians, and other terrestrial small animals such as mustelidae, are not only excluded from human diet, but also regarded as unclean.

> And these are they which are unclean unto you among the creeping things that creep upon the earth: the weasel, and the mouse, and the great lizard after its kind, and the gecko, and the land-crocodile, and the lizard, and the sand-lizard, and the chameleon. These are they which are unclean to you among all that creep: whosoever doth touch them, when they are dead, shall be unclean until the even. (Lev. 11: 29-31)

In Leviticus, every kind of reptile is also excluded from diet: "And every creeping thing that creepeth upon the earth is an abomination; it shall not be eaten." (Lev. 11:41)

Further, all aquatic creatures, either marine, riverine or lacustrine, other than fishes are condemned in Leviticus and Deuteronomy:

> And of all that have not fins and scales in the seas, and in the rivers, of all that move in the waters, and of all the living creatures that are in the waters, they are an abomination unto you. (Lev. 11:10)

> Of all that live in water you may eat these: whatever has fins and scales you may eat. And whatever does not have fins and scales you shall not eat; it is unclean for you. (Deut. 14:9-10)

The consumption of insects and of other small flying creatures is likewise excluded in Leviticus, paralleled by Deuteronomy:

> All winged creeping things that go upon all fours are an abomination unto you. (Lev. 11:20)
>
> But all winged creeping things, which have four feet, are an abomination unto you. (Lev. 11:23)
>
> And all winged insects are unclean for you; they shall not be eaten. (Deut. 14:19)

With respect to small flying animals, an interesting exception to dietary prohibition is represented by arthropods. Following Leviticus (not paralleled by Deuteronomy), these animals may be eaten:

> Yet these may ye eat of all winged creeping things that go upon all fours, which have legs above their feet, wherewith to leap upon the earth. Even these of them ye may eat: the locust after its kind, and the bald locust after its kind, and the cricket after its kind, and the grasshopper after its kind. (Lev. 11: 21–22)

The absence of clear archaeological data for relics of small animals in relation to human food consumption precludes from establishing whether the population in Israel really followed the prescriptions listed in the Biblical books. It is reasonable to suppose that most people piously followed the religious dietary prescriptions. But it is also reasonable to suppose that in daily life there were disobediences and exceptions, the latter especially on occasions of urgent necessities such as famines, due either to natural events or provoked by human emergencies such as wars or protracted sieges.

A similar set of data regarding religious dietary prohibitions is missing for other Near Eastern cultures. No hint of a detailed, fixed dietary regimen sanctioned in prescriptive texts like that in Israel can be found in Mesopotamian, Hittite or Syrian documents. However, some scant data may be drawn from hemerologies, a category of texts in which each day of the year is associated with some specific duty or prohibition. Unfortunately, scholars have paid little attention to these texts and hence few of them have been published and edited to date. Most dietary obligations or prohibitions concern the consumption of large mammalians or birds; however we can learn from a Mesopotamian hemerology that on the first and fourth days of the month Tašritu (the seventh month of the Mesopotamian calendar, coinciding roughly with October), the Mesopotamians "should not eat dormouse" (Scurlock, 2002b citing Labat, 1939, nos. 168: 11-15; 170: 38–41; 176: 1–3, in Collins 2002: 393; Hulin, 1959, no. 48:8, cf. no. 52:38). From this prohibition we may obviously deduce that the human consumption of dormouse was a common and widespread practice.

Consumption of small animals on special occasions is also attested in some Mesopotamian medical texts, which list recipes for the preparation of various drugs. For example, a recipe from Nineveh prescribes the sick person to drink beer and to eat a kind of mouse or a shrew (Küchler, 1904: 2 and 35). However, ancient Near Eastern medicine is a peculiar field in which magic has an important role. Therefore, the consumption of small animals such as mouse (or shrew) was certainly prescribed because of the magical or symbolic power attributed to those animals.

As mentioned above, sacrificial animals might be considered, albeit very cautiously, indicative of human consumption on the premise that gods might eat what humans ate. Thus we find that barley-fattened bandicoot rats (a rodent living in the southern marshes) were offered in the regular daily meal given to Marduk, chief of the Babylonian gods, during the Seleucid period (IVth-IInd centuries BC) (Scurlock, 2002b, citing Englund, 1995, 37–55, in Collins, 2002: 390). At the end of the IIIrd millennium BC, the southern Mesopotamian calendar included a month which seems to have been named "porcupine-eating" (Sallaberger, 1993: I: 195; however, Cohen, 1993: 145–147 suggests that the month's name should be translated "month of the bitter herbs").

In general, scant data stemming from written texts preclude tracing a methodologically correct comparison between the dietary prescriptions of ancient Israel and those of neighboring cultures. The few elements pointed out so far seem to suggest that there were some important differences with regard to small animals, the Mesopotamian diet being a little more "liberal" than that of Israel; however, it is not possible to prove that this difference was not due to some peculiar development in some specific periods or regions. On the other hand, the attention duly paid to accurately listing animals excluded from human consumption in the Biblical books could well be a resolute determination to differentiate Israel from neighboring populations. This would allow the conclusions that in Israel the dietary conventions of other Near Eastern peoples were totally or partly dismissed insofar they were considered religiously unclean, and that such uncleanliness may have derived, at least in part, from the permission to eat some selected small animals such as the above mentioned dormouse.

However that may be, from both written documents and monumental iconography it can be deduced that Hebrew, Syrian, and Mesopotamian dietary conventions with regard to small animals agreed at least on one point. Consumption of locusts appears not only to have been allowed in all three cultures from the religious point of view, but even systematically practiced in daily life, at least in Syria and Mesopotamia.

As already seen, the dietary prescriptions in Leviticus (11:21) did allow consumption of locusts, bald locusts, grasshoppers, and crickets, which are grouped under the general label of "all winged creeping things that go upon all fours, which have legs above their feet, wherewith to leap upon the earth." These are clearly separated from the "winged creeping things that go upon all fours" (Lev. 11:20) which are excluded from the diet, and which should be identified almost certainly with all species of insects: the number of legs higher than two helps in distinguishing them from birds, and reaffirms the variable number of legs of invertebrates. The characteristic marking the difference seems to have been their ability to jump rather than their being winged, e.g. crickets (if the translation of the Hebrew word is correct). On the other hand, it seems that permission to eat these animals among all other flying creatures other than birds depends on their ability to damage and destroy cultivated fields. In other parts of the Bible, locusts like caterpillars are depicted as a dangerous pest, similar to famine, pestilence, and war ("If there be in the land famine, if there be pestilence, blasting, mildew, locust, [or] if there be caterpillar; if their enemy besiege them in the land of their cities; whatsoever plague, whatsoever sickness [there be]" (1Kings 8:37), and "for this reason they are one of the seven plagues inflicted upon Egypt" (Exodus 10:12-18).

The harmful activity of locusts is rather well attested in Mesopotamian documents of the IInd and Ist millennium BC. In some letters written in the XVIIIth century BC to the king of Mari, the capital of a Syrian state on the midcourse of the Euphrates, the royal governors report to the king about the damages caused by impressive invasions of locusts, and take various measures to destroy them, or at least limit their desctrution of the harvest (such as filling canals with water, or massive killing effected by the local population; fire, the expected most effective defense, is not reffered to in these texts) (Heimpel, 1996; Lion and Michel, 1997: 710–716). The governors mention damage to the early spring barley crops.

In Akkadian, locusts are usually designated *erbu*; in the Mari texts, however, other designations appear such as *sansar* or *sarsar* (cf. Hebrew *ṣelāṣal* in Deut. 28:42; very curiously, the Syrian Akkadian term *ṣanṣar/ṣarṣar* represents the origin of the Italian word *zanzara*, however designating the mosquito: the intermediation of some local Arabian dialect may be taken into account), *erhizzu/irgiṣu*, or *ergilatum* (Lion and Michel, 1997: 708–710). These terms may well designate different kinds of insects or arthropods; but perhaps they designate either locusts of different sizes, or various phases in the physical development of locust nymphs (Geller, 2003).

In other letters, the Mari governors mention the capture of locusts in the field and their delivery to court. The governor of the town Saggaratum informs the king of Mari that locusts caught in a field have been sent to the capital: "There were *ṣanṣar*-locusts in the valley... . The locusts covered [everything]. Now, I send these locusts to my lord." (Jean, 1941: no. 107, 22–27).

Similarly, the governor of the town Terqa informs the king of Mari that locusts had invaded his district, but could not install themselves in the fields because of hot weather; then, he adds that some locusts had been caught (either by the population or by the governor's servants) and that he was delivering them to the capital: "When I was to send this tablet to my lord, *ṣanṣar*-locusts arrived to Terqa. When they arrived, it was hot, and they did not land. Thus, now I am sending all the locusts which have been caught." (Kupper, 1948: no. 62, 5-18).

In neither letter is it stated whether the locusts sent to court had been killed in the field or collected live and then preserved; however, the latter alternative seems more probable in light of another letter, written by the governor of the town Qattunân to the king of Mari. *Erhizzu*-locusts caught in the fields had been sent to the king in reed cages, but had died because of the long distance traveled; consequently, the governor replaced them with another kind of locusts.

About the *erhizzu*-locusts of which my lord wrote to me. Here, where *ergilatum*-locusts can be caught, there are no *erhizzu*-locusts. I sent five nomads and they picked up *erhizzu*-locusts at [the town] Musilanum of the Talhayum district. The distance being long, these *erhizzu* locusts died in their reed cages. I have herewith placed 38 *erhizzu*-locusts under my own seal and conveyed them to my lord...." (Birot, 1993; 64, 4–18; I quote the translation offered by Geller, 2003).

Destruction and collection of locusts are also attested to in letters of the Ist millennium BC, stemming from the royal archives of Kalhu, the capital of the Neo-Assyrian empire from the 9th century BC to 709 BC, and of Nineveh, the capital from 705 BC. From Kalhu stems a letter written by an official (probably a governor) to the Assyrian king reporting that in accordance with a royal order, the whole population of the province had killed the locusts in the field, thereby preserving the harvest from destruction:

The bodyguard [o]f the king, my lord, wrote to me: 'Kill the locusts!'. We put down as many as there were, none of them even touched the harvest. All the subjects of the king my lord destroyed them. The crops in the entire territory of the king my lord, are very well. (Saggs, 1952: no. CIII: 6-r.17)

Also the governor of Assur, the ancient religious capital of Assyria, wrote to the king about the killing of locusts in his province, following orders given by the king himself (Parpola 1987: no. 104: 5–9: "As to what the king, my lord, wrote to me 'In order to kill locusts go to the villages Mê-tabūte and Amante as far as the town Kasappa.'" The letter, found in Nineveh, is unfortunately fragmentary).

It seems that fields believed to be liable to an abnormal influx of locusts' numbers were marked off to facilitate a prompt intervention at the crucial moment of the locusts' intensive reproduction, as stated in another letter to the king of the governor of Assur:

[As to the lo]custs [concerning which the king, my lord], wrote to me: 'Send word that fields infested with locusts be marked off with boundary stones! They must be knocked out (?) [the moment] they are about to shed!'. (Parpola, 1987, no.103:6-13; letter found in Nineveh).

Destruction and collection of locusts in the field are mentioned in a letter written by a lesser official to a high dignitary in the Syrian region of the Lower Khabur river:

They wrote to me from the Palace [*i.e.* the central administration in the Assyrian capital]: 'Write down all the locusts that you collect and all that you kill, and send to the Palace!' Now then I am writing it down and sending it to my lord."

After listing various quantities of locusts (measured in weight) arranged by villages (where the locusts had been taken or killed), the official states that the killing was started when the locust number increased excessively:

> When they were few, we collected them, ... pushed them into a seah measure and measured them with it; when they became oppressive, we just killed them in the middle of the field. (Parpola, 1987, no. 221:4-9, rev. 2-6)

It is clear that both in the IInd millennium Syria and the Ist millennium Assyria, locusts were collected before they started to multiply in the field; killing them took place only when their number became excessive.

As for the possible identification of the locust species mentioned in the Syrian and Mesopotamian texts, in a detailed study the Assyriologist W. Heimpel (Heimpel, 1996) has convincingly demonstrated that it was the *Dociostaurus maroccanus* (subfamily Gomphocerinae), commonly known as the Moroccan locust, a species which is still at home in the Syrian and Mesopotamian Jazirah, and which severely affected the environs of Jerusalem in April 1915 and southwestern Turkey from 1910 to 1918 (Heimpel, 1996: 103–104). Heimpel duly noted that the Mari texts which mention the killing and the collection of locusts are dated only to spring months, while no text can be assigned to summer months; furthermore, in the texts there is no mention of damage to typical summer crops such as wheat. This excludes the possibility that the Mari locusts might have been the *Schistocerca gregaria* (subfamily Cyrtacanthacridinae), known as the Desert locust, which swarms in peak summer.

Both in the Mari and Neo-Assyrian texts, the king appears clearly to be directly involved in the taking of measures for killing locusts in due time. This may reasonably be interpreted as a routine preoccupation for the maintenance of field productivity in the provincial territories. However, in many texts the killing of locusts is linked with a delivery of animals to the royal court requested by the king or the royal administrators themselves. Even a lesser provincial official located in far-off central Syria might be requested by the central administration to carefully report the number of taken and killed locusts. As seen from the Mari texts, it is also clear that locusts were to be delivered to the royal court live and that dead locusts were discarded. The dispatch of locusts to the royal court depended certainly on the demand of refined culinary items for the royal table, viz. locusts for consumption. This is attested in a letter from Mari, where a specialized cook (Lion and Michel, 1997: 718–719) in need of locusts is sent to court to collect the requisite number.

> The *lurakkûm*-cook who is in [the town] Saggaratum has no locusts; and this is what he told me: 'My lord had locusts be taken from [the town] Šubat-Šamaš'. Now, I will send the *lurakkûm*-cook to my lord: let them give him locusts [for] his duty. (Jean, 1941: no. 136, 4–13)

The consumption of locusts was common in the Mesopotamian world, since they appear along with other edible items in various texts, such as some old-Babylonian letters requesting the delivery of food ("send me 100 locusts and ..." Stol, 1981: no. 15, 27–28; "Send me locusts, cress, ..., chickpeas, lentils, and garlic," Stol, 1981, no. 152, 24–26. Both letters from Larsa in central Babylonia). From these letters, it appears that locusts were considered ingredients of complex dishes, some of which are cursorily mentioned in texts of various genres. In the so-called "Dream Book" (an Assyrian text listing omens to be understood from dreams), a soup of locusts is mentioned along with soups of barley, lentils, etc. (Oppenheim, 1956: 315, Sm. 2073+ iii, 9; cf. Lion and Michel, 1997: 718). Locusts seem to have been indispensable in the preparation of a special soup, named *šiqqu*, as attested in an old Babylonian letter from Tell Harmal (central Babylonia):

> A second matter. We are going to prepare a *šiqqum*-sauce. [We] do not [have] any *sansarni*-locust. Since you have at your disposal 1 *pi*-measure and 4 *qu*-measure of *sansarni*-locusts, and they were promised to us, buy 3 [more] *ban*-measure of locusts wherever there is any and send them to us. (Goetze, 1958: no. 17, 15-22)

The locust *šiqqu*-soup was considered a medicine to be taken together with pomegranate juice (*ibid.*). Details of the recipe of the *šiqqu*-soup are not known but the term *šiqqu* seems to designate a kind of soup whose preparatory procedure was equal in importance to the soup *per se*, since a *šiqqu*-soup of fish is also frequently mentioned in the texts (Lion and Michel, 1997: 718–719). Perhaps the *šiqqu*-soup may be compared with the Roman *garum*-sauce, which was prepared following various recipes which had different, dominant ingredients. In describing specific recipes for preparing a beverage including locusts, Herodotus (IV, 17, 2) relates that a Libyan population used to dry and mince locusts, then mix them with milk. The association of locusts with liquids seems thus a customary mode of consumption.

The delivery of locusts to the royal court, attested to both in Mari and in Assyria, was clearly for the purpose of preparing elitist dishes. No recipe including locusts as an ingredient has been preserved in the extant documentation. In any case, discarding locusts that died during transport, attested to in the Mari letter, indicates that recipes, like that for the *šiqqu*-soup, required live locusts to start with. In fact there is no written document suggesting that killed locusts may have been subjected to some special treatment in order to preserve them (e.g., pickling).

Some slight indications about the preparation and presentation of dishes based on locusts can be deduced from a vivid representation of a ceremonial banquet hosted by the Assyrian king Sennacherib perhaps in 692 or 691 BC for the inauguration of his "Palace without rivals" in Nineveh (archaeologically known as the "Southwest Palace"). The walls of a long ceremonial corridor were decorated with a series of carved stone slabs which depicted a long procession of men transporting multivarious foods to be served at the royal banquet, or displayed on tables, or enclosed in vessels, or carried in the hands (part of the procession is depicted in Matthiae, 1996, Fig. 8.2). Two of these slabs bear the so far unique, strikingly detailed representation of a set of locusts (or large grasshoppers) ready to be served at the royal table (Figs. 9.1 and 9.2).

The locusts are fixed on long spits, each bearing opposed rows of animals, arranged as though stationary on the ground, their legs resting on the spit, the articulations of their bent legs at a level higher than their bodies. The bodies of the animals are depicted as completely and perfectly preserved, even their antennae and wings. A comparison of the size of the men bearing the spits and that of the transported locusts, shows the latter to be large adult animals, excluding any possibility of nymphs in one of the five stages of development (for the growth stages of locusts, see Gilbert, 2002: 41). Since in the detailed representation there is no trace of any kind of string attaching the locusts to the spit, one may safely deduce that they were not brought live to the royal table. So, either they had been killed and cooked in the royal kitchen before the ceremonial entrance, or merely killed (and perhaps prepared in some way) to be cooked in the presence of guests at the royal table. Had they been cooked before the ceremonial presentation, the cooking process must have been extremely delicate given the perfect preservation of their body and above all their antennae and wings. The cooking procedure may have been delicate sauteing in oil or butter, as done in modern Near East (Geller, 2003). While on the other hand, had they been freshly killed in

Fig. 9.1: Slabs from Corridor LI in the Southwest Palace of Sennacherib, Nineveh (702–692 B.C.). (from A.A. Layard, 1853. Discoveries in the Ruins of Nineveh and Babylon. London, 339 pp.).

Fig. 9.2: Slab from Corridor LI in the Southwest Palace of Sennacherib, Nineveh (702–692 B.C.). (from A.A. Layard, 1853. Discoveries in the Ruins of Nineveh and Babylon. London, 339 pp.).

the kitchen, preparation at the royal table can only be conjectured. It may be that the locusts *per se* were fried or cooked in some other manner at the table. But the left slab depicts a couple of servants directly behind those carrying the spitted locusts, clutching in both hands what seem to be two strings of fruits or vegetables, either garlic, small onions or, more likely, small pomegranates. As noted above, two of these items appear with locusts in texts dealing with food preparation. Medical texts prescribe association of pomegranate juice and locust-*šiqqu*-soup; Old Babylonian letters requesting food list garlic along with locusts. Thus, we may assume that either locust-*šiqqu*-soup was mixed with pomegranate juice, or some other special dish mixing garlic and locusts was prepared at the table to entertain the royal personages.

Inclusion of locusts in the menu prepared for, or at, the royal table on ceremonial occasions shows clearly that at court such animals were considered a luxury item. However, we may take for granted that common people too killed and collected locusts or grasshoppers for direct home consumption. Certainly, following recipes poorer than those of the royal court, they might have offset, at least in part, the shortage of cereal products caused by locust swarms. As for the royal table, it cannot be excluded that the serving of locusts may have also had a symbolic and ideological significance. The public display of the corpses of the most celebrated enemy of harvest, although transformed into luxury food, might efficaciously represent the king's ability to take care of the harvest in his country and his constant concern for his subjects' welfare.

References

Birot M. 1993. Correspondance des gouverneurs de Qattunân. Archives royales de Mari, vol. XXVII. Editions Recherche sur les Civilisations, Paris.

Cohen M.E. 1993. The Cultic Calendars of the Ancient Near East. CDL Press, Bethesda.

Collins B.J. (ed.). 2002. A History of the Animal World in the Ancient Near East. Handbook of Oriental Studies, Section One. The Near and Middle East, vol. 64. EJ Brill, Leiden.

Englund R.K. 1995. There's a rat in my soup! Altorientalische Forschungen 22: 37–55.

Geller M.J. 2003. Diet and Regimen in the Babylonian Talmud. *In:* Grottanelli C. and Milano L. (eds.). Food and Identity. S.A.R.G.O.N., Padova (in press).

Gilbert A.S. 2002. The Native Fauna of the Ancient Near East. *In:* Collins B.J. (ed.), 2002, pp. 3–75.

Goetze A. 1958. Fifty Old Babylonian Letters from Tell Harmal. Sumer 14: 3–78.

Heimpel W. 1996. Moroccan Locusts in Qattunan. Revue d'Assyriologie 90: 101–120.

Hesse B. and Wapnish P. 1997. An archaeozoological perspective on the cultural use of mammals in the Levant. *In:* Collins B.J. (ed.), 2002, pp. 457–491.

Hulin P. 1959. A Hemerological Text from Nimrud. Iraq 21: 42–53.

Jean Ch.-F. 1941. Lettres Archives Royales de Mari, Lupzimezie Nationale, Paris, vol. II.

Küchler F. 1904. Beiträge zur Kenntnis der assyrisch-babylonischen Medizin. Leipzig.

Kupper J.-R. 1948. Lettres Archives royales de Mari, Lupzimezie Nationale, Paris, vol. III.

Labat R. 1939. Hémérologies et ménologies d'Assur. Paris.

Lion B. and Michel C. 1997. Criquets et autres insectes à Mari. Ann. Recherches Interdisciplinaires Mari 8: 707–724

Matthiae P. 1996. L'arte degli Assiri. Cultura e forma del rilievo storico. Laterza, Bari, Roma.

Oppenheim A.L. 1956. The Interpretation of Dreams in Ancient Near East. Trans. Amer. Phil. Soc. NS 46/III, Philadelphia, PA.

Parpola S. 1987. The Correspondence of Sargon II, Part I. Letters from Assyria and the West. State Archives of Assyria I. Helsinki University Press, Helsinki.

Saggs H.W.F. 1952. The Nimrud Letters, 1952, Part IX. Iraq 36, 199-221.

Sallaberger W. 1993. Der kultische Kalender der ur III-Zeit. Untersuchungen zur Assyriologie und Vorderasiatischen Archäologie 7. Walter De Gzuyter, Berlin.

Scurlock J. 2002a. Animals in Ancient Mesopotamian Religion. *In:* Collins B.J. (ed.), 2002, pp. 361-403.

Scurlock J. 2002b. Animal Sacrifice in Ancient Mesopotamian Religion. In: Collins B.J. (ed.), 2002, pp. 390-403.

Smith B.D. 1995. The Emergence of Agriculture. Scientific American Library, Washington DC.

Stol M. 1981. Letters from Yale. Altbabylonische Briefe in Umschrift und Übersetzung 9. EJ Brill, Leiden.

Zohary D. and Hopf M. 1993. Domestication of Plants in the Old World: Clarendon Press, Oxford, UK.

10

Human Consumption of Lepidoptera, Termites, Orthoptera, and Ants in Africa

François Malaisse

Laboratoire d'Ecologie, Faculté Universitaire des Sciences Agronomiques de Gembloux, Passage des Déportés, 2, B-5030 Gembloux Belgium. E-mail: malaisse.f@fsagx.ac.be

The use of insects as food for humans deserves greater recognition from governments and from donor agencies. Edible insects are a neglected resource to alleviate malnutrition and hunger (van Huis, 1996).

Abstract

Human consumption of Lepidoptera, Isoptera, Orthoptera, and Formicidae in Africa is reviewed based on published literature and recent inquiries carried out at the author's request. This information is analyzed according to species diversity, ethnolinguistic groups concerned, chemical composition, as well as some ecological aspects such as caterpillar food plants, diversity of items eaten in termitaria, etc.

At least 100 caterpillar species provide human food in Africa, of which 85 have been evaluated. Values are respectively known for 25–30 ethnospecies and 18 taxa for Isoptera, 89 of about 120 species of Orthoptera, and at least a tenth of the edible ant species—more than 250 species altogether.

From a chemical point of view, large diversities exist at various levels according to taxa involved (at family as well as species level).

Discussion focuses on the importance of uptake, sustainable management, preliminary knowledge needed, and choice of valuable species. It has been noted that caterpillars are the most convenient wild edible food entering sustainable production. The future of large termitaria and *Macrotermes* species is dark, even if the present biomass remains tremendous. More accurate studies are needed.

Key Words: entomophagy, Africa, caterpillars, termites, Orthoptera, ants

Introduction

Entomophagy has received considerable interest these last decades. With about 2,000 edible insect species listed throughout the world (Ramos-Elorduy, 2004), more and more interest has recently been devoted to both diversity and nutritional aspects. Human consumption of insects considers some 104 families distributed in 14 orders (Table 10.1). Of these, four orders take a dominant position regarding the number of species involved, namely Coleoptera, Hymenoptera, Lepidoptera, and Orthoptera; as far as values regarding number of families are concerned, Lepidoptera takes first place with 21 families.

This paper focuses on the consumption of Lepidoptera (mainly caterpillars), Isoptera (termites), Orthoptera (locusts, grasshoppers, crickets) and Formicidae (ants). Practices pertaining to caterpillars are referred to as "campeophagy"—derived from Greek "καμπη", meaning caterpillar and "φαγειν", meaning to eat (Malaisse, 2003). Eating practices for the other groups are similarly termed termitophagy, acridophagy and myrmecophagy.

While consumption of locusts is recorded in the Bible [Mark 1:6] (Southwood, 1977), the interest of peoples inhabiting tropical Africa in other edible insect groups is also ancient, as indicated in comments made in manuscripts from the colonial period, some of which are cited below.

Some articles and books have reviewed information pertaining to entomophagy on a worldwide scale, such are Bergier (1941), Bodenheimer (1951), and DeFoliart (2002). Concerning Africa, information on edible insects is given by Tango Muyay (1981), Bahuchet (1985), Nonaka (1996), Seignobos et al. (1996), Malaisse (1997), and Roulon-Doko (1998).

Materials and Methods

Several levels of knowledge may be distinguished within the literature dealing with entomophagy. One set of articles simply reports that certain insects or insect groups are consumed with no mention of the species but indicating countries, regions or ethnolinguistic groups. A second set of publications lists ethnospecies when several species are involved, and sometimes provides primary information about their ecology. In a third set, Linnean species names are

Table 10.1: Diversity of families with edible insects in the world

Orders	Families	Orders	Families	Orders	Families
Anisoptera	Aeshnidae	**Hemiptera**	Belastomatidae		Hesperiidae
	Libellulidae		Cicadidae		Lasiocampidae
Anoplura	Peliculidae		Coreidae		Limacodidae
Coleoptera	Anobiidae		Corixidae		Lymantriidae
	Bruchidae		Naucoridae		Mimallonidae
	Buprestidae		Nepidae		Momphidae
	Carabidae		Notonectidae		Noctuidae
	Cerambycidae		Pentatomidae		Notodontidae
	Cicindelidae		Scutelleridae		Nymphalidae
	Chrysomelidae	**Homoptera**	Aphididae		Papilionidae
	Curculionidae		Cicadellidae		Pieridae
	Dynastidae		Cicadidae		Psychidae
	Dytiscidae		Coccidae		Pyralidae
	Elateridae		Dactylopidae		Saturniidae
	Histeridae		Flatidae		Sphingidae
	Hydrophilidae		Fulgoridae	**Neuroptera**	Corydalidae
	Lucanidae		Membracidae	**Orthoptera**	Acrididae
	Melolonthidae		Psyllidae		Blattidae
	Passalidae	**Hymenoptera**	Apidae		Catantopidae
	Tenebrionidae		Cynipidae		Gryllacrididae
Diptera	Calliphoridae		Formicidae		Gryllidae
	Chaoboridae		Halictidae		Gryllotalpidae
	Conopidae		Megachilidae		Hemiacrididae
	Culicidae		Thynnidae		Hymenopodidae
	Ephydridae		Vespidae		Mantidae
	Leptidae		Xylocopidae		Mecopodidae
	Muscidae	**Isoptera**	Hodotermitidae		Pamphagidae
	Rhagionidae		Rhinotermitidae		Phaneropteridae
	Sarcophagidae		Termitidae		Phasmidae
	Stratiomyidae	**Lepidoptera**	Arctiidae		Pyrgomorphidae
	Syrphidae		Brahmaeidae		Tettigoniidae
	Tipulidae		Castniidae	**Plecoptera**	Perlolidae
Ephemeroptera	Baetidae		Cossidae		Pteronarcyidae
	Ephemeridae		Geometridae	**Trichoptera**	Hydropsychidae
			Hepialidae		Stenopsychidae

given. A fourth set deals with the chemical composition and nutritional value of edible insects.

As far as materials are concerned, two main problems arise. Old scientific names of insects need to be updated on the basis of recent systematic revisions on the one hand, and correct Latin spelling has to be checked on the other, and their nominator has also to be provided.

With respect to the methodology of collecting information, three approaches were used. First, old literature was reviewed. Second, edible insects were listed and collected by Kimba (Bas-Congo and Tabwa), Kisimba (Katanga), Latham (Bas-Congo), Ouedraogo (Burkina Faso), and Saracco-Lanata (Gabon). Third, some collected material was sent for identification.

Most information is currently cited by country. I opted to use an ethnolinguistic division as human groups are considered the most important in the present study.

A map showing the distribution of ethnolinguistic groups for Africa was collated from previous essays (Murdock, 1959), as well as from more local information dealing with Angola (Milheiros, 1967), Bas-Congo (Boone, 1973), Burkina Faso (Grimes, 1996), Southeast Congo (Boone, 1961), Tanzania (Thomas, 1975), and Zambia (Anon., 1998). Some 1,032 groups were retained and their territorial zone of influence mapped.

Campeophagy in Africa

Introduction

As far as we know, the oldest comment dealing with human consumption of caterpillars in Africa occurs in a manuscript relating the Simon van der Stel expedition in 1685–86 in Namaland, south of Namibia (Waterhouse, 1932; Palmer and Pitman, 1972). A drawing by Claudius (Fig. 10.1), presumably of *Imbrasia tyrrhea* (Oberprieler, *in litt.*), appears in the manuscript, which is housed in Dublin and was edited in 1932 by Waterhouse. According to van der Stel: "This caterpillar is called Aroube by the Namaquas and is found in their country. The monster is regarded by them as a delicacy and a dainty dish, for when they have first squeezed out of it all the green ordure, they impale it on a wooden spit and lay it on the embers until it is baked hard, and then they consign it with gusto to their eager bellies".

Diversity of Information

Distribution and diversity of campeophagy has been reported in explorer accounts as well as travel stories (de Almeida, 1946; Bloomhill, 1958), local dictionaries of the colonial period (Courtois, 1900; Dinkelacker, 1914; Baião, 1939; White Fathers, 1949; Hannan, 1959), ethnographic studies, ethnozoological reports (Gaerdes, 1959), and lastly in local notes and surveys. Synthesis of the information gathered is expressed in two different ways: (a) a table in which each source is cited according to the ethnolinguistic group involved and (b) a map of distribution of ethnolingusitic groups practicing campeophagy (Fig. 10.2).

Fig. 10.1: Drawing of *Imbrasia tyrrhea*, an edible caterpillar, by Claudius in Simon van der Stel's expedition manuscript (after Palmer and Pitman, 1972).

Table 10.2 presents the ethnolinguistic groups for which we have been able to cite human consumption of caterpillars (and pupae), all together some 109 sources. Some 121 ethnolinguistic groups practice campeophagy in Africa, constituting 12% of the ethnic groups recognized for the continent. Interest in campeophagy (Fig. 10.3) has increased, with about 51 papers published during the last decade.

Diversity of Edible Species

According to present knowledge regarding comestible caterpillars in tropical Africa, some 87 species of 11 families (Table 10.3) have been determined. In addition to these, at least some 30 other species are consumed elsewhere according to Adriaens (1953), Bouyer (1999), DeFoliart (1991a), McGregor (1995), Roulon-Doko (1980, 1998), and Tango Muyay (1981). They belong to families Agaristidae, Cossidae, Geometridae, Nymphalidae (Acraeineae and Nymphalineae), and Papilionidae.

Fig. 10.2: Distribution map of campeophagy in Africa.

In total, some 15 families are concerned, Saturniidae ranks first with some 51 different species or 50%, followed up by Notodontidae (16 species, 16%), and Sphingidae (10 species, 10%).

Color photographs of African edible caterpillars can be found in Malaisse (1997) (24 species), Latham (1999a-b, 2000, 2002) (12 Linnean species, 8 ethnospecies); Oberprieler (1995) has provided color plates of 22 species of Saturniidae. Malaisse and Parent (1980) have set up a key of determination of 31 edible caterpillars of Katanga.

Other caterpillar families worldwide also contain edible species, as indicated in papers dealing with Central America (Aldasoro Maya, 2003; Ramos-Elorduy, 1991); e.g. Castniidae, Ceratocampidae, Hepialidae, Megathymidae, Nycteolidae, and Pyralidae.

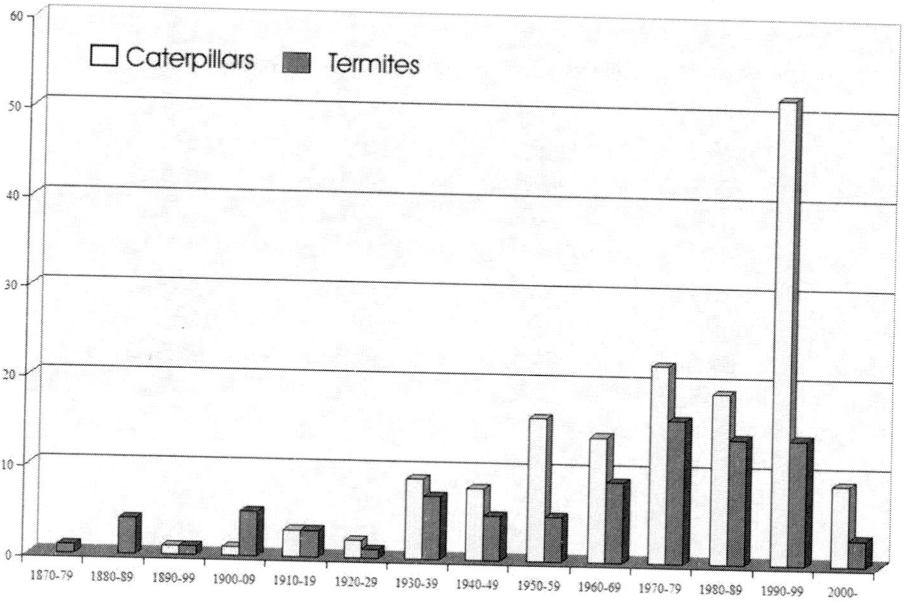

Fig. 10.3: Number of publications dealing with edible caterpillars and termites in Africa according to decades.

Food of Edible Caterpillars

Table 10.3 contains preliminary information on plants eaten by the diverse edible caterpillars of Africa (Latham, 2000, 2002; Malaisse, 1983, 1997; Oberprieler, 1995; Pinhey, 1972, 1975; Takeda, 1990; Turk, 1990). A study of host plants concerning 163 caterpillar species of Upper Katanga (Malaisse, 1983) states various strategies, from monophagous caterpillar species to polyphagous ones, some of which live on 15 different plant species.

The information has been categorized into large territorial entities, from West Africa to South Africa. *Cirina forda* provides a good example of the variation in diversity of host plants according to the five territories concerned. This caterpillar feeds on *Vitellaria paradoxa* or karité in West Africa, *Autranella congoensis* in the Central African Republic, *Crossopteryx febrifuga* in Lower Congo, *Albizia antunesiana* in Katanga, and *Burkea africana* in South Africa. The biological cycle is monovoltine in Katanga (Malaisse, 1997) but bivoltine in South Africa (Van den Berg et al., 1973). Such information is of interest to rearers of caterpillars as a source of animal proteins.

Table 10.2: Ethnolinguistic groups (or territories) of Africa for which campeo-, termito-, acrido- or myrmecophagy has been indicated.

N°	Ethnolinguistic unit (Territory)	Country	Cam	Ter	Acr	Myr	Reference(s)
			E/L	*	E/L	E/L *	
(1)	(2)	(3)	(4)	(5)	(6)	(7)	(8)
N5	Kabyle	Algeria			+		Bergier 1941
N7	Moroccan	Morocco			+		Lucas *in* Bergier 1941
N45	Chaamba	Algeria			+		Künkel d'Herculais 1891
N63	Ahaggaren (Kel)	Algeria			1/1	1/1	Gast 2000
N66	Teda = Toubou	Chad		I			Carbou 1912
8	Peul	Burkina Faso		X			Ouedraogo et al. 2003
8bis	Barani	Burkina Faso		I			Ouedraogo et al. 2003
28bis	Nyabwa	Ivory Coast			-/1		Zézé-Béké 1989
29	Kono	Guinea-Conakry	+				Villiers 1957
37	Dan (Man)	Ivory Coast		I			Villiers 1947
37a	Toura=Win	Ivory Coast		I			Iroko 1996
39	Sénoufo	Burkina Faso	1/1	I			Ouedraogo et al. 2003
39a	Natioro	Burkina Faso		QI			Ouedraogo et al. 2003
39b	Blé	Burkina Faso		I			Ouedraogo et al. 2003
42	Bété	Ivory Coast		I			Paulian 1946
46	Grebo Gbaepo	Liberia		I			Saimon *in* DeFoliart 2002
47	Kru (E Monrovia)	Liberia		QI			Landry *in* DeFoliart 2002
47	Kru	Ivory Coast		I			Holas 1980
48	Sapo (E Monrovia)	Liberia		QI			Landry *in* DeFoliart 2002
50	Dioula = Dyula	Burkina Faso	1/1	X			Ouedraogo et al. 2003
50bis	Dzùùngoo	Burkina Faso		I			Ouedraogo et al. 2003
66	Bambara	Mali	+	I			Binger 1892
69	Dogon	Burkina Faso		I			Ouedraogo et al. 2003
70	Samo = Sana	Burkina Faso		QI			Ouedraogo et al. 2003

Table 10.2: (Contd.)

Table 10.2: (Contd.)

(1)	(2)	(3)	(4)	(5)	(6)	(7)	(8)
71	Songhai = Songoy	Burkina Faso		I			Ouedraogo et al. 2003
75	Hinyanka	Mali		I			Binger 1892
77	Marka	Burkina Faso		QI			Ouedraogo et al. 2003
78	Bobo Madaré	Burkina Faso	+				Sawadogo *in* DeFoliart 2002
78	Bobo Madaré	Burkina Faso		I			Ouedraogo et al. 2003
78bis	Bomu = Bore	Burkina Faso		I			Ouedraogo et al. 2003
79	Bwamu	Burkina Faso		QI			Ouedraogo et al. 2003
80	Tusya(n)	Burkina Faso		I			Ouedraogo et al. 2003
81	Mosi=Mooré	Burkina Faso		I	+		Bertrand *in* DeFoliart 2002
81	Mosi=Mooré	Burkina Faso		I	+		Sawadogo *in* DeFoliart 2002
81	Mosi =Mooré	Burkina Faso		QI			Ouedraogo et al. 2003
82	Gurmanche	Burkina Faso		QI			Ouedraogo et al. 2003
82bis	Yana	Burkina Faso		QI			Ouedraogo et al. 2003
83	Lilse = Lyélé	Burkina Faso		I			Ouedraogo et al. 2003
84	Ko = Kipirsi	Burkina Faso		I			Ouedraogo et al. 2003
85	Nuni = Nunuma	Burkina Faso		QI			Ouedraogo et al. 2003
86	Busansi = Bissa	Burkina Faso		QI			Ouedraogo et al. 2003
87	Lobi	Burkina Faso	1/1	I			Ouedraogo et al. 2003
90	Dyan	Burkina Faso		QIE			Ouedraogo et al. 2003
91	Karaboro	Burkina Faso	1/1	I			Ouedraogo et al. 2003
92	Wara Wara	Burkina Faso		I			Ouedraogo et al. 2003
94	Bolon = Boka	Burkina Faso		I			Ouedraogo et al. 2003
96	Birifor	Burkina Faso		QI			Ouedraogo et al. 2003

Table 10.2: (Contd.)

Table 10.2: (Contd.)

(1)	(2)	(3)	(4)	(5)	(6)	(7)	(8)
97	Dagara = Dagari	Burkina Faso	1/1	QIE			Ouedraogo et al. 2003
97b	Bobo Madaré	Burkina Faso		I			Ouedraogo et al. 2003
97d	Seeku	Burkina Faso		QI			Ouedraogo et al. 2003
100	Kassena = Kasem	Burkina Faso		I			Ouedraogo et al. 2003
131	Moba	Burkina Faso		I			Ouedraogo et al. 2003
154	Bariba	Nigeria	+/1	QI	+/3		Fasoranti and Ajiboye 1993
155	Zerma = Dyerma	Burkina Faso		QI			Ouedraogo et al. 2003
157	Hausa = Haoussa	Burkina Faso		I			Ouedraogo et al. 2003
173	Nupe	Nigeria	+/1	QI	+/3		Fasoranti and Ajiboye 1993
174	Yoruba	Nigeria	+				Golding 1942
174	Yoruba	Nigeria	+				Ene 1963
174	Yoruba	Nigeria	+/1				Ashiru 1988
174	Yoruba	Nigeria	+/1	QI	+/3		Fasoranti and Ajiboye 1993
210	Soga (Namwenda)	Uganda		I			Osmaton 1951
301	Fang = Pangwe	Gabon		+			Largeau 1901
301	Fang = Pangwe	Gabon	21/-				Tessmann 1913
316	Nzebi	Gabon	+				Rougeot 1955
316	Nzebi	Gabon	12/-				Saracco-Lanata (in litt.)
318	Kota	Gabon	4/-				Saracco-Lanata (in litt.)
335	Boma = Buma	D.R. Congo		I			Hochegger 1973
335	Boma = Buma	D.R. Congo		I			Hochegger 1975
339	Teke	Gabon	17/-				Sarraco-Lanata (in litt.)
339	Teke	Congo		I			Savorgnan de Brazza *in* Ney 1984
339	Teke	Congo			+		Guiral *in* Bergier 1941
339	Teke	Congo		I			Adzou *in* Iroko 1996
339	Teke	D.R. Congo			44/3		Kisimba et al. 2003
342	Kongo	D.R. Congo		+			Baumann 1887
342	Kongo	D.R. Congo		+			Laman 1936

Table 10.2: (Contd.)

Table 10.2: (Contd.)

(1)	(2)	(3)	(4)	(5)	(6)	(7)	(8)
342	Kongo	D.R. Congo			4/-		Kisimba et al. 2003
342.1	Kongo [Bwende=Lari]	Congo	13/1	I	13/11		Nkouka 1987
342.8	Kongo[Mboma]	D.R. Congo		+			Hegh 1922
342.10	Kongo [Ndibu]	D.R. Congo	27/-	+			Kimba Kasalwe (in litt.)
342.12	Kongo[Ntandu]	D.R. Congo		+			Hegh 1922
342.12	Kongo[Ntandu]	D.R. Congo	22/12				Latham 1999a-b, 2000, 2002
342.17	Kongo[Ambriz]	Angola		I			Monteiro 1875
356a	Mongo (Ngando)	D.R. Congo	11/4	QIS			Takeda 1990
356b	Mongo (Ntomba)	D.R. Congo	+	2/-	1/1		Pagezi 1985, 1988
358	Yans = Yanzi	D.R. Congo		I			de Plaen 1974
358	Yans = Yanzi	D.R. Congo	33/-	QISN	7/-		Tango Muyay 1981
359	Mbala	D.R. Congo		+			Torday and Joyce 1905
359	Mbala	D.R. Congo		+			Hegh 1922
359	Mbala	D.R. Congo	9/4	I	1/1		Adriaens 1953
359	Mbala	D.R. Congo	24/-		11/-		Mbemba and Remacle 1992
359bis	Pelende	D.R. Congo	>8/1	I	+		Adriaens 1951
359bis	Pelende	D.R. Congo	4/-		3/-		Mbemba and Remacle 1992
360	Suku	D.R. Congo			+		Mbemba and Remacle 1992
360bis	(T)Samba	D.R. Congo	+				Mbemba and Remacle 1992
361	Songo	D.R. Congo	+		+		de Beaucorps 1941
362	Sondi = Sonde	D.R. Congo	4/2				Leleup and Daems 1969
363	Pende	D.R. Congo	7/3				Leleup and Daems 1969
363	Pende	D.R. Congo	18/-				Daems *in* Leleup and Daems 1969
363	Pende	D.R.Congo	+	I	+		Kirinich *in* DeFoliart 2002
369	Nzing=Dzing	D.R. Congo		+			Mertens 1935
375	Tetela	D.R. Congo			3/-		Kisimba et al. 2003
376	Songye	D.R. Congo	+		47/3		Kisimba et al. 2003
378	Kanyok	D.R. Congo	+	+	42/3		Kisimba et al. 2003
379	Luba Katanga	D.R. Congo		I			Van Avermaet and Mbuya 1954
379	Luba Katanga	D.R. Congo		+			Mahauden 1965
379	Luba Katanga	D.R. Congo	+	+	40/3		Kisimba et al. 2003

Table 10.2: (Contd.)

Table 10.2: (Contd.)

(1)	(2)	(3)	(4)	(5)	(6)	(7)	(8)
379	Luba Katanga	D.R. Congo				1/-,Q	Malaisse, this volume
380	Luba Kasai	D.R. Congo		7/-			Callewaert 1922
380	Luba Kasai	D.R. Congo		+			De Clercq 1960
380	Luba Kasai	D.R. Congo	5/4	+	32/-		Kisimba et al. 2003
380a	Kuba	D.R. Congo		I			Torday 1925
380a	Kuba	D.R. Congo		I			Vansina 1964
380a	Kuba	D.R. Congo	+	IS	+	+	Boyle *in* DeFoliart 2002
380b	Lulua	D.R. Congo	9/-	IS	1/-	1/-	Katya Kitsa 1989
380b	Lulua	D.R. Congo	5/4		3/-		Kisimba et al. 2003
380c	Kabinda	D.R. Congo	8/5		47/3		Kisimba et al. 2003
381	Kete	D.R. Congo	+	IS	+	+	Boyle *in* DeFoliart 2002
385	Lunda	Zambia	6/-		1/-		Mbata 1995
385	Lunda-Ruund	Zambia		IS			Malaisse (unpubl. data)
385	Lunda-Ruund	Zambia	9/7		41/3		Kisimba (in litt.)
386	Cokwe=Tshokwe	Angola		+			Pogge 1880
386	Cokwe=Tshokwe	Angola		+			MacJannet 1947
386	Cokwe=Tshokwe	Angola	+				Roeges *in* Leleup and Daems 1969
386	Cokwe=Tshokwe	Angola	+	+			Silow 1983
386	Cokwe=Tshokwe	D.R. Congo	12/8		10/-		Kisimba et al. 2003
388	Holo	D.R. Congo	> 6/-				Denis 1964
390	(U)Mbundu North	Angola		I			Serpa Pinto 1881
390bis	Gingas (Malanje)	Angola		I			Santos Oliveira et al. 1976
391	(U)Mbundu South	Angola	17/-	3/-	9/-		Bossard 1996
392bis	Bienos=Bihénos	Angola		I			Serpa Pinto 1881
395	Nyaneka	Angola		I			Lang and Tastevin 1937
396	Herero	Angola		X			Silow 1983
396	Herero	Botswana		I			Anderson *in* von Berensberg 1907
396	Herero	Namibia		I			Steinhardt 1922
396	Herero	Namibia	2/2	I	1/1		Marais et al. [n.d.]
396	(Ochi)Herero	Namibia	-/6				Marais 1996
397	Ganguella	Angola	+	+			Serpa Pinto 1881
398	Luimbi	Angola	+				Baum 1903
399	Luc(h)azi	Zambia	+	+			Silow 1983
399	Luc(h)azi	Zambia	1/-	1/-			Mbata 1995
400	Kwanyama=Ambo	Namibia		+			Tönjes 1910

Table 10.2: (Contd.)

Table 10.2: (Contd.)

(1)	(2)	(3)	(4)	(5)	(6)	(7)	(8)
400	Kwanyama	Namibia	2/2	+	1/1		Marais et al. [n.d.]
400bis	Kwanyama= Owambo	Namibia	-/5				Marais 1996
401bis	Ndonga=Kavango	Namibia	-/2				Marais 1996
402	Cimba=Zemba	Namibia	2/2	+	1/1		Marais et al. [n.d.]
403	Lwena=Luvale	Zambia		+			Horton 1953
403	Lwena=Luvale	Zambia	18/-	+			White 1959
403	Lwena=Luvale	Zambia		+			Silow 1983
403	Lwena=Luvale	Zambia	4/-	1/-	4/-		Mbata 1995
404	Ndembo	Zambia	+	+			Silow 1983
404	Ndembo	D.R. Congo	6/6		6/-		Kisimba et al. 2003
405	Kaonde	Zambia	5/-	1/-	2/-		Mbata 1995
405	Kaonde	Zambia	20/15				Kisimba et al. 2003
406	Mbunda	Zambia	31/-	I			Silow 1983
406bis	Yauma = Iauma	Angola	+	I			Silow 1983
407	Nkoya	Zambia	12/-	X			Silow 1983
407bis	Mashasha	Zambia	+	X			Silow 1983
409	Lozi	Zambia		X			Silow 1983
409	Lozi	Zambia			1/-		Mbata 1995
410	Nyengo	Zambia	+	+			Silow 1983
413	Mbukushu	Zambia		+			Silow 1983
421a	Hechware	Zimbabwe	+				Hobane 1995
421a	Hechware = Sarwa	Botswana	+				Bryden *in* Bodenheimer 1951
421f	Tswana (Kgalagadi)	Namibia	-/1				Marais 1996
421g	Tswana Kwena	Botswana	1/1	+	+		Livingstone 1857
421h	Tswana Tlokwa	Botswana	5/3	1/1	6/1	+	Grivetti 1979
423	(Si)Ndebele	Zimbabwe	1/-	+			Pelling 1966
423	(Si)Ndebele	Zimbabwe		IS			Gomez 1989
423	(Si)Ndebele	Zimbabwe	+	+	+	+	Steele (in litt.)
424	Gwembe Tonga	Zambia	+	I			Scudder 1962
424	Gwembe Tonga	Zambia		I			Reynolds 1968
424	Tonga	Zambia				1/-	Torrend 1931
424	Tonga	Zambia		I	7/-		Mbata 1995
425	Ila	Zambia		IG		X	Smith and Dale 1968
427	Lenje	Zambia	+	I			Torrend 1931
427	Lenje	Zambia		I	1/-		Mbata 1995
429	Lala	Zambia		I			Madan 1913
429	Lala	Zambia	+	I			Thomson 1954
429	Lala	Zambia		I			Silow 1983
429	Lala	Zambia	1				Mbata 1995

Table 10.2: (Contd.)

Table 10.2: (Contd.)

(1)	(2)	(3)	(4)	(5)	(6)	(7)	(8)
429	Lala	D.R. Congo	17/14		6/-		Kisimba et al. 2003
430	Nsenga	Zambia		I			Silow 1983
430	Nsenga	Zambia	10/-	I	4/-		Mbata 1995
431	Kunda	Zambia	1/-		1/-		Mbata 1995
432	Cewa	Zambia		I			Stannus 1910
432	Cewa	Zambia		I			Hodgson 1933
432	Cewa	Zambia		I			Rita-Ferreira 1966
432	Cewa	Zambia		I		1/-	Silow 1983
432	Cewa	Zambia		I	4/-		Mbata 1995
432	Cewa	Malawi	14/-	I			Clarke et al. 1996
433	Nyanja-Ngosi	Zambia	3/-	I	11/-		Mbata 1995
433	Nyanja	Zambia	13/-				Mkanda and Munthali 1994
435	Bisa	Zambia		I			Silow 1983
435	Bisa	Zambia	8/6				Mbata and Chidumayo 1999
435	Bisa	Zambia	8/5				Mbata and Chidumayo 2003
436	Aushi	Zambia	+/1	I	1/-		Kisimba and Muzinga (in litt.)
437	Lamba	Zambia		I			Madan 1913
437	Lamba	Zambia	+	I			Doke 1931
437	Lamba	Zambia	21/18				Demesmaecker 1997
437	Lamba	D.R. Congo			1/-		Kisimba et al. 2003
438	Sanga	D.R. Congo	1/1	I			Lambrechts and Bernier 1961
438	Sanga	D.R. Congo			+	1/-,Q	Malaisse, this volume
440	Shila	D.R. Congo		IS			Kisimba and Muzinga (in litt.)
441	Tabwa	D.R. Congo	+	I			Bergier 1941
441	Tabwa	D.R. Congo	+	QIS	2/1		Kimba (in litt.)
442	Bemba	Zambia	6/-	I			Richards 1939
442	Bemba	Zambia		I			White Fathers 1954
442	Bemba	D.R. Congo	35/26				Malaisse and Parent 1980
442	Bemba	Zambia	26/-	I	5/-		Mbata 1995
442	Bemba	D.R. Congo	38/24	QIS	1/25		Malaisse 1997
442	Bemba	D.R. Congo	40/27	QIS	1/25		Malaisse, this volume
442	Bemba	D.R. Congo			3/-		Kisimba et al. 2003
443	Lungu	Zambia		I			Silow 1983

Table 10.2: (Contd.)

Table 10.2: (Contd.)

(1)	(2)	(3)	(4)	(5)	(6)	(7)	(8)
445	Mambwe	Zambia		I			Silow 1983
448	Nyakyusa	Malawi	+				Makungwa et al. 1997
449bis	Malila	Tanzania	1/-				Latham (in litt.)
454	Nyamwezi	Tanzania		I	+		Spellig 1929
454	Nyamwezi	Tanzania		I			Blohm 1931
454	Nyamwezi	Tanzania	3/-	3/1	-/5		Harris 1940
458	Tumbwe	D.R. Congo		IS			Kisimba and Muzinga (in litt.)
458bis	Holoholo	D.R. Congo	+	+	+		Schmitz 1912
459	Hemba	D.R. Congo		I			Colle 1913
459	Hemba	D.R. Congo	3/-		45/3		Kisimba et al. 2003
459	Hemba	D.R. Congo		IS			Malaisse (unpubl. data)
459bis	Zela	D.R. Congo	2/-				Kisimba et al. 2003
461	Bangu Bangu (Nonda)	D.R. Congo			6/-		Kisimba et al. 2003
464	Lega-Shabunda	D.R. Congo		I			Delhaise 1909
464	Lega-Shabunda	D.R. Congo	+	I			Bergier 1941
474	Bwa (Ababwa)	D.R. Congo		I			De Calonne 1909
474	Bwa (Ababwa)	D.R. Congo		I			Costermans 1955
477	Bali	D.R. Congo		I			Schebesta 1934
482	Bira	D.R. Congo		I			Brisson no date
483	Nande	D.R. Congo	+	+	+		Kakule (in litt.)
485	Nyoro	Uganda		I			Casati 1892
485	Nyoro	Uganda		I			Johnston 1902
485	Nyoro	Uganda		I			Osmaton 1951
487	Ganda	Uganda		I	+		Johnston 1902, 1904
487	Ganda	Uganda		IS			Roscoe 1911
487	Ganda	Uganda		I			Osmaton 1951
487	Ganda	Uganda		I			Snoxall 1967
487	Ganda	Uganda		QIS			Menzel and d'Aluisio 1998
488	Nkore	Uganda		I			Davis 1938
490	Ziba	Tanzania	5/3	3/1	6/1		Harris 1940
494	Shi	D.R. Congo			2/-		Cuypers 1970
495	Rwanda	Rwanda		I			d'Hertefelt 1962
496	Rundi	Burundi		I			Meyer 1916
496	Rundi (Hutu)	Burundi		I	+		Trouwborst 1962
497	Ha	Tanzania		I			Scherer 1959

Table 10.2: (Contd.)

Table 10.2: (Contd.)

(1)	(2)	(3)	(4)	(5)	(6)	(7)	(8)
501bis	Isanzu	Tanzania		I			Kohl-Larsen 1943
512	Kamba	Kenya		I			Lindblom 1969
517	Caga	Kenya		I			Müller 1947
520	Pare = Asu	Tanzania		I			Kotz 1922
531	Gogo	Tanzania	+				Cole 1902
531bis	Sandawe	Tanzania	+	+			Newman 1975
543	Pogoro	Tanzania		I			Hendle 1907
544	Ngindo	Tanzania	+		+	+	Crosse-Upcott 1958
550	Yao	Mozambique		I			Sanderson [n.d.]
550	Yao	Mozambique	+/1				Harris 1940
551	Ngoni	Zambia	+	I	+		Chakanga *in* DeFoliart 2002
554bis	Pangwa	Tanzania	9/-	I	>5/2		Stirnimann 1976
558	Tumbuka	Zambia	+	I	3/-		Mbata 1995
558	Tumbuka	Malawi	+				Makungwa et al. 1997
570	Nyungwe	Mozambique		I			Courtois 1900
571	Venda	Zimbabwe	+	I			Wessmann 1908
571	Venda	Zimbabwe	+	I	+		Stayt 1931
571	Venda	Zimbabwe	9/1				Jackson 1954
571	Venda	R.S.A	+	I			McCallum 1993
571	Venda	R.S.A.		I			Crafford 1991
571	Venda	R.S.A.		QIS			van der Waal (in litt.)
571	Venda	R.S.A.	+		-/19		van der Waal 1996, 1999
573	Tsonga	R.S.A.		I			Junod 1927
573	Tsonga	R.S.A.		I		1/-,If	Jackson 1954
574	Thonga	R.S.A.	+				Junod 1927
574	Thonga	R.S.A.	+				Silow 1976
575	Ronda	R.S.A.	+				Silow 1976
576	Swazi	Swaziland	1/-	I	+		Beemer 1939
577	Sotho	R.S.A.		I			Endemann 1911
577bis	Pedi (Sotho central)	R.S.A.	5/5	I	+	1/1,If	Quinn 1959
579	Zulu	R.S.A.		I			Kranz 1880
579	Zulu	R.S.A.		I			Bryant 1905
580	Xhosa	R.S.A.	+				Kropf 1899
582	Shona-Ndau	Zimbabwe	1/1	I	+	+	Benhura and Chitsaku 1990
583	Shona-Karanga	Zimbabwe	13/7	IS	10/5	1/1	McGregor 1995
583	Shona-Karanga	Zimbabwe	+	+	+	+	Steele (in litt.)
584	Shona	Zimbabwe				1/1	Hannan 1959

Table 10.2: (Contd.)

Table 10.2: (Contd.)

(1)	(2)	(3)	(4)	(5)	(6)	(7)	(8)
584	Shona	Zimbabwe	6/-	I	+	1/1,If	Duncan 1933
584	Shona	Zimbabwe	7/1		+		Gelfand 1971
584	Shona	Zimbabwe	5/3	IS	8/8	1/1	Chavanduka 1975
584a	Shona (Zezuru)	Zimbabwe	+/1				Biehler 1950
584a	Shona (Zezuru)	Zimbabwe	+/1				Hannan 1959
584a	Shona (Zezuru)	Zimbabwe	+	+	+	+	Steele (in litt.)
584b	Shona (Manyika)	Zimbabwe	+	+	+	+	Steele (in litt.)
585	Shona-Korekore	Zimbabwe			+	+	Steele (in litt.)
587bis	Sandawe	Tanzanie		+			Newman 1975
590	Haya = Ziba	Tanzania		I			Stuhlmann 1894
590	Haya = Ziba	Tanzania		I			Rehse 1910
590	Haya = Ziba	Tanzania		I			Césard 1936
591	Nandi	Kenya	+	+			Yagi 1997
629bis	Ik = Teuso	Uganda		I			Turnbull 1972
629bis	Ik = Teuso	Uganda		I			Turnbull 1973
630	Teso (Kumi, Ngora)	Uganda		+			Osmaton 1951
632	Acholi	Uganda		IM			Osmaton 1951
684	Zande	D.R. Congo	+				Emin-Pacha 1888
684	Zande	Sudan		IO			Colrat de Montrozier 1902
684	Zande	D.R. Congo		W			Costermans 1950
684	Zande	Sudan	+	I			de Schlippé 1956
684	Zande	D.R. Congo	+	I			De Smet 1972
685	Abarambo	D.R. Congo		W			Costermans 1950
688	Mangbetu	D.R. Congo		I			Emin-Pacha 1888
688	Mangbetu	D.R. Congo		+			Bequaert 1921
689	Medje	D.R. Congo	+				Bequaert 1921
699	Alur	Uganda		S			Osmaton 1951
700	Madi	Sudan		+			Costermans 1950
705	Logo	D.R. Congo		ILNOW			Costermans 1955
705bis	Avokaya	D.R. Congo		IOW			Costermans 1955
775	Masa (Bugudum)	Cameroon	2/1	2/1 I	+/14		Mignot 2003
789	Bongo	Sudan		IO			Schweinfurth 1875
794	Mandja = Manza	Central Afr. R.		I			Vergiat 1937
795	Nzakara	Central Afr. R.		I			Retel-Laurentin 1986
800	Gbaya (Bodoe)	Central Afr. R.	59/13	12/-QIS	+	3/-,If E	Roulon-Doko 1998
804	Bwaka (Babinga)	Central Afr. R.		IE			Trilles 1932
804	Bwaka (Babinga)	Central Afr. R.		I			Schebesta 1940
804	Bwaka (Babinga)	Central Afr. R.		I			Bahuchet 1972
804	Bwaka (Babinga)	Central Afr. R.		I			Sevy 1972

Table 10.2: (Contd.)

Table 10.2: (Contd.)

(1)	(2)	(3)	(4)	(5)	(6)	(7)	(8)
804	Bwaka (Babinga)	Central Afr. R.	+	I	+		Bahuchet 1975
804a	Bwaka = (Y)Aka	Central Afr. R.	11/8	4/4	3/11		Bahuchet 1985, Hladik 1995
804a	Bwaka = (Y)Aka	Congo	>6/-				Kitanishi 1995
804b	Issongo	Central Afr. R.	12/7				Hladik 1995
804c	Bofi	Central Afr. R.	17/9	+			Malaisse, this volume
821	Yangere	Central Afr. R.	21/-				Masseguin and Antonini 1938
824	Baka	Sudan		W			Costermans
825	Mündü	D.R. Congo		W			Costermans 1955
826	Ndjem=Badjoué	Cameroon		I			Abe'Ele 1998
831	Béti	Cameroon	1/-	QIW	4/-	3/-	Ndounga *in* Bergier 1941
831	Béti	Cameroon		I			Tsala and Vincent 1969
838	Bamileke	Cameroon		I			Jacques-Félix 1948
841	Bafia	Cameroon		I			Tessmann 1934
846	Duala	Cameroon		I			Dinkelacker 1914
849	Nso = Nsaw	Cameroon		I			Banadzem 1980
849	Nso = Nsaw	Cameroon		I			Warnier 1985
900	Fali	Cameroon		I			Lebeuf 1961
900	Fali	Cameroon		I			Valentin 1968
901	Gisiga	Cameroon		I			Seignobos et al. 1996
909	Mofu-Diamaré	Cameroon	3/-	I	+		Seignobos et al. 1996
909bis	Mofu-Gudur	Cameroon			55/29		Barreteau 1999
979	Ibo = Igbo	Nigeria			14/5		Barker et al. 1977
986	Ebira	Nigeria	+/1	QI			Fasoranti and Ajiboye 1993
990	Bassa	Cameroon	3/-				Merle 1958
992	!Kung	Namibia		I			Marshall 1976
992	!Kung	Namibia		I	+/1	1/1	Green 1998
993	San	Namibia	3/3				Oberprieler 1995
993	San	Botswana	+				Imamura-Hayaki 1996
993	San [ǀGui]	Botswana		1/-,I			Silberbauer 1972
993	San [ǀGui]	Botswana	7/1	3/1,IN	+	2/-	Nonaka 1996
993bis	San (Bushman)	Namibia	-/1				Marais 1996
994	Nama(n)	Namibia		I	+		von Berensberg 1907
994	Nama(n)	Namibia	4/4				Oberprieler 1995

Table 10.2: (Contd.)

Table 10.2: (Contd.)

(1)	(2)	(3)	(4)	(5)	(6)	(7)	(8)
994	Nama(n)	Namibia	-/3				Marais 1996
994bis	Damara	Namibia	4/4				Oberprieler 1995
994bis	Damara	Namibia	-/5				Marais 1996
1005	Antankarana	Madagascar	+		+		Decary 1937
1005	Antankarana	Madagascar	+				Bergier 1941
1007	Antandroy	Madagascar			+		Decary 1937
1008	Bara	Madagascar	5/2		3/1		Gade 1985
1009	Betsileo	Madagascar	+				Ellis 1838
1009	Betsileo	Madagascar			+		Catat *in* Bergier 1941
1009	Betsileo	Madagascar	5/2		3/1		Gade 1985
1013	Merina	Madagascar	+/1		+		Decary 1937
1013	Merina	Madagascar	5/2		3/1		Gade 1985
1015	Sakalava	Madagascar	+/1		+		Decary 1937

Cam = campeophagy, Ter = termitophagy, Acr = acridophagy, Myr = myrmecophagy.

+ = at least one species, amount not precised; E/L = ratio ethnospecies/linnean species; - = not available

* E = eggs, G = geophagy, I = imago, If = alate female, L = larvae, M = mushroom bed, N = nymph, O = termite oil, Q = queen, S = soldier, W = worker.

X : confirmation of no consumption of an order.

Seasonality

Information concerning gathering periods of edible caterpillars available to the Gbaya of the Central African Rep. (Roulon-Doko, 1980, 1998), Binga of Lobaye (Bahuchet, 1975), Ntandu of Low Congo (Latham, 1999a), Yansi of Bandundu (Tango Muyay, 1981), Luluwa of West Kasaï (Katya Kitsa, 1989), Bemba of Katanga (Malaisse and Parent, 1980) and Zambia (Richards, 1939), Gwembe Tonga of Middle Zambezi (Scudder, 1962) as well as southern Mozambique (de Almeida, 1946), North Transvaal (Velcich, 1963), and South Africa (Van den Berg et al., 1973). The publications indicate strong seasonality, differences according to sites, and a strong correlation with rainfall patterns.

Chemical Composition

Some twenty papers deal with the chemical composition of edible caterpillars from Africa: (Chinn, 1945; Tihon, 1946a; Adriaens, 1951, 1953; Dartevelle, 1951; Nunes, 1960; Paulian, 1963; Malaisse et al., 1969; Heymans and Evrard, 1970; Le Clerc et al., 1976; Santos Oliveira et al., 1976; Malaisse and Parent, 1980; Dreyer

Table 10.3: Species diversity of edible caterpillars in Africa and their host plants*

Families	Species	St	References	Host plants
Brahmaeidae				
	Dactylocerus lucina Drury	c	Malaisse and Lognay 2003	
Hesperiidae				
	Coeliades libeon Druce	c	Dartevelle 1951	
Lasiocampidae				
	Bombycomorpha pallida Distant	c	Ramos-E1orduy 1991	
	Borocera madagascariensis Boisduval	c, p	Gade 1985	*Uapaca bojeri* [Mad]
	Borocera sp.1	c	Gade 1985	
	Catalebeda jamesoni Bethune-Baker	p	DeFoliart 2002	
	Gonometa postica Walker	c, p	Quin 1959	
	Libethra cajani Vinson	p	DeFoliart 2002	
	Pachymeta robusta Aurivillius	c	Malaisse and Roulon-Doko 2003	*Julbernardia paniculata* [SC]
	Pachypasa bilinea Walker	p	DeFoliart 2002	
	Rombyx radama Coquillet	p	DeFoliart 2002	
Limacodidae				
	Hadraphe ethiopica (Bethune-Baker)	c	Malaisse et al. 2003	*Julbernardia paniculata* [SC], *Brachystegia spiciformis* [SC], *Baphia obovata* [SC]
Lymantriidae				
	"nziozu" (kintandu)	c	Latham 2002	*Leptoderris congolensis* [C]
	Rhypopteryx poecilanthes Colenette	c	Malaisse and Lognay 2003	*Symphonia globulifera* [C]
Noctuidae				
	Helicoverpa armigera (Hübner)	c	Mignot 2003	

Table 10.3: (Contd.)

Table 10.3: (Contd.)

Families	Species	St	References	Host plants
	Prodenia sp.1	c	Adriaens 1953	
	Sphingomorpha chlorea Cr.	c	Silow 1976	
	Spodoptera exempta (Walker)	c	Mbata 1995	
	Spodoptera exigua Hübner	c	Mbata 1995	
Notodontidae				
	Anaphe infracta Walsingham	c	Godwey 1912	
	Anaphe panda (Boisduval)	c	Harris 1940, Le Clerc et al. 1976	*Bridelia micrantha* [E], *Combretum* sp. [SC]
	Anaphe reticulata Walker	c	Steele (in litt.)	*Dombeya rotundifolia* [S]
	Anaphe venata Butler	c	Bahuchet 1985, Ashiru 1988, 1989	*Bridelia micrantha* [W], *Cola gigantea* [W], *C. lateritia* [NC], *Triplochyton scleroxylon* [W]
	Antheua insignata Gaede	c	Malaisse and Parent 1980	*Brachystegia boehmii* [SC], *Hymenocardia ulmoides* [C], *Julbernardia paniculata* [SC]
	Antheua sp. 1 "malomba loka"	c	Latham 2002	*Milletia eetveldeana* [C]
	Antheua sp. 2 "mfundi"	c	Latham 2002	*Milletia eetveldeana* [C]
	Antheua sp. 3 "mfundi"	c	Latham 2002	*Leptoderris congolensis* [C]
	Busseola fusca Hmps.	i	DeFoliart 2002	
	Cerurina cf. *marshalli* (Hampson)	c	Malaisse and Roulon-Doko 2003	
	Desmeocraera sp.	c	Silow 1976	
	Drapedites uniformis Swinhoe	c	Malaisse and Parent 1980	*Pterocarpus tinctorius* [SC]
	Elaphrodes lactea (Gaede)	c	Seydel 1939	*Albizia antunesiana* [SC], *A. ferruginea* [C], *Brachystegia boehmii* [SC], *B. longifolia* [SC], *B. microphylla* [SC], *B. spiciformis* [SC], *B. utilis* [SC], *Irvingia grandifolia* [NC], *Julbernardia paniculata* [SC], *Lophira alata* [NC], *Piptadeniastrum africanum* [NC], *Pterocarpus angolensis* [SC], *Pycnanthus angolensis* [NC], *Zanha golugensis* [NC]

Table 10.3: (Contd.)

Table 10.3: (Contd.)

Families	Species	St	References	Host plants
	Ipanaphe carteri Walsingham	c	Latham in Malaisse and Lognay 2003	
	"lusambwa" (chibemba)="nsindi" (ntandu)	c	Malaisse and Parent 1980, Latham 1999	*Eriosema psoraloides* [C], *Hyparrhenia diplandra* [C,SC]
	"miengeti"	c	Latham 2002	*Milletia eetveldeana* [C]
Nymphalidae				
	Cymothoe caenis (Drury)	c	Malaisse (unpubl.)	*Caloncoba welwitschii* [C]
Papilionidae				
	cf. *Papilio* sp.	c	Malaisse and Roulon-Doko 2003	
Psychidae				
	Deborrea malagassa Heylaerts	c	Decary 1937	*Acacia* sp. [Mad]
	Eumeta cervina Druce	c	Steele (in litt.)	
	Eumeta rougeoti Bourgogne	c	Steele (in litt.)	
	Eumeta sp.	c	Pagezy 1975	
	"náà-so-?ínì"	c	Roulon-Doko 1998	
Saturniidae(*)				
	Antherina suraka (Boisduval)	c	Gade 1985	*Uapaca bojeri* [Mad]
	Argema mimosae (Boisduval)	c	Kropf 1899	*Commiphora mollis* [SC], *Sclerocarya birrea* [S], *Spirostachys africana* [S]
	Athletes gigas (Sonthonnax)	c	Malaisse and Parent 1980	*Brachystegia taxifolia* [SC], *Julbernardia paniculata* [SC]
	Athletes semialba Sonthonnax	c	Malaisse and Parent 1980	*Acacia campylacantha* [SC], *Brachystegia globifera* [SC], *B. spiciformis* [SC]
	Bunaea alcinoe Stoll (1)	c	Silow 1976	*Balanistes aegyptiaca* [SC], *Bauhinia reticulata* [S], *Cussonia* sp. [SC], *Ekebergia benguellensis* [SC], *Entandrophragma angolense* [NC], *Piliostigma thonningii* [SC], *Sarcocephalus latifolius* [C]

Table 10.3: (Contd.)

Table 10.3: (Contd.)

Families	Species	St	References	Host plants
	Bunaea aslauga Kirby	c	Harris 1940	
	Bunaeopsis aurantiaca (Rothschild)	c	Malaisse and Parent 1980	*Hyparrhenia diplandra* [SC]
	Cinabra hyperbius (Westwood)	c	Silow 1976; Malaisse and Parent 1980	*Brachystegia spiciformis* [SC], *B. venosa* [SC], *Julbernardia paniculata* [SC]
	Cirina butyrospermi Vuillet	c	Bergier *in* Merle 1958	*Vitellaria paradoxa* [W]
	Cirina forda (Westwood)	c	Adriaens 1953, Oberprieler 1993	*Albizia antunesiana* [SC], *Autranella congolensis* [NC], *Bridelia ferruginea* [NC], *Burkea africana* [SC,S], *Carissa edulis* [S], *Crossopteryx febrifuga* [C], *Erythrophleum africanum* [SC], *E. guineense* [NC], *E. suaveolens* [SC], *Julbernardia globiflora* [SC], *Pappea capensis* [S]
	Epiphora bauhiniae (Guérin-Méneville)	c	Pagezy 1975	*Ziziphus jujuba* [W], *Z. mauritiana* [SC], *Z. mucronata* [S], *Z. spinachristii* [W]
	Gonimbrasia alopia Westwood	c	Latham 2002	*Albizia ferruginea* [C], *Chaetocarpus africanus* [C], *Manotes expansa* [C]
	Gonimbrasia anthina Karsch	c	Latham 1999	*Aframomum albo-violaceum* [C,SC], *Antidesma venosum* [C], *Manotes expansa* [C], *Strychnos pungens* [C]
	Gonimbrasia belina (Westwood)	c	Velcich 1963, Oberprieler 1993	*Colophospermum mopane* [SC,S], *Carissa grandiflora* [S], *Diospyros mespiliformis* [SC], *Ozoroa crassinervia* [S], *O. longipes* [S], *Sclerocyara birrea* [SC]
	Gonimbrasia cytherea (Fabricius)	c	Smit 1964	*Euclea daphnoides* [S], *Protea mellifera* [S]
	Gonimbrasia dione (Fabricius) (2)	c	Silow 1976	*Antidesma membranaceum* [C], *Bauhinia petersiana* [S], *Brachystegia laurentii* [C], *Bridelia duvigneaudii* [SC], *Diplorynchus condylocarpon* [SC], *Khaya nyasica* [SC], *Milicia excelsa* [C], *Musanga cecropioides* [C], *Ricinodendron heudelotii* [C], *Uapaca guineensis* [C], *U. nitida* [SC], *Vitex madiensis* [C]
	Gonimbrasia hecate Rougeot	c	Malaisse and Parent 1980	*Combretum psidioides* [SC], *Terminalia mollis* [SC]

Table 10.3: (Contd.)

Table 10.3: (Contd.)

Families	Species	St	References	Host plants
	Gonimbrasia macrothyris (Rothschild)	c	Malaisse and Parent 1980	*Brachystegia spiciformis* [SC], *Chaetocarpus africanus* [C], *Julbernardia paniculata* [SC], *Manotes expansa* [C], *Ochna afzeli* [C]
	Gonimbrasia melanops (Bouvier)	c	Latham 2002	*Entandrophragma angolense* [NC], *E. cylindricum* [NC], *Macaranga monandra* [C]
	Gonimbrasia rectilineata Sonthonnax	c	Malaisse and Parent 1980	*Erythrophleum africanum* [SC], *Julbernardia paniculata* [SC], *Ochna schweinfurthiana* [SC], *Syzygium guineense* [SC], *S. owariense* [SC], *Uapaca kirkiana* [SC]
	Gonimbrasia rhodina Rothschild	c	Latham 1999	
	Gonimbrasia zambesina (Walker)	c	Malaisse and Parent 1980	*Diospyros mespiliformis* [SC], *Piliostigma thonningii* [SC], *Strychnos potatorum* [SC]
	Goodia kuntzei (Dewitz)	c	Malaisse and Parent 1980	*Acacia sieberiana* [S], *Bauhinia galpinii* [S], *Brachystegia venosa* [SC], *Dichrostachys cinerea* [SC], *Julbernardia globiflora* [SC], *J. paniculata* [SC]
	Gynanisa ata Strand	c	Malaisse and Parent 1980	*Brachystegia boehmii* [SC], *B. longifolia* [SC], *B. spiciformis* [SC], *B. taxifolia* [SC], *B. venosa* [S], *Julbernardia paniculata* [SC]
	Gymnanisa maja (Klug)	c	Oberprieler 1993	*Acacia erioloba* [S], *A. karroo* [S], *Colophospermum mopane* [S]
	Heniocha apollonia (Cramer)	c	Marais 1996	*Acacia karroo* [S], *A. mollissima* [S]
	Heniocha dyops (Maassen and Weymer)	c	Oberprieler 1993, Marais 1996	*Acacia erubescens* [S], *A. hereoensis* [S], *A. mellifera* [S], *A. nigrescens* [S]
	Heniocha marnois (Rogenhofer)	c	Marais 1996	*Acacia burkei* [S], *A. erubescens* [S], *A. nigrescens* [S], *A. polyacantha* [SC]
	Imbrasia epimethea (Drury)	c	Adriaens 1953	*Afzelia quanzensis* [SC], *Erythrophleum africanum* [SC], *E. suaveolens* [SC], *Funtumia africana* [C], *Holarrhena*

Table 10.3: (Contd.)

Table 10.3: (Contd.)

Families	Species	St	References	Host plants
				floribunda [C], *J. globiflora* [SC], *J. paniculata* [SC], *Petersianthus macrocarpus* [C], *Ricinodendron heudelotii* [C]
	Imbrasia ertli Rebel	c	Oberprieler 1993	*Brachystegia glaucescens* [SC], *B. spiciformis* [SC], *Funtumia auriculiformis* [C], *Holarrhena floribunda* [C], *Julbernardia globiflora* [SC], *J. paniculata* [SC], *Petersianthus macrocarpus* [C], *Ricinodendron heudelotii* [C]
	Imbrasia obscura (Butler)	c	Bahuchet 1985	*Albizia ferruginea* [C], *Pentaclethra macrophylla* [C], *Tetrapleura tetraptera* [NC], *Triplochyton scleroxylon* [NC]
	Imbrasia truncata Aurivillius	c	Bahuchet 1985	*Entandrophragma angolense* [NC], *Petersianthus macrocarpus* [NC], *Sorindeia zenkeri* [C]
	Imbrasia tyrrhea (Cramer) (3)	c	Malaisse and Parent 1980	*Acacia karroo* [S], *Diospyros lycioides* [S], *Euclea pseudebenus* [S], *Ozoroa concolor* [S], *Uapaca kirkiana* [SC], *U. nitida* [SC], *U. pilosa* [SC], *U. sansibarica* [SC]
	Imbrasia cf. *wahlbergii* (Boisduval)	c	Latham 2002	*Maesobotrya vermeulenii* [C]
	Lobobunaea goodii (Holland)	c	Takeda 1990	*Piptadeniastrum africanum* [C], *Uapaca guineensis* [C]
	Lobobunaea phaedusa (Drury) (4)	c	Seignobos et al. 1996	*Anona senegalensis* [C], *Crossopteryx febrifuga* [C]
	Lobobunaea saturnus (Fabricius)	c	Malaisse and Parent 1980	*Diplorynchus condylocarpon* [SC], *Julbernardia paniculata* [SC], *Syzygium guineense* [SC]
	Melanocera menippe (Westwood)		DeFoliart 2002	
	Melanocera nereis (Rothschild)	c	Latham 1999	*Albizia ferruginea* [C]
	Melanocera parva Rothschild	c	Malaisse and Parent 1980	*Brachystegia microphylla* [SC], *Julbernardia paniculata* [SC], *Ochna schweinfurthiana* [SC]
	Micragone ansorgei Rothschild (5)	c	Malaisse (unpubl.)	*Brachystegia spiciformis* [SC], *Julbernardia paniculata* [SC]

Table 10.3: (Contd.)

Table 10.3: (Contd.)

Families	Species	St	References	Host plants
	Micragone cana Aurivillius	c	Malaisse and Parent 1980	*Diplorynchus condylocarpon* [SC], *Syzygium owariense* [S]
	Micragone herilla (Westwood)	c	Ramos-Elorduy 1991	
	Pseudantheraea discrepans Butler	c, p	Malaisse and Lognay 2003	*Entandrophragma angolense* [NC], *E. candollei* [NC], *Macaranga monandra* [C], *Uapaca guineensis* [C]
	Pseudobunaea irius (Fabricius)	c	Marais 1996	*Brachystegia spiciformis* [SC], *Julbernardia globiflora* [SC], *Pilostigma thonningii* [S, SC]
	Rohaniella pygmaea (Maassen and Weyding)	c	Marais 1996	*Burkea africana* [S]
	Tagoropsis natalensis Felder	c	Malaisse and Parent 1980	*Allophyllus africanus* [SC]
	Tagoropsis sp.1	c	Gade 1985	*Uapaca bojeri* [Mad]
	Urota sinope (Westwood)	c	Malaisse and Parent 1980	*Erythrina abyssinica* [SC]
	Usta terpsichore (Maassen and Weyding)	c	Malaisse and Parent 1980	*Commiphora* spp. [S], *Monotes caloneurus* [SC], *Sclerocarya birrea* [S],
	Usta wallengrenii (Felder)	c	Gaerdes 1959; Oberprieler 1993; Marais 1996	*Commiphora* spp. (*C. africana, C. angolensis, C. glandulosa, C.glaucescens, C. saxicola, C. tenuipetiolata, C. virgata*) [S]
Sphingidae				
	Acherontia atropos (Linn.)	c	Latham 1999	*Solanum macrocarpum* [C]
	Coelonia fulvinotata (Butler)	c	Malaisse and Roulon-Doko 2003	
	Herse convolvuli (Linn.)	c	McGregor 1995	*Ipomoea batatas* [SC,S]
	Hippotion eson (Cramer)	c	Malaisse and Lognay 2003	*Amorphophallus abyssinicus* [SC]
	Hippotion sp. 1	c	Malaisse and Roulon-Doko 2003	

Table 10.3: (Contd.)

Table 10.3: (Contd.)

Families	Species	St	References	Host plants
	Lophostethus demolini (Angas)	c	Malaisse and Roulon-Doko 2003	*Hibiscus tiliaceus* [W]
	Nephele comma Hopffer	c	Silow 1976	
	Platysphinx stigmatica (Mabille)	c	Malaisse and Lognay 2003	
	Platysphinx sp. 1 "munsona"	c	Latham 1999	*Milletia versicolor* [C]
	Platysphinx sp. 2 "kienga"	c	Latham 1999	*Sarcocephalus latifolius* [C]

Host plants : [W]=West, [Mad]=Madagascar, [NC]=North Central, [C]=Central, [E] East, [SC]=South Central, [S]=Southern Africa.

St = stade : c=caterpillar, i=imago, p=pupae.

*Nomenclature according to Bouyer (1999). Synonymies for some taxa quoted in previous publications: (1) = *Bunaea caffraria* Stoll, (2) = *Imbrasia petiveri* (Guérin-Méneville), (3) = *Imbrasia rubra* (Bouvier), (4) = *Lobobunaea christyi* (Sharpe), (5) = *Micragone bilineata* Rothschild.

and Wehmeyer, 1982; Kodondi et al., 1987a; Ashiru, 1988; Sekhwela, 1989; Mbemba and Remacle, 1992; Glew et al., 1999). One study deals with proteins (Landry et al., 1986), two with lipids (Motshegwe et al., 1998; Zinzombe and George, 1994), and even one with vitamins (Kodondi et al., 1987b).

Tables 10.4 to 10.7 synthesize according to families involved, the main data published. Energetic values comprise 350 to 500 kcal, proteins 45 to 80% dry matter with a higher mean value for Notodontidae, fats 8 to 27%, and ash 4 to 14% according to species involved. Glutamic acid, phenylalanine and tyrosine are the main amino acids, linolenic, palmitic and stearic acids the main lipids; as for vitamins, B_2 and P.P. are high and B_1 and B_6 low compared with nutritional requirements (Kodondi et al., 1987b).

Table 10.4: Composition of edible caterpillars of Tropical Africa (values per 100 g dry matter)

Composition	Saturniidae N = 20	Hesperiidae N = 1	Limacodidae N = 1	Notodontidae N = 6
Proteins (g)	(44.1) 63.7 +/- 10.4 (79.6)	51.2	69.6	(45.6) 53.7 +/- 5.4 (61.0)
Fats (g)	(8.1) 13.8 +/- 4.5 (21.5)	12.4	9.2	(10.1) 21.7 +/- 8.3 (26.0)
Carbohydrates (g)	(3.7) 13.8 +/- 9.2 (29.4)	15.6	12.7	(13.1) 18.4 +/- 7.0 (24.1)
Ashes (g)	(3.8) 6.7 +/- 2.6 (14.4)	11.6	8.5	(3.7) 5.3 +/- 1.4 (7.7)
Ca (mg)	(50) 148 +/- 124 (500)	220	1600	(20) 108 +/- 86 (200)
P (mg)	(500) 1099 +/- 680 (2300)	1160	900	(450) 710 +/- 445 (1500)
Fe (mg)	(10) 81 +/- 81 (300)		20	(10) 42 +/- 31 (80)
E.v. (Kcal)	(371) 449 +/- 36 (504)	348	397	(397) 463 +/- 49 (485)

E.v. = energetic value.
(*Sources:* Demesmaecker 1997; Kodondi et al. 1987a; Malaisse 1997, 2003; Malaisse and Lognay 2003; Santos Oliveira et al. 1976).

Termitophagy

Introduction

Consumption of termites by various peoples of Africa is noted in numerous publications, old as well as recent. Already Labat (1732: 183) reported that Cavazzi found that the alates were a delicacy for the populations of the kingdoms of Congo, Angolle, and Matamba: "What a treat to have them roasted!"

In general, local people often consider termites delicacies. Techniques involved in collecting termites (alates or soldiers) have been described from time to time. Some, in particular the collection of soldiers, are similar to those used by chimpanzees (Roulon-Doko and Joulian, 1994). Silow (1983) did an extensive study of trapping methods, and mentioned hand-picking of sunset termites, cage-

Table 10.5: Amino acid composition of edible African caterpillars (in % of proteins)

Amino acids	Attacidae N=14	Notodontidae N=1 to 3
aspartic acid	(8.5) 8.8 +/- 0.4 (9.3)	0.6
glutamic acid	(13.6) 14.5 +/- 0.8 (15.0)	0.4
alanine	(4.0) 4.4 +/- 0.4 (4.7)	1.8
arginine	(5.6) 6.2 +/- 0.6 (6.6)	0.3
cystine	(1.3) 1.6 +/- 0.3 (2.0)	
glycine	(3.7) 3.8 +/- 0.2 (4.1)	1.4
histidine	(1.7) 2.8 +/- 0.6 (3.4)	(0.8) 2.3 +/- 1.4 (3.4)
isoleucine	(2.4) 4.5 +/- 2.0 (10.9)	(2.2) 3.9 +/- 1.5 (4.9)
leucine	(3.7) 6.6 +/- 1.4 (9.1)	(1.3) 4.7 +/- 3.0 (6.7)
lysine	(3.9) 6.9 +/- 1.2 (9.1)	(0.9) 4.8 +/- 3.4 (6.8)
methionine	(1.1) 1.9 +/- 0.5 (2.4)	
(methionine + cystine)	(0.8) 1.5 +/- 0.5 (2.1)	(1.5) 2.2 +/- 0.9 (2.8)
phenylalanine	(1.7) 5.2 +/- 2.0 (6.5)	2.2
(phenylalanine + tyrosine)	(8.9) 11.3 +/- 2.0 (14.7)	(13.2) 14.2 +/- 1.3 (15.1)
proline	(2.0) 2.1 +/- 0.1 (2.2)	1.9
serine	(4.5) 4.7 +/- 0.2 (4.9)	
threonine	(3.9) 4.4 +/- 0.3 (5.1)	(0.4) 3.1 +/- 2.3 (4.5)
tryptophan	(0.7) 1.3 +/- 0.5 (1.7)	
tyrosine	(1.3) 5.5 +/- 3.0 (7.7)	2.5
valine	(4.2) 6.6 +/- 2.0 (10.2)	(1.8) 4.4 +/- 2.2 (5.7)

Sources: Ashiru 1988; Demesmaecker 1997; Kodondi et al. 1987b; Santos Oliveira et al. 1976.

trapping of sunset fliers, trapping with cupola-shaped cages, net-basket, grass-cornet or torchlight, and digging out "nymphs" and winged termites.

Some exhaustive studies deal with ethnological and ecological aspects: termites and termitaria by Hegh (1922) and Iroko (1996); ethnological studies devoted to the Aka (Bahuchet, 1985) or the Gbaya (Roulon-Doko, 1992, 1998); ethnozoological approach of the Ngangela and the Nkoya (Silow, 1983), the ethnoecological work concerning Upper Katanga (Malaisse, 1997) as well as a study devoted to termite consumption in Burkina Faso (Ouedraogo et al., 2003).

Termitophagy: Stages Consumed

The ethnolinguistic groups consuming termites are listed in Table 10.2. A distinction was attempted between consumption of the queen, alate adults, soldiers, workers, eggs, "nymphs", mushroom beds, oil and soil (Fig. 10.4). Some 163 sources dealing with human consumption of termites were consulted. The

Table 10.6: Lipid composition of edible African caterpillars (in % of total fatty acids)

Fatty acids		Attacidae N=12	Notodontidae N=1 to 2
lauric acid	C12:0	(0.10) 0.17 +/− 0.06 (0.20)	
myristic acid	C14:0	(0.10) 0.56 +/− 0.61 (2.30)	0.90
pentadecanoic acid	C15:0	(0.10) 0.23 +/− 0.12 (0.50)	0.20
palmitic acid	C16:0	(8.74) 21.47 +/− 5.53 (28.45)	(11.61) 20.86 (30.10)
palmitoleic acid	C16:1	(0.10) 0.39 +/− 0.21 (0.87)	1.00
heptadecanoic acid	C17:0	(0.11) 3.70 +/− 9.14 (29.70)	(1.37) 1.44 (1.50)
stearic acid	C18:0	(1.00) 18.99 +/− 7.46 (33.42)	(5.40) 12.42 (19.44)
oleic acid	C18:1	(1.70) 5.83 +/− 3.61 (8.40)	
	C18:1 cis-9	(4.85) 9.36 +/− 2.63 (12.90)	(8.37) 9.14 (9.90)
	C18:1 isom.*	(0.23) 0.32 +/− 0.07 (0.42)	0.10
linoleic acid	C18:2ω–6	(4.40) 9.15 +/− 6.11 (27.20)	(6.10) 8.71 (11.31)
linolenic acid	C18:3ω-3	(2.80) 34.14 +/− 10.78 (45.12)	(41.70) 41.72 (41.73)
arachidic acid	C20:0	(0.20) 1.51 +/− 2.43 (7.50)	(0.12) 0.79 (1.46)
eicosadienoic acid	C20:2	(0.10) 0.25 +/− 0.21 (0.40)	
other fatty acids		(1.15) 2.43 +/− 1.15 (4.35)	(4.08) 4.40 (4.71)

*double link isomere position
Sources: Demesmaecker 1997; Kodondi et al. 1987b; Malaisse and Lognay 2003.

Table 10.7: Vitamin composition of edible African caterpillars (values per 100 g dry matter)

Vitamin		Attacidae N=3	Hesperiidae N=1
folic acid	µg	(6.3) 21.1 +/− 15.4 (37.0)	
nicotinic acid	mg		22.7
pantothenic acid	mg	(7.3) 8.8 +/− 1.5 (10.2)	3.8
biotin	µg	(23.0) 32.7 +/− 11.2 (45.0)	92.0
cholecalciferol	µg	22.2 (N=1)	
cyanocobalamin	µg		6.0
niacin	mg	(9.4) 10.4 +/− 0.9 (11.0)	
pyridoxin	µg	(50.0) 90.0 +/− 45.8 (140.0)	252.0
retinol	µg	(30.0) 35.0 +/− 7.8 (44.0)	
beta-carotene	µg	(6.3) 6.8 +/− 0.7 (7.6)	
riboflavin	mg	(3.2) 4.1 +/− 1.0 (5.1)	3.4
thiamin	mg	(0.2) 0.2 +/− 0.1 (0.3)	0.6
alpha-tocopherol	mg		51.0

Sources: Kodondi et al. 1987a; Paulian 1963.

179 ethnolinguistic groups practicing termitophagy constitute 17% of the groups recognized for Africa. Thirteen papers pertaining to this subject have been published during the last decade (Fig. 10.3). Relevant comments have been published in travel stories of earlier explorers dating from the end of the 18th century (Smeathman, 1781), the 19th (Drummond, 1888), and beginning of the 20th.

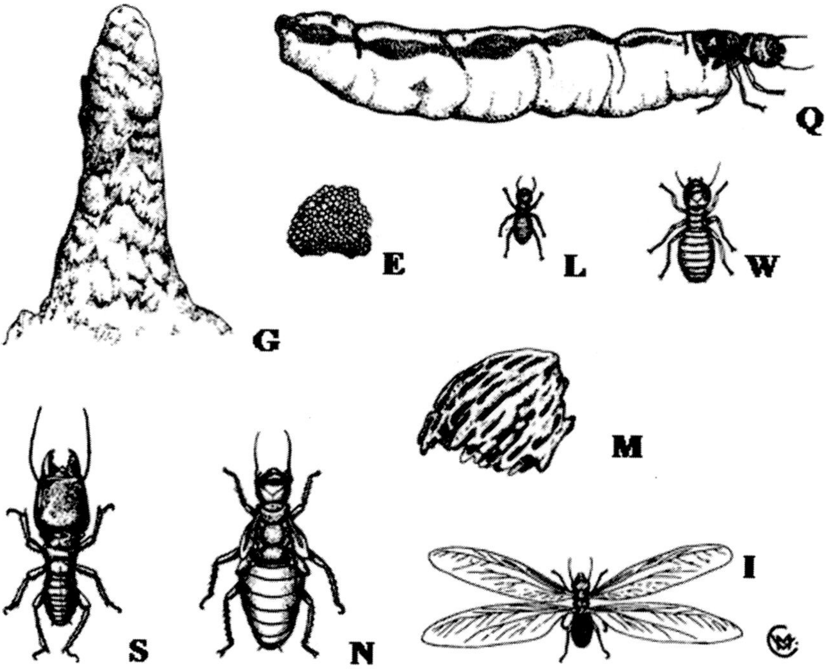

Fig. 10.4: Diversity of items in termitophagy (E = eggs, G = geophagy, I = imago, L = larva, M = mushroom bed, N = nymph, Q = queen, S = soldier, W = worker, M = Signature of drought woman).

Mapping Ethnolinguistic Groups Practicing Termitophagy

Silow (1983) made a distribution map of consumption of winged termite sexuals in Bantu-speaking Africa (Figure 10.5A). This information is complemented by an original map of termitophagy in Africa (Fig. 10.5B).

In areas with strongly contrasting dry and wet seasons, where *Macrotermes* species construct large termite mounds, consumption of termites appears to be of greater importance than in the evergreen rainforest areas (Bouillon and Mathot, 1965; Ruelle, 1970).

Diversity of Edible Species

At least 18 different species belonging to two families and four subfamilies are consumed (Table 10.8). The number of ethnospecies eaten habitually by the Zande is 9 (De Smet, 1972); Luba Kasai (Callewaert, 1922) and Logo-Avokaya 7 (Costermans, 1955); Mangbetu 5 (Emin Pacha, 1888); Aka 4 (Bahuchet, 1985); and Gbaya (Roulon-Doko, 1998), Yansi (Tango Muyay, 1981) and Bemba 3 (Malaisse, 1997). Recognition of termites by local people or termite

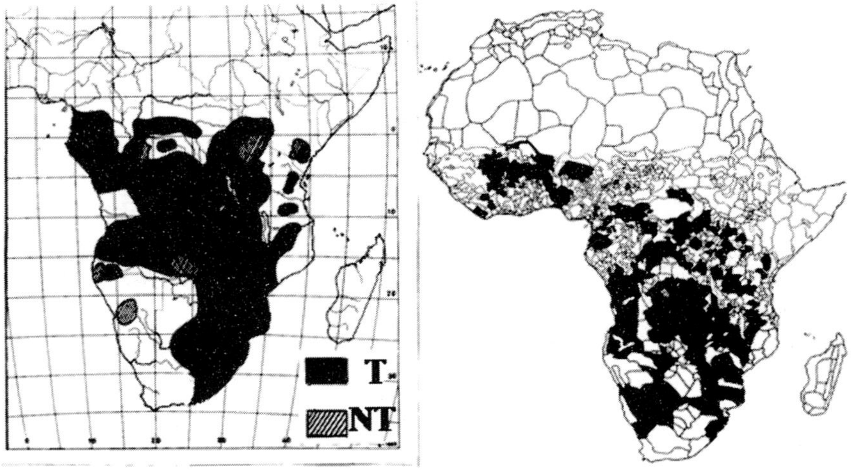

Fig. 10.5: Distribution map of termitophagy in Africa. (A) Left: Flying termites in Bantu-speaking Africa (after Silow, 1983): T = human consumption of termites; NT = no human consumption of termites. (B) Right: Termitophagy according to ethnolinguistic groups (original).

Table 10.8: List of Termite species eaten in Africa

Families subfamilies	Genera	Specific epithets	Ethno-linguistic group (Country)	Reference
Hodotermitidae				
	Hodotermes	*mossambicus* (Hagen)	Tswana (Botswana)	Grivetti 1979
	Hodotermes	*mossambicus* (Hagen)	San (Botswana)	Nonaka 1996
	Hodotermes	sp.[1]	? (South Africa)	Fuller 1918
	Microhodotermes	*viator* (Latreille)	? (South Africa)	Bodenheimer 1951
	Microhodotermes	spp.[2]	?	Logan 1992
Termitidae Macrotermitinae				
	Acanthotermes	*acanthothorax* (Sjöstedt)[3]	?W and S Lake Victoria (Uganda, Tanzania)	Harris 1940
	Macrotermes	*bellicosus* (Smeathman)	? (Nigeria)	Ukhun and Osasona 1985
	Macrotermes	*falciger* (Gerstäcker)[4]	Shona (Zimbabwe)	Chavanduka 1975
	Macrotermes	*mossambicus* (Hagen)	? (Zambia, Zimbabwe)	Silow 1983
	Macrotermes	*mossambicus* (Hagen)[5]	?	Logan 1992
	Macrotermes	*muelleri* (Sjöstedt)[6]	? (D.R. Congo)	DeFoliart 2002
	Macrotermes	*natalensis* (Haviland)	Masa (Cameroun)	Mignot 2003

Table 10.8: (Contd.)

Table 10.8: (Contd.)

Families Subfamilies	Genera	Specific epithets	Ethno-linguistic group (Country)	Reference
	Macrotermes	*natalensis* (Haviland)	? (D.R. Congo)	Bequaert 1921
	Macrotermes	*subhyalinus* (Rambur)	Mofu (Cameroun)	Seignobos et al. 1996
	Macrotermes	*subhyalinus* (Rambur)	Gingas (Angola)	Santos Oliveira et al. 1976
	Macrotermes	*vitrialatus* (Sjöstedt)	Mbunda(Angola)	Silow 1983
	Microtermes	sp.	Nandi (Kenya)	Yagi 1997
	Odontotermes	*badius* (Haviland)[7]	Pedi (R.S.A.)	Quin 1959
	Odontotermes	*magdalense* Grassé and Noirot	Mofu (Cameroun)	Seignobos et al. 1996
	Odontotermes	sp.	Sanga (Katanga, D.R. Congo)	Malaisse 1997
	Pseudacanthotermes	*militaris* Hagen	S Lake Victoria (Tanzania)	Harris 1940
	Pseudacanthotermes	*militaris* Hagen	?Ganda (Kampala, Uganda)	Pomeroy 1976
	Pseudacanthotermes	*spiniger* Sjöstedt	? (D.R. Congo)	Bequaert 1921
	Pseudacanthotermes	*spiniger* Sjöstedt	?Ganda (Kampala, Uganda)	Pomeroy 1976
	Pseudacanthotermes	*spiniger* Sjöstedt	Mbunda (Angola)	Silow 1983
	Protermes	sp.	Aka (Central Afr. Rep.)	Bahuchet 1985
Apicotermitinae				
	Apicotermes	sp.	Aka (Central Afr. Rep.)	Bahuchet 1985
Termitinae				
	Megagnathotermes	*katangensis* Sjöstedt	Sanga (D.R. Congo)	Malaisse 1997
Nasutermitinae				
	Trinervitermes	sp.	Luba Kasai (D.R. Congo)	Callewaert 1922

[1]Fuller (1918) mentions a *Hodotermes* species. Two taxa are presently recognised: *H. erithreensis* Sjöstedt and *H. mossambicus*. It is logical to suppose that the taxon referred to by Fuller regarding southern Africa is *H. mossambicus*, as stated by Coaton (1949).

[2]Logan (1992) considered several species of *Microhodotermes*, but recent lists retain only three taxa, namely *M. maroccanus* (Sjöstedt), *M. viator* (Latreille), and *M. wasmanni* (Sjöstedt).

[3]Harris (1940) considered several species of *Acanthotermes* spp., including *A. militaris*. This last species is regarded here as *Pseudacanthotermes militaris*, since only one taxon is known to exist for the genus *Acanthotermes*.

[4]*Macrotermes swaziae* Fuller, cited by Bodenheimer (1951), is a synonym of *M. falciger* (Ruelle 1970).

[5]*Macrotermes michaelseni* (Sjöstedt), cited by Logan (1992), is a synonym of *M. mossambicus* (Ruelle 1970).

[6]*Termes gabonensis* Sjöstedt, cited by DeFoliart (2002), is a synonym of *M. muelleri* (Ruelle 1970).

[7]Quin (1959) cites the consumption of *Termes badius*, considered here as *Odontotermes badius*.

ethnotaxonomy is based on termite biology (season and hour of swarming), termite mound aspects, termite body color, as well as the fungi (*Termitomyces* species) associated with them.

Chemical Composition

Some 12 papers deal with the chemical composition of edible termites from Africa (Tihon, 1946b; Auffret and Tanguy, 1947–1948; Cemlik, 1969a,b; Heymans and Evrard, 1970; Phelps et al., 1975; Santos Oliveira et al., 1976; Platt, 1980; Ukhun and Osasona, 1985; Malaisse, 1997; Malaisse and Parent, 1998). One study deals in particular with proteins (Hladik, 1977). Termite castes eaten have been differentiated into alate adults, soldiers, and queens (Tables 10.9 to 10.11). Indeed, soldiers are rich in proteins and, unlike alates, poor in fats. Fat body lipids of soldiers differ markedly in composition from those alates or queens. In queens they consist predominantly of oleic, palmitic, and stearic acids, while in alates linoleic acids predominate.

The foregoing results agree with data available for South America (Redford and Dorea, 1984) and Asia (Gope and Prasad, 1983).

Table 10.9: Composition of edible Termites and Orthoptera of Tropical Africa (values per 100 g dry matter)

Composition	Termite Alate adults	Termite Soldier	Orthoptera N=11
Proteins (g)	39.4	61.3	(13.7) **40.4** +/– 23.8 (69.5)
Fats (g)	42.7	5.3	(5.3) **23.7** +/– 22.2 (68.8)
Carbohydrates (g)			(11.3) **12.4** +/– 1.6 (13.5)
Ashes (g)	8.8	6.8	(1.1) **4.5** +/– 2.8(10.0)
Ca (mg)	390	170	(43) **260** +/– 182 (600)
P (mg)			(60) **594** +/– **533** (1600)
Fe (mg)			(15) **29** +/– 13 (40)
E.v. (kcal)			(196) **423** +/– 208 (713)
N (g)	6.3	9.8	

Sources: Heymans and Evrard 1970; Mbemba and Remacle 1992; Malaisse and Parent 1998.

Acridophagy in Africa

Introduction

No synthesis at regional or African levels is available regarding acridophagy. Nevertheless some studies are pertinent and need to be pointed out, namely those of Barker et al. (1977), Roulon-Doko (1998), Seignobos et al. (1996), Barreteau (1999), van der Waal (1999), Gast (2000), and Kisimba et al. (2003). In eastern

Table 10.10: Amino acid composition of edible African termites (in % of proteins)

Amino acids	**Termites** **Alates (*Macrotermes* spp.)** **N = 4**
aspartic acid	(7.6) 9.0 +/- 1.4 (10.4)
glutamic acid	(8.8) 9.8 +/- 0.9 (10.5)
alanine	(5.9) 7.0 +/- 1.7 (9.0)
arginine	(5.1) 5.9 +/- 0.9 (6.9)
cystine	(0.1) 0.9 +/- 0.7 (1.9)
glycine	(3.9) 4.4 +/- 0.5 (4.8)
histidine	(2.7) 3.6 +/- 1.3 (5.1)
isoleucine	(3.2) 4.0 +/- 0.8 (5.1)
leucine	(6.8) 7.5 +/- 0.5 (7.8)
lysine	(3.5) 5.3 +/- 1.3 (6.6)
methionine	(0.7) 1.3 +/- 0.4 (1.6)
(methionine + cystine)	
phenylalanine	(4.1) 4.4 +/- 0.2 (4.6)
(phenylalanine + tyrosine)	
proline	(4.8) 5.2 +/- 0.6 (5.6)
serine	(1.2) 2.9 +/- 1.5 (4.0)
threonine	(2.7) 3.7 +/- 0.7 (4.0)
tryptophan	0.8
tyrosine	(3.0) 5.0 +/- 1.9 (6.9)
valine	(5.1) 5.9 +/- 1.0 (7.3)

Sources: Phelps et al. 1975; Santos Oliveira et al. 1976; Hladik 1977; Ukhun and Osasona 1985.

Nigeria, the Igbo people recognise 13 species of grasshoppers (including *Zonocerus variegatus, Actractamorpha aurivilii, Gymnobothrus flaviventris, Eyprepocnemis ibandana*), as well as the praying mantis and the locust *Locusta migratoria*. These insects were reported as food in the Alayi, Umushahia, and Aba areas (Barker et al., 1977). For the Mofu also, green grasshoppers, various crickets, praying mantis, and several gryllids are considered delicacies and mostly captured by children (Seignobos et al., 1996). Barreteau (1999), in an ethno-zoological study of the crickets and mantis consumed by the Mofu-Gudur of northern Cameroon, recorded respectively 74 and 11 ethnospecies with vernacular names, of which 56 crickets and 7 mantis are eaten. Van der Waal (1999) with respect to the Venda of northern South Africa recorded more than 150 local names of grasshopper species and 42 common species, of which a large number is eaten. Roulon-Doko (1998) listed 28 ethnospecies of crickets and grasshoppers eaten by the Gbaya of the Central Africa Republic and provided several scientific names. Kisimba et al. (2003) listed 46 edible ethnospecies with vernacular names for 18 ethnolinguistic groups of the D.R. Congo.

Consumption of acridids in the 19th century was of importance, in northern Africa (Camboué, 1886; Künckel d'Herculais, 1891); colorful comments have been found in these old texts.

Table 10.11: Lipid composition of edible African Termites and Orthoptera (in % of total fatty acids)

Fatty acids		Termites Alates	Termites Queen[1] N = 2	Orthoptera N = 1
lauric acid	C12:0	0.1	n.d.	0.4
myristic acid	C14:0	0.9	(1.0) 1.4 ± 0.6 (1.9)	2.1
tsuzuic acid	C14:1	0.5		
pentadecanoic acid	C15:0	0.6		0.3
palmitic acid	C16:0	33.0	(17.2) 18.9 + 2.4 (20.6)	24.0
palmitoleic acid	C16:1	0.5	1.6	0.6
margaric acid	C17:0	2.6		0.7
heptadecanoic acid	C17:1	0.7		
stearic acid	C18:0	1.4	(6.3) 8.8 ± 3.5 (11.3)	15.7
oleic acid	C18:1	9.5	(61.3) 66.5 ± 7.4 (71.8)	
	C18:1 cis-9			33.3
	C18:1 isom.*			
linoleic acid	C18:2	43.1	(1.6) 3.0 ± 2.0 (4.4)	9.9
linolenic acid	C18:3	3.0	0.5	6.9
arachidic acid	C20:0	0.4		1.2
eicosadienoic acid	C20:2	3.7		
behenic acid	C22:0	0.1		
other fatty acids				5.3

* double link isomere position
[1]fat bodies.
Sources: Cemlik 1969b; Kodondi et al. 1987b; Malaisse 1997; Malaisse and Lognay 2003.

Mapping Ethnolinguistic Groups Practicing Acridophagy

All together some 67 sources relating human consumption of Orthoptera were consulted. Some 90 ethnolinguistic groups practiced acridophagy, constituting 9% of the groups recognized in Africa.

Figure 10.6 provides a map of acridophagy in Africa. Greater importance is observed in territories dominated by savanna ecosystems.

Diversity of Edible Species

Table 10.12 is a preliminary list of species of Orthoptera eaten in Africa. Not only crickets and locusts, but also praying mantis, cockroaches, gryllidae, etc. are evidenced. According to present knowledge some 96 species of 13 families have been identified. Besides these species at least, some forty others are pertinent according to Barreteau (1999) and van der Waal (1999).

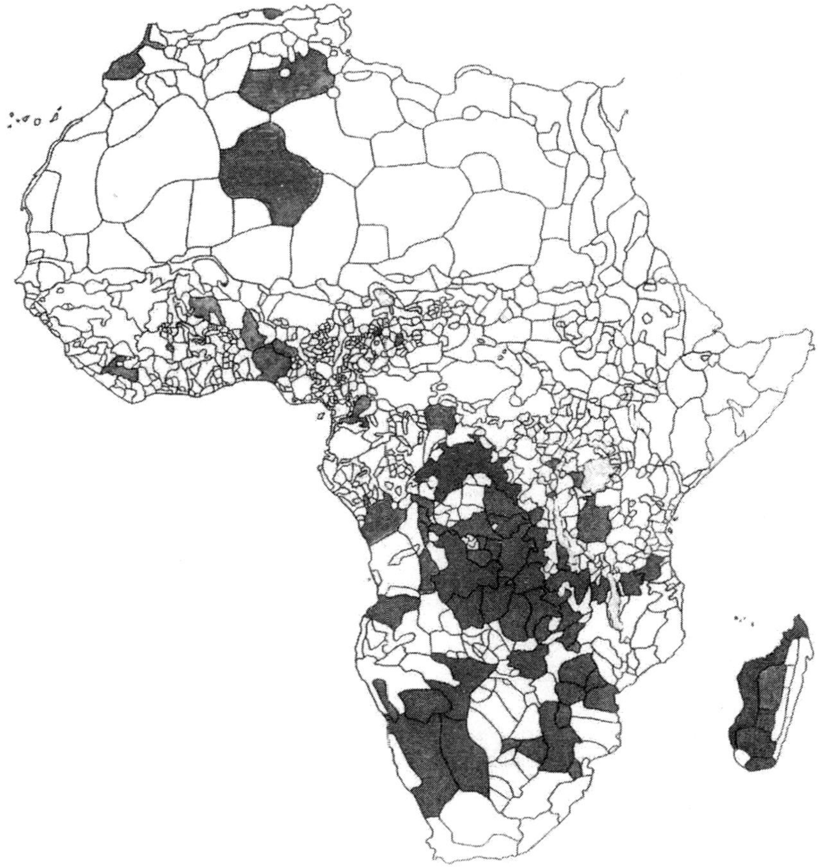

Fig. 10.6: Preliminary distribution map of acridophagy in Africa.

Acrididae, with 59 species or 52% of the total known distributed in 9 subfamilies, ranks first, followed by Catantopidae (10 species, 10%), and Pyrgomorphidae (6 species, 6%).

Color drawings of African Acrididae, including edible species, are presented by Mestre (1988) and Launois-Luong and Lecoq (1989), while color photographs of some edible Orthoptera (4 species) are reproduced by Malaisse (1997).

Chemical Composition

Six papers deal with chemical composition of edible Orthoptera from Africa (Adler, 1934; Lapp and Rohmer, 1937; Heymans and Evrard, 1970; Mbemba and Remacle, 1992; Malaisse, 1997; Malaisse and Parent, 1998). According to these authors, differences between the species identified are significant (Table 10.9).

Table 10.12: Preliminary list of Orthoptera eaten in Africa*

Families / Subfamilies	Species	References
Acrididae		
Acridinae	*Acrida bicolor* Thunberg	Chavanduka 1975
Acridinae	*Acrida sulphuripennis* (Gerstaecker)	Mbata 1995
Acridinae	*Acrida turrita* L.	van der Waal 1999
Acridinae	*Chirista compta* (Walker)	van Huis 1996
Acridinae	*Locustana pardalina* Walker	Barreteau 1999
Acridinae	*Orthochtha venosa* (Ramme)	Barreteau 1999
Acridinae	*Roduniella insipida* Karsch	Bahuchet 1985
Acridinae	*Sherifura hanoingtoni* (Uvarov)	Barreteau 1999
Calliptaminae	*Acorypha clara* (Walker)	Barreteau 1999; Mignot 2003
Calliptaminae	*Acorypha glaucopsis* (Walker)	Barretau 1999
Calliptaminae	*Acorypha picta* (Krauss)	Barreteau 1999
Calliptaminae	*Caloptenopsis nigrovariegata* (I. Bolívar)	Mbata 1995
Coptacridinae	*Poecilocerastis tricolor* I. Bolívar	Kisimba et al. 2003
Cyrtacanthacridinae	*Acanthacris ruficornis* (Fabricius)	Shaxon et al. 1974
Cyrtacanthacridinae	*Acridoderes strenuus* (Walker)	van Huis 1996
Cyrtacanthacridinae	*Anacridium burri* Dirsh	Malaisse 1997
Cyrtacanthacridinae	*Anacridium melanorhodon* (Walker)	van Huis 1996
Cyrtacanthacridinae	*Anacridium moestum* Serville	van der Waal 1999
Cyrtacanthacridinae	*Anacridium wernerellum* (Karny)	van Huis 1996
Cyrtacanthacridinae	*Cyrtacantharis aeruginosa* Stoll	Mbata 1995
Cyrtacanthacridinae	*Cyrtacantharis tatarica* (L.)	Mbata 1995
Cyrtacanthacridinae	*Kraussaria angulifera* (Krauss)	van Huis 1996
Cyrtacanthacridinae	*Nomadacris septemfasciata* (Serville)	Chavanduka 1975
Cyrtacanthacridinae	*Ornithacris cyanea* Stoll	Ramos Elorduy 1987
Cyrtacanthacridinae	*Ornithacris magnifica* I. Bolívar	Mbata 1995
Cyrtacanthacridinae	*Ornithacris turbida* (Walker)	Bani 1995
Cyrtacanthacridinae	*Orthacanthacris humilicrus* (Karsch)	Mignot 2003
Cyrtacanthacridinae	*Schistocerca gregaria* (Forskål)	Mbata 1995
Euryphyminae	*Amblyphymus* sp.	van der Waal 1999
Eyprepocnemidinae	*Cataloipus cognatus* (Walker)	van der Waal 1999
Eyprepocnemidinae	*Cataloipus cymbiferus* (Krauss)	Barreteau 1999
Eyprepocnemidinae	*Cataloipus fuscocoeruleipes* Sjöstedt	van Huis 1996
Eyprepocnemidinae	*Cyathosternum* spp.[2]	Gelfand 1971
Eyprepocnemidinae	*Eyprepocnemis* cf. *plorans* Charpentier	Roulon-Doko 1998
Eyprepocnemidinae	*Heteracris coerulescens* Stål	Bahuchet 1985
Eyprepocnemidinae	*Heteracris guineensis* Krauss	van Huis 1996
Eyprepocnemidinae	*Heteracris* sp.	van der Waal 1999
Eyprepocnemidinae	*Tylotropidius didymus* Thunberg	Roulon-Doko 1998
Eyprepocnemidinae	*Tylotropidius gracilipes* (Brancsik)	Barreteau 1999
Gomphocerinae	*Brachycrotaphus tryxalicerus* (Fischer)	Barreteau 1999
Gomphocerinae	*Krausella amabile* (Krauss)	Barreteau 1999, Mignot 2003
Gomphocerinae	*Mesopsis abbreviatus* P. de Beauvois	Roulon-Doko 1998
Gomphocerinae	*Stenohippus mundus* (Walker)[1]	Mignot 2003
Oedipodinae	*Aiolopus thalassinus* (Fabr.)	van der Waal 1999
Oedipodinae	*Gastrimargus africanus* (Saussure)	van Huis 1996
Oedipodinae	*Gastrimargus determinatus* (Gerstäcker)	Barreteau 1999

Table 10.12: (Contd.)

Table 10.12: (Contd.)

Families / Subfamilies	Species	References
Oedipodinae	*Humbe tenuicornis* (Schaum)	van der Waal 1999
Oedipodinae	*Locusta migratoria* L.	Shaxon et al. 1974
Oedipodinae	*Oedaleus carvalhoi* I. Bolívar	van der Waal 1999
Oedipodinae	*Oedaleus flavus* L.	van der Waal 1999
Oedipodinae	*Oedalus nigeriensis* (Uvarov)	Barreteau 1999
Oedipodinae	*Oedaleus nigrofasciatus* (De Geer)	Mbata 1995
Oedipodinae	*Oedaleus senegalensis* (Krauss)	Mignot 2003
Oedipodinae	*Paracinema tricolor* (Thunberg)	Barreteau 1999
Oedipodinae	*Pycnodictya flavipes* Miller	van der Waal 1999
Tropidopolinae	*Afroxyrrhepes procera* (Burm.)	Roulon-Doko 1998
Tropidopolinae	*Homoxyrrhepes punctipennis* (Walker)	Mbata 1995; Barreteau 1999
Tropidopolinae	*Tristria* cf. *conops* Karsch	Roulon-Doko 1998
Tropidopolinae	*Tristria* cf. *discoidalis* I. Bolívar	Roulon-Doko 1998
Truxalinae	*Truxalis burtti* Dirsh	van der Waal 1999
Truxalinae	*Truxalis johnstoni* (Dirsh)	Barreteau 1999
Truxalinae	*Truxaloides constrictus* (Schaum)	van der Waal 1999
Blattidae		
	genus 1	Boyle *in* DeFoliart 2002
	genus 2	Kimba (in litt.)
Catantopidae		
Catantopinae	*Catantops melanostichus* Schaum	van der Waal 1999
Catantopinae	*Catantops quadratus* Walker	Bahuchet 1985
Catantopinae	*Catantops stramineus* (Walker)	Barreteau 1999
Catantopinae	*Diabolocatantops axillaris* (Thunberg)	van Huis 1996
Catantopinae	*Exopropacris modica* (Karsch)	Barreteau 1999
Catantopinae	*Harpezocatantops stylifer* (Krauss)	Barreteau 1999; Mignot 2003
Catantopinae	*Oxycatantops congoensis* (Sjöstedt)	Bani 1995
Catantopinae	*Oxycatantops spissus* (Walker)	Bani 1995; Barreteau 1999
Catantopinae	*Parapropacris notata* (Karsch)	Roulon-Doko 1998
Catantopinae	*Phaeocatantops decoratus* (Gerst.)	van der Waal 1999
Gryllacrididae		
	Gryllacris africana[2]	Bahuchet 1985
Gryllidae		
	Acheta spp.	Mbata 1995
	Brachytrupes membranaceus (Drury)	Shaxon et al. 1974
	Gryllus bimaculatus De Geer	Mbata 1995
	Gymnogryllus leucostictus Burm[3]	
Gryllotalpidae		
	Gryllotalpa africana P. de Beauvois	Chavanduka 1975
Hemiacrididae		
	Acanthoxia gladiator Westwood	Roulon-Doko 1998
	Hieroglyphus daganensis Krauss	Mignot 2003
	Mazaea granulosa Stål.	Roulon-Doko 1998
Hymenopodidae		
	Pseudoharpax virescens (Gerst.)	Seignobos et al. 1996

Table 10.12: (Contd.)

Table 10.12: (Contd.)

Families / Subfamilies	Species	References
Mantidae		
	Epitenodera houyi (Werner)[4]	Roulon-Doko (1998)
	Hoplocorypha garuana (Giglio-Tos)	Barreteau 1999
	Mantis religiosa L.[5]	Quin 1959
	Miomantis paykullii Stål	Seignobos et al. 1996
	Sphodromantis centralis Rehn	Kisimba et al. 2003
	Tarachodes saussurei (Giglio-Tos)	Barreteau 1999
Mecopodidae		
Mecopodinae	*Anoedopoda erosa*[2]	Bahuchet 1985
Mecopodinae	*Leprocristus* sp.[2]	Roulon-Doko 1998
Pamphagidae		
Porthethinae	*Lamarckiana bolivariana* (Sauss.)	van der Waal 1999
Porthethinae	*Lamarckiana punctosa* (Walker)	van der Waal 1999
Pyrgomorphidae		
	Chrotogonus senegalensis (Krauss)	Barreteau 1999
	Phymateus viridipes Stål	Chavanduka 1975
	Pyrgomorpha cognata (Krauss)	Barreteau 1999
	Pyrgomorpha vignaudii (Guérin-Méneville)	Bahuchet 1985
	Zonocerus elegans Thunberg	Quin 1959
	Zonocerus variegatus (L.)	Roulon-Doko 1998
Tettigoniidae		
Conocephalinae	*Conocephalus* spp.	Bahuchet 1985
Conocephalinae	*Lanista* sp.	Roulon-Doko 1998
Conocephalinae	*Ruspolia differens* Serville[6]	Mbata 1995, Malaisse 1997
Copiphorinae	*Pseudorhynchus lanceolatus* Fabr.	Roulon-Doko 1998
Decticinae	*Anabrus simplex* Haldeman	Egan 1917

*Taxonomy according to Mestre (1988) and Dirsh (1965).

[1]*Stenobothrus epacromioides* Krauss, cited by Mignon (2003), is treated as a synonym of *Stenohippus mundus* (Walker).

[2]We were unable to establish which taxon is under consideration. Genus *Leprocristus* could not be traced but *Leptoscirtus* exists.

[3]*Gymnogryllus elegans* Guér. is treated as a synonym of *G. leucostictus* Burm.

[4]*Epitenodera gambiense* (Beier), cited by Roulon-Doko (1998), is regarded as a synonym of *E. houyi* (Werner).

[5]Quin (1959) cited *Mantis religiosa*, a species not occurring in southern Africa. We were unable to establish which taxon was considered; might be a *Sphodromantis* sp.

[6]*Homorocoryphus nitidulus vicinus* (Walker), cited by Shaxon et al. (1974), is regarded as a synonym of *Ruspolia differens* Serville.

Myrmecophagy in Africa

Introduction

Silow (1983) devoted some 68 pages to the ethnomyrmecological knowledge of Ngangela and Nkoya in Mid western Zambia. Of the thirteen species recognized by them, only one is eaten, the edible winged thief ant. Roulon-Doko (1998) in a chapter on ants as perceived by the Gbaya, indicated that this ethnic group eats one ethnospecies as a condiment and the eggs of two others. After the eggs have been gathered from the center of the ant hill, those of one species are eaten fresh after exposure to the sun and those of the other cooked (Roulon-Doko, 1998). The Ju/'hoansi of western Kalahari and San | Gui of central Kalahari likewise use the golden ant *Camponotus fulvopilosus*, which sprays formic acid, as a seasoning (Nonaka, 1996; Green, 1998). Even the vicious red ants are used by the Ngindo to provide a dish, after first scorching the marching column with a firebrand then roasting and pulverizing (Crosse-Upcott, 1958). The reference of Grivetti (1979) to consumption by Tswana Tlokwa of wasp larvae, related to the family of Formicidae is not clear.

Diverse ethnolinguistic groups consider adults of some ant species, namely winged thief ants, as delicacies. The species concerned belong to genus *Carebara* (*C. junodi* Forel and *C. vidua* Sm.) (Silow, 1983). The ecology of *Carebara vidua* is not well known. This Myrmicinae taxon lives in a termite nest and purportedly feeds on malnourished termites. The biomass of the queen is 3,400 to 4,600 times greater than that of the workers (Girard and Lepage, 1991).

In some (sub)arid countries, nests of ants, namely of *Megaponera* sp. (Seignobos et al., 1996) and *Messor aegyptiacus* (Gast, 2000), are a source of various seeds, providing food during gap periods. Gast (2000) listed some 16 different edible seeds for Kel Ahaggar.

Ethnolinguistic Groups Practicing Myrmecophagy

Altogether some 20 sources mentioning human consumption of ants were consulted (see Table 10.2). So we can cite some 20 ethnolinguistic groups practicing myrmecophagy, i.e. 1.9% of the groups recognized for Africa.

Diversity of Edible Species

Table 10.13 lists some ants used for food as well as their uses.

Chemical Composition

No information is available regarding the chemical composition of edible ants of Africa.

As to palatability, *C. vidua* is reported as an appetising and fruity meat by Quin (1959).

Discussion

Insects have been and still are eaten by a large group of people belonging to diverse ethnolinguistic groups of Africa.

Table 10.13: Ants used as a food source in Africa[1]

Subfamilies	Species	Use	Reference
Formicinae	*Camponotus* sp.	condiment (formic acid)	Nonaka 1996
Formicinae	*Camponotus fulvopilosus* De Geer	condiment (formic acid)	Green 1998
Formicinae	*Oecophylla smaragdina* Fabricius	larva; imago	Chinn 1945
Myrcineae	*Carebara junodi* Forel	alate female	Silow 1983
Myrcineae	*Carebara vidua* F. Smith	alate female	Quin 1959
Myrcineae	*Messor aegyptiacus* (Emery)	seeds from nest	Gast 2000
Ponerinae	*Pachycondyla* sp.[2]	seeds from nest	Seignobos et al. 1996

[1]Nomenclature after Bolton (1995).
[2]*Megaponera* Mayr, referred to by Seignobos et al. (1996), is regarded as a junior synonym of *Pachycondyla* F. Smith.

In towns, for reasons such as foreign life style influence and sensibility of various foreign religions, the practice of eating insects is diminishing (Katsia Kitsa, 1989; Abdullah, 1975). However, accessibility to food in general, and animal products in particular, enhances the significance of edible insects (Plate V, 1 to 4; Fig. 10.7).

Estimations of daily, monthly or annual uptake of caterpillars have still to be made for some ethnolinguistic groups. Gomez et al. (1961) published results

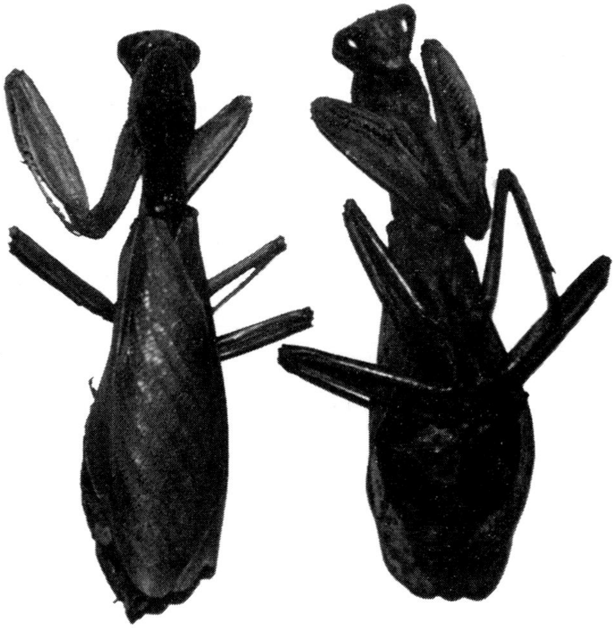

Fig. 10.7: Some edible African insects. *Sphodromantis centralis* (environs of Lubumbashi) (photo F. Malaisse).

of a survey conducted in all territories of the Congo. Insect consumption was calculated and represents more than 40% of animal proteins in some parts of the country. Around Brazzaville, 40 g dry weight smoked caterpillars are consumed per day (Paulian, 1963), in the vicinity of Kananga about 15 g a day during five months (Katya Kitsa, 1989), and in other parts of the continent values of 30–50 dry weight of insects per day, mainly caterpillars, are not at all rare for several months.

Regarding chemical composition and energy values, insects considered here are able to supply a valuable amount of energy, animal proteins, lipids and vitamins. Several authors (Munthali and Mughogho, 1992; Cunningham, 1996; Malaisse, 2001; DeFoliart, 2002) consider that the current use and sustainable intensification of edible insects should be pursued. But, any intensification program would necessitate a preliminary ethnoecological study (Malaisse, 2001). Only such an approach would enable, after an open dialogue with the local people to be involved, a judicious and effective project.

Caterpillars hold a dominant position (DeFoliart 1991a,b, 1995; Holden, 1991; Munthali and Mughogho, 1992; Cunningham, 1996). Selection of valuable species should be based on preliminary studies concerning their occurrence, seasonality, energy balance (Sholtz, 1982), and abundance of their food plants. Leguminous trees are frequently considered important forage plants for cattle; Turk (1990) provided a list of 42 species. These food plants have to be carefully considered in terms of their several uses, i.e., timber, firewood, edible products, fungal symbiosis, domestic, medicinal or cultural uses, as emphasized by Malaisse (2001) (Fig. 10.8).

Research carried out on *Elaphrodes lactea* in Katanga (Malaisse et al., 1974; Malaisse-Mousset et al., 1970) and a Symposium held in Gaborone on *Imbrasia belina* or "phane" (Brandon, 1987 a,b; Ferreira, 1995; Gashe et al., 1996; Gashe

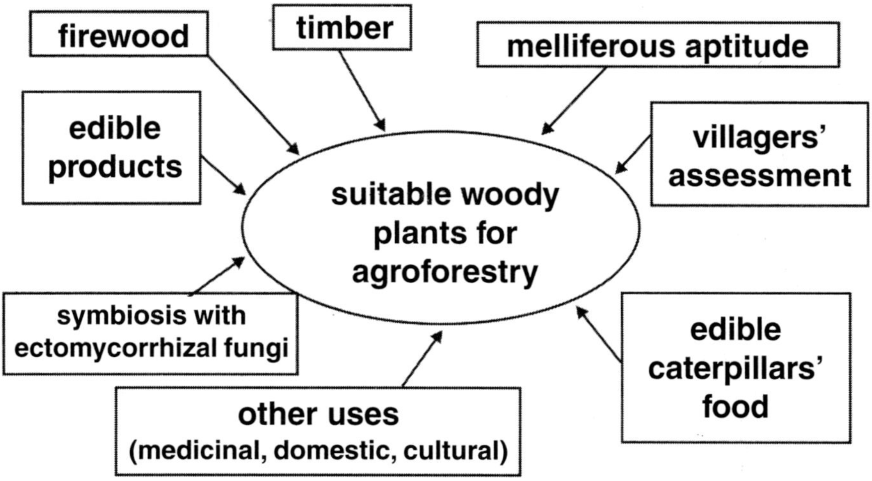

Fig. 10.8: Strategy of selection of trees for agroforestry.

et al., 1997; Illgner and Nell, 2000; Letsie, 1996; Marais, 1996; Moruakgomo, 1996; Mpuchane et al., 1996; Ohiokpehai et al., 1996; Siame et al., 1996; Styles, 1995, 1996) have led to a good basic knowledge for the management of these two species. More than 1,500 tons of *I. belina* are sold each year. A bag containing 80,000 caterpillars sold for 120 to 150 rands (US $75 in 1995). Gatherers are able to earn some 2,500 rands (US $ 1,250) in seven weeks (Styles, 1995; Cunningham, 1996).

Unfortunately, the future of high termitaria and *Macrotermes* spp. is bleak (DeFoliart, 2002). In suburban regions and towns the flora and fauna of have been destroyed. Termite hills have become the main source of brick-clay and a favourable site for growing maize crops (Aloni et al., 1993). The decrease in number of large hills began a long time ago, following the destruction of surrounding vegetation, which resulted in a more empty hills. It is difficult to retain termites as a sustainable food source, especially since many agronomists consider them a nuisance (Pearce, 1997).

Nevertheless, in some countries, termite hills are under ownership, assuring good management, so long as neither soldiers nor queen are harvested.

Making a rough estimation of the termite edible castes biomass is very difficult. It was done in Upper Katanga and data have been provided on the

Table 10.14: Rough estimation of *Macrotermes falciger* biomass in Upper Katanga (square degree of Lubumbashi, Congolian part)[1]

Main vegetation units	Vegetation cover (ha)		High mounds density per ha	Percentage active	Active mounds[3]	
	Original[2]	Nowadays			Original	Nowadays
Open forest	884,675	727,258	3.4	22	661.7	544.0
Wooded savanna	40,347	15,925	2.5	8	8.1	3.2
Tree and shrub savanna	40,348	15,925	2.3	2	1.9	0.7
Dambo	39,429	26,559	0.2	0	0	0
Alluvial savanna	34,920	28,200	0.6	0	0	0
Dense riparian forest	7,877	162	0.0	—		
Marshy vegetation	7,747	4,027	0.0	—		
Cupriferous steppe	565	565	0.0	—		
Dry evergreen forest	260	115	4.7	27	0.33	0.15
Derived savanna	—	214,821	2.7	0	0	0
Urban area	—	11,960	—	—		
Cultivation	—	5,777	2.7	2		0.31
Man-made reservoir	—	3,720	—	—		
Afforestation	—	1,154	2.7	2		
Total	1,056,168				672	548

[1]After Goffinet (1976), Malaisse (1985), Malaisse and Bizangi (1985).
[2]Beginning of 20th century.
[3]Values in thousand (10^3).
Relative dry weights for one *Macrotermes falciger* colony are respectively of 50.9 % alate imagos, 1.1% for big soldiers, 29.2 % for big workers, 16.5% for small workers, with a great total of 19.3 kg for one colony. Rough estimation of edible termites' dry weight still available for Lubumbashi square quadrat (Congolian part) is of 10,576 tons, with about one half of alates adults produced each year! (Malaisse 1997).

relative importance of the diverse vegetation unit covers (at the beginning of the 20th century and nowadays), on density of high termite hills according to the respective vegetation units, and percentage of active mounds in each unit. Moreover, estimations of relative importance of castes, population size, and biomass have been given (Table 10.14). Locusts were long ago condemned a pest or applauded as manna. They are now monitored and controlled, which is why swarms occur less frequently. The cricket *Brachytrupes* has a large distribution and is regarded a delicacy by several populations inhabiting tropical Africa. No information is available regarding their population dynamics.

Lastly ants are a food rarely eaten in Africa, contrary to some South American territories. *Carebara* spp., the most interesting ants, are hosted in termitaria, a subecosystem whose management, as mentioned above, is threatened.

Acknowledgments

I am indebted to Mrs. A. Roubinkova for tremendous support in obtaining books as well as papers dealing with this subject, and to Mrs. C. Van Marsenille for pen-and-ink drawings.

I appreciate the support given in Katanga by Kisimba Kibuye Emile and Muzinga Yumba Laurent for field observations and material collection. I received valuable local information from several friends. Special thanks to J. Kimba (D.R. Congo), P. Latham (Bas Congo, Tanzania), G. N'Gasse (Ecofac-Project, Bangui, Central African Rep.), R. Oberprieler (formerly Namibia), M. Ouedraogo (Burkina Faso), F. Sarraco-Lanata (Gabon), and A. Steele (Zimbabwe).

Checking of insect Latin spelling and their nominator was done in collaboration with Dr. U. Dell' Asta (Lepidoptera), Dr. J. Ruelle (Isoptera), Dr. E. Deconinck (Heteroptera) and J. Debecker—all of the Musée d'Afrique centrale at Tervuren (Belgium).

D. Cohen provided assistance with the English expression.

References

Abdullah M. 1975. Recopilatión de noticias sobre insectos comestibles con comentarios personales y recetas culinarias. Graellsia, 29: 225–238.

Abe'Ele M.P. 1998. La pêche chez les Badjoué. Interaction entre les techniques, le temps, les terroirs et les ressources prélevées à Doumo (Périphérie de la Réserve de Faune du Dja, Est-Cameroun). Second rapport semestriel, septembre 1998. Rapport Projet Forêts Communautaires/Fac. Univ. Sci. Agron. Gembloux, Belgium.

Adler E. 1934. Nutritive value of locusts. Farming South Africa 9: 29.

Adriaens E.L. 1951. Recherches sur l'alimentation des populations du Kwango. Bull. Agric. Congo Belge,. 42: 227–270, 473–552.

Adriaens E.L. 1953. Note sur la composition chimique de quelques aliments mineurs indigènes du Kwango. Ann. Soc. Belg. Méd. Trop. 33: 531–544.

Aldasoro Maya E.M. 2003. Etude ethnoentomologique dans la communauté hñahñu "El Dexthi" de la vallée du Mezquital (Hidalgo, Mexique). Colloque "Les Insectes dans la tradition orale", Villejuif (France). (in press)

Aloni K., Dikumbwa N., Malaisse F., and Pirotte A. 1993. Termites: insectes nuisibles, friandises ou bâtisseurs de cathédrales et de villes? Défis-Sud 12: 36–37.

Anonymous. 1998. Zambia, Map of Tribes.

Anonymous. 2003. http://www.utoronto.ca/forest/termite/speclist.htm

Ashiru M.O. 1988. The food value of the larvae of *Anaphe venata* Butler (Lepidoptera: Notodontidae). Ecol. Food Nutr. 22: 313–320.

Ashiru M.O. 1989. Silkworms as money spinners. Nigerian Field 54 (1–2): 3–10.

Auffret Ch. and Tanguy F. 1947–48. Note sur la valeur alimentaire des termites. Bull. Médical de l'A.O.F., pp. 395–397.

Bahuchet S. 1972. Etude écologique d'un campement de Pygmées Babinga (région de la Lobaye, République centrafricaine). J. Agric. Trop. Bot. Appl. XIX (12): 509 –559.

Bahuchet S. 1975. Ethnozoologie des pygmées Babinga de la Lobaye, République centrafricaine. *In:* Pujol R. (éd.). L'homme et l'animal. Premier colloque d'ethnozoologie. Inst. Inte. d'Ethnosci. Paris, pp. 53–61.

Bahuchet S. 1985. Les pygmées Aka et la forêt Centrafricaine. Selaf, Paris, 640 pp.

Baião D.V. 1939. Dictionario Ganguella-Portuguêz. Lisboa,

Banadzem J. 1980. Les notions de temps et d'espace chez les Nso du Cameroun. Mémoire de maîtrise, Univ. Paris, vol. I.

Bani G. 1995. Some aspects of entomophagy in the Congo. Food Insects Newsl. 8 (3): 4–5.

Barker D., Oguntoyinho J., and Richards P. 1977. The utility of the Nigerian peasant farmer's knowledge in the monitoring of agricultutral resources. Monitoring and Assessment Research Centre (M.A.R.C.), Chelsea College, Univ. of London, London. Report 4, 55 pp.

Barreteau D. 1999. Les Mofu-Gudur et leurs criquets. *In:* L'homme et l'animal dans le bassin du Lac Tchad. Actes du Colloque du réseau Mega-Tchad, Editions IRD, collections Colloqies et Séminaires, n°00/354, Univ. Nanterre, Nanterre, France, pp. 133–169.

Baum H. 1903. *Kunene—Sambesi-Expedition*. Verl. Kolonial-Wirschaftlichen Komitees, Berlin, xi+593 pp.

Baumann O. 1887. Beiträge zur Ethnographie des Congo. *Mitteilungen der Anthropologischen Gesellschaft in Wien*, vol. 17.

Beemer H. 1939. Notes on the diet of the Swazi in the protectorate. Bantu Studies, XIII (3): 199–236.

Benhura M.A.N. and Chitsiku I.C. 1990. Food consumption patterns in the Mutambara District of Zimbabwe. Central African J. Med. 36 (5): 120–128.

Bequaert J. 1921. Insects as Food; How they have augmented the food supply of mankind in early and recent times. J. Amer. Mus. Nat. Hist. XXI (2): 191–200.

Bergier E. 1941. Peuples entomophages et Insectes comestibles. Etude sur les mœurs de l'Homme et de l'Insecte. Rullières Frères, Avignon, France, 231 pp.

Biehler E. 1950. A Shona Dictionary with an Outline of Shona Grammar. Longmans, Green and Co., Cape Town, R.S.A., 337 pp. (4th ed.).

Binger L.G. 1892, 1894. Du Niger au golfe de Guinée par le pays de Kong et le Mossi 1887–1889. Hachette, Paris, 2 vols., S.7 515 + 416 pp.

Blohm W. 1931. Die Nyamwezi. Land und Wirtschaft. de Gruyter and Co., Hamburg, Germany, xii + 182 pp.

Bloomhill G. 1958. Africa à la carte. Some favourite dishes of the African gourmet. African World 1958 (Nov.): 9–10.

Bodenheimer E.S. 1951. Insects as Human Food. A Chapter of the Ecology of Man. W. Junk, The Hague, The Netherlands, 352 pp.

Bolton B. 1995. A New General Catalogue of the Ants of the World. Harvard Univ. Press, Cambridge, Massachusetts, MA, USA, 504 pp.

Boone O. 1961. Carte ethnique du Congo. Quart sud-est. Musée Royal Afr. Centrale, Ann., série in-8°, sc. humaines 37: 271 pp.

Boone O. 1973. Carte ethnique du Congo. Quart sud-ouest. Musée Royal Afr. Centrale, Ann., série in-8°, sc. humaines 78: 406 pp.

Bossard E. 1996. Quelques notes sur l'alimentation et les apports traditionnels occultes en Angola. Garcia de Orta, Sér. Bot., Lisboa 13 (1): 7–41.

Bouillon A. and Mathot G. 1965. Quel est ce termite africain? *Zooléo*, Ed. de l'Université, Léopoldville, 1: 115.

Bouyer T. 1999. Catalogue des Saturniidae africains. Entomologia Africana, hors série 1: 73 pp.

Brandon H. 1987a. The snack that crawls. Int. Wildl. March–April, pp. 16–21.

Brandon H. 1987b. Revoilà le temps des chenilles. Géo, 103: 136–145.

Brisson, R. (n.d.). Vocabulaire Bira (mimeo).

Bryant A.T. 1905. A Zulu-British Dictionary, Marianhill Mission Press, Durban, R.S.A. 778 pp.

Bukkens S.G.F. 1997. The nutritional value of edible insects. Ecol. Food Nutr. 36 (2–4): 287–319.

Callewaert R. 1922. Witte mieren in Kasai. Congo, 2 (3): 366–380.

Camboue R.P. 1886. Les sauterelles à Madagascar. Bull. Soc. nat. d'Acclimatation France 33: 168–172.

Carbou H. 1912. La région du Tchad et du Ouadaï. Tome I: "Etudes ethnographiques. Dialecte toubou". Ernest Leroux Edit., Paris, 380 pp.

Casati G. 1892. Dix années en Equatoria (trad. de Hessem L.). Librairie Firmin-Didot, Paris, xi + 502 pp.

Cemlik S.H.W. 1969a. The neutral lipids from various organs of the termite *Macrotermes goliath*. J. Insect Physiol. 15: 839–849.

Cemlik S.H.W. 1969b. Composition of the neutral lipids from termite queens. J. Insect Physiol. 15: 1481–1487.

Césard E. 1936. Le Muhaya. Anthropos, 31(1–2): 97–114, (3–4): 489–508, (5–6): 821–849.

Chavanduka D.M. 1975. Insects as a source of protein to the African. Rhodesia Sci. News 9: 217–220.

Chinn M. 1945. Notes pour l'étude de l'alimentation des indigènes de la Province de Coquilhatville. Ann. Soc. Belg. Méd. Trop. 25: 57–149.

Coaton W.G.H. 1949. Notes on some South African species of the families Hodotermitidae and Kalotermitidae. J. Entom. Soc. Southern Afr. 12: 13–77.

Colle H. 1902. Notes on the Wagogo of German East Africa. J. Royal Anthropological Inst. vol. 32.

Colle H. 1913. Les Baluba (Congo Belge) I–II. Collection de Monographies ethnographiques. Albert Dervit, Institut Int., Bibliographie, Bruxelles, 10–11: 918 pp.

Colrat de Montrozier R. 1902. Deux ans chez les anthropophages et les sultans du centre africain. Librairie Plon, Paris, 326 pp.

Costermans B.J. 1950. Termieten-larvenoogst bij de Logo. Kongo-Overzee XVI: 185–197.

Costermans J.B. 1955. Het Termieten-stoken bij de Logo-Avokaya. Aequatoria 18 (1): 14–25.

Courtois V.J. 1900. Diccionario Cafre-Tetense-Portuguez. Imprensa da Universidade, Coimbra.

Crafford J.E. 1991. Insects as a source of food, folklore and folk taxonomy in Venda. Proc. Eighth Entomological Cong., Entomological Society of Southern Africa, Bloemfontein. 12

Crosse-Upcott A.R.W. 1958. Ngindo famine subsistence. *Tanganyika Notes, Records*, Dar es Salaam, 50: 1–20.

Cunningham T. 1996. Saturniid subsidy: cash and protein from edible caterpillars of Zambesian woodlands. *In:* Campbell B. (ed.). The Miombo in Transition: Woodlands and Welfare in Africa. Center Int. Forestry Res. (CIFOR), Bogor, Indonesia, pp. 107–108.

Cuypers J-B. 1970. L'alimentation chez les Shi. Musée royal de l'Afrique centrale. Tervuren (Belgique). Ann., série in 8°, Sciences humaines, 67: xii + 249 pp.

Dartevelle E. 1951. Sur un Hespéride des environs de Léopoldville et remarques sur la valeur alimentaire des chenilles de Lépidoptères. Lambillionea 51: 12–16, 18–20.

de Almeida A. 1946. Cronica de etnografia colonial. Carne de invertebrados—Tabu alimentar dos nativos das colonias portuguesas. Bol. Geral Colon. 32 (249): 107–115.

de Beaucorps R. 1941. Les Basongo de la Luniungu et de la Gobari. Inst. Royal Colonial Belge, Section Sc. morales et politiques, Mémoires, Coll. 8°, X (3): 188.

De Calonne A. 1909. Les Ababua. Mouvement Sociologique Int. (Bruxellles), vol. 10.

De Clercq A. 1960. Dictionnaire Tshiluba-Français. Soc. Missionnaire de Saint-Paul, Léopoldville, nouv. éd., 392 pp.

DeFoliart G.R. 1991a. Edible caterpillars: A potential agroforestry resource ? Food Insects Newsl. 4 (2): 3–4.

DeFoliart G.R. 1991b. Forest management for the protection of edible caterpillars in Africa. Food Insects Newsl. 6: 1–2.

DeFoliart G.R. 1995. Edible insects as minilivestock. Biodiv. Cons. 4: 306–321.

DeFoliart G.R. 2002. http://www.food-insects.com/book7_31/Chapter
de Plaen G. 1974. Les structures d'autorité des Bayanzi. "Encyclopédie universelle", Editions universitaires, Paris, 336 pp.
de Schlippé P. 1956. Shifting Cultivation in Africa. The Zande System of Agriculture. Routledge and Kegan Paul, London, 313 pp., 96 pl.
De Smet R.E. 1972. Enquête de Fuladu, 1959: L'emploi du temps du paysan dans un village Zande du nord-Est du Zaïre. CEMUBAC, Bruxelles. 89, 396 pp.
Delhaise C.G.F.F. Cdt. 1909. Les Warega (Congo Belge). Coll. monographies ethnographiques, sociologie descriptive. Albert De Wit, Institut Int. Bibliographie, Bruxelles 5: 376 pp.
Demesmaecker A. 1997. Contribution à l'écologie: Les chenilles comestibles du Copperbelt, Zambie. Travail de fin d'Etudes, Fac. Univ. Sc. Agron. Gembloux, 117 pp.
Denis J. 1964. Ngombe Ya Tumba. Village Hoto du Haut-Kwango. Bull. Soc. Belge Etudes géographiques (SOBEG) 29 (1): 133–167.
d'Hertefel M 1962. Le Rwanda. *In:* d'Hertefelt M., Trouwborst A.A., and Schefer J.H. (eds.) Les anciens royaumes de la zone interlacustre méridionale, Rwanda, Burundi, Buka. Musée royal d'Afrique centrale, Tervuren (Belgium). Monographies ethnographiques 6: 11–111.
Dinkelacker E. 1914. Wörterbuch der Duala-sprache. Abhandlungen des Hamburgischen Kolonialinstituts, Hamburg (Germany), vol. 16.
Dirsh V.M. 1965. The African genera of Acridoidea. Anti-Locust Research Centre, Cambridge Univ. Press, Cambridge (U.K.) 579 pp.
Doke C.M. 1931. The Lambas of Northern Rhodesia. A Study of Their Customs and Beliefs. G. Harrap and Co., London 408 pp.
Dreyer J.J. and Wehmeyer A.S. 1982. On the nutritive value of mopane worms. S. Afr. J. Sci. 8: 33–35.
Drummond H. 1888. Tropical Africa. Hodder and Stoughton, London vii + 228 pp., 2 maps.
Ducan J.R. 1933. Native Food and Culinary Methods. Native Affairs Dept. Ann., Rhodesia 11: 101–106.
Dugast I. 1955–59. Monographie de la tribu des Ndiki (Banem du Cameroun). Travaux et Mémoires de l'Institut d'Ethnologie, Paris 58: xxiv + 823 pp.
Ellis W. Rev. 1859. Three Visits to Madagascar during the Years 1853–1854–1856. John Potter and Co., London, UK, 428 pp.
Emin-Pacha. 1888. Die Tagebücher von Dr Emin Pacha. G. Westermann, Hamburg, 448 pp. (2nd ed., 1927).
Endemann K. 1911. Wörterbuch der Sotho-Sprache. Abhandlungen des Hamburgischen Kolonialinstituts, Hamburg (Germany), vol. 7.
Ene J.C. 1963. Insects and Man in West Africa. Ibadan Univ. Press, Ibadan, Nigeria.
Fasoranti J.O. and Ajiboye D.O. 1993. Some edible insects of Kwara State, Nigeria. Amer. Entomol. 39: 113–116.
Ferreira A. 1995. Saving the Mopane Worm. South Africa's wiggly protein snack in danger. Food Insects Newsl. 8: 6.
Fuller C. 1918. Notes on white ants. On the behaviour of true ants towards white ants. Bull. South Afr. Biol. Soc. 1 (1): 16–20.
Gade D.W. 1985. Savanna woodland, fire, protein and silk in highland Madagascar. J. Ethnobiol. 5 (2): 109–122.
Gaerdes F. 1959. Schädliche une nützliche Insekten in Südwestafrika. South West Afrika Ann. 1959, pp. 71–77.
Gashe B.A., Leggett K., and Mpuchane S.F. (eds.). 1996. Phane. Proc. First Multidisciplinary Symp. on Phane (18 June 1996), Dept. Biol. Sc., Gaborone, 139 pp. ISBN 99912-913-3-4
Gashe B.A., Mpuchane S.F., Siame B.A., Allotey J., and Teferra G. 1997. The microbiology of phane, an edible caterpillar of the emperor moth, *Imbrasia belina*. J. Food Protect. 60 (11): 1376–1380.
Gast M. 2000. Moissons du désert. Ibis Press, Paris, 160 pp.
Gelfand M. 1971. Diet and Tradition in an African Culture. E. and S. Livingstone, London, UK, 248 pp.
Girard C. and Lepage M. 1991. Vie et mort des termitières cathédrales. Insectes 82: 3–6.

Glew R.H., Jackson D., Sena L., VanderJadt D.J., Pastuszyn A., and Millson M. 1999. *Gonimbrasia belina* (Lepidoptera : Saturniidae): a nutritional food source rich in protein, fatty acids, and minerals. Amer. Entomol. pp. 250–253.

Godwey C.C. 1912. On the utilisation of an indigenous silkworm (*Anaphe infracta* Wism.) in Uganda. Ann. Ent. Res. 3: 269–274.

Golding F.D. 1942. The silkworms of Nigeria. Farm Forest 3: 35–40.

Gomez M.I. 1989. A Resource Inventory of Indigenous and Traditional Food. Univ. of Harar Press, Harar, Ethiopia.

Gomez P.A., Halut R., and Collin A. 1961. Production de protéines animales au Congo. Bull. agric. Congo 52 (4): 698–815.

Gope B. and Prasad B. 1983. Preliminary observations on the nutritional value of some edible insects in Manipur. Journ. Advanced Zool. 4: 55–61.

Green S.V. 1998. The Bushman as an Entomologist. Antenna-London 22 (1): 4–8.

Grimes B.F. (ed.). 1996. Ethnologue. Summer Institute of Linguistics, Dallas, TX (USA) 288 pp. (13th ed.)

Grivetti L.E. 1979. Kalahari agro-pastoral-hunter-gatherers : the Tswana example. Ecol. Food Nutr. 7: 235–256.

Hannan M. 1959. Standard Shona Dictionary. Macmillan and Co., NewYork, NY, xvii + 825 pp.

Harris W.V. 1940. Some notes on insects as food. Tanganyika Notes Records, Dar es Salaam 9: 45–48.

Hegh E. 1922. Les termites. Partie générale. Description.— Distribution géographique.— Classification.— Biologie.— Vie sociale.— Alimentation.— Constructions.—Rapports avec le monde extérieur. Impimerie Industrielle et Financière, Bruxelles, Belgium, 756 pp.

Hendle P.J. 1907. Die Sprache der Wapogoro. Archiv für das Studium Deutscher Kolonialsprachen, Berlin, vol. 6, 171 pp.

Heymans J-C. and Evrard A. 1970. Contribution à l'étude de la composition alimentaire des insectes comestibles de la Province du Katanga. Probl. Soc. Congolais (Bull. Trimestr. C.E.P.S.I.) 90–91: 333–340.

Hladik A. 1995. Valorisation des produits de la forêt dense autres que le bois d'œuvre. Rapport de mission. République centrafricaine, Projet Ecofac, groupement AGRECO/CIRAD-forêt, Paris, 57 pp.

Hladik C.M. 1977. Chimpanzees of Gabon and chimpanzees of Gombe. Some comparative data on the diet. *In:* Clutton-Brock T.H. (ed.). Primate Ecology. Acad. Press, New York, NY, pp. 481–503.

Hochegger H. 1973. La division du travail entre l'homme et la femmechez les Buma. *In:* Rapport sur le Vème Colloque de Bandundu sur l'agriculture et l'élevage dans l'entre Kwango-Kasai (Rép. du Zaïre). Ceeba Publ., Bandundu (Zaïre), pp. 21–43.

Hochegger H. 1975. *Normes et pratiques sociales chez les Buma (Rép. du Zaïre).* Ceeba Publ., Bandundu (Zaïre), sér. II, 26: 519 pp.

Hodgson A.G.O. 1933. Notes on the Achewa and Angoni of the Dowa District of the Nyasaland Protectorate. J. Roy. Anthropol. Inst. 63: 123–164.

Holas B. 1980. Tradition Krou. Fernand Nathan, Paris, 573 pp.

Holden S. 1991. Edible caterpillars—a potential agroforestry Resource? Food Insects Newsl. 4 (2): 3–4.

Horton A.E. 1953. A Dictionary of Luvale. Rhan Brothers, El Monte, CA, iv + 434 pp.

Hulstaert G. 1961. Les Mongo, aperçu général. Archives d'Ethnographie. Musée roy. Afr. centrale, Tervueren 5: 66 pp.

Illgner P. and Nell E. 2000. The geography of edible insects in sub-Saharan Africa: a study of the mopane caterpillar. Geogr. J. 166 (4): 336–351.

Imamura-Hayaki K. 1996. Gathering activity among the central Kalahari San. African Study Monog., suppl. 22: 47–66.

Iroko A.F. 1996. L'homme et les termitières en Afrique. Karthala Press, Paris, 298 pp.

Jackson A.P. 1953. Ample Food without Ploughing. Native Affairs Dept. Ann. (NADA) 31: 62–66.

Jacques-Félix H. 1948. Ignames sauvages et cultivées du Cameroun. Etudes camerounaises, 1 (21–22): 13–18.

Johnston H. 1902, 1904. The Uganda Protectorate. Hutchinson and Co., London, UK, vol. I: 1–470: vol. II: 471–1018.

Joulian F. and Roulon-Doko P. 1994. Comparaison d'une activité technique chez les hommes et chez les chimpanzés : la collecte des termites. Techniques et culture 23–24: 29–62.

Junod H.A. 1927. The Life of a South African Tribe. Macmillan, London, UK, 2 vols. 557 + 660 pp.

Kassner T. 1911. My Journey from Rhodesia to Egypt, Including an Ascent of Ruwenzori and a Short Account of the Route from Cape Town to Broken Hill and Lado to Alexandria. Hutchinson and Co., London, UK. xiv + 309 pp.

Katya Kitsa. 1989. Contribution des aliments comestibles à l'améliorationde la ration alimentaire au Kasai-Occidental. Zaïre-Afrique 239: 511–519.

Kisimba K.E., Lumbu S., and Malaisse F. 2003. About acridophagy in Katenga. Geo-Eco-Trop., 26 (in press).

Kitanishi K. 1995. Seasonal changes in the subsistence activities and food intake of the Aka hunter-gatherers in Northeastern Congo. African Study Monogr. 16 (2): 73–118.

Kodondi K.K., Leclercq M., and Gaudin-Harding F. 1987a. Vitamin estimations of three edible species of attacidae caterpillars from zaïre. Intern. J. Vitam. Nutr. Res. 57: 333–334.

Kodondi K.K., Leclercq M., Bourgeay-Clausse M., Pascaud A., and Gaudin-Harding F. 1987b. Intérêt nutritionnel de chenilles d'attacidés du Zaïre: composition et valeur nutritionnelle. Cah. Nutr. Diét. 22 (6): 473–477.

Kohl-Larsen L. 1943. Auf der Spuren des Vormenshen. Deutsche Afrika-Expedition 1934–1936 und 1937–1939. Strecker and Schröder, Stuttgart, Germany, 358 pp.

Kotz E. 1922. Im Banne der Furcht. Sitten und Gebräucher der Wapare in Ostafrika. Advent-Verlag, Hamburg, Germany, 245 pp.

Kranz W. 1880. Natur- und Kulturleben der Zulus. Sitten und Gebräucher der Wapare in Ostafrika. Advent-Verlag, Wiesbaden, Germany.

Kropf A.A. 1899. A Kaffir-English Dictionary. Lovedale pp. U.K.

Künckel d'Herculais J. 1891. Note sur les populations acridophages en extrême sud de l'Algérie. Bull. Soc. Entomol. France, pp. 24–26.

Labat J-B. (Rév. P.). 1732. Relation historique de l'Ethiopie occidentale contenant la description des royaumes de Congo, Angolle, et Matamba, traduite de l'italien du P. Cavazzi et augmentée de plusieurs relations pratique des auteurs. Charles Jean-Baptiste Delespine, Paris, vol. I, 495 pp.

Laman K.E. 1936. Dictionnaire Kikongo-Français, avec une étude phonétique décrivant les dialectes les plus importants de la langue dite kikongo. Mém. Inst. Roy. Col. Belge, Sect. sci. morales politiques, Bruxelles, 8°, II: 1183 pp.

Lambrechts A. and Bernier G. 1961. Enquête alimentaire et agricole dans les populations rurales du Haut-Katanga (1957–1958). Centre d'Etude des Problèmes sociaux indigènes, coll. Mémoires 11: 236 pp.

Landry S.V., DeFoliart G.R., and Sune M.L. 1986. Larval protein quality of six species of Lepidoptera (Saturniidae, Sphingidae, Noctuidae). J. Econ. Entomol. 79: 600–604.

Lang A. and Tastevin C. 1937. La tribu des Va-Nanyeka. Mission Rohan-Chabot, Angola et Rhodesia. Ethnographie, vol. 5. Imprimerie Crété, Corbeil, France, 213 pp.

Lapp C. and Rohmer J. 1937. Composition et valeur alimentaire du pèlerin (*Schistoceca gregaria*). Bull. Soc. Chimie Biol. 19: 321–324.

Largeau V. 1901. Encyclopédie pahouine, Leroux, Paris.

Latham P. 1999a. Edible caterpillars and their food plants in Bas Congo. Latham P. (ed.). Forneth, U.K., 38 pp.

Latham P. 1999b. Edible caterpillars of the bas Congo region of the Democratic Republic of the Congo. Antenna-London, 23 (3): 134–139.

Latham P. 2000. Chenilles comestibles et leurs plantes hôtes de la Province du Bas-Congo. Latham P. (ed.), Forneth, UK, 40 pp.

Latham P. 2002. Some edible caterpillars and their food plants in Bas-Congo, Democratic Republic of Congo. Latham P. (ed.), Forneth, UK, 49 pp.

Launois-Luong M.H. and Lecoq M. 1989. Vade-mecum des criquets du Sahel. Min. Affaires étrangères des Pays-Bas et Cirad/Privas, coll. Acridologie opérationnelle 5: 125 pp.

Le Clerc A.M., Ramel P., and Acker P. 1976. Note au sujet de la valeur nutritionnelle d'une chenille amimentaire : Anathepanda infracta. Ann. Nutr. Aliment. 21: 69–72.

Lebeuf J-P. 1961. L'habitation des Fali, montagnards du Cameroun spetentrional; technologie, sociologie, symbolisme. Guides bleus, Librairie Hachette, Paris, 608 pp.

Leleup N. and Daems H. 1969. Les chenilles alimentaires du Kwango. Causes de leur raréfaction et mesures préconisées pour y remédier. Agric. Trop. Bot. Appl. 16 (1): 1–21.

Letsie L. 1996. A gendered socio-economic study of Phane. *In:* Gashe B.A. Leggett K., and Mpuchane S.F. (eds.). Phane, Proc. 1st Multidisciplinary Symp.. Dept. Biol. Sci. Gaborone, pp 104–121. ISBN 99912-913-3-4.

Lindblom G. 1969. The Akamba in British East Africa: an Ethnological Monograph. Negro Univ. Press, New York, NY, 607 pp.

Livingstone D. 1857. Missionary Travels and Researches in South Africa, Including a Sketch of Sixteen Years Residence in the Interior of Africa, and a Journey from the Cape of Good Hope to Loanda, on the West Coast: Thence Across the Continent, Down the River Zambezi, to the Eastern Ocean. John Murray, London, UK, 687 pp.

Logan J.W.M. 1992. Termites (Isoptera), a pest or resource for small farmers in Africa. Trop. Sci., 32, 71–77.

Lomer C.J. and Prior C. (eds.) 1992. Biological control of locusts and grasshoppers. C.A.B. Int./ I.I.T.A., 394 pp. ISBN 0 85198 779 6.

MacJannet M.B. 1947. Chokwe-English, English-Chokwe Dictionnary and Grammar Lessons. Vila Luzo, Angola. (2nd ed).

Madan A.C. 1913. Lala-Lamba-Wisa and English, English and Lala-Lamba-Wisa Dictionary. Oxford Univ. Press, Oxford, UK.

Mahauden C. 1965. Kisongokimo (Chasse et magie chez les Balubas). Flammarion, Paris, 253 pp.

Makungwa S.D., Crafter S.A., Awimbo J., and Broekhoven J. 1997. Assessment of grass, honey and edible caterpillars in some forest reserves in Malawi. *In:* Crafter S.A. and Awimbo J. (eds.). Non-timber Forest Products: Value, Use and Management Issues in Africa, Including Examples from Latin America. I.U.C.N., Gland, Switzerland, pp. 37–40.

Malaisse F. 1983. Trophic structure in Miombo (Zambezian tropical woodland). Ann. Fac. Sc. Lubumbashi 3: 119–162.

Malaisse F. 1997. Se nourrir en forêt claire. Approche écologique et nutritionnelle. Presses agronomiques de Gembloux/ Centre Techn. Coop. Agricole et Rurale (Wageningen), 384 pp.

Malaisse F. 2001. Lessons from the Past for a better Future: Ethnolcology, a promising Link between Tradition and Science regarding Biodiversity Management. *In* Science and Tradition: Roots and Wings for Development. A.R.S.O.M./Unesco, Brussels, 89–106.

Malaisse F. 2003. Campeophagy: a state of knowledge report. Geo-Eco-Trop . 26(1): 37–56.

Malaisse F. and Parent G. 1980. Les chenilles comestibles du Shaba méridional (Zaïre). Les Nat. Belges 61 (1): 2–24.

Malaisse F. and Parent G. 1998. Chemical composition and energetic value of some edible products provided by hunting or gathering in the open forest (Miombo). Geo-Eco-Trop, 21 (1–4): 65–71.

Malaisse F. and Lognay G. 2003. Les chenilles comestibles d'Afrique tropicale. *In:* Motte-Florac E. and Thomas J.M.C. (eds): Les Insectes dans la Tradition orale, Ethnoscience, Peeters-Selaf, Paris, 5: 271–295.

Malaisse F. and Roulon-Doko P. 2003. Les chenilles consommées par les Gbaya de Centrafrique (in press.)

Malaisse F., Malaisse-Mousset M., and Evrard A. 1969. Aspects forestiers et sociaux des pullulations de Tunkubiu. Faut-il détruire ou protéger Elaphrodes lactea (Gaede) (Notodontidae) ? Bull. Trim. C.E.P.S.I. 86: 27–36.

Malaisse-Mousset M., Malaisse F., and Watula C. 1970. Contribution à l'étude de l'écosystème forêt claire (Miombo). Note 2: Le cycle biologique d'Elaphrodes lactea (Gaede) et son influence sur l'écosystème "Miombo". Trav. Serv. Sylv. Univ. Off. Congo 8: 3–10.

Malaisse F., Verstraeten C. and Bulaimu T. 1974. Contribution à l'étude de l'écosystème Forêt claire (Miombo). Note 3: Dynamique des populations d'Elaphrodes lactea (Gaede) (Lep. Notodontidae). Rev. Zool. Afr. 88 (2): 286–310.

Malaisse F., Demesmaecker A., Matera J., Wathelet B., and Lognay G. 2003. Enfin "Tubambe" dévoile son identité! *Hadraphe ethiopica* (Bethune-Baker) (Limacodidae), une chenille comestible des forêts claires zambéziennes. Biotechnol. Agron. Soc. Environ. 7 (2): 67–77.

Marais E. 1996. Omaungu in Namibia: *Imbrasia belina* (Saturniidae: Lepidoptera) as a commercial resource. *In:* Gashe B.A., Legett K., and Mpuchane S.F. (eds.) Phane. Proc. 1st Multidisciplinary Symp.. Dept. Biol. Sci. Gaborone, 23–31. ISBN 99912-913-3-4.

Marais E., Kainge P., Muzambani E., and Eiseb S. (n.d.). Insects and Culture in Namibia. Nat. Museum of Namibia, Windhoek (poster).

Marshall L. 1976. The Kung of Nyae Nyae. Harmand Univ. Press, Cambridge, UK, 433 pp.

Masseguin (méd.) and Antonini (pharm.) 1938. Les chenilles comestibles dans la Haute-Sangha. Bull. Soc. Rech. congolaises 25: 133–145.

Mbata K.J. 1995. Traditional Uses of Arthropods in Zambia. Food Insects Newsl. 8 (3): 1, 5–7.

Mbata K.J. and Chidumayo E.N. 1999. Emperor moth caterpillars (Saturniidae = Attacidae) for a snack: Traditional processing of edible caterpillars for home consumption and for sale in the Kopa area of Zambia. Food Insects Newsl. 12 (1), 1–5.

Mbata K.J. and Chidumayo E.N. 2003. Traditional values of caterpillars (Order Lepidoptera) among the Bisa people of Zambia (submitted).

Mbemba F. and Remacle J. 1992. Inventaire et composition chimique des aliments et des denrées alimentaires traditionnels du Kwango-Kwilu au Zaïre. Presses universitaires de Namur, Namur, Belgique, 80 pp.

McCallum T.G. 1993. White woman witch Doctor. Tales from the African life of Rae Graham. Miami, FL, USA, 247 pp.

McGregor J. 1995. Gathered produce in Zimbabwe's communal areas changing resource availability and use. Ecol. Food Nutr. 33: 163–193.

Ménard. 1909. Dictionnaire Français-Kirundi et Kirundi-Français. Roulers, Belgium.

Menzel P. and D'Aluisio F. 1998. Man-eating Bugs. The Art and Science of Eating Insects. Ten Speed Press, Berkeley, CA, USA, 192 pp. ISBN 1-58008-022-7

Merle M. 1958. Des chenilles comestibles. Notes africaines, 77: 20–23.

Merriweather A. 1968. Molepolole mission history. Botswana Notes Records 1: 15–17.

Mertens J. 1935. Les Ba-Dzing de la Kamtsha. Inst. roy. col. belge, Bruxelles, mém. in 8°, Sect. Sci. mor. pol., IV (1): 381 pp.

Mestre J. 1988. Les Acridiens des formations herbeuses d'Afrique de l'Ouest. CIRAD-PRIFAS, Montpellier France 330 pp.

Meyer H. 1916. Die Barundi., Leipzig, Germany.

Mignot J-M. 2003. Les Arthropodes selon les Masa Bugudum (Nord Cameroun) : classificatins, savoirs et usages. Une introduction. Colloque Les "insectes" dans la tradition orale. Villejuif, France (in press).

Milheiros M. 1967. Notas de Etnografia angolana. Instituto de Investigaçao cientifica de Angola, Lunda, 342 pp.

Mkanda F.X. and Munthali S.M. 1994. Public attitudes and needs around Kasungu National Park, Malawi. Biod. Cons., 3: 29–44.

Monteiro, J.J. 1875. Angola and the River Congo. Macmillan and Co., London, UK, I:ix + 305 pp.

Moruakgomo M.B.W. 1996. Commercial utilisation of Botswana's veld products—the economics of phane: the dimensions of phane trade. *In:* Gashe B.A., Legett K., and Mpuchane S.F. (eds.): Phane, Proc. 1st Multidisciplinary Symp. Dept. Biol. Sci., Gaborone, pp. 32–45, ISBN 99912-913-3-4

Motshegwe S.M., Holmback J., and Yeboah S.O. 1998. General properties and the fatty acid composition of the oil from the mophane caterpillar, *Imbrasia belina*. J. Amer. Oil Chem. Soc. 7 (6): 725–728.

Mpuchane S., Taligoola K. and Gashe B.A. 1996. Fungi associated with *Imbrasia belina*, an edible caterpillar. Botswana Note Records 28: 193–197.

Mughogho D.E.C. and Munthali S.M. 1995. Ecological interaction between *Gonimbrasia belina* (L.) and *Gynanisa maia* (L.), and th e impact of fire on forage preferences in a savanna ecosystem. African J. Ecol. 33: 84–87.

Müller E. 1947. Wörterbuch der Djaga-Sprache (Madjame-Mundart) gesprochen am Kilimandjaro in Ostafrika. Beihefte zur Zeitschrift für Eingeboren-Sprachen, Hamburg 25: 411 pp.

Munthali S.M. and Mughogho D.E.C. 1992. Economic incentives for conservation : bee-keeping and Saturniidae caterpillar utilization by rural communities. Biod. Cons. 1: 143–154.

Murdock G.P. 1959. Africa. Its Peoples and Their Culture History. McGraw-Hill Book Co., New York, NY, xiii+456 pp.

Myles T.G. 1998. Proposed Taxonomy of the Order Isoptera, 7 pp. http://www.utoronto.ca/forest/termite/taxon.htm

Newman J.L. 1975. Dimensions of Sandawe diet. Ecol. Food Nutr., 4: 33–39.

Ney N. (éd.). 1984. Conférence et lettres de P. Savorgnan de Brazza sur ses trois explorations dans l'Ouest Africain, de 1875 à 1886. Brazzaville/Heidelberg, Allemagne, 463 pp.

Nkouka E. 1987. Les insectes comestibles dans les sociétés d'Afrique centrale. Muntu 6: 171–178.

Nonaka K. 1996. Ethnoentomology of the Central Kalahari San. African Study Monogr., suppl. 22: 29–46.

Nunes José Antonio Pereira. 1960. A alimentaçao na Baixa de Cassange. Anais Inst. Med. trop., Lisboa, XVII (1/2): 283–435.

Oberprieler R. 1995. The Emperor Moths of Namibia. Ekogilde, Hartbeespoort, R.S.A., ix + 91 pp.

Ohiokpehai O., Bulawayo B.T., Mpotokwane S., Sekwati B., and Bertinuson A. 1996. Expanding the uses of phane, a nutritionally rich local food. *In:* Gashe B.A., Legett K., and Mpuchane S.F. (eds.). Phane, Proc. 1st Multidisciplinary Symp. Dept. Biol. Sc., Gaborone, 84–103. ISBN 99912-91-3-4.

Osmaton H.A. 1951. The termite and its uses for food. Ugandan J. 15: 80–83.

Ouedraogo M., Lognay G., and Malaisse F. 2003. Chenilles comestibles du Burkina-Faso. Geo-Eco-Trop (in press).

Pagezy H. 1975. Les interrelations homme-faune de la forêt du Zaïre. In: Pujol R. (éd.) L'homme et l'animal. Premier colloque d'ethnozoologie. Inst. Int. Ethnosci., Paris, pp. 63–88.

Pagezy H. 1988. Contraintes nutritionnelles en milieu forestier équatorial liées à la saisonnalité et la reproduction: réponses biologiques et stratégies de subsistance chez les Ba-Oto et les Ba-Twa du village Nzalekenga (Lac Tumba, Zaïre). Thèse de doctorat, Univ. de Droit, d'Economie et des Sciences d'Aix-Marseille, Fac. des Sciences et Techniques de Saint Jérôme, 257 pp.

Palmer E. and Pitman N. 1972. Trees of Southern Africa covering all known indigenous species in the Republic of South Africa, South-West Africa, Botswana, Lesotho and Swaziland. A.A. Balkema, Cape Town, R.S.A., vol. 1: 703.

Paulian R. 1946. Eléments d'un vocabulaire zoologique des dialestes forestiers de basse Côte d'Ivoire. J. Soc. Africanistes XVI: 23–28.

Paulian R. 1963. Coeliades libeon Druce, chenille comestible du Congo. Bull. Inst. rech. Sci. Congo 2: 5–6.

Pearce M.J. 1997. Termites. Biology and Pest Management. CAB Int. Oxon UK, 172 pp. + 32 plates. ISBN 0 85199 130 0

Pelling J.N. 1966. A Practical Ndebele Dictionary. Daystar Publications, Bulawayo, Rhodesia.

Phelps R.J., Struthers J.K., and Moyo S.J.L. 1975. Investigations into the nutritive value of *Macrotermes falciger* (Isoptera: Termitidae). Zool. Afr. 10: 123–132.

Pinhey E. 1972. Emperor Moths of South and South Central Africa. Struik, Cape Town, R.S.A. 150 pp. ISBN 0 86977 016 0

Pinhey E. 1975. Moths of Southern Africa. Tafelberg, Cape Town, 273 pp. R.S.A., ISBN 0 624 00784 7

Pinto Serpa. 1881. Comment j'ai traversé l'Afrique depuis l'Atlantique jusqu'à l'Océan indien à travers des régions inconnues. Librairie Hachette et Cie, Paris, I:xxx + 456 pp.

Platt B.S. 1980. Tables of representative values of food commonly used in tropical countries. HMSO, Mecial Res. Council, London, Special rept. ser. 302: v + 46 pp.

Platt E.E. 1921. List of foodplants of Some South African lepidopterous larvae. S. Afr. J. Nat. Hist. 3 (1): 65–138.

Pogge P. 1880. Im Reiche des Muata Jamwo. Tagebuch meiner im Auftrage der Deutschen Gesellschaft zur Erforschung Aequatorial-Afrika's in die Lunda-Staaten unternommenen Reise. Dietrich Reiner, Berlin, 247 pp.

Pomeroy D.E. 1976. Studies on a population of large termite mounds in Uganda. Ecol. Entomol. 1: 49–61.
Quin P.J. 1959. Foods and feeding habits of the Pedi with special reference to identification, classification, preparation and nutritive value of the respective foods. Witwatersrand Univ. Press, Johannesburg, R.S.A. xvii + 278 pp, 134 plates.
Ramos-Elorduy (de Conconi) J. 1991. Los insectos como fuente de proteínas en el futuro. Editorial Limusa, Mexico, 148 pp.
Ramos -Elorduy J. 2004. Insects: A hopeful food (this volume).
Redford K.H. and Dorea J.G. 1984. The nutritional value of invertebrates with emphasis on ants and termites as food for mammals. J. Zool. 203: 385–395.
Rehse H. 1910. Kiziba, Land und Leute. Strecker und Schröder, Stuttgart, Germany.
Retel-Laurentin A. 1986. Contes du pays Nzakara (Centrafrique). Karthala, Paris, 310 pp.
Reynolds B. 1968. The material culture of the peoples of the Gwembe valley. *In:* Kariba Studies. Manchester Univ. Press, Manchester, UK, III: 262 pp.
Richards A.I. 1939. Land, labour and diet in Northern Rhodesia: An economic study of the Bemba tribe. Oxford University Press, Int. African Inst., Oxford, UK, 425 pp.
Rita-Ferreira A. 1966. Os Cheuas de Macanga. Memórias do Instituto de Investigação cientifíca de Moçambique, Lourenco Marques, Mosambique, 8 (c): 332 pp.
Roscoe J. 1911. The Baganda, an Account of Their Native Customs and Beliefs. MacMillan, London, UK, xix + 547 pp.
Rougeot P-C. 1955. Les Attacides (Saturnidae) de l'Equateur Africain Français. MackMillan, Encyclopédie entomologique. Ed. Paul Lechevalier, Paris, XXXIV: 116 pp. + xii plates.
Roulon-Doko P. 1980. Le savoir des Gbaya de Centrafrique sur les chenilles comestibles, leur comportement et leur habitat. Bull. de liaison de la SEZEB (Paris), 8: 20–25.
Roulon-Doko P. 1992. La collecte des termites chez les Gbaya 'bodoe de la savane Centrafricaine. Ecol. Humaine (Laboratoire d'Ecologie humaine, Aix-en-Provence, France), X (2): 41–54.
Roulon-Doko P. 1998. Chasse, cueillette et cultures chez les Gbaya de Centrafrique. L'Harmattan, Paris, 540 pp.
Roulon-Doko P. and Joulian F. 1994. Comparaison d'une activité technique chez les hommes et chez les chimpanzés: la collecte des termites. Tech. Cult. 23–24: 29–62.
Ruelle J.E. 1970. A revision of the termites of the genus *Macrotermes* from the Ethiopian Region (Isoptera : Termitidae). Bull. Br. Mus. nat. Hist. (Ent.) 24: 365–444.
Sanderson G.M. (n.d.). A Dictionary of the Yao Language. Government Printer, Zomba, Nyassaland.
Santos Oliveira J.F., Passos de Carvalho J., Bruno de Sousa R.F.X,. and Madalena Simao M. 1976. The nutritional value of four species of insects consumed in Angola. Ecol. Food Nutr. 5: 91–97.
Schebesta P. 1940. Les Pygmées. Traduit de l'allemand par F. Berger. Gallimard, Paris, 199 pp.
Scherer J.H. 1959. The Ha of Tanganyika. Anthropos 54: 841–904.
Schmitz R. 1912. Les Baholoholo (Congo belge). Collection Monographies ethnographiques, Albert De Wit, Institut Int. Bibliographie, Bruxelles, Belgium IX: xxxii + 605 pp. + ix plates + 1 map.
Scholtz C.H. 1982. Trophic ecology of Lepidoptera larvae associated with woody vegetation in a savanna ecosystem. S. Afr. Nat. Sci. Prog. Rept. 55: v + 25 pp.
Schweinfurth G. 1875. Au cœur de l'Afrique: 1868–1871. Voyages et découvertes dans les régions inexplorées de l'Afrique. Trad. H. Loreau, Hachette et Cie, Paris, 508 pp.
Scudder T. 1962. The ecology of the Gwembe Tonga. *In:* Kariba Studies, Manchester Univ. Press, Manchester, UK, 2: 274 pp.
Seignobos C., Deguine J-P. , and Aberlenc H-P. 1996. Les Mofu et leurs insectes. J. d'Agric. trad. et de Bota. Appl. 38 (2): 125–187.
Sekhwela M.B.M. 1989. The nutritive value of Mophane bread – Mophane insect secretion (Mophote or Maboti). Botswana Notes, Records 20: 151–154.
Sevy G.V. 1972. *T*erre ngbak : étude de l'évolution de la culture matérielle d'une population forestière centrafricaine. SELAF, Paris, 416 pp.
Siame A.B., Mpuchane S.F., Gashe B.A., Allotey J., and Teferra G. 1996. Nutritional qualities of mophane worms, *Imbrasia belina* (Westwood) and the microorganisms associated with the

worms. *In:* Gashe B.A., Legett K., and Mpuchane S.F. (eds.). Phane, Proc. 1st Multidisciplinary Symp. Dept. Biol. Sc., Gaborone, pp. 80–83. ISBN 99912-913-3-4.

Silberbauer G.B. 1972. The G/wi Bushmen. *In:* Bicchieri M.G. (ed.). Hunters and Gatherers Today. Holt, Rinehart and Wilston, New York, NY, pp. 271–326.

Silow C.A. 1976. Edible and other insects of Mid-Western Zambia—Studies in Ethno-entomology II. (Sweden): Institutionen för Allmän och Jämförande Ethnografi, Uppsala, Sweden, Occ. Papers V: 223 pp.

Silow C.A. 1983. Notes on Ngangela and Nkoya Ethnozoology: ants and termites. Etnografiska Musuem, Göteborg, Sweden, Ethnol. Stud. 36: vii+177 pp.

Smeathman H. 1781. Some account of the termites which are found in Africa and other hot climates. Phil. Trans. Roy. Soc. LXXI: 139–192.

Smith E.W. and Dale A.M. 1968. *The Ila-speaking Peoples of Northern Rhodesia*. University Books, New-Hyde Park, New-York, 423 + 433 pp.

Snoxall R.A. 1967. Luganda-English Dictionary. Clarendon Press, Oxford UK, xxxvi + 357 pp.

Southwood T.R.E. 1977. Entomology and mankind. Amer. Scient. 65: 30–39.

Spellig F. 1927. Die Wanjamwesi. Ein Beitrag zur Völkerkunde Ostafrikas. Zeitschrift Ethnol. 59: 201–252.

Stannus H.S. 1910. Notes on some tribes of British Central Africa. J. Roy. Anthropol . Inst. 40: 285–336.

Stayt H.A. 1931. The Bavenda. Oxford Univ. Press, London, UK, 392 pp.

Steinhardt D. 1922. Von wehrhaften Reisen und seinem Reiche. Hamburg, Germany (3rd ed.).

Stirnimann H. 1976. Existenzgrundlagen und traditionelles Handwerk der Pangwa von SW.-Tansania. Studia Ethnographica Friburgensia 4: 311 pp.

Styles C. 1994. There's big value in mopane worms. Farmer's Weekly, July 22: 1999 20–23 pp.

Styles C.V. 1995. The mopane worm: too profitable for its own good? Our Living WorldVol. 28.

Styles C.V. 1996. The biological ecology of *Imbrasia belina* (Saturniinae, Saturniidae) with reference to its behaviour, physiology, population dynamics, impact within mopane veld and utilisation within South Africa. *In:* Gashe B.A., Legett K., and Mpuchane S.F. (eds.). Phane, Proc. 1st Multidisciplinary Symp.. Dept. Biol. Sc., Gaborone, pp. 9–13. ISBN 99912-913-3-4.

Takeda J. 1990. The dietary repertory of the Ngandu people of the tropical rain forest : an ecological and anthropological study of the subsistence activities and food procurement technology of a slash-and-burn agriculturist in the Zaire River basin. Afric. Study Monogr., Suppl. 11: 1–75.

Tango Muyay. 1981. Les insectes comme aliment de l'homme. CEEBA Publ., Série II **69**, Bandundu, Zaïre, 177 pp.

Tessmann G. 1913. Die Pangwe.Völkerkundliche Monographie eines west-afrikanischen Negerstammes. E. Wasmuth, Berlin, Band I : xxi + 275 pp., Band II : 422 pp.

Tessmann G. 1934. Die Bafia, und die Kultur der Mittel Kamerun-Bantu. Strecker and Schröder,, Stuttgart, 270 pp.

Thomas I. 1975. Tribes. *In:* Berry L. (ed.). Tanzania in Maps, University of London Press, London, 112–113, 168.

Thompson B.P. 1954. Two studies in African nutrition. An urban and a rural community in Northern Rhodesia. Rhodes-Livingstone Pap. 24: 1–57.

Tihon L. 1946a. Contribution à l'étude du problème alimentaire indigène au Congo Belge. Bull. Agric. Congo Belge 37 (4): 829–868.

Tihon L. 1946b. A propos des termites au point de vue alimentaire. Bull. Agric. Congo Belge 37 (4): 865–868.

Tönjes H. 1910. Wörterbuch der Ovambosprache. Lehrbucher des Seminars für Orientalishische Sprachen, Berlin.

Torday E. 1925. On the trail of the Bushongo. Seeley, Service and Co. London, UK. 286 pp.

Torday E. and Joyce T.A. 1905. Notes on the Ethnography of the Ba-Mbala. J. Roy. Anthropol. Inst., vol. 35.

Torrend J. 1967. An English-Vernacular Dictionary of the Bantu-Botatwe Dialects of Northern Rhodesia. Gregg Press, Farnborough, UK. 649 pp.

Tosi E. and Daccordi M. 1983. Insetti commestibili. Slurp! Nigrizia, 101 (6): 49–52.

Trilles R.P. 1932. Les Pygmées de la forêt équatoriale. Librairie Bloud et Gay, Paris, 521 pp.

Trouwborst A.A. 1962. Le Burundi. Musée royal Afrique Centrale. Ann., Sc. hum., Tervuren, Monogr. ethnogr. 6: 113–169.

Tsala T. and Vincent J-F. 1969. Mille et un proverbes béti: Minkana beti, minted Awon Ai Fus, ou la société béti à travers ses proverbes. Yaoundé, 413 pp. (mimeo)

Turk D. 1990. Leguminous trees as forage for edible caterpillars. Nitrogen-Fixing-Tree-Research-Reports 8: 75–77.

Turnbull C. 1973. Les Iks: survivre par la cruauté au Nord-Ouganda. In: Terre humaine, Plon, Paris, 389 pp.

Turnbull C.M. 1972. The Mountain People. Jonathan Cape, London, UK, 309 pp.

Ukhun M.E. and Osasona M.A. 1985. Aspects of the nutritional chemistry of Macrotermes bellicosus. Nutr. Repts. Int. 32 (5): 1121–1130.

Valentin P. 1968. Termiten—geröstet oder as Suppe. Kosmos, Bild unsener Welt, 10 (68): 24–25.

Van Avermaet E. and Mbuya B. 1954. Dictionnaire Kiluba-Français. Ann. Musée Royal Congo Belge, Linguistique 7: 838 pp.

Van den Berg M.A. 1974. Biological studies on Cirina forda (West.) (Lepidoptera : Saturniidae), a pest of wild seringa trees (*Burkea africana* Hook.). Phytophylactica 6: 61–62.

Van den Berg M.A., Catling H.D., and Vermeulen J.B. 1973. The distribution and seasonal occurrence of Saturniidae (Lepidoptera) in Transvaal. Phytophylactica 5: 111–114.

van der Waal B. 1996. Importance of grasshoppers as traditional food in villages in northern Transvaal, South Africa. *In:* Jain S.K. (ed.). Ethnobiology in Human Welfare. Deep Publ., New Delhi, pp. 35–41.

van der Waal B. 1999. Ethnobiology and uses of grasshoppers in Venda, Northern Province, South Africa. S. Afr. J. Ethnol. 22 (2): 103–109.

Van Dyk J. 2000. Nutritional value of various insects per 100 grams. >http://www.ent.iastate.edu/misc/insectnutrition.html<

Van Huis A. 1996. The traditional use of arthropods in sub-Saharan Africa. Proc. Exper. Appl. Entomol., N.E.V. Amsterdam 7: 3–20.

Vansina J. 1964. Le royaume Kuba. Musée royal Afrique centrale, Ann. in 8°, Sci. Humaines, Monographies ethnographiques, 49: ix + 196 pp.

Velcich G. 1963. Mopani worms. BaNtu, 10: 604–605.

Vergiat A-M. 1937. Mœurs et coutumes des Manjas. Payot, Paris, 323 pp.

Villiers A. 1947. Une manne africaine: les termites. 239–240.

Villiers A. 1957. Les Lépidoptères de l'Afrique Noire Française. I.F.A.N., Dakar, Initiations africaines, XIV (1): 84 pp.

von Berensberg H.P. 1907. Uses of insects as food, delicacies, medicines or in manufactures. Natal Agricultural J. and Ministry Records, 10: 757–762.

Walker N.J. 1995. Late Pleistocene and Holocene huner-gatherers of the Matopos. Societas Archaeologia Upsaliensis, Uppsala, Studies in African Archaeology 10: 284 pp.

Warnier J.P. 1985. Echanges, développement et hiérarchies dans le Bamenda précolonial (Cameroun). Studien zür Kulturkund, 76: 323 pp. Franz Steiner Verlag, Wiesbaden, Germany.

Waterhouse G. 1932. Simon van der Stel's Journal of Expedition to Namaqualand 1685–1686. Green, Dublin, Ireland, xxviii + 85 pp.

Wessmann R. 1908. The Bawenda of the Spelonken (Transvaal): a Contribution towards the Psychology and Folklore of African Peoples. London, 154 pp. (rev. ed. 1969, New York, NY)

White C. 1959. A preliminary survey of Luvale rural economy. Rhodes-Livingstone Papers 9: 1–58.

White F. 1983. The Vegetation of Africa. A descriptive memoir to accompany the Unesco/AEFAT/UNISO vegetation map of Africa, 356 pp. ISBN 92-3-101955-4

White Fathers. 1954. Bemba-English Dictionary. Longmans, Green & Co., London, 1505 pp.

Yagi S. 1997. Case study of edible insects in Enzaro Village, Kenya. >http://ss.jircas.affrc.go.jp/engpape/annualreprot/1997/divisions/cpphdivision1.htm<consulted ii.2001

Zézé-Béké P. 1989. Les interdits alimentaires chez les Nyabwa de Côte-d'Ivoire. J. des Africanistes 59 (1–2): 229–238.

Zinzombe, I.M. and George S. 1994. Larval lipid quality of Lepidoptera: Gonimbrasia belina. Botswana Notes/Records 26: 167–173.

11

Insects Eaten in Africa (Coleoptera, Hymenoptera, Diptera, Heteroptera, Homoptera)

Arnold Van Huis

Laboratory of Entomology, Wageningen University, P.O. Box 8031, 6700 EH Wageningen, The Netherlands. E-mail: Arnold.vanHuis@wur.nl

Abstract

In sub-Saharan Africa 246 insect species are eaten, of which 30% belong to Lepidoptera, 29% to Orthoptera, and 6% to Isoptera. The other 35% include Coleoptera (19%), Homoptera (7%), Hymenoptera (5%), Heteroptera (3%), Diptera and Odonota (1%). The most important beetle species eaten is the larvae of the palm weevil (*Rynchophorus phoenicis*), which is considered a delicacy throughout humid Africa. Harvesting of the beetle larvae can be timed by cutting palms or trees several weeks carlier. From Hymenoptera, queens of the ant *Carebara vidua* are consumed. Other ant species are often used to flavor dishes. Bee larvae are often eaten together with the harvested honey. Pregnant women in Africa consume clay not only from termite mounds but also from nests of the mud-dauber wasp, *Synagris* sp. This geophagy provides them with the minerals and trace elements necessary for fetal growth. From Diptera, the swarming lake fly *Chaoborus edulis* is made into a cake, which is very rich in protein and iron. From Heteroptera, the pungent *Natalicola delegorguei* is eaten in southern Africa and *Agonoscelis versicolor* in Sudan from which an edible oil is derived. Of the Homoptera, a number of cicada species are eaten. Excretions from the psyllid *Arytaina mopane* are known as mopane bread or lerp, which is collected as food in southern Africa. The waxy substance of another homopteran, the flattid *Phromnia rubra* is eaten in Madagascar. Possibilities are mentioned to stimulate the consumption of edible insects in Africa.

Key Words: Africa, ants, beetles, bugs, edible insects, food insects, entomophagy, flies, geophagy, honey, Madagascar, nutrition, Africa, weevils.

Introduction

The use of traditional food is sustainable and has economic, nutritional and ecological benefits for rural communities in sub-Saharan Africa. A total of 246 insect species (most at species and only a few at genus or family level) are edible in sub-Saharan Africa (Van Huis, 2003A). Of these, the majority are Lepidoptera (30%), Orthoptera (29%), and Coleoptera (19%); below 10% are Homoptera (7%), Isoptera (6%), Hymenoptera (5%), Heteroptera (3%), Diptera and Odonota (1%) (Fig. 11.1). Here only those insect orders are discussed in which the following number of edible species were found: Coleoptera (48), Homoptera (17), Hymenoptera (13), Hemiptera (7), Diptera (2) and Odonata (1), and a few other arthropod species. All representatives from Africa encountered in these insect orders (or taxa) are listed in Table 11.1.

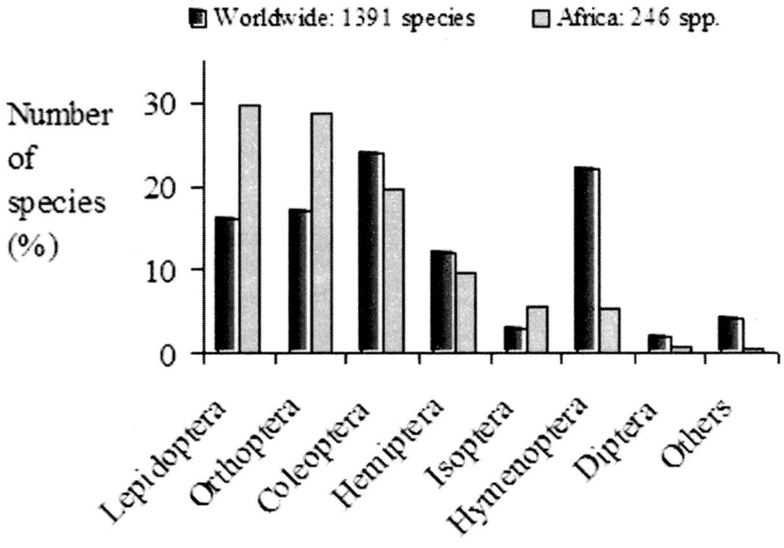

Fig. 11.1: Percentage of insect species per order eaten worldwide (Source: Ramos-Elorduy, 1997) and in sub-Saharan Africa (van Huis, 2003A).

Caterpillars, termites, and lake flies have a high protein content, and termites and beetle larvae a high fat content (Table 11.2). Among micronutrients, Fe is very high in lake fly cake and moderately so in caterpillars, ants, and palm weevil larvae. Vitamin B_1 and B_2 are highest in lake fly cake and palm weevil larvae. The bioefficiency of micronutrients, notably iron, is markedly higher in insects than in vegetables, which is important given the fact that 50% of the African children suffer from anemia (ACC/SCN, 2000).

Table 11.1: List of arthropod species consumed in different countries of sub-Saharan Africa (adapted from Van Huis, 2003)

Order/Family	Species	Countries and Reference
Coleoptera		
Curculionidae	*Eugnoristus monachus* Oliv.	Madagascar (Bodenheimer, 1951)
	Rhyna sp.	Madagascar (Bodenheimer, 1951)
	Rhynchophorus sp.	Madagascar (Bodenheimer, 1951)
	Rhynchophorus phoenicis Fabr.	Angola (Santos Oliveira et al., 1976), Cameroon (Bodenheimer, 1951), Congo (Bani, 1995; Nkouka, 1987; Takeda, 1990), Nigeria (Fasoranti and Ajiboye, 1993)
	Polyclaeis equestris Boheman	South Africa (Quin, 1959)
	P. plumbeus Guerin	South Africa (Quin, 1959)
	Sipalinus aloysii-sabaudiae Camerano	Tanzania (Bodenheimer, 1951; Harris, 1940)
Carabidae	*Scarites* sp.	Angola (Bergier, 1941), Madagascar (Decary, 1937)
Buprestidae	*Sternocera funebris* Boheman	Zimbabwe (Chavanduka, 1976; Gelfand, 1971)
	S. interrupta Oliv.	Cameroon (Seignobos et al., 1996)
	S. orissa Buq.	Botswana (Nonaka, 1996), South Africa (Bodenheimer, 1951; Quin, 1959), Zimbabwe (Chavanduka, 1976; Gelfand, 1971)
Cerambycidae	*Acanthophorus maculatus* Lameere	Zambia (Mbata, 1995)
	A. capensis White	Zambia (Mbata, 1995)
	A. confinis Laporte	Zambia (Mbata, 1995)
	Ancylonotus tribulus Fabr.	Gabon and Senegal (Netolitzky, 1919), West Africa (Bergier, 1941)
	Ceroplesis burgeoni Breuning	southern Africa (Malaisse, 1997)
	Dorysthenes forficatus Fabr.	North Africa (Ghesquière, 1947)
	Mallodon downesi Hope	Central Africa (Bergier, 1941), South Africa (Bodenheimer, 1951)
	Macrotoma edulis Karsch	Sao Tomé and Principe (Netolitzky, 1919)
	M. natala Thomson	Botswana (Roodt, 1993)
	Petrognatha gigas Fabr.	Gabon (Bergier, 1941), Senegal (Netolitzky, 1919)
	Plocaederus frenatus Fåhraeus	Central Africa (Bergier, 1941)
	Pycnopsis brachyptera Thomson	D.R. Congo (Malaisse, 1997)
	Sternotomis itzingeri katangensis Allard	D.R. Congo (Malaisse, 1997)
	Zographus aulicus Bertolini	D.R. Congo (Malaisse, 1997)
Dytiscidae	*Cybister hova* Fairm.	Madagascar (Decary, 1937)
Elateridae	*Tetralobus flabellicornis* Linn.	Central Africa (Bodenheimer, 1951)
Lucanidae	*Cladognathus serricornis* Latr.	Madagascar (Decary, 1937)
Passalidae	Unknown sp.	Madagascar (Bodenheimer, 1951)
Scarabaeidae		
Cetoniinae	*Goliathus cacicus* Voet	Central Africa (Bergier, 1941)
	G. regius Klug	Central Africa (Bergier, 1941)

Table 11.1: (Contd.)

Table 11.1: (Contd.)

Order/Family	Species	Countries and Reference
	G. cameronensis	Central Africa (Bergier, 1941)
	G. goliathus Drury	Central Africa (Bergier, 1941)
Dynastinae	*Augosoma centaurus* Fabr.	Cameroon (Bodenheimer, 1951), Congo (Bani, 1995; Nkouka, 1987), D.R. Congo (Takeda, 1990)
	Oryctes boas Fabr.	Congo (Bani, 1995; Nkouka, 1987), Nigeria (Fasoranti and Ajiboye, 1993), South Africa (Bergier, 1941; Netolitzky, 1919)
	O. monocerus Oliv.	South Africa (Bergier, 1941; Netolitzky, 1919)
	O. nasicornis Linn.	Madagascar (Bergier, 1941)
	O. owariensis Palisot	Congo (Bani, 1995; Nkouka, 1987), South Africa (Bergier, 1941; Netolitzky, 1919)
Melolonthinae	*Lepidiota mashona* Arrow	Zimbabwe (Chavanduka, 1976; Gelfand, 1971; Weaving, 1973)
	L. anatina Brenske	Zimbabwe (Chavanduka, 1976)
	L. nitidicollis Kolbe	Zimbabwe (Chavanduka, 1976)
	Proagosternus sp.	Madagascar (Decary, 1937)
	Tricholespis sp.	Madagascar (Decary, 1937)
Rutelinae	*Popillia femoralis* Klug	Cameroon (Bodenheimer, 1951)
Scarabaeinae	*Pachylomera femoralis* Kirby	Zambia (Mbata, 1995)
Trichiinae	*Platygenia* spp.	Africa (Ghesquière, 1947)
	P. barbata Afzelius	D.R. Congo (Adriaens, 1951)
Tenebrionidae	xylophagous insects	Congo (Nkouka, 1987)
Diptera		
Chaoboridae	*Chaoborus edulis* Edwards	East African lakes (Bergier, 1941; Owen, 1973), Tanzania (Bodenheimer, 1951; Harris, 1940), Uganda, Kenya (Van Huis, this volume)
	C. pallidipes Theob.	Uganda (Bergier, 1941)
	Chaoborus sp.	East African lakes (Bergier, 1941)
Hemiptera/Heteroptera		
Belastomatidae	*Belastoma* sp.	Congo (Bani, 1995; Nkouka, 1987)
Coreidae	*Petascelis remipes* Signoret	Zimbabwe (Chavanduka, 1976; Weaving, 1973)
	P. wahlbergi Stål	Zimbabwe (Chavanduka, 1976)
Nepidae	*Nepa* sp.	Madagascar (Decary, 1937)
Pentatomidae:	*Agonoscelis versicolor* Fabr.	Sudan ((Van Huis, this volume), Delmet , 1975)
Tessaratomidae	*Natalicola delegorguei* Spin.	South Africa (Faure, 1944), Zimbabwe (Chavanduka, 1976; Bodenheimer, 1951)
	N. pallidus Westwood	Zimbabwe (Van Huis, this volume; Weaving, 1973)
Hemiptera/Homoptera		
Cicadidae	*Afzeliada* sp.	Congo (Nkouka, 1987)
	Afzeliada afzelii Stål	D.R. Congo (Malaisse, 1997)
	Afzeliada duplex Diabola	D.R. Congo (Malaisse, 1997)
	Ioba horizontalis Karsch	D.R. Congo (Malaisse, 1997)
	I. leopardina Distant	D.R. Congo (Malaisse, 1997), Zambia (Mbata, 1995), Zimbabwe (Chavanduka, 1976; Malaisse, 1997)

Table 11.1: (Contd.)

Table 11.1: (Contd.)

Order/Family	Species	Countries and Reference
	Monomatapa insignis Distant	Botswana (Roodt, 1993)
	Munza furva Distant	D.R. Congo (Malaisse, 1997)
	Orapa sp.	Botswana (Roodt, 1993)
	Platypleura adouma Distant	Congo (Nkouka, 1987)
	P. stridula Linn.	Zambia (Mbata, 1995)
	Sadaka radiata Karsch	D.R. Congo (Malaisse, 1997)
	Ugada limbalis Karsch	Congo (Nkouka, 1987), D.R. Cong (Malaisse, 1997), Zambia (Mbata, 1995)
	U. giovanninae Boulard	Congo (Nkouka, 1987)
	U. limbimaculata Karsch	Congo (Nkouka, 1987), D.R. Congo (Malaisse, 1997)
Flatidae	*Phromnia rubra* Signoret[1]	Madagascar (Decary, 1937)
Fulgoridae	*Pyrops tenebrosa* Fabr.	Madagascar (Decary, 1937)
Psyllidae	*Arytaina mopane* Pettey[1]	Botswana (Sekhwela, 1988), Zimbabwe (Weaving, 1973)
Hymenoptera		
Apidae	*Apis adansoni* Latr.	D.R. Congo (Takeda, 1990), Tanzania (Harris, 1940), Zambia (Mbata, 1995)
	Apis mellifira Linn.	Senegal (Gessain and Kinzler, 1975), Zambia (Mbata, 1995)
	Dactylurina staudingeri Gribodo	D.R. Congo (Takeda, 1990)
	Trigona sp.	D.R. Congo (Takeda, 1990)
	T. ferruginea gambiensis Moure	Senegal (Gessain and Kinzler, 1975)
	T. occidentalis Darchen	Senegal (Gessain and Kinzler, 1975)
	T. ruspolii Mag.	Senegal (Gessain and Kinzler, 1975)
	T. senegalensis Darchen	Senegal (Gessain and Kinzler, 1975)
Formicidae	*Anomma nigricans* Illiger (eggs)	Cameroon (Van Huis, this volume)
	Oecophylla spp.	Cameroon, Congo (Bani, 1995; Nkouka, 1987)
	Camponotus sp.	Botswana (Nonaka, 1996)
	Carebara vidua Smith	South Africa (Bodenheimer, 1951; Quin, 1959), Zambia (Mbata, 1995; Silow, 1983), Zimbabwe (Chavanduka, 1976; Jackson, 1954 *in* Gelfand, 1971)
	C. lignata Westwood	southern Africa (Van Huis, this volume)
Odonata	larvae	Madagascar (Decary, 1937)
	adults	Nigeria (Bodenheimer, 1951)
	Trithemis arteriosa Burmeister	D.R. Congo (Malaisse, 1997)
Arachnida	*Epeira nigra* Vinson	Madagascar (Decary, 1937)
	Nephila madagscariensis Vinson	Madagascar (Decary, 1937)

For explanations: see Materials and Methods.

[1] From these homopterans the secreted product is eaten (that of *Arytaina mopane* is called "lerp").

Table 11.2: Protein, fat, iron, and vitamins B_1 and B_2 content of some edible insects in Africa

Insect	Protein	Fat g/100 g	Fe g/100 g	Vitamins (mg/100 g) B1	B2
Lepidoptera					
Mopane worm (dried)[1]	57	16	31	0.6	2.0
Isoptera					
Mature alates	38	46	8	0.1	1.2
Coleoptera					
Palm weevil larvae[2]	20	42	15	3.4	2.5
Hymenoptera					
Ants[3]	17	4	22	0.4	1.0
Diptera					
Lake fly cake[4]	49	10	78	1.5	4.1

[1]*Gonimbrasia belina*; [2] *Rhynchophorus phoenicis*; [3] *Oecophylla* spp.; [4] *Chaoborus edulis*
Source: Bukkens, 1997.

Materials and Methods

Information was collected from published literature and through collation of personal interviews. Interviews were conducted in 1995 and 2000 by and large with scientifically trained personnel. Some of the results obtained in 1995 have been published (Van Huis, 1996). In total, 308 persons from 27 countries in West, East and Southern Africa were interviewed. To avoid misunderstandings about the identity of the arthropod species or taxa, most of the people interviewed had necessarily to be scientists or technicians trained in entomology. Doubts were resolved with pictures from books or insect museum collections.

A number of insects are utilized for medicinal purposes. These practices are not included here although certain insect products which are eaten (honey, termite soil, lerp, vespid clay), are occasionally mentioned. The use of insects as food for livestock has also not been included.

For the taxonomic status (order, family) of the insect species mentioned here see Table 11.1. Bergier (1941), Bodenheimer (1951), Netolitzky (1919) and Silow (1976) are referenced when they cite very early authors from difficult-to-access sources. Valid (most recent) species names are used, and only for some well-known species, are old names added in parentheses. Contrary to the convention of the International Code of Zoological Nomenclature (ICZN, 1999), author names are not parenthesized to indicate a species and genus combination different from that used for the original description. To ascertain the historical status of names and accurate use of parentheses would require an inordinate amount of bibliographic searching. For some insect species encountered in the literature an authority name could not be found. Subfamilies are indented in the table.

Entomophagy Per Insect Order

The various practices of entomophagy are treated per insect order, starting with Coleoptera, then Hymenoptera and Diptera, followed by Heteroptera (Hemiptera and Homoptera). Dragonflies and some spider species are consumed sporadically.

Coleoptera

Larvae of the palm weevil *Rynchophorus* spp. (Col.: Curculionidae) are eaten in Asia, Africa, and the Americas (Plate V, 5). The African species is *R. phoenicis*. The main host plant of the weevil is the oil palm (*Elaeis guineensis*) but it can also reproduce in other palms (*Borassus, Calamus, Hyphaene, Phoenix, Raphia*) (Malaisse, 1997). The female beetle lays her eggs in wounds made by human activity in the traditional tapping of palms for wine, or those made by other insects, in particular *Oryctes* spp. Trees under stress and fallen palms also serve as breeding sites. Because the larvae taste so good, there is a danger of indiscriminate felling of trees, which would deprive the community of palm products. Therefore, in Nigeria, children may be discouraged from eating larvae by taboos invoked by elders (Fasoranti and Ajiboye, 1993). The larvae enter the trunk and make tunnels. Mature larvae are collected, washed and fried or grilled without adding oil, as they are high in fat. Condiments include onion, pepper and a little salt.

People in the D.R. Congo have a unique way of ascertaining the appropriate moment for harvesting edible larvae of weevil, long-horned, and scarab beetles, which occur in standing or rotting palm trees of *Elaeis, Raphia, Chamaerops* and *Cocos nucifera* trees (Ghesquière, 1947). They lay their ears against the palm tree, listening for sounds of a nibbling beetle. In Cameroon my informants indicated that this method is used to harvest the correct instar of *R. phoenicis* larvae from the palm tree. Roulon-Doko (1998) for the Central African Republic, Santos Oliveira et al. (1976) for Angola, and Takeda (1990) for the D.R. of Congo have also mentioned gatherers listening at tree trunks for collection of beetle larvae.

Larvae of *Oryctes* spp. breed in decomposing organic matter, including dunghills. Upon collection the last four abdominal segments, containing most of the gut contents, are removed before processing (Fasoranti and Ajiboye, 1993). The larvae are thoroughly washed before frying and condiments added to taste. The Gbaya in the Central African Republic earlier collected dung beetles by searching the excreta of buffaloes. However, in the 1970s cows replaced buffaloes and the beetles are no longer collected (Roulon-Doko, 1998).

Roulon-Doko (1998) mentioned 16 edible beetle larvae collected by the Gbaya in the Central African Republic. However, only five are deliberately looked for by adults, another five by children, and the rest collected when accidentally found. Decary (1937) mentioned that in Madagascar beetle larvae are eaten fried. Apparently they were very costly and affordable only by the rich. To preclude endangering irrigation, searching for them in the bunds of rice fields was prohibited and hence people had to go long distances to collect them.

Roulon-Doko (1998) indicated that to collect cerambycids, *Burkea africana* or "trees of *Burkea Africana*" trees around the village are felled. A few days later the cerambycids arrive to attack the tree, and the person waiting near the trunk gathers them. From flower beetles or cerambycids, both the larvae and the adults are eaten, often grilled. From adult beetles the wings and the legs are removed (Nonaka, 1996). Larvae are also prepared by macerating in a mortar and pestle with various condiments and cooking them (Bergier, 1941: p. 61).

The San in the central Kalahari collect the buprestid beetle *Sternocera orissa* in January during outbreaks from the leaves of *Acacia mellifera* and *Kyllinga alba* (Nonaka, 1996). The quarry is roasted in hot ash and sand and the hind wings then removed. The heads are picked off if eaten directly. The insect may also be pulverized and mixed with fruits or wild plants to form a paste. Females containing eggs are particularly relished.

Hymenoptera

In southern and East Africa the queens of the thief ant *Carebara vidua* are eaten. These ants live in the earthen walls separating the roomy galleries and chambers of termites. The workers, just one mm in length, can slip unobtrusively in and out of the termite nest chambers, feeding on the termite eggs and larvae. The queens are vastly larger than the workers and measure up to 25 mm, the weight differential being a factor 4,000 (Wheeler, 1927; Hölldobler and Wilson, 1990). Because of her size she cannot care for her eggs and feed her tiny young when she begins a new colony. As compensation for this handicap, she is equipped with dense tufts of hair on her legs, to which during her nuptial flight the minute workers cling. The fellow travellers aid her in founding the new colony (Skaife, 1979: 253). The queens are relished as a delicacy in Africa and caught in large numbers during their diurnal nuptial flight. Their gasters are torn off, and eaten raw or fried with salt as snacks or a garnish. Oil is not necessary as the ants already contain enough fat for frying. The unsects can be preserved by salting and sun-drying.

Gbaya in the Central African Republic use ants as a condiment (Roulon-Doko, 1998). The women pierce the nest in a tree and brush the insects into a container. Roulon-Doko (1998) also indicated that children eat ant eggs. For one species they collect and eat just the raw eggs carried outside the nest by the ants. From another ant species they collect the stored eggs by digging out the nest. These eggs are fried before eaten. Several interviewees in Cameroon also indicated that dug-out eggs are fried and eaten. The San in central Kalahari collect *Camponotus* sp. by poking a nest with a digging stick and tapping the ground around the nest with the hand (Nonaka, 1996). They then pound wild plants at home and near the end of the process mix in the ants. This adds a sweet-sour flavor to the mixture, which enhances the taste of the plants. To alleviate thirst the pygmies in the D.R. Congo seek ant nests (probably an *Oecophylla* sp.) in trees and press the nymphs found to obtain a sweet-sour liquid (Bergier, 1941: 66).

Nests of wild bees are exploited for honey. The bouquet of the honey of the stingless bee (*Trigona* spp.) is preferred over that of the true honeybee (*Apis* spp.). Honey is used everywhere in Africa to solve numerous health problems (Van Huis, 2003B). The bee larvae are collected for food (Owen, 1973: 135). Bodenheimer (1951: 172) mentioned that the Hottentots consider the eggs, larvae, and combs of bees a delicacy. In Tanzania bee larvae are generally eaten raw together with combs containing honey (Harris, 1940). They may also be shaken out of the comb and added with honey to porridge. In the D.R. Congo, the Ngandu prefer the honeycomb with honey and live bee larvae (Takeda, 1990). In other parts of the country bee larvae and pupae are grilled (Bergier, 1941: 76). The Greater Honeyguide *Indicator indicator* leads hunters to bee nests where it feeds on adult and larval bees and on beeswax (Maclean, 1993: 408–9). On the other hand, the San in Kalahari locate bees by running after the swallow-tailed bee-eater (*Merops hirundineus*) or tracing the footprints of the honey badger (*Mellivora capensis*) (Nonaka, 1996).

Decary (1937) mentioned that larvae of wasps were eaten in Madagascar but already in the 1930s the custom was disappearing. Takeda (1990) also mentioned earlier consumption of hymenopteran wasps along with the honeycomb as a side dish for cassava.

Hunter (1984) reported that pregnant women in Sierra Leone eat the clay of mud-dauber wasps. Geophagy is a widespread practice of pregnant women in all continents, aside from Africa. In every country visited my interviewees mentioned this practice as pregnancy "craving". Different sorts of clay are used. In Sierra Leone, 50% of the women regularly consume clay from termite mounds (about 40–140 g a day). However, they also use clay from the nest of mud-daubers but less often, probably due to the smaller quantities available. There are many species of mud-daubers in Africa. The genus *Synagris* occurs in most of tropical Africa and builds its nest suspended from poles supporting the thatched roof of a house. In this way it is domestically accessible. Hunter (1984) concluded that geophagy is probably a response to the frequency of pregnancy, hard physical labor of women, anemia, calcium deficiency, and other dietary insufficiencies. Insect geophagy provides women with a valuable supply of essential minerals and trace elements in response to the critical needs of fetal development.

Diptera

The East African lake fly *Chaoborus edulis* producing the "Kungu" cake already mentioned by Livingstone (see also Cotterril, 1968: 415–418), drifts en masse over the lake in accordance with lunar cycles (Owen, 1973; MacDonald, 1956).

The clouds of these flies move to the lakeside where they are collected by whirling baskets attached to long handles (Armitage et al., 1995; Harris, 1940; Owen, 1973: 135). They are squashed and dried in the sun and the resultant cake is an important source of protein in Uganda and elsewhere in East Africa. Narrative Press (2001: 274) reports that Livingstone wrote in 1861: "The people gather these minute insects by night, and boil them into thick cakes, to be used as a relish—millions of midges in a cake. A kungo cake, an inch thick .... was offered to us; it was very dark in colour and tasted not unlike caviar, or salted locusts".

Heteroptera

Local inhabitants of certain areas of Drakenberg in Eastern Transvaal of South Africa and in eastern parts of Zimbabwe consider the heteropteran *Natalicola delegorguei* a great delicacy. Adults congregate in large numbers at the end of the rainy season on certain kinds of bushes and trees (Bodenheimer, 1951). They excrete a pungent fluid (Bodenheimer, 1951; Faure, 1944) that can cause severe pain and even temporary blindness if it comes into contact with the eyes (Scholtz, 1984). Gatherers must turn their eyes away when collecting them at sundown or early morning. The head is removed by rubbing it against a hard object. The thorax is then pressed to remove the secretion. Subsequently they are then eaten roasted, cooked or raw, alone or with other food dishes. These stinkbugs contain large fat reserves. Although large numbers are often collected, small quantities are eaten at a sitting, partly because they are very rich and partly because they are considered a preeminent delicacy.

The pentatomid *Agonoscelis versicolor*, a pest of rainfed sorghum and sesame in Sudan, is collected during the dry season, when they mass in rock cracks in the Nuba mountains, about 500 km south of Khartoum (Delmet, 1975). The insects are fried (without oil), crushed, and pressed. The extracted edible oil can be conserved for a long time. This oil is considered of better quality then sesame oil. It can also be used for veterinary purposes, treating scab diseases of horses, sheep and camels. It is used industrially to smoothen sheep skins. Seignobos et al. (1996) have mentioned such a heteropteran insect in northern Cameroon.

Homoptera

Quite a number of species of cicadas are eaten in Africa; just in Katanga, D.R. of Congo, eight species are mentioned (Malaisse, 1997). Adults alight on the trunks of trees emit a shrieking sound at the beginning of the rainy season. In southern Africa children trap cicadas by climbing trees or using a long pole whose end has been dipped in glue fabricated from certain plant galls and tree resins. The Gbaya in the Central African Republic use sticks from *Lantana rhodesianus* or grass stems of *Andropogon gayanus* with glue at the end, to collect edible flower beetles (Cetoniinae), cicadas, and grasshoppers (Roulon-Doko, 1998). The Mofu-Gudur in northern Cameroon use sap from *Diospyros mespiliformis* or mistletoe as glue (Barreteau, 1999). The cicadas may be immersed in hot water, after which the wings are pulled off. They are then boiled in enough water to allow evaporation for 10 to 30 minutes, after which they are fried in a little oil. They can be preserved by salting and sun-drying after boiling.

The psyllid *Arytaina mopane* Pettey is a phloem-feeder of the tree *Colophospermum mopane* (Leguminosae: Caesalpiniaceae) in southern Africa, which can be completely defoliated by the edible mopane worm *Gonimbrasia belina* Westwood (Lepidoptera: Saturniidae). The psyllid larva constructs a scutcheon-like cover, the lerp, which protects the developing insect. These lerps covering the leaf surface are collected by the local people in large quantities for food. It is only available during the dry season and is usually washed away by rain (Sekhwela, 1988). The chemical composition of the lerps is given by Ernst and Sekhwela (1987). Lerps have a high concentration of monosaccharides, which probably explains the marked palatability of mopane leaves for cattle and the preference of the African population for the lerp.

Another homopteran, the flattid *Phromnia rubra* lives on Combretaceae in Madagascar. The nymphs cluster on twigs and stems in the dry forest of the south and west. Their wispy tail consists of a waxy substance, which is thought unpalatable to predators (Preston-Mafham, 1991). However, Decary (1937) mentioned that this secretion drips on branches where it can obtain the thickness of a fist. The people are fond of this sugary white substance and they call it "cicadellid honey".

Other Insect Groups and Arthropods

The larvae of several species of Odonota are collected in abandoned rice fields or in marshes in the highlands of Madagascar (Decary, 1937). However, when fried they are apparently not very tasteful. The Pangwe in Cameroon fish the larvae because diuretic qualities are attributed to the eaten insect (Bergier, 1941: 76).

In Madagascar two spider species are eaten, both fried in oil or fat (Decary, 1937). Ground-dwelling bird spiders are eaten by children in the Central African Republic (Roulon-Doko, 1998).

Conclusions

Promotion of consumption of edible insects offers an affordable meat source for low-income situations. This possibility has largely been overlooked by both nutritionists and agronomists. However, the African culture itself recognizes the value of insects as food and this minilivestock is eaten throughout the continent. In central and southern Africa in particular edible insects are an important protein and micronutrient source for the rural poor and a delicacy for the urban.

However, their importance as a food source is limited because insects can only be harvested from wild vegetation and agricultural crops during short periods of time.

Can we promote edible insects as a food resource in sub-Saharan Africa? Wild populations of edible insects in open accessory areas could be better exploited were more attention given to ecological and social aspects. Harvesting techniques could probably be improved as well. Studies are needed to domesticate edible insects and rear them onfarm with low costs and simple techniques. Lastly, finding ways to longer conserve comestible insects would certainly stimulate commercialization of this source of human nutrition.

Acknowledgments

Wageningen University, the Netherlands, provided the funds for traveling to a number of African countries and several European libraries in 1995 and 2000. Yde Jongema of our Laboratory spent considerable effort in checking the correct scientific names of the arthropods mentioned in Table 11.1 and his assistance is very much appreciated. Those in Africa who provided assistance and/or information are far too numerous to be mentioned here, but their contributions are gratefully acknowledged.

References

ACC/SCN 2000. Fourth report on the world nutrition situation. UN Administrative Committee on Coordination, Sub-Committee on Nutrition (ACC/SCN), in collaboration with IFPRI, Geneva, 121 pp.

Adriaens E.L. 1951. Recherches sur l'alimentation des populations au Kwango. Bull. agric. Congo belge 62: 473–550.

Armitage P.D., Cranston P.S., and Pinder L.C.V. (eds). 1995. The Chironomidae: Biology and Ecology of Non-biting Midges. Chapman & Hall, London, UK. 572 pp.

Bani G. 1995. Some aspects of entomophagy in the Congo. *Food Insect Newsl.* 8: 4–5.

Barreteau D. 1999. Les Mofu-Gudur et leurs criquets. *In:* Baroin C. and Boutrais J. (eds.). L'homme et l'animal dans le bassin du lac Tchad. Actes du colloque du réseau Mega-Tchad, Orléans 15-17 octobre 1997. Collection Colloques et Séminaires, no. 00/354. Université Nanterre, Paris, pp. 133–169.

Bergier E. 1941. Peuples entomophages et insectes comestibles: étude sur les moeurs de l'homme et de l'insecte. Imprimerie Rullière Frères, Avignon, Vancluse, France.

Bodenheimer F.S. 1951. Insects as Human Food: a Chapter of the Ecology of Man. Dr. W. Junk Publ., The Hague 352 pp.

Bukkens S.G.F. 1997. The nutritional value of edible insects. Ecol. Food. Nutr. 36: 287–319.

Chavanduka D.M. 1976. Insects as a source of protein to the African. Rhod. Sci. News 9: 217–220.

Chitsiku I.C. 1989. Nutritive value of foods of Zimbabwe. Zambezia 16: 67–97.

Cotterril H.B. 1968. Natural history appendix on the Kungu fly, pp. 415–417. *In:* Cotteril H.B. (ed.). Travels and Researches among the Lakes and Mountains of Eastern and Central Africa from the Journals of J. Frederic Elton. Frank Cass & Co. Ltd., London, UK, 417 pp. (1st publ. 1879).

Decary R. 1937. L'entomophagie chez les indigènes de Madagascar. Bull. Soc. ent. Fr. (9 juin 1937), pp. 168–169.

Delmet C. 1975. Extraction d'huile comestible d'*Agonoscelis versicolor* Fabricius, Hétéroptère Pentatomidae, au Djebel Guli, Soudan. *In:* Pujol R. (ed.). L'Homme et l'Animal. Premier Colloque d'Ethnozoologie, Paris, pp. 255–258.

Ernst W.H.O. and Sekhwela M.B.M. 1987. The chemical composition of lerps from the mopane psyllid, *Arytaina mopane* (Homoptera, Psyllidae). Insect Biochem. 17(6): 905–909.
Fasoranti J.O. and Ajiboye D.O. 1993. Some edible insects of Kwara State Nigeria. Amer. Entomol. 39: 113–116.
Faure J.C. 1944. Pentatomid bugs as human food. J. Ent. Soc. S. Afr. 7: 111–112.
Gelfand M. 1971. Insects. *In:* Diet and Tradition in African Culture, E&S Livingstone, Edinburgh, pp. 163–171.
Gessain M. and Kinzler T. 1975. Miel et insectes à miel chez les Bassari et d'autres populations du Sénégal Oriental. *In:* Pujol R. (ed.). L'Homme et l'Animal. Premier Colloque d'Ethnozoologie, Paris, pp. 247–254.
Ghesquière J. 1947. Les insectes palmicoles comestibles. In: Lepesme P. and Ghesquiere J. (eds.) *Les Insectes des Palmiers* Lechevalier, Paris. Appendice II, pp. 791-793.
Harris W.V. 1940. Some notes on Insects as Food. Tanganyika Notes Rec. 9: 45–48.
Hölldobler B. and Wilson E.O. 1990. The Ants. Springer-Verlag, Berlin, 732 pp.
Hunter J.M. 1984. Insect clay geophagy in Sierre Leone. J. Cultural Geography 4 (2): 2–13.
ICZN 1999. International Code of Zoological Nomenclature. International Trust for Zoological Nomenclature. Natural History Museum, London, 306 pp.
Narrative Press (ed.). 2001. A Popular Account of Dr. Livingstone's Expedition to the Zambesi and Its Tributaries and the Discovery of Lakes Shirwa and Nyassa 1858–1864 by David and Charles Livingstone. Narrative Press, Santa Barbara, CA, 433 pp.
MacDonald W.W. 1956. Observations on the biology of chaoborids and chironomids in Lake Victoria and on the feeding habits of the 'elephant-snout fish' (*Mornyrus kannume* Forsk.). J. Anim. Ecol. 25: 36–53.
Maclean G.L. 1993. Robert's Birds of Southern Africa. John Voelcker Bird Book Fund, Cape Town, Rep. S. Africa, 871 pp.
Malaisse F. 1997. Se nourrir en forêt claire africaine: approche écologique et nutritionelle. Les Presses Agronomiques de Gembloux Gembloux, Belgium, 384 pp.
Mbata K.J. 1995. Traditional uses of arthropods in Zambia. Food Insect Newsl. 8(3), 1, 5–7.
Netolitzky F. 1919. Käfer als Nahrungs und Heilmittel. Koleopt. Rdsch. 8 (1–3), 21–26, 47–60.
Nkouka E. 1987. Les insectes comestibles dans les sociétés d'Afrique Centrale. Rev. Sci. Cult. CICIBA, Muntu 6: 171–178.
Nonaka K. 1996. Ethnoentomology of the central Kalahari San. African Study Monogr. Suppl. 22: 29–46.
Owen D.F. 1973. Man's Environmental Predicament: An Introduction to Human Ecology in Tropical Africa. Univ. Press, Oxford, UK, 214 pp.
Preston-Mafham K. 1991. Madagascar: A natural history. Facts on File Ltd., Oxford, UK.
Quin P.J. 1959. Food and Feeding Habits of the Pedi with Special Reference to Identification, Classification, Preparation and Nutritive Value of the Respective Foods. Witwatersrand Univ. Press, Johannesburg, Rep. S. Africa, 278 pp
Ramos-Elorduy J. 1997. Insects: a sustainable source of food. Ecol. Food. Nutr. 36(2–4): 247–276.
Roodt V. 1993. The Shell Field Guide to the Common Trees of the Okavango Delta and Moremi Game Reserve. 113 pp.
Roulon-Doko P. 1998. Les activités de cueillète (Chap. V, Part 2 "Les Invertébrés"), pp. 247–342. *In:* Chasse, cueillette et culture chez les Gbaya de Centrafrique. Éditions L'Harmattan, Paris, 540 pp.
Santos Oliveira J.F.S., Passos de Carvalho J., Bruno de Sousa R.F.X., and Madalena Simão M. 1976. The nutritional value of four species of insects consumed in Angola. Ecol. Food. Nutr. 5: 91–97.
Scholtz C.H. 1984. Useful Insects. De Jager-HAUM publishers, Pretoria, Rep. S., 48 pp.
Seignobos C., Deguine J.P. and Aberlenc H.P. 1996. Les Mofu et leurs insectes. J. Agric. Trad. Bota. Appl. 28(2): 125–187.
Sekhwela M.B.M. 1988. The nutritive value of Mophane bread—Mophane insect secretion (maphote or maboti). Botswana Notes Rec. 20: 151–153.

Silow C.A. 1976. Edible and other insects of mid-westen Zambia. Studies in Ethno-entomology II. Occasional Papers V, Institutionen för Allmän och Jämförande Etnografi vid Uppsala Universitet. Almqvist & Wiksell, Uppsala, Sweden, 223 pp.

Silow C.A. 1983. Notes on Ngangela and Nkoya Ethnozoology: ants and termites. Etnol. Studier, no. 36. Etnografiska Museum, Göteborg, Sweden, 177 pp.

Skaife S.H. 1979. African Insect Life. Hamlyn Publ. Group Ltd., London, 279 pp.

Takeda J. 1990. The dietary repertory of the Ngandu people of the tropical rain forest: an ecological and anthropological study of the subsistence activities and food procurement technology of a slash-and-burn agriculturist in the Zaire river basin. African Study Monographs, Suppl. 11: 1–75..

Van Huis A. 1996. The traditional use of arthropods in sub-Saharan Africa. Proc. Exper. Appl. Entomol. N.E.V. Amsterdam 7: 3–20.

Van Huis A. 2003A. Insects as food in sub-Saharan Africa: Review. Insect Science and Its Application 23: 163–185.

Van Huis, A. 2003B. Medical and stimulating properties ascribed to arthropods and their products in sub-Saharan Africa, pp. 367–382. *In*: Motte-Florac E. and Thomas J.M.C. (eds.). *Insects in oral literature and traditions*. Ethnosciences 11. Société d'études linguistiques et anthropologiques de France (Series): 407. Peeters, Paris, 633 pp.

Van Huis A. 2004. Insects eaten in Africa (Coleoptera, Hymenoptera, Diptera, Heteroptera, Homoptera) (this volume).

Weaving A. 1973. Insects: A Review of Insect Life in Rhodesia. Irwin Press Ltd., Salisbury, Wiltshire, England.

Wheeler W.M. 1927. Ants: Their Structure, Development and Behavior. Columbia Univ. Press, Columbia, NY.

12

Notes on Edible Insects of South Benin: A Source of Protein

SEVERIN TCHIBOZO[1], ARNOLD VAN HUIS[2], AND MAURIZIO G. PAOLETTI[3]

[1]*Laboratoire d'Ecologie Appliquée, Faculté des Sciences Agronomiques, Université d'Abomey-Calavi 04 B.P., 0385 Cotonou, Bénin. Tel. and Fax +229 303084, E-mail: Tchisev@avu.org*
[2]*Dépt. des Sciences de Production Végétale, Laboratoire d'Entomologie B.P. 8031, 6700 EH Wageningen, Les Pays-Bas.*
E-mail: arnold.vanhuis@users.ento.wag-ur.nl and arnold@vanhuis.com
[3]*Dept. Biology, Padova University, 35100-Padova, Italy*
E-mail: paoletti@civ.bio.unipd.it

Abstract

Insects have been and still are consumed in South Benin. They are a very important source of animal protein able to successfully substitute some meats and improve the health of badly nourished children. Four kinds of insects are principally collected in South Benin—*Oryctes* spp., *Rhynchophorus phoenicis* (Fabr.); *Brachytrupes membranaceus* (Drury), and *Macrotermes falciger*. The species mostly eaten in Benin are *Macrotermes falciger* and *Oryctes* spp. (Tchibozo, 2002). The various aspects investigated here are: the species eaten, techniques of gathering, culinary usages, communities consuming them, and their economic importance.

Key Words: entomophagy, Benin, *Rhynchophorus phoenicis, Brachytrupus membranaceous, Macrotermes falciger*.

Introduction

Terrestrial invertebrates have played several roles during the course of human development. In the Old Testament, Moses recommended to the Hebrews: "So, here are those that you can eat: the different species of grasshopper, cricket, grigs, and locusts" (Lev.11,22). The practice of entomophagy compensates shortages in proteins and lipids vis-à-vis chicken and pork meat (Bizé, 1997). However, in many cities of the D.R. Congo, edible insects cost more than other sources of protein such as beef and fish (Munyuli Bin Mushambanyi, 2000). Caterprillars also constitute a relatively important source of unrefined animal proteins, ranging from 54.5% to 72.5%, lipids around 16%, and a caloric value around 460 cal/100 g (BEDIM, 1994; Malaisse, 1997). In the region of Tanganyika in Congo-Zaire, termites are eaten raw, often live, but also dried (BEDIM, 1992). The larvae of bees, sweet and melting in the mouth, are very rich in Vitamin D and should be considered as a means of combating rachitis, which afflicts most of the ill-nourished children of the continent and severly compromises their growth (Munyuli Bin Mushambanyi, 2000).

Materials and Methods

We first targeted in 2000 through prospection in South Benin those localities in which insects are eaten (Pobè, Dasso, Saketé and Kétou during May–July, the wet season, and during November, the dry season). We stayed in those areas to assess the different stages in which these edible insects are handled (species eaten, techniques of gathering, culinary uses, and the communities consuming them).

Species Consumed

In South Benin, the species consumed in particular are the larvae of *Oryctes* spp., *Rynchophorus phoenicis* F., the grig *Brachytrupes membranaceus* (Drury), winged termites and the queen of *Macrotermes falciger* (Gerstäcker) (Table 12.1).

Techniques of Gathering

Consumers as well as a few middlemen (especially in Pobè) collect the insects wanted.

Larvae of *Oryctes* spp. are collected from rubbish heaps and the rotting trunk of palm trees (Plate VI, 1 and 2). Those of *Rhynchophorus phoenicis* are collected in the stipes of palm trees in which the palm wine has already been extracted. The hunters ascertain their presence by listening at the stipes beforehand, for the clatter made by the mandibles during insect alimentation. Grigs (*Brachytrupes membranaceus*) are more often collected by children from their burrows or while on the move. Winged termites (*Macrotermes falciger*) are collected after the rain in a large pan containing water placed under an electric light (Fig. 12.1). Inhabitants of the localities lacking electric lights, place lanterns in large empty pans

Table 12.1: Different species of insects eaten in Sud-Bénin

Insects eaten (order, family, type and species)	**Name in French**	**Name in local language**		**Status**
		Nagot ethnic group	**Fon ethnic group**	
Coleoptera				
Dynastidae *Oryctes* spp.	Larve de Dynaste	Woiwo	Tran	Non threatened
Curculionidae *Rhynchophorus phoenicis* (F.)	Larve du stipe de palmier à huile	Woiwo	Hli	Non threatened
Orthoptera				
Gryllidae				
Brachytrupes membranaceus (Drury)	Grillon	Hyrè	Abosaklé	Non threatened
Isoptera				
Termitidae				
Macrotermes falciger (Gerstäcker)	Termite	Iba	Toutou	Threatened

Fig. 12.1: Winged termites (*Macrotermes falciger*) on a white pan used as a trap under an electric lamp in Benin (photo S. Tchibozo).

to trap the winged termites. The queen is captured after the termitarium has been completely vacated (Plate VI, 3, 4, 5).

The most favorable season for the gathering and collecting insects is the wet one, i.e. the months of May, June, and July. Larvae of *Rhynchophorus phoenicis*, however, are exceptionally collected year round.

Culinary Uses

Larvae of *Oryctes* spp. are first thoroughly cleansed of excreta and dirt, parboiled at low heat, then fried (preferably in "red oil") and served as a meat dish in tomato or some other sauce. *Rhynchophorus phoenicis* larvae, well cleansed after gathering, are slightly perforated with a sharp knife, then spread in a pan and hoasted. When the larvae have turned transparent, the oil is drained from the pan, leaving just a small quantity into which condiments (tomato, chili, onions) are added. All the ingredients are simmered, then served with rice, or a pasta made from corn flour, but especially with "Piron" (*Manihot exculenta* product).

Grigs (*Brachytrupes membranaceus*) on the other hand are grilled, and then peeled just before eating. Like grilled fish, they can be peeled and consumed at any time. In the case of a very successful collection, grilled grigs are deskinned and served in a tomato or palm-nut sauce. As for winged termites, the wings are removed and the insect then fried in "red oil" (palm oil). Or they are slightly grilled intact to facilitate removal of the wings, then returned to the pan, some salt added, and fried/grilled to the right consistency. They may be consumed as such or served in a tomato sauce. The queen of the termitarium is grilled in her skin and then deskinned or sometimes sucked out directly from her envelope.

Communities Consuming Edible Insects

The consumption of insects in Benin is still not common and is even taboo in certain families. People are often surprised to see their neighbors consuming insects; sometimes they are curious and some tempted into tasting or eating them. The biggest consumers are the communities Nagots, natives of Pobè, Kétou, and a few Fons (most of whom have lived in Nigeria).

Insects as Animal Protein

Insects hold an important role in the nutrition of millions of individuals. Around one billion people nowadays suffer from malnutrition. In Asia and Africa, animal proteins are generally lacking in their diets. The minimum daily protein requirement is 35 g for a male weighing 70 kg (FAO norms) (Mignon, 2002). Since insects constitute a very important source of animal protein, their direct consumption and use in the food products, such as biscuits, chocolates, babyfoods, etc. should be encouraged. In fact, why not even substitute meat with insect flavored cubes! Communities suffering from protein deficiencies need to be educated about the important nutritional values of insects. The magazine Spore no. 77 (1988), cited by (Munyuli Bin Mushambanyi, 2000), states that in Oriental Kasai, D.R. Congo, insect consumption in a group of 2000 people, reaches 50 g fresh insects per day per person.

Economic Importance

In the 1980s, a woman originating from Hozin (Vally of Oueme, South Benin) collected wild larvae of *Oryctes* spp. and commercialized them in Nigeria. Her flourishing activity ceased in 1997 due to insufficiency of larvae. Members of local communities could by means of adequate technological information start breeding various insects in order to diversify their economic resources. In Pointe Noire, D.R. Congo, the price of 30 or 40 legless larvae is five times that of 1 kg beef (Bizé, 1997).

As indicated by Mignon (2002), citing BEDIM, ethnic groups of high demographic development must be aware of the very fragile equilibrium of nature and hence develop the means for substituting/supplementing wild insect collections with bred insects.

Conclusion

There are far more edible species in the tropical and subtropical regions of Benin than are actually eaten (Van Huis, 1997). Entomophagy can seriously offset the limitation and lack of proteins for human inhabitants of developing countries. Breeding of edible insects must be considered if correct management of the natural local resources is to be secured.

These and other initial data will spur chemical analysis of various insects in order to determine the percentage rate in proteins, lipids and caloric value. Already ethnic groups have begun insect breeding without such analyses, in order to compensate protein shortages through consumption of insects.

References

BEDIM. 1992. Bulletin Semestriel d'Information sur le Mini-Elevage, 1 (2): 18 pp.

BEDIM. 1994. Bulletin Semestriel d'Information sur le Mini-Elevage, 3 (2): 18 pp.

Bizé V. 1997. Les insectes, une ressource alimentaire d'avenir. *Insectes,* 106–107 (3): 10–13.

DeFoliart G. R. 1995. Food Insects Newsl. 8 (1): 12 pp.

Malaisse F. 1997. Se nourrir en forêt claire africaine. Approche écologique et nutritionnelle. Les presses agronomiques de Gembloux et CTA, Belgique, 384 pp.

Mignon J. 2002. L'entomophagie: une question de culture? Tropic., 20 (3): 151–155.

Munyuli Bin Mushambanyi T. 2000. Etude préliminaire orientée vers la production des chenilles consommables par l'élevage des papillons (*Anaphe infracta*: Thaumetopoeidae) à Lwiro, Sud-Kivu. République Démocratique du Congo. Tropic. 18 (4): 208–211.
Tchibozo S. 2002. Rapport d'étude préliminaire sur les invertébrés terrestres (arthropodes) utiles (comestible, médicinale et mythologique) du Bénin. 9 pp. + annexes.
Van Huis A. 1997. The traditional use of arthropods in Sub-Saharan Africa. Proc. Exper. Appl. Entomol., N.E.V. Amsterdam, 7: 20.

13

Edible Insects in Japan

Jun Mitsuhashi

Dept. of Bioscience, Tokyo University of Agriculture, 1-1 Sakuragaoka 1-chome, Setagaya-ku, Tokyo 156-8502, Japan. E-mail: junmths@nodai.ac.jp

Abstract

In Japan, the most popular edible insect is a grasshopper, *Oxya yezoensis* or *O. japonica*, an insect rich in proteins. It is consumed in large amounts even today. The mixtures of river-living insect larvae are particularly relished. The larvae and pupae of a wasp, *Vespula lewisi*, are consumed in considerable amounts, while pupae and female adults after oviposition are the preferred stages of *Bombyx mori*; the pupae are particularly rich in nutrients. All of these insects are cooked with soy sauce and sugar, and are sold as canned foods. In addition to these insects, the larvae of cerambycid beetles are preferentially eaten in the countryside. Larvae of the dobsonfly, *Protohermes grandis* (Neuroptera), has been consumed as a traditional medicine.

Key Words: edible insects, traditional insect foods, Japanese insect foods, food insects.

Introduction

Numerous insect species were consumed in the past in Japan, especially in montane regions. According to a survey by Miyake (1919), 55 species were recognized as food and 123 used for medicinal purposes. In Mexico, 303 species have been reported as foods mainly by indigenous people (Ramos-Elorduy, 1992). In the survey done by Miyake, the insect species were not well identified, and the recorded species could well be a group of species, for example, grasshoppers containing several species of Acrididae. The number of species would probably increase to levels comparable with Mexico were the insects properly identified. Traditional insect foods are *inago, zaza-mushi, hachinoko, kaiko, magotaro-mushi,* and *teppo-mushi*. In the past these insect foods were important sources of protein, especially in districts situated far from the coast, because people could not obtain sufficient meat and fish.

Insects are still eaten today but mostly as favorite foods; their nutritional value is overshadowed by an abundant supply of modern preparations.

Some insects are currently sold as canned foods and are comparatively expensive. Such foods are probably eaten mainly for nostalgic reasons. Detailed descriptions of each are given below.

Inago

Inago is a kind of grasshopper, mostly *Oxya yezoensis* (Plate VI, 1) but sometimes another species, *O. japonica* is also consumed (Acrididae, Orthoptera). It is polyphagous but prefers rice plants and has been a significant pest in paddy fields. Nevertheless, as a valuable protein source it was served as a side dish at table. The *inago* population decreased drastically after World War II between 1950–1970 due to excessive use of insecticides in paddy fields. After insecticidal pollution became a problem leading to restrictions on insecticides use *inago* populations gradually recovered, and people were again able to collect considerable amounts. *Inago* are collected by untrained as well as professional collectors. *Inago*s are pushed through a bamboo tube (a cut stem without nodes) to which the bag is attached; once bagged, the *inago* cannot escape. People also catch *inago* by hand in the paddy fields, after the rich harvest, and keep them in a bag. Adults are usually eaten. Primary and junior high school pupils have collected large numbers of *inago* as an extracurricular activity and sold them to professional collectors. Reportedly, in 1974 a junior high school in Miyagi Prefecture earned ¥2,800,000 through the sale of *inago*; part of the money went toward improving school facilities for extracurricular activities (Sankei Newspaper Co., 1977).

Collectors also purchase *inago* and set up temporary facilities near collection sites for their pretreatment. The *inago* caught are starved overnight to ensure gut evacuation, then boiled in a large cauldron, dried, frozen, and sent back to the home factory for final cooking, in which soy sauce and sugar are generally added. Many Japanese people like *inago* and Japanese paddy field collections are unable to meet the demand. Some years ago a company imported a grasshopper, *Melanoplus differentialis*, from Canada to replace *inago*. The collected grasshoppers were boiled and shipped to Japan in refrigerated containers.

For private use, *inago* is boiled for three to four minutes, cooled, and dried in the sun for one or two days. The hind legs and wings are removed, the insects laid in a pan and seasoned with 200 g sugar and 150 g soy sauce per 500 g *inago*. They are heated in the pan over a medium flame until most of the liquid evaporates. About 750 g cooked *inago* are obtained from 1,000 g fresh *inago*.

Cooked *inago* appear dark and retain their original shape and pattern so that the species can still be identified if the *inago* is immersed in water to remove the soy sauce and sugar. Once an orthopteran taxonomist found a rare species of *inago* in Japan, *O. ninpoensis*, in a package of commercialized cooked *inago* (Fukuhara, 1986).

Some cooked *inago* are packed in plastic containers for immediate consumption and others canned for preservation. The former can be seen in kiosks in some railway stations and also in some department stores and supermarkets in autumn. About 150,000 kg fresh *inago* are processed annually (Kaneman Co., pers. comm.). The price of 100 g bagged *inago* was ¥ 600 and ¥ 1,000 for canned in 1996–1997.

Inago can also be deep-fried and as in all cases must be bagged overnight to allow gut evacuation, then killed by immersion in boiling water for three to four minutes. The boiled *inago* is sun-dried for one day, then slipped into boiling oil, removed and seasoned with salt. Roasted *inago* is also eaten, in which case, it is more often may be skewered before roasting.

Zaza-mushi

Zaza-mushi is the common name of larvae that live in river shallows where the sound of the water flow can be linguistically expressed as *za-za*. The common name originated from this sound.

So, *zaza* is a coined onomatopoetic word, and *mushi* the base word for insects. Therefore *zaza-mushi* means insects that live in river shallows. The term usually includes larvae and nymphs of Ephemeroptera, Odonata, Neuroptera, Plecoptera, Trichoptera, and sometimes even Hemiptera.

The species composition of *zaza-mushi* varies depending on location and time. In the past nymphs of stoneflies (Plecoptera) predominated, but more recently larvae of caddisflies (Trichoptera) have displaced them. The Tenryu River in Nagano Prefecture is noted for *zaza-mushi*. Several decades ago, many people collected *zaza-mushi*, and consequently the insect populations decreased. The local government then restricted the period of *zaza-mushi* collection and concomitantly a license was required. Presently, only licensed persons can collect *zaza-mushi* during winter. The river fisher association receives all applications for licenses and obtains permission to collect *zaza-mushi* from the Ministry of Land, Infrastructure and Transport, which is responsible for managing rivers. Although *zaza-mushi* occurs in almost all rivers of Japan, that from Tenryu River is said to be the best. Even in Tenryu River, a very small section in Ina city is said to be the very best. This is because the water is extremely clean and contains appropriate nutrients for *zaza-mushi*. Although collection is restricted to winter, it is said that December to February are the best months. *Zaza-mushi* collected in other seasons is said to be less flavorful.

During winter a professional can collect 2 kg *zaza-mushi* per day. There are currently about forty such collectors. To collect *zaza-mushi*, they go into the river wearing wire shoes on their feet. They set up a net downstream and kick over

Fig. 13.1: Collecting *zaza-mushi* in Tenryu River near Ina city in January, Japan (photo J. Mitsuhashi).

stones on the bottom upstream, releasing larvae hidden between and under them into the flow (Fig. 13.1). The flowing larvae are trapped in the net downstream. The collected *zaza-mushi* sells at a price of ¥9,000 ($90) per kg. Out of season most collectors earn income by gathering wild resources such as edible plant shoots, mushrooms or wasps.

Although, *zaza-mushi* consists of various insect species, the majority belong to Trichoptera. There are two types of trichopteran larvae: some species make cases with small stones or sunken plant materials, while others make soft cases. The latter type such as *Stenopsyche griseipennis* (Plate VI, 7), *Parastenopsyche sauteri*, and *Hydropsycheodes brevilineata* is caught by collectors. Oviposition takes place in July to September and the larvae overwinter in a mature stage. Before overwintering, larvae accumulate fats and carbohydrates in their bodies which gives them a good taste.

Although cooked in a manner similar to *inago*, with soy sauce and sugar, *zaza-mushi* retains a distinctive flavor. A single company in Nagano Prefecture started producing canned *zaza-mushi* in 1956. At that time, the price of canned *zaza-mushi* was ¥55 per 100 g, but it has now increased to ¥3,000. About 4,000 kg of canned *zaza-mushi* are prepared annually (Kaneman Co., pers. comm.). Deep-frying is said to be another excellent way of cooking it. Once fried *zaza-mushi* was exported to the US as a canned product (Torii, 1957).

Hachinoko

Hachinoko means larvae of the wasp, *Vespula lewisi* (Vespidae). The food *hachinoko* is a mixture of cooked larvae, pupae and sometime adults of the wasp. The wasp makes a nest underground and in autumn, worker wasps busily collect food.

These wasps are carnivorous and like frog meat. To locate a nest, people first catch a frog, take a small piece of meat, and attach small silk flocks. The remainder of the frog's body is placed where wasps will find it easily. When a wasp comes to the frog carcass and cuts out a small piece of meat, people replace the meat cut by the wasp with the piece bearing flocks. When the wasp flies carrying the flocked frog piece, the people follow it easily because the flocks are white and glossy. When the collectors locate the nest, they put a firecracker (sometimes specially made for this purpose) in the nest, and light it. The smoke paralyzes all the wasps inside and the nest is then dug out with a shovel or spade. The caps of the cells are cut off and the larvae and pupae picked out. These are washed and cooked with soy sauce and sugar (Fig. 13.2) and eaten as a delicacy or a side dish. Wasp rice is also made by mixing cooked wasps with cooked rice. This was a favorite dish of the late Emperor Hirohito (Mitsuhashi, 1988). It is said that he ate wasp-rice even when sick and unable to eat other foods.

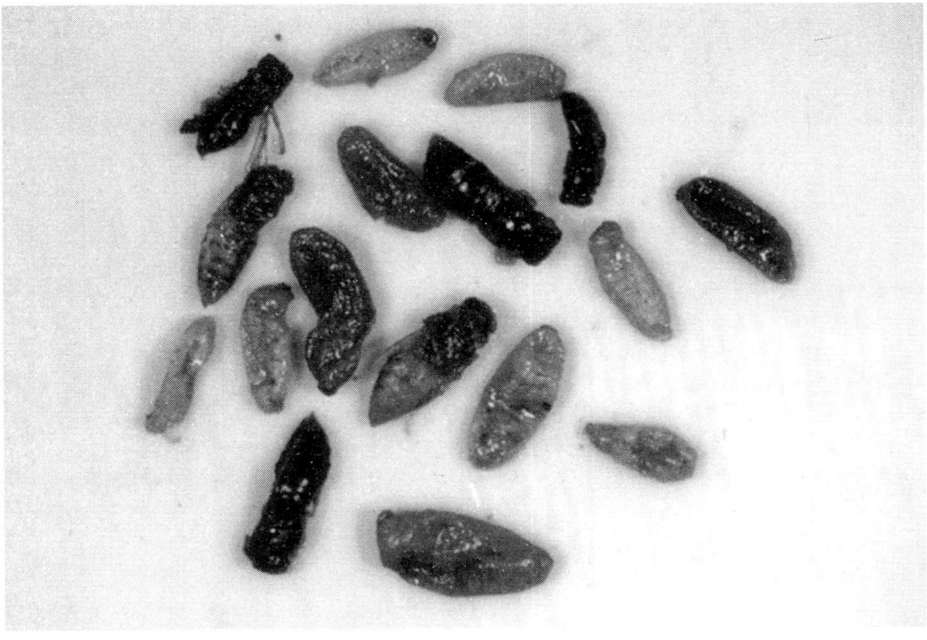

Fig. 13.2: Cooked *hachinoko* (*Vespula lewisi*) Japan (photo J. Mitsuhashi).

Around 1985, some companies sold canned larvae of drones of *Apis mellifera* since the majority of male bees are useless to the beekeeper. However, honeybee larvae did not last long on the market, because the taste is inferior to that of *hachinoko*. Wasp larvae are collected by both amateurs and professionals. Professional collectors travel for great distances to find wasp nests, collecting tons of wasps. However, during the course of processing, the weight of the wasps is reduced to half at the time of cooking, and the final product weighs only a third of the original weight.

Recently, an amateur group has been trying to conserve the wasp. They catch the queens in autumn and overwinter them at home, releasing them the following spring. Canned or bottled *hachinoko* is available from several companies in Nagano Prefecture (Fig. 13.3). The cost is currently ¥1,500 per 100 g. About 40,000 kg of *hachinoko* are collected for processing every year (Kaneman Co., pers. comm.). Some people, especially children, relish live wasp larvae which, they say, taste sweet.

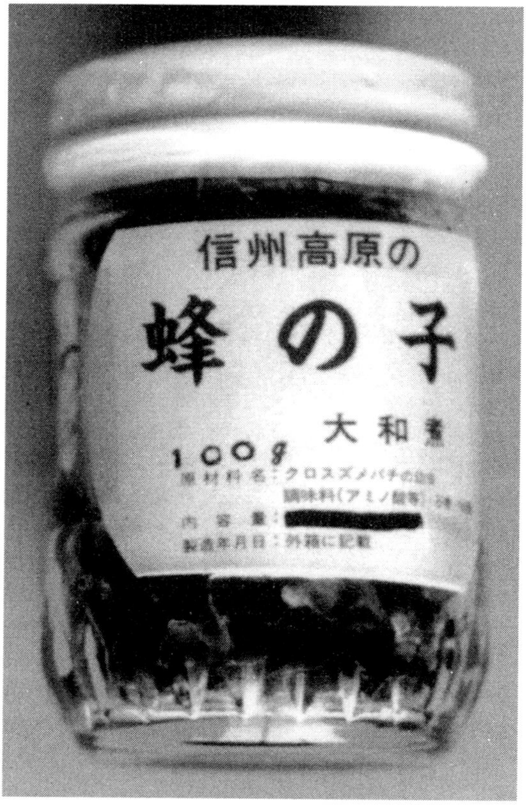

Fig. 13.3: Bottled *hachinoko*, Japan (photo J. Mitsuhashi).

Kaiko

Kaiko means commercial silkworm, *Bombyx mori*. In Asia, the habit of eating silkworm pupae is common where sericulture is popular. When silk is spun from cocoons, people first kill the pupae within the cocoons by heat treatment. Since pupae are no longer of value in sericulture after they have spun, they are usually used as foods for domestic animals or fish, or as fertilizer. However, people cook some of the pupae with soy sauce and sugar for human consumption. They are eaten as a delicacy or a side dish. Cooked silkworm pupae are sold as canned food. The moths are also eaten. In Japan, silkworm growers usually obtain silkworm eggs from special companies, who supply them disease-free; the moths are not useful after oviposition. One food company was inspired to cook the moths with soy sauce and sugar. Cooked adult moths are also available canned. Fried silkworm pupae are eaten after seasoning with salt.

During and after World War II when food supplies were scarce, girls who worked in the silk mills ate silkworm pupae as they emerged from the cocoons, without further cooking. It is also said that in olden times silkworm growers ate live silkworms during rearing as a refreshment. In some districts the larvae attacked by nuclear polyhedrosis virus are supposedly good to eat, probably because diseased larvae do not have well-developed silk glands, which in healthy larvae contain abundant sticky gelatinous material. Cooked silkworms have a strong mulberry taste that some people dislike.

The pupae of wild silkworms such as *Antheraea yamamai* and *A. pernyi* are also eaten. However, cultivation of these wild silkworms is not widespread in Japan. In China, consumption of pupae of *A. pernyi* is supposedly common.

Silkworm has also been consumed for medicinal reasons.

Silkworms are rich in nutrients, and were often used as a tonic (Umemura, 1943). According to the famous old Chinese pharmacological text "Honzo-komoku"(Suzuki, 1976), adult moths can be used as an aphrodisiac. Perhaps this is based on the behavior of the moths, which copulate immediately upon emergence from cocoons. In some districts, larvae and their feces have been used as an antifebrile (Miyake, 1919), larvae and pupae as a medicine for sore throat, and pupae for nephritis (Umemura, 1943). Larval exuviae were believed to cure hemorrhage while hatched eggshells purportedly caused permanent infertility (Suzuki, 1976). Silkworm feces are known to contain rich nutrients and are used as fertilizers and foods for domestic animals and fish. They have also been used medicinally and are said to act as a spasmolytic, an anodyne, a tranquilizer for rheumatism, head ache, arthralgia, neuralgia, and skin ulcer (Suzuki, 1976). There are also some incredible stories wherein, according to "Honzo-komoku", a pregnant woman can change the sex of her embryo from female to male by taking a single fecal pellet of summer brood larvae three times a day (Suzuki, 1976).

Magotaro-mushi

Magotaro-mushi is the larvae of the dobsonfly, *Protohermes grandis* (Neuroptera) (Plate VI, 8). Sometimes they are found among *zaza-mushi*. Used primarily as a traditional medicine for children, Saigawa River in Fukushima Prefecture is famous river for *magotaro-mushi*. Collected *magotaro-mushi* are skewered and dried (Plate VI, 9), and in this state packed and sold in small boxes made of paulownia wood. When used, they are roasted and eaten. It is said that *magotaro-mushi* is effective in tranquilizing nervous babies and children. Roasted *magotaro-mushi* is said to be effective in treating intestinal worms, tuberculosis and stomach and intestinal troubles (Umemura, 1943). *Magotaro-mushi* is known to contain essential and nonessential amino acids, sterols, and pantothenate (Watanabe, 1982). However, specific components responsible for medicinal effects are not known and it is not recognized as a medicine by the Japanese Pharmacopoeia (Yasue, 1987).

Teppo-mushi

Teppo-mushi is the larvae of long-horned beetles (Coleoptera: Cerambycidae) which live in tree trunks. They first bore tunnels in the inner bark, then in the xylem. They are found by the frass they push out of the tunnel opening. The larvae of various species are eaten, among which *Batocera lineolata* is the largest in Japan. Full-grown larvae reach a length of about 10 cm. Some people prefer them raw, saying that the live larvae taste sweet.

Larvae in living or dried wood are microbiologically safe to eat raw. Eating those found in rotten wood, raw may cause parasitoses. Generally, larvae are pulled out of the tunnel using a hook or cutting the log and roasted over a flame or in a frying pan. Most people who eat insects say that *teppo-mushi* is the most delicious among the edible preparations. Unfortunately, collection of a large number of larvae is difficult. Hence at present there are no commercially available preparations of *teppo-mushi*.

Nutritional Evaluation of Traditional Insect Foods

In Japan, edible insects are currently consumed mainly as favored delicacies. In the past or during wartime when foods were scarce, however, edible insects were mainstays in people's nutrition.

Fresh *inago* contains about 66% water (Table 13.1). In dry matter, crude protein is the most abundant constituent and accounts for 77% of the dry weight; however, the real protein content is about 70% of the crude. The main proteins are alkali soluble, and albumin, globulin and prolamine have been found in considerable amounts.

Table 13.1: Chemical composition of *inago* (in %)

	Fresh *inago* (Ichikawa, 1936a)	Dried *inago* (Korikawa, 1934)	Anhydrous *inago* (Korikawa, 1934)
Water	65.85	11.89	—
Total nitrogen	4.08	10.86	12.32
Crude protein	25.51	67.86	77.00
Crude lipid	1.96	4.52	5.12
Carbohydrate	1.38	—	—
Glycogen	—	0.0 1	0.01
Chitin	—	9.25	10.37
Crude ash	2.23	3.75	4.26

Table 13.2: Distribution of nitrogen to protein amino acids in *inago*, in % (Korikawa, 1934)

Total nitrogen	100
Amide nitrogen	7.05
Melanin nitrogen	5.51
Diamino nitrogen	27.70
Monoamino nitrogen	59.00
Arginine nitrogen	8.91
Cystine nitrogen	0.88
Histidine nitrogen	11.45
Lysine nitrogen	7.57
Tryptophan nitrogen	0.88
Tyrosine nitrogen	7.11

As for amino acids, *inago* is rich in histidine and lysine but poor in cystine and tryptophan (Table 13.2). The nutritional value of *inago* protein has been evaluated by supplying it to mice as a protein source. Since 10% dried *inago* powder added to the mouse diet as the sole protein source supported normal mouse growth, the *inago* protein was evaluated as qualitatively good (Korikawa, 1934).

Inago contains rather less lipids and carbohydrates (Table 13.1) than other edible insects. These lipids contain various unsaturated fatty acids. About 16% of the lipids is unsaponifiable, of which 44% is sterols. *Inago* lipids are generally not considered good in quality (Tazaki, 1940–1945).

Minerals such as sodium, potassium, and phosphate are abundant (Table 13.3). Further, compared with other foods such as cereals, *inago* is rich in Cu, Fe, and Mn (Ichikawa, 1936a). Contrarily it contains little calcium, resulting in a low CaO/MgO ratio (0.246), a value considered unfavorable for proper nutrition. On the other hand, *inago* contains high Fe, a content comparable to that of bovine liver (Tazaki, 1940–1945).

With respect to vitamins, *inago* contains a considerable amount of vitamin A but is poor in vitamin B_1 (Tazaki, 1940–1945).

Table 13.3 Ash composition of *inago*, in % (Ichikawa, 1936b)

SiO_2	15.352
Al_2O	1.990
Fe_3O_3	0.816
CaO	0.833
MgO	2.097
K_2O	21.148
Na_2O	17.920
P_2O_5	36.948
MnO	0.158
TiO_2	0.108
CuO	0.191

The aforesaid nutritional contents fluctuate from year to year (Korikawa, 1934), and season to season, with the highest contents attained at the end of August (Ichikawa, 1936b). These contents may also vary between females and males and it was claimed that females contained twice as much protein as males. Some minerals were also found to differ between the sexes (Ichikawa, 1937). The food energy value of fresh *inago* is 107 kcal/100 g, comparable to that of low fat beef (Tazaki, 1940–1945).

Silkworm pupae, easily obtained in large numbers in sericultural areas, were routinely used as a side dish at hungry times or a protein source for production of soy sauce (Esaki, 1958). Three pupae have generally been considered to have the same nutritional value as one chicken egg (Esaki, 1958). According to Ishimori (1944), dried pupae still contain 7.2% water. The most abundant constituent is protein (48.9%), followed by lipid (29.6%). Among other constituents, glycogen is 4.65%, chitin 3.73%, and ash 2.19%. Silkworm pupae contain 7–8 times as much vitamin B_2 as yeast.

As mentioned above, silkworm feces are rich in nutrients. Dried feces contain 83.77–90.44% organic acids, 9.56–16.23% ash, and 1.91–3.60% total nitrogen. The amino acids histidine, leucine, and lysine predominate. The feces are also rich in sterols such as ß-sitosterol, cholesterol, and ergosterol. Lastly, feces also contain uric acid, phosphoric acid, potassium, calcium, vitamin A and B groups, plant growth hormones, and chlorophyll (Namba, 1980).

Chemical analyses have not been done on *hachinoko*. However, data on bees are available (Hocking and Matsumura, 1960). Protein is the major component and lipid less compared with other edible insects. From Table 13.4, it can be seen that the bee is an outstanding vitamin source. Bee larvae contain as much as twice vitamin A as egg yolk, and 10 times as much vitamin D as cod liver oil.

Dried *magotaro-mushi* is rich in protein (57.0%). It also contains 11.3% crude lipids, 4.9% ash, and 10.8% water. In the hydrolyzed protein of *magotaro-mushi*, alanine, aspartic acid, glutamic acid, isoleucine, leucine and methionine are rich, and arginine, glycine, proline, serine, taurine, tryptophan, and tyrosine found

Table 13.4: Chemical composition of Bee Brood (Hocking and Matsumura, 1960)*

	Mature larvae (%)	Pupae (%)
Water	77.00	70.20
Protein	15.40	18.20
Lipid	3.71	2.39
Ash	3.02	2.17
Glycogen	0.14	0.75
Vitamin A#	106.70	51.30
Vitamin D#	6863.30	5165.00

*Values are average percentages of wet weights
#IU/g wet weight.

in considerable amounts. The amounts of cystine, histidine, lysine and phenylalanineare are significantly lower. For vitamins, *magotaro-mushi* is rich in pantothenate (brochure of Hoshina Co., Fukushima, Japan).

Conclusion

At present it is not necessary to consume insects as nutritional sources in Japan. However, utilization of some insects as favorite foods will continue. *Inago* are the most popular edible insects and the amount collected in Japan cannot meet consumer demand. It is said that considerable amounts of *inago* are imported from other countries such as China and Korea.

When nutritional shortage occurs, *kaiko* will be a promising candidature for nutrient sources, because it is rich in proteins, lipids, and vitamins, and is easily obtained in large quantity.

Most Japanese, however, have no desire to eat insects and generally dislike them. It is not necessary to eat insects with their shape intact. When used as common foods or as an important nutrient source, they can be pulverized dry. This might eliminate the refusal reaction against insects by the majority of people.

In Japan, many insects have also been used medicinally. In most cases, however, the effective components are not clear. Some insects were used as medicine apparently out of superstition. However, some medicinally used insects were certainly rich in nutrients and may have helped recovery from disease by improving the nutritional condition of the sick, especially when foods were lacking due to natural disaster or war.

References

*Esaki, C. 1958. A collection of T. Esaki's essays. Hokuryukan, Tokyo, 345 pp.
*Fukuhara N. 1986. Story of rediscovery of *Oxya ninpoensis*. Insectarium 23: 262–269.
Hocking B. and Matsumura F. 1960. Bee brood as food. Bee World, 41, 113–120.
*Ichikawa C. 1936a. Chemical investigation on *inago*-1. J. Agr. Chem. 12: 408–411.

*Ichikawa C. 1936b. Chemical investigation on *inago*-2. J. Agr. Chem. 12: 412–414.
*Ichikawa C. 1937. Chemical investigation on *inago*-3. J. Agr. Chem. 13: 710–712.
*Ishimori N. 1944. Agriculture, Insects, and Silkworm. Kyoiku Kagaku Sha, Tokyo, 119 pp.
*Korikawa K. 1934. Chemical investigation on *inago*. 1. General components and proteins. J. Zymol. Jpn. 12: 361–365.
Mitsuhashi J. 1988. Rice with cooked wasps: Emperor Hirohito's favorite dish. Food Insect Newsl. 1 (2): 2.
*Miyake T. 1919. Surveys on food and medicinal insects in Japan. Agr. Exp. Sta. Spec. Rept. 3l: 203 pp.
*Namba T. 1980. Iconography of Japanese and Chinese Medicinal Materials in Color. Hoikusha, Osaka, vol. 2, 521 pp.
Ramos-Elorduy J. 1992. Biogeographical aspects of some edible insects from Mexico. III Intern. Congr. Ethnobiol., Mexico City, 1992, abstract p. 141.
*Sankei Newspaper Co. (ed.). 1977. Revive Small Organisms (Part II). Hokuyosha, Tokyo, 232 pp.
*Suzuki S. 1976. Honzo-Komoku (Translation) Shunyo-do, Tokyo, vol. 10, 649 pp.
*Tazaki T. (during World War II, 1940–1945). Nutritional value of *inago* and methods for cooking it. All Japan Side Job Assoc. Mimeo. 32 pp.
*Torii T. 1957. Notes on *zaza-mushi* in Ina-Tenryu. Shinkonchu 10: 26–29.
*Umemura J. 1943. Konchu-Honzo. Shobunkan, Nagoya, 209 pp.
*Watanabe T. 1982. Cultural history of medicinal insects. Tokyo-Shoseki, Tokyo, 210 pp.
*Yasue Y. 1987. The magotaromushi show gallery. Insectarium 24: 242–245.

---

*Asterisked entries in Japanese.

14

Insects: A Hopeful Food Source

JULIETA RAMOS-ELORDUY

Instituto de Biologia UNAM. Aptdo. postal 70-153. 04510 México D.F. MEXICO.
E-mail: relorduy@ibunam.ibiologia.unam.mx

Abstract

Insects as a major animal group possess an enormous biodiversity and form a colossal biomass in nature that is generally wasted. Insects offer us many benefits, including use in human and animal nutrition, medicine, religion, art and handicrafts. They are also efficient recyclers of organic matter and provide a source of economic gain for the poor through their sale.

Even if developed nations consider insects to be an emergency food or non-conventional food of low prestige, they are part of the daily diet of the greater part of humanity, where they are regarded as healthy, nutritious and tasty food, at times constituting the only significant source of quality protein for these people. To date, around 2,000 edible insect species have been identified throughout the world.

Because of their high nutritive value not only in proteins but also in fats, minerals and vitamins, and their ubiquitous presence, insects are a "potential" source of sustainable food for humans. Once we select suitable species and develop appropriate breeding methods, insects would be able to provide a reliable and sustainable source of high-quality animal protein.

Key Words: edible insects, entomophagy, nutritive value, commercialization, recycling, sustainability.

Introduction

Insects in the Human Diet: Tradition and Acculturation

Diet is a reflection of a population's socioeconomic conditions. Dietary habits and taste perceptions are closely bound to a population's history and geographic origins and evolve in relation to its life style, tradition, and education. This may explain why in some developed countries insects are considered a primitive food, whereas other cultures consider them a valuable and integral part of their diet.

Human development is linked to alimentation and for this reason nutrition is a key to human progress and a productive life. With 2/3 of the population in third world countries suffering from hunger or malnutrition (De Castro, 1973), sustainable food alternatives are needed that will be accepted by these people and will fit in with their already deeply rooted dietary habits. For example, according to Zubiran et al., in Mexico 39 million people suffered from malnutrition in 1974. In this context, edible insects are an important and promising food resource to develop in the near future.

Undoubtedly, it is generally people living in tropical rural areas who consume the largest variety and quantity of insects. However, many edible insects are now being eaten by town people of a higher social status, as they have become "gourmet" dishes or delicacies (Ramos-Elorduy and Conconi, 1994; Ramos-Elorduy, 1998; Ramos-Elorduy et al., 2003).

Food consumption patterns are increasingly shaped by mass media. Large-scale and often inspiring advertisings evoke a desire to eat new and different products that do not constitute part of the traditional diet, even in the most remote societies. There is a marked tendency in countries undergoing economic and sociocultural expansion to search for novelties and to blindly adopt the "prestigious" western way of life, which is considered "convenient", even though it may not be compatible with the local agriculture and economy, nor good for the health. This acculturation process carries the risk of losing knowledge of collecting, preparing, eating, preserving, and using edible insects, a knowledge that has often been carried down through the generations by word of mouth. In this way, we risk the depreciation or loss of a nutritious food source that is practically cost free and constitutes a natural renewable resource of wide distribution. Ironically, in many advanced countries, there is a tendency among the middle and upper social classes to "return" to natural and therapeutic foods, including insects (Evans, 1993).

Entomophagy: An Ancestral Nutritional Habit

Entomophagy is a deeply rooted nutritional habit among the majority of the world's populations of the tropical area. Various insect species are eaten according to local values, beliefs, credence systems and cosmogony, following the ecological opportunities and limitations concerning the availability of the resource (Ramos-Elorduy, 1999).

Evidence of enthomophagy exists dating back to the earliest phase of human evolution and can be traced throughout all of human prehistory (Sutton, 1990; McGrew, 1981; Lanfranchi, this volume; Tommaseo-Ponzetta, this volume).

Entomophagy still persists because humans have found insects at every step of their progress, especially those people who live in harmony with nature. Various ethnic groups named and classified them and were very aware of the right time to capture them, in order not to extinguish this resource.

Insects as a Code of National Identity

Consumption of insects serves as an identification factor among different people.

In warmer parts of the planet, edible insects have been in constant supply, whereas in temperate or colder zones with fewer species, invertebrates could not represent a consistent nutritional option.

Countries in temperate zones often conquered many tropical ones, as a rule also imposing their alimentary habits. In general they considered edible insects a low prestige food, and for many people entomophagy was looked upon as a last option, even if—due to their exquisite flavor—some insect species were accepted by the conquerors as delicacies or a rare food (Ramos-Elorduy, 1996a).

In indigenous languages the names given to insect species may assume a particular meaning, which can be mystical, religious or even magical, or allude to a special ceremony or festivity. Some edible species are associated with fellowship, love, protection, care or affinity, so people of different cultures make mental associations, which are deeply rooted in their cultural background (Ramos-Elorduy, 2000).

Migrations compelled people to adapt and adjust to new cultures and new nutritional diets, accepting new general rules, with new perspectives, and establishing transitional forms of behavior. These transformations brought about decisive psychological change in individuals, who concomitantly tried to conserve some of their traditions and social customs—in which food held a special role (Paoletti et al., 2001)—as a means of expressing themselves, communication, and self-identity. Some edible insects, being part of this "special" food, are sent to the people living in a new country by their relatives or brought to them by friends or compatriots (Ramos-Elorduy, 1996a). Therefore, the demand for some insect species comes principally from people who emigrate to other regions who currently constitute a real and potential market.

Ways of Obtaining Food Resources

The poor nutritional status of the greater part of humanity is associated with the degradation of our natural and cultural patrimony. Most modern agrarian policies do not take into account the impact on the environment and generally favor new technological input and highly productive practices. Such policies contrast with the conditions required to conserve and preserve ecosystem resources so as to guarantee sustainability in food production. Modern agrarian policies often neglect the potential benefits of diverse environmental units and their existent biodiversity, thus missing the opportunity of extracting and using various resources that could help provide basic necessities for the population. Also, most agricultural production is directed toward exportation, even at the expense of local food demand. Worldwide, the agricultural picture is characterized by land exhaustion, fertilizer and pesticide pollution, and deforestation. This, combined with high demographic pressure and the spread of poverty and malnutrition, obliges us to pay serious attention to other natural and renewable food resources.

There are two basic approaches to farming, one the traditional approach based on extraction of existing resources, and the second, termed here the agroindustrial approach, in which appropriation of resources is based on a profound and continuous alteration of the ecosystem by humans. This is a relatively modern product of the scientific and industrial revolution and is dependent on the industrial world for a continuous supply of required inputs such as fossil energy sources for fertilizers and pesticides (Toledo, 1995). In the traditional approach the farmer views nature as a self-sufficient holistic system; contrarily the agroindustrial approach considers nature a separate entity to be subjected to manipulation, exploitation, and domination (Toledo, 1995).

Importance of Traditional Knowledge

Traditional knowledge (habits, beliefs, values, behaviors, myths, rites, etc.) is often ignored, as currently happens when—due to irrational management of the environment—the biota and numerous natural ecosystems may disappear or their impoverishment provoke an extinction of a species. On the other hand, industrialized societies are not the only ones to change: indigenous cultures are also changing, but at a different rate (Motte-Florac, 1995).

All these facts influence people to emigrate thereby weakening their traditional knowledge base. Also, young people are losing interest in ancient wisdom, so indigenous people and knowledge are threatened with an imminent elimination, which would be an irreplaceable loss (Motte-Florac and Ramos-Elorduy, 2002).

Preservation of entomological knowledge in terms of biological diversity, aside from aiding indigenous people, would assist the western world in widening the base of ideas for solving some of today's vital problems.

The western world would benefit from an awareness of the positive value of insects. Traditional knowledge could provide profitable alternatives which would avoid a systematic extermination of species through the large-scale use of pesticides and insecticides.

Traditional knowledge of insects sheds light on their remarkable nutritional values as well as on new ways of using them in organic agriculture, medicine, technology, and the arts.

Clement (1993) stated that the potential importance of insects in human and animal diets is widely ignored by those attempting to plan for sustainable development in the world.

Insects from a Biological and Agronomic Point of View

Insects are the major animal group on the earth. They provide an enormous biodiversity and constitute a huge amount of biomass. The actual number of insect species is not known but has been estimated to be circa three to four million (Mittenmeier, 1988). Insects are found everywhere because they colonize any type of habitat in the aquatic and terrestrial milieu, playing a vital role in an area's ecology (Sutton, 1988; Chauvin, 1968). Insects are mostly primary consumers and due to their high rate of reproduction tend to dominate all sources of energy because of competitive exclusion. The many benefits insects offer us are often overlooked and underestimated. For instance, they can be used in human and animal nutrition, in medicine, and also as recyclers of organic matter.

Many insect species constitute pests in a large variety of crops under different conditions and in different countries. The biomass of insect pests is significant and enormous quantities of toxic substances are used to kill them. Indeed, many insect pests are exterminated at a significant cost, whereas they are perfectly edible and could be used as a food source (Ramos-Elorduy and Pino, 1994). In fact, insect pests are eaten by several ethnic groups (Ramos-Elorduy and Pino, 1993a) and many species are preserved and stored for consumption, e.g. locusts and caterpillars of different species in Africa and *Sphenarium* grasshoppers and many bugs in Mexico. Indeed, it is ironical that many international and nongovernmental organizations try to save crops that contain no more than 14% protein by killing another food source (insects) that may contain up to 75% high-quality protein.

Biodiversity and Biogeography of Edible Insect Species

Perhaps because of their ubiquitous presence, insects have been part of the human diet since ancient times throughout the world. As written in the Bible or the Koran, they are chosen because of their kindness, harmlessness, and flavor, even though many Westerners still consider entomophagy a sign of "barbarism" (Ramos-Elorduy, 1990). Many different insect species are eaten in different countries around the world. To date the author and her colleagues have recorded 1,681 edible insect species consumed by people in five continents (Ramos-Elorduy and Conconi, 1994; Ramos-Elorduy, 1998) (Table 14.1): if we add to this the 230 unclassified edible insect species reported by Paoletti et al. (2001), the number reaches 1,911 species; but perhaps it is lower since some of the 230 unclassified species may have been reported earlier.

Table 14.2 shows the geographic distribution of edible insect species recorded to date as well as the number of known entomophagic countries in each continent: tropical and subtropical populations are the greatest consumers of edible

Table 14.1: Number of edible insect species recorded in the world

Order	Common name	No. of species
Thysanura	Silverfish	1
Anoplura	Lice	3
Ephemeroptera	Mayflies	19
Odonata	Dragonflies	29
Orthoptera	Grasshoppers, Cockroaches, and Crickets	267
Isoptera	Termites	61
Hemiptera	Bugs	102
Homoptera	Cicada and Leafhopper, Mealybugs	78
Neuroptera	Dobsonflies	5
Lepidoptera	Butterflies and Moths	253
Trichoptera	Caddishflies	10
Diptera	Flies and Mosquitoes	34
Coleoptera	Beetles	468
Hymenoptera	Ants, Bees, Wasps	351
Total		**1681**

Source: Ramos-Elorduy J., 2004.

Table 14.2: Consumption of insect species in the world

Continent	No. of species	Percentage	No. of entomophagic countries	Relative index of entomophagy	Rank
Africa	524	30.03	35	14.97	3
America	679	38.91	23	29.52	1
Asia	349	20.00	18	19.38	2
Australia	152	8.71	14	10.85	4
Europe	41	2.35	12	3.42	5
World	**1,745**	**100**	**102**		

Modified from Ramos-Elorduy, 1997a.
Relative Index of Entomophagy determined by dividing the total number of species in each continent by the number found in entomophagic countries.

insects. The species numbers are greater than those of Table 14.1 because some species are eaten in more than one continent.

From the biodiversity point of view, it is interesting to compare the total number of known edible insect species with the number of vertebrate spicies generally used to obtain animal proteins. For example, in the Mezquital Valley, Hidalgo, Mexico (Ramos-Elorduy and Pino, 1979), 61 edible insect species are eaten compared to 15 vertebrate species. Note that the number of edible insect species "officially" recorded to date probably constitutes just a small fraction of their total number because few scientific studies have been undertaken in this field. Table 14.3 lists the major studies on edible insects throughout the world.

Table 14.3: Principal studies on edible insects throughout the world

Author(s)	Year(s)	No. of species	Country according to date
AFRICA			
Quinn	1959	15	South Africa (Pedi Culture)
Leleup and Daems	1969	7	Zaire (Kwango)
Gelfand	1971	16	Zimbabwe (Shona Culture)
Bahuchet	1972–1978	41	African Central Rep. (Aka, Mbuti)
Chavanduka	1975	20	Various countries
Santos Oliveira et al.	1976	4	Angola
Malaisse and Parent	1980	35	Zaire
Muyay	1981	51	Zaire (Bandundu, Yansi, Nsalaba)
Kodondi	1984–1987	4	Zaire
Nkuoka	1987	26	African Central Rep.
Pagezy	1990–1993	31	Zaire (Twa)
Takeda	1990	19	Zaire (Ngandu)
Malaisse and Parent	1991	33	Zambia
Roulon-Doko	1991	185	African Central Rep.
Malaisse	1997	112	Zaire
Latham	2002	12	Zaire, Democratic Republic of Congo
ASIA AND AUSTRALIA			
Nguyen-Caong-Tieu	1928	29	Tonkin, China
Bristowe	1932	36	Thailand
Reim	1962	40	Australia
Van Der Meer	1965	48	Indonesia (Karo Batak)
Meyer-Rochow	1973–1997	61	New Guinea (Papua), Australia and northeast India
Mitsuhashi	1980–1999	119	Japan
Luo Zhi Yi	1997	18	China
Tommaseo and Paoletti	1997	11	Irian Jaya, Indonesia
Yhoung-Aree, Puwuastien, Attig	1997	19	Thailand
Xiaoming Chen and Ying Feng	1999	177	China
AMERICA			
Essig	1931	23	USA
Ruddle	1973	13	Colombia (Yukpa)
Pereira	1974	52	Brazil (Amazonas)
Ramos-Elorduy	1974–2002	504	Mexico (46 ethnic groups)
Ebeling	1986	15	USA
Dufour	1987	19	Colombia and Venezuela (Tatuyo)
Sutton	1988	20	USA (Great Basin)
Onore	1997	93	Ecuador
Paoletti	2000–2002	230	Amazonia (Amerindian groups)

Modified from Ramos-Elorduy, 1997a.

The studies cited here include only a major research in which the species were taxonomically identified. Many other studies have been done but the authors simply mention that people consume grasshoppers, ants, wasps, crickets, worms or honey, without classifying or identifying the species ingested. Recently,

Smith (1992) of the World Conservation Monitoring Center reported 19 edible insect species. Perhaps this low number is due to westerners only now acknowledging insects as an alimentary resource.

Biomass of Edible Insect Species

Insects in nature constitute a nonnegligible biomass, as exemplified by insect pests. In particular, grasshoppers represent tons of wasted edible insect protein. Examples of the biomass provided by some edible insect species are listed in Table 14.4.

Table 14.4: Quantity of various edible insect species collected in different countries

Insect group	Biomass	Country	Species (reference)
Grasshoppers	9 tons/year	Algeria	*Locusta and Schistocerca* (Gunn, 1960)
	10 tons/year	Thailand	*Oxya velox* (I.F.N.L., 1987)
	> 10 tons/year	Mexico	Oaxaca, *Sphenarium genus* (Ramos-Elorduy, 1996a)
	5–6 tons/year	USA	*Anabrus simplex* (DeFoliart, 1989)
Termites	1,331 kg/month	Zaire	*Macrotermes* spp. (Phelps, 1988: in I.F.N.L., 1988)
	large quantities	Philippines	*Macrotermes gilvus* (I.F.N.L., 1988)
	large quantities	African Central Rep.	(Roulon-Doko, 1999)
	Zillions, twice a year	Indonesia: Java	(Wasousky, 1993)
Bugs	3 tons/family/year	Mexico: Tulancalco, Hidalgo	*Thasus gigas* (Ramos-Elorduy and Conconi, 1996)
	5 tons/family/year	Mexico: Cuautla, Morelos	*Euschistus* spp. and *Edessa* spp. (Ramos-Elorduy and Conconi, 1996)
Butterflies and Moths	280–300 kg/year	African Central Republic, Zaire: Kwango, Shaba Meridional	*Anaphe, Bunaea, Lobobunaea, Nudaurelia, Gynanisa maja ata* Strand (Bahuchet, 1972–78; Leleup and Daems, 1969; Malaisse and Parent, 1980, 1991; Kodondi, 1984–87; Pujol, 1990)
	5 tons/year	Zaire: Kwango	*Anaphe, Bunaea, Lobobunaea, Nudaurelia* (Adriaens, 1951)
	35 kg/year	South Africa, Bapedi ethnos	*Gonimbrasia belina* (Quinn, 1959)
	20 kg/year	South Africa, Pedi ethnos	*Gonimbrasia belina* (Quin, 1959)
	140 kg/person/year	Zambia	Mopanie worms (*Gonimbrasia belina*) (I.F.N.L., 1991)
	1600 kg/year	Zimbabwe, Botswana	(I.F.N.L., 1988)
	2000 kg/year	India	*Bombyx mori* (Ichponani and Malek, 1971)
	3000 kg/year	Mexico: Chiapas	*Latebraria amphypirioides* (Ramos-Elorduy and Conconi, 1996)
	many tons	USA	*Coloradia pandora* (Blake and Wagner, 1987)
	183 metric tons	India, Assam	Eri silkworm (Chowdhury, 1982)

Table 14.4: (Contd.)

Table 14.4: (Contd.)

Beetles	5,000 samples/day	Cameroon	*Popillia japonica* (DeFoliart, 1989)
	2000 samples/h	Colombia, Venezuela: Tatuyo Ethnos	*Rhyncophorus palmarum* (Dufour, 1987)
	1.5–2 kg/h	New Guinea	*Rhyncophorus palmarum* (Bergier, 1941)
Ants	39 tons/year/ family	Mexico: Arriaga Chiapas	*Atta* spp. (Ramos-Elorduy and Conconi, 1996)
Wasps	many kg/year	Japan	*Vespula lewisi* (Mitsuhashi, 1995)
Aquatic insects	16 tons/year	Zaire	Kasai (Kitsa, 1989)
Mayflies	1 kg/day/season	France	Mayflies (Fontaine, 1959)
Insect Consumption	12,000 tons/year	Zaire	Kananga (Kitsa, 1989)

I.F.N.L.: Insect Food Newsletter, Ramos-Elorduy 1997a.

Insect Consumption Patterns

Since insect consumption is part of a population's cultural heritage, knowledge of how to find, gather, prepare, and conserve insect resources is handed down through generations by word of mouth. Children learn from their parents how to find or collect insects through imitation or instruction. Most insect species are captured casually but in some cases people actively search for or trap them. Insect species are selected for consumption according to seasonal presence and abundance, so as to economize on the energy and time used in catching them (Ramos-Elorduy, 1997b).

Consumption of insects is virtually always part of the "normal" diet. However local celebrations in honor of some insect species are held in several places (Bodenheimer, 1951; Ramos-Elorduy and Pino, 1988, 1989).

Insects are eaten in all developmental stages, i.e., eggs, larvae, pupae, and as adults. However, most are eaten in immature stages when the exoskeleton is reduced and softer (Ramos-Elorduy 1982). Among the principal orders, the species consumed in greater quantities are the larger and more succulent ones. Insects generally have a crunchy consistence, often enhanced by frying or roasting, so that when they are ingested salivation and mastication signal "I've had enough to eat".

Insects may be eaten daily and in quite large numbers in many different areas. Often a sequence exists in the consumption of different edible species in the course of the year. A recent study in Kananga, Zaire, recorded the quantity of insects consumed among 2,006 people: an average ingestion of 663 g of insects per person per month was found or 22.1 g dried insects/day (50 g fresh insects/day). Kitsa (1989) calculated an insect consumption in the region of about 12,000 tons/year. Paoletti et al. (2000) determined the annual consumption of invertebrates in the Tukanoan Village of Yapu, Colombia, with a population of about 100 people, as: *Atta* soldiers 100 kg/y, *Syntermes* soldiers 133 kg/y, wasps larvae and pupae (3 species) 2 kg/y, stingless bees (one species) 1.5 kg/y, palm worm 6 kg/y, woodborer beetle larvae 2.5 kg/y (Paoletti et al., 2000, reports as kg y^{-1}).

Thus, even if we do not know precisely the quantity of insects eaten daily, nor their amount in weight, it can be seen that they form a significant part of the diet of many people

Storage and Preservation

In some areas insects are a major staple food and during periods of abundance, when insect collection occurs, peasants preserve and store them for later consumption. Preservation of insects is generally done by drying them in the sun, over hot ashes, or in earthen ovens. Boiling and preserving them in salt is also common.

Preparation

Insects are roasted, fried or boiled and generally eaten in combination with other food items. Only a very few species are eaten live, e.g. some "*Jumiles*" or honey ants. However in the Amazon many insect larvae, such as palm worms are eaten raw on the spot (Dufour, 1987; Paoletti et al., 2000). In the Amazon, peasants have used wood worms as a source of fat for frying (Ramos-Elorduy 1997a).

Palatability

The organoleptic characteristics of insects, such as their crunchy texture, absence of odor, and white coloration of most of their larvae, as well as their flavor, could favor their acceptance among Western people. The flavors of edible insects are not unknown to us. Some insect species are said to have a flavor reminiscent of apples, almonds or pine nuts. Others have no special flavor and take on that of the oil in which they are fried or the relish dressing (e.g. garlic, onion, lemon or chili). Aquatic insects often have a fish flavor when eaten fresh, and a flavor similar to that of dried shrimps when preserved. People who traditionally consume insects know how to prepare them to make a good and tasty meal.

Gathering

Methods for capturing insects are usually fairly simple. Often they are collected by hand or netted. Sometimes big sticks are used as well as scoops or barretas, machetes or axes. It is of course necessary to handle tools of this kind efficiently.

Seasonal Capture

Some species, such as wasps, are caught year round; others are caught only in one or two seasons of the year when their abundance curve peaks. Peasants do not catch them when their abundance declines, which permits them to fully mature mate, and thus ensure survival of the species (Ramos-Elorduy et al., 1997, 1998, 2002).

Nutritive Value of Edible Insects

Insects generally contain a high content of good quality protein, vitamins, and minerals and also contribute a significant amount of energy. Protein content generally ranges from 30% (wood worms) to 81% (*Polybia* wasp). Only honey ants have a low protein content of 9.63 %. Figure 14.1 shows the variation in protein content in species of different orders.

In Figure 14.1 it should be noted that the variation in protein quantity of edible insects is high, the highest seen in Hymenoptera and the lowest variation in order Coleoptera, even though the highest number of species is seen in this order. It may also be observed in Figure 14.1 that fish is the only food with a proteinic content higher than the majority of insect species, but equal to those of *Polybia* wasps from La Mixteca, Oaxaca (81%) (Ramos-Elorduy, 1997a).

The biological quality of insect protein is good, having chemical scores from amino acid profiles comparable to the WHO/FAO/UNU (1985) pattern, that ranges from 10% to 96%. However, some insect species have low values of lysine and tryptophan, in relation to those recommended by WHO/FAO/UNU (1985) for preschoolers and adults: for adults their values are higher than those

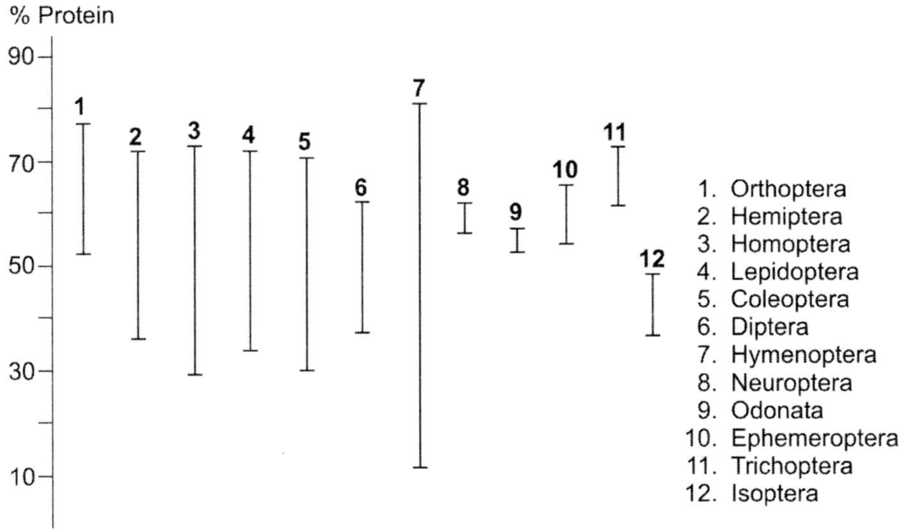

Fig. 14.1: Rank of various orders of edible insects of Mexico, according to the protein content (%) in each species in each order compared with different conventional alimentary products (From Ramos-Elorduy J., 1997a). A—Beans, B—Lentils, C—Soybean, D—Chicken, E—Egg, F—Cattle, G—Fish.

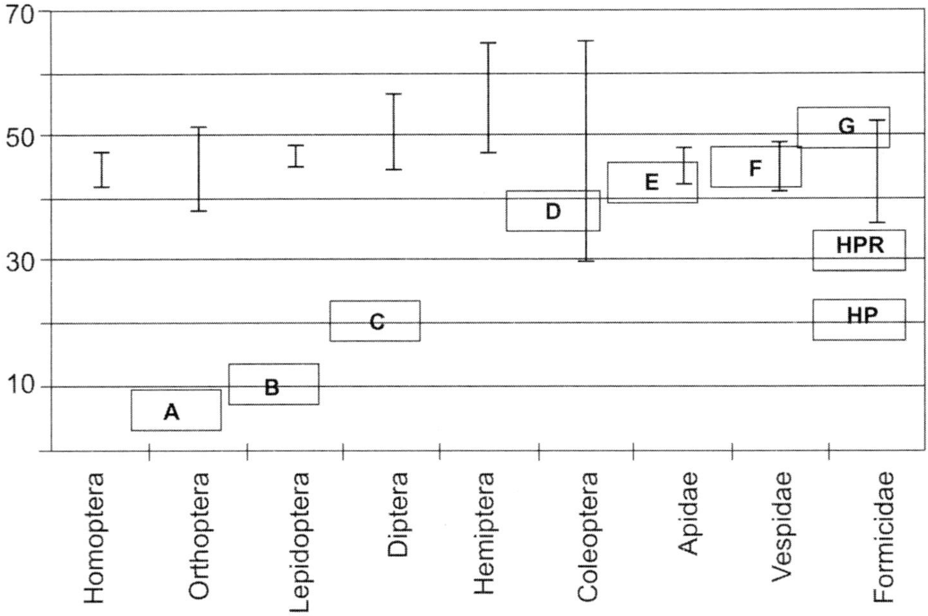

Fig. 14.2: Rank of various orders of edible insects of Mexico. According to essential amino acid total content, compared with different conventional alimentary products (modified from Conconi, 1993). A—Beans, B—Lentils, C—Soybean, D—Fish, E—Poultry, F—Cattle, G—Egg. HPR—WHO/FAO/UNU Pattern (1985) for preschoolers and adults.

recommended but for preschoolers they are deficient in some values (Ramos-Elorduy et al., 1982; Conconi, 1993; Ladrón de Guevara et al., 1995; Ramos-Elorduy, 2004). In Africa, some insect proteins show a serious deficiency of isoleucine (Kodondi et al., 1984). Bukkens (1997, and this volume) has given a very good summary of the amino acids found in edible insects and their values.

It is important that foods contain high quantities of proteins but even more important than the quantity is their quality, which depends on their amount of essential amino acids. Variations within and between the various orders have been observed. Lepidoptera and Hymenoptera (Apidae) show a minimal variation, while the highest variability is found in order Coleoptera (Fig. 14.2). Nonetheless compared with the same conventional products mentioned before and the quantities recommended by WHO/FAO/UNU (1985), it is apparent that fish which have the highest protein content are surpassed here by many insect species, because their amount of essential amino acids is lower. For this reason, edible insects are a good nutritional food in terms of the great importance they have in the optimization of the body's functions.

Fats

The quantity of fats edible insects contain varies according to the species analyzed and their developmental stages. The larval stage of orders Lepidoptera, terrestrial Coleoptera, Hymenoptera (wasps and ants), Pentatomidae, Hemiptera, and holometabolous species in general, which have different alimentary habits, are high in fatty acids.

The type of fats found in these insects are mostly polyunsaturated, so they do not harm the body. If we compare the different fat types of edible insects with other conventional animal organisms, we could say that insects mostly supply the saturated and monounsaturated types (Table 14.5).

Table 14.5: Proportion of fatty acids (%) of diverse organisms for obtention of animal protein (Adapted from National Research Council, 1988)

Organisms	Saturated acid	Monounsaturated acid	Polyunsaturated acid
Cattle	52.0 (28.1)	44.2	3.2
Pig	44.1 (24.3)	44.3	11.6
Poultry	35.5 (20.2)	40.8	22.7
Fish	29.6 (22.6)	39.6	30.8
Insects	11.0–43.4 (0.1–9.1)	55.9	57–100

For cattle, average of 27 cuts; pig 16; poultry 8; fish 3 types with 2 products of each type (e.g. Halibut, Tuna); insects maximum and minimum values from 27 analyzed species (percentage of stearic acid given in parentheses). Modified from DeFoliart G., 1991.

Bukkens (this volume) analyzes and compares results obtained by different authors and mentions that "there is little similarity in the fatty acid composition of related insect species (from the same taxonomic family) collected in different locations. However, the fatty acid composition of related species collected in the same location is similar. This suggests that the fatty acids composition is largely influenced by the host plant on which they feed. All food insects analyzed are a significant source of the essential fatty acids linoleic acid (C18: 2n-6) and linolenic acid (C18:3 n-3).

Energy

Most edible insects are rich in fat and supply ample food energy (Ramos-Elorduy and Pino, 1990a).

Fats are the compounds that bring the highest amount of energy to the body, i.e., twice the amount given by carbohydrates or proteins. Proteins cannot be efficiently assimilated if energy is insufficient in the diet, so human beings need a certain amount of calories every day in order to utilize proteins properly. The caloric value of insects varies, every order being ranked differently according to

Table 14.6: Rank of calories of diverse orders of edible insects in Mexico

Insect Order	Kcal/100 g	Difference
Odonata	431–520	89
Ephemeroptera	354–355	1
Orthoptera	336–438	102
Isoptera	347–508	161
Hemiptera	329–629	300
Homoptera	394–469	75
Neuroptera	332–366	34
Lepidoptera	293–777	484
Coleoptera	283–653	370
Diptera	217–499	282
Hymenoptera	380–561	181
Conventional products		
Cereals	330–370	40
Vegetables	308–352	44
Legumes	388–421	33
Meats	165–705	540

Modified from Ramos-Elorduy, 2000.

the species type involved. Those belonging to the aquatic environment have a lower amount, but a substantial variation exists in the terrestrial environment, with 76.13% of the species studied (Table 14.6) ranging from 370 to 570 kcal/100 g and those insect orders with the highest fat content having the most calories, as mentioned above.

Almost all species have energy values higher than those of everyday foods of plant or animal origin (Table 14.6); only pork contains more calories than the majority of edible insects. This is an important fact because insects provide the energy needed by the body to carry out all its various functions and activities.

Minerals

Minerals are indispensable inorganic elements because the body is not able to synthesize them. They have an impressive variety of metabolic functions, as they build, activate, regulate and control many chemicals, e.g. iron in the hemoglobin, zinc in insulin. They are divided into three groups: macronutrients (Ca, P, K, Mg, Cl, S), micronutrients (Fe, Cu, I, Mn, Co, Zn, Mo), and ultramicronutrients (F, Cd, Li, Cr, Se, B) (Ramos-Elorduy et al., 1998a).

The sodium, potassium, calcium, zinc, iron, and magnesium contents among the various orders are shown in Table 14.7. Orders that have the lowest variation are Orthoptera, Lepidoptera, and Hymenoptera.

Edible insects are low in sodium and sometimes in calcium, high in zinc, iron, and potassium, and rich in magnesium.

Table 14.7: Range of content of diverse mineral salts of some orders of edible insects in Mexico compared with conventional products (g/100 g)

	Na	K	Ca	Zn	Fe	Mg
Insects						
Orthoptera	0.066–0.609	0.044–0.574	0.051–0.120	0.016–0.078	0.016–0.044	0.352–0943
Hemiptera	0.020–0.572	0.014–0.256	0.075–0.104	0.024–0.112	0.012–0.130	0.744–2.550
Lepidoptera	0.048–0.544	0.048–2.912	0.048–0.088	0.022–0.040	0.017–0-054	0.384–1.628
Hymenoptera	0.063–1.608	0.063–1.030	0.040–0.224	0.016–0.050	0.014–0.046	0.348–1.129
Conventional Products						
Other animals						
Cattle	0.060	0.370	0.01	0.00042	0.028	0.025
Poultry	0.086	0.321	0.02		0.015	0.023
Fish	0.104	0.256	0.01	0.0025	0.0302	0.023
Turkey				0.00296		
Milk			0.12	0.00334	0.0001	0.01
Egg			0.05	0.00144	0.023	0.01

Modified from Ramos-Elorduy, Muñoz and Pino, 1998a.

But to make a true comparison among the species, they must be at the same stage of development and if dried need to have undergone the same method or type of preservation.

According to these results, the majority of edible insects have an adequate proportion of total ashes and a very high proportion of K, Ca, Fe, Mg. Generally their content of the different minerals studied for conventional consumption, shows that edible insects can supply the daily mineral requirements of humans, depending on age, sex, activity, and physiological state.

Some insect species contain important quantities of some of the mineral salts. For example, termites are rich in calcium and sulphur, and grasshoppers are rich in iron and zinc. The magnesium content of Mexican edible insects is higher than that of all other food products studied. Calcium, iron, and potassium contents are higher than those of most food products of plant and animal origin. None of the species provides lithium (Ramos-Elorduy et al., 1998a).

Vitamins

Vitamins are micronutrients indispensable in the human diet because, as happens with minerals, the body is not able to synthesize them properly. They control many metabolic functions such as the efficient working of the essential enzymatic system. Almost all of them are water soluble (e.g. vitamin C and those of the B group), or liposoluble (as A, D, E, and K); their sources are of plant or animal origin (Ramos-Elorduy and Pino, 2001a).

Insects are rich in B group vitamins such as niacin, riboflavin, and thiamine (Ramos-Elorduy et al., 1988) that are often very scarce in tropical diets. Kodondi et al. (1987) reports rich vitamin contents in some Attacidae, edible caterpillars, of Zaire.

Insect Consumption

Insects thus have the potential for improving people's diet by significantly contributing to their protein intake and reducing deficiencies in minerals and vitamins (Ramos-Elorduy et al., 1984). For instance, in a study done by Belgian economists in Zaire (formerly Belgian Congo), it was found that in some villages insects constitute 81% of the total animal protein ingested by people. Unfortunately, only caterpillars and grasshoppers were reported and without classification or specifying their nutritive value (Gomez et al., 1961). In other parts of Africa insects likewise constitute an important part of the daily diet (Gelfand 1971; Kodondi et al., 1984; Oliveira et al., 1976; Owen, 1973; Quin, 1959; Takeda, 1990). In the Central African Republic, caterpillars (Attacidae in Europe = Saturniidae in America), *Pseudantherea discrepans,* may provide up to 50% of the total proteins consumed by the Babinga dwarfs (Bahuchet, 1972) and over 50% among the Aka dwarfs (Bahuchet, 1978).

Commercialization of Edible Insect Species

Commercialization of edible insects is practiced in many countries (Table 14.8). It is done at different levels with insects sold in small or large quantities and either at the national or international level (Ramos-Elorduy, 1997b). They are sold live, fresh, boiled, roasted, or cooked, and almost always in different measures. Both indigenous and foreign species are frequently canned by transnational American or Japanese companies (Taylor, 1975; Mitsuhashi 1980, 1999). In Venda, South Africa, commerce with the indigenous species has reached a business worth one million dollars per year (Van der Waal, 1994).

Examples of this commercialization are given in Table 14.9; undoubtedly, many more examples exist.

Table 14.8: Countries that commercialize edible insects

Asia		Africa		Americas		Europe		Australia
Turkey	•Δ	Burkina Faso	•Δ	Canada	Δ	Belgium	Δ	Not yet recorded
China	•	R.C.A.	•Δ	Mexico (95 sp.)	•Δ	France	Δ	
Japan	•Δ	Zaire	•Δ	Brazil	•Δ	Spain	Δ	
Thailand	•Δ	Congo	•Δ	Guatemala	•	Germany	Δ	
		Botswana	•	Colombia	Δ			
		Zimbabwe	•	Peru	•Δ			
		Tanzania	•	Venezuela	•Δ			
		South Africa	•	Ecuador	•			
				United States	•Δ			

Level: • National, Δ International (from Ramos-Elorduy, 1997a).

Table 14.9: Commercialization and prices of some edible insect species

	Quantity	Cost*	Reference—Country
Canned products			
White Agave worm	30 g	50 Cdlls	Ramos-Elorduy and Conconi, 1996—Mexico
Baby Bees	2 ounces	220 dlls	Hocking and Matsumura, 1960—Canada
Atta Ants	453 g	20 dlls	Contessi, 1993—Colombia
Locusta Grasshoppers	6.5 g	8 dlls	I.F.N.L., 1988—USA
Bombyx mori pupae	35 g	50 Yens	Mitsuhashi, 1995—Japan
Vespula lewisi larvae	40 g	50 Yens	
Oxya japonica larvae	35 g	50 Yens	
Hepialus armoricanus	20 worm	35 dlls	I.F.N.L., 1993—China
Liometopum apiculatum	1/2 K	100 dlls	Ramos-Elorduy and Conconi, 1996—Mexico
Aegiale (A) *hesperiaris*	1/2 K	150 dlls	
Euchistus taxcoensis	1/2 K	80 dlls	
Euchistus strennus	1/2 K	65 dlls	
Sphenarium purpurascens	1/2 K	35 dlls	
Prepared food			
Tenebrio molitor	5 g	45 dlls	I.F.N.L. 1993—USA
Acheta domestica	10 g	65 dlls	
Liometopum apiculatum	30 g	20 dlls	Ramos-Elorduy et al. 2003—Mexico
Aegiale (A) *hesperiaris*	30 g	25 dlls	
Xyleutes redtembacheri	30 g	21 dlls	

Price variations over time are common. For example, price variation for different (85) edible species in Mexico ranged from 108% to 6500% (13 years) (Ramos-Elorduy and Conconi, 1996); price variation for rice grasshopper in Thailand: 2,333% to 5000% (9 years) I.F.N.L.: Insect Food Newsletter, 1992 (modified from Ramos-Elorduy, 1997a).

Commercialization is stimulated by the deeply rooted habit of eating insects by some immigrant groups; edible insects are not a common and generalized product (Ramos-Elorduy, 1996a). Indeed, most edible insect species belong to particular habitats or ecosystems in a certain country and can only be obtained elsewhere by means of commerce and trade.

Edible insect products are subject to price changes as are many other alimentary products. No product fetches a high price if demand is nil. For the people of rural areas, edible insects often have no monetary cost but offer the possibility of some cash income if sold (Ramos-Elorduy and Conconi, 1996). According to Evans (1993), commercialization would aid in the preservation of many insect species.

Breeding Edible Insect Species

Breeding Techniques

Bee rearing is well known (*Apis* spp., *Melipona* spp., *Trigona* spp., and *Lestrimelitta* spp.) in various countries (Brazil, Mexico, China) but we can assert that the culture of other edible insect species is not as well established as that of native stingless bees raised in ceramic pots (*Trigona* species) (Ramos-Elorduy, 1982) or in empty tree trunks (*Melipona* species) (Weaver and Weaver, 1981) and that of honeybees in baskets or big hollow trunks covered with dung. The existing culture of some species, such as the "*escamoles*" ants (*Liometopum apiculatum* and *Liometopum occidentale* var. *luctuosum*) or the "madroño butterfly" (*Eucheria socialis*), is more a matter of taking care of the nest than actually culturing and various wasp species of *Polybia* were termed a "protoculture" by Ramos-Elorduy et al. (1997).

Cultures of edible insects generally do not need a complex infrastructure and their care is simple. For instance, in some cases (ants, bees, wasps) they feed themselves, while in other cases residues of organic matter of plant and animal origin can be utilized, which translates into reduced costs and a tidy profit. Kok (1983) proposed a model to produce insects for human consumption on a large scale.

Another type of culture for human or animal nutrition occurring in premises with controlled conditions, recycles organic wastes for, say mealworms, crickets, cockroaches or flies (Ramos-Elorduy, 2000).

Insects constitute an important part of the trophic chain in various ecosystems. They largely nourish birds, amphibians, fish, and/or certain reptiles. They represent approximately 70 to 80% of bird diets. Thus insect cultures can also be used indirectly for human nutrition, through their use in animal nutrition.

Conversion Efficiency

Insects show a good conversion efficiency, which is an important fact for the commercial production of animal protein. The average conversion ratio of some insect species studied is 4 to 5 : 1. In comparison, the ratio for poultry is 2.6 : 1, for sheep 19 : 1 and for livestock 20 : 1 (Taylor, 1975). This good conversion efficiency is partly due to the poikilothermal nature of insects which implies that they do not have to allocate a large portion of their food to maintaining their body temperature. For instance, crickets may be smaller than cows, but they convert plants into biomass five times faster.

Here are some examples of conversion efficiency of different animals nourished with different insects (Table 14.10).

Recycling and Animal Nutrition

Insects are essential agents feeding on organic matter in nature. Insects can act as efficient biotransformers (Ramos-Elorduy, 1996b) converting organic wastes into animal biomass rich in proteins and suitable for use in animal nutrition. These organic wastes are constant, abundant, and have a very low or zero cost.

Table 14.10: Conversion efficiency obtained when animals were fed with insects coming from organic waste recycle in Mexico

% inclusion	C.E.	Animals
Tenebrio molito		**Poultry**
0	1.37	Arbor Acress × Vantres
5	1.39	(Ramos-Elorduy. Avila, Rocha, and Pino, 2002)
10	1.34	
0	1.36	Isa Brown
5	1.27	(Ramos-Elorduy. Avila, Pino, and Martinez, 2002)
10	1.32	
20	1.20	
0	1.90	Arbor Acress
2.5	1.98	(Perdomo, 1999)
5	2.37	
10	2.33	
Musca domestica		
0	1.63	Indian River
50	1.38	(Ramos Calcaneo et al., 2002)
10	1.45	
25	1.37	
Cochlyiomia hominivorax		
0	1.71	Indian River
5	1.29	(Ramos Calcaneo et al., 2002)
10	1.26	
25	1.24	
30	1.30	
Sphenarium purpurascens		
0	1.41	Indian River
30	1.47	(Ramos Calcaneo et al., 2002)
Axayacatl		
0	1.59	Arbor Acress
2.5	1.78	(Barrera, Ramos-Elorduy, Avila, and Pino, 2000)
5	1.90	
10	1.93	
Tenebrio molitor		**Ostrich**
0	1.63	(*Strutio camelus* L.) race African Black
5	1.305	(Ramos-Elorduy, Avila, Medina Sánchez, and Pino, 2001d).
Periplaneta americana		**Fishes**
0	2.6	*Carassius auratus* L.
10	2.5	(Hernández, Ramos-Elorduy, and Pino, 2000)
20	2.8	
30	2.2	
Musca domestica		
0	1.38	*Salmo gardnieri* (Rainbown trout)
100	1.28	(Ramos-Elorduy, Villegas, and Pino, 1988)
Tenebrio molitor		**Pig**
Tenebrio 8%	1.67	York Landrace White to weaning 23 days of age
Pig Plasma 8%	1.59	(Ramos-Elorduy, Gamboa, Pino, Borbolla, and Avila, 1997)
Soja 31.37%	1.90	

C.E. = Conversion efficiency (modified from Ramos-Elorduy, 2000).

In fact, insects tested for their capacity to recycle organic waste developed well in most of the substrata offered, generally showing good conversion efficiency and producing more biomass and weight than those checked in optimal conditions (Ramos-Elorduy, 1996b) (Table 14.11). The nutritive value of the insects raised varied according to the media used but was generally slightly superior to controls. Time to recycle wastes varied depending on the insect species and on the substrate employed. With substrates that provide a balanced diet, from 92 to 95% of the media was consumed and transformed into insect tissue (Ramos-Elorduy and Pino, 1990b). The degree of transformation of proteins ranged from 5 to 8% bad-quality proteins in the substrate, to 43–61% good-quality proteins in the insects (Ramos-Elorduy et al., 1988). All these facts show the possibilities of recycling organic matter as a culture medium for insects, for the purpose of obtaining nutritious insect biomass.

Table 14.11: Important characteristics of insects

1. Large biodiversity.
2. Short life cycle.
3. Found in every ecosystem, colonizing different habitats.
4. Localization of a large quantity in the same place.
5. Numerous and acceptably edible.
6. High population and biomass.
7. High reproductive degree.
8. Enormous variety of diets and capable of utilizing many indigenous resources not used by man.
9. Form part of the traditional diet of many cultures.
10. Possess good organoleptic characteristics.
11. Their consumption constitutes in some cases a psychic reward because of sweet flavor.
12. Good nutritive value.
13. Contain a large quantity of good quality proteins.
14. Provide a significant quantity of energy.
15. Possess ample good and known flavor variations; also allow versatile preparations.
16. Generally have a crunchy consistence.
17. Easy conservation and preservation.
18. Most species are innocuous.
19. Have an enormous potential in human and animal nutrition and in medicine.
20. Their profit is integral, so total profit of their biomass.
21. Good digestibility of dry matter and proteins.
22. Breeding is generally easy.
23. Breeding of some species is controlled.
24. Breeding could be increased easily and quickly.
25. Breeding could utilize many resources not used by humans.
26. Breeding costs are generally low.
27. Cultures can use waste as inputs.
28. Culture gives an alimentary and economical return.
29. Have a better conversion efficiency than most other animals.
30. Have good resource return rates (Jones and Madsen 1991).
31. No generation of waste occurs.
32. Excreta can be used as fertilizers.
33. Good transformers of low nutritive elements into much richer compounds.
34. Can be exploited for inclusion in different products of modern food technology.

Trials performed with insects in animal nutrition show that they can be substituted for soybean or fish flour in the feeding of poultry or fish with equal or better results (DeFoliart et al., 1982; Ramos-Elorduy et al., 1988; Ramos-Elorduy, 1996b; Ramos-Elorduy et al., 2002), and in ostrich (Ramos-Elorduy et al., 2001d) or pig nutrition (Ramos-Elorduy et al., 1997) compared to pig plasma that is a very expensive fodder. All these experiments were undertaken in the initial phases of their life, i.e., when animals need the highest quantity and best quality of proteins.

Why Insects could be a Potential Food

Because of their inherent characteristics, insects are a promising food source. Therefore, efforts should be made to know them better and make a selection of species suitable to raise as food for animal and human nutrition.

Popular interest must be promoted in traditional foods based on wild insect species, semiwild species (protected and semicultivated species in situ) and semidomesticated species (ex situ in nurseries and solar vegetable gardens) that are not under threat of extinction, and their rational exploitation encouraged through culturing to assure their sustainability.

An activity or process is sustainable when the permanent availability of the elements necessary for satisfying the present needs without endangering those of future generations can be guaranteed. The concept of "using to conserve" has had success in various programs of natural resource conservation (Evans, 1993).

Regional culturing of insects will give rise to an agroindustry that can lead to many socioeconomic benefits (Ramos-Elorduy, 1974), among which are absorption of rural labor, supply of the home market, foreign currency caption, and an improved standard of living of the rural population.

There are no official records of the magnitude of the benefits insect agroindustries have effected worldwide. However, in Mexico for instance, we know that honey (not including all the national honeybee cultures) reached a currency caption of US $52 million in 1991 (Anonymous, 1992). Other Mexican insect species are already in demand by restaurants in the United States, Finland and Germany, and in Colombia and Venezuela *Atta* are exported to Santander, Spain, with huge profits (Ramos-Elorduy, 1998).

Conclusion

True, the ideas put forward in this paper may appear unconventional and difficult to realize, but some of them have been tried with good results. Very successful agroindustries of insect species rearing are those of silkworms, shell or grana mealybugs, and several wild stingless bee species. These ancient insect breeding techniques have proven both economically rewarding and sustainable over time, which may explain the striking resemblance between existing insect cultures throughout the world (Ramos-Elorduy and Conconi, 1993). The most amazing example is that of the honeybee culture "exported" to practically all countries of the earth. Honeybee culture has provided meals and an income to people in these countries, showing that an exotic insect species can be domesticated and exploited in a sustainable manner.

The examples of insect cultivations discussed in this paper are all run by rural people. They are able to recognize, study, increase and exploit the local edible insect species in a systematic way. We can help them by providing recent information on these and other insect species, and by making the necessary arrangements to implement insect breeding on a wider scale, using the natural resources that each ecosystem possesses, be it a terrestrial or aquatic species. It is important to realize that this activity will provide skills and income to many unemployed persons. For instance, the eri silk culture in Assam, India, hires no less than 40,000 families (Anonymous, 1992). A rustic anonymous said that a cultivation generally requires low investment and provides high yields over a short period of time (Ramos-Elorduy, Pino and Cuevas, 1998b, Ramos-Elorduy and Pino, 2001b).

We should recognize the value of the traditional approach of various ethnic groups to agricultural systems in the formation of new sustainable forms of exploiting natural resources. "Small is beautiful" said Schumacher (Gibbons, 1991) and that it can also be sustainable has been proven by many ethnic groups throughout the world that collect and cultivate edible insects in small defined areas. In some parts of Africa there is even legislation for the collection of edible caterpillars, as is the case in Zambia (Anonymous, 1991).

Given their organoleptic characteristics insects are an attractive food source. They can be used in existing food technologies and some are presently used in the production of candy, cookies, dressings, sausages, etc., or even constitute a complete meal (Ramos-Elorduy, 1998). Thus insects already play a significant role in nutrition and the economy (Ramos-Elorduy, 1997b) and the vista to an economical entomology beckons. Of course, this is a gigantic task and challenge. It implies the selection of suitable species that can be readily raised in small spaces or in the laboratory, elaboration of a design to manage and control a selected species culture, and exploration of the possibility of using organic waste as input. Finding an economic and nutritional use for insect species provides an important means for precluding species extinction.

Insects are a promising food alternative, but international laws are mandatory because of the need to regulate and standardize the exploitation and commercialization of edible insects in different countries of the world (Ramos-Elorduy et al., 2003).

References

Adriaens E.L. 1951. Recherches sur l'alimentation des population au Kwango. Bull. Agric. Congo Belge XLII (3): 470–814.

Anonymous 1987, 1988, 1991, 1992, 1993. *In:* The Food Insects Newsl. II (1), IV (2), V (2), VI (1, 2).

Anonymous 1992. Reporte sobre producción y exportación de miel de abeja. Ed. SARH, México p. 52.

Arana F. 1991. Comer Insectos. Ed. Planeta, México D.F. 1ª Ed. 102 p.

Arnaud D.F., Ilbert F., Montgruel L., and Tubiana L. 1994. Biodiversity: Enjeu économique ou enjeu pour les générations futures. *In*: Biodiversity: Le Fruit Convoité, Ed. Arnaud, Ilbert and Montgruel eds. FPH/Solagral, Montpellier, Francia, p. 7–10.

Bahuchet S. 1972. Etude écologique d'un campement de pygmées Babinga (Région de la Lobaye R.C.A.). J. D'Agric. Tropic. Bot. Appliq. XIX (12): 509–559.

Bahuchet S. 1978. Introduction à l'ethnoécologie des pygmées Aka de la Lobaye Empire Centre Africain. Thèse Ecole des Hautes Etudes en Sciences Sociales. 348 pp. Paris, France.

Barrera S.C., Ramos-Elorduy J., Avila G.E., and Pino J.M. 2000. The "Axayacatl" (Insecta-Hemiptera), an edible insect as an example in animal nutrition. VIIth Int. Cong. Ethnobiology. p. 41, Ed. ISE University of Georgia, Athens Georgia U.S.A.

Beckerman S. 1979. La a bundancia de proteinas en la Amazonia: una respuesta a Gross. *Amazonia Peruana* vol. III (6): 91–126.

Bergier E. 1941. Insectes Comestibles et Peuples Entomophages. Edit. Rullières Fréres Avignon, France, 209 pp.

Beutelspacher C.R. 1989. Las Mariposas Entre Los Antiguos Mexicanos. Ed. Fondo de Cultura Económica, México, D.F. 102 p.

Blake E.A. and Wagner M.R. 1987. Collection and consumption of Pandora moth (*Coloradia pandora lindsleyi*) (Lepidop. Saturniidae) larvae by Owens Valles and Mono Lake Paiutes. Bull. Ent. Soc. Amer. 33 (1): 139–141.

Bodenheimer E.S. 1951. Insects as Human Food. Dr. W. Junk Publishers The Hague, Printed in Cd. Netherlands by Drukkerijen RBK Haarlem. 352 pp.

Bongaarts J. 1980. Does malnutrition, affect fecundity? A summary of the evidence. *Science* 208: 564–569.

Breuil H., Gómetz P.S., and Aguilo J.G. 1912. Les Peintures Ruprestres d'Espagne, Les Abris del Bosque It Alpera (Albacete), Ed. L'Antropologie, Paris, France 23, p. 629.

Bristowe M.A. 1932. Insects and otehr invertebrates for human consumption in Siam. *Trans. Ent. Soc.* Part II: 387–405.

Bukkens S.G.F. 1997. The nutritional value of edible insects. Ecol. Food Nutr., 36 (2–4): 287–319.

Bukkens S.G.F. 2004. Insects in the human diet: Nutritional aspects (this volume).

Cambefort Y. 1991. Le Scarabéeet les Dieux Ed. Boubée, Paris France 273 p.

Cartmill L. 1972. Arboreal adaptations and the origin of the order Primates in The Functional and Evolutionary Biology of primates. Ed. R. Tuttle, Chicago Illinois U.S.A., Aldine, p. 97–122.

Carrasquedo U.R., M. Flores M., J. López T., and M.E. Vázquez H. 1988. Insectos comestibles en el Estado de Tlaxcala, Tésis Prof. Univ. Autonom. Tlaxcala. México, Depto de Ciencias de la Educación 102 p.

Chauvin R. 1968. Traité De L´Abeille. Tome V. Histoire, Ethnographie, Folklore. Edit. Masson, Paris France, 152 pp.

Chavanduka D.M. 1975. Insects as a source of protein to the African. *Rhod. Sci. News* 9 (7); 217–220.

Chowdhury S.N. 1982. Eri Silk Industry. Ed. Directorate of Sericiculture and Weaving Govt. of Assam, Assam.

Clement C.R. 1993. Native Amazonian fruits and nuts: composition, production and potential use for sustainable development. *In:* Tropical Forest, People and Food: Biocultural Interactions and Applications to Development Eds Hladik C.M., Hladik A, Linares O., Pagezy H., Semple A., and Hadley M. MAB/UNESCO, and The Parthenón Publishing Group, Paris, France, pp. 139–152.

Conconi E. 1997. Estudio de mercadeo de gusano blanco de maguey. Tésis Prof. ITAM (Instituto Tecnológico Autónomo de México), México, 87 pp.

Conconi M. 1993. Estudio comparativo de 42 especies de insectos comestibles con alimentos convencionales en sus valores nutritivo, calórico, proteínico y de aminoácidos haciendo énfasis

en la aportación de los aminoácidos esenciales y su papel en el metabolismo humano. Tésis Prof; Fac. Ciencias Univ. Nal. Autón. Méx., México, D.F.: 71 pp.
Contessi S. 1993. *In:* Insect Food Newsl. VIII (2): 4.
De Castro J. 1973. Geopolitica del hambre. Guadarrama, Madrid, Spain, vol. I: 339 pp.; vol. II: 331 pp.
DeFoliart G.R. 1989. The use of insects as food and as animal feed. Bull. Ent. Soc. Amer. 35: 22–35.
DeFoliart G.R., Finke M., and Sunde M.L. 1982. Protein value of the mormon cricket (Orthoptera-Tetigoniidae) harvested as a high protein feed for poultry. J. Econ. Ent. 75: 848–852.
Dufour D. 1987. Insects as food. A case study from the northwest Amazon. Amer. Anthropol. 89 : 383–397.
Ebeling W. 1986. Handbook of Indias Food and Fibers of Arid America. Univ. of Calif. Press, U.S.A. 971 p.
Essig E.O. 1931. The value of insects to the California Indians. *Scient. Mon.* 38(2): 181–186.
Evans M.I. 1993. Conservation by commercialization. *In:* UNESCO. Food and Nutrition in the Tropical Forest Biocultural Interactions and Applications to Development. UNESCO, The Parthenon Publishing Group, Paris, France, pp. 815–828.
Fairhead J.R. and Leach, M. 1998. Termites, society and ecology: Perspectives from West Africa. *In*: Ecology and Post-Modernity, Posey Darell (editor). Cup, Cambridge, Great Britain.
Fontaine J. 1959. Un problème biologique intéressant. Les chutes de manne. Science et Nature 32: 9–15.
Gelfand M. 1971. Diet and Tradition in an African Culture. G.F. Brock Publ., Edinburg, Scotland, 248 pp.
Gibbons A. 1991. Small is beautiful: Microlivestock for the third world. Science, pp. 253–378.
Gomez P., Halut R., and Collin A. 1961. Production de protéines animales au Congo. Bull. Agric. Congo 52(4): 689–815.
Gunn D.L. 1960. The biological background of locust control. Ann. Rev. Entomol. 5: 279–300.
Hernández-Pacheco, F. 1921. Escena pictórica con representaciones de insectos de época paleolítica. *Bull. Rsc. Span Hist Nat.*, Madrid Spain 50: 62–77.
Hernández M.M., Ramos-Elorduy J., and Pino J.M. 2000. Supervivencia y desarrollo de la carpa japonesa alimentada con diferentes inclusiones de Periplaneta americana L. proveniente de reciclaje de desechos orgánicos. Mem. XXXV Cong. Natl. Entomol. Méx. pp. 176–186.
Heyden D. 1974. La Diosa Madre: Itzpálotl. *Bull. del INAH*, México Segunda época, No. 11 p. 13–14.
Holden, S. 1991. Edible Caterpillars-a Potential Agroforestry Resource?, *The Food* Insects Newsletter IV (2): 3–4.
Hocking B. and Matsumura F. 1960. Bee brood as food. Bee World 41 (5): 113–120
Hunn E.S. 1977. The Clasification of Discontinuities in Nature, Tzeltzal Folk Zoology, Academic Press, New York, U.S.A., 368 p.
Ichponani J. and Malek N. 1971. Evaluation of de-oiled silkworm pupae meal as protein source in chicken rations. Br. Poul. Sci. 12: 231–234.
Jones K.T. and D.B. Madsen. 1991. Further experiments in native food procurement. *Utah Archaeol.* 91: 68–77.
Kitsa K. 1989. Contribution des insectes comestibles à l'amélioration de la ration alimentaire au Kasai-Occidental au Zaïre. Zaire-Afrique 239: 511–519.
Kodondi K., Leclercq G., and Gaudin-Harding F. 1987. Vitamin estimations of three edible species of Attacidae caterpillars from Zaire. Int. J. Vit. Nutr. Res. 57: 333–334.
Kodondi K., Leclercq M., Bourgeay-Causse M., Pascaud A., and Gaudin-Harding F. 1984. Intérêt nutritionnel de chenilles d'Attacides du Zaïre. Composition et valeur nutritive. Cah. Nutr. Diet. XXII (6) : 478–485.
Kok R. 1983. The production of insects for human food (*Stegobium paniceum*). J. Can. Inst. Food Sci. Technol. V (16): 5–18.
Ladrón de Guevara O., Padilla P., García L., Pino J.M., and Ramos-Elorduy J. 1995. Amino acid determination in some edible Mexican insects. Amino Acids: 9: 161–173.

Lanfranchi G.B. 2004. Minilivestock consumption in the ancient Near East: The case of locusts (this volume).

Leigh, E.G. 1993. Epilogue: comments on the Paris Symposium by a concerned biology. Chapter 94 pp. 829–838 *In:* Tropical Forest, People and Food bioculturel Interactions and Applications to Development, Ed. Hladik C.M., Hladik A., Linares O., Pagezzi H., Semple A., and Hadley M. Ed. MAB/UNESCO and The Parthenon Publishing Group. Paris, France.

Leleup P. and Daems A. 1969. Les Chenilles Alimentaires du Kwango. J. Agric. Trop. et Bot. Appl. XVI (1) : 1–21.

Luo Zhi-Yi. 1987. Insects as food in China. *J. Ecol. of Food and Nutr.* 36: (2–4): 201–207.

Malaisse F. 1997. Se nourrir en forêt claire, Ed. Presses Agronomiques de Glemboux, Belgique, 237 p.

Malaisse F. and Parent G. 1980. Les Chenilles Comestibles du Shaba Méridional (Zaïre). Nat. Belges 61 (1): 2–24.

Malaisse F. and Parent G. 1991. Disponibilité des produits sauvages comestibles de la région Zambézienne: Ecologie et valeur alimentaire des insectes et des reptiles. Résumés Int. Symp. UNESCO (Alimentation en Forêt Tropicale) p. 20. Paris, France.

McGrew W.C. 1981. The female chimpanzee as a human evolutionary prototype. *In:* F. Dhalberg (ed.). Woman the Gatherer, Columbia Univ. Press, New Haven, CT, pp. 35–73.

Meyer-Rochow V.B. 1973. Edible insects in three different ethnic groups of Papua and New Guinea. Amer. J. Clinical Nutr. 26: 663–677.

Meyer-Rochow V.B. and Changkija S. 1995. Uses of insects as human food in Papua, New Guinea, Australia and North-east India in *Abstracts Int. Symposium Biodiversity in Agriculture*, p. 24. Pekin, China.

Mitsuhashi J. 1980. Edible Insects in the World. Kokinshoin. Kanda, Tokio, Japan, 270 pp. (in Japanese).

Mitshuhashi J. 1995. Personal communication.

Mitshuhashi J. 1999. People Who Eat Insects 400 (in Japanese). Heibon-Sha, Tokyo, Japan.

Mittenmeier R.E. 1988. Several working papers for the Biodiversity Task Force of the World Bank. The World Bank, Washington, DC. USA.

Motte-Florac E. 1995. Evolution de la thérapeutique chez les P'urhépecha de la Sierra Tarasca (Michoacán, Mexique), pp. 55–70. *In:* Grenand F. and Randa V. (eds.). Transitions Plurielles: Exemples dans Quelques Societés Amériques. Peeters no. 349, París.

Motte-Florac E. and Ramos-Elorduy J. 2002. Is the traditional knowledge of insects important? *In:* Stepp J.R., Wyndham F.S., and Zarger R.K. (eds.). Ethnobiology and Biocultural Diversity. Proc. 7th Int. Cong. Ethnobiol., Univ. Georgia Press, Athens, GA, 717 pp.

Muyay, T. (1991). Les insects comme aliments de l'homme. CEEBA Public. Serie II Vol. 69. 177 p. Bandundu Republique Du Zaire.

Nerlove, S.B. 1974. "Woman's workload and infant feeding practices: a relationship with demographic implications" Ethnol 13 (2): 207–214.

Nguyen-Caong-Tieu. 1928. Notes sur les insectes comestibles au Tonkin. Bull. Econ. Indochine. 31: 735–744.

Nkuoka E. 1987. Les insectes comestibles dans les societés d'Afrique Centrale Muntu. Rev. Sci. Cult. du CICIBA 6: 171–178

Oliveira S., Passos de Carvalho J., Bruno de Sousa R., and Sinao M. 1976. The nutritional value of four species of insects consumed in Angola. Ecol. Food Nutr. 5: 91–97.

Onore G. 1997. A brief note on Edible Insects in Ecuador. J. Ecol. of Food and Nutr. 5: 91–97.

Owen D. 1973. Man's environmental predicament. In: An Introduction to Human Ecology in Tropical Africa. Oxford Univ. Press, London, UK.

Pagezy H.S. 1990. Seasonal variation of food supply in the Lake Tumba region of Zaire in Hladik, M. Bauchet, S. and de Garinel. Eds. Food and Nutrition in the African Rain Forest pp. 37–42. UNESCO-CNRS. Paris, France.

Pagezy H.S. 1993. Importance of natural resources in the diet of the young child in a flooded tropical forest in Zaire, Chapter 30: 365–380. in Tropical Forests People and Food Biocultural Interactions and Applications to Development. Ed. MAP UNESCO and The Parthenon Publishing Group. Paris, France.

Paoletti M.G., Buscardo E., and Dufour D.L. 2001. Edible invertebrates among Amazonian Indians. A disappearing knowledge. Environ., Devel., Sustainability 2: 195–225.

Paoletti M.G., Dufour D.L., Cerda H., Torres F., Pizzoferrato L., and Pimentel D. 2000. The importance of leaf- and litter-feeding invertebrates as sources of animal protein for the Amazonian Amerindians. Proc. R. Soc. London B 267: 2247–2252.

Perdomo H. 1999. Evaluación del efecto de la suplementación con larvas de *Tenebrio molitor* L. (Coleoptera: Tenebrionidae) a una dieta de iniciación de pollos de engorda. Tésis Fac. de Ciencias, Univ. Nal. Autón. Méx., México, D.F., 59 pp.

Pereira, N. 1974. Panorama Da Alimentaçao Indigena: Comidas Bebidas y Toxicos na Amazonia Brasileira. Ed. Livraria Sao Jose, Rio de Janeiro, Brasil 442 p.

Phelps J. 1988. *In:* Insect Food Newsl. vol VI (1): 6.

Peters, M.T. 1993. Insects and Human Society. Ed. T.M. Peters 393 Bay Road Belchertown, MA. U.S.A. 450 p.

Pimentel, D. 1980. Energy and land constraints in food protein production. Science 190: 754–761.

Posey, D.A. 1990. Introduction to Ethnobiology: Its Implications and Applications, in Posey D.A. and Overall L. Eds. Ethnobiology: Implications and Applications. Proc. First Int. Congress of Ethnobiology Belem, Brazil, Vol. I, pp. 1–8.

Pujol, R. (1990). Personal Communication.

Quinn P. 1959. Foods and Feeding Habits of the Pedi. These Witwatersrand Univ. Johannesburg, South Africa, 30 pp.

Ramos Calcaneo B., Ramos-Elorduy J., and Pino M.J.M. 2002. Estudio del efecto de tres insectos en la alimentación de pollos de la variedad Indian River. (submitted to Rev. Mex Vet.)

Ramos-Elorduy F. 1974. Fundamentos Económicos para el fomento de industrias rurales en México. Thesis Esc. Nal. Econ. Univ. Nal. Autón, México, D.F., 79 pp.

Ramos-Elorduy J. 1982. Los Insectos Como Fuente de Proteínas en El Futuro. Edit. Limusa, México, D.F., 149 pp. (2nd ed.).

Ramos-Elorduy J. 1990. Edible insects barbarism or solution to the hunger problem. *In:* Posey D.A. (ed.) Ethnobiology Implications and Applications. Proc. First Int. Cong. Ethnobiology, Belem, Brazil. vol. I, pp. 151–158.

Ramos-Elorduy J. 1993. Insects in the diet of tropical forest peoples in mexico, Chap. 17 pags. 205–212 in Tropical Forest People and Food Biocultural Interactions and Applications to Development. Ed. MAB UNESCO and The Parthenon Publishing Group Paris, France.

Ramos-Elorduy J. 1996a. Insect consumption as a means of national identity. *In:* Ethnobiology in Human Welfare. Editor S.K. Jain, Deep Publication, New Delhi, India, pp. 9–12.

Ramos-Elorduy J. 1996b. Insects as intermediate biotransformers. *In:* Human Ecology.

Ramos-Elorduy J. 1997a. Insects: A sustainable source of food? Ecol. Food Nutr. 36 (2–4): 247–276.

Ramos-Elorduy J. 1997b. Importance of edible insects in the nutrition and economy of the people living in the rural areas. Ecol. Food Nutr. 36 (1): 347–366.

Ramos-Elorduy J. 1998. Creepy Crawly Cuisine. The Gourmet Guide to Edible Insects. Inner Traditions Intern. Rochester Vermont Park St. Press, 150 pp.

Ramos-Elorduy J. 1999. El consumo de insectos como un hábito ancestral. *In:* Chalchihuite. Homenaje a Doris Heyden. Ed. Instituto Nacional Antropología e Historia, México, D.F., pp. 275–305.

Ramos-Elorduy J. 2000. La etnoentomología actual en México, en la alimentación humana. In: La Medicina Tradicional y en el Reciclaje y Alimentación Animal. Mem. XXXV Cong. Natl. de Entomol. México, pp. 3–46.

Ramos-Elorduy J. 2003. La Entomofagia, Etnoentomomedicina y Etnoentomoreciclaje. *In:* Biodiversidad, Taxonomía y Biogeografía de Artrópodos de México: Hacia una Síntesis de su Conocimiento, Mexico vol. IV (in press).

Ramos-Elorduy, J. 2004. La Etnoentomología en la alimentación, la medicina y el reciclaje, In: Eds. J.E. Llorente B., J.J. Morrone, O. Yánez O., I. Vargas F. Biodiversidad, taxonomiya y biogeografia de artropodos de mexico: Hacia una sintesis de su conocimiento Editorial: Conabio, Facultad de Ciencias UNAM, Instituto de Biología UNAM, Las Prenas de Ciencias UNAM, Vol. IV: 329.413. Mèxico, D.F.

Ramos Elorduy J., Motte-Florac E. and Conconi M. 2004. Entomophagie au Mexique. perspectives et avenir. Ed. Peeters Louvain Paris., France 223 p. (in press).

Ramos-Elorduy J. and Conconi M. 1992. Resemblance of the techniques for exploit some edible insect species in different ethnic groups all over the world. II Int. Cong. Ethnobiol. Abstracts Ed. ISE, Mexico D.F. p. 141.

Ramos-Elorduy J. and Conconi M. 1993. Resemblance of the techniques for exploiting some edible insect species in different ethnic groups all over the world. II Int. Cong. Ethnobiol. Abstracts Ed. ISE, Mexico, D.F., p. 141.

Ramos-Elorduy J. and Conconi M. 1994. Edible Insectes of the world. (Liste des espèces, lieux de consommation et ethnies qui les consomment). Fourth Int. Cong. Ethnobiol. Abstracts. Lucknow, UP, India, p. 141.

Ramos-Elorduy J. and Conconi M. 1996. Conservación y Mercadeo de los Insectos Comestibles. In: Etnobiología en El Conocimiento y Conservación de los Recursos Naturales y Culturales Eds. Rodriguez and Vargas UNAM (In press). México, D.F.

Ramos-Elorduy J. and Pino M.J.M. 1979. Insectos comestibles del Mezquital y su valor nutritivo. Ann. Inst. Biol. UNAM Ser. Zool. 50: 563–574.

Ramos-Elorduy J. and Pino M.J.M. 1988. Influencia de algunos insectos comestibles en la vida mística de diversos grupos étnicos de México. 11 th. Ann. Ethnobiol. Conf. Abstracts Ed. Soc of Ethnobiology USA, Mexico, D.F., p. 33–34.

Ramos-Elorduy J. and Pino J.M. 1989. Los Insectos Comestibles Entre Las Antiguas Culturas De México. AGT, 139 pp.

Ramos-Elorduy J. and Pino J.M. 1990a. Contenido calórico de algunos insectos comestibles de México. Rev. Soc. Quim. Mex. 34(2): 56–68.

Ramos-Elorduy J. and Pino J.M. 1990b. Variation de la valeur nutritive de *Tenebrio molitor* l élévé sur differentes substrats. *In:* Fleurat-Lessard F. and Ducom P. (eds.). Proc. 5th Int Work Conf. on Stored Product Protection. vol. I, pp. 201–210. Bordeaux, France.

Ramos-Elorduy J. and Pino J.M. 1993a. Biogeographical aspects of some edible insects from Mexico. II Int. Cong. Ethnobiol. Abstracts. p. 43.

Ramos-Elorduy J. and Pino J.M. 1994. Aprovechamiento en la alimentación, de los insectos plaga como un método natural de control. V Cong. Int. Manejo Integr. Plagas Costa Rica Abstracts, Ed. Soc. Ent. Costa Rica, p. 43.

Ramos-Elorduy J. and Pino J.M. 2001a. Contenido de vitaminas de algunos insectos comestibles de México. Rev. Soc. Quím México 45 (2): 66–76.

Ramos-Elorduy J. and Pino J.M. 2001b. Insectos comestibles de Hidalgo, México Ann. Inst. Biol. UNAM Ser. Zool. 72 (1): 43–84.

Ramos-Elorduy J. and Pino J.M. 2002. Edible Insects of Chiapas, Mexico. Ecol. Food Nutr. 41: 271–279.

Ramos-Elorduy J., Avila E., Medina H., and Pino J.M. 2001d. Uso de insectos en la alimentación primaria de aves de criadero. Mem. XXXVI Cong. Natl. Entomol. Mex. Queretaro, México, p. 125.

Ramos-Elorduy J., Avila E., Pino J., and Martínez V.H. 2002. Effect of Tenebrio molitor coming from recycled organic wastes on the nutrition of broilers Isa Brown. (submitted to J. Anim. Sci.) USA.

Ramos-Elorduy J., E. Avila G., Rocha H., and Pino J.M. 2002. Use of *Tenebrio molitor* (Coleoptera: Tenebrionidae) to recycle organic wastes and as feed for broiler chickens. J. Econ. Entomol. 95 (1): 214–220.

Ramos-Elorduy J., Pino J.M., and Bourges H. 1982. Valor nutritivo y calidad de la proteína de tres insectos comestibles de México. Fol. Entomol. Mex. 53: 111–118.

Ramos-Elorduy J., Pino J.M., and Cuevas S.C. 1998. Insectos comestibles del Estado de México y determinación de su valor nutritivo. Ann. Inst. Biól. UNAM. Ser. Zool. 69 (1): 65–104.

Ramos-Elorduy J., Pino J.M. and Villegas J. 1988. The efficiency of the insect *Musca domestica* in recycling organic wastes as a source of proteins. *In:* Biodeterioration. Elsevier, Amsterdam, Netherlands, vol. III, pp. 805–810.

Ramos-Elorduy J., Motte-Florac E., and Conconi M. 2003. Entomophagie au Mexique. Perspectives et avenir. Peeters Louvain Ed., Paris., 223 pp.

Ramos-Elorduy J., Muñoz J.J., and Pino J.M. 1998a. Determinación de minerales en algunos insectos comestibles de México. Rev. Soc. Quím México 42(1): 18–33.

Ramos-Elorduy J., Morales J., Pino J.M., and Nieto Z. 1988. Contenido de Tiamina, Riboflavina y Niacina en algunos insectos comestibles de México. Rev. Tech. Alim. 22: 76–81.

Ramos-Elorduy J., Pino J.M., Escamilla E., Alvarado M., Lagunes J., and Ladrón de Guevara O. 1997. Edible insects of Oaxaca State Mexico and their nutritive value. J. Food Comp. Anal. 10: 142–157.

Ramos-Elorduy J., Pino J.M., Márquez C., Rincón F. and O., Alvarado M., Escamilla E., and Bourges H. 1984. Protein content of some edible insects in Mexico. J. Ethnobiol. 4: 61–72.

Ramos-Elorduy J., Pino M.J.M. and Gonzalez M.O. 1981. Digestibilidad in vitro de algunos insectos comestibles de México. Fol. Ent. Mex. 49: 141–154.

Ramos-Elorduy J., Pino M.J.M., Corona C.R., and Medina D.V. 1985. Estudio de los insectos comestibles de Guerrero y su valor nutritivo. Res. VII. Cong. Nac. de Zoología Vol. II: 1107–1126.

Ramos-Elorduy J., Pino M.J.M., and Romero S.L.A. 1988. Determinación del valor nutritivo de algunas especies de insectos comestibles del Estado de Puebla. An. Inst. de Biol. Univ. Nal. Autón Méx. Ser. Zoología 58 (1): 355–372.

Ramos-Elorduy J. and Pino M.J.M. 1989. Los insectos comestibles en el Mexico antiguo (estudio etnoentomologico). Ed. A.G.T. México, D.F. 108 p.

Ramos-Elorduy J., Lévieux J., and Lenoir A. 1992. Possibilités de renforcement des fondations chez deux especès de formis d'intêret économique. Premiers resultats (*Liometopum*. Hymenoptera. Formicidae) Annls. Soc. Ent. Fr. (N.S.) 28 (2): 215–219.

Ramos-Elorduy J. and Pino M.M. (1992). Biogeographical aspects of some edible insects from Mexico. II Int. Cong. Ethnobiol. Abstracts Ed. ISE, Mexico, D.F. p. 143.

Ramos-Elorduy J. and Pino M.J.M. (1993). Algunos insectos comestibles del Estado de Veracruz. Primera Reunión de Investigadores sobre Fauna Veracruzana. Xalapa Veracruz, México, Res. p. 78–79.

Ramos-Elorduy J., Pino M.J.M., and Hernández B.S. 1994. Edible Hymenóptera of Yucatan Península. 12th Int. Cong. of IUSSI. Les insectes sociaux. París-Sorbonne France p. 350.

Ramos Edorduy J., Gamboa B. Pino M.J.M., Borbolla G., and Avila E. 1997. Los insectos en la alimentación animal, un ejemplo. *Mem. XXXII Cong. Nal. Ent. Mexico* D.F., p. 42–43.

Ramos-Elorduy J. and Pino M.J.M. 2001c. Insectos comestibles del Estado de Michoacán. An Inst. Biol. Univ. nal. Autón. Méx. Serie Zool. (In Press).

Ramos-Elorduy J., Pino M.J.M., and Morales de L.J. 2002. Análisis Químico Proximal, Vitaminas y Nutrimentos Inorgánicos de Insectos consumidos en el Estado de Hidalgo, México. Folia Entomol. Mex. 41(1): 15–29.

Ratcliffe, B.C. 1990. The significance of scarab beetle in the ethnoentomology of non-industrial indigenous people. In Possey D.A. and Overall L. Eds; Ethnobiology Implications and Applications, proc. First Int. Congress of Ethnobiology Belem, Brazil Vol. I, pp. 159–185.

Redford, K.H. and Dorea J.G. 1984. The nutritional value of invertebrates with emphasis on ants and termites as food for mammals. *J. Zool.* 203: 385–395.

Reim, H. 1962. Die Insektennahrung der Australischen Ureinwohner Ed. Akademie-Verlag-Berlin 165 p.

Roulon-Doko P. 1999. Chasse, Cueillette et Culture Chez les Gbaya de Centrafrique. Ed. L'Harmattan, Paris, France, pp. 539.

Ruddle, K. 1973. The Human use of insect: examples for Yupka, Biotropica 5: 94–110.

Schurr, K. 1972. Insects as a Major protein source in sewage lagoon biomass useable as animal food Proc. North Cent. Branch. Entomol. Soc. Am. 27: 135–137.

Smith H.S. 1992. Global Biodiversity: Status of the Earth's Living Resources. Chapman and Hall, London, UK, 594 pp.

Sutton M. 1988. Insects as food: Aboriginal entomophagy. *In:* The Great Basin. Anthropological, Ed. Ballena Press Publishing, Novato California, U.S.A., 33: 115 pp.

Sutton M. 1990. Insects resources and Pliopleistocene hominid evolution. *In:* Ethnobiology: Implications and Applications, Eds. Posey D. and Overall L. Proe First Int. Congress of Ethnobiology. Belem, Brazil. vol. 1, pp. 195–207.

Takeda J. 1990. The dietary repertory of the Ngandu people of the tropical rain forest and ecological and anthropological study of the subsistence activities and food procurement technology of slash-and-burn agriculturists in the Zaire river basin. African Study Monograph. Supplem. issue no. 11, 75 pp. Ed. Koto Univ. Press, Japan

Taylor R. 1975. Butterflies in My Stomach. Edit. Woodbridge Press Publishing Co. Santa Barbara, California, U.S.A., 134 pp.

Toledo V. 1995. Etnobiología para un desarrollo sustentable: Hacia una nueva etapa de la investigación en México. *In:* Etnobiologia en El Conocimiento y Conservación De Los Recursos Naturales y Culturales, Eds. Rodriguez and Vargas UNAM (In press) México, D.F.

Tommaseo-Ponzetta M. and Paoletti M.G. 1997. Insects as food of the Irian Jaya Populations J. Ecol. of Food and Nutr. 36 (2–4): 321–346.

Tommaseo-Ponzetta M. 2004. Insects: Food for human evolution (this volume).

Van der Meer M.J.C. 1965. Insects eaten by the Karo-Batak people. (A contribution to emtomobromatology) Entomologische Berichten 25(6): 101–107.

Van der Waal B.C.W. 1994. The importance of grasshoppers (Fam. Acrididae) as traditional food in villages in northern Transvaal, South Africa. Fourth Int. Cong. Ethnobiol. Abstracts. Lucknow, UP, India, p. 140.

Wasosky G. 1993. In: Insects Food Newsl. V (2): 8.

Weaver N. and Weaver E.C. 1981. Beekeeping with the stingless bee *Melipona beeckei* by the Yucatan Maya. Bee World 62: 7-19.

WHO/FAO/UNU. 1985. Necesidades De Energia Y De Proteinas. Serie de informes técnicos 724. WHO, Gèneve, Switzerland, 220 pp.

Wilson, E.O. 1985. The Biological diversity crisis: A challenge to science. *Issues Sci. Technol* 2: 20–29.

Xiaoming Chen & Ying Feng. 1999. The Exible Insects of China. Ed. (in chinese) 181 p. China

Yhoung-Aree J., Puwastien P., and Attig G.A. 1997. Edible Insects in Thailand: An unconventional protein source? *Ecol. of Food and Nutr.* 36 (2–4): 133–149.

Zubiran S., Chavez A., Bonfil G., Aguirre G., Cravioto J., and De la Vega D. 1974. La Desnutricion Del Mexicano. Ed. Fondo de Cultura Económica, Mexico, D.F., 62 pp.

15

Edible Invertebrates among Amazonian Indians: A Critical Review of Disappearing Knowledge

M.G. Paoletti[1*] and D.L. Dufour[2]

[1]*Dept of Biology, Padova University, Via U. Bassi 58/b, Padova, Italy.*
[2]*Dept. of Anthropology, University of Colorado, Boulder, CO 80309, USA. E-mail: Darna.Dufour@colorado.edu*
**Author for correspondence, E-mail: paoletti@civ.bio.unipd.it*

Abstract

For the indigenous populations of Amazonia, invertebrates constitute an important component of the diet. Information on entomophagy for 39 ethnic groups (and three other post-Columbian settlers) or about 21.4% of the 182 groups known in the Amazon Basin is presented here, but utilization of this non-conventional food resource is surely much more widespread. A database is given of all the information available for each ethnic group regarding the species included in the diet, scientific and the ethno name if known, stage of life cycle consumed, manner of preparation and, when known, host plant. This database lists 209 scientifically identified species. Information on an additional 426 species and ethno names, with an insecure link to Linnean taxonomy suggest that local knowledge is very extensive. The database represents not only an easy-to-consult resource, but also a support for further research. Caterpillars, termites, leafcutter ants, bees, wasps, and Coleoptera seem to be the more collected items, together with a few aquatic ones. The most intensively collected are those dependent on forest leaves and litter, representing in general the higher biomass, so much work needs to be done for other groups, including caterpillars, aquatic insects, grasshoppers, snails, and spiders. Knowledge of the relations between indigenous populations and ecosystems is indeed the base for the preservation of natural and cultural biodiversity. We are at the beginning of a survey that has to be expanded.

Key Words: Amazon food database, entomophagy, edible invertebrates, sustainable food procurement, biodiversity, Amazonian Indians, minilivestock, insects as food

Introduction

An update of a previously published list (Paoletti et al., 2001) of the invertebrate species consumed by different groups of Amazonian Indians assesses which groups are currently or were involved in the consumption of insects and other invertebrates (Crustacea, Arachnida, Insecta, Anellida, Mollusca), and which organisms are/were used, and why others were neither consumed nor used as medicine.

Reports of the alimentary use of invertebrates have been recorded since the first Western conquistadors, travelers, and naturalists visited the Amazon Basin (Bodenheimer, 1951). In most cases these are reports of single observations that focus on the exotic nature of the food sources, assumed to be types of emergency foods only. Good examples are the observations of A.R. Wallace on the consumption of earthworms collected from bromeliads in the Rio Negro or of palm worms (Wallace, 1853, 1889). Also evident in these early reports, and a few later ones, is the general sense of repulsion felt by observers unaccustomed to eating small animals. Giacone did little to hide his repulsion when he wrote: "when Indians feed on reptiles and bugs it produces a *fetor nauseante*" (Giacone, 1949). A missionary once told us: "in that particular village they are so poor they eat insects". In most cases, visitors and people that have been in contact with Amerindians, including several anthropologists, have considered the consumption of small invertebrates to be a strange practice associated with times of scarcity. Most Amerindians are aware of the Western distaste for invertebrates and hence do not like to show Westerners the kind of invertebrates they eat. When asked about the use of insects as food, they typically respond that "some neigbors eat insects" or "these old people once ate insects". For instance, M.G.P. found that Guajibo in the Alto Orinoco were only willing to discuss edible insects after he noticed that the floor was littered with wings (Plate VII, 1) of *Tropidacris cristata* (Plate VII, 2 and 3). The Guajibo are very fond of *Tropidacris cristata,* and remove the wings before roasting them. *Tropidacris cristata*, however, are not eaten by other Amerindian groups in the same region.

In contrast, our experience showed invertebrates to be perfectly good food sources and included in the human diet throughout the year, when available as well as during special seasons. Amerindians value many of these organisms for their flavour as well as for their nutritional value (Ruddle, 1973; Coimbra, 1984, 1985; Dufour, 1987; DeFoliart, 1992; Cerda et al., 2000; Marconi et al. 2002; Paoletti et al., 2000, 2001; 2003). The Yukpa Indians (Colombia) reportedly prefer some of their traditional insect foods to fresh meat (Ruddle, 1973). Fresh and smoked earthworms, motto, and kuru, are better appreciated by the Ye'Kuana of the

Alto Rio Padamo, Venezuela and highly prized compared to other food such as fish, game, pork or beef. Further, insects and other small invertebrates can make a significant contribution to the diet. For example, among the Tukanoan Indians (Colombia), insects and other small invertebrates provide up to 12% of the animal protein in men's diets and 26% in women's during the months of May and June (Dufour, 1987). Some data are summarized in Table 15.1 regarding the amount of invertebrates eaten in the Amazon. Much work still needs to be done to better appreciate and document this use.

Table 15.1: Invertebrates eaten by the American Indians as percentage of animal food

Ethnic group	**Period**	**Invertebrates as a percent of total meat consumed**	**Reference**
Tukanoans, Colombia	May–June	12% men, 26% women	Dufour, 1987
Piaroa, Cuao river, Venezuela	yearly	8%	Zent, 1992
Guajibo, Alcabala de Guajibo, Venezuela	Rainy season	60%–70%	Paoletti et al., 2001

Materials and Methods

All the references to entomophagy in the Amazon Basin found in the literature were compiled and added to our unpublished observations and fieldwork during the period 1994–2002 in Alto Orinoco, Venezuela, and Ecuador. Data from classical works such as Wallace (1853, 1889) and Bodenheimer (1951) were also included but the focus was on more recent literature and personal communications from anthropologists actively working in the region (Paoletti et al., 2001). The references included met the following criteria: (1) the author(s) provided taxonomic identifications for the invertebrates discussed, or a sufficiently detailed description of the morphology and ecology of the organisms that they could be tentatively identifed at least to order and family; (2) the author(s) identified the group of people involved. Not included was the literature containing only general or superficial reports of entomophagy (innumerable), or references to entomophagy providing only local names (ethno names) and not the associated Linnean taxonomic names or the Amerindian group involved. With regard to the latter we made several exceptions. One was the larvae of the many wood-boring insects consumed by Amerindians. These proved very difficult to identify and in some cases are locally given a few collective names. For instance, *makoia*, for the Yanomamo refers to the larvae of potentially hundreds of xilophagous beetles (Scarabaeidae, maybe also Buprestidae and Cerambycidae). The second exception was caterpillars, consumed in abundance by Amerindians, but for which we had much difficulty in obtaining taxonomic identifications. For example, from Yanomamo interviewees we obtained a list of 25 edible caterpillars and their host plants, but even with the assistance of expert taxonomists, have only been able to properly identify one (Table 15.2). Interestingly, it appears that Yanomamo caterpillar taxonomy is more advanced than Western. At present, most of the caterpillars have only ethno names and in a few cases family names. The third exception was cases in which only honey, or propolis and wax are the products gathered, and the consumption of brood and pupae is not specifically mentioned.

Table 15.2: Ethnonames of caterpillars consumed by the Yanomamo, Alto Orinoco, Venezuela, July 1997

Caterpillars (Lepidoptera)	Host plants
Magna (this species is possible appreciated by Piaroa)	*Humisci*
Irokiri	*Mokhe*
Wagio wagio	**Wagiowagio nato* vine
Casha	*Kanaiiha and hawari* vine
Parima	*Tahinani* tree
Mamocorisina	*Canainini* tree
Iro krukuiu	*Atari hahi* tree
Pascou	*Toho toto* vine
Wero wero	*Scawara curimi* tree
Opomoschi	*Penahe' and Ohata waco* tree
Penahe nosi	**Peena peena and Curatasci cote* tree
Wateoma	*Warapahi* tree
Paruri hesikaki	?
Hewakema	?
Niyã	?
Kirakirami umo	?
Shakukumi	?
Kraya	?
Maya	?
Yakureto	*Kumicihenati* tree
Mapayawa	**Mapayawa hena* tree
Hewekewekemorewe	*Pitahamu* tree
Wawa hena	*Maecotoma* tree
Shiyahumi	*Atarihia* tree
Pore mapuusiki	*Pitaha* tree

*The only available similarities in the folk name between host plant and caterpillars.

Results

Data on the entomophagy in 39 Amerindian groups and a few non-native villagers are presented, constituting only some 22% of the 182 groups known in the Amazon Basin (Bahuchet, 1993). One ethnic group was found, the Matis of Brazil, who do neither collect nor use insects or other invertebrates as food; they even avoid eating deer (Milton, 1997; pers. comm., 1999). Presumably most other Amazonian Indians gather and consume insects and/or other invertebrates but information is not available, as least to us. In reviewing the literature we found that of the many papers reporting on entomophagy and other invertebrates eaten, only a few carefully identified and recorded the amount of the invertebrates consumed (Ruddle, 1973; Coimbra, 1984; Dufour, 1987; Zent, 1992; Cerda et al., 2000; Cerda et al., this volume).

The database (Appendix 15.1) lists the current state of our knowledge of the invertebrates consumed by Amerindians (and for a few other peoples) in the Amazon. We have listed the species eaten by stage of life cycle (larvae, pupae, or adult stage), and how consumed (sometimes raw on the spot, sometimes roasted in leaf packages, more often fried, baked or roasted, in some cases smoked). For those cases insects whose taxonomic identification could not be ascertained, we have listed only the ethno names collected. In such cases however, we are confident that, based on discussions with local people as to details of color, size, morphology, host plant, etc., the different ethno names really indicate different Linnean species.

Types of Invertebrates Utilized as Food

Finkers (1986) mentioned about 142 ethnonames for insects possibly eaten by Yanomamo for Venezuela but however without species identification. Dufour and Sander (2000) counted 56 species of insects consumed for all of South America but Onore (2004) mentioned 82 species for Ecuador alone. We present here 209 invertebrate species potentially eaten in Amazonia and consider our current knowledge far from exhaustive. Indeed, we are at the very beginning of systematic knowledge of the consumption of invertebrates in Amazonia. Local people know not only the species eaten, but also closely related species not considered resources or considered poisonous (e.g. some caterpillars). For instance, the Yanomamo name crickets and grasshoppers (at Motorema: *seki sekima* and *rain reim*) but do not eat them. The same is true for cicadas (at Motorema: *rororomi cona*) of which they know the larval, nymphal, and adult stages!). Other ethnogroups that do not eat grasshoppers, such as the Ye'Kuana and Piaroa, nevertheless have names for the large species eaten by the Guajibo (*Tropidacris cristata*)—*masiseniene* and *poracheneuwe*. In most cases we found no terrestial snails mentioned as edible among the Piaroa, Guajibo and Yanomamo albeit they have ethnonames for several kinds. Apparently only Ye'kuana eat terrestrial snails (at least at Alacran, Alto Rio Padamo). The Guajibo suggested that terrestrial snails are poisonous. However they use the shell of a large terrestrial species to prepare and smoke *jopo* (a drug made with a leguminous seed of *Piptadenia peregrina* at Sabaneta Guaiabal, Puerto Ayacucho; Schultes and Raffauf, 1990). Among myriapods only one large chilopod (*Scolopendra gigantea*) has been mentioned as edible, by the Guajibo.

Table 15.3 shows our estimation of the number of species consumed in Amazonia and gives a general estimation of currently known species for the Amazon and for South America. For Isoptera the number of species exceeds 100

Table 15.3: Estimation of species in South America and/or Amazonia (e.n. = ethnonames)

Order, Family	Tribe, Genus	Amazon known species	South American species	Edible species reported here	Ethnic groups	Source
Anellida Glossoscolecidae		350?	1500?	4	3	Pers. eval.
Gastropoda				3	3	Fred Thompson
Araneae Theraphosidae	larger terrestrial	27		3	2	Rogerio Bertani, 2002 pers. comm.
Myriapoda, Chilopoda				1	1	Gordon,1989
	Avicularia	33	54	1	2	Rogerio Bertani, 2002 pers. comm.
Crustacea, Decapoda	Shrimps	16	30??	2	7	Carlo Froglia, 1999 pers. comm.
Crustacea, Decapoda	Crabs		146	4 ?	7	Rodriguez, 1982
Anoplura	Louses			1	3	
Isoptera	Arboreal nests	35 (only Brazil)	79	4	4	Reginaldo Constantino, 1999 pers. comm.
Isoptera	*Syntermes*		23	4	7	Reginaldo Constantino, 1999 pers. comm.
Hymenoptera Formicoidea	*Atta*		15	3	16	Holldobler and Wilson, 1990
Hymenopt., Apidae & Meliponini[1]			250–350	25 (104 e.n.)	15	David Roubik, 1999 pers. comm.
Hymenoptera Vespidae			535	33 (62 e.n.)	12	James Carpenter, 1999 pers. comm.
Hemiptera & Homoptera				5	4	
Trichoptera				2	2	
Odonata		300		6	2	Jourg De Marmels
Neuroptera, Megaloptera				3	2	Atilano Contreras-Ramos

Table 15.3: (Contd.)

Table 15.3: (contd.)

Orthoptera Tettigoniidae				2	2	
Orthoptera Acrididae				8	3	
Coleopt. Curcul. Rhynchopho-rinae	Largest ones	6		7	22	Pers. eval.
Coleopt. Curcul. Rhynchopho-rinae	*Metamasius*	30??	40	6	4	Vaurie, 1966
Coleoptera Bruchidae				6	6	
Coleoptera Scarabaeidae		4,000??		17	16	Pers. eval.
Coleoptera Lucanidae		100??		3	3	Pers. eval.
Coleoptera Buprestidae		2,000??		1	2	Pers. eval.
Coleoptera Cerambycidae		4,000??		7	6	Pers. eval.
Coleoptera Passalidae		10??		2	2	Pers. eval.
Diptera				3	3	
Lepidoptera		30,000??		42 (50 e.n.)	14	Pers. eval.
TOTAL		—	—	209 (428)		(parentheses incl. ethnonames)

but for Hymenoptera and Coleoptera rises to about 600 in the first (estimation for South America) thousands in the second (for Amazonia). Still this taxonomic information, probably represents an underestimation of the actual number of species that live in the Amazon Basin and have probably been tested locally as potential food items. Not all insects are eaten by all ethnic groups and different Amerindian groups have different preferences as well.

The variety of invertebrates adopted as food is impressive: leaf and litter consumers, lignivores, pollinivores, scavengers, predators, and generalists, as well as aquatic forms of both detritivores (Diptera Sratiomyidae, Megaloptera, Trichoptera) and predators such as Odonata and Megaloptera. In most cases immature forms such as larvae, caterpillars, and pupae are preferred, but adults are also collected and eaten, for instance the large *Rhynchophorus* and some scar-abs among most Amerindians. In most Indian groups it is difficult to assess which items are eaten fresh on the spot and those eaten only after some sort of cooking. For example, we know that grasshoppers and caterpillars are eaten "only" after cooking. Our data on 39 Amerindian groups suggests that even if actual knowledge is far from complete, different groups have developed different preferences. To give one example, the Yukpa and Guajibo (Hiwi) consume grasshoppers while other Amerindians living nearby and/or in similar environments (Piaroa, Curripaco, Ye'kuana or Yanomamo around Puerto Ayacucho, Venezuela) do not. Apparently only Piaroa and Ye'Kuana consume earthworms nowadays while according to Wallace (1853; 1889), Tukanoans did consume them in days gone by. In the early 1980s the latter were collecting them for fishing bait but not eating them (D.L.D. pers. obs.). The Ye'kuana are specialists in earthworm recognition; to wit, they have 16 ethnonames for earthworms (Glossoscolecidae) Table 15.4. We consider that their very prominent nomenclature related to the fact that they actively consume earthworms, at least two species (Paoletti et al., 2003; Moreno and Paoletti, 2004).

Apparently only two Indian groups, the Piaroa and the Yanomamo, consume large terrestrial spiders, such as *Theraphosa apophysis* and *T. blondi* which

Table 15.4: List of earthworms (Glossoscolecidae) ethnonames from only two Ye'Kuana villages in the Alto Rio Padamo area, Amazonas, Venezuela

Guatamo: group of persons and Luis Garcia; at Alacran Angel Garcia 1998– 2002	Earthworm ethnonames	Characteristics	USE
?	SCICIU	White-pink, 6 cm, upper river banks	Only for fishing
?	CATASU'	Red-brown, 6 cm, upper river banks	Only for fishing
?	MAWADA	?	Only for fishing
?	CANAJE'	?	Only for fishing
Andiorrhinus (Amazodrilus) motto n. sp.	*MOTTO*	White, lower river banks	Edible
	DAICIK	White, small, lower river banks	Edible
	VEJAJ	White, medium size	Edible
	TOCCAMO	White	Edible
	TAEGIC	White	Edible
	MODOIDDI	White	Edible
	MOUATO	?	Edible
	ARAITO'	?	Edible
	CETOKA	?	Edible
?	KURUJICETTE	30 cm, dark brown, in forest	Only for fishing
Andiorrhinus (Amazodrilus) kuru n. sp.	KURU	Red-brown, 40–60 cm , in forest	Edible
?	SARIDI	40–50 cm dark brown, in forest	Edible

live in terrestrial burrows (Plate VII, 5) and *Avicularia avicularia*, a web spider species.

Most groups eat a variety of species of caterpillars but unfortunately these species are very poorly known (see Table 15.2). A Ye'Kuana village leader told us (M.G.P., November, 2002) that he grew up eating a caterpillar called *masamasadi* in the way we consume pork or chicken. Most groups also use the larvae of palm Curculionidae (Rhynchophorinae) and a few other seed beetles such as Bruchidae. Some species of wood-boring Coleoptera are also eaten, but limited data exist as to the types and quantities consumed. It is known, for instance, that some Tukanoan women eat several different larvae because logs of felled trees in their gardens show considerable evidence of larvae extraction; this resource has never been systematically investigated however (D.L.D. pers. obs.). The Yanomamo at Motorema (Alto Orinoco, Venezuela) also collect *makoia* larvae [*Pelidnota* (*Chalcoplethis*) sp.] from seasoned logs in their gardens.

Many groups consume ants of the genus *Atta* and several termites of the genus *Syntermes*. *Syntermes* construct nests very deep in the soil (Constantino, 1995) and collect leaves and litter on the soil surface during the night. Other genera of smaller arboreal termites (genera *Nasutitermes* and *Labiotermes*) are exploited by Yanomamo, Tukanoans, Lecos, etc. Some groups such as Curripaco, Piaroa, and Guajibo (Hiwi) near Puerto Ayacucho prepare a sauce called *catara* (Fig. 15.9) for sale in the Puerto Ayacucho Indian market made with liquids extracted from cassava (*Manihot esculenta*), hot peppers (*Capsicum* sp.), and soldiers of *Atta cephalotes* and/or soldiers of *Eciton burchelli*. In the same market, termite soldiers of genus *Syntermes* are occasionally sold. In Tena and Quito, Ecuador, insects coming from the Amazon and the Andes are frequently marketed (G. Onore, 1999, pers. comm. and this book).

The Caboclos and other recent colonizers in the Amazon such as some black communities, sometimes consume certain invertebrates, among which *Atta* ants and palm worms (*Rhynchophorus* sp.) are reported more often.

Strategies Developed

We have shown that leaf- and litter-feeding invertebrates (Paoletti et al., 2000) are the most frequently exploited in Amazonian forested areas. This is transparently a strategy of using invertebrates that thrive on leaves and litter—the most renewable resource crop in the tropical forest. For instance up to 96.5% of the invertebrates consumed by the Tukanoans are leaf- and litter-feeders (Table 15.5).

The key taxa are:

—leaffeeders: caterpillars, *Atta* ants;

—litterfeeders: *Syntermes* and some Glossoscolecidae earthworms.

In most cases the invertebrates exploited form aggregations (social or gregarious communities) are of larger body size. However, small termites of genera *Nasutitermes* and *Labiotermes*, less than one cm long, are gathered in the

Fig. 15.1: *Shuru; Euryrhynchus amazoniensis* collected at Alacran, Alto Rio Padamo, Venezuela (photo M.G Paoletti).

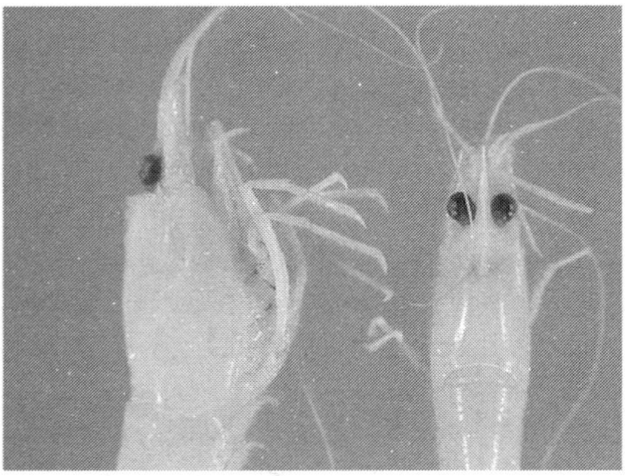

Fig. 15.2: *Macrobrachium* sp. from Vaupes, Colombia (photo M.G. Paoletti).

Fig. 15.3: *Pomacea* sp., a highly valued aquatic snail in Alto Orinoco (photo M.G. Paoletti).

Fig. 15.4: *Memu: Plekocheilus* sp. collected at Alacran, Alto Rio Padamo, Venezuela (photo M.G. Paoletti).

Nutritional Aspects

In addition to amino acids, the nutrients, oligoelements, vitamins, and fatty acids—especially PUFA (polyunsaturated fatty acids)—have proven of great interest in most of the invertebrates eaten in the Amazon. As a source of essential linolenic, linoleic, and arachidonic acids, they seem to somehow represent the need for fat, which for many reasons is a limited resource in the Amazon (Paoletti et al., 2003, 2003a and Paoletti and Dreon this book).

It is very important to maintain the existing folk wisdom about such traditional resources and to promote collection of further developments.

What is Needed for Further Research in Amazonia

In spite of some excellent research, our current knowledge of invertebrates as food resources is still rudimentary. Identification of the invertebrates eaten, their host plants, their composition and biology, including timing of collection, remain largely unknown. Very limited information is available for many invertebrates, such as the Rhynchophorinae weevils complex living on palms (Cerda et al., 2000; Cerda et al., this book), caterpillars, locusts, termites and others. Moreover, little is known of how the termite (*Syntermes*) digests litter or of the density of *Atta* in different areas of Amazonia. A more extended survey of Amerindian groups could greatly improve our knowledge of these organisms.

Further, very limited data are available on the quantity of invertebrates eaten by the Amerindians (Dufour, 1987; Zent, 1992; Paoletti et al., 2000) and how their nutritional compositions compare with other available protein sources.

To promote and maintain the use of insects and small invertebrates as food resources without destroying the forest, practical rearing plans are also required and small and large-scale experiments are needed at the village level. Designing minilivestock systems that could sustain local villagers (Hardouin and Stievenart, 1993; Paoletti and Bukkens, 1997; DeFoliart, 1999; Cerda et al., 2000; Paoletti et al., 2000) is a way of reducing forest destruction, providing innovative tools for local communities, and potential ecotouristic activities.

Acknowledgements

We are indebted to many researchers who shared with their useful information. We particularly thank all local communities involved in our investigation. We are grateful to Gene De Foliart, Katerine Milton, Danilo Lopez Hernandez, Prudencio Chacon, Felix Moretto, Giuseppe Bortoli, Juan Finkers, Alejandro Signi, and Gary Dawson for supporting us with stimuli, suggestions, and practical help. Fieldwork was expedited by Paul Patmore, Hugo Cerda, Franz Torres, and Yelinda Araujo. INIA, FONACIT in Venezuela and CNR Italy provided some support for fieldwork in Alto Orinoco, Amazonas, Venezuela. Many people assisted in identification of the materials listed here, including host plants, and locally providing useful information. Carlo Froglia, Guido Pereira, J. Vincente Garcia, Fred Thompson, Nestor J. Cazzaniga, Reginaldo Constantino, Rogerio Bertani, Rick C. West, William Goitia, Tim Myles, Paul Eggleton, David Nickle, Hugh Rowell, Ken Tennesen, Jurg Demarmels, James Carpenter, David Roubik, Stephane Boucher, Robin M Giblin-Davis, Richard C. Vogt, Alan Lane de Melo, Thomas K. Wood, Giovanni Onore, Boris Kondrateff, Charles O'Brien, Keith Brown, Willy Kuschel, Carlos Bordon, and Cornelius C. Berg. We thank them for their great support and help.

References

Bahuchet S. 1993. Situation des Populations indigènes des Forêts Denses Humides, European Commission, Dg XI, Brixellen, 511 pp.

Beckerman S. 1977. The use of palms by the Bari Indians of the Maracaibo Basin, Principes 22 (4): 143–154.

Bodenheimer F.S. 1951. Insects as Human Food. Dr. W. Junk Publications. The Hague, The Netherlands.

Cerda H., Martinez R., Briceno N., Pizzoferrato L., Manzi P., Tommaseo Ponzetta M., Marin O., and Paoletti M.G. 2000. Palm worm: (*Rhychophorus palmarum*) traditional food in Amazonas, Venezuela: Nutritional composition, small scale production and tourist palatability, Ecol. Food Nutr. 40(1): 13–32.

Chagnon N.A. 1968. Yanomamo: The Fierce People. Holt, Reinhart and Winston, Inc., New York, NY, pp. 30–32.

Clastres P. 1972. The Guayaki, *In:* Bicchieri M.G. (ed.). Hunters and Gatherers Today. A Socio-economic Study of Eleven Such Cultures in the Twentieth Century. Holt, Reinhart and Winston, Inc., New York, NY pp.139–159.

Cocco L. 1982. Iyewei-teri. Quince Anos entre los Yanomamos. Escuela técnica popular. Don Bosco, Caracas.

Coimbra C.E.A. Jr. 1984. Estúdios de ecologia humana entre os Surui do Parque Indígena Aripuana, Rondónia. 1. O uso de larvas de Coleopteros (Bruchidae e Curculionidae) na alimentação, Revista Brasileira de Zoologia 2: 35–47.

Coimbra C.E.A. 1985. Estúdios de ecologia humana entre os Surui do parque Indígena Aripuana, Rondónia. Aspectos alimentares, Bol. Museu Paramense Emílio Goeldi Antropol., 2 (1): 57–87.

Constantino R. 1995. Revision of the neotropical genus *Syntermes* Holmgren (Isoptera: Termitidae). Univ. Kansas Sci. Bull. 55 (13): 455–518.

DeFoliart G.R. 1992. Insects as human food. Crop Protection 11: 395–399.

DeFoliart G.R. 1999. Insects as food: why the Western attitude is important. Ann. Rev. Entomol. 44: 21–50.

Denevan W.M. 1971. Campa subsistence in the grand Pajonal, eastern Peru, Geog. Rev. 61, 496–518.

Dufour D.L. 1987. Insects as food: a case study from the Northwest Amazon. Amer. Anthropol. 89: 383–397.

Dufour D. and Sander J. 2000. Insects *In:* Kiple K.F. and Ornelas K.C. (eds.). The Cambridge World History of Food. Cambridge Univ. Press, New York, NY, pp. 546–554.

Finkers J. 1986. Los Yanomamos y su Sistema Alimenticio, Vic. Ap. P. Ayacucho Venezuela, 262 pp.

Giacone A. 1949. Os Tucanos e outras Tribos do Rio Uaupes, Afluente do Negro-Amazonas, Impressa Oficial do estado de São Paulo, Brasil.

Gordon D. G. 1998. The Eat a Bug Cookbook. Ten Speed Press, Berkeley CA, USA

Hardouin J. and Stievenart C. 1993. Invertebrates (Minilivestock) Farming. Proc. Seminar La Union, Philippines, November 1992. EEC-DGXII/CTA/IFS/ DMMMSU/ITM, Philippines.

Hawkes K., Hill K., and O'Connell J.F. 1982. Why hunters gather: optimal foraging and the Achè of the eastern Paraguay, Amer. Ethnol. 9: 379–398.
Hill K. and Kaplan H. 1983. Field report and research summary: The Mashco-Piro of Manu National Park, Peru, Manuscript Dept. Anthropology, Univ. Utah, Salt Lake City, UT, USA.
Hill K., Kaplan H., Hawkes K., and Hurtado A.M. 1987. Foraging decision among Achè hunter-gatherers: new data and implications for optimal foraging models, Ethnol. Sociobiol. 8: 1–36.
Hölldobler B. and Wilson E.O. 1990. The Ants. Harvard Univ. Press, Cambridge, MA, 732 pp.
Hurtado A.M., Hawkes K., Hill K., and Kaplan H. 1985. Female subsistence strategies among Achè hunter-gatherers in Eastern Paraguay, Human Ecol. 13: 1–28.
Lizot J. 1988. Los aborígenes de Venezuela, pt. III. *In:* J. Lizot (ed.). Etnología Contemporánea, vol. II. Fundación La Saille, Caracas, pp. 479–513.
Marconi S., Manzi P., Pizzoferrato L., Buscardo E., Cerda H., and Paoletti M.G. 2002. Nutritional evaluation of terrestrial invertebrates as traditional food in Amazonas. Biotropica 34 (2): 273–280.
Medeiros Costa-Neto E. 1998. Folk taxonomy and cultural significance of 'abeja' (Insecta, Hymenoptera) to the Pankarare, north-eastern Bahia state, Brazil. J. Ethnobiol. 18 (1): 1–13.
Menzel P. and D'Aluisio F. 1998. Man Eating Bugs. Ten Speed Press, Berkeley, CA, 192 pp.
Milton K. 1984. Protein and carbohydrate resources of Maku Indians of north-western Amazonia. Amer. Anthropol. 86: 7–27.
Milton K. 1997. Real men don't eat deer. Discover, June, pp. 46–53.
Moreno A. and M.G. Paoletti 2004. *Andiorhinus kuru* n. sp. one giant earthworm (Oligochaeta: Glossoscolecidae) food resource for Makiritare Indians of the alto rio Padamo, Amazonas, Venezuela. *Canadian Journal of Zoology*, 82(3): 1–14 (in press).
Onore G. 1997. A brief note on edible insects in Ecuador, Ecol. Food Nutr. 36 (2–4): 277–285.
Onore G. 2004. Edible insects in Ecuador (this volume)
Paoletti M.G. and Bukkens S.G.F. 1997. Minilivestock, Ecol. Food Nutr. 36 (2–4), 95–346. (spec. issue).
Paoletti M.G., Buscardo E., and Dufour D.L. 2001. Edible invertebrates among Amazonian Indians, a disappearing knowledge. *In:* Begissi A. and Hens L. (eds.). Local Knowledge in the Tropics: Relevance to Conservation and Management. Environment, Development and Sustainability 2: 195–225.
Paoletti M.G., Dufour D., Cerda H., Torres F., Pizzoferrato L., and Pimentel D. 2000. Leaf- and litter-feeding invertebrates are important Amerindian sources of animal protein in Amazonia. Proc. Roy. Soc., London B 267: 2247–2253.
Paoletti M.G., Buscardo E., VanderJagt D.J., Pastuszyn A., Pizzoferrato L., Huang Y-S., Chuang L-T., Millson M., Cerda H., Torres F., and Glew R. 2003. Nutrient content of Earthworms consumed by Yekuana Amerindians of the Alto Orinoco of Venezuela. Proc. Roy. Soc. London B. 270: 249–257.
Paoletti M.G., Buscardo E., Vanderjagt D.J., Pastuszyn A., Pizzoferrato L., Huang V., Chung T., Millson M., Cerda H., and Glew R.H. 2003a. Nutrient content of termites (*Syntermes* soldiers) consumed by Makiritare Amerindians of Alto Orinoco of Venezuela. Ecol. Food Nutr. 42: 173–187.
Posey D.A. 1979. Ethnoentomology of the Gorotire Kayapo of Central Brazil. PhD. Univ. Georgia, Athens, 177 pp.
Posey D.A. 1987. Ethnoentomological survey of Brazilian Indians. Entomol. Gener. 12 (2–3): 191–202.
Ramos-Elorduy J. 2004. Insects: A hopeful food source (this volume)..
Rodriguez G. 1982. Les crabes d'eau douce d'amerique. ORSTOM, Paris, 223 pp.
Ruddle K. 1973. The human use of the insects: examples from the Yukpa, Biotropica 5 (2): 94–101.
Schlenker H. 1974. Makiritare (Venezuela, Orinoco-Quellgebiet) sammeln, zubereiten und essen von Würmern. *In:* Wolf G. (ed.). Encyclopaedia Cinematographica, Institut für den Wissenschaftlichen Film, Gottingen, pp. 1–11.
Schultes R.E. and Raffauf R.F. 1990. The Healing Forest. Dioscorides Press, Portland, OR, USA.

Smole W.J. 1976. The Yanoama Indians. A Cultural Geography, Univ. Texas Press, Austin TX, 272 pp.

Vaurie P. 1966. A revision of the neotropical Genus *Metamasius* (Coleoptera Curculionidae, Rhynchophorinae). Bull. Amer. Mus. Nat. Hist. 131 (3): 211–338.

Wallace A.R. 1853. On the insects used for food by the Indians of the Amazon. Royal Entomol. Soc. London 2: 241–244.

Wallace A.R. 1889. A Narrative of Travels on the Amazon and Rio Negro, Dover Publ., Inc., New York, NY.

Zent S. 1992. Historical and Ethnographic Ecology of the Upper Cuao River Votiha: Clues for an Interpretation of Native Guianese Social Organisation, PhD Diss., Columbia University, New York, NY.

Appendix 15.1: Database of invertebrates consumed by different Amerindians.

Family	Local name	Species[1]	Ethnic group[2]		Host plant	Eaten[3]	Reference
Anellida Glossoscolecidae			TUKANOANS, ?		Bromeliaceae	bo, y	Wallace, 1853, 1889
Anellida Glossoscolecidae	*Gua* or *Wua*	*Andiorrhinus (Amazonidrilus) motto* Righi and Araujo	PIAROA, Ven.	A		bo, y	Paoletti, 1995–1999
Anellida Glossoscolecidae	*Motto*	*Andiorrhinus (Amazonidrilus) motto* Righi and Araujo	YE'KUANA, Ven.	W		bo, y	Schlenkner, 1974
Anellida Glossoscolecidae	*Wiae*	*Andiorrhinus (Amazonidrilus) motto* Righi and Araujo ?	PIAROA (Cuao river) Ven.				Zent, 1992[4]
Anellida Glossoscolecidae	*Kuru*	*Andiorrhinus kuru* Moreno and Paoletti	YE'KUANA, Ven.	W		bo, y	Paoletti, 1998–1999[5]
Myriapoda, Chilopoda		*Scolopendra gigantea* L.	GUAJIBO, Venezuela			ra	Gordon, 1998
Anoplura		*Pediculus* sp.	MAKU, Colombia-Brazil			r	Milton, pers. comm., 1999
Anoplura		*Pediculus* sp.	YANOMAMO, Ve			r	Smole, 1976[6]
Anoplura		*Pediculus* sp.	YUKPA, Ven.				Ruddle, 1973[7]
Araneae	*Eikia*		PIAROA, Ven.			ra	Paoletti, 1996–1998
Araneae	6 ethno names		YANOMAMO, Ven.	A		ra	Paoletti coll. 1997
Araneae		*Avicularia* sp.	YANOMAMO, Ven.			ra	Lizot, 1988
Araneae	*Aukyae*	*Theraphosa* sp.??	PIAROA, Cuao river, Ven.				Zent, 1992
Araneae	*Jojo*	*Theraphosa* sp.??	YANOMAMO, Ven. (Boca Ocamo)				Cerda and Torres collection, 1999

Appendix 15.1: (Contd.)

Appendix 15.1: (Contd.)

Family	Local name	Species[1]		Ethnic group[2]		Host plant	Eaten[3]	Reference
Araneae Theraphosidae	?	*Avicularia avicularia* L.		YANOMAMO (Rio Siapa) Ven.		Near gardens	Roasted in fireplace	Rick C. West, pers. comm. 2003
Araneae Theraphosidae	*Hajo*	*Theraphosa apophysis* (Tinter 1991)		YANOMAMO (Achacoa, Motorema) Ven.			Roasted in fireplace	Expedition XI.2002
Araneae Theraphosidae	*Moriri*	*Theraphosa* sp. ??		YE'KUANA (Guatamo) Ven.			ra	Expedition XI.2002
Araneae Theraphosidae	*Waikoshi-nemi*	*Theraphosa apophysis* (Tinter 1991) ??		YANOMAMO (Motorema) Ven.				Expedition XI.2002
Araneae Theraphosidae		*Theraphosa blondi*		YANOMAMO (Rio Siapa) Ven			ra	R. West
Coleopt. Bruchidae				PARAKANA, Brazil		*Orbignya* spp.	r	Milton, pers. comm. 1999
Coleopt. Bruchidae	*Kadeg*	*Caryobruchus* sp.	L	SURUI, Brazil	W	*Orbignya martiana* Barb. Rodr.	f, r	Coimbra, 1984
Coleopt. Bruchidae	*Eteme*	*Caryobruchus* sp.	L	YUKPA, Ven.	W	*Scheelea* spp.	r, ra	Ruddle, 1973
Coleopt. Bruchidae	*Kadeg*	*Pachymerus cardo* Fahreus	L	SURUI, Brazil	W	*Orbignya martiana* Barb. Rodr.	f, r	Coimbra, 1984
Coleopt. Bruchidae		*Pachymerus nucleorum* (F.)	L	TIMBIRA, Brazil		*Orbignya martiana* Barb. Rodr.		cit. in Coimbra, 1984
Coleopt. Bruchidae?				MASHCO-PIRO, Peru		Larvae on palm nuts		Hill and Kaplan, 1983

Appendix 15.1: (Contd.)

Appendix 15.1: (Contd.)

Family	Local name	Species[1]		Ethnic group[2]		Host plant	Eaten[3]	Reference
Coleopt. Buprestidae	*Mae Do Dol*	*Euchroma gigantea* L.	L	KAYAPO, Brazil				Posey, 1987
Coleopt. Buprestidae	*Boopika*	*Euchroma gigantea* L.	L, A	TUKANOANS, Colombia	W, C			Dufour, 1987
Coleopt. Cerambycidae	*Pikoroa*	*Acrocinus longimanus*	L	TUKANOANS, Colombia	W, C			Dufour, 1987
Coleopt. Cerambycidae	*Chaacho*	*Macrodontia cervicornis*	L	ACHE, Paraguay				Hurtado et al., 1985[8]
Coleopt. Cerambycidae		*Oncideres* sp.	A, L	AWA, Ecuador				Onore, 1997
Coleopt. Cerambycidae		*Oncideres* sp.	L, A	SARAGUROS, Ecuador				Onore, 1997
Coleopt. Cerambycidae		*Psalidognathus atys* White	L	Ecuador				Onore, 2004
Coleopt. Cerambycidae		*Psalidognathus cacicus* White	L	SARAGUROS, Ecuador			f	Onore, 1997
Coleopt. Cerambycidae		*Psalidognathus erithrocerus* Reiche	L	Ecuador				Onore, 2004
Coleopt. Cerambycidae		*Psalidognathus modestus* Thomson	L	Ecuador				Onore, 2004
Coleopt. Curculionidae	*Poatask*	*Anthonomus* sp.	A	YUKPA, Ven.	C	*Zea mays* L.	ra	Ruddle, 1973
Coleopt. Curculionidae		*Cosmopolites sordida* Germar	A	ASHUARAS, Ecuador				Onore, 1997
Coleopt. Curculionidae		*Cosmopolites sordida* Germar	A	SHUARAS, Ecuador				Onore, 1997
Coleopt. Curculionidae		*Dynamis borassi* F.		GUAJIBO (HIWI), Ven.				Paoletti, 1996–1998[9]

Appendix 15.1: (Contd.)

Appendix 15.1: (Contd.)

Family	Local name	Species[1]		Ethnic group[2]	Host plant	Eaten[3]	Reference
Coleopt. Curculionidae		*Dynamis borassi* F.		PIAROA, Ven.			Paoletti, 1996–1998[9]
Coleopt. Curculionidae		*Dynamis nitidula* Guérin	L, P	ASHUARAS, Ecuador			Onore, 1997
Coleopt. Curculionidae		*Dynamis nitidula* Guérin	L, P	COFANES, Ecuador			Onore, 1997
Coleopt. Curculionidae		*Dynamis nitidula* Guérin	L, P	HUAORANIS, Ecuador			Onore, 1997
Coleopt. Curculionidae		*Dynamis nitidula* Guérin	L, P	QUICHUAS, Ecuador			Onore, 1997
Coleopt. Curculionidae		*Dynamis nitidula* Guérin	L, P	SECOYAS, Ecuador			Onore, 1997
Coleopt. Curculionidae		*Dynamis nitidula* Guérin	L, P	SHUARAS, Ecuador			Onore, 1997
Coleopt. Curculionidae		*Dynamis perryi* Waterhouse	L, P	ASHUARAS, Ecuador			Onore, 1997
Coleopt. Curculionidae		*Dynamis perryi* Waterhouse	L, P	COFANES, Ecuador			Onore, 1997
Coleopt. Curculionidae		*Dynamis perryi* Waterhouse	L, P	HUAORANIS, Ecuador			Onore, 1997
Coleopt. Curculionidae		*Dynamis perryi* Waterhouse	L, P	QUICHUAS, Ecuador			Onore, 1997
Coleopt. Curculionidae		*Dynamis perryi* Waterhouse	L, P	SECOYAS, Ecuador			Onore, 1997
Coleopt. Curculionidae		*Dynamis perryi* Waterhouse	L, P	SHUARAS, Ecuador			Onore, 1997
Coleopt. Curculionidae		*Metamasius cinnamominus* Perty	A	ASHUARAS, Ecuador			Onore, 1997

Appendix 15.1: (Contd.)

Appendix 15.1: (Contd.)

Family	Local name	Species[1]		Ethnic group[2]	Host plant	Eaten[3]	Reference
Coleopt. Curculionidae		*Metamasius cinnamominus* Perty	A, L, P	PIAROA, Ven. Caño Tigre	*Socratea exorrhiza* (Mart.) Wendland		Govoni, 1998, det. by O'Brien[10]
Coleopt. Curculionidae		*Metamasius cinnamominus* Perty	A	SHUARAS, Ecuador			Onore, 1997
Coleopt. Curculionidae		*Metamasius cinnamominus* Perty		YANOMAMO, Ven.			Paoletti coll. 1998
Coleopt. Curculionidae		*Metamasius dimidiatipennis* Jekel	A	ASHUARAS, Ecuador			Onore, 1997
Coleopt. Curculionidae		*Metamasius dimidiatipennis* Jekel	A	SHUARAS, Ecuador			Onore, 1997
Coleopt. Curculionidae		*Metamasius hemipterus* L.	A	ASHUARAS, Ecuador			Onore, 1997
Coleopt. Curculionidae		*Metamasius hemipterus* L.	A	SHUARAS, Ecuador			Onore, 1997
Coleopt. Curculionidae		*Metamasius sericeus* Olivier	A	ASHUARAS, Ecuador			Onore, 1997
Coleopt. Curculionidae		*Metamasius sericeus* Olivier	A	SHUARAS, Ecuador			Onore, 1997
Coleopt. Curculionidae		*Metamasius* sp.	L	PIAROA, Ven.			Govoni, 1998[10]
Coleopt. Curculionidae		*Rhinostomus barbirostris* F.	L, P	ASHUARAS, Ecuador			Onore, 1997
Coleopt. Curculionidae		*Rhinostomus barbirostris* F.	L, P	AWA, Ecuador			Onore, 1997

Appendix 15.1: (Contd.)

Appendix 15.1: (Contd.)

Family	Local name	Species[1]		Ethnic group[2]		Host plant	Eaten[3]	Reference
Coleopt. Curculionidae		*Rhinostomus barbirostris* F.	L, P	COFANES, Ecuador				Onore, 1997
Coleopt. Curculionidae	*Maquanato* [adult]	*Rhinostomus barbirostris* F.		GUAJIBO (HIWI), Ven.	L, A			Paoletti, 1996–1998[9]
Coleopt. Curculionidae		*Rhinostomus barbirostris* F.	L	HOTI, Ven.			r, ra	Zent collection, 2000, pers. comm.
Coleopt. Curculionidae		*Rhinostomus barbirostris* F.	L, P	HUAORANIS, Ecuador				Onore, 1997
Coleopt. Curculionidae		*Rhinostomus barbirostris* F.	L, P	NEGR. ESMERALD, Ecuador				Onore, 1997
Coleopt. Curculionidae	*Ciuve*	*Rhinostomus barbirostris* F.		PIAROA, Ven.				Paoletti, 1996–1998[9]
Coleopt. Curculionidae		*Rhinostomus barbirostris* F.	L, P	QUICHUAS, Ecuador				Onore, 1997
Coleopt. Curculionidae		*Rhinostomus barbirostris* F.	L, P	SHUARAS, Ecuador				Onore, 1997
Coleopt. Curculionidae		*Rhinostomus barbirostris* F.	L, P	SIONAS, Ecuador				Onore, 1997
Coleopt. Curculionidae	*Mayora*	*Rhinostomus barbirostris* F.		SURUI, Brazil	A	*Jessenia bataua* (Mart.) Burret	f, r	Coimbra, 1984
Coleopt. Curculionidae	*Picu*	*Rhynchophorus palmarum* L.	L	ACHE, Paraguay				Hawkes et al., 1982[11]
Coleopt. Curculionidae		*Rhynchophorus palmarum* L.	L, P	ASHUARAS, Ecuador				Onore, 1997
Coleopt. Curculionidae		*Rhynchophorus palmarum* L.	L, P	AWA, Ecuador				Onore, 1997
Coleopt. Curculionidae		*Rhynchophorus palmarum* L.		BANIWA, Ven.		Many different ones	ra	Paoletti, 1996–1998[12]

Appendix 15.1: (Contd.)

Appendix 15.1: (Contd.)

Family	Local name	Species[1]		Ethnic group[2]	Host plant	Eaten[3]	Reference
Coleopt. Curculionidae		*Rhynchophorus palmarum* L.		BARI, Ven.	*Jessenia* spp.	ra	Beckerman, 1977[13]
Coleopt. Curculionidae	*Boreu*	*Rhynchophorus palmarum* L.		BORORO, Brazil	*Scheelea phalerata* (Mart. ex Spreng.) Burret	dr	Albisetti and Venturelli, 1962[14]
Coleopt. Curculionidae		*Rhynchophorus palmarum* L.		CAINGUA, Brazil	Palmae apposit. fallen		Metraux, 1948[14]
Coleopt. Curculionidae		*Rhynchophorus palmarum* L.	L, P	COFANES, Ecuador			Onore, 1997
Coleopt. Curculionidae	*Muddi*	*Rhynchophorus palmarum* L.	A	CURRIPACO, Ven.			Paoletti, 1998[15]
Coleopt. Curculionidae	*Mutzubuto*	*Rhynchophorus palmarum* L.	A	GUAJIBO (HIWI), Ven.			Paoletti, 1996–1998[9]
Coleopt. Curculionidae	*Alerito*	*Rhynchophorus palmarum* L.	L	GUAJIBO (HIWI), Ven.			Paoletti, 1996–1998[9]
Coleopt. Curculionidae	*Mynda*	*Rhynchophorus palmarum* L.	L	GUAYAKI, Paraguay			Clastres, 1972
Coleopt. Curculionidae		*Rhynchophorus palmarum* L.	L	HOTI, Ven.		r, ra	Zent collection, 2000, pers. comm.
Coleopt. Curculionidae		*Rhynchophorus palmarum* L.	L, P	HUAORANIS, Ecuador			Onore, 1997
Coleopt. Curculionidae		*Rhynchophorus palmarum* L.	L	MAKU, Cobia-Brazil		s, y	Milton, 1984[16]
Coleopt. Curculionidae		*Rhynchophorus palmarum* L.	L, P	NEGR. ESMERALD, Ecuador			Onore, 1997

Appendix 15.1: (Contd.)

Appendix 15.1: (Contd.)

Family	Local name	Species[1]		Ethnic group[2]		Host plant	Eaten[3]	Reference
Coleopt. Curculionidae	*Signue*	*Rhynchophorus palmarum* L.	A	PIAROA (Gavilan) Ven.		Palmae varia		Paoletti, 1996–1998[9]
Coleopt. Curculionidae		*Rhynchophorus palmarum* L.	A, L, P	PIAROA, Ven.		*Socratea exorrhiza* (Mart.) Wendland		Govoni, 1998[10]
Coleopt. Curculionidae	*Irikã* or *Irike*	*Rhynchophorus palmarum* L.	L	PIAROA, Ven.		Palmae varia		Paoletti, 1996–1998[17]
Coleopt. Curculionidae		*Rhynchophorus palmarum* L.	L, P	QUICHUAS, Ecuador				Onore, 1997
Coleopt. Curculionidae		*Rhynchophorus palmarum* L.	L, P	SECOYAS, Ecuador				Onore, 1997
Coleopt. Curculionidae		*Rhynchophorus palmarum* L.	L, P	SHUARAS, Ecuador				Onore, 1997
Coleopt. Curculionidae	*Mojojoi*	*Rhynchophorus palmarum* L.	L	SIONA SECOYA, Ecuador		Palmae varia	f, r	Vicklers, pers. comm., 1999
Coleopt. Curculionidae		*Rhynchophorus palmarum* L.	L, P	SIONAS, Ecuador				Onore, 1997
Coleopt. Curculionidae	*Mayora*	*Rhynchophorus palmarum* L.		SURUI, Brazil	A	Palmae and *Jacaratia dodecaphylla*[18]	r	Coimbra, 1984, 1985
Coleopt. Curculionidae		*Rhynchophorus palmarum* L.	L, P	TSACHILAS, Ecuador				Onore, 1997
Coleopt. Curculionidae	Waraa	*Rhynchophorus palmarum* L.	L	TUKANOANS, Colombia	M			Dufour, 1987[19]
Coleopt. Curculionidae	*Ou* [larv.]; *Jora* [ad]	*Rhynchophorus palmarum* L.	L, A	YANOMAMO, Ven.	A	Palmae varia	r, ra	Paoletti coll. 1998

Appendix 15.1: (Contd.)

Appendix 15.1: (Contd.)

Family	Local name	Species[1]		Ethnic group[2]	Host plant	Eaten[3]	Reference
Coleopt. Curculionidae		*Rhynchophorus palmarum* L.		YE'KUANA, Ven.			Paoletti, 1998–1999[5]
Coleopt. Lucanidae		*Sphaenognathus feisthamelii* Guérin-Menéville	L	QUICHUAS, Ecuador			Onore, 1997
Coleopt. Lucanidae		*Sphaenognathus feisthamelii* Guérin-Menéville	L	SARAGUROS, Ecuador			Onore, 1997
Coleopt. Lucanidae		*Sphaenognathus lindenii* Murray	L	QUICHUAS, Ecuador			Onore, 1997
Coleopt. Lucanidae		*Sphaenognathus metallifer* Bomans and Lacroix	L	CANARIS, Ecuador			Onore, 1997
Coleopt. Passalidae		*Passalus interruptus* L.	L	Colombia, Paraguay			De Foliart, web
Coleopt. Passalidae	*Yayaru*	*Veturius sinuosus* (Drapiez)	L	TUKANOANS, Colombia			Dufour, 1987
Coleopt. Scarabaeidae	*Makoia*	*Pelidnota (Chalcoplethis)* sp.	L	YANOMAMO, Ven.	A	ra	Paoletti coll. 1998[20]
Coleopt. Scarabaeidae		*Ancognatha castanea* Erichson	L	CANARIS, Ecuador			Onore, 1997
Coleopt. Scarabaeidae		*Ancognatha castanea* Erichson	L	OTAVALOS, Ecuador			Onore, 1997
Coleopt. Scarabaeidae		*Ancognatha castanea* Erichson	L	PILAHUINES, Ecuador			Onore, 1997
Coleopt. Scarabaeidae		*Ancognatha castanea* Erichson	L	QUICHUAS, Ecuador			Onore, 1997

Appendix 15.1: (Contd.)

Appendix 15.1: (Contd.)

Family	Local name	Species[1]		Ethnic group[2]	Host plant	Eaten[3]	Reference
Coleopt. Scarabaeidae		*Ancognatha castanea* Erichson	L	SALAZACAS, Ecuador			Onore, 1997
Coleopt. Scarabaeidae		*Ancognatha castanea* Erichson	L	SARAGUROS, Ecuador			Onore, 1997
Coleopt. Scarabaeidae		*Ancognatha jamesoni* Murray	L	QUICHUAS, Ecuador			Onore, 1997
Coleopt. Scarabaeidae		*Ancognatha vulgaris* Arrow	L	CANARIS, Ecuador			Onore, 1997
Coleopt. Scarabaeidae		*Ancognatha vulgaris* Arrow	L	OTAVALOS, Ecuador			Onore, 1997
Coleopt. Scarabaeidae		*Ancognatha vulgaris* Arrow	L	QUICHUAS, Ecuador			Onore, 1997
Coleopt. Scarabaeidae		*Clavipalpus antisanae* Bates	L	QUICHUAS, Ecuador			Onore, 1997
Coleopt. Scarabaeidae		*Coelosis biloba* L.	L	AWA, Ecuador			Onore, 1997
Coleopt. Scarabaeidae		*Coelosis biloba* L.	L, P	TSACHILAS, Ecuador			Onore, 1997
Coleopt. Scarabaeidae		*Democrates burmeisteri* Reiche	L	SALAZACAS, Ecuador			Onore, 1997
Coleopt. Scarabaeidae		*Democrates burmeisteri* Reiche	L	QUICHUAS, Ecuador			Onore, 1997
Coleopt. Scarabaeidae		*Dynastes hercules* L.	L, A	HUAORANIS, Ecuador			Onore, 1997
Coleopt. Scarabaeidae	*Moa Hioro*	*Dynastes hercules* L.	L	TUKANOANS, Colombia			Hugh-Jones, 1999 pers. comm.
Coleopt. Scarabaeidae		*Golopha aeacus* Burmeister	L	OTAVALOS, Ecuador			Onore, 1997

Appendix 15.1: (Contd.)

Appendix 15.1: (Contd.)

Family	Local name	Species[1]		Ethnic group[2]	Host plant	Eaten[3]	Reference
Coleopt. Scarabaeidae		*Golopha aeacus* Burmeister	L	QUICHUAS, Ecuador			Onore, 1997
Coleopt. Scarabaeidae		*Golopha aegeon* Drury	L	OTAVALOS, Ecuador			Onore, 1997
Coleopt. Scarabaeidae		*Golopha aegeon* Drury	L	QUICHUAS, Ecuador			Onore, 1997
Coleopt. Scarabaeidae		*Heterogomphus bourcieri* Guérin	L	QUICHUAS, Ecuador			Onore, 1997
Coleopt. Scarabaeidae		*Leucopelaea albescens* Bates	A, L	OTAVALOS, Ecuador			Onore, 1997
Coleopt. Scarabaeidae		*Leucopelaea albescens* Bates	L, A	QUICHUAS, Ecuador			Onore, 1997
Coleopt. Scarabaeidae		*Leucopelaea albescens* Bates	L, A	SALAZACAS, Ecuador			Onore, 1997
Coleopt. Scarabaeidae		*Megasoma acteon* L.	L, A	KAYAPO, Brazil			Posey, 1987
Coleopt. Scarabaeidae		*Pelidnota nigricauda* Bates	L, A	QUICHUAS, Ecuador			Onore, 1997
Coleopt. Scarabaeidae		*Platycoelia forcipalis* Ohaus	L	?, Ecuador			Onore, 1997
Coleopt. Scarabaeidae		*Platycoelia parva* Kirsch	L	?, Ecuador			Onore, 1997
Coleopt. Scarabaeidae		*Platycoelia rufosignata* Ohaus	L	?, Ecuador			Onore, 1997
Coleopt. Scarabaeidae	*Poxta*	*Podischnus agenor* Oliv.	L, A	YUKPA, Ven.		r, ra	Ruddle, 1973
Coleopt. Scarabaeidae		*Praogolofa unicolor* Bates	A, L	OTAVALOS, Ecuador			Onore, 1997

Appendix 15.1: (Contd.)

Appendix 15.1: (Contd.)

Family	Local name	Species[1]		Ethnic group[2]		Host plant	Eaten[3]	Reference
Coleopt. Scarabaeidae		*Praogolofa unicolor* Bates	L, A	PILAHUINES, Ecuador				Onore, 1997
Coleopt. Scarabaeidae		*Praogolofa unicolor* Bates	L, A	QUICHUAS, Ecuador				Onore, 1997
Coleopt. Scarabaeidae		*Praogolofa unicolor* Bates	L, A	SARAGUROS, Ecuador				Onore, 1997
Coleopt. Scarabaeidae		*Strateagus* sp.	L, A	KAYAPO, Brazil				Posey, 1987
Coleopt. Scarabaeidae		Unidentified larva	L	MAKU, Colombia-Brazil			s, y	Milton, 1984
Coleopt.?	*Poshori*	Larvae?	L	CAMPA, Gr. Pajomal, Peru		Maize cob piles		Denevan, 1971[21]
Coleopt. Cerambycidae		*Macrodontia cervicornis* L.	L	Ecuador				Onore, 2004
Crust. Decap. Palaemonidae	*Acorsosoto*	*Macrobrachium* sp.		GUAJIBO (HIWI), Ven.			ra	Paoletti, 1996–1998[9]
Crust. Decap. Palaemonidae		*Macrobrachium* sp.		TUKANOANS, Colombia	A		ra	Dufour, coll.
Crust. Decap. Palaemonidae	*Suju imgra-sciscirimi*	*Macrobrachium* sp.?		YANOMAMO, Ven.	A		ra	Paoletti coll. 1997
Crust. Decap. Palaemonidae	*Suju*	*Macrobrachium* sp.?		YANOMAMO, Ven.	A		ra	Paoletti coll. 1997
Crust. Decapoda				CURRIPACO, Ven.				Paoletti, 1998[15]
Crust. Decapoda				PIAROA, Ven.				Paoletti, 1996–1998[9]
Crust. Decapoda				SIONA SECOYA, Ecuador				Vicklers, pers. comm., 1999
Crust. Decapoda				SIONA SECOYA, Ecuador				Vicklers, pers. comm., 1999

Appendix 15.1: (Contd.)

Appendix 15.1: (Contd.)

Family	Local name	Species[1]		Ethnic group[2]		Host plant	Eaten[3]	Reference
Crust. Decapoda	*Gorpa*			SURUI, Brazil	A		R, f	Coimbra, 1985[22]
Crust. Decapoda	*Gorpa-Sakap*			SURUI, Brazil	A		R, f	Coimbra, 1985
Crust. Decapoda				YANOMAMO, Ven.				Smole, 1976[6]
Crust. Decapoda				YE'KUANA, Ven.				Paoletti, 1998–1999
Crust. Decapoda	*Churu*	*Euryrhynchus amazoniensis* Tiefenbacher, 1978		YE'KUANA (Alacran)		Small streams	ra	Expedition XI.2002
Crust. Decapoda	*Caluwey*	*Fredius* sp.?		GUAJIBO (HIWI), Ven.			ra	Paoletti, 1996–1998[9]
Crust. Decapoda	*Jora orajesci*	*Fredius* sp.?		YANOMAMO, Ven.	A		ra	Paoletti coll. 1997
Crust. Decapoda	*Oko*	*Fredius* sp.?		YANOMAMO, Ven.	A		ra	Paoletti coll. 1997
Crust. Decapoda	*Oko okojesci*	*Fredius* sp.?		YANOMAMO, Ven.	A		ra	Paoletti coll. 1997
Diptera Culicidae				YANOMAMO, Ven.			r	Smole, 1976[6]
Diptera Sratiomyidae		Gen. sp.?	L, P	TUKANOANS, Colombia	A	*Manihot esculenta* Crantz		D.L. Dufour coll.
Diptera Stratiomyidae	*Minu*	*Chryschlorina* sp.	L	YUKPA, Ven.	A		ra	Ruddle, 1973
Gastropoda	*Sunama*			YANOMAMO, Ven.	A	Aquatic species	ra	Paoletti coll. 1997
Gastropoda, Ampullariidae		*Pomacea* sp.		PIAROA, Ven.		River Orinoco	??	Inigo Narbaiza, FUDECI, 2003
Gastropoda, Ampullariidae		*Pomacea* sp.		YANOMAMO, Ven.		River Orinoco	??	Inigo Narbaiza, FUDECI, 2003

Appendix 15.1: (Contd.)

Appendix 15.1: (Contd.)

Family	Local name	Species[1]		Ethnic group[2]		Host plant	Eaten[3]	Reference
Gastropoda, Bulimulidae	*Memu*	*Plekocheilus* sp.		YE'KUANA (Alacran) Ven.		Gardens	ra	Expedition XI.2002
Hemiptera, Belastomatidae	*Rijo*	*Belastoma* sp.		YE'KUANA (Alacran) Ven.			ra	Expedition XI.2002
Hemiptera, Naucoridae	*Rijo*	*Ambrysus* sp.		YE'KUANA (Alacran) Ven.			ra	Expedition XI.2002
Homopt. Membracidae		*Umbonia spinosa* F.	L, A	QUICHUAS, Ecuador				Onore, 1997
Homoptera, Cicadidae		*Carineta fimbriata* Walker	A	QUICHUAS, Ecuador				Onore, 1997
Homoptera Cicadidae?	*Kaae*		A	PIAROA, Cuao river, Ven.				Zent, 1992
Homoptera, Membracidae	*Mere in am*	*Umbonia spinosa* F.[34]		TUKANOANS, Colombia	C	*Inga* spp.	r, t	Dufour coll. 100[23]
Hymenopt. Apidae		*Apis mellifera* (L.)	L, P, H	NEGR. ESMERALD, Ecuador				Onore, 1997
Hymenopt. Apidae	*Mynga*	*Apis mellifera* (L.)?		ACHE, Paraguay				Hill et al., 1987
Hymenopt. Apidae		*Bombus atratus* Franklin	L	CANARIS, Ecuador				Onore, 1997
Hymenopt. Apidae		*Bombus atratus* Franklin	L	OTAVALOS, Ecuador				Onore, 1997
Hymenopt. Apidae		*Bombus ecuadorius* Meunier	L	?, Ecuador				Onore, 1997
Hymenopt. Apidae		*Bombus funebris* Smith	L	CANARIS, Ecuador				Onore, 1997
Hymenopt. Apidae		*Bombus funebris* Smith	L	OTAVALOS, Ecuador				Onore, 1997

Appendix 15.1: (Contd.)

Appendix 15.1: (Contd.)

Family	Local name	Species[1]		Ethnic group[2]		Host plant	Eaten[3]	Reference
Hymenoptera Apidae		*Bombus robustus* Smith	L	?, Ecuador				Onore, 1997
Hymenopt. Apidae		*Tetragonisca angustula* Latreille	L	?, Ecuador				Onore, 1997
Hymenopt. Apidae		*Tetragonisca angustula* Latreille	L	CANARIS, Ecuador				Onore, 1997
Hymenopt. Eumenidae	*Tptimi*	*Eumenes canaliculata oliv.*	Tabu	YUKPA, Ven.	A			Ruddle, 1973
Hymenopt. Formicidae	*Kidakue*	*Atta cephalotes* L. (det. W. Goitia 2002)		YE'KUANA, (Guatamo) Ven.			soldiers	Expedition XI.2002
Hymenopt. Formicidae		*Atta cephalotes* L.	A	ASHUARAS, Ecuador				Onore, 1997
Hymenopt. Formicidae		*Atta cephalotes* L.	A	AWA, Ecuador				Onore, 1997
Hymenopt. Formicidae		*Atta cephalotes* L.	A	COFANES, Ecuador				Onore, 1997
Hymenopt. Formicidae		*Atta cephalotes* L.	A	HUAORANIS, Ecuador				Onore, 1997
Hymenopt. Formicidae	*Sauva*	*Atta cephalotes* L.		KAYAPO, Brazil				Posey, 1979
Hymenopt. Formicidae		*Atta cephalotes* L.	A	NEGR. ESMER-ALD., Ecuador				Onore, 1997
Hymenopt. Formicidae	*Anangù*	*Atta cephalotes* L.	A	QUICHUAS, Ecuador				Onore, 1997
Hymenopt. Formicidae		*Atta cephalotes* L.	A	SECOYAS, Ecuador				Onore, 1997
Hymenopt. Formicidae		*Atta cephalotes* L.	P, L	SHUARAS, Ecuador				Onore, 1997
Hymenopt. Formicidae		*Atta cephalotes* L.	A	SIONAS, Ecuador				Onore, 1997

Appendix 15.1: (Contd.)

Appendix 15.1: (Contd.)

Family	Local name	Species[1]		Ethnic group[2]		Host plant	Eaten[3]	Reference
Hymenopt. Formicidae	*Makaiyaa*	*Atta cephalotes* L.	S, al	TUKANOANS, Colombia	W, C		ra	Dufour, 1987[24]
Hymenopt. Formicidae	*Kidakue*	*Atta cephalotes* L.		Ye'Kuama (Guatamo) Ven.		Soldiers for *catara*		Exp. XI.2002
Hymenopt. Formicidae	*Meka*	*Atta cephalotes* L.?	al	SIONA SECOYA, Ecuador				Vicklers, pers. comm., 1999
Hymenopt. Formicidae	*Ruhaa*	*Atta laevigata* Smith	S, al	TUKANOANS, Colombia	W, C		ra	Dufour, 1987[25]
Hymenopt. Formicidae		*Atta sexdens* L.	A	ASHUARAS, Ecuador				Onore, 1997
Hymenopt. Formicidae		*Atta sexdens* L.	A	AWA, Ecuador				Onore, 1997
Hymenopt. Formicidae		*Atta sexdens* L.	A	COFANES, Ecuador				Onore, 1997
Hymenopt. Formicidae		*Atta sexdens* L.	A	HUAORANIS, Ecuador				Onore, 1997
Hymenopt. Formicidae	*Sauva*	*Atta sexdens* L.		KAYAPO, Brazil				Posey, 1979
Hymenopt. Formicidae		*Atta sexdens* L.	A	NEGR. ESMERALD., Ecuador				Onore, 1997
Hymenopt. Formicidae		*Atta sexdens* L.	A	QUICHUAS, Ecuador				Onore, 1997
Hymenopt. Formicidae		*Atta sexdens* L.	A	SECOYAS, Ecuador				Onore, 1997
Hymenopt. Formicidae		*Atta sexdens* L.	P, L	SHUARAS, Ecuador				Onore, 1997
Hymenopt. Formicidae		*Atta sexdens* L.	A	SIONAS, Ecuador				Onore, 1997

Appendix 15.1: (Contd.)

Appendix 15.1: (Contd.)

Family	Local name	Species[1]		Ethnic group[2]	Host plant	Eaten[3]	Reference
Hymenoptera Formicidae	*Biapuna*	*Atta sexdens* L.	S, al	TUKANOANS, Colombia	W, C	ra	Dufour, 1987[26]
Hymenopt. Formicidae	*Coqui*	*Atta* sp.	al	CAMPA, Gr. Pajonal, Peru			Denevan, 1971[21]
Hymenopt. Formicidae	*Petabuto*	*Atta* sp.	al	GUAJIBO (HIWI), Ven.			Paoletti, 1996–1998[27]
Hymenopt. Formicidae	*Welito*	*Atta* sp.	al	GUAJIBO (HIWI), Ven.			Paoletti, 1996–1998[27]
Hymenopt. Formicidae	*Sauva*	*Atta* sp.		KAYAPO, Brazil			Posey, 1979
Hymenopt. Formicidae		*Atta* sp.		LECOS (Bolivia)		ra	Stefano Vanzin, 2003
Hymenopt. Formicidae	*Huaeti*	*Atta* sp	al	PIAROA, Cuao river, Ven.			Zent, 1992
Hymenopt. Formicidae	*Weneru*	*Atta* sp.	al	PIAROA, Cuao river, Ven.			Zent, 1992
Hymenopt. Formicidae	*Wiriya terae*	*Atta* sp.	al	PIAROA, Cuao river, Ven.			Zent, 1992
Hymenopt. Formicidae	*Akue*	*Atta* sp.	al	PIAROA, Ven.			Paoletti, 1996–1998[9]
Hymenopt. Formicidae	3 ethno names	*Atta* sp.	S	PIAROA, Ven.			Paoletti, 1996–1998[9]
Hymenopt. Formicidae	*Mohra*	*Atta* sp.		SURUI, Brazil	A	r, f	Coimbra, 1985
Hymenopt. Formicidae	*Sauba*	*Atta* sp.	al	YANOMAMO, Ven.		r, ra	Smole, 1976[6]
Hymenopt. Formicidae	*Oshe ana*	*Atta* sp.	al, S	YANOMAMO, Ven.	A	r, ra	Paoletti coll. 1998
Hymenopt. Formicidae		*Atta* sp.		YE'KUANA, Ven.			Paoletti, 1998–1999
Hymenopt. Formicidae	*Kiavu*	*Atta* sp.	al	YUKPA, Ven.	A	r, ro	Ruddle, 1973
Hymenopt. Megachilidae		*Megachile* sp.	L	OTAVALOS, Ecuador			Onore, 1997

Appendix 15.1: (Contd.)

Appendix 15.1: (Contd.)

Family	Local name	Species[1]		Ethnic group[2]		Host plant	Eaten[3]	Reference
Hymenoptera Megachilidae		*Megachile* sp.	L	PILAHUINES, Ecuador				Onore, 1997
Hymenopt. Megachilidae		*Megachile* sp.	L	QUICHUAS, Ecuador				Onore, 1997
Hymenopt. Megachilidae		*Megachile* sp.	L	SALAZACAS, Ecuador				Onore, 1997
Hymenopt. Megachilidae		*Megachile* sp.	L	SARAGUROS, Ecuador				Onore, 1997
Hymenopt. Melaponinae	55 ethno names		HLP	KAYAPO, Brazil				Posey, 1979
Hymenopt. Meliponinae				ACHE, Paraguay				Hill et al., 1987
Hymenopt. Meliponinae	*Cheti*			ACHE, Paraguay				Hill et al., 1987
Hymenopt. Meliponinae	4 ethno names			CURRIPACO, Ven.				Paoletti, 1998[15]
Hymenopt. Meliponinae	4 ethno names			GUAJIBO (HIWI), Ven.			r, ra	Paoletti, 1996–1998
Hymenopt. Meliponinae				PANKARARE, Brazil				Medeiros Costa-Neto, 1998[29]
Hymenopt. Meliponinae	4 ethno names			PIAROA, Ven.				Paoletti, 1996–1998
Hymenopt. Meliponinae	33 ethno names			YANOMAMO, Ven.	A		r, ra	Paoletti coll. 1997
Hymenopt. Meliponinae	4 ethno names			YE'KUANA, Ven.				Paoletti, 1998–1999
Hymenopt. Meliponinae	*Okosji*	*Frieseomelitta varia* (Lepeletier)		YANOMAMO, Ven.		Mavaca, on beehive J. Finkers		Expedition XI 2002

Appendix 15.1: (Contd.)

Appendix 15.1: (Contd.)

Family	Local name	Species[1]		Ethnic group[2]		Host plant	Eaten[3]	Reference
Hymenopt. Meliponinae		Gen. sp. ?	L, P	TUKANOANS, Colombia	A			Dufour coll.[30]
Hymenopt. Meliponinae	*Waloya-ey*	*Melipona grandis* Guerin	HLP	SURUI, Brazil	A		r	Coimbra, 1985
Hymenopt. Meliponinae	*Warwarkuy -ey*	*Melipona schwarzi* Moure	HLP	SURUI, Brazil	A		r	Coimbra, 1985
Hymenopt. Meliponinae	*Yoidmeb-ey*	*Nannotrigona bipunctata polystica* (Moure)	HLP	SURUI, Brazil	A		r	Coimbra, 1985
Hymenopt. Meliponinae	*Wakabpe-ey*	*Nannotrigona* n. sp.	HLP	SURUI, Brazil	A		r	Coimbra, 1985
Hymenopt. Meliponinae	*Yoiditer-ey*	*Nannotrigona xanthotricha* (Moure)	HLP	SURUI, Brazil	A		r	Coimbra, 1985
Hymenopt. Meliponinae		*Partamona testacea* (Klug)		YANOMAMO, Ven.				Paoletti coll. 1998
Hymenopt. Meliponinae	*Yape-ey*	*Plebeia* n. sp.	HLP	SURUI, Brazil	A		r	Coimbra, 1985
Hymenopt. Meliponinae	*Kirun-ey*	*Ptilotrigona lurida* (Smith)	HLP	SURUI, Brazil	A		r	Coimbra, 1985
Hymenopt. Meliponinae		*Scaptotrigona xanthotricha* (Moure)		YANOMAMO, Ven.				Paoletti coll. 1998
Hymenopt. Meliponinae	*Waptirir-ey*	*Tetragonisca angustula* Latr.		SURUI, Brazil	A		r	Coimbra, 1985
Hymenopt. Meliponinae	*Arma-ey*	*Tetragonisca branneri* Cocker.	HLP	SURUI, Brazil	A		r	Coimbra, 1985
Hymenopt. Meliponinae	*Gosereyb-ey*	*Tetragonisca dorsalis* (Smith)	HLP	SURUI, Brazil	A		r	Coimbra, 1985
Hymenopt. Meliponinae	*Wano*	*Trigona (Tetragona) clavipes* F.	HLP	YUKPA, Ven.	M		y, ro	Ruddle, 1973

Appendix 15.1: (Contd.)

Appendix 15.1: (Contd.)

Family	Local name	Species[1]		Ethnic group[2]		Host plant	Eaten[3]	Reference
Hymenoptera Meliponinae	*Wikisa*	*Trigona amulthea* (Oliver)	HLP	YUKPA, Ven.	M		y, ro	Ruddle, 1973
Hymenopt. Vespidae				CURRIPACO, Ven.				Paoletti, 1998[15]
Hymenopt. Vespidae	8 ethno names			GUAJIBO (HIWI), Ven.			r, ra	Paoletti, 1996–1998[31]
Hymenopt. Vespidae	39 ethno names		HLP	KAYAPO, Brazil				Posey, 1979
Hymenopt. Vespidae			HLP	PANKARARE, Brazil				Medeiros Costa-Neto, 1998[32]
Hymenopt. Vespidae	*Paehu Acyu*		P	PIAROA, Cuao river, Ven.				Zent, 1992
Hymenopt. Vespidae	5 ethno names			PIAROA, Ven.				Paoletti, 1996–1998
Hymenopt. Vespidae				YANOMAMO, Ven.				Lizot, 1988
Hymenopt. Vespidae	10 ethno names			YANOMAMO, Ven.	A		r, ra	Paoletti coll. 1997[33]
Hymenopt. Vespidae	*Copina*			YANOMAMO (Guarapana) Ven.			r, ra	Ex. XI.2002
Hymenopt. Vespidae				YE'KUANA, Ven.				Paoletti, 1998–1999
Hymenopt. Vespidae	*Mamisici-acorin*	*Agelaia pallidiventris* (Richards)	L, P	YANOMAMO, (Motorema) Ven.	A		r, ra	Paoletti coll. 1998[34]
Hymenopt. Vespidae		*Angiopolybia paraensis* Spinola	P, L	ASHUARAS, Ecuador				Onore, 1997
Hymenopt. Vespidae		*Angiopolybia paraensis* Spinola	P, L	SHUARAS, Ecuador			ra	Onore, 1997

Appendix 15.1: (Contd.)

Appendix 15.1: (Contd.)

Family	Local name	Species[1]		Ethnic group[2]	Host plant	Eaten[3]	Reference
Hymenoptera Vespidae		*Apoica pallens* F.	P, L	ASHUARAS, Ecuador			Onore, 1997
Hymenopt. Vespidae		*Apoica pallens* F.	P, L	SHUARAS, Ecuador		ra	Onore, 1997
Hymenopt. Vespidae		*Apoica pallida* Olivier	P, L	ASHUARAS, Ecuador			Onore, 1997
Hymenopt. Vespidae		*Apoica pallida* Olivier	P, L	SHUARAS, Ecuador		ra	Onore, 1997
Hymenopt. Vespidae		*Apoica strigata* Richards	P, L	ASHUARAS, Ecuador			Onore, 1997
Hymenopt. Vespidae		*Apoica strigata* Richards	P, L	SHUARAS, Ecuador		ra	Onore, 1997
Hymenopt. Vespidae		*Apoica thoracica* R. de Buysson	P, L	ASHUARAS, Ecuador			Onore, 1997
Hymenopt. Vespidae		*Apoica thoracica* R. de Buysson	P, L	SHUARAS, Ecuador		ra	Onore, 1997
Hymenopt. Vespidae	*Utia*	*Apoica thoracica* R. de Buysson	L, P	TUKANOANS, Colombia	A	r, ra	Dufour, 1987
Hymenopt. Vespidae		*Brachygastra lecheguana* Latr.	L, H	COLONOS DE LA COSTA, Ecuador			Onore, 1997
Hymenopt. Vespidae		*Brachymenes wagnerianus* Saussure	P, L	ASHUARAS, Ecuador			Onore, 1997
Hymenopt. Vespidae		*Brachymenes wagnerianus* Saussure	P, L	SHUARAS, Ecuador		ra	Onore, 1997
Hymenopt. Vespidae		*Mischocyttarus rotundicollis* Cameron	P	?, Ecuador			Onore, 1997

Appendix 15.1: (Contd.)

Appendix 15.1: (Contd.)

Family	Local name	Species[1]		Ethnic group[2]	Host plant	Eaten[3]	Reference
Hymenopt. Vespidae		*Mischocyttarus tomentosus* Zikan	P	?, Ecuador			Onore, 1997
Hymenopt. Vespidae		*Montezumia dimidiata* Saussure	P, L	ASHUARAS, Ecuador			Onore, 1997
Hymenopt. Vespidae		*Montezumia dimidiata* Saussure	P, L	SHUARAS, Ecuador		ra	Onore, 1997
Hymenopt. Vespidae		*Polistes bicolor* Lepeletier	P, L	ASHUARAS, Ecuador			Onore, 1997
Hymenopt. Vespidae		*Polistes bicolor* Lepeletier	P, L	SHUARAS, Ecuador		ra	Onore, 1997
Hymenopt. Vespidae		*Polistes deceptor* Schultz	P, L	ASHUARAS, Ecuador			Onore, 1997
Hymenopt. Vespidae		*Polistes deceptor* Schultz	P, L	SHUARAS, Ecuador		ra	Onore, 1997
Hymenopt. Vespidae		*Polistes occipitalis* Ducke	P, L	ASHUARAS, Ecuador			Onore, 1997
Hymenopt. Vespidae		*Polistes occipitalis* Ducke	P, L	SHUARAS, Ecuador		ra	Onore, 1997
Hymenopt. Vespidae		*Polistes testaceicolor* Bequaert	P, L	ASHUARAS, Ecuador			Onore, 1997
Hymenopt. Vespidae		*Polybia aequatoriales* Zavattari	P, L	SHUARAS, Ecuador		ra	Onore, 1997
Hymenopt. Vespidae		*Polybia dimidiata* F.	P, L	SHUARAS, Ecuador		ra	Onore, 1997
Hymenopt. Vespidae		*Polybia emaciata* Lucas	P, L	SHUARAS, Ecuador		ra	Onore, 1997

Appendix 15.1: (Contd.)

Appendix 15.1: (Contd.)

Family	Local name	Species[1]		Ethnic group[2]		Host plant	Eaten[3]	Reference
Hymenopt. Vespidae		*Polybia flavifrons* Smith	P, L	SHUARAS, Ecuador			ra	Onore, 1997
Hymenopt. Vespidae	*Utia*	*Polybia rejecta* (F.)	L, P	TUKANOANS, Colombia	A		r, ra	Dufour, 1987
Hymenopt. Vespidae		*Polybia testaceicolor* Bequaert	P, L	SHUARAS, Ecuador			ra	Onore, 1997
Hymenopt. Vespidae	*Toti Utia*	*Stelopolybia angulata* (F.)	L, P	TUKANOANS, Colombia	A		r, ra	Dufour, 1987
Hymenopt. Vespidae		*Stelopolybia baezae* Richards	L	CANARIS, Ecuador				Onore, 1997
Hymenopt. Vespidae		*Synoeca corneliana* Richards	L	?, Ecuador				Onore, 1997
Hymenopt. Vespidae		*Synoeca lobipleura* Richards	L	?, Ecuador				Onore, 1997
Hymenopt. Vespidae		*Synoeca ornata* Ducke	L	?, Ecuador				Onore, 1997
Hymenopt. Vespidae		*Synoeca virginea* F.	P	?, Ecuador				Onore, 1997
Hymenopt. Vespoidea Polistinae	*Wanacana*	*Mischocyttarus* sp.	L, P	YUKPA, Ven.	A		y, ra	Ruddle, 1973
Hymenopt. Vespoidea Polistinae	*Mist Koruka*	*Polistes canadensis* (L.)	L, P	YUKPA, Ven.	A		y, ra	Ruddle, 1973
Hymenopt. Vespoidea Polistinae	*Nonawu*	*Polistes pacificus* (F.)	L, P	YUKPA, Ven.	A		y, ra	Ruddle, 1973
Hymenopt. Vespoidea Polistinae	*Mist*	*Polistes versicolor* (Oliv.)	L, P	YUKPA, Ven.	A		y, ra	Ruddle, 1973
Hymenopt. Vespoidea Polistinae	*Piowara*	*Polybia ignobilis* (Haliday)	L, P	YUKPA, Ven.	A		y, ra	Ruddle, 1973

Appendix 15.1: (Contd.)

Appendix 15.1: (Contd.)

Family	Local name	Species[1]		Ethnic group[2]		Host plant	Eaten[3]	Reference
Insecta	142 ethno names			YANOMAMO, Ven.				Finkers, 1986
Insecta	6 ethno names			YANOMAMO, Ven.				Cocco, 1982
Insecta		All insects are taboo to them	not	MATIS, Panoan, Brazil				Milton, 1997[35]
Isoptera Termitidae	*Bucoa*	*Labiotermes labralis* (Holmgren)	al, P	TUKANOANS, Colombia	A	nests on trees	r	Dufour coll. N.107[36]
Isoptera Termitidae	*Arepo*	*Labiotermes* sp. cfr . *labralis*	L, P, al	YANOMAMO, Ven.	A		r, ra	Paoletti coll. 1998[37]
Isoptera Termitidae				KAYAPO, Brazil				Posey, 1979
Isoptera Termitidae				YANOMAMO, Ven.				Smole, 1976[6]
Isoptera Termitidae	*Jajoma*	*Labiotermes* sp. cfr . *labralis*	P, al	YANOMAMO, Ven.	A			Paoletti coll.1998
Isoptera Termitidae		*Nasutitermes* sp. *corniger* (Motschul.)		LECOS, Bolivia			r, ra	Candelaria, leg.S. Vanzin XI.2002
Isoptera Termitidae	*Hajara asip*	*Nasutitermes* sp.	al, P	YANOMAMO (Hachacoa) Ven.			ra	Ex. XI. 2002
Isoptera Termitidae	*Haoma*	*Nasutitermes* sp.	al, P	YANOMAMO (Hachacoa)			ra	Ex. XI. 2002
Isoptera Termitidae	*Arepo*	*Nasutitermes ephrateae* (Holmgren)	al, P	YANOMAMO, Ven.			ra	Cerda and Torres collection, 1999[38]
Isoptera Termitidae	*Arepo*	*Nasutitermes ephrateae* (Holmgren)	al, P	YANOMAMO, (Hachacoa) Ven.			ra	Ex. XI. 2002
Isoptera Termitidae	*Bupena*	*Syntermes aculeosus* Emerson	S	TUKANOANS, Colombia	W, C			Dufour, 1987
Isoptera Termitidae	*Ofo*	*Syntermes* sp.	S	GUAJIBO (HIWI), Ven.				Paoletti, 1996–1998[9]

Appendix 15.1: (Contd.)

Appendix 15.1: (Contd.)

Family	Local name	Species[1]		Ethnic group[2]	Host plant	Eaten[3]	Reference
Isoptera Termitidae		*Syntermes* sp.	S	MAKU, Cobia-Brazil		s, y	Milton, 1984
Isoptera Termitidae	*Aekua*	*Syntermes* sp.	al	PIAROA, Cuao river, Ven.			Zent, 1992
Isoptera Termitidae		*Syntermes* sp.	S, he	PIAROA, Ven.		bo	Paoletti, 1996–1998[9]
Isoptera Termitidae	*Wapaacashi*	*Syntermes* sp.	S	YANOMAMO, Ven.	A	r, ra	Paoletti coll. 1998
Isoptera Termitidae		*Syntermes* sp.	S, he	YANOMAMO, Ven.		r	Lizot, 1988
Isoptera Termitidae	*Meka Bupuara*	*Syntermes spinosus* Latr.	S	TUKANOANS, Colombia	W, C		Dufour, 1987
Isoptera Termitidae		*Syntermes tanygnathus* Constantino	S	TUKANOANS, Colombia	W, C		Dufour, 1987
Isoptera Termitidae	*Osce Ana*	*Synthermes territus* Emerson	S, he	YANOMAMO, (Motorema) Ven.		ra	Expedition XI.2002
Isoptera, Termitidae	*Seri*	*Syntermes aculeosus* Emerson	S, he	YE'KUANA, Ven.		ra	Paoletti, 1998–1999
Lepidopt.	*Kaenia*		L	PIAROA (Cuao river), Ven.			Zent, 1992
Lepidopt.	*Kanaeru*		L	PIAROA (Cuao river), Ven.			Zent, 1992
Lepidopt.	*Raesaenia*		L	PIAROA (Cuao river), Ven.			Zent, 1992
Lepidopt.	*Ruaenia*		L	PIAROA (Cuao river), Ven.			Zent, 1992
Lepidopt.	*Uhuminu*		L	PIAROA (Cuao river), Ven.			Zent, 1992
Lepidopt.	4 ethno names		L	PIAROA, Ven.	A	ra	Paoletti, 1996–1998

Appendix 15.1: (Contd.)

Appendix 15.1: (Contd.)

Family	Local name	Species[1]		Ethnic group[2]	Host plant	Eaten[3]	Reference
Lepidopt.	*Pitiusip*		L	SHUARAS, Ecuador		f, bo	Onore, 1997
Lepidopt.	*Tampidura*		L	SHUARAS, Ecuador		f, bo	Onore, 1997
Lepidopt.	*Yankinia*		L	SHUARAS, Ecuador		f, bo	Onore, 1997
Lepidopt.				SURUI, Brazil	*Castilla ulei*		Coimbra, 1985
Lepidopt.				SURUI, Brazil	*Tetragastris altissima*		Coimbra, 1985
Lepidopt.	*Kasha*		P	YANOMAMO, Ven.			Smole, 1976[6]
Lepidopt.	*Kishipinan-oma*		L	YANOMAMO, Ven.			Cerda and Torres coll., 1999[38]
Lepidopt.	*Masamasadi*		L	YE'KUANA (Guatamo) Ven.	*Deocjc' Pourouma guianensis*		Ex. XI.2002
Lepidopt.		(Black and red)	L	MAKU, Colombia-Brazil		s, y	Milton, 1984
Lepidopt. Brassolidae		*Brassolis astyra* Godman and Salvin	L	HUAORANIS, Ecuador			Onore, 1997
Lepidopt. Brassolidae		*Brassolis astyra* Godman and Salvin	L	QUICHUAS, Ecuador			Onore, 1997
Lepidopt. Brassolidae		*Brassolis sophorae* L.	L	HUAORANIS, Ecuador			Onore, 1997
Lepidopt. Brassolidae		*Brassolis sophorae* L.	L	QUICHUAS, Ecuador			Onore, 1997
Lepidopt. Castnidae		*Castnia daedalus* Cramer	L	HUAORANIS, Ecuador			Onore, 1997
Lepidopt. Castnidae		*Castnia daedalus* Cramer	L	QUICHUAS, Ecuador			Onore, 1997

Appendix 15.1: (Contd.)

Appendix 15.1: (Contd.)

Family	Local name	Species[1]		Ethnic group[2]		Host plant	Eaten[3]	Reference
Lepidopt. Castnidae		*Castnia licoides* Boisduval	L	HUAORANIS, Ecuador				Onore, 1997
Lepidopt. Castnidae		*Castnia licoides* Boisduval	L	QUICHUAS, Ecuador				Onore, 1997
Lepidopt. Castnidae		*Castnia licus* Drury	L	HUAORANIS, Ecuador				Onore, 1997
Lepidopt. Castnidae		*Castnia licus* Drury	L	QUICHUAS, Ecuador				Onore, 1997
Lepidopt. Hepialidae		*Hepialus* sp.	L	NEGR. ESMERALD, Ecuador				Onore, 1997
Lepidopt. Hepialidae		*Hepialus* sp.	L	QUICHUAS, Ecuador				Onore, 1997
Lepidopt. Hesperioidea	*Une gawa*	*Hesperidae* sp.	L	TUKANOANS, Colombia		*Pupunha* spp. leaves		Hugh-Jones, 1999 pers. comm.
Lepidopt. Mimallonidae?	*Outunima*			TUKANOANS, Colombia	A	*Anacardium* spp.		Dufour, 1987
Lepidopt. Noctuidae	*Batyia*			TUKANOANS, Colombia	A	*Erisma japura* Spruce ex Warming		Dufour, 1987
Lepidopt. Noctuidae	*Hadi ia*		L	TUKANOANS, Colombia		*Erisma japura* Spruce ex Warming leaves		Hugh-Jones, 1999 pers. comm.
Lepidoptera Noctuidae	*Sona-saa*		L	TUKANOANS, Colombia				Hugh-Jones, 1999 pers. comm.

Appendix 15.1: (Contd.)

Appendix 15.1: (Contd.)

Family	Local name	Species[1]		Ethnic group[2]		Host plant	Eaten[3]	Reference
Lepidopt. Noctuidae?	*Mamneg-ey*		L	SURUI, Brazil		*Bertholletia excelsa* Hum. and Bon.		Coimbra, 1985
Lepidoptera Noctuidae Catocalinae	*Yupuna*		L	YUKPA, Ven.	A	*Zea mays* L.?	ra	Ruddle, 1973
Lepidopt. Noctuidae Hadeninae	*Yupuna*		L	YUKPA, Ven.	A	*Zea mays* L.?	ra	Ruddle, 1973
Lepidopt. Noctuidae Hadeninae	*Yupuna*	*Laphygma frugiperda* (Smith)	L	YUKPA, Ven.	A	*Zea mays* L.?	ra	Ruddle, 1973
Lepidopt. Noctuidae Hadeninae	*Yupuna*	*Mocis repanda* F.	L	YUKPA, Ven.	A	*Zea mays* L.?	ra	Ruddle, 1973
Lepidopt. Notodontidae	*Menehaia*		L	TUKANOANS, Colombia	A	*Inga* spp.		Dufour, 1987
Lepidopt. Saturnidae	*Ia hoa kuna*		L	TUKANOANS, Colombia				Hugh-Jones, 1999 pers. comm.
Lepidopt. Saturnidae	*Mene besuwu-sa*		L	TUKANOANS, Colombia		*Inga* spp.		Hugh-Jones, 1999 pers. comm.
Lepidopt. Saturnidae		*Automeris* sp.		TUKANOANS, Colombia				Hugh-Jones, 1999 pers. comm.
Lepidopt. Saturnidae	*Hutia*	*Dirphia* sp.?	L, P	TUKANOANS, Colombia	A	secondary forest		Dufour, 1987
Lepidopt. Saturnidae?	*Jebakemi*		L	YANOMAMO, Ven.				Cerda and Torres coll., 1999[38]
Lepidopt. Saturnidae?	*Tikatikamo*		L	YANOMAMO, Ven.				Cerda and Torres coll., 1999[38]

Appendix 15.1: (Contd.)

Appendix 15.1: (Contd.)

Family	Local name	Species[1]		Ethnic group[2]		Host plant	Eaten[3]	Reference
Lepidopt. Sphingidae	*Kiinamono*	*Erinnyis ello* (L.)	L	TUKANOANS, Colombia		*Manihot esculenta* Crantz		Dufour, 1987
Lepidopt. Sphingidae	*Kiinamono*	*Erinnyis ello* (L.)	L, P	TUKANOANS, Colombia	A	*Manihot esculenta* Crantz		Dufour, 1987
Lepidoptera Sphingidae	*Opomoschi*	*Erinnyis ello* (L.)	L	YANOMAMO, Ven.	A	*Manihot esculenta* Crantz		Paoletti coll., 1998
Lepidopt. varia	3 ethno names		L	CURRIPACO, Ven.	A		ra	Paoletti, 1998[15]
Lepidopt. varia	4 ethno names		L	GUAJIBO (HIWI), Ven.			ra	Paoletti, 1996–1998
Lepidopt. varia	25 ethno-names		L	YANOMAMO, Ven.	A			Paoletti coll., 1997
Lepidopt. varia	3 ethno names?		L	YE'KUANA, Ven.				Paoletti, 1998–1999
Lepidoptera, Nymphalidae	*guachanzo-curo*	*Panacea prola* Doubleday	L	Napo district, Ecuador		*Cariodendron orinocense*	ra	Onore, 2004
Neuroptera Megalopt. Corydalidae		*C. armatus* and *C. peruvianus*		Cordillera, Peru				Menzel and D'Aluisio, 1998
Neuropt. Megalopt. Corydalidae	*Stpaykt*	*Corydalus* sp.	L, A	YUKPA, Ven.	W, C		ra	Ruddle, 1973
Odonata Aeschnidae		*Aeschna brevifrons* Hagen	L	QUICHUAS, Ecuador			r	Onore, 1997
Odonata Aeschnidae		*Aeschna marchali* Rambur	L	QUICHUAS, Ecuador			r	Onore, 1997

Appendix 15.1: (Contd.)

Appendix 15.1: (Contd.)

Family	Local name	Species[1]		Ethnic group[2]		Host plant	Eaten[3]	Reference
Odonata Aeschnidae		*Aeschna peralta* Ris	L	QUICHUAS, Ecuador			r	Onore, 1997
Odonata Aeschnidae		*Coryphaeschna adnexa* Hagen	L	OTAVALOS, Ecuador				Onore, 1997
Odonata Aeschnidae		*Coryphaeschna adnexa* Hagen	L	QUICHUAS, Ecuador				Onore, 1997
Odonata Gomphidae	*Wahuhe'*	*Zonophora* sp.		YE'KUANA (Alacran) Ven.			r	Ex. XI.2003
Odonata Libellulidae	*Wahuhe'*	*Dasythemis* sp.		YE'KUANA (Alacran) Ven.			r	Ex. XI.2002
Orthopt.	3 ethno names			GUAJIBO (HIWI), Ven.		Sabaneta Guaiabal		Paoletti, 1996–1998[9]
Orthoptera Acrididae	*Pisatpi*	*Aidemona azteca* Saussure	A	YUKPA, Ven.	W, C		ra	Ruddle, 1973
Orthopt. Acrididae	*Pisatpi*	*Orphulella* sp.	A	YUKPA, Ven.	W, C		ra	Ruddle, 1973
Orthoptera Acrididae	*Pisatpi*	*Osmilia flavolineata* De Geer	A	YUKPA, Ven.	W, C		ra	Ruddle, 1973
Orthopt. Acrididae	*Kosopina*	*Osmilia* sp.	A	YUKPA, Ven.	W, C		ra	Ruddle, 1973
Orthopt. Acrididae	*Kosopina*	*Schistocerca* sp.	A	YUKPA, Ven.	W, C		ra	Ruddle, 1973
Orthopt. Acrididae	*Sakaramo*	*Tropidacris latreillei* (Pty.)	A	YUKPA, Ven.	W, C		ra	Ruddle, 1973
Orthopt. Acrididae	*Tgbuto*	*Rhammatocerus* sp.		GUAJIBO (HIWI), Ven.			ra	Paoletti, 1996–1998[9]
Orthopt. Acrididae		*Schistocerca* sp.	A	QUICHUAS, Ecuador				Onore, 1997
Orthopt. Acrididae	*Servato*	*Tropidacris cristata* L.	A	GUAJIBO (HIWI), Ven.	C		ra	Paoletti, 1996–1998[9]

Appendix 15.1: (Contd.)

Appendix 15.1: (Contd.)

Family	Local name	Species[1]		Ethnic group[2]	Host plant	Eaten[3]	Reference
Orthopt. Tettigonidae	*Pistina*	*Conocephalus angustifrons* Redt.	A	YUKPA, Ven.	W, C	ra	Ruddle, 1973
Orthopt. Tettigoniidae	*Nyimia*	*Neoconcephalus* sp.	A	TUKANOANS, Colombia			Hugh-Jones, 1999 pers. comm.
Trichoptera		Gen. sp. ?	L	YE'KUANA (Alacran) Ven.			Exp. XI.2003
Trichoptera	-	-	L	YE'KUANA, Ven.	W, C	ra	Ex. XI.2002
Trichoptera Hydropsychidae	*Misipsi*	*Leptonema* sp.	L	YUKPA, Ven.	W, C	ra	Ruddle, 1973

[1]A = Adult; L = larvae or caterpillar; P = pupa, neanid, nymph; H = honey; LPH = larvas, pupas and honey; S = soldiers; he = only the head; al = swarming alates

[2]A = All group collects; W = only women; C = children; M = mostly men

[3]bo = soup; ra = roasted; r = raw; f = fry; t = toasted; dr = drink; s = sauce (catara); y = yearly

[4]Zent has recorded one year's collection June 1986–July 1987.

[5]Interviews and collections at Toki, Guatamo and Buena Vista (Alto Rio Padamo, Amazonas, Venezuela).

[6]He made his observations for Barafiri, he reports data of Valero collected by Biocca.

[7]The Yukpa kill the lice with a bite and always discard it (Ruddle, 1973).

[8]Ken Hill (pers. comm. 10.08.1999) argues that this name is based on a determination by a nonspecialist studying larvae that could not be properly identified.

[9]Alcabala de Guajibo.

[10]Collected by Guido Goyoni, near Babilla de Pintado, Puerto Ayacucho, Venezuela, as edible by Piaroa. The genus *Metamasius* includes 40 species from South America (Vaurie, 1966).

[11]Cite 1.32 hours per kg of palm larvae.

[12]Personal data collection from speaking with Baniva peoples in Puerto Ayacucho.

[13]Observed that the Bari, for providing *Rhynchophorus palmarum* L., fell only *Jessenia* spp. palms.

[14]from Coimbra, 1984.

[15]Interview especially at Cucurital, near Puerto Ayacucho, Amazonas, Venezuela, with Capitan Eleasar Torquato and family.

[16]She mentions only four insects without local or scientific names.

[17]Interviews and collections in Gavilan, Cano Tigre, Babilla de Pintado near Puerto Ayacucho, Venezuela.

[18]*Jacaratia dodecaphylla* (Caricaceae) is felled to be attacked by *Rhynchophorus palmarum* L. After 2–3 months the larvae are eaten only fried. The larvae are obtained also from other palms.

[19]Palms are felled to grow the palm worms.

[20]Collected in Mavaka, Alto Orinoco at Motorema village from logs in fallow garden 1998. Ignacio says that only old people eat it after having discarded the gut. Det. C. Ratcliffe on adults obtained from the larvae.

[21]He mentions 5–6 *Atta* species eaten. He mentions larvae and beetles and says that children are actively involved with this sort of collection.

[22]Personal information 4.VII.99.

[23]Yapu 25.VII.1986 on guama tree *Inga* spp.

[24]The alatae are called *mëkaiyaa liara* which swarms early morning in the first half of the rainy season.

[25]The alatae are called *ruhaa liara;* swarms about 3 pm in the first half of the rainy season.

[26]The alatae are called *biapuna liara;* swarms about 4 pm in the first half of the rainy season.

[27]Interview and collections in Sabaneta Guaiabal, Alcabala Guajibo and Coromoto near Puerto Ayacucho, Venezuela. They produce a spicy sauce, Catara sold on the Market, containing soldiers of *Atta* spp.

[28]Interview and collections in Coromoto near Puerto Ayacucho, Venezuela.

[29]He cites 15 Meliponinae handled (not eaten?).

[30]Acaricunra June 1977 (not numbered).

[31]Interview and collections in Sabaneta Guaiabal, Alcabala Guajibo and Coromoto near Puerto Ayacucho, Venezuela.

[32]He cites 7 wasps Polistinae (not eaten?).

[33]Interviews in August 1997 at Puerto Ayacucho, Venezuela, with the following groups of Yanomamo:

*Renato Moi, 22, Comunidad Mawaka, Alto Orinoco, Venezuela, Amazonas

*Carlo Gomez, 22, Comunidad Mawaka, Alto Orinoco, Venezuela, Amazonas

*Porfirio Miguel, Comunidad de Platanal, Alto Orinoco, Venezuela, Amazonas

*Maxeopewe Camacho, 30, Comunidad Mawaka, Alto Orinoco, Venezuela, Amazonas

*Simon Silva, 25, Comunidad de Arata, Alto Orinoco, Venezuela, Amazonas

*Gustavo Pairawa, Comunidad Motorema, Alto Orinoco, Venezuela. Amazonas

*Pompeio Nuñez, 32, Ocama, Alto Orinoco, Venezuela, Amazonas

*Santiago Mareño, 28, Mawaca, Alto Orinoco, Venezuela, Amazonas

Technique used: the 8 interviewees answering as a team, most of the time all together, for 7 total hours of interview on three different days. After a transcription into typed pages, the two literates Carlo Gomez and Renato Moi made corrections of spellings and added a few ethno names. For each ethno name we made accurate descriptions to avoid synonymies.

[34]Collected in Mavaka, Alto Orinoco at Motorema village from a nest from a forest tree after smoking it, 1998 (det. J. Carpenter).

[35]Milton, 1997 says that deer, Agouti paca, different species of monkeys and all insects are taboo for this group.

[36]Nest on trees: Yapu 18.XI.86

[37]*Motorema*, aerial nests on trees, only white forms eaten.

[38]Collected by H. Cerda and F. Torres on May 28, 1999 at Boca de Ocamo, Alto Orinoco, Venezuela.

16

Edible Insects in Ecuador

GIOVANNI ONORE

Departamento de Biologia, Pontificia Universidad Católica del Ecuador, Apartado 17-01-2184, Quito, Ecuador. E-mail: GONORE@puce.edu.ec

Abstract

In Ecuador the ancestral tradition of entomophagy still exists, particularly in the countryside where the native population is relatively isolated from technological progress. Eighty-two (82) edible species are listed for the country; none is a main dish but many are used to complement other animal protein sources in the diet. The most common edible insects belong to orders Coleoptera and Hymenoptera and are consumed at either the larval or the adult stage.

Key Words: Ecuador, edible insect species, entomophagy, *Panacea prola* biology.

Introduction

In their early history, humans obtained most of their protein from vertebrates through hunting and fishing. However, in natural ecosystems when humans were short of protein, collecting insects proved to be an interesting and permanent alternative (Bodenheimer, 1951). In fact, insects represent a wholesome food. For example, 100 g of dried termites contain on average 53% protein, 15% fat and 3.5% carbohydrates and significant amounts of phosphorus, iron, sodium, and potassium (Tosi and Daccordi, 1983). Insects can be variously consumed: raw, fried, boiled, salted and dried, smoked, and as ingredients in sauces.

In Ecuador the tradition of entomophagy is very old. This can be seen from archaeological records in the museums of Quito, in which edible insects from orders Coleoptera and Lepidoptera are mentioned. Also, reports from the early colonization period mention the entomophagy of Ecuadorian populations (Velasco, 1789). The tradition is still alive today especially in the remote areas of the country (Figueroa and Albuja, 1983; Gonzales and Ortiz de Villalba, 1985; Smith and Paucar, 2000).

Involuntary entomophagy refers to cases of ingestion of insects without intent or awareness. This form of entomophagy is very frequent. Virtually all fruit and vegetables and many other types of food contain insects or fragments of insects that are ingested along with their hosts, without consumer awareness. The most common examples are fruit fly larvae, *Anastrepha* spp. (Diptera), in guava fruits (*Psidium guajava* L.) and *Ceratitis capitata* (Wiedemann) (Diptera) in citrus.

A peculiar example of entomophagy is the habit that mothers in Ecuador have of picking lice (*Pediculus hominis capitis* De Geer) out of their children's hair and often eating instead of discarding them.

In general, however, the main scope of entomophagy is to provide protein or add variety and taste to the diet. The latter type of entomophagy is focussed here.

Inventory of Edible Insects in Ecuador

To obtain a preliminary inventory of edible insects, I gathered information during field trips from 1980 to 2003 all round Ecuador. My sources of information were from the natives and from personal observations during meals, hunting expeditions, agricultural activities, and market transactions with the native populations. Specimens of edible insects were collected wherever possible, both as imago and larvae. The larvae were then reared to adult for identification. The available edible insects were also tasted for palatability. The insects were prepared and labeled in the field, brought to the Quito Catholic Zoology Museum (QCAZ) where they were stored and identified. I followed the classification of Smart (1981); Borror et al. (1989); D'Abrera (1984, 1987); Kristensen (1998). The host plants were classified following Gentry (1993) and Patzelt (2002).

The inventory of edible insect species in Ecuador so obtained is given in Table 16.1. In three cases, the native *Shuaras* interviewees could not provide samples of the caterpillars consumed in the area due to seasonal unavailability. For these cases, only the local name has been reported in Table 16.1.

Insects currently consumed in Ecuador belong to these orders: Coleoptera, Hymenoptera, Lepidoptera, Homoptera, Orthoptera, and Odonata (Table 16.1).

Coleoptera

Coleoptera comprises beetles, which in Ecuador are called "catsos" in the adult stage and "cuzo" in the larval stage. These insects can be served boiled with salt, cooked with onions, or eaten live, such as *Metamasius* spp. The taste of raw *Metamasius* spp. could definitely compete with that of peanuts.

Around the time of the "finados" (All Souls' Day), the white beetle *Platycoelia lutescens* ("catso blanco") (Plate VIII, 1) emerges in significant numbers and is actively sought in the area around Quito. In the early morning hours between 3.00–4.00 a.m., native hunters explore the slopes of the Andes in the vicinity of Quito. According to tradition, the success of the harvest is directly related to the amount of rain and noise of thunder. On arrival back home, the elytra and legs of the beetles are removed, the beetles fried in pork fat with a local variety of long, white, leeklike onion, salted and served with "mote", a kind of soft maize. The taste of *Platycoelia lutescens* is similar to pork scraps (Fig. 16.1).

In the "paramo" (grassy and bushy areas in the high Andes) of Antizana, farmers harvest the green beetle *Pelidnota nigricauda* ("catso verde"), a species closely related to the white beetle found around Quito. Inhabitants of the inter-Andean valley consume various other species of Coleoptera closely related to the beetles mentioned above, but definitely less tasty. In the latter region, fried balls of mixed beetles may include several species of insects. In Salcedo, Calderón and Otavalo, the "catsos blancos" used as food are bigger in size than those from the Quito region and belong to species *Praogolofa unicolor*. Very often farmers in the highlands compete with chickens and dogs in hunting for "cuzos" or white grubs, which they store in "shigras" (a traditional bag) to add some variety to their dinner. In Cotopaxi and Tungurahua on the border of the Llanganati forest, farmers carrying out farming practices find the fat larvae of *Psalidognathus*

Fig. 16.1: *Platycoelia lutescens*. ("Catsos blanco") Fried adult specimens mixed with onion. Quito, XXIV de Mayo market, All Souls' Day (photo G. Onore).

Table 16.1: Edible insect species from Ecuador

Order	Scientific name	Family	Edible phase
Coleoptera	*Ancognatha castanea* Erichson	(Scarabaeidae)	–
	Ancognatha jamesoni Murray	(Scarabaeidae)	–
	Ancognatha vulgaris Arrow	(Scarabaeidae)	–
	Clavipalpus antisanae Bates	(Scarabaeidae)	–
	Coelosis biloba L.	(Scarabaeidae)	–
	Cosmopolites sordida Germar	(Curculionidae)	+
	Democrates burmeisteri Reiche	(Scarabaeidae)	–
	Dynamis nitidula Guérin	(Curculionidae)	–, @
	Dynamis perryi Waterhouse	(Curculionidae)	–, @
	Dynastes hercules L.	(Scarabaeidae)	–
	Golopha aeacus Burmeister	(Scarabaeidae)	–
	Golopha aegeon Drury	(Scarabaeidae)	–
	Heterogomphus bourcieri Guérin	(Scarabaeidae)	–
	Macrodontia cervicornis L.	(Cerambycidae)	–
	Metamasius cinnamominus Perty	(Curculionidae)	+
	Metamasius dimidiatipennis Jekel	(Curculionidae)	+
	Metamasius hemipterus L.	(Curculionidae)	+
	Metamasius sericeus Olivier	(Curculionidae)	+
	Oncideres sp.	(Cerambycidae)	+,–
	Pelidnota nigricauda Bates	(Scarabaeidae)	+, –
	Platycoelia forcipalis Ohaus	(Scarabaeidae)	–
	Platycoelia lutescens Blanchard	(Scarabaeidae)	+, –
	Platycoelia parva Kirsch	(Scarabaeidae)	–
	Platycoelia rufosignata Ohaus	(Scarabaeidae)	–
	Proagolofa unicolor Bates	(Scarabaeidae)	+,–
	Psalidognathus atys White	(Cerambycidae)	–
	Psalidognathus cacicus White	(Cerambycidae)	–
	Psalidognathus erithrocerus Reiche	(Cerambycidae)	–
	Psalidognathus modestus Thomson	(Cerambycidae)	–
	Rhynchophorus palmarum L.	(Curculionidae)	–, @
	Sphaenognathus feisthamelii Guérin-Menéville	(Lucanidae)	–
	Sphaenognathus lindenii Murray	(Lucanidae)	–
	Sphaenognathus metallifer Bomans et Lacroix	(Lucanidae)	–
Hymenoptera	*Angiopolybia paraensis* Spinola	(Vespidae)	+
	Apis mellifera (L.)	(Apidae)	–, @
	Apoica pallens Fabricius	(Vespidae)	+
	Apoica pallida Olivier	(Vespidae)	+
	Apoica strigata Richards	(Vespidae)	+
	Apoica thoracica R. de Buysson	(Vespidae)	+
	Atta cephalotes L.	(Formicidae)	+
	Atta sexdens L.	(Formicidae)	+
	Bombus atratus Friese	(Apidae)	–
	Bombus ecuadorius Meunier	(Apidae)	–
	Bombus funebris Smith	(Apidae)	–
	Bombus robustus Smith	(Apidae)	–
	Brachygastra lecheguana Latreille	(Vespidae)	–
	Brachymenes wagnerianus Saussure	(Vespidae)	+
	Megachile sp.	(Megachilidae)	–

Table 16.1: (Contd.)

Table 16.1: (Contd.)

Order	Scientific name	Family	Edible phase
	Mischocyttarus rotundicollis Cameron	(Vespidae)	+
	Mischocyttarus tomentosus Zikan	(Vespidae)	+
	Montezumia dimidiata Saussure	(Vespidae)	+
	Polistes bicolor Lepeletier	(Vespidae)	+
	Polistes deceptor Schultz	(Vespidae)	+
	Polistes occipitalis Ducke	(Vespidae)	+
	Polistes testaceicolor Bequaert	(Vespidae)	+
	Polybia aequatorialis Zavattari	(Vespidae)	+
	Polybia dimidiata (Fabricius)	(Vespidae)	+
	Polybia emaciata Lucas	(Vespidae)	+
	Polybia flavifrons Smith	(Vespidae)	+
	Stelopolybia baezae Richards	(Vespidae)	–
	Stelopolybia corneliana Richards	(Vespidae)	–
	Stelopolybia lobipleura Richards	(Vespidae)	–
	Stelopolybia ornata Ducke	(Vespidae)	–
	Synoeca virginea Fabricius	(Vespidae)	+
	Tetragonisca angustula Latreille	(Apidae)	–
Lepidoptera	*Brassolis astyra* Godman et Salvin	(Nymphalidae)	–
	Brassolis sophorae L.	(Nymphalidae)	–
	Castnia daedalus Cramer	(Castniidae)	–
	Eupalamides cyparissias Fab.	(Castniidae)	–
	Castnia licoides Boisduval	(Castniidae)	–
	Castnia licus Drury	(Castniidae)	–
	Hepialus sp.	(Hepidalidae)	–
	Panacea prola Doubleday	(Nymphalidae)	–
	'Pitiusip'	(Unknown)	–
	'Tampidura'	(Unknown)	–
	'Yankinia'	(Unknown)	–
Homoptera	*Carineta fimbriata* Walker	(Cicadidae)	+
	Umbonia spinosa Fabricius	(Membracidae)	+, –
Orthoptera	*Schistocerca* sp.	(Acrididae)	+
Odonata	*Aeschna brevifrons* Hagen	(Aeschnidae)	–
	Aeschna marchali Rambur	(Aeschnidae)	–
	Aeschna peralta Ris	(Aeschnidae)	–
	Coryphaeschna adnexa Hagen	(Aeschnidae)	–

Edible phase: (–) larva; (@) pupa; (+) imago.

erithrocerus and *P. modestus* crawling between grasses after heavy rainfall. These larvae are collected and added to soups.

Probably the best-known Coleoptera used as food is the palm weevil *Rhynchophorus palmarum*, known locally as "mayón" "gualpa" or "chontacuro". An interesting anecdote from the Napo area concerns this insect. In the 1950's, the Capuchin missionary Angel de Ucar, a great lover of plants and animals, introduced the first African oil palm (*Elaeis guineensis* Jacq.) from neighboring Colombia to the Coca Catholic Mission, probably for decorative reasons. The native Quichuas, living in the area of the Mission, traditionally cut the living trees of *Mauritia flexuosa* L., a local palm species growing in swampy forests ("morete"), and *Bactris gasipaes* H.B.K., a spiny cultivated palm ("chonta"), in order to attract the palm weevil for oviposition. After several weeks, the natives returned to the felled trunks to harvest the fat weevil larvae. Given their traditional use of the palm trees, the natives were very intrigued about the use of the newly introduced exotic palm and decided that the missionary must be using the African oil palm to rear fatter weevils as a delicacy for the bishop. Therefore they baptized the African oil palm as "chontacuro-del-obispo" which translates literally as: palm for rearing weevils for the bishop.

One of the resources collected for food by the Shuaras, native hunters of the forest in the central-eastern zone of Ecuadorian Amazon, are the small weevils associated with Arecaceae (palms and relatives), which are consumed live as an appetizer. In the Saraguro area, the fat long-horned beetle larvae of *Psalidognathus cacicus* and *Pexteuso atys* are fried and used as food and medicine.

Hymenoptera

Among Hymenoptera, bees, wasps and ants are used as food throughout the country. Although virtually everyone is familiar with the taste of honey from bees (*Apis mellifera* L.), few know the delicious taste of the fat larvae inside the honeycomb. These killer bees arrived in Ecuador in the 1970s, occupying the entire country. The native community of S. Lorenzo promptly learned how to harvest the honey and consume the honeycombs full of larvae and pupae.

In Portoviejo and many other places, nests of the wasp *Brachygastra lecheguana* ("belton") are collected to obtain the honey, which is eaten together with the larvae and pupae. The Shuaras in Macas and Sevilla Don Bosco use bags to capture the wasp nests and then roast the nest inhabitants for consumption.

Regular consumers of wasps guarantee that they taste very good but some cases of ingestion of wasps have reportedly led to hearing impairment. I have carefully tried every single species of edible wasp without experiencing this problem. Quite likely hearing impairment is related to the wasp's nutrition, since at particular times of the year wasps visit toxic flowers.

Our Colombian neighbors are very fond of flying ants as food, known to them by the local name "culonas". In Santander, this is a typical dish served at the local airport. The same tradition of eating ants has survived in eastern parts of Ecuador where the species *Atta cephalotes* and *Atta sexdens* are consumed by the local population and are known by the native Quichuas as "añangu". *Atta* spp. have a unique and fatty taste.

In Ecuador there are at least a hundred species of stingless bees (Meliponidae), which are valued for their lemonlike honey. During honey harvesting, the hunter often also eats the larvae mixed with the honey and pollen. I observed the same custom in the Valladolid area with tiny *Tetragonisca angustula* brood combs.

Children in the inter-Andean valley search earnestly for nests of the leafcutter bee *Megachile* sp. in the holes and cracks of bamboo stems. They eat the nest contents, larvae and all.

Lepidoptera

Larvae of various species of Lepidoptera (butterflies and moths) are eaten on both sides of the Andes, in particular those of families Hepialidae, Nymphalidae, and Castniidae. Larvae of Castniidae are considered by patrons to be a variety of weevil species because they resemble palm weevils. The Shuaras in Sucúa boil and fry unidentified caterpillars which they call "tapindura", "yankinia" and "pitiusip".

In Napo province I had an opportunity to investigate the larvae of Brush-Footed Butterflies eaten by native Indian people and known as "guachanzocuro". I carefully reared the caterpillars in a laboratory and obtained the imago of *Panacea prola* Doubleday 1948 (Lepidoptera: Nymphalidae) (Plate VIII, 2, 3 and 4). The adult of this brightly colored butterfly is active during the hottest hours of the day and alights along roads and in settlers farms as it is attracted by domestic animal dung and leftover kitchen food. When disturbed, it flies to nearby branches but soon returns to the same pabulum. Dorsally the wings appear mainly bluish, with whitish and black stripes while the under part of the fore-wings is reddish. The female lays eggs in small groups under *Caryodendron orinocense* (Euphorbiaceae) leaves (Fig. 16.2).

Just hatched larvae are 2.5–3.5 mm long, live in communities, and prefer the underside of leaves. At the last instar they are 50–60 mm long, orange with bluish spots and transverse strips of black hairs. During the last instar they become typically hairy, very voracious and consume all the foliage on a tree, leaving it bare.

Apparently final-stage larvae show no preference and attach to either male or female trees. When ready to pupate, the caterpillars descend to the surrounding vegetation although some pupae may hang in dry leaves of the same host plant.

In the town of Archidona LW 77 50 00 LS 00 58 00 (I.G.M., 1978–1982) and the Napo region, Quichua natives (in Peru the same Indians are called Quechua) anxiously await the arrival of the larvae harvesting season in March and April, with slight variations according to the local microclimate.

Harvesting of mature larvae is done in the morning. Women and children usually gather the caterpillars and store them in *Calathea* spp. (Marantaceae) leaves, baskets or buckets. They are left for two days so the intestines can be voided. Now they are boiled in salt water, wrapped in Marantaceae leaves, fried in oil, drained and ground into thin paste, mixed with vegetable sauce, and combined either with cassava, palm hearts or boiled fruit of "chonta" (*Bactris gasipaes*) Arecaceae.

Fig. 16.2: *Caryodendron orinocense* (Euphorbiaceae). Small trees hosting the edible *Panacea prola*. Ecuador, Archidona (photo M.G. Paoletti).

Caterpillar preparation is done by the women who, following native tradition, are the family pillars and manage the kitchen, while the men dedicate themselves to hunting, fishing, preparing the fields and cutting down trees in the jungle.

The larvae are known in Quichua as "guachanzocuro", "guachanzo" or "huachanzo", the common name of the host plant, and "curo", worm or caterpillar.

The trees sprout new buds and leaves after complete defoliation with no apparent signs of stress. However, when the affected tree is felled, the pest can be traced back through the years; a transverse section shows a ring that corresponds exactly to the period of defoliation, which is similar to those indicating the winter rest period in the Northern Hemisphere. In other words, it is possible to trace the cycle of the *Panacea* outbreak over the years by "reading" the rings in the trunk.

So far the only known host plant is *Caryodendron orinocense* Karsten, a dioecious Euphorbiaceae with wide neotropical distribution: throughout Venezuela, Colombia, Ecuador, Peru, Bolivia and Brazil. Gara and Onore (1989) were the first to mention *Caryodendron orinosense* as a host plant of *Panacea prola*.

Following Jiménez and Bernal (1992) this tree is known in Ecuador, depending on the area, as "kofán", "namampí", "nambi", "tocay". Gentry (1993) cited it with the Peruvian names of "meto huayo" or "inchi".

Caryodendron orinosense known to the Ecuadorian farmers as "maní de árbol", reaches a height of 30 m and higher in the jungle canopy. This useful plant provides valuable wood for carpentry and its nuts are rich in oil, similar to the common edible nuts. Recently several enterprises have shown a considerable amount of interest in establishing agroindustrial states of "huachanzo" in order to produce valuable edible oil.

Velasco (1789) mentioned the plant as "dindi" and defined the oil extracted from its seeds as "sweet and delicious, similar to almond oil".

Homoptera, Orthoptera, and Odonata

Among the Homoptera, Orthoptera, and Odonata there are also edible insects. The bean tree (*Inga edulis* Mart.) is the host of colonies of the treehopper *Umbonia spinosa*; the juvenile insects of this species are highly prized for their taste. They are consumed when their spines are still tender. In later stages, the spines can cause serious problems at the moment of ingestion. I suffered a sore throat when I tried the hoppers. Within order Orthoptera, grasshoppers are highly valued by natives throughout the Andes, with those related to genus *Schistocerca* preferred.

Above the surface of swamps, one frequently finds large green-and-brown dragonflies (Odonata) that have aquatic larvae. In the vicinity of Latacunga, these aquatic larvae of genus *Aeschna* are consumed in the same way as freshwater shrimps.

Conclusion

In Ecuador, the tradition of entomophagy can be found throughout the territory, but especially in the high Andes and Amazon where the natives have less contact with western "civilization". The most consumed exapods are definitively beetles, followed by ants, wasps, caterpillars and other minor groups. A first inventory resulted in 82 edible insect species. The author guarantees from his own experience that the organoleptic characteristics of edible insects vary, but in general the taste is excellent. In conclusion, the most valued delicacy is the larvae of *Rhynchophorus palmarum*, followed by adults of *Platycoelia lutescens*.

Acknowledgment

I acknowledge with appreciation the following persons, whose tremendous contribution made this paper possible: Dr. Laura Arcos Terán and Dr. Patricio Ponce for the Pontificia Universidad Católica del Ecuador (PUCE) institutional support; Dr. Jaime Jaramillo for plant identification; Dr. Tjitte De Vries for logistic support in the Cuyabeno area; the Mamallacta family in Archidona (Napo province, Ecuador) for sharing their ancestral native knowledge; my PUCE students for their collaboration during field trips; Florencio Maza for preparing the specimens at QCAZ; Dr. Sandra Bukkens for preparing the manuscript for publication; unknown referees for critical revision of the text; and finally Maurizio G. Paoletti for encouraging the undertaking.

References

Bodenheimer N. 1951. Insects as Human Food. W. Junk, The Hague, The Netherlands.

Borror D.J., Triplehorn C.A., and Johnson N.F. 1989. An Introduction to the Study of Insects. Saunders College, Philadelphia, PA. (6th ed.)

D'Abrera B. 1984. Butterflies of South America. Hill House, Victoria, Australia.

D'Abrera B. 1987. Butterflies of the Neotropical Region, Part 3. Brassolidae, Acraeidae and Nymphalidae (Partim). Hill House, Victoria, Australia.

Figueroa S.S. and Albuja L.V. 1983. Importancia y conservación de la vida silvestre ecuatoriana: Mamíferos ecuatorianos considerados raros o en peligro de extinción. M.A.G. Galaxy, Quito, Ecuador.

Gara R. and Onore G. 1989. Entomología forestal, Proyecto DINAF, AID, apoyo al sector forestal. Ministerio de Agricultura y Ganadería, Quito, Ecuador.

Gentry A.H. 1993. Woody Plants of Northwest South America (Colombia, Ecuador, Peru) with Supplementary Notes on Herbaceous Taxa. Conservation International, Washington, DC, USA.

Gonzales A. and Ortiz de Villalba J.S. 1985. Rio Napo, Realidad amazónica ecuatoriana. CICAME, Vicariato Apostólico del Aguarico, Pompeya, Napo, Ecuador.

I.G.M. 1978–1982. Índice toponímico de la República del Ecuador, 1–8 (A-Q). Instituto Geográfico Militar, Quito, Ecuador.

I.G.M. 2002. Mapa del Ecuador. Inst. Geog. Militar, Quito, Ecuador.

Jiménez L.C. and Bernal H.Y. 1992. El inchi *Caryodendron orinocense* Karsten, (Euphorbiaceae). Secretaria Ejecutiva del Convenio Andrés Bello. SECAB-CAF, Santa Fé de Bogotá, D.C. Colombia.

Kristensen N.P. (ed.). 1998. Lepidoptera: Moths and Butterflies, 1. Walter de Gruyter GmbH & Co., Berlin, Germany.

Patzelt E. 2002. Flora del Ecuador. Imprefepp, Quito, Ecuador.

Smart F.P. 1981. The Illustrated Encyclopedia of the Butterfly World. Corgi Books. London, UK.

Smith A.B.T. and Paucar A. 2000. Taxonomic review of *Platycoelia lutescens* (Scarabaeidae: Rutelinae: Anoplognathini) and a description of its use as food by the people of the Ecuadorian highlands. Ann. Entomol. Soc. of Amer. 93(3):408–414.

Tosi E. and Daccordi M. 1983. Insetti commestibili. Nigrizia (101) 6: 49–52.

Velasco J. De. 1789. Historia del Reino de Quito en la América Meridional, Tomo I, Parte 1. Rev. ed. 1946 by El Comercio, Quito, Ecuador.

17

Palm Worm (Coleoptera, Curculionidae: *Rhynchophorus palmarum*) A Traditional Food: Examples from Alto Orinoco, Venezuela

H. CERDA[1], Y. ARAUJO[2], R.H. GLEW[3] AND M.G. PAOLETTI[4]

[1] *Simon Rodriguez University, Apartado Postal 47.925, Caracas 1041-A, Venezuela. E-mail: hcerda@reacciun.ve*
[2] *INIA-Mérida, Edo. Mérida, Venezuela*
[3] *Dept. of Biochemistry and Molecular Biology, University of New Mexico School of Medicine, Albuquerque, NM, USA*
[4] *Dept. of Biology, Padova University, Via U. Bassi 58/b, 35121, Padova, Italy. E-mail: paoletti@civ.bio.unipd.it*

Abstract

Current knowledge about the so-called palm worms, weevil beetles, and Curculionidae widely used as food in the Amazon is summarized. The Indians gather the palm worms from damaged or fallen palm stems and eat them raw or roasted. We analyzed the nutrient composition of the palm worm and found that it is an excellent source of protein, fat, vitamins A and E, and minerals. Development of a local, controlled, small-scale palm worm production system implemented by the Indians in the Amazonas is described. Larvae are bred using wild palm materials and traditional Indian plants. Larval survival and density in each palm substrate were analyzed together with their nutrient composition. These data were compared with the mother palms *cucurito* (*Maximiliana maripa*), *seje* (*Jessenia bataua*), and *moriche* (*Mauritia flexuosa*). Finally, the palatability of the palm worm to non-Amerindian tourists is assessed. The nutrient composition of the palm worm, the simplicity of a more controlled local production system, and the acceptability of the palm worm to tourists make this nonconventional resource promising, both as a nutritional food and as a source of cash income for the Indians.

Key Words: palm worms, *Rhynchophorus palmarum*, edible insects, Jivi, Guajibo, Amazon Indians, nutritional composition, palms, minilivestock, amino acids, minerals, fatty acids.

Introduction

Palm weevil larvae and in some cases adults (Coleoptera, Curculionidae, especially *Rhynchophorus* and the smaller *Metamasius*) are appreciated as a delicious food throughout the tropical world; those of *R. ferrugineus papuanus* are even consumed on ritual occasions by the Asmat of Irian Jaya, Indonesia (Sowada, 1995; Tommaseo and Paoletti, 1997, this volume). In Africa, palm worms (*Rhynchophorus phoenicis*) are appreciated by many different populations (Malaisse, 1997). The taste of palm worms—fat, legless larvae—of *R. palmarum* has been appreciated in South America for centuries (DeFoliart, 1993).

Worldwide, there are about 24 species of large Rhynchophorinae and many more species of *Metamasius* (Wattanapongsiri, 1966; Wibmer and O'Brien, 1986) that are potentially edible. However, their practical interest relates to the damage they inflict on palm plantations, especially through the transmission of parasitic diseases (Howard et al., 2001). We found evidence from Vietnam that some other large weevils living inside bamboo shoots are especially appreciated by children and eaten raw. Those weevils (Plate 1.10) possibly belong to several genera (as suggested by C. Obrien, personal communication 2003: *Otidocephalus, Cyrtotrachelus,* and *Macrocheirus*).

Although palm worms are the most common insects eaten by Amerindians in the Amazon and play an important role as a source of protein and fat, data pertaining to their nutrient composition are scarce (Sanchez et al., 1997). Personal records exist for the Piaroa, Makiritare, Curripaco, Yanomamo and Guajibo in Alto Orinoco, Venezuela and the Lecos in Bolivia. For instance, Chagnon (1968) reported that the Yanomami Indians deliberately cut down palm trees to provide fodder for adult insects, and some weeks later the palm stems harbor numerous large fat larvae. The same holds true for *R. ferrugineus* in New Guinea where sago grub production is consistent (Townsend, 1974; Sowada, 1995; Meyer-Rochow and Changkija, 1997; Tommaseo and Paoletti, 1997 and current volume; Meyer-Rochow this volume); these fatty larvae are sold—raw or cooked—in the local markets (Mercer, 1997). Trials of breeding have been reported for Thailand (see Yhoung-Aree and Viwatpanich, this volume).

Several authors (Chagnon, 1968; Beckerman, 1977; DeFoliart, 1993; Dufour, 1987; Sanchez et al., 1997), including ourselves, have observed the strategy these people have developed to optimize yield of larvae. Target palms are cut down; in the Venezuelan Amazonian region, larvae are collected mainly from the *moriche* palm (*Mauritia flexuosa*). In Bolivia, the Lecos collect palm worms in particular

Mollusks

Early records reveal that land snails were cooked and eaten (Giles, 1889; Basedow, 1925). In Arnhem Land and Cape York, jungle land snails (*Xanthomelon pachystylum*) are collected even now from beneath leaf litter at the base of trees and roasted on coals for consumption (Isaacs, 1987). Shellfish (freshwater mussels), high in protein and low in fat, are roasted (Gillon and Knight, 1986).

Oligochaeta (Earthworms)

Smyth (1878) suggested that Aborigines could have utilized the Giant Gippsland Earthworm (*Megascolides australis*) simply on the basis of its large body size, but there is no evidence that Australian Aborigines ate earthworms. Earthworms were collected for fishing (Duncan-Kemp, 1961).

Crustacea

Some of the 140 species of freshwater crayfish in Australia were sought as food (Johnston, 1943). These included species of *Euastacus* (Kohen and Merrick, 1998) and *Cherax* (Horwitz and Knott, 1995). Species belonging to the *Cherax destructor* complex, better known by its common name of "yabby" are widely distributed in southeastern and southwestern Australia (Horwitz and Knott, 1995). It has been suggested that Aborigines assisted the spread of *Cherax* by translocating them to inland water bodies by keeping them in moist soil (Horwitz and Knott, 1995). They were cooked over ashes or boiled, and are high in protein and unsaturated fats (Gillon and Knight, 1986).

Blattodea (Cockroaches)

In Arnhemland, several different species (*Cosmozosteria* spp.) are used as a topical anesthetic. They are crushed and rubbed onto bites and stings or heated in hot ashes, then squeezed to release a liquid onto injuries (ACNTA, 1988).

Isoptera (Termites)

In the western deserts, women collected termites and cleaned them by winnowing and rocking them in dishes with hot ashes to kill and cook them. Sometimes live termites were pounded and kneaded into a raw oily cake (Tindale, 1966). Winged termites were also collected and eaten (Cleland, 1966). The mounds of certain species of termites were used for medicinal reasons; parts of the mound were crushed and eaten, or mixed with liquid and drunk to stop diarrhoea and to replace electrolytes (ACNTA, 1988).

Hemiptera

The two main groups of Hemiptera utilized for food are psyllids and coccids (scale insects), although Meyer-Rochow (1975) reported that hemipterans from other families were consumed by Walbiri and Pintupi. Considerable confusion in the use of common names exists because a large proportion of Australian psyllids that feed on *Eucalyptus* have immature stages that build a sugary covering called a "lerp". Lerps are collected as a source of sugar but loose usage of the names "lerps" and "scales" (a product of scale insects), compounded with misapplication of the terms "manna" and "honeydew", has resulted in inconsistencies in descriptions (Yen, 1984).

Several hundred species of lerp-building psyllids are found on *Eucalyptus* spp. in Australia (Yen, 2002). One of the more widespread genera is *Glycaspis* with over 100 species and it is quite likely that it is these conical-shaped lerps that are eaten over much of the continent. The lerps are produced by the immature stages of the psyllids (Fig. 18.1). Reports of lerps from different species of *Eucalyptus* used as food are numerous (Irvine, 1947; Cleland and Johnston, 1939; Latz, 1995; Bryce, 1998). In Victoria, lerps were collected and either eaten raw or mixed in a wooden vessel with gum from *Acacia* trees dissolved in water (Oates

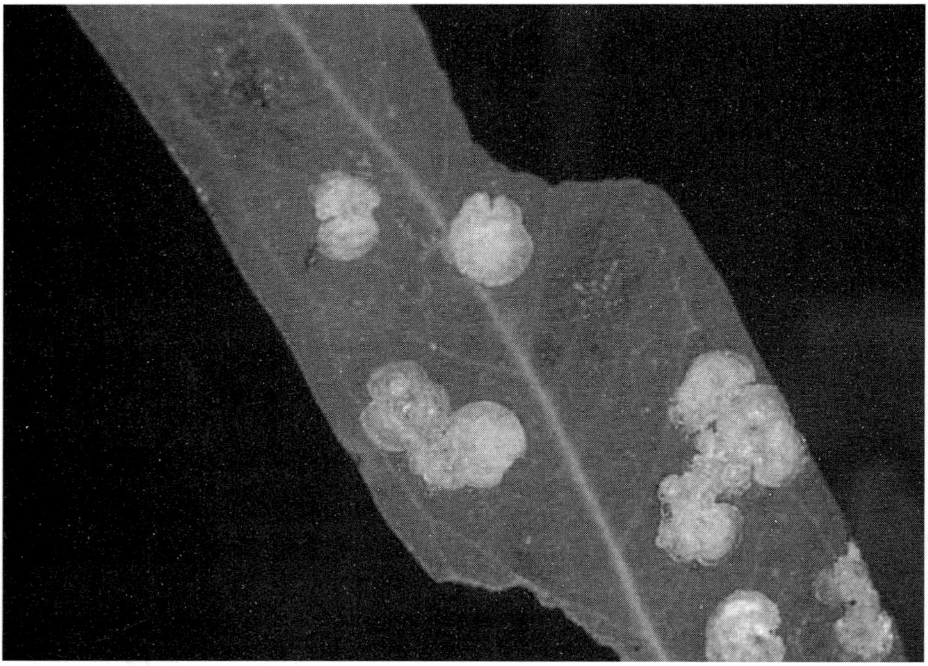

Fig. 18.1: Psyllid lerps. These are the sugary cases produced by the immature stages of some species of psyllids (Hemiptera: Psylloidea) feeding on the leaves of *Eucalyptus* species, Australia (photo A.L. Yen).

and Seeman, 1979). In arid Australia, *Eucalyptus* branches were collected and put on a hard surface to dry the lerps which, after being shaped into balls, were eaten at leisure (Goddard and Kalotas, 1995). Meagher (1974) describes three types of "scale insects" from Western Australia, one of which is sometimes found on the ground beside ant nests; it is most likely that these are psyllid lerps as ants have been observed collecting them (Yen, pers. obs.)

The use of fire by Aborigines to flush out larger animals when hunting and to regenerate plants is well known (Jones, 1969); one result of this burning is that the young growth (coppice) produced by mallee eucalypt after fire is conducive to the build-up of psyllid lerps, and could be viewed as a deliberate use of fire to cultivate lerps.

Several species of scale insects are used as food. The more obvious are the gall-forming species such as *Cystococcus pomiformis* which forms the so-called bush coconut or bloodwood apple on *Corymbia opaca* (Desert Bloodwood) (Sweeney, 1947; Cleland, 1966; Isaacs, 1987; Latz, 1995); the woody gall is cracked open and the white flesh inside eaten (Goddard and Kalotas, 1995). A smaller gall from *Acacia aneura* (Mulga), called the Mulga apple, is also eaten; the identity of the gall-forming species is uncertain but thought to be a coccid (Cleland, 1966; Isaacs, 1987; Goddard and Kalotas, 1995) although Latz (1995) has stated that it is the larva or pupa of a gall-forming wasp.

Scale insects, along with the honeydew they produce, from at least three species of *Acacia* and one of *Hakea* are eaten (Latz, 1995; Bryce, 1998). The best-known scale is *Austrotachardia acaciae* from Mulga. It is directly sucked from the plant or dissolved in water to make a sweet drink (Cleland, 1966; Goddard and Kalotas, 1995).

Several reports mention cicadas being roasted and eaten in central Australia (Schulz, 1891; Chewings, 1936; Meggitt, 1962; Meyer-Rochow, 1975). There is doubt as to the accuracy of the first two reports because they both describe insects emerging from pupal cases in the ground at the base of river Red Gum trees (*Eucalyptus camaldulensis*) (Schulz, 1891; Chewings, 1936); it is more likely that these are goat or ghost moths.

Orthoptera (Grasshoppers, Crickets, Locusts)

The general opinion is that Orthoptera were not a major dietary component in Australia, and were either eaten occasionally (Meyer-Rochow, 1975) or not at all. Reports exist of giant crickets collected by women, roasted on cooking stones and eaten by men (Duncan-Kemp, 1961); roasted grasshoppers pulverized and kept as flour when traveling (Duncan-Kemp, 1961); plague locusts toasted over fire (Tindale, 1966); and children collecting and eating Cylindrachetidae (sand-gropers) after rain (Tindale, 1972). Tindale (1981) discussed the role of grasshoppers in the seasonal food cycles of Aborigines in the desert and the southern arid coastal region. Schulz (1891) reported that the Aborigines around Finke River region would not eat locusts.

Mantodea (Mantids)

There is one report of a species of mantid roasted on hot stones and eaten (Johnston, 1943); the Pintupi, however, considered them poisonous (Meyer-Rochow, 1975).

Diptera (Flies)

Unidentified insects, described as "a fly", frequented desert bloodwood trees (*Corymbia opaca*) and made "singing noises" with their wings. On dying, they fell to the ground into spinifex, from which they were collected and eaten raw. These "flies" also made a sweet substance that adhered to bloodwoods and was also used as food (Chewings, 1936; Sweeney, 1947). It is doubtful that these insects were in fact Diptera.

Coleoptera (Beetles)

The larvae of many species of beetles developing in trunks or logs were eaten. These were generally larvae of jewel beetles (Buprestidae) or longicorn beetles (Cerambycidae), although beetles from other families may have also been used.

In southwestern Western Australia, edible larvae were extracted from *Xanthorrhoea, Acacia, Eucalyptus,* and *Banksia* (Meagher, 1974). In *Xanthorrhoea,* up to 100 grubs of the beetle *Bardistus cibarius* (Buprestidae) could be found in decayed or rotting trees, and tops of trees were deliberately lopped to promote increase in beetle numbers (Meagher and Ride, 1979).

Lepidoptera (Moths)

In the case of moths, larvae were the life history stage usually used as food, although adults of some species were also utilized. Tindale (1932) recorded two species of hepialid moths eaten by Aborigines: *Trictena argentata* and *Abantiades marcidus* (Plate VIII, 5). Tindale (1938) better detailed how Aborigines in the Warburton Ranges dug up larvae and pupae of *T. argentata* from the soil at the base of river Red Gums (*Eucalyptus camaldulensis*) or caught adults as they emerged, which were then cooked on hot ashes. The larvae of *Endoxyla leucomochla* were eaten raw or cooked only to the tender stage (Tindale, 1962). *Acacia kempeana* roots with grubs inside were broken off and carried fresh to eat later (Fig. 18.2). Other edible grubs discussed by Tindale (1966) include *Endoxyla biarpiti, Catoxophylla cyanauges* and *Endoxyla amphiplecta*. Witchetty grubs in the trunks of river Red Gum were extracted by means of a grass hook made from plants such as the curly windmill grass (*Enteropogon acicularis*) (Low, 1990). Witchetty grubs were also used in treatment of burns and major wounds; they

Fig. 18.2: Witchetty grubs (exposed and still contained within sealed root sections) on a wooden carrying dish. The grubs still sealed within the roots remain alive and can be transported for eating later, Australia (photo A.L. Yen).

were crushed or pounded and spread thickly over the affected area (ACNTA, 1988). There are about 70 species of cossid moths in Australia; unlike the species described by Tindale from the roots of an *Acacia* species, other species live in the trunks of species of *Acacia* and *Eucalyptus* (Nielsen et al., 1996).

Latz (1995) differentiated between edible grubs found in the root or trunk of plants from those that feed on plant foliage in central Australia. The latter are considered less tasty than the root- or trunk-dwellers. The white-lined hawk moth, *Hyles lineata* (Sphingidae) has edible green caterpillars (known as *anumara*) found on Tar Vine (*Boerhavia coccinea*) or Long-leaf Emu bush (*Eremophila longifolia*) (Tindale, 1972; Low, 1989; Bryce, 1998). The larvae are numerous at irregular intervals after rain and were collected and kept alive until they had voided all the *Boerhavia* material from their bodies. They were cooked by shaking in hot ashes to separate them from the frass (Tindale, 1972). There are several references to eating foliage-feeding caterpillars but the identity of the species cannot be ascertained (Smyth, 1878; Schulz, 1891; Bodenheimer, 1951). Chewings (1936) recorded green caterpillars feeding on "grasses" collected in large numbers. The heads were pulled off, and the body contents squeezed out with the fingers. They were then dried in hot ashes and either eaten or stored as much. Later they were pounded with stones, kneaded into a paste, and baked on coals. These could have been the *muluru* caterpillar (Kimber, 1984). The preparation was undertaken by the Wangkangurru and the Yarluyandi people of Lake Eyre Basin (Hercus, 1989). The caterpillar involved (called *muluru* in the Diyari language) is the grass witchetty grub whose scientific identity is not yet known but might be the *anumara* of the Arrente people. *Muluru* only occurred during very restricted periods and its preparation was a special ceremony undertaken by older women (Hercus, 1989). These grass witchetty grubs were dried and stored because considered inedible unless pulverized (Hercus, 1989). Another comestible species is the processionary caterpillar (*Ochrogaster lunifer*), eaten only in times of extreme hardship (Latz, 1995). There are unsubstantiated reports of pupae of processionary caterpillars being dug up and eaten (Isaacs, 1987). The larvae of the processionary caterpillar spin a large silken bag as a communal shelter. In central Australia, the bags are collected, thoroughly cleaned, and the underlying layer of silk used as a protective dressing over open sores or burns (ACNTA, 1988).

Consumption of Bogong moths (*Agrotis infusa*) was first described by Bennett (1834) and the story of Bogong moths detailed by Flood (1980). In the alpine areas of southeastern Australia, adults of the moth *Agrotis infusa* congregate during late spring and in summer in the rocky recesses of mountains. Aboriginal men visited the congregation sites during the moth period (November–January) to collect and roast the moths. Moths were smoked out and caught on an outstretched kangaroo skin or a net made out of plant fiber. The moths are very fatty and were carefully roasted to remove scales, legs and wings. The roasted bodies were either eaten as such or made into "cakes" to be smoked and eaten later. The surviving adult moths stayed on until about March, then set off on their migratory flight to oviposition sites up to 2,000 km away. Moth collection was undertaken by men and as to whether women were permitted to feed on them is debatable. Several different groups of Aborigines were involved in moth hunting.

"Edible Grub" Category

The most widespread insect food in Australia falls into the category of "edible grubs". They are generally larvae of Coleoptera (beetles) or Lepidoptera (moths) that feed in the trunks, branches or roots of plants as mentioned above. They have been given common names such as bardi/bardee grubs, or witchetty/witchety/witjuti grubs. Nomenclatural problems arose because many observers had to assume that any large white grub was either a bardi or a witchetty grub (since no clear distinction between the two grubs had been attempted). The problem of proper naming was further confused by the fact that not until 1953 did Tindale formally describe a species of cossid moth as *Xyleutes leucomochla* (now *Endoxyla leucomochla*) that fed on the roots of *Acacia ligulata* as the witjuti grub (Fig. 18.3). Later, Tindale (1962) assigned the term witchetty grub to several species of stem-boring and root-feeding members of Hepialidae (ghost

Fig. 18.3: Witchetty grub exposed in section of root of *Acacia* plant, Australia (photo A.L. Yen).

moths), Cossidae (goat moths), Buprestidae (jewel beetles), and Cerambycidae (longicorn beetles). In so doing, the term 'witjuti' or 'witchetty' grub was equated to a large group of edible groups from taxa other than *Endoxyla leucomochla*, and Nielsen et al. (1996) concluded that the actual cossid species to which the name "witjuti grub" was assigned is not known. However, there was probably geographical variation in the use of these names; for example. Hercus (1989) listed the grub names in the Wangkangurru language of the Simpson Desert, and the term used by them for buprestid beetle and cossid moth larvae as well as a general term for caterpillars is *pardi* (= *bardi*).

Latz (1995) listed 24 plant species from central Australia harboring edible grubs. The Great and Little Sandy Deserts have 25 plant species with edible grubs (Walsh, 1990). Other records of edible grubs from various host plants include Sweeney (1947), Irvine (1957), Cleland and Johnston (1933), Cleland and Tindale (1959), Greenway (1973), Thomson (1975), Chapman et al. (1995), and Goddard and Kalotas (1995).

Hymenoptera (Bees, Wasps and Ants)

Honey ants, *Camponotus inflatus*: Several species of honey ants exist in Australia but *Camponotus inflatus* is the one favored for food, albeit *Melophorus bagoti* is sometimes used (Conway, 1990, 1992). *C. inflatus* lives in underground colonies beneath Mulga (*Acacia aneura*) trees. The workers collect honeydew from scale insects and plant secretions on Mulga to feed the repletes (Fig. 18.4). Honey ants were generally collected by women while available (Sweeney, 1947; Devitt, 1989), with a ban on collection during certain times to let the nests recover (Duncan-Kemp, 1933). Traditionally, honey ants were used as a sweetener in seedcakes made from woollybutt grass (*Eragrostis eriopoda*) and inland pigweed (*Portulaca oleracea*) (Bryce, 1998). Honey ants were also mixed with diluted nectar from bauhinias and fermented in the Channel country of Queensland (Duncan-Kemp, 1933).

Green tree ants, *Oecophylla smaragdina*: Green tree ants build nests of leaves tied together with larval silk. They are widely used in northern Australia as an expectorant and an antiseptic. The nest is either placed in cold water and squeezed or boiled in water. The resultant liquid is drunk to cure coughs, colds, mouth and lip sores, and sometimes for gastrointestinal disorders. The abdomen of queens is sometimes eaten for colds or simply as a snack, or crushed and spread on the skin for scabies or itch. The ants are sometimes crushed and mixed in water, then drunk as an emetic (Crawford, 1982; ACNTA, 1988). Green tree ants are also used to make the sour root bulbs of *Microstemma* more palatable by adding squashed ants that give them a lemony taste (Isaacs, 1987).

Fig. 18.4: Honey ant repletes in Australia (photo A.L. Yen).

Other ants: There are references to consumption of workers of other ant species, but all seem to refer to a single observation (Bodenheimer, 1951) and its authenticity is questionable. More reliable are the various versions of eating ant "eggs", which in fact are ant larvae and pupae of larger species of *Myrmecia* and *Camponotus* (Smyth, 1878; Meagher, 1974). They were collected and "sieved" by tossing up and down on bark or wood and then roasted.

Sugar bags: "Sugar bag" is the common name given to nests built by native bees of genus *Trigona*. They are more widespread in northern Australia and are a valued component of the diet, as evidenced by Aboriginal stories associated with them (McKeown, 1936). These native bees are stingless and build their nests in tree hollows or crevices in the ground. The honeycombs are collected for their sweet honey (McKeown, 1936; Thomson, 1949). They are less abundant in arid regions but do occur on desert bloodwood trees (*Corymbia opaca*) (Sweeney, 1947; Latz, 1995). Men collected sugar bags (Schulz, 1891; Meyer-Rochow, 1975), which were located by catching native bees, gluing a tiny piece of feather or spider's web on them, and following them back to the hive (Gillon and Knight, 1986; Isaacs, 1987). Hive establishment by sugar-bag bees was encouraged by women draping ghost gum with bunches of yellow-flowering punjilla. The bees hovered and buzzed persistently over a tree hollow, and women successfully tempted them to settle there (Duncan-Kemp, 1961). Smyth (1878) gave an account of an intoxicating drink made from old honeycombs in New South Wales. In Queensland, Bauhinia flowers were pounded and the liquid drained into a coolamon and mixed with sugar bag or ant honey and set aside to ferment, which takes 8–10 days (Duncan-Kemp, 1933). In Arnhem Land and in the Kimberleys, the Aborigines differentiate types of bees according to location of their nests, behavior, and sweetness of honey produced (Isaacs, 1987). There are six species in Australia, but as the genus has only recently been revised (Dollin et al., 1997), there is no information in these observations as to the identities of the species involved.

Wasps: Low (1989) recorded smoking of Vespidae larvae in their nests and subsequent consumption.

Seeds harvested by ants: Aborigines collected these seeds and made them into a paste or flour for cooking. There are several records of seeds collected by Aborigines around the entrances of seed-harvesting ants (Duncan-Kemp, 1933; Tindale, 1981; Lowe and Pike, 1990; Latz, 1995). Seedcakes were sometimes made and buried for emergencies (Duncan-Kemp, 1933).

Other Insects

Observations on consumption of other insects are often vague. Meggitt (1962) recorded that the Walbiri had four types of edible grubs from *Acacia*, cerambycid larvae, cossid larvae, honey ants, lerps, and sugar bags as preferred foods, and termites, grasshoppers, crickets, cicadas, weevils, and lice as occasional food.

Meggitt (1962) also noted insects not eaten by Walbiri: sugar ants, mole crickets, houseflies, bushflies, bullants, adult moths and butterflies, scorpions, mosquitoes, centipedes, march flies, black beetles, mantids, stick insects, wasps, as well as spiders. The Martu in the Great and Little Sandy Deserts ate witchetty grubs, wasp galls, lerps, winged termites and ant pupae (Walsh, 1990). Pintubi ate termites, occasionally grasshoppers and locusts, some Hemiptera, some beetles, honey ants, and moth larvae (Meyer-Rochow, 1975).

Vertebrates

Australian Aborigines ate almost all species of nonmarine vertebrates and many marine species. The larger-bodied species included macropods (kangaroos and wallabies), dugongs, and crocodiles, while smaller-bodied species included possums, bandicoots, echidnas, flying foxes, goannas, lizards, snakes, and turtles (Isaacs, 1987). Many species of birds were eaten; the larger-bodied species included emus, ducks, magpie geese, and native pigeons (Isaacs, 1987). Fish were an important source of protein and many species were speared, netted or trapped (using logs as barriers in streams and rivers) (Isaacs, 1987). Recent unpublished research indicates that Aborigines in southeastern Australia farmed eels (Heather Builth, pers. comm.). Since European settlement, Aborigines have hunted, introduced mammals such as rabbits, pigs, and water buffalo (Isaacs, 1987).

Importance of Insects in Aboriginal Diet

Nutritional Value

The nutritional value of Australian insects has only been studied in an *ad hoc* manner and only at the broad taxonomic level (e.g. cossid larvae of uncertain species identity, lerps without reference to species). The nutritional content of edible grubs is 7–50% protein, 14–47% fat, and 7–10% sugar (Meyer-Rochow, 1978; Peterson, 1978; Cherikoff, 1989; Latz, 1995). Some edible grubs weigh up to 30 g or more, so are capable of providing significant nutrients (Tindale, 1981). Lerp is low in protein and high in carbohydrates (73.3 g/100 g) and minerals (16.5–75 mg/100 g of sodium, potassium, and calcium; Peterson, 1978). The composition of honey ants (*Camponotus inflatus*) is 37% carbohydrate and 60% fat (Devitt, 1986, 1989). The average fat content of male Bogong moths is 61% and females 51% of dry weight; they are rich in fat but not in protein (Flood, 1980).

An overlooked factor in hunter/gatherer societies is the problem of weaning children in the total absence of sources of milk other than the mother's. In desert areas, children can be maintained on the breast for up to three years and for as long as five years on a supplemental basis. There is increased use of accessory foods after year one and the most important appears to have been edible moth and beetle larvae rich in animal fats. Absence of such supplemental foods led to malnutrition among children and a form of scurvy among adults (Tindale, 1981).

There has been speculation about the importance of insects in the diet of Australian Aborigines. Calaby (1971) suggested that invertebrates were of little interest except in times of food shortage. Gould (1969) found insects such as honey ants, termites and two types of edible grubs classified as *kuka* (staple food), while Mulga galls, lerps and scale insects were *mirka* (supplementary food). Tindale (1953), referring to the desert environment, considered insects an important part of the diet in this harsh environment. They are an important buffer food against short-term drought, since many have a two-year life cycle, often extended to three or more years during unfavorable times. Such buffer foods may be especially important against bad seasons. At such times, cossid larvae that feed in roots of several species of shrubs are vitally important (Tindale, 1981). Lerps are often overlooked as important dietary items, but they are highly nutritious and tasty, available in dry periods, and readily collected (Walsh, 1990).

The importance of insects was probably not restricted to times of food shortage. Certainly food items such as edible grubs, honey ants, and sugar bags are desirable food items regardless of environmental conditions. Tindale (1981) described the food year in two contrasting environments: the desert and the arid southern coastal region. In the desert, the main perennial foods tend to be lizards and small mammals and sometimes termites and edible grubs. This diet is seasonally supplemented by fully mature gravid grasshoppers and witchetty grubs (May), lerp (June), honey ants (October), adult hepialid and cossid moths (February, if rains have fallen) and early-laying grasshoppers (April). In the southern coastal region, the seasonal supplementary invertebrates are grasshoppers (January-February), edible grubs (March), hepialid moth adults (April, following rain), *Unio* shellfish in lagoons and river shallows (November), and developing grasshopper nymphs as well as termites (December). In northern coastal Australia, the rich marine resources are more important and use of insects, except for sugar bags, may not have been so widespread.

Aboriginal Culture and Insects

Aborigines believe that each food type was created by an ancestral spirit and in some cases, such as the honey ants of Papunya, the ancestral spirits metamorphosed (Isaacs, 1987). The animals are linked totemically with the Aborigines and rites and rituals are often observed in relation to food. Some animals are totemic relatives and there may be taboos among one group of Aborigines (but not another adjacent group), while other food is permitted only during certain times of the year. There are totems for various edible grubs, honey ants, lerps and sugar bags, but there are also totems for insects not of edible status (Meyer-Rochow 1978).

Ceremonies for increase of insect foods have been reported, generally held shortly before the usual breeding season, for edible grubs, the *anumara* caterpillar, and honey ants (Tindale and George, 1971; Tindale, 1972, 1981).

Information Limitations

Considerable fragmentary information exists about insect foods of Australian Aborigines. It is not possible to comprehensively list all the insect taxa utilized due to the vacuity of some historical records and the uncertain identity of many of the observations. The reasons for this situation include:

1. Availability of information. In many Aboriginal tribes, information was often maintained by individuals and sometimes was dependent upon the sex of the informant ("men's business" and "women's business"). One potential outcome of this differentiation of knowledge is that male anthropologists may not always have had access to knowledge maintained by women and vice versa.

2. Sacred information. There are public and restricted realms of knowledge in Aboriginal cultures, and information obtained may depend upon the sex and age of those interviewed (Yen et al., 1997).

3. Language. Linguistics is very important in recording anthropological information. Few of the early explorers or anthropologists in Australia were conversant with even one of the 270 distinct Aboriginal languages (and as many as 600–700 dialects (Australian InFo International, 1989), and the lack of standard uniform phonetics caused many problems. One example is the reported *Arrente unirringita* or witchetty grub ceremony recorded by Spencer and Gillen (1899) near Alice Springs; *udnirringita* means "caterpillar" and probably refers to one of the plant-feeding species rather than the root-feeding cossid larva (Gill, 1998). Another issue is the use of the same word by Aborigines to describe two very different taxonomic identities based on nontaxonomic characters. For example, in Pitjantjatjara, *wanka* refers to both spiders and silk-spinning caterpillars, and *mirin-mirinpa* refers to both crickets and cicadas (Yen et al., 1997).

4. Scientific misidentification by recorders. Many of the records have been confused by incorrect identifications. This may simply be due to incorrect naming, such as confusing psyllid lerps with coccid scales, or caused by identification problems associated with immature larvae of beetles and moths. The use of names such as bardi and witchetty grubs to describe any large white grub has further compounded the problem.

5. Taxonomic issues. Many species of insects in Australia are still undescribed. As taxonomy is based on adults, it is not surprising that identification of immature stages is difficult. The lack of life history information in conjunction with the need for taxonomic research leaves western science behind the Aboriginal naming system. There are many plants that harbor edible grubs, either in the branches, trunk or roots. Most Aborigines classify edible grubs by their own binomial naming system indicating "edible grub" from "name of plant species".

It is likely that there are many different species of edible grubs as they involve families with large numbers of species such as the Hepialidae (goat moths), Cossidae (ghost moths) and Buprestidae (jewel beetles).

The main insects used as food by Australian Aborigines can be categorized into six types: (1) edible grubs of different taxa in the Coleoptera and Lepidoptera; (2) honey ants; (3) scale insects; (4) lerps; (5) sugar bags, and (6) Bogong moths. Of these, edible grubs are the group most widely eaten across Australia. Lerps and scale insects are also widespread. Honey ants and sugar bags are restricted to large parts of Australia, while Bogong moths are restricted to the southeastern Australian alps.

In reality, the terms "edible grubs", "lerp" etc. hide the fact that each of these groups constitutes many different species, so the number of insect species eaten is considerably larger. Only the first five types listed are still utilized in any major way today. Bogong moths are no longer collected as food. With further erosion of traditional ways of life among Australian Aborigines, further knowledge about ethnoentomology may be lost.

Insect and Invertebrate Foods Today

Invertebrate foods are not considered a major food by Australian Aborigines nowadays because of their adoption of the western practice of purchasing prepared foods in retail outlets. While some Aboriginal communities in more remote regions of Australia still collect invertebrates for food, the practice is not so common now. As Menzel and D'Aluisio (1998) were told by one interviewee, it is easier to purchase food in the local supermarket.

Bush foods have become part of the food scene in Australia, but it is a small niche market to supply restaurants rather than a general adoption by the general community (Menzel and D'Aluisio, 1998). There have been attempts to use indigenous Australian ingredients (both plants and animals) in introduced cuisines (Cherikoff, 1989). Australian vertebrates such as native ducks, kangaroos, crocodiles and fish, have, unless protected by wildlife conservation legislation, been readily utilized by non-Aboriginal Australians since European settlement. Except for freshwater and marine invertebrates, integration of Australian insects into contemporary cuisine is relatively uncommon. When it does occur, it is generally in forms not traditionally practiced by Aborigines (e.g. witjuti grub soup or moth damper) (Cherikoff, 1989).

Commercial production of invertebrates for human consumption is primarily associated with freshwater crustaceans. They are a major commodity sought by recreational fishers today. The result is that some species have become threatened and others require regulations to control the timing of fishing, numbers caught, and minimum size caught (Merrick, 1997; Morey, 1998). Three species, *Cherax destructor* (yabby), *C. quadricarinatus* (red claw), and *C. cainii* (marron) are used in commercial aquaculture (Huner, 1994). Artificial nests have been used to encourage honey production by native bees (Cherikoff, 1989). There are no records regarding attempts to commercialize endemic Australian insects for human consumption.

Conclusion

There is no doubt that insects constituted an important part of the diet of Australian Aborigines in the past and are still important in parts of Australia. The relative importance of insects varied according to geographic location and seasonal factors. There was probably less reliance on insects in areas with access to rich marine resources and the greatest diversity of insect foods was found in the semiarid and arid parts of Australia. In listing edible Australian plants, Cherikoff (1989) recorded 16 species of plants with edible grubs in "dryland", four plant species with edible grubs in the temperate Sydney region, and only two plant species with edible grubs from rainforests. There is still much to be learned about the use of insects as food in Australia, especially correct identification of the different species utilized and the life history stages of many of these species. Any attempts to commercially breed these insect foods in future will be made easier if their correct identities are known.

Acknowledgments

I thank Maurizio G. Paoletti for requesting this article and members of the Mutitjulu Community at Uluru-Kata Tjuta National Park for teaching me about insect foods. David Cheal provided useful comments on the manuscript.

References

ACNTA, Aboriginal Communities of the Northern Territory of Australia. (1988). Traditional Bush Medicines. Greenhouse Publ. Pty Ltd., Darwin, N. Terr., Australia.

Australian InFo International. (1989). Australian Aboriginal Culture. Australian Government Publ. Service, Canberra, NSW.

Basedow H. (1925). *The Australian Aborigines*. F.W. Preece and Sons, Adelaide, S. Australia.

Bennett G. 1834. Wanderings in New South Wales, Batavia, Pedir Coast, Singapore, and China: Being the Journal of a Naturalist in Those Countries during 1832, 1833, and 1834. Richard Bentley, London, UK.

Bodenheimer F.S. 1951. Insects as Human Food. Dr. W. Junk, The Hague, The Netherlands.

Bryce S. (comp). 1998. Women's Gathering and Hunting in the Pitjantjatjara Homelands. Inst. Abor. Develop., Alice Springs, N. Terr., Australia. (Rev. 1998 reprint).

Calaby J.H. 1971. Man, fauna, and climate in Aboriginal Australia. *In:* Mulvaney D.J. and Golson J. (eds.). Aboriginal Man and Environment in Australia. Australian Natl. Univ. Press, Canberra, NSW, pp. 80–93.

Campbell T.G. 1926. Insect foods of the Aborigines. Austr. Mus. Mag. 2: 407–410.

Chapman A., Hadfield B., Douglas W., Gardner J., Hutchison P., and Rolland G. 1995. Aboriginal names for some biota from the Great Victoria Desert, Western Australia. CALMScience 1: 349–361.

Cherikoff V. 1989. The Bushfood Handbook. Cherikoff Pty Ltd.: Boronia Park, New South Wales.

Cherry R. 1993. Use of insects by Australian Aborigines. Cult. Entomol. Digest 1: 14–15.

Chewings C. (1936). Back in the Stone Age. Angus and Robertson, Sydney, NSW, Australia.

Cleland J.B. 1966. The ecology of the Aboriginal in South and Central Australia. *In:* Cotton B.C. (ed.). Aboriginal man in South and Central Australia. Part I. Government Printer, Adelaide, S. Australia, pp. 111–158.

Cleland J.B. and Johnston T.H. 1933. The ecology of the Aborigines of Central Australia: botanical notes. Trans. Roy. Soc. S. Austr. 57: 113–124.

Cleland J.B. and Johnston T.H. 1939. Aboriginal names and uses of plants in the northern Flinders Ranges. Trans. Roy. Soc. S. Austr. 63: 172–179.

Cleland J.B. and Tindale N.B. 1959. The native names and uses of plants at Haast Bluff, Central Australia. Trans. Roy. Soc. S. Austr. 82: 123–140.

Conway J.R. 1990. Copping it sweet: the honey ant in Aboriginal culture. *Geo* 12: 54–61.

Conway J.R. 1992. Notes on the excavation of a nest of *Melophorous bagoti* Lubbock in the Northern Territory, Australia (Hymenoptera: Formicidae). J. Austr. Entomol. Soc. 31: 247–248.

Crawford I.M. 1982. Traditional Aboriginal plant resources in the Kalumburu area: aspects in ethno-economics. Rec. W. Austr. Mus. Suppl. 15: 1–86.

Devitt J. 1986. A taste for honey: Aborigines and the collection of ants associated with mulga in Central Australia. *In:* Sattler P.S. (ed.). The Mulga Lands. Roy. Soc. Queensland, Brisbane, Queensland, Australia, pp. 40–44.

Devitt J. 1989. Honeyants: A desert delicacy. Austr. Nat. Hist. 22: 588–595.

Dollin A.E., Dollin L.J., and Sakagumi S.F. 1997. Australian stingless bees of the genus *Trigona* (Hymenoptera: Apidae). Invert. Taxon. 11: 861–896.

Duncan-Kemp A.M. 1933. Our Sandhill Country. Angus and Robertson Ltd., Sydney, NSW, Australia.

Duncan-Kemp A.M. (1961). Our Channel Country. Angus and Robertson Ltd., Sydney, NSW, Australia.

Flood J. (1980). The Moth Hunters. Austr. Inst. Abor. Studies, Canberra, NSW, Australia.

Giles E. (1889). Australia Twice Traversed, Vols 1, 2. Sampson, Low, Marston, Searle and Rivington Ltd., London, UK.

Gillon A. and Knight E. 1986. Bourke Bush Food. Develop. Advisory Publ. N.S.W., Dubbo, NSW, Australia.

Goddard C. and Kalotas A. (comp. and ed). 1995. Punu: Yankunytjatjara Plant Use. IAD Press, Alice Springs, N. Terr., Australia.

Gould R.A. 1969. Subsistence behaviour among the Western Desert Aborigines of Australia. Oceania 39: 253–274.

Greenway J. 1973. Down among the Wild Men. Hutchinson, Sydney, NSW, Australia.

Hercus L.A. 1989. Preparing grass witchetty grubs. Rec. S. Austr. Mus. 23: 51–57.

Horwitz P. and Knott B. 1995. The distribution and spread of the yabby *Cherax destructor* complex in Australia: speculations, hypotheses and the need for research. Freshwater Crayfish 10: 81–91.

Huner J.V. 1994. Freshwater crayfish aquaculture in North America, Europe, and Australia. Food Products Press, New York, NY.

Irvine F.R. 1957. Wild and emergency food of Australian and Tasmanian Aborigines. Oceania 28: 113–142.

Isaacs J. 1987. Bush food: Aboriginal food and herbal medicine. Weldons Pty Ltd., Sydney, NSW, Australia.

Johnston T.H. 1943. Aboriginal names and utilization of the fauna in the Eyrean region. Trans. Roy. Soc. S. Austr. 67: 244–311.

Jones R. 1969. Fire-stick farming. Austr. Nat. Hist. 16: 224–228.

Kimber R.G. 1984. Resource use and management in central Australia. Austr. Abor. Studies 1984/2: 12–23.

Kohen J.L. and Merrick J.R. 1998. Limited use of freshwater crayfishes (genus *Euastacus*) by Aborigines in eastern New South Wales: records and comments. Proc. Linn. Soc. New South Wales 119: 101–105.

Latz P. 1995. Bushfires and bushtucker: Aboriginal Plant Use in Central Australia. IAD Press, Alice Springs, N. Terr., Australia.

Low T. 1989. Bush tucker: Australia's Wild Food Harvest. Angus and Robertson, Sydney, NSW, Australia.

Low T. 1990. Witchetty grubs. Austr. Nat. Hist. 23: 284–285.

Lowe P. and Pike J. 1990. Jilji: Life in the Great Sandy Desert. Magabala Books, Broome, W. Australia.

McKeown K.C. 1936. Insect Wonders of Australia. Angus and Robertson Ltd., Sydney, NSW, Australia.

Meagher S.J. 1974. The food resources of the Aborigines of the south-west of Western Australia. Rec. W. Austr. Mus. 3: 14–65.

Meagher S.J. and Ride W.D.L. 1979. Use of natural resources by the Aborigines of south-western Australia. *In:* Berndt R.M. and Berndt C.H. (eds.). Aborigines of the West: their past and their present. Univ. Western Australia Press, Nedlands, W. Australia, pp. 66–80.

Meggitt M.J. 1962. Desert People—a Study of the Walbiri Aborigines of Central Australia. Angus and Robertson, Sydney, NSW, Australia.

Menzel P. and D'Aluisio F. (1998). Man Eating Bugs. Ten Speed Press: Berkeley, CA.

Merrick J.R. 1997. Conservation and field management of the freshwater crayfish, *Euastacus spinifer* (Decapoda: Parastacidae), from the Sydney region, Australia. Proc. Linn. Soc. New South Wales 118: 217–225.

Meyer-Rochow V.B. 1975. Local taxonomy and terminology for some terrestrial arthropods in five different ethnic groups of Papua New Guinea and Central Australia. J. Roy. Soc. W. Austr. 58: 15–30.

Meyer-Rochow V.B. 1978. The diverse use of insects in traditional societies. Ethnomedizin 5: 287–300.

Meyer-Rochow V.B. and Changkija S. 1997. Uses of insects as human food in Papua New Guinea, Australia, and North-East India: cross-cultural considerations and cautious conclusions. Ecol. Food Nutr. 36: 159–185.

Morey J.L. 1998. Growth, catch rate and notes on the biology of the Gippsland Spiny Freshwater Crayfish, *Euastacus kershawi* (Decapoda: Parastacidae), in West Gippsland, Victoria. Proc. Linn. Soc. New South Wales 119: 55–69.

Nielsen E.S., Edwards E.D., and Rangsi T.V. (eds). 1996. Checklist of the Lepidoptera of Australia. CSIRO, Collingwood, Victoria, Australia.

Oates A. and Seeman A. 1979. Victorian Aborigines: Plant Foods. National Museum of Victoria, Melbourne, Victoria, Australia.

Peterson N. 1978. The traditional pattern of subsistence to 1975. *In:* Hetzel B.S. and Frith H.J. (eds.). The Nutrition of Aborigines in Relation to the Ecosystem of Central Australia. CSIRO, East Melbourne, Victoria, pp. 25–35.

Reim H. 1962. Die Insektennahrung der Austraslischen Ureinwohner. Akademie-Verlag, Berlin.

Schulze L. 1891. The Aborigines of the upper and middle Finke River: their habits and customs, with introductory notes on the physical and natural-history features of the country. Trans. Roy. Soc. S. Australia 14: 210–245.

Smyth R.B. 1878. The Aborigines of Victoria. Government Printer, Melbourne, Victoria.

Spencer B. and Gillen F.J. 1899. The Native Tribes of Central Australia. Macmillan and Co. Ltd., London, UK.

Sweeney G. 1947. Food supplies of a desert tribe. Oceania 17: 289–299.

Thomson D.F. 1949. Economic Structure and the Ceremonial Exchange Cycle in Arnhem Land. MacMillan and Co. Ltd., Melbourne, Victoria, Australia.

Thomson D. 1975. Bindibu Country. Thomas Nelson, Melbourne, Victoria, Australia.

Tindale N.B. 1932. Revision of the Australian Ghost Moths (Lepidoptera Homoneura, Family Hepialidae). Rec. S. Austr. Mus. 4: 497–536.

Tindale N.B. 1938. Ghost Moths of the Family Hepialidae. S. Austr. Natural. 19: 1–6.

Tindale N.B. 1953. On some Australian Cossidae including the moth of the witjuti (witchety) grub. Trans. Roy. Soc. S. Australia 76: 56–65.

Tindale N.B. 1962. Witchety grub. Australian Encyclopaedia 9: 339–340. Grolier Soc. Australia, Sydney, NSW.

Tindale N.B. 1966. Insects as food for the Australian Aborigines. Austr. Nat. Hist. 15: 179–183.

Tindale N.B. 1972. The Pitjandjara. *In:* Bicchieri, M.G. (ed). Hunters and Gatherers Today. Holt, Rinehart & Winston, New York, NY, pp. 217–268.

Tindale N.B. 1981. Desert aborigines and the southern coastal peoples: Some comparisons. *In:* Keast A. (ed). Ecological Biogeography of Australia. Dr. W. Junk Publ., The Hague, The Netherlands, pp. 1853–1884.

Tindale N.B. and George B. 1971. The Australian Aborigines. Lloyd O'Neil Pty Ltd., Melbourne, Victoria.

Walsh F.J. 1990. An ecological study of traditional Aboriginal use of "country": Martu in the Great and Little Sandy Deserts, Western Australia. Proc. Ecol. Soc. Australia 16: 23–37.

Yen A.L. 1984. Honeydew, manna, scale and lerp. Victorian Natural. 100: 244–245.

Yen A.L. 2002. Short-range endemism and Australian Psylloidea (Insecta: Hemiptera) in the genera *Glycaspis* and *Acizzia* (Psyllidae). Invert. System. 16: 631–636.

Yen A.L., Gillen J., Gillespie R., Vanderwal R. and the Mutitjulu Community. 1997. A preliminary assessment of Anangu knowledge of Central Australian invertebrates. Mem. Mus. Victoria 56: 631–634.

19

Traditional Food Insects and Spiders in Several Ethnic Groups of Northeast India, Papua New Guinea, Australia, and New Zealand

V.B. Meyer-Rochow

Dept. of Biology, University of Oulu, SF-90014 Oulu, P.O. Box 3000, Finland and School of Engineering & Science, International University Bremen, D-28725 Bremen, P.O. Box 750561, Germany. E-mail: b.meyer-rochow@iu-bremen.de

Abstract

Taxonomic identifications and vernacular names of some insects and spiders, consumed by the following ethnic communities, are presented: Ao-Naga (Northeast India: Nagaland), Meeteis (Northeast India: Manipur), Chimbu (Central Highlands: Papua New Guinea), Onabasulu (Southern Highlands: Papua New Guinea), Kiriwina (Trobriand Islands: Papua New Guinea), Walbiri (Central Australia), Pintupi (Central Australia), and Maori (New Zealand/Aotearoa). Differences and similarities of entomophagous habits among the groups are briefly examined with regard to cultural, ethnic, and linguistic relationships. It is postulated that prehistorically two centres in the region under discussion existed where entomophagy evolved and from where the practice spread: Southern India and Southeast Asia. It is further postulated that not protein-rich, but sugar- and fat-containing insects were the first species to find a place in the regular diet of prehistoric man. Species containing mostly protein were added to the local food spectrum later. According to this scenario, the food insect preferences of the Australian Aborigines reflect those of the earliest insect-eating humans. Evidence is also presented that domestication of the silkworm and use of its product (i.e., silk) could have arisen in Northeast India as early as 4000 years BC as a consequence of eating wild silkworm larvae and pupae.

Key Words: Entomophagy, Ethnoentomology, Cultural Entomology, Austronesia, Naga, Meeteis, Papua New Guinea, Walbiri, Pintupi, Maori.

Introduction

The nutritional behaviour of humans has accompanied the evolution of mankind throughout the ages and it is impossible to divorce health derived from food and diet from well-being (Teuteberg 1991; Meyer-Rochow 1998). Insects have played a far greater role in the cultural histories of the peoples of the earth than is generally recognized. An excellent example are the medium-size Bogong moths (*Agrotis infusa*) and the Australian Aborigines. Annual aggregations of millions of individuals of the moth in the mountains of New South Wales and Victoria, have lured scores of Aboriginal men to the area every year and led to weeks, if not months of feasting on the scorched and roasted bodies of this noctuid (McKeown, 1944). In the New Guinea highlands Bulmer (1968) investigated "mysteries of croaking worms" among the Kalam people, a study which led him to an appreciation of the folkloristic significance of singing insects among many tribes of that great island. Finally, more than a hundred years ago Spencer and Gillan (1899) made us aware of the position of insects in Aboriginal myth and belief, when they listed some 30 insect totems, among them spider-man, the sawfly clan, and totemic groups of maggot, honeybee, blowfly, bushfly, marchfly, to name but a few.

Although these and other observations were made by researchers with a Western cultural background, it would be fair to say that investigators brought up under the influence of Western civilizations, do frequently have a problem in that they conceive of insects as nothing more than incidental pests (Meyer-Rochow et al. 2000). Consequently, they tend to distance themselves from these—in their view—ugly and unpleasant creatures (Hall, 1969). Lyon (1974) saw this attitude as a basic weakness in the overwhelming majority of all anthropological studies and Posey (1976) spoke of an "ethnocentric bias". Although one would have hoped for a change since the recognition of the problem, there is apparently still considerable room for improvement.

Mountford (1971) in his study of the Aborigines of Ayers Rock, their beliefs and their art, could perhaps be forgiven for almost completely ignoring insects, but even in 1989 Oliver in an otherwise comprehensive review of the native cultures of Australia and the Pacific Islands (including Papua New Guinea) barely touched upon insects. And modern tomes on historical, sociological, and cultural aspects of food also still frequently fail to mention insects as a separate food category. Yet, insects (including their edible products) are exactly that: a genuine category of food!

Some insects are ingested involuntarily and unintentionally by everyone at some time, whether of vegetarian disposition or not, and may be important contaminants of food, harbouring vectors of ill health and disease (Gorham, 1976). But other insects have been sought after, deliberately collected, and eaten more than willingly since ancient times by large sections of the world's population. As long as humans have been around, insects have been the dominant group of animals on earth, both in number of species and individuals (Raven et al., 1971).

There is hardly a group of insects that did not have a fancier among the peoples of this world (Bodenheimer, 1951) and there is ample evidence for the high regard in which insects were held by our forebears.

Cave drawings of grasshoppers, executed by Cro-Magnon man more than 10,000 years ago, are known from the grotto "Les Trois Frères" in Ariège (Kühn, 1952); rock drawings of the honey-ant totem by Australian Aborigines have been reported (Spencer and Gillen, 1899) and body art and sand drawings have incorporated insects; illustrations on shards of Chinese pottery, 2,000 years old, clearly show bees (Zhou, 1979), and even antique Greek coins bear the image of a honey-producing insect or a grasshopper on one side (Schimitschek, 1968). Human consumption of insects was undoubtedly once widespread and clearly dates back to the dawn of mankind (locusts as a permitted, kosher food item are specifically mentioned in the Bible: Leviticus, Ch. 11: 21; Lanfranchi, this volume).

What led to the relatively recent decline in the use of insects as food among non-Western societies stems, in our view, from the misguided belief that aping the Western way of life is the fastest way for technologically less advanced societies to be accepted by Europeans. While this may have been the case in colonial times, it no longer holds true and countries in which entomophagy is still practiced or given up only recently, should take note of the fact that since the 1970s traditional and regional foods have become trendy and shown a remarkable renaissance in Europe (Köstlin, 1975) and to some extent eastern Asia as well (Pemberton, 1994).

Indeed, I can see no sound reason why in the Western world oysters, crabs, or a Scottish "haggis" should be considered delicacies, while flour beetles, fried caterpillars or roasted locusts are viewed with revulsion. As scientists, it seems to me not only our duty to record as much as possible in many societies about the uses of insects as food, but also to actively encourage local people in cultures in which such traditions still exist to continue the practice of entomophagy. Toward this end I have repeatedly pointed out in German (Meyer-Rochow, 1973a), English (Meyer-Rochow, 1975a; 1976), Japanese (Meyer-Rochow, 1982a), Finnish (Meyer-Rochow, 1988), and even Estonian (Meyer-Rochow, 1990) that it simply makes no sense to fight a chemical battle against insects, to destroy them in biological warfare, and to eliminate them from what is considered an edible food item, when in reality the insects embody valuable and delectable nourishment. To corroborate the latter I arranged in February 1990 a 5-course insect dinner for well-known public figures in Hamilton (New Zealand). The banquet received TV-coverage and was copied a year later in the USA. Szent-Ivany (1958) had also drawn attention to the widespread use of insecticides in Papua New Guinea, underscoring the extreme risk to health on observing that locals collected and ate the poisoned pests. Why kill these insects at all, when they per se are nutritious? One is reminded of a parallel from the world of plants: 80% of what is identified in Nebraska as "weeds" and targeted for destruction, actually consists of perfectly edible, and often even outright delicious species (Welsch, 1975).

Materials and Methods

All investigations, except for the Meeteis of Manipur, were based on personal visits to the regions studied and interviews between the author and local folk. Communication difficulties, when such arose, were usually overcome through interpreters or use of sign language. Where locals were asked to collect food insects, they would often return with large numbers of individuals belonging to one or two of the most common taxa. They would furthermore preferentially catch the larger and easier targets. To preclude this unwanted "selection", we ourselves frequently collected insects, taking care to include representatives of taxonomically different groups, i.e. orders and families.

The animals were then shown to persons regarded locally as knowledgeable by their companions. The responses of the people interviewed were recorded on tape or written down phonetically and later transcribed phonemically. Occasionally people were shown drawings of insects in the book "The Insects of Australia" (CSIRO, 1970) but, like Waldron and Gallimore (1973), we too found that untrained people had difficulty recognizing line drawings of insects. The problem of picture recognition is thought to be threefold: a) book figures of insects and spiders are not usually drawn to size, b) most of the line drawings lack colour, and c) all figures including colour photographs are two-dimensional representations.

Some of the insect and spider species were identified on the spot, others were preserved in 50% ethanol or air-dried and identified at the Australian National University, University of Western Australia, University of the Waikato, and Zoological Survey of India in Calcutta. For entomological classification of the insects into orders, the well-established system used by Borror and DeLong (1964) was followed.

All the inquiries were conducted over a limited period of time (no more than approx. 2 weeks were available for each ethnic group). Consequently, conclusions about possible seasonal changes in the abundance of food insects could not be drawn, but this was never an objective of this preliminary survey anyway.

Results

Northeast India

Northeast India is an area of astonishing ethnic, linguistic, and cultural diversity and being familiar with only a few peoples of the region, it would be preposterous to assume presentation of a comprehensive list of the edible insect species of that part of the world. A brief account of some from the Mizo Hill area, how, for example, the bug *Ochrophora montana* is collected, prepared for human consumption, and used for oil extraction, is given by Sachan et al. (1987).

A recent summary of edible Orthoptera, Isoptera, Hemiptera, Lepidoptera, and Hymenoptera in Northeast India is provided by Pathak and Rao (2000) and the nutritional value of some edible insects from Manipur has been examined by Gope and Prasad (1983). The Ao-Naga, Meeteis, and Khasi are discussed below.

About 20 Naga languages, divided into 5 subgroups (Campbell, 1991), are spoken by approximately 1.5 million people living in Nagaland (a state within the Indian Union) and in Northwest Myanmar. All Naga languages are classified as Tibeto-Burmese. Nagaland enjoys a monsoon climate with an average annual rainfall of 200–250 cm and an average humidity of 85%; summer temperatures lie in the mid-twenties, while winter temperatures are approximately 10 degrees lower. Nagaland's population density is approximately 82 km^{-2}. Results of our inquiry are given in Table 19.1. Photographs of a variety of edible insects for sale at a local market in Kohima, are shown in Plate IX, 1, 2, and 3.

Meeteis and Khasi

The other northeastern ethnic groups for whom data on insects as food (Table 19.2) can be provided are the Meeteis of Manipur and the Khasi of Meghalaya (Table 19.3). The Meeteis of Manipur, a state of 2.2 million inhabitants, live in a very hilly area south of Nagaland and west of Myanmar. They represent one of the many peoples of Manipur, inhabiting mostly the valleys, while 29 major tribes live in the hills. The total area of Manipur is 22,327 km^{-2}, of which only 2,238 km^{-2} constitute valleys, the remainder being hilly. Summer temperatures of Manipur are generally above those of its northern neighbour.

The climate of Meghalaya (home of the Khasi) occupies a position between that of Manipur East and Nagaland. Meghalaya covers an area a little over 22,000 km^2 and is a hilly stretch of land north of Bangladesh, divided into five districts. The dominant tribal entities are the Khasis, the Jaintias, and the Garos. The Garos inhabit western Meghalaya, the Khasis central Meghalaya, and the Jaintias the eastern regions.

Papua New Guinea

In Papua New Guinea and neighboring islands, a bewildering variety of insects (Meyer-Rochow, 1973b; Meyer-Rochow and Changkija, 1997; Tommaseo-Ponzetta and Paoletti, 1997 and this volume), and even spiders (e.g., Bulmer, 1974), has found acceptance as nutrition for humans. My own research was originally conducted among Central Highlanders (the Chuave), Southern Highlanders (the Onabasulu), and Trobriand Islanders (Kiriwinians), but very recently (November 2002) inhabitants of coastal New Britain were also visited.

The Chuave (Table 19.4) are a constituent of the Chimbu people whose population density reaches 250 km^{-2}. Despite the tropical latitude, seasonal variations in the weather are distinct and winter nights, because of the altitude at

Table 19.1: Some edible insects and spiders of the Ao-Nagas

Taxon /Scientific name	English name	Vernacular name	Remarks
Odonata			
Acisoma parnorpaides Rambar, 1842	dragonfly	atsü-kumbo	both species collected from streams
Aeschna nymphs	dragonfly	anga-mechep nymphs/larvae	
Orthoptera			
Acheta bimaculatus (DeGeen, 1775)	field cricket	chokokza	all species abundant and edible
Brachytrupes achatinus (Stoll)	common cricket	shati-chokok	"
Liogryllus bimaculatus De Geer	spotted cricket	mesü-chokok	"
Gryllodes melanocephalus	blackhead cricket	chokok	"
Gryllotalpa africana Palisot-Beauvois	mole cricket	chokok	"
Acrida gigantea (Herbst)	grasshopper	chupong	"
**Acridium melanocorne* Linn.	brown locust	koropong changkok	"
**Acridium perigrinum*	migratory locust	wara serapong	seasonally abundant
Mecopoda elongata L.	green locust	serapong changkok	edible
**Thylotropides ditymus*?	nocturnal grasshopper	aya changkok	"
**Lima cordid*?	long-horned grasshopper	alu changkok	"
**Holochlora albida* Brunner	long-horned leaf grasshopper	aya changkok	"
Hierodula coarctata Saussure	praying mantis	aei changkok	predatory, but edible
Isoptera			
Odontotermes obesus (Rambur)	termites, white ants	along, aneung	fried edible
Hemiptera			
**Bagrada picta* Fabr.	painted bug	tsüngi	*Osbickia* spp. and *Schima wallichii*,
**Dolycoris indicus* Stall.	bamboo bug	pollo	*Dendrocalamus hamiltoni* are hosts
Lethocerus indicus (Lepeletier et Serville)	giant water bug	atsü leplo	collected from ponds
Lohita grandis Gray	giant red bug	alu tsüngi	"
Gerris spinole	water strider	tsümeroki	"
Gerris sp.	water strider	tsümeroki	"
Homoptera			
**Cicada viridis* L.	cicada	chang changkok	seasonally abundant
**Cicada* sp.	cicada	loyang	seasonally abundant
Coleoptera			
Batocera rubra	woodborer	arulangtang	considered delicious
**Xysterocera globosa Oliv.*		arulangtang tasula	

Table 19.1: (Contd.)

Table 19.1: (Contd.)

Taxon /Scientific name	**English name**	**Vernacular name**	**Remarks**
**Xysterocera* sp.	pink woodborer	angami tsükha	*Butea minor* (host plant)
Neocerambyx paris (Wiedemann, 1821)	woodborer	tsükha	*Albizia* sp. (host plant)
**Coelosterma scabrator* F.	woodborer	tsükha	*Alnus nepalensis* and *Calicarpa* spp.
**Coelosterma* sp.	woodborer	khuro tsükha	*Morus* spp. (host plants)
Rhynchophorus ferrugineus (oliver)	palm weevil	morong	considered delicious
Balaninus album	banana weevil	merong	"
Xylotrupes gideon (Linnaeus)	hercules beetle	lessepo	"
Trichoptera			
aquatic nymphs	caddisfly larvae	tsü-longben	in ponds and streams
Lepidoptera			
**Diacrisia obliqua* walker	woolly bear caterpillars	wakak	edible
**Phyllosomia ricini* (Boisduval)	erriworm	errimesen	very common food
Antheraea assamensis Heilten	muga silkworm (emperor)	mugamesen	"
Antheraea roylei (Moore)	tusar oakworm (emperor)	sarang mesen	"
Bombyx sp.	banana leaf roller	mango longpen	"
Bombyx mori L.	silkworm	mugamesen	commonly marketed
Attacus cynthia Drury	tusar oakworm	mugamesen	"
Attacus ricini (Boisduval)	erriworm	allishi mesen	"
Malacosoma sp.	tent caterpillars	mesang-long	*Schima wallichii*
Hymenoptera			
Crematogaster dohni	red ant	kongja-za	food and medicinal
Oecophylla smaragdina (Boisduval) (Fabricius)	weaver ant	kongja	edible
Icaria artifex (De Saussere)	common wasp	kibang ninang	larvae consumed
Vespa orientalis L.	giant hornet	nadi	common insect food
Vespa sp.	small hornet	sünteng-nadi	"
Apis sp.	honeybee	longnang	less common, but prized
Arachnida			
Avansa sp.	semicolonial spider	tsüksen	both spider species, legs tied, marketed
Nephila clavata L. Koch	striped, giant orb-weaver	tsüksen tasula	

* = names as provided by the Zoological Survey of India.

which the Chuave live, outright cold. The Onabasulu (Table 19.5) inhabit the area north of Mt. Bosavi in the Southern Highlands and the first-ever census of 1966 yielded a total population of 200 people, who spoke their own distinct language. The population density is low (12 km^{-2}) and the climate characterized

Table 19.2: Some edible insects of the Meeteis people (N. Biraj Singh and Abhik Gupta, pers. comm.)

Taxon /Scientific name	English name	Vernacular name	Remarks
Odonata			
Anisoptera:			
Cordulegaster sp.	dragonfly larvae	maikhumbi	fried in oil
Other Anisoptera	dragonfly larvae		"
Isoptera Pterygote forms	termites/white ants winged individuals		roasted without oil and eaten after removal of wings
Orthoptera (some species of Tettigonidae and Acrididae)	some species of long- and short-horned grasshoppers		fried with or without oil
Hemiptera			
Nepidae: *Lacotrephes* sp.	water scorpion	naushek	eaten steamed, roasted or as flavoring agent for sauces
Ranatra sp.	water stick insect	esing mapi	fried in oil
Notonectidae:			
Anisops sp.	backswimmer	long khajing	fried in oil
	other aquatic bugs		fried in oil
Coleoptera Dytiscidae: *Laccophilus* sp.	water beetle	tharaikokpi	fried in oil
Hymenoptera Some larvae and pupae of Formicidae and Apidae	larvae and pupae of red ants and bee larvae		fried in oil or turned into a hot curry

by almost constant daily rains throughout the year. Especially during certain festivals (Plate IX, 4), a considerable amount of insects, e.g. sago palm grub, is consumed (Meyer-Rochow, 1983). The Kiriwinians (Table 19.6) are an anthropologically well-studied Melanesian people with an Austronesian language, inhabiting the Trobriand Islands off the coast of Southeast New Guinea. The climate is warm and pleasant throughout the year; distinct rainy seasons do not exist and the population density is approximately 120 km^{-2}.

In Papua New Guinea the existence of highly seasonal food insects limited to certain areas (e.g. Ephemeroptera as food among the locals of the Sepik river village of Malu; Szent-Ivany and Ujházy, 1973) can cause the entomophagy researcher to miss important species or to come to the wrong conclusions. Nonetheless, it is quite clear that Orthoptera and Coleoptera form the "backbone" of the local food insects and provide most of the edible species. It was reported earlier (Meyer-Rochow, 1973b) that both immature and adult forms of numerous species of beetles belonging principally to families Cerambycidae,

Table 19.3: Some edible insects from the Khasi Hill region (D. Khathing, pers. comm.), unfortunately not zoologically identified nor sighted by the author. The prefixes "ka" and "u" in vernacular names denote female and male respectively

Taxon/Scientific name	English name	Vernacular name	Remarks
Orthoptera	grasshoppers or locusts	niangphlang	caught in baskets by the millions in autumn (Nov.) during harvest time
	cricket	ka sharaij	has 4 different names according to age and sound
Isoptera	termites/white ants	ka kber	usually swarms in the autumn afternoon
Hemiptera or Coleoptera	"beetle"?	ka niangsbai	feeds on oak leaves; appears by the millions in autumn (2 types)
	"big beetles"	ka niangsohriew	bigger than sp. above and with lateral spots
	"giant beetle"	ka nianglyhur	largest of all the flying "beetles" (very loud buzz)
Unidentifiable (maybe cicada).	?	Ka niangkrai	ground insect; 2 types, appears in autumn, but doesn't feed; lives 2–3 months
Coleoptera	grubs/beetle larvae	eri-worm	favorite food; cultivated
Hymenoptera Some larvae and pupae of Vespidae and Apidae	common wasp ground hornet	u dkhew u kyieng	nest on trees, roofs, etc. "king of all wasps", nest underground
	rock bee	u lwai	nest on high rocks
Lepidoptera	caterpillar	u niang kseh	large caterpillar, mostly on stem of pine trees
Unidentifiable	unidentifiable	u shalyngur	comes in reddish, brownish, black varieties

Curculionidae, Scarabaeidae, Passalidae, and Tenebrionidae were consumed and that long- as well as short-horned grasshoppers and locusts, crickets, mole-crickets, praying mantids, and stick insects (Fig. 19.1) were also components of many a local diet. Special attention was paid to head lice (Plate IX, 5), which were not only eaten but also considered living indicators of the state of health of a child.

Differences between the three ethnic groups in relation to the species of insects consumed did, of course, exist and were mentioned in Meyer-Rochow (1973b), but the then-postulated correlation between number of insects consumed and human population density can no longer be upheld.

Table 19.4: Some edible insects of the Chuave people with Kinuku dialect (Thurman, pers. comm.) in parentheses

Taxon/Scientific name	English name	Vernacular name	Remarks
Mantodea	Praying mantis	keikabu (keikapu)	eaten
Orthoptera			
Tettigonidae	long-horned grasshoppers	weriwawa (sirikine)	eaten
Acrididae	short-horned grasshoppers	giba (siname)	eaten
Grylloidea	crickets	keko (ekera)	eaten
Gryllotalpa sp.	mole cricket	wiwi, wewi (kuoko)	eaten
Phasmatodea	stick insects	kumatoru, komatoru	sometimes eaten
Homoptera			
Cicadidae	cicadas	giuro (gioro)	some regularly eaten
Hemiptera			
Mictis sp.	leaf and vegetable bugs	ga(d)raniba (garanipa)	eaten, but sprays poison in human eyes
Phthiraptera	lice	numan (numan)	perhaps eaten
Coleoptera	beetles		some grubs and larvae
Passalidae	sugar beetles	gomuna (gomina)	regularly eaten
Scarabaeidae: *Xylotrupes gideon, Oryctes* sp.	rhinoceros beetles	wawe (waw gomina)	eaten as grub
Lucanidae (spp.)	stag beetles	gomuna (gomina)	grub probably eaten
Cerambycidae (spp.)	longicorn beetles	emeiba (emeipa)	eaten (at least grub)
Curculionidae (spp.)	weevils	eimeiba (emeipa)	eaten (at least grub)
Lepidoptera and Coleoptera	any woodboring larvae	omun (omon)	many are eaten
		konokono (topatopa)	general term
Lepidoptera	moths and butterflies	monsumuna	stuffed in bamboo, eaten
	social caterpillars		
Hymenoptera	ants and their pupae	sin (sin)	some eaten
	honeybee	dum (ipa dum)	honey and larvae eaten
Scorpiones/Uropygi	scorpions/whip scorpions?	wiwi (ekera)	like mole cricket (some possibly eaten)
Diplopoda/Peripatidae	millipedes/walking worms	onobanugan (onopamukan)	said to be eaten (identified from book)
	honey pot ant	ngari	

Australia

Still the most thorough and comprehensive treatise on insects as edible food among Australian Aborigines is that of Reim (1962). The book, however, rarely gives vernacular names, nor does it deal with the insect food of Central Australian

Table 19.5: Some edible insects of the Onabasulu people

Taxon/Scientific name	English name	Vernacular name	Remarks
Odonata			
Anisopetra and Zygoptera	damsel- and dragonflies	wodien	larvae like small crayfish, eaten
Orthoptera			
Tettigonidae (e.g., *Valanga* sp.)	long-horned grasshopper	sak(g)e	eaten
Coleoptera			
Curculionidae:			
Rhynchophorus bilineatus (Montr.)	sago palm beetle	yagi	grubs highly esteemed, eaten
Rhynchophorus ferrugineus (Olivier)	"musical weevil"	hugu	classified according to host tree, some undoubtedly eaten
Various families	hardwood borers	waba	
Hymenoptera			
Oecophylla smaragdina	weaver ant	yesi	eaten

Pintupis. Certain aspects of the Walbiri (termed 'Walpari' by Reim, 1962) insect food are mentioned, but Reim focuses on the most common Aboriginal food insects and various geographic regions of Australia. More recently Cherry (1991) presented a brief review of the contemporary and historical use of insects for a range of purposes, including food, by the Aborigines of Australia.

The two ethnic groups dealt with here can be considered hunter-gatherer societies of Central Australia, who speak mutually unintelligible, but nevertheless closely related languages of the Southwest group of the Pama Nyungan language family (Wurm, 1970). Traditionally, the population density of both groups has been extremely low. Like other Australian Aborigines, Walbiri and Pintupi (Tables 19.7 and 19.8) made use of a variety of insects and their products and, in particular, ate fatty lepidopteran and coleopteran larvae extracted from tree trunks and plant roots; termites and sugar-containing insects such as honeypot ants (Plate IX, 6) and bee larvae (Fig. 19.2) were also consumed, however.

New Zealand

The New Zealand Maoris are a Polynesian people with Austronesian linguistic affinities, who settled in the islands, known to them as 'Aotearoa', approximately 1,000 years prior to the arrival of the Europeans. The only insect still regularly consumed by them (if available) is the larva of a cerambycid beetle (*Prionoplus reticularis*) known locally as the "huhu" grub. The term is also used for the pupa (which is eaten) and the adult beetle (which is not normally used as food). The

Table 19.6: Some edible insects of the Kiriwina people

Taxon/Scientific name	English name	Vernacular name	Remarks
Mantodea:			
Tenodera sp., *Hierodula sternosticta*	praying mantis	tataya	eaten
Orthoptera,			
Tettigonidae:	long-horned	dila	it chirps: not eaten
Caedicia sp.,	grasshoppers	pwewesa	it doesn't chirp: eaten
Valanga sp.		nipawa	(big forms) some
Caelifera	short-horned	gagata	eaten
	grasshoppers	kilili	(small forms) some eaten
	small green hoppers		all edible
Grylloidea: *Metioche* sp., *Teleogryllus* sp.	crickets house cricket	sigwa, kinaneita sigwapolu	some bush crickets eaten
Phasmatodea:	stick insects		
Phasmatinae		kwapu	several spp. eaten
Eurycantha horrida		kidoka	some eaten
Boisduval			eaten
Phthiraptera	lice	kutu	eaten
Coleoptera spp.?	beetles	kim (general term for typical beetle shape)	some grubs are eaten
Hymenoptera			
Oecophylla smaragdina (Fabricius)	weaver ant	siboyeki	eaten
Arachnida			
Araneidae: *Nephila* sp.	giant orb weaving spiders	kapari	in olden days eaten

larvae can be found in a variety of rotting timbers, including those of introduced trees.

According to unconfirmed reports (Houia, pers. comm.), cicadas, locally known as "kihikihi", may have been eaten in some parts of New Zealand or could have found some use in indigenous medicines. The manuka beetle *Pyroneta festiva* and puriri moths (*Aenetus virescens*) are also said to have found occasional acceptance as human food.

Some expressions (see Broughan and Reed, 1987) referring to maggots, caterpillars, and grubs exist in the Maori language (to witness: *Iro te iro, homei kia kainga, ka kai hoki ta i a au*, which means in English "Fly-blown or not, give it to me. I'll eat the maggots now, for ultimately they'll eat me"), but to conclude from this and similar idiomatic references to insects that the speaker actually consumed them is presumptuous.

Fig. 19.1: Ten cm long edible stick insect, *Eurycantha horrida* in Australia (photos B. Meyer- Rochow).

Discussion

A critical assessment of the results can take a variety of forms. Lists of edible insect species can be discussed in the context of health and nutritional status of the ethnic groups involved, but they can also provide support for studies dealing with cultural and linguistic affinities among various peoples. They may even allow insights into the psychology of an ethnic group, its artistic expressions, and its metaphysical beliefs (Fig. 19.3).

A comprehensive picture of the contribution that insects make to a local diet can only be obtained if qualitative results are supplemented by quantitative data, but even to obtain reliable and complete *qualitative* information would require continuous observations throughout the year. To collect quantitative measurements of the amounts and percentages of each insect species consumed over the span of one year would compound the difficulties of such research. Nonetheless, real progress in the quest to understand the impact and importance of entomophagy in societies in which the custom is still an integral part of the local culture, can only come from the combined qualitative and quantitative approach. Without it any conclusion must remain unsubstantiated guesswork, however

Table 19.7: Some edible insects of the Walbiri people with terms in brackets after Meggitt (1962)

Taxon/Scientific name	English name	Vernacular name	Remarks
Isoptera	termites, white ants		some eaten
Orthoptera			
Teleogryllus commodus (Walker)	cricket	djabalari djabalari	rarely eaten
Acridoidea	grasshoppers and locusts	tindilga (djindilga)	rarely eaten
Homoptera			
Cicadidae	cicadas	lirinba	occasionally eaten
Coccoidea	scale insects	manda	some regularly eaten
Psylloidea	plant lice (lerp)	(jiljalbu)	regularly eaten
Coleoptera			
Euryscaphus sp.	carabid beetle	niedi niedi	eaten
Cerambycidae	longicorn beetles and bardies	mija mija	regularly eaten
Curculionidae	weevils	(lodu)	occasionally eaten
Lepidoptera			
Cossidae	witjuti or witchetty grub	malguri; (ngalgari)	regularly eaten
unidentified	caterpillars	waiburi; waioupi; (ladjul)	some eaten
Hymenoptera			
Camponotus inflatus (Lubb.)	honey-pot ant	ingurani; (jirambi), but other species (bingi) not eaten	highly prized food (under mulga *Acacia aneura* tree)
Trigona sp.	native honeybee	djolala	highly prized food

much food for thought and incentives as well as guidelines for future work such a conclusion may provide.

The nutritional value of various insect species has been examined and found to be remarkably high (e.g., reviews by Bodenheimer, 1951; DeFoliart, 1989; Bukkens, 1997; Feng and Chen, 1999; Bukkens, this volume). Insects generally consist of easily digestible proteins and fats, and small but significant amounts of carbohydrates and vitamins (Meyer-Rochow, 1978, 1979, 1982a; Meyer-Rochow and Changkija, 1997). If the fat content of an insect equals or exceeds the protein component, we speak of "f"-species; when protein amounts are greater than the lipid content, we deal with "p"-species. The term "s"-species designates an insect in which carbohydrates ("s" = sugar) dominate. Because of the high amounts of protein that make up many an insect's body, it has frequently been thought that insect consumption arose out of a need to supplement protein-intake. Another, less popular, idea is that the need to acquire salt and minerals made people search for and consume insects (Ichinose, 1989). However, in the light of my own research both hypotheses may be wrong (see below).

The chitin of the insect cuticle has been credited with cancer-preventive properties (Goodman, 1989) and grasshoppers (= "inago") have been praised by

Table 19.8: Some edible insects of the Pintupi people

Taxon/Scientific name	English name	Vernacular name	Remarks
Isoptera	termites, white ants	longurlma	eaten
Orthoptera			
Acridoidea	grasshoppers and locusts	djindilga	rarely eaten
Homoptera			
Cicadelloidea	leafhoppers	jugri jugri	some regularly eaten
Hemiptera			
various spp. of various families, e.g. Lygaeidae, Reduviidae, Pentatomidae	vegetable and stink bugs	patana	eaten
Coleoptera	beetles	nidinidi; niriniri	some grubs and adults regularly eaten
Lepidoptera			
Lymantridae	moths	binda binda (flying species that approach light)	grubs and adults eaten
Cossidae	witjuti or witchetty grubs	maku	regularly eaten
Hymenoptera			
Camponotus inflatus (Lubb.)	honey-pot ant	ngari	highly prized food
Trigona sp.	native honeybee	djorata: tjurratja	delicacy

Japanese nutritionists for their mineral content for years (Nagano, pers. comm.). Based on dry weight, Australian witchetty grubs (different spellings exist—"witchety, witjuty, or witjuti"—but refer to the same animal) possess 38% protein and 39.8% fat and are thus, nutritionally speaking, "f"-species. The corresponding figures for the desert locust, a "p"-species, are 75% and 20%, while those of termites, members of the "f"-category, amount to 36% and 44%, respectively (Table 19.9). Unpalatable and outright poisonous insects do exist (Berenbaum, 1993; Blum, 1994) but the majority appear wholesome and safe, to such an extent that a New Zealand survival book recommends consumption of insects in emergency situations (Hildreath, 1974).

However, as Reim (1962) has quite emphatically stated, insects need not merely be a food source in times of emergency and starvation, but can represent a genuine food category as in the Australian Aboriginal societies. The same could be said for the Nagas, numerous Papua New Guinea tribes, Thai peasants (Watanabe and Satrawaha, 1984; Chen et al., 1998)), and many other peoples of Southeast Asia (Watanabe, 1983). However, even where a great variety of insect

Fig. 19.2: Native Australian, stingless *Trigona* sp. honeybee, whose "honeybag" in a hollow tree together with the larvae is considered a delicacy among Central Australian Aborigines (photos B. Meyer-Rochow).

Table 19.9: Composition of five kinds of edible insects (after Meyer-Rochow, 1982a)

	100 g dry matter			100 g fresh matter		
	Desert locust	**Termite**	**Witjuti grub**	**Witjuti grub**	**Grasshopper**	**Bee larva**
Protein	75	36	38	16.5	64.2	20.3
Carbohydrate	traces	8	5.7	2.5	traces	19.7
Fat/Lipid	20	44	39.8	17.2	2.4	7.9
Ashes	5	12	16.2	7.1	3.3	9.5
Water	-	-	-	56.7	30.1	42.6

Witjuti grub is the Aboriginal name of edible root- and stem-boring larvae of certain moths (e.g. Cossidae) and longicorn beetles (Cerambycidae). In addition to the contents given in this Table, witjuti grubs contain 400 international units (i.u.) vitamin D, 100 i.u. vitamin A per 100 g, 6 ppm iron, 5 ppm copper, and 19 ppm zinc. 100 g of grasshopper (the Japanese 'inago') contain approx. 300 i.u. vitamin A, 920 i.u. carotenes, 20 mg vitamin C, 7 mg of vitamin $B_1 + B_2$, and healthy amounts of calcium, as well as niacin. Data combined from different sources according to Meyer-Rochow (1982a).

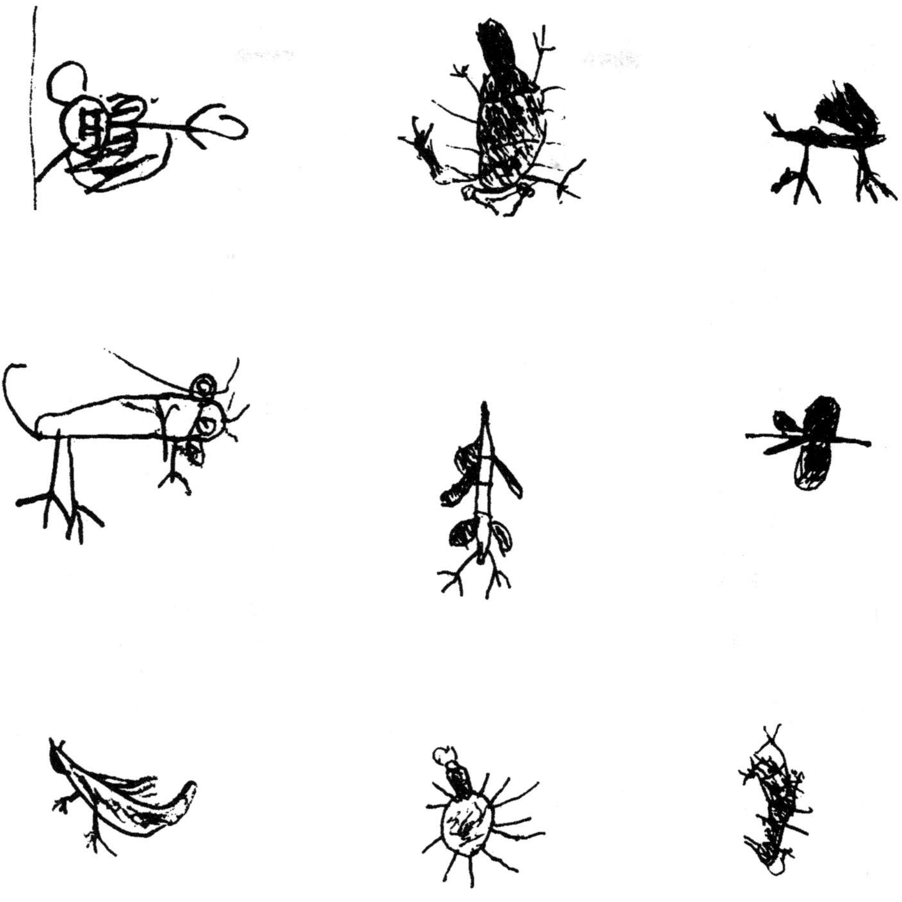

Fig. 19.3: Drawings of arthropods by ca. 19-year-old male Onabasulu youth, who had never before used pencil and paper, depicting from top left to bottom right: beetle, cockroach, cricket, ant, dragonfly, butterfly, grasshopper, spider, scorpion. Note that several insects were given claws of the cassowary bird and that fast running species had more than 6 legs (photos B. Meyer-Rochow).

species are utilized as food, consumption of insects is never random, haphazard, or unselective. Food taboos, known so well for other kinds of food stuffs and present in one form or another in every society on Earth (Meyer-Rochow and Moro, 1995), also operate when it comes to insects. For example, except for of a few northern coastal Australian tribes, the Aborigines never consume locusts, although the latter are widely appreciated as food in other parts of the world, including nearby the Papua New Guinea. Another example from the region involves the Kiriwina Islanders. They express a very differentiated view on the acceptability of these insects as food (those that chirp and have long antennae are not eaten and among those with short antennae only some species are considered "fit for consumption" (Meyer-Rochow 1975b). One is reminded of the Japanese, who will eat grasshoppers ("inago") but not locusts ("bata"), and the Maoris of New Zealand who, despite the occurrence of locusts there, are not known to touch them as food while happily ingesting the beetle larvae known as "huhu" grubs.

A people, tribe, or ethnic group never makes use of the full potential of edible items in its surroundings and a statement such as "early hunter-gatherers must have experimented with nearly all" the food items (Heiser, 1990) is clearly without scientific foundation. An especially well-studied example in this context are the Ache of the Paraguayan jungle (Hill and Hurtado, 1989). In spite of being surrounded by hundreds of edible mammalian, avian, reptilian, amphibian, and piscine species, they exploit only 50 of them. Turning to the plants, fruits, and insects, the situation is no different: only 40 in total are exploited. Ninety-eight percent of the calories in the diet of the Ache are supplied by only 17 different food sources.

The reasons for such selectivity are, of course, complex and have been examined for five ethnically and/or religiously different human societies (Meyer-Rochow and Moro, 1995). The availability of the species in question and the ease with which certain specimens can be procured are certainly involved, but traditions and beliefs may also dictate which species are/are not acceptable. It is clear that a child's choice of food is influenced by its parents' dietary habits (Koivisto-Hursti 1999). However, no correlation between the number of insect species used as food and human population density appears to exist (this paper), and harshness or productivity of a region (although assessable only with severe limitations) also appear to provide no distinct clues. Even if climatic differences and the total number of insect species available to choose from in a given environment were to be considered, one would still encounter some difficulties in explaining why some tribes use more (or less) insects as food as others.

Among the Australian Aborigines grubs of numerous species of beetles and moths (collectively and interchangeably termed witchetty grubs or bardies: Reim 1962) form a favorite food. Some termite species also find acceptance. This liking of "fatty", timber-boring insect larvae is shared with peoples as far away as Northeast India (Nagas) and the South Pacific (the Maoris of New Zealand). Termites, however, a food item also known to have been used for medicinal purposes, link the Aborigines more to South Indian tribals such as the Kurubas for instance (Moro, pers. comm.). The diet of traditionally living Aborigines was not high in fat (O'Dea, 1991) and it is possible that some of their fat needs were met by consuming lipid-rich insects. Incidentally, the fat composition of witchetty or witjuti grubs resembles that of olive oil (Naughton et al., 1986). But do fat-rich species represent the earliest arthropods that the human palate found acceptable? How far back in time are entomophagous practices rooted anyway?

We cannot precisely answer these questions but can assume that even our hominoid precursors in all likelihood would have tasted some insects. Monkeys, as Nickle and Heymann (1996) have shown in a thorough study, extensively consume locally abundant insect species. The fact that we possess taste receptors that respond to "sweet" but none that specifically inform us of lipids or proteins, perhaps indicates that the earliest insects to be accepted by humans were sweet species. This would also accord with the view that our immediate hominoid ancestors were predominantly fruit-eating primates (Dudley, 2000). I therefore suggest that historically the search for and acceptance of fatty insects followed the acceptance of sugar-containing, sweet forms and that the consumption and specific search for protein-rich insect species came later.

It is now known that Aborigines have lived in Australia for at least, 40,000–50,000 years (Flood, 1983). But we do not know how long present-day Aboriginal populations have occupied the territories they now regard as their "home turf" and whether there have been multiple colonizations. Our observations have documented a strong similarity in the insect diets of the two Central Australian Walbiri and Pintupi tribes and even some of the insect terminology (e.g. the respective terms "djolala" and "djorata or tjurratja" for native honeybee) is similar (Meyer-Rochow, 1975b, 1982b). Aboriginal insect food in central Australia differs from that of the northern coastal tribes and Papuan, Austronesian, and Asian neighbors in that it is largely made up of "f" and "s" species, whereas that of the others is dominated by "p" insects. The antiquity of Australian Aboriginal colonization therefore receives a "boost" from results in entomophagy research.

Linguistic affinities between the Aboriginal insect terms and those used in Papua New Guinea or Southern and Southeast Asia cannot, of course, be expected because the isolation of the Australian Aborigines has simply been too long. But Kovalainen (1975), Valonen (1975), and Härkönen (1998) have demonstrated that food terminology and eating habits can indeed often provide valuable clues about ethnic relationships (although in their cases, i.e., the Karelian people, shorter periods and geographic distances were involved). That food habits, even among a people dispersed around the globe, can persist for over thousands of years is a fact, especially when linked to some cultural identity or religion (the Jewish "kashrut" is a case in point, Meyer-Rochow and Moro, 1995). Our limited food insect material can, thus, be helpful in the greater context of understanding the cultural histories and origins of the societies from which the information on insect food habits and insect terminologies was gleaned.

According to the data presented in this paper, the greatest variety of terrestrial arthropods is consumed by the Nagas, whose present-day territory lies in Northeast India and Northwest Myanmar. The Tibeto-Burmese-speaking Nagas and other "Trans-Ganges tribals" are, according to Schmidt (1977), culturally, if not linguistically related to inhabitants of the Malay peninsula and the Sunda Islands (Ghosh, 1982). In the broadest sense and according to currently accepted views, we find Tibeto-Burmese at one end of a linguistic continuum and Austronesian at the other (Ruhlen, 1987). The Trans-New Guinea phylum of Papuan languages has only recently been recognized as a more or less coherent group (Oliver, 1989) and finds itself placed in the greater Indo-Pacific phylum, predating those few Austronesian languages (for instance Kiriwina) that were able to establish a foothold in some coastal areas of the island of New Guinea.

If the various wood-boring grubs, leaving aside honey-pot ants, native bees, and manna-producing "s" insects, represent some of the historically older sources of edible insects, one might expect to find linguistic similarities in the terminology of this food item. And indeed the terms "morong", "muga", and "mesen" in various combinations employed by Nagas in relation to edible grubs and caterpillars, are not all that different from the Thai "malaeng", commonly part of the name of edible insect species (Watanabe and Satrawaha, 1984). Quite similar too are the terms of "megoma", "monsumuna", and "mugan" used by Chuave to describe wormlike arthropods such as caterpillars and millipedes and even the Polynesian "mongomonga" for a wide variety of cockroaches, scuttling beetles, and bugs (Pond, 1984) does not sound too different. The Maori word "huhu" for edible grubs and beetles is almost identical to "hugu", an edible weevil in the language of the Onabasulu of the Southern Highlands of Papua New Guinea—coincidence? Other similarities are present. To Thai peasants of Khon Kaen, bees are "pung" (Watanabe and Satrawaha, 1984); in Chuave (Papua New Guinea) the same insect is "dum", and to the Kalam of the Kaironk Valley, also Papua New Guinea, edible insects generally are mostly "jong" (Bulmer, 1974), while Nagas refer to the edible termites as "along or aneung" (this paper).

Although there is as yet little conclusive evidence for the following scenario, I believe it to be likely that practices of entomophagy in Southern Asia and beyond benefited from tribal migrations (voluntary or forced) as well as trade along ancient routes. If we accept that a tropical setting provides humans with a better opportunity to forage for insects than a colder and entomologically impoverished environment, then Southern India and the Burmese/Thai/Malay regions would have been the most likely ones in which insects could have become a regular type of human food. From these areas the habit could have spread first, in most ancient times, to Australia from India via the Sunda Islands and then in separate migratory waves from the Thai/Burma/Malay region to the Northwest (Naga), North (China) and Northeast (Taiwan, Japan). Southward a more recent dispersal, once again via the Sunda Islands, reached Papua New Guinea and most recently Oceania as far as New Zealand.

It has been suggested, originally by Reim (1962), that Australian Aborigines retained from the repertoire of edible insects principally those that provided them with fat (termites, witchetty grubs) and carbohydrates in the form of sugar (honey-pot ant, sugary lerp, and scale insects). An alternative view is that at the time of the first Aboriginal immigration to Australia the habit of using "p"-species of insects in the human diet had not yet "caught on" and thus allowed the tradition of using "f"-species to become firmly established. For Naga tribes people, Thai peasants, and Papua New Guinea highlanders, which were all less isolated than the Australian Aborigines, the situation may have been different and allowed a quick acceptance of "p"-species, facilitated through human migrations, trade contacts, etc. We feel that this hypothesis ought to be given some consideration.

Quite a different aspect in relation to insects as food in Southern Asia, Oceania, and Australia could be brought into the discussion, namely the various "spin-offs" of entomophagous practices. Insect products, such as fermented refuse honeycombs came into use and provided a mildly intoxicating drink called "bull" in certain parts of Aboriginal Australia (McKeown, 1944). Insect medicines and remedies for a wide variety of ailments became common (in northern Queensland, for instance, a liquid made from the bodies of green tree ants was used against headache and fatigue, McKeown, 1944) and cossid larvae found use as "pacifiers" for Aboriginal babies during weaning (Tindale, 1953). There must have been hundreds more such uses of insects but as with the significance of insects (and spiders) in local folklore, myth, legend, song, and dance (Meyer-Rochow, 1978, 1979), it is now often too late to obtain such information.

From what little information is available on insects in folk medicine, it appears that such medications were particularly popular in places where the total consumption of insects as food (although still practiced) and the number of species consumed declined (Meyer-Rochow and Changkija, 1997). In the traditional medicines of Japan (Umemura, 1943, cited in Schimitschek, 1968), China (Chen and Akre, 1994), and Korea (Pemberton, 1999), almost every insect taxon was represented and to this day some species are being used as remedies against a variety of ailments. In Finland and Germany, where no insect species is considered edible these days, some species still retain a folk-medicinal or magical use, especially in the treatment of arthritis, rheumatism, and warts (Meyer-Rochow, unpubl.).

A final paragraph must address insect husbandry and domestication. In Europe the honeybee was domesticated because of its products honey and wax. But given the fact that in so many cultures the larvae were also appreciated as food, it may well be that ancient European bee domesticators ate them. Be that as it may, it is widely accepted that the oldest fully domesticated insect, the silkworm, was known to the Chinese at least 4,000 years ago (Zhou, 1979). Yet how was the gathering of silk discovered? Placed by some researchers in the area of Northeast India ("...references in Sanskrit literature suggest that a silk industry may have existed in India . . . possibly even as early as 4,000 BC..."; Cloudsley-Thompson, 1976), silkworm domestication and the subsequent use of silk by humans, could well be rooted in entomophagy. Suzuki (cited in Nunome, 1988), talking about a "multiorigin of silkworm use in ancient times", appears to harbour the same sentiment, and after all, the use of silkworm pupae as food is still widespread. Conceivably during alimentary preparation of wild silkworms and their relatives (*Antheraea* spp. are, for example, even now consumed by the Nagas) the discovery was made that hot water softens the pupal case and allows the silk to be separated and wound.

Actions that approached "cultivation" or "domestication" of another insect, the sago-palm grub, were reported by Townsend (1970), cited in Oliver (1989) and Mercer (1994). Furthermore, Reim (1962) has stated that the Aborigines of Southwest Australia were engaged in "Bardy Zucht" and actually rearing the grubs of the cerambycid *Bardistus cibarius* by deliberately damaging host plants and allowing colonizations with the beetle to take place. Conservation of such insects for use at a later date requires drying and/or roasting and storage in a safe place, which is exactly what some Aboriginal tribes did (Reim, 1962).

My survey ends here with a certain amount of sadness. So much remains to be done, so much is still unknown, and so much will forever remain unknown, because we have badly neglected the interrelationship between humans and insects in the past. Was, for example, the dearth of all kinds of large insects from the Moluccan Island of Ceram, which the famous 19th century zoologist Alfred Russel Wallace in 1883 called "a zoological desert" surrounded by islands on which there was no shortage of spectacular insects, the result of overexploitation of a resource by the Melanesian inhabitants of this sizable land mass or were other causes responsible? We still have no answer to that and many other specific questions, but let it be noted that avenues for a responsible environmental policy are now earnestly being sought (Rolston, 2000) and that wildlife, which includes insects and other invertebrates, has at last been recognized as valuable in its own right (Chardonnet et al., 2002). We are now dealing with questions that have begun to receive greater attention and recognition worldwide. In the words of Saito (2002) "we need to cultivate an ecologically informed nature aesthetics...to maintain...ecological health and well-being". Ethnoentomological and cultural entomological studies have finally come of age. Having inspired anthropologists, social scientists, linguists, zoologists, and historians, may they be an inspiration for our employers and funding agencies as well!

Acknowledgments

I thank my companions, helpers, guides, and informants in the field, Dr. Abhik Gupta and N. Biraj Singh of Assam University/Silchar (India) for information on the Meeteis, Professor D. Khathing of North East Hill University/Shillong (India) for information on edible insects from the Khasi Hill district and Dr. Sapu Changkija of Nagaland University/Meziphima (India) for photographs 1–3, Plate IX. Dr. Jozsef Gal assisted with digitizing the slides and preparing the CD.

References

Berenbaum M.R. 1993. Sequestered plant toxins and insect palatability. Food Insects Newsl. 6 (3): 1–9.

Bible (Torah): Leviticus 11, 1–31.

Blum M.S. 1994. The limits of entomophagy: a discretionary gourmand in a world of toxic insects. Food Insects Newsl. 7 (1): 1–11.

Bodenheimer F.S. 1951. Insects as Human Food. Dr. W. Junk, The Hague, The Netherlands.

Borror D.J. and DeLong D.M. 1964. An Introduction to the Study of Insects. Holt, Rinehart, and Winston, New York, NY.

Broughan A.E. and Reed A.W. 1987. Maori Proverbs (reviewed and edited by T.S. Karetu). Reed-Methuen, Auckland, North Island, NZ.
Bukkens S.G.F. 1997. The nutritional value of edible insects. Ecol. Food Nutr. 36: 287–319.
Bulmer R.N.H. 1968. Worms that croak and other mysteries of Karam natural history. Mankind 6: 621–639.
Bulmer R.N.H. 1974. Folk biology in the New Guinea Highlands. Soc. Sci. Inform. 13: 9–28.
Campbell G.L. 1991. Compendium of the World's Languages. Routledge, London, UK.
Chardonnet P., des Clers B., Fischer J., Gerhold R., Jori F., and Lamarque F. 2002. The value of wildlife. Revue Sci. 21: 15–51.
Chen P.P., Wongsr S., Jamyanya T., Rinderer T.E., Vongsamanode S., Matsuka M., Sylvester H.A., Oldroyd B.P., and Chen-Ping P. 1998. Honeybees and other edible insects used as human food in Thailand. Amer. Entomol. 44: 24–29.
Chen Y. and Akre R.D. 1994. Ants used as food and medicine in China. Food Insects Newsl., 7 (2): 1–10.
Cherry R.H. 1991. Use of insects by Australian Aborigines. Amer. Entomol. 37: 8–13.
Cloudsley-Thompson J.L. 1976. Insect and History. Weidenfeld and Nicolson, London, UK.
CSIRO (Commonwealth Scientific and Industrial Research Organization). 1970. The insects of Australia. Univ. Press, Melbourne, Melbourne, Victoria.
DeFoliart G.R. 1989. The human use of insects as food and as animal feed. Entomol. Soc. Amer., Spring 1989: 22–35.
Dudley R. 2000. Evolutionary origins of human alcoholism in primate frugivory. Quart. Rev. Biol. 75: 3–15.
Feng Y. and Chen X.M. 1999. Review on nutritive value of edible insects. Forest Res. (Beijing), 12: 662–668 (in Chinese with English summary).
Flood J. 1983. Archaeology of the Dreamtime. Collins, Sydney, NSW, Australia.
Ghosh B.B. 1982. History of Nagaland. Chand, New Delhi, India.
Goodman W.G. 1989. Chitin: a magic bullet? Food Insect Newsl. 2: 2.
Gope B. and Prasad B. 1983. Preliminary observation on the nutritional value of some edible insects of Manipur. J. Adv. Zool. 4: 55–61.
Gorham J.R. 1976. Insects as food. Bull. Soc. Vector Ecol. 3: 11–16.
Hall E.T. 1969. The hidden dimension. Anchor, Garden City.
Härkönen M. (1998). Uses of mushrooms by Finns and Karelians. Int. J. Circumpolar Health 57: 40–56.
Heiser C.B. 1990. Seed to Civilization. Harvard Univ. Press, Cambridge, MA.
Hildreath B. 1974. How to Survive in the Bush, on the Coast, and in the Mountains of New Zealand. Shearer Governmint Printers, Wellington, North Island, NZ.
Hill K. and Hurtado A.M. 1989. Hunter-gatherers of the New World. Amer. Scientist 77: 436–443.
Ichinose K. 1989. More insect eating. Nature (London) 337: 513–514.
Koivisto-Hursti U.K. 1999. Factors influencing children's food choice. Annales medicini 31 suppl. 1: 26–32.
Köstlin K. 1975. Die Revitalisierung regionaler Kost. *In:* Valonen N. and Lehtonen J.U.E. (eds.). Ethnological Food Research. Vammalan Kirjapaino Oy, Vammala, Finland, pp. 159–166.
Kovalainen P. 1975. Die Nahrung als ein Gruppenmerkmal der karelischen Minderheiten. *In:* Valonen N. and Lehtonen J.U.E. (eds.). Ethnological Food Research. Vammalan Kirjapaino Oy, Vammala, Finland, pp. 147–158.
Kühn H. 1952. Die Felsbilder Europas. Franck, Stuttgart, West Germany.
Lyon P. 1974. Native South Americans: Ethnology of the Least Known Continent. Little, Brown and Company, Boston, MA.
McKeown K.C. 1944. Insect Wonders of Australia. Angus and Robertson, Sydney, New South Wales, Australia.
Meggitt M.J. 1962. Desert People—a study of the Walbiri Aborigines of Central Australia. Angus and Robertson, Sydney, NSW, Australia.

Mercer C.W.L. 1994. Sago grub production in Labu swamp near Lae—Papua New Guinea. Klinkii 5: 30–34.

Meyer-Rochow V.B. 1973a. Ernährung: Insekten gefällig? Selecta 15: 1396.

Meyer-Rochow V.B. 1973b. Edible insects in three different ethnic groups of Papua New Guinea. Amer. J. Clinical Nutr. 26: 673–677.

Meyer-Rochow V.B. 1975a. Can insects help to ease the problem of world food shortage? Search 6: 261–262.

Meyer-Rochow V.B. 1975b. 2. Local taxonomy and terminology for some terrestrial arthropods in five different ethnic groups of Papua New Guinea and Central Australia. J. Roy. Soc. Western Australia 58: 15–30.

Meyer-Rochow V.B. 1976. The use of insects as human food. Food Nutr. 33: 151–152.

Meyer-Rochow V.B. 1978/79. The diverse uses of insects in traditional societies. Ethnomedicine 5: 287–300.

Meyer-Rochow V.B. 1982a. Konchu to minzokugaku: konchu no iroiro na riyou. Insectarium 19: 174–179 (in Japanese).

Meyer-Rochow V.B. 1982b. Klima und Sprachenentwicklung. Naturwissenschaftliche Rundschau 35: 203–204.

Meyer-Rochow V.B. 1983. Bosabi no hitobito. Minzokugaku 22: 52–58 (in Japanese).

Meyer-Rochow V.B. 1988. Maanmainiot ötökät. Tiede 8: 58–61 (in Finnish).

Meyer-Rochow V.B. 1990. Need suureparased suuparased putukad. Horisont 6: 35–40 (in Estonian).

Meyer-Rochow V.B. 1998. Ethnic identities, food, and health. Inter. J. Circumpolar Health 57: 2–3.

Meyer-Rochow V.B. and Moro S.D. 1995. Food taboos. 1992. Proc. Jamaican Soc. Science Technology 6: 1–8.

Meyer-Rochow V.B. and Changkija S. 1997. Uses of insects as human food in Papua New Guinea, Australia, and North-East India: cross cultural considerations and cautious conclusions. Ecol. Food Nutr. 36: 159–187.

Meyer-Rochow V.B., Okui K., Henshall K., and Brook D.L. 2000. The mental concept of "worm/bug" in Japanese and English idioms. J. Finnish Anthropol. Soc. 25: 29–40.

Mountford C.P. 1971. Ayers Rock: Its People, Their Beliefs, and Their Art. Pacific Books, Sydney, NSW, Australia.

Naughton J.M., O'Dea K., and Sinclair A.J. 1986. Animal food in traditional Aboriginal diets: polyunsaturated and low in fat. Lipids 21: 684–690.

Nickle D.A. and Heymann E.W. 1996. Predation on Orthoptera and other orders of insects by tamarin monkeys, *Sanguinus mystax* and *Sanguinus fuscicollis nigrifrons* (Primates: Callitrichidae) in northeastern Peru. J. Zool. (London) 239: 799–819.

Nunome J. 1988. Archaeology of silk and cloth. Yuzan-kaku, Tokyo.

O'Dea K. 1991. Traditional diet and food preferences of Australian Aboriginal hunter-gatherers. Phil. Trans. Roy. Soc. (London) B 333: 233–241.

Oliver D.L. 1989. Oceania: the native cultures of Australia and the Pacific Islands. Univ. Hawaii Press, Honolulu.

Pathak K.A. and Rao K.R. 2000. Insect pests as human food of tribals in North Eastern region of India. Indian J. Entomol. 62: 97–100.

Pemberton R.W. 1994. The revival of rice-field grasshoppers as human food in South Korea. Pan-Pacific-Entomol. 70: 323–327.

Pemberton R.W. 1999. Insects and other arthropods used as drugs in Korean traditional medicine. J. Ethnopharmacology 65: 207–216.

Pond W. 1984. Polynesian classification of small land animals: a description from Niuatoputapu Island in the Kingdom of Tonga. Weta 7: 93–107.

Posey D.A. 1976. Entomological considerations in Southeastern Aboriginal demography. Ethnohistory 23: 147–160.

Raven P.H., Berlin B., and Breedlove D.E. 1971. The origins of taxonomy. Science 174: 1210–1213.

Reim H. 1962. Die Insektennahrung der australischen Ureinwohner. Akademie Verlag, Berlin.

Rolston H. 2000. The land ethic at the turn of the millennium. Biodiv. Conserv. 9: 1045–1058.
Ruhlen M. 1987. A guide to the world's languages, vol. 1: Classification. E. Arnold, London, UK.
Sachan J.N., Das B.B., Gangwar S.K., Pathak K.A., and Katiyar J.N. 1987. Insects as food in north-eastern hill region of India. Bull. Entomol. 28: 67–68.
Saito Y. 2002. Ecological designs: promises and challenges. Environ. Ethics 24: 243–261.
Schimitschek E. 1968. Insekten als Nahrung, in Brauchtum, Kult und Kultur. *In:* Helmcke J.-G., Starck D. and Wermuth H. (eds.). Handbuch der Zoologie - eine Naturgeschichte der Stämme des Tierreichs. Akademie Verlag, Berlin, vol. 4, pp. 1–62.
Schmidt P.W. 1977. Die Sprachenfamilien und Sprachenkreise der Erde. H. Buske Verlag, Hamburg.
Spencer B. and Gillen F.J. 1899. The native tribes of Central Australia. Macmillan, London, UK.
Szent-Ivany, J.J.H. 1958. Papua New Guinea Insects. Proc. Tenth Int. Cong. Entomol. 1956, pp. 427–437, University Press, Montreal.
Szent-Ivany J.J.H. and Ujházy E.I.V. 1973. Ephemeroptera in the regimen of some New Guinea people and in Hungarian folk songs. Eatonia 17: 1–3.
Teuteberg H.J. 1991. Food patterns in the European past. Ann. Nutr. Metab. 35: 181–190.
Tindale N.B. 1953. On some Australian Cossidae including the moth of the Witjuti (Witchety) grub. Trans. Roy. Soc. South Australia 76: 56–65.
Tommaseo-Ponzetta M. and Paoletti M.G. 1997. Insects as food of the Irian Jaya populations. Ecol. Food Nutri. 36: 321–346.
Townsend P.K.W. 1970. Cited in Oliver (1989).
Umemura J. 1943. Cited in Schimitschek (1968).
Valonen N. (1975). Ueber die Methoden und Probleme der finnischen Volksnahrung. *In:* Valonen N. and Lehtonen J.U.E. (eds.). Ethnological Food Research. Vammalan Kirjapaino Oy, Vammala, Finland. pp. 302–318.
Waldron L.A. and Gallimore A.J. 1973. Pictorial depth perception in Papua New Guinea, Torres Strait, and Australia. Austr. J. Psycho. 25: 89–92.
Wallace A.R. 1883. The Malay Archipelago—the land of the orangutan and the bird of paradise. Macmillan, London, UK.
Watanabe H. (1983). Minami no dobutsu shi. Uchida Rokakuho, Tokyo (in Japanese).
Watanabe H. and Satrawaha R. 1984. A list of edible insects sold at the public market in Khon Kaen, Northeast Thailand. Southeast Asian Studies 22: 316–325.
Welsch R.L. 1975. American plains Indian ethnogastronomy. *In:* Valonen N. and Lehtonen J.U.E. (eds.). Ethnological Food Research. Vammalan Kirjapaino Oy, Vammalan, Finland, pp. 319–326.
Wurm S. 1970. Classification of Australian languages including Tasmanian. *In:* Sebeok T.A. (ed.). Linguistics in Oceania, Mouton, The Hague, vol. 8, pp. 721–778.
Zhou R. 1979. Insect studies of ancient China. Kunchong Zhishi, 16: 237–240 (in Chinese).

20

Edible Insects in the Laos PDR, Myanmar, Thailand, and Vietnam

JINTANA YHOUNG-AREE[1] AND KANVEE VIWATPANICH[2]

[1]*Institute of Nutrition, Mahidol University, Puthamonthon 4 Road, Nakhon Pathom 73130, Thailand. E-mail: nujya@mahidol.ac.th*
[2]*Dept. of Community Medicine, Faculty of Medicine, Thammasat University, Pathumthani 12120, Thailand. E-mail: kanvee@alpha.tu.ac.th*

Abstract

The contribution that insects make to life is enormous. During consultant missions to the Laos PDR, Myanmar, and Vietnam, the authors conducted an anthropological study of edible insects in these countries. Informal conversations were held with food insect vendors and consumers. Field notes pertaining to these missions and those obtained from rapid appraisals in Thailand were compiled and analyzed. The results demonstrate that the selected countries share common forest and watershed resources as they all form part of the Mekong River Basin. Due to comparable ecological and environmental conditions, as well as certain sociocultural practices, people living in these countries have similar preferences regarding insect foods. Approximately 164 insect species are eaten: these include beetles (61 species), termites (2 species), bugs (11 species), moths (47 species), cicadas (11 species), dragonflies (4 species), bees and ants (16 species) as well as crickets and locusts (22 species). The study confirmed that 44 species of insects are eaten in northeastern Thailand alone. The diversity of insect foods is greater in urban areas than in rural areas. Rural folk eat insects as their main dish, whereas urban people eat insects either as a main dish or snack, or both. Moreover, although insects are a nutritious food source, cooking methods may alter their quality. Rural consumers or indigenous migrants may obtain fewer calories from insect foods as they prefer meals that are low in fat. Urban consumers, however, have several different ways of preparing insect foods using fat (frying, frittering) and can obtain a greater variety of insects. At present, insect foods are increasingly in demand and the marketing of edible insects is becoming more lucrative. Insect food distributors may earn a net profit of approximately 800–1,200 Baht/day (19–30 US $) for those selling in slum areas and about 2,000 Baht/night (49 US $) , if they operate in tourist zones. On a commercial scale, entrepreneurs may earn as much as 600–700 million Baht per year (14.8–17.3 million US $ y^{-1}). However, the significant quantity of edible insects collected is having an adverse effect on the ecosystem and food chain, resulting in a biodiversity crisis. Nevertheless, the farming of insects as a human food source has real potential. For instance, studies are underway in Thailand on the farming of certain crickets, bamboo moths, and sago beetles. Modified technologies are needed, however, to yield the same quality of insects as those obtained from the natural environment. Financial support for edible insect research is one important means of maintaining biodiversity on the planet.

Key Words: edible insect, insect eater, food habits, insect breeding, income, the Laos PDR, Myanmar, Thailand, Vietnam

Introduction

Insects are vitally important in maintaining a balanced ecosystem. Being cold-blooded, they are more efficient in transforming plant biomass into animal biomass (Lindroth, 1993). Yet insects contribute to human life in ways that go beyond this function. For example, those who enjoy wearing silk can thank silkworms for their thread, and those who are partial to honey should be grateful to bees. Insects are also used as animal feed, particularly in poultry farming (WREN media, 2003; www.szgdocent.org/ff/f-arth3c.htm). Moreover, several insect species are now recognized for their medicinal properties, especially in China, Japan and Thailand (WREN media, 2003; Viwatpanich, 1999; and Zimian et al., this volume) and Korea (Pemberton, this volume). In poor rural families living in communities where insects abound, the family members often collect them in large quantities (e.g. locusts, ants and crickets) and sell them. In Thailand people use insects and their products for cosmetics and to produce novelty items such as brooches, which are made from dried metallic wood-boring beetles. Insects are also used in games (Meyer-Rochow, 1978, 1979) and as a means of predicting the future (Viwatpanich, 2003). Finally, in some countries, insects also confer social status. For instance, a man who specializes in harvesting honey is recognized as being extremely courageous (Tembo, 1993).

As a source of food, insects are nutritious and high in protein (Meyer-Rochow, 1976; Bukkens, 1997; Institute of Nutrition, Mahidol University, 2002). Of the 800,000 species of known, approximately 2,000 edible ones have been recorded (WREN media, 2003). Further, in Thailand, for example, edible insects can be prepared in 15 different ways (Viwatpanich, 1999).

Objectives of this Study

The people of Southeast Asia are traditionally renowned as insect eaters. Due to the importance of insects both as an alternative food source and a means of increasing income levels in poorer families, anthropological methodologies were used in this study to document the availability, use, value and demand for edible insects in four selected Southeast Asian countries, i.e., the Laos People's Democratic Republic (Laos PDR), Myanmar (erstwhile Burma), Thailand, and Vietnam. In each of these countries information was obtained through consultant missions and office visits.

Data was collected through in-depth interviews, observations, and informal discussions with village members, insect vendors and distributors, as well as consumers who frequent insect stores and outlets in these countries. Photographs were taken mainly at insect food outlets, as the opportunity arose. Fifteen insect distributors and 30 insect eaters participated in interviews. Informal discussions were conducted with two Burmese vendors, a Vietnamese hawker, a seller at an insect shop (Plate X, 1 and 2) in northeastern Thailand (Khon Kaen province), two Bangkok economically disadvantaged hawkers, and nine Bangkok street vendors. To determine income and related variables, discussions were also held with insect distributors in northeastern Thailand and Bangkok (in both rich areas and slums). Investigators also visited folk healers in the Kanchanaburi province in western Thailand to collect information about the medicinal properties of insects. All the data collected were synthesized based on content analyses.

The scope of the study followed the conceptual framework shown in Figure 20.1. Information was collected on the availability of insects in terms of whether farmed or still abundantly found in nature (mainly in rural areas), or whether obtained through wholesale or retail outlets (in urban areas). Data were gathered on the specific uses of insects (for instance, as food sources or in folk medicine) and their economic potential, along with firsthand information on how people consume insects and cooking methods. The nutritive values of insects were compiled from the Food Composition Database (Institute of Nutrition, Mahidol University, 2002).

Commonalities in the Laos PDR, Myanmar, Thailand, and Vietnam

The Laos PDR, Myanmar, Thailand, and Vietnam are Southeast Asian countries renowned for being rich in flora and fauna. Geographically, these nations share common fringing mountains and plateaus (Hirsch and Cheong, 1996). These countries and China share the Mekong River along a common border of 1,400 km. The Mekong Basin is a major source of natural resources, particularly for Thailand and the Laos PDR, as they share a common border of 900 km or 64% of

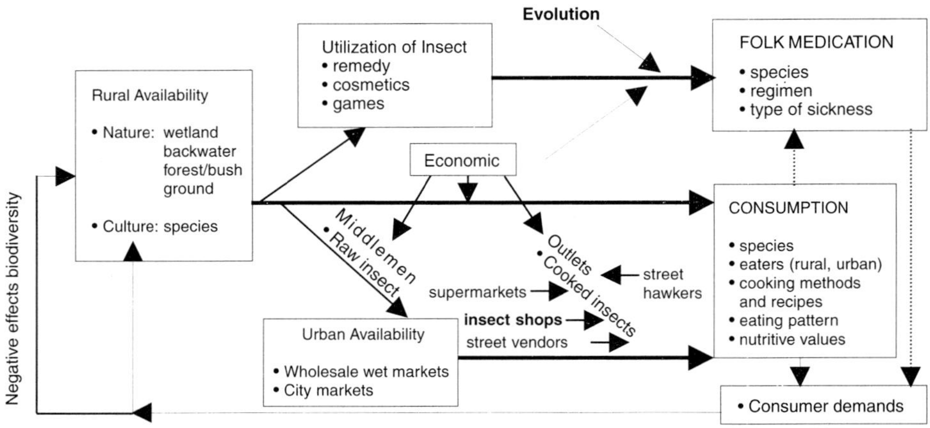

Fig. 20.1: Edible insects in Laos PDR, Myanmar, Thailand and Vietnam.

the Mekong's total length. It is estimated that about 35% of the water in the Mekong River comes from watersheds located in the Laos PDR. Approximately 202,000 km^2 out of a total land area of 236,800 km^2 in the Laos PDR territorial area (or 85% of its total land) is within the Mekong Basin. Likewise, about 183,000 km^2 of the Mekong River or one-third of Thailand is located in the Mekong Basin. It lies mostly in Thailand's northeastern provinces together with a portion of the nation's eastern and northern regions. In Vietnam, the Basin covers 29% or 96,000 km^2 of the country's total area. The Mekong Delta is crucial to Vietnam as it produces half of the nation's supply of rice and 40% of its total agricultural outputs. For Myanmar, only 24,000 km^2 of eastern Shan State is found within the Basin's boundaries, accounting for 3% of the total Basin area. In addition to the Mekong River, both the Irawadee and Sarawin Rivers provide resources for inhabitants along the border of Myanmar and Thailand.

Forests and watersheds are crucial in sustaining the ecosystem in the region. The Laos PDR and Myanmar are relatively rich in forests and watersheds compared to Thailand and Vietnam. In the Laos PDR, 45% or 11.6 million hectares of total land area are covered in forest. Approximately 70% of the nation's southern provinces and 25% of its northern provinces are covered in forest. In Vietnam, only 2 million hectares or about 26% of the total land area is forested.

The Basin's wetland areas serve as important habitats for migratory birds and resident wildlife as well as a variety of fish species. For instance, a survey by Meusch and Yhoung-aree (2003) identified approximately 200 species of aquatic animals found in the Southern province of the Laos PDR. This evidence suggests that the aquatic biodiversity of the Mekong Basin is one of the richest in Asia. Because of the ecological conditions found in these countries, the areas are home to a wide variety of insects, some of which are utilized as human food sources. Due to a lack of written documentation, it is difficult to determine when the local populations started eating insects. Nonetheless, poems, verses, and folk medicine manuscripts found in Thailand, the Laos PDR, and Myanmar indicate that insects have been traditionally valued sources of food and medicine (Sing, 1981; Washirarungsi and Sitimanodhamma 1982; pers. comm. with folk healers).

The prominence of insects in the local diet is related to the availability of other food sources. For example, in areas rich in such conventional foods as fish, frogs and other animals, insects serve only as an alternative food source, as occurs in some rural areas of the Laos PDR, Myanmar, and Thailand. In areas of chronic food scarcity such as in northern and northeastern Thailand, insects are a common food item.

Types of Edible Insects, Their Availability, and Acquisition

In the four countries studied, folk classifications of insects are based on their size, mobility, and habitat (Viwatpanich, 1999). Insects are considered small in size and dependent on wings or legs for mobility. Those who eat scorpions classify it as an insect because it lives in cryptic places. In folk classification, an aquatic organism with legs is considered an insect; it differs from such creatures as fish, snakes, snails, eels, etc, which are not classified as insects. Therefore folk classifications of insects include animals that can live or hide in the air, on or under the ground, in branches/leaves of trees, bushes, shrubs, and in water. These folk classifications determine collecting methods and techniques.

Insects are eaten at different stages of their life cycle, the stages differing among species (Jamornman, 2003). For instance, the eggs of ants, giant water bugs, and cicadas are commonly eaten, while more developed stages such as caterpillars, grubs, pupae and larvae of ants, dragonflies, moths, beetles, bees, hornets, silkworm, and weevils may be preferred. Adult insects such as beetles, crickets, ants, bees, and hornets are also commonly consumed (Table 20.1). Jamornman (2003) compiled a list of approximately 164 species of edible insects in Thailand: we found that most of these species are also commonly consumed by people living in the other Southeast Asian countries (Meyer-Rochow, this volume). These insects include beetles (61 species), termites (2 species), bugs (11 species), moths (47 species), cicadas (11 species), dragonflies (4 species), bees and ants (16 species) and crickets and locusts (22 species). A study by Viwatpanich (1999) confirmed that 44 species of insects are eaten in the Ubon Ratchathani province (in northeastern Thailand) alone.

Aquatic Insects

In general, these insects live in still water, swamps, and river shallows. They survive among the aquatic plants in swamps, depending on algae and small organisms for food. Some species of insects are found in clear water, e.g. the

Table 20.1: Edible insects based on adult habitats

Aquatic habitat		Ground habitat		Tree/Branch habitat	
Common name	**Scientific name**	**Common name**	**Scientific name**	**Common name**	**Scientific name**
Backswimmer	*Anisops* sp.	Dung beetle	*Copris nevinsoni* Waterhouse	Band-wing grasshopper	*Ducetia japonica* Thunburg
Crawling water beetle	*Eretes sticticus* L.	Dung beetle	*Paragymnopleurus aethiops* Sharp	Bombay locust	*Patanga succincta* L.
Damselflies	*Ceriagrion* sp.	Dung beetle	*Heliocopris bucephalus* F.	Cicada	*Meimuna opalifera* Walker
Giant water bug	*Lethocerus indicus* Lep.-Serv.	Dung beetle	*Heliocopris dominus* Bates	Cone-headed grasshopper	*Euconocephalus* sp.
Predaceous diving beetle	*Cybister rugosus* Macl.	Field cricket	*Gryllus bimaculatus* De Geer	Long-horned grasshopper	*Pseudophyllus titan* White
True water beetle	*Cybister limbatus* F.	Ground cricket	*Acheta confirmata* Walker	Long-horned beetle	*Batocera rubus* L.
River skimmer	*Crocothemis* sp.	House cricket	*Acheta testacea* Walker	Long-horned beetle	*Threnetica lacrymans* Thoms.
Water scavenger beetle	*Hydrous cavistanum* Bedel	Mole cricket	*Gryllotalpa africana* Beauvois	Mantid	*Hierodura* sp.
Water scorpion beetle	*Laccotrephes ruber* L.	Short-tailed cricket	*Brachytrupes portentosus* Lichtenstein	Licht	Honeybee *Apis florae* F.
		Scarab beetle	*Holotrichia* sp.	Hornet	*Polistes stigma* F.
		Scarab beetle	*Anomala cupripes* (Hope)	Metallic wood-boring beetle	*Sterocera acquisignata* Saunders
		Spider	*Melophaeus* sp.	Moths (bamboo caterpillars)	*Omphisa fuscidentalis* Hampson
		Termites	*Termes* sp.	Red ant	*Oecophylla smaragdina* F.
		Silkworm pupae*	*Bombyx mori* L.	Rhynoceros beetle	*Xylotrupes gideon* L.
				Skipper	*Erionata thrax thrax* L.

Table 20.1: (contd.)

Table 20.1: (contd.)

Aquatic habitat		Ground habitat		Tree/Branch habitat	
Common name	Scientific name	Common name	Scientific name	Common name	Scientific name
				Slant-faced grasshopper	*Acrida willemsei* Pirsh
				Spur-throated grasshopper	*Chondracris rosea* DeGeer
				Stink bug	*Tesseratama javanica* Thunberg
				Tortoise beetle	*Sagra femorata* Drury
				Wasp	*Vespa* sp.
				Weevil	*Hypomeces squmosus* F.
				Weevil	*Rhynchophorus ferrugineus* Olivier

* Produced by culture

backswimmer, giant water bug, predaceous diving beetle, and river skimmers. Others live in eutrophic water, e.g. the crawling water beetle, damselflies, water scavenger beetle, and the water scorpion beetle. Although people generally prefer adult insects for food, premature/young insects are also consumed. Aquatic insects are collected using simple tools such as the single holder dip-net ("Sa-whing" in both the Thai and Lao language) and the dual holder dip-net ("Yhang" in both languages). The net used depends on the size of the swamp and backwater, as well as the gender of the hunter. The dual holder dip-net is more suited to large backwaters, while the single holder dip-net is preferred in small swamps. Moreover, the dual holder dip-net is generally used by men, whereas women who collect insects use the single holder dip-net. Skillful collectors can also collect insects by hand. Of all aquatic insects, the largest are the giant water bugs. As these appear at night, a blue fluorescent light is used to lure them into the net: they are then stored in closed baskets.

Ground Insects

Ground insects usually thrive in wetland climates with high humidity and muddy soils. Insects living underground have stronger legs; parts of their legs have a harrow-like shape, enabling them to dig deep holes in which to live. Mole crickets can dig a hole 10–20 cm deep. Some crickets are able to dig relatively deeper burrows (approximately 40–60 cm in depth) to avoid sunlight. Dung beetles breed and grow under piles of cattle dung and human wastes. Spiders make hollows which are 5–8 cm wide and 30–50 cm deep around grass roots or shrubs and cover the opening with silk. Ground insects normally obtain food from the roots of trees, young plants, tubers, and young leaves. Thus in certain circumstances they are termed pests by plantation owners as they destroy plants such as rice seedlings, young trees, and nursery crops, among others.

To collect these types of insects, it is necessary to understand their habitats and behavior. Generally, a spade can be used to dig them out of their burrows. Insects such as crickets, beetles, and winged termites can be lured out at night using lamps. They can be collected by hand or caught using water-filled bins or closed baskets under lamps.

Tree/Shrub/Bush Insects

Insects in this group often have strong wings to protect them from predators. Viwatpanich (1999) reported that villagers in northeastern Thailand are able to collect 22 types of edible insects from trees. When it rains and young leaves begin to emerge, in particular the leaves of tamarind, mango, sago, and coconut trees, these insects usually emerge from their ground dwellings at night to feed on them. Grasshoppers and locusts also belong to this group and are common pests that eat young rice and corn plants.

In general, these edible insects are large enough to be caught by hand. However, simple instruments (e.g., bamboo sticks, torches, resin, buckets, closed baskets) can also be used to collect them. Although such insects can be caught either during the day or at night, most are caught during the day. A stick tipped with glue or resin is used to hunt cicadas and stinkbugs. When the hunting is done at night a torch is needed. A common method used to collect tree-dwelling insects is to shake the trees so that the insects drop into water buckets placed priorly on the ground; the insects are then stored in closed baskets.

Caterpillars of the nocturnal moth *Omphisa fuscidentalis* are commonly eaten by ethnic groups residing along the borders of Thailand, Myanmar, and the Laos PDR. Various ethnic groups call the bamboo caterpillars (*Omphisa fuscidentalis*) by different names, e.g. "Dae," "Mangmae", or "Nae." Thai people call it "Rode Duan" or *express train* due to its trainlike appearance. This species is mostly found in the deep jungles in northernmost Thailand, especially in the provinces of Tak, Chiangrai, Mae Hongson, Nan and Prae. These areas are adjacent to the Shan State of Myanmar. The Karen, one of the minority groups residing along the border of Thailand and Myanmar, are known to be especially partial to bamboo caterpillars. Their indigenous hunting wisdom has taught them that these caterpillars live between the bamboo nodes and that they make a distinctive sound. They know that when the bamboo leaves turn yellow, caterpillars are to be found inside the stems. This occurs as the caterpillars eat the inner layer and suck nutrients from the bamboo. To harvest the caterpillars, the entire bamboo node is cut into sections. Once the node is opened, the caterpillars are extracted and boiled for a short time to preserve them.

Fierce insects such as red ants, bees, hornets, and wasps normally live in or on large trees located in less disturbed areas. For red ants, hunters use a bamboo stick to stir up the nest which causes the ants to drop into either a bucket of water or onto the ground (where a cloth or leaves have been laid). After chasing or winnowing the adult ants out, the pupae and young ants are collected. To harvest bees, hornets, and wasps, hunters should take care to protect themselves from the stings of angry swarms. In some cases, hives are burned by lighting a quick fire and the white grubs collected and eaten raw immediately.

On the whole, edible insects are available in season (Fig. 20.2) and various species can be obtained from different habitats to provide a regular food source for an entire year. They are collected mostly from natural sources by using simple techniques passed down through generations.

Consumption of Edible Insects

Cooking Methods and Recipes

Asian food is well known throughout the world and many Asian culinary cultures have incorporated insects in their menus. Yhoung-aree et al. (1997) have

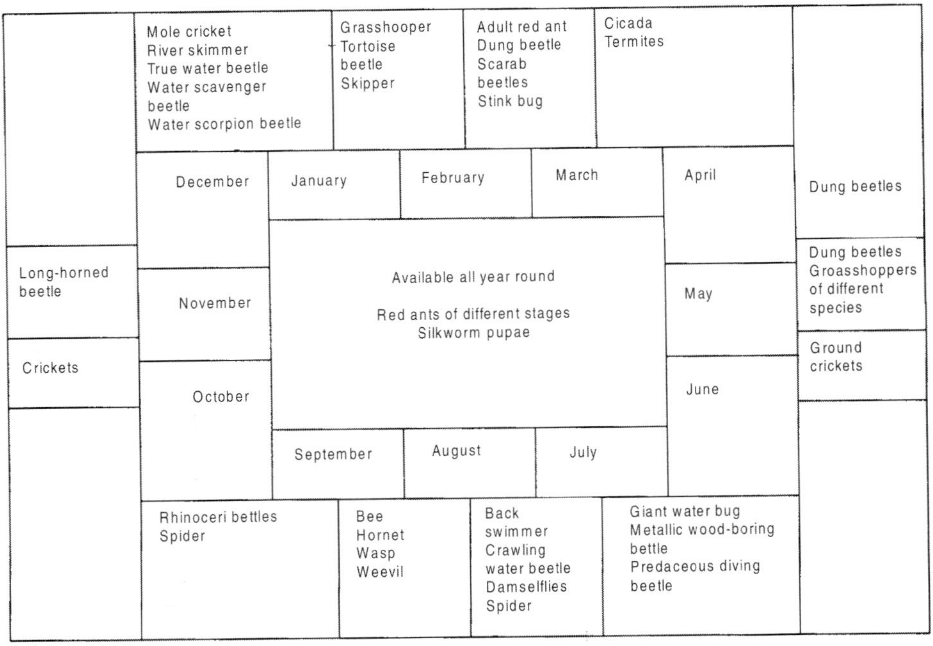

Fig. 20.2: Peak availability of edible insects.

Fig. 20.3: Fried scarabaeid beetles (photo K. Viwatpanich).

Fig. 20.4: Crickets (*Gryllus mimaculatus*) (photo B. Mukthabhan).

Fig. 20.5: Caterpillars, *Omphisa fuscidentalis* (photo B. Mukthabhan).

documented cooking methods and recipes for edible insects in Thailand. To date, most of these recipes are still popular in the Laos PDR and Thailand. Some dishes are common in Myanmar (e.g., fried crickets, roasted bamboo caterpillars) and Vietnam (e.g., fried crickets and fried scorpions). However, recipes such as insect burgers and insect sandwiches are becoming a specialty in Thailand. Some of the cooking methods used to prepare edible insects are given below.

Barbecuing, smoking: These methods can be used to cook crickets, giant water bugs, and beetles. In the process of hunting bees, hornets and wasps, smoking is used to chase these insects away. As a result, the hives containing the grubs are precooked and ready to eat.

Burger: This western style of food preparation is used for silkworm and bamboo caterpillars. These insect burgers are available in tourist areas in Bangkok. The introduction of this food to Thai society is an indication of the diversification of food varieties to which local ingredients and eating habits can readily be adapted.

Chili sauce/paste: Minced insects are mixed with chili, garlic, onion and salt (or fish sauce) and commonly eaten by indigenous Thais and Laotians and ethnic groups of Burma. Fermented fish is often added in Thai-Laotian recipes. The insects suitable for chili-insect sauces are beetles, cicadas, crickets and male giant water bugs (male bugs exude a unique smell; Plate X, 2). Bamboo caterpillars can also be cooked as a paste.

Deep-frying: This is the main cooking method used for bamboo caterpillars, beetles, bugs, crickets, grasshoppers, locusts and scorpions. Pepper and sometimes spices are added during frying, thereby enhancing the distinctive flavor of the insects. Some cooks may decorate the dishes with fried leaves of bitter lime (Bai makrood in Thai) or pandanus (Bai-tuey in Thai) to enhance the aroma. This dish goes well with beer and is especially popular in urban outlets.

Frittering: Cricket and locust fritters are commonly available in urban food shops. However, this recipe is said to be less appealing than those which require deep-frying.

Homoke (steaming/charcoal baking in a banana leaf): Insects are mixed with chili-based ingredients, then wrapped in a banana leaf and cooked either by steaming or baking. In rural areas, people prefer charcoal baking because of its aroma and taste. Red ants, bees, hornets, silkworm, wasps, and premature insects, as well as aquatic insects taste better baked or steamed.

Poaching: This method is commonly used to cook ant eggs/pupae. Spices are added to improve the taste of this dish which is sometimes prepared in a manner similar to scrambled eggs. The technique is also known as "oil free-frying".

Roasting: All kind of insects, especially beetles, can be prepared by roasting, except bees and wasps. To prepare delicious roasted insects, a pan is heated with a little oil, the insects added and sprinkled with salt and pepper, then baked.

Salad: An excellent insect salad can be made using immature red ants and their eggs/pupae, cicadas, and premature aquatic insects mixed with finely chopped lemon grass, onion, fresh chili, lemon juice, sugar and fish sauce (or salt) to taste. Fermented fish makes this dish even more delicious and the addition of a garnish such as mint, chives and coriander renders it unique. A native recipe found in the northern and northeastern regions of Thailand and the Laos PDR adds roasted ground rice to insect salads.

Sandwich: This dish can be found in Thailand. Like burgers, insect sandwiches have recently been introduced in Thailand. The preparation of an insect sandwich is similar to that of other types of sandwiches found in western countries. Insects are used instead of chicken, beef or ham. Silkworm and bamboo caterpillars can also be used in this manner.

Sautéeing: This is not a common method of cooking insects, except in the case of wasps and a mixture of immature insects, which are sautéed with cabbage.

Soup/curry: This method is appropriate for ant eggs, and beetles, locusts, and wasps. Some people add coconut milk to improve taste and increase the nutritive value of the dish.

It is obvious that insects can be used to make a variety of delicious dishes. Indigenous food on sale in urban areas is often served to local migrants. Tourists will find that they can choose insect dishes from imported recipes, such as burgers and sandwiches, as well as from a variety of local, indigenous recipes.

Eating Patterns

City and rural people prepare insects in different ways, depending on their preferred eating habits. In rural areas, indigenous diets are usually low in fat. The most popular insect recipes involve roasting, poaching, addition of chili paste, and a spicy soup without coconut milk. Other foods are likewise cooked without oil. In general, insects are served as the main dish at family meals in rural areas. Because the availability of insects varies seasonally, the quantity eaten may not be high. As reported by Yhoung-aree et al. (1997), people in rural areas consume insects as food more during the dry season than the rest of the year. In vulnerable groups in particular, edible insects assist in sustaining household food supplies.

Insect eating in urban areas is not the same today as in the past (Yhoung-aree et al., 1997). Urban eating patterns show an increasing trend in the demand for insects as food and in the quantity eaten. Insect foods are now available in most supermarkets, specialized insect food shops, and from insect food hawkers and street vendors.

In urban areas, insect eaters are generally migrants who come from various regions of Thailand as well as from other countries. Thailand is an example of a country with a heavy rural-to-urban migration flow. An estimated 3,159,069 migrants (excluding illegal migrants) had moved to urban areas by 1997 (National Statistics Office, URL: http://www.nso.go.th/eng/stat/migrant/migtab1.htm).

The Thai Dept of Immigration estimated that illegal migrant workers reached 525,480 in 1994. The majority had arrived from neighboring countries, mainly Cambodia, the Laos PDR, Myanmar and Vietnam. The reasons for their immigration to Thailand concerned matters of trade, political conflict, natural or man-made disasters, and religious cultural traditions (Asia Pacific Migration Research Network, URL: www.unesco.org/most/apmrnw14.htm).

Migrants who eat insects can be divided into three groups. First, those who are usually indigenous insect eaters in their native areas. The largest group are people originating from the northern and northeastern provinces of Thailand and the Laos PDR who commonly eat insects as a main course, cooked with chili paste (crickets, giant water bugs), roasted (beetles, silk pupae) or fried (bamboo caterpillars, beetles, crickets, mole crickets, and locusts). These dishes can also be bought from street vendors who most likely come from the same regions. In slum areas, only fried locusts distributed by hawkers are available. Street vendors may also sell precooked hives containing grubs of bees, hornets and wasps at affordable prices.

Secondly, in Thai insect food sellers can be found in areas frequented by local migrants and tourists such as nightclubs, night bazaars, and bus stations (the last especially in Vietnam). In these areas, several varieties of fried insects are available, generally bamboo caterpillars, beetles, crickets, mole crickets, female giant water bugs, locusts, silk pupae, and scorpions. They are normally eaten as snacks and accompanied by beer. Insects sold in these areas are therefore rather costly (Table 20.2). For instance, fried female giant water bugs cost 5 Baht per piece, whereas one fried scorpion costs 10 Baht (40–41 Baht = one US $).

Lastly, permanent migrants who marry and settle in cities also introduce their insect eating customs into the diets of their new families. They usually eat insect dishes in their main meals, e.g. as insect-chili paste, or as snacks such as fried bamboo caterpillars, crickets, and locusts. Urban insect consumers usually buy insect foods in supermarkets. The most common types available from these outlets are chili paste with male giant water bugs, and individual fried bamboo caterpillars, beetles, crickets, mole crickets, male giant water bugs, locusts, silk pupae, and scorpions.

In conclusion, people living in rural areas eat insects as main dishes, whereas urban dwellers eat insect foods either as a main dish, a snack, or both (e.g. among indigenous insect-eating migrants). Insect dishes are more diverse in urban areas than in rural. Nutritionally, rural insect eaters or indigenous migrants to urban areas may obtain fewer calories from insect foods than most urban eaters, as they prefer recipes low in fat. Urban dwellers contrarily consume a wider variety of insects and follow more diverse methods of cooking, as well as recipes with more fat.

Table 20.2: Nutrient content and price of insect foods sold in urban areas of Bangkok (Khoa San Road and Or-Tor-Gor) and Chiang Mai (40.5 Baht = 1 US $)

			Nutrient content in 100 g			Baht/100 g	
Vernacular name	English name	Scientific name	Kcal	Protein (g)	Fat (g)	Khoa San Road	Or-Tor-Gor Market
1. Fried insects:							
Mang Tub-tao	Predaceous diving beetle	*Cybister limbatus* F.	301	24.3	18.6	33.25	–
Mang Ki-noon	Scarab beetle	*Holotrichia* sp.	215	15.5	12.9	34.50	–
Jing-rheed	Crickets	*Acheta confirmata* Walker	465	14.9	17.0	38.50	51.28
Mang-dana	Giant water bugs	*Lethocerus indicus* Lep.-Serv.	303	22.9	19.8	78.00	–
Tukkatan Patunkka (or Esaan)	Bombay locusts	*Patanga succincta* L.	221	16.6	14.8	66.75	69.00
Tukkatan E-Moh	Spur-throated grasshoppers	*Chondracris rosea* DeGeer	290	23.8	17.6	–	62.50
Rod-duan	Bamboo caterpillars	*Omphisa fuscidentalis* Hampson	644	25.5	55.3	90.90	133.30
Mod daeng	Red ants	*Oecophylla smaragdina* F.	231	16.1	15.0	37.00	–
Non-Mhai	Silkworm pupae	*Bombyx mori* L.	241	14.1	18.5	30.30	34.00
Mang-pong	Scorpions	Keterometrus longimanus	NA	NA	NA	294.00	127.39

			Nutrient content in 100 g			Chiang Mai market (Baht/100 g)
			Kcal	Protein (g)	Fat (g)	
2. Chili paste containing...						
Namprig-Jigoong	Crickets	*Acheta confirmata* Walker	79	4.6	1.3	20.00
Namprig Mangda	Giant water bugs	*Lethocerus indicus* Lep.-Serv.	90	4.3	1.3	33.00
Namprig-Mangjon	Mole crickets	*Gryllotalpa africana* Beauvois	101	11.7	4.3	20.00
Namprig-Toh	Wasps	*Vespa* sp.	108	12.1	2.6	35.50

Value of Edible Insects

Nutritive Values of Insects

The Thai Food Composition Database provides nutritive values of common edible insects in Thailand, which are summarized in Table 20.3. Common food items are presented in this Table as references to compare the nutrients found in edible insects. The protein content of conventional animal food sources are used: 3.3 g per 100 ml milk, 12.8 g per 100 g for eggs, and 18–28 g per 100 g for common meat (pork and chicken). Processed foods such as fermented fish are common to city dwellers and recognized as a secondary staple food for a number of ethnic groups. The protein content of ham is 32 g while fermented fish has only 6 g protein. Compared with these figures, insects are a good source of protein. As evaluated by Sirichakwal and Sungpuag (1982) and Ngunboonsri et al. (1989), amino acid contents for the protein quality of scarab beetles, dung beetles, cicadas, locusts and bamboo caterpillars were about 40–80% of reference proteins (Joint FAO/WHO Expert Committee, 1973). With respect to calories, levels found in insects are comparable to those found in conventional animal sources. Table 20.3 also shows that insects are good sources of minerals, especially phosphorus, potassium, iron, and vitamins (such as vitamin B_1, B_2 and niacin). Moreover, the addition of oil during frying, roasting, and sautéeing (with vegetables) increases the nutrient contents of insects as food. If they become acceptable to people living at poverty levels, these dishes could be extremely valuable contributions to their food and nutritional requirements.

Recorded Health Attributed to Edible Insects

Insects and their by-products have been used in folk medicines in many cultures throughout the world. In Brazil, 42 insects are reportedly used in folk medicine (Costa-Neto, 2002). In Indo-Chinese countries, indigenous ethnic groups have also used insects as medicine. In Thailand, anecdotal evidence exists on the use of insect therapy among rural people. Folk medicine probably evolved before modern or conventional medical services and provides options to people, especially the poor. The Institute of Thai Traditional Medicine has put much effort into documenting research findings of folk regimens. For instance, Subchareon et al. (2001) conducted a survey on the Thai folk system of medicine in relation to insects in four regions, covering 14 provinces. The insects used by folk healers belong to 11 orders. The most common species used are bees, carpenter bees, stingless bees, sphecid wasps, wasps, large red ants, termites, cockroaches, Mod-Rii, the cultivated lac scale (*Kerria lacca*), hornets, bedbugs, giant water bug, grasshoppers, water scorpion beetles, various subgroups of dung beetles, and fluted scales.

Some of the therapeutic uses of some common medicinal insects are the following. The juice of red ants supposedly aids detoxification of the blood. In women who have just given birth, red ant juice is said to be useful in restoring the uterus and jettisoning any aftermath in the uterine canal. The juice is also used to stimulate pulse and heartbeat. Red ant juice is likewise useful in arresting hemorrhage quite common in miscarriages. A mixture of red ants and mint leaves puportedly relieves asthmatic symptoms. In many cases, the adult red ants are squeezed and their aroma inhaled to alleviate feelings of dizziness. Some people have even been known to chew them (Jongleebhan, 2003).

It is believed that the water scorpion bug (*Laccotrephes ruber* L.) helps children recover from urinary incontinence. Insect therapy may also prove effective in chronic illnesses. For instance, consumption of roasted dung beetles is said to cure children who suffer from "saang," the folk term for malnutrition. This word is commonly used among rural Laotians and Thais. In addition to prescribed drugs, eating cockroaches reputedly assists in recovery from kidney diseases (Viwatpanich, 1999).

Economic Value

Edible insects may also provide economic benefits to humans, either directly or indirectly. Indirectly, country folk who eat insects as their main dishes reduce family expenses on food. Farmers whose crops are attacked by grasshoppers may recover some of their losses by selling the insect pest or using it as food. There is also evidence of direct economic benefits for insect collectors/middlemen. In times of scarcity, native collectors may earn 120 Baht kg^{-1} for crickets, 270 Baht kg^{-1} for locusts and 300–370 Baht kg^{-1} for bamboo caterpillars. Naturally, middlemen may earn even more. As cold storage is required during the peak season, the business is most advantageous to those who can provide cold storage facilities.

In any case, the edible insect business is a lucrative one. According to fresh insect distributors at the Klong Tuey market and other boat markets in Bangkok, there is an increasing demand for edible insects in the urban marketplace. Middlemen supply two deliveries a day, at 3:00 a.m. and 1:00 p.m. Up to a total of 2,000 kg of insects per day are supplied to sellers at the Klong Tuey (Bangkok) market alone. Their profit from selling fresh insects is estimated to be around 10%. Even though this amount is said to be low, the goods have a high turnover rate.

Distributors of urban insect foods have commented on the positive aspects of this business. Small-scale street vendors and insect hawkers in a Bangkok slum revealed that they earn about 800–1,200 Baht per day or about 50–60% profit. Those found in Bangkok's tourist night spots can earn a similar profit. A rough estimate of their net profit would be about 2,000 Baht per night. In an insect food shop in northeasterm Thailand, bamboo caterpillars are imported from the province of Chiang Rai (northern Thailand), and grasshoppers, crickets, and mole crickets are brought in from the Ubon Ratchathani province

Table 20.3: Nutritive values of edible insects cooked by blanching (per 100 g insects).

Common name	Scientific name	Water	Macronutrient				Micronutrient									
		Water g	Kcal g	Protein g	Fat g	CHO g	Vit A RE	B_1 mg	B_2 mg	Niacin mg	Ca mg	P mg	Fe mg	Na mg	K mg
Reference items															
Milk, sterilized		89.7	61	3.3	3.4	4.2	35	0.03	0.24	_	114	82	0.2	52	_
Eggs (fresh), hen		72.8	155	12.8	10.8	1.6	368	0.15	0.61	0.4	38	230	3.1	178	111
Pork (minced)		63.0	229	18.2	17.0	0.7	2	0.96	0.10	4.6	5	170	0.9	160	300
Chicken (meat)		60.8	193	28.6	8.7	0.1	5	0.06	0.22	6.7	7	171	0.6	80	101
Ham (pork)		61.6	168	32.2	3.6	1.7	NA	0.62	0.29	2.8	9	170	2.0	79	230
Fermented fish (flesh)		60.5	97	6.1	3.7	7.7	NA	NA	NA	NA	683	400	4.4	7,180	175
1. Aquatic insects															
Giant water bug	*Lethocerus indicus* Lep.-Serv.	63.2	182	19.8	8.3	7.1	NA	0.09	1.5	3.9	44	226	13.6	84	192
Predaceous diving beetle	*Cybister limbatus* F.	61.2	180	21.0	7.1	7.9	NA	0.31	3.51	6.9	37	205	6.4	62	198
2. Ground insects															
Dung beetle	*Paragymnopleurus aethiops* Sharp	68.4	136	17.2	4.3	7.2	NA	0.19	1.09	3.4	31	158	7.7	293	288
Ground cricket	*Acheta confirmata* Walker	67.1	188	17.5	12.0	2.4	NA	0.11	0.8	3.2	NA	143	NA	NA	NA
House cricket	*Acheta testacea* Walker	71.4	134	12.9	5.5	8.1	NA	0.36	1.91	3.1	76	185	9.5	87	306
Mole cricket	*Gryllotalpa africana* Beauvois	71.2	136	15.5	6.3	4.4	NA	0.2	1.89	4.8	76	254	41.7	97	268
Short-tailed cricket	*Brachytrupes portentosus* Licht	73.3	125	12.8	5.7	5.7	NA	0.26	1.78	2.3	88	163	14.4	57	277

Table 20.3: (contd.).

Table 20.3: (contd.)

			Macronutrient				Micronutrient								
Common name	Scientific name	Water g	Kcal g	Protein g	Fat g	CHO g	Vit A RE	B_1 mg	B_2 mg	Niacin mg	Ca mg	P mg	Fe mg	Na mg	K mg
Scarab beetle	*Holotrichia* sp.	74.1	98	13.4	1.4	7.9	NA	0.29	1.19	4.0	23	207	6.4	465	463
Winged termites*	*Termes* sp.	71.6	NA	65.5	16.0	7.1	NA	NA	NA	NA	NA	NA	NA	NA	NA
3. Tree/bush insects															
Red ants (young)	*Oecophylla smaragdina* F.	66.1	194	12.7	12.5	7.7	NA	0.31	0.71	3.3	23	173	3.4	50	168
Bombay locust	*Patanga succincta* L.	61.1	169	20.6	6.1	7.9	NA	0.23	1.86	4.6	35	238	5.0	267	237
Spur-throated grasshopper	*Chondracris rosea* DeGeer	76.7	105	14.3	3.3	4.6	NA	0.19	0.57	6.7	28	150	3.0	32	217
Hornet pupae	*Polistes stigma* F.	72.8	140	14.8	6.8	4.8	48	0.02	0.89	3.0	46	120	0	0	0
Hornet, grubs		71.7	141	13.7	6.3	7.4	34	0.01	0	3.0	61	142	0	0	0
Moths (Bamboo caterpillars)—fried	*Omphisa fuscidentalis* Hampson	4.5	644	25.5	55.3	11.0	0	0	0	0	41	356	2.7	609	674
4. Culture insect															
Silkworm pupae	*Bombyx mori* L.	75.3	127	12.2	7.0	4.0	NA	0.12	1.05	0.9	42	167	1.8	14	139

Source: Food Composition Database for INMUCAL Program, Institute of Nutrition, Mahidol University, 2002.

* Klinhome et al., 1984.

NA: no data available.

(northeast), the Laos PDR and Cambodia. The price of these insect foods is similar to those sold in Bangkok.

Vendors of edible insects are pleased with their urban business as it brings in a good return. The nutritive values obtained from insects, however, are costly. For instance, 100 kcal obtained from bamboo caterpillars cost 14.1 Baht (see Table 20.2). Due to the high demand, the supply of fresh insects from domestic sources is inadequate. In past years, farmers suffered from insect pests that destroyed their crops (e.g. rice, corn). Some farmers offset this problem by including some of the insects in their diet. Nowadays, due to the lucrative business of the edible insect market, these farmers prefer to grow corn to actually feed the pests, e.g. grasshoppers, because the income earned from selling grasshoppers is much higher than that from corn, according to the owner of the Malang Esaan Insect Shop.

Perspectives for Edible Insects

Edible Insect Markets

Insect foods are important not only for country dwellers, but also for urban populations. The demand for insects for domestic consumption has increased substantially, especially in Thailand. In a workshop "Edible Insects" held on National Agriculture Day, February 3, 2003, four entrepreneurs revealed how the edible insect business is now a lucrative one in Thailand.

The first entrepreneur stated that he started his business in 1987 when an infestation of grasshoppers began destroying rice farms in Thailand's central provinces. As a result, he became a street vendor of fried grasshoppers (Plate X, 3). He built up a good business, although he had to keep moving from place to place in order to follow the significant grasshopper infestations. During the first six years of business, he was able to sell grasshoppers initially at 25 Baht kg^{-1}, an amount that later increased to 70 Baht kg^{-1} in the 6th year. At that time, his business was located in the western province of Kanchanaburi. Immediately following this, 5-year long infestations of grasshoppers occurred in Prachinburi province (in the east) to which he moved. He continued his business in this province and then moved to Lopburi (upper central) and Pitsanulok (lower north) respectively. During the 10th year in Pitsanulok, he earned an average of 150 Baht kg^{-1} for grasshoppers and 300 Baht kg^{-1} for bamboo caterpillars, his best-selling insect. Starting out on a small scale, his business has grown so much that he is now using cold storage to store insects of different species (giant water bugs, crickets, mole crickets, predaceous water beetles). Moreover, he pointed out that the supply of insects from domestic sources is so inadequate that insects have to be imported from the Laos PDR, Myanmar, and Cambodia. During the peak season approximately 100,000 kg y^{-1} crickets are stored. They are obtained from collectors who capture them from natural sources. The investment cost for crickets is between 270–300 Baht kg^{-1}. Bamboo caterpillars imported from Laotian and Burmese collectors cost about 300 Baht kg^{-1}. The quantity of bamboo caterpillars supplied is approximately 200,000 kg y^{-1}. He also noted that insects collected from the wild are preferred because they are believed to be healthier and taste better. Hence, they are more expensive.

A second entrepreneur shared information regarding his insect franchise. From the consumer market he was able to develop eight insect food products that he packages attractively. For the first 15 days after he started his business, he earned 50,000 Baht d^{-1}. Moreover, he began with 60 outlets, a number that expanded in the second month to 90 outlets. His products are now readily found in supermarkets and tourist spots throughout the country. They can also be delivered to consumers by post. Canned insect foods are exported to mainly Asian countries. The business has been encouraged by the government's policy for Small and Medium Entrepreneurs (SME). He estimates that his business requires more than one million kilograms of crickets per year, which are collected from the wild or cultured. Ten middlemen are involved. They collect the insects along the borders of the Laos PDR and Myanmar in northern Thailand and the border of Cambodia in eastern Thailand. The cost for these insects is 80 Baht kg^{-1}. The domestic supply, especially from insect farms, is inadequate due to low productivity. The wholesale cost for insects from farms is about 130 Baht kg^{-1}, slightly higher than that supplied by the border hunters. Overall, his company earns 600–700 million Baht per annum.

The third entrepreneur breeds crickets. His farm is situated on the outskirts of Bangkok (province of Pathumthani). He feels that crickets are in high demand, both for human consumption and for animal feed. Although he is knowledgeable about cricket breeding, disease and pests have caused a high mortality rate. Consequently, he is advocating further research to improve the productivity of crickets.

The last speaker worked in the marketing area for a famous silk company in Thailand. He shared information on the economic crisis that his company faced in 1997. At that time, market prices of silk textiles dropped dramatically. His company's crisis also affected at least 4,000 contract silk farmers in the northeastern provinces. The company modified its production strategy: instead of harvesting silk cocoons for textile purpose, the farmers were trained to harvest them for human food. During their life cycle, there is a stage at which silkworms produce thread to develop cocoons. If the worms are harvested about 5 days before that stage, they are tastier when cooked. The silkworms at this stage look like "bamboo caterpillars" and are therefore called "Rod Dum Esaan,", which roughly translates as "a truck of hard-working capacity".

Edible Insect Supplies

The use people make of forests as a food source, which includes edible insects, has put significant pressure on the environment. The Royal Government of

Thailand is aware of this problem and of the urgent need to conserve the country's insect natural resources. Hence, the SME policy and research encourage inclusion of both edible insect breeding and product development. For instance, native farmers in the south have tried to cultivate sago grubs because they believe these insects provide them with a good income. Much effort has been made with research and development to raise bees, crickets, and bamboo caterpillars because they are already popular foods and assure a lucrative business.

Bees and honey: In 1998, an estimated 20,777 kg of edible insects were collected from a conservation forest (*Dong Phu*) in the Sisaket province (Northeast, Thailand) with an area of 13,282 rai (or 2,125 hectares) (Subansenee et al., 1995). Bees and honey are important products obtained from forest areas. In Thailand, four species of bees were reported to provide a good yield. In 1990, the total annual honey production was about 2,000,000 kg. This natural honey is used for both domestic consumption and export. The volume of exports increased from 745,000 kg (11 million Baht) in 1987 to 1,194,400 kg (26.9 million Baht) in 1994 (Subansenee, 1995). Since beekeeping has become an increasingly important economic activity in Thailand, the government encourages people with suitable land to farm bees. It is expected that in 2003 there will be 2,500 apiaries; the honey from these will be exported to the world food market (News aired via Radio, 100.6 MHz, July 20, 2003).

Likewise, the Vietnamese government also considers bees and honey to be a lucrative business. Pure forest honey commands a better price on the domestic market due to its superior taste and reputed medicinal qualities. Vietnam is exporting honey to Japan, the United States, Taiwan, and other Asian countries (Department of Agriculture and Rural Development of Bac Kan Province, 1997). During the first half of 2001, Vietnam's Central Bee Company (VCBC) exported more than 4,000,000 kg of honey.

Crickets (*Gryllus bimaculatus*) farming: The cricket is one of the most common edible insects in Asia which people regard as a good source of income. In Thailand in 2002, 53 of the 76 provinces had cricket farms. Kaniknant (2002) documented a simple technique for breeding crickets requiring a large, well-sealed container, about 80 cm in diameter. Inside this container, materials such as crevice stone, hollow logs, grass and leaves are placed, providing areas where the crickets can live and hide. The crickets are placed in this container at a ratio of one male per two females. They are fed on pet food, grains, vegetable scraps (e.g., cabbages, kale, morning glory) and grass. It is believed that crickets have poor eyesight and that the males use sound to attract the females. Egg laying commences two days after mating. Crickets can lay eggs for 40 consecutive days. The eggs laid in the first 5 days peak at about 1,000 to 2,000 eggs per female. After hatching, 20 days are required for the crickets to develop into adults. In 31 to 43 days they are fully grown and able to enter another breeding cycle and produce a new generation. Their life span is between 73 and 96 days.

Cricket breeders commented that neither the yield nor the quality of insects obtained by farming are as high as those of insects collected from a natural environment. Crickets collected from nature are healthy, tastier and bring in a good income. As for body size, those collected from nature are relatively larger, as observed in rural areas of the Laos PRD and Myanmar. In comparison, crickets farmed are thinner and lighter in weight. Apparently the environment provided on farms is not entirely adequate.

Breeding bamboo caterpillars: Breeding bamboo larvae as a source of income is being promoted by the Thai Department of Forestry, Ministry of Agriculture and Cooperatives. Therefore, scientists are being encouraged to conduct research into breeding these caterpillars. Kaniknant (2003) spent 3 years determining various methods of breeding bamboo caterpillars.

Bamboo caterpillars, *Omphisa fuscidentalis*, are an important source of cash for the mountain tribes and native gatherers in the northern provinces of Thailand, which include Chiangrai, Nan, Prae, Maehongson, and Tak. According to Kaniknant (2003), the adult *O. fuscidentalis* is a brownish nocturnal moth. Five to to 7 days after mating, the female oviposits in the sheathing bract of the bamboo. In 12 days the white eggs change to light brown and hatch. Swarms of young larvae migrate inside the bamboo nodes enriched by the tender inner layers of the plant. Surprisingly, they are able to chew holes and enter into the nodes within a day. Holes are also made in the partitions throughout a stem so that the caterpillars can move from section to section to feed on the inner bamboo surfaces. During this stage, the larvae molt. It takes about ten months for the young larvae to develop into mature larvae. They feed on the bamboo layers for 40 to 60 days. The fully grown larvae then migrate downward to the nodes at the base of the stem. They encapsulate themselves in a cocoon for 8 months and develop into fully grown adults. Subsequently, they fly out from the bamboo home at night through the holes made earlier. Life outside their bamboo home can be short since the adults live only 5–7 days. Males die after mating and females die after ovipositing. The total life span is thus approximately 10–11 months.

The worms are harvested upon migrating to the bottom of the bamboo stems where they form cocoons. In nature, this stage occurs between December and March. Hence, the larvae at this stage bring in a good price for the vendors.

Bamboo caterpillars can be farmed in two ways: the first consists of creating a man-made environment using a nylon net that covers the young nodes of living bamboo stems. Adult moths are kept inside at a ratio of four males to four females. One female produces about 1–7 clusters of eggs and in general, 80–100 larvae can be collected from one bamboo stem. The second method makes use of a natural process by which the moths are allowed to live in an open bamboo plantation without being disturbed. Such a plantation should be situated at about 400 meters above sea level or higher to ensure that the environment is humid.

Conclusion

The findings from this study show that collecting insects for food is a strongly entrenched custom among ethnic groups in the Laos PDR, Myanmar, Thailand, and Vietnam. In these countries, about 164 species are considered edible. Forty-four species are commonly eaten in the Mekong Basin provinces, especially in Thailand. Although insects have traditionally been popular among rural populations, rapid socioeconomic changes in these regions have favored a rural-to-urban migration, leading to the subsequent adoption of insect eating among urban dwellers.

The insects used as food in urban areas are more diverse than in rural areas. Rural people eat insects as main dishes, whereas urban people eat insects either as a main dish, a snack or both. Moreover, although edible insects are a nutritious food source, cooking methods may alter their quality. Rural eaters or indigenous migrants may obtain fewer calories from insect food sources because they prefer meals lower in fat. Urban eaters, however, have several different ways of preparing insects using fat (frying, frittering) and they also consume a greater variety of insects. At present, the demand for insects as food is increasing, and the business of marketing edible insects is becoming more lucrative. Insect food distributors may earn a net profit of about 800–1,200 Baht d^{-1} if selling in slum areas, and about 2,000 Baht/night if businesses are located in tourist areas. On a commercial scale, entrepreneurs may earn as much as 600–700 million Baht per year. A consequence of this increase in the quantity of edible insects collected is that the ecosystem and food chain have been adversely affected, resulting in a biodiversity crisis. However, it would be possible to raise (farm) these insect as a human food source. In Thailand, for instance, studies are underway on farming particular crickets, bamboo moths, and sago beetles. However, higher levels of technology are needed to provide farmed insects of a quality equal to that of insects obtained from the natural environment. In this respect, an important means of maintaining biodiversity on the planet would be to provide financial support for research into edible insects.

Acknowledgments

A number of interviewees (insect eaters and vendors, folk healers), friends and colleagues gave us morale support while writing this article and we are indeed grateful to them. A special note of appreciation is given to Associate Professor Benja Mukthabhan and her family who helped in photographing at the Insect Shop in Khon Kaen province, Northeast Thailand. Special thanks are given to Ms Duangkamol Pomsuwan and Wetchayan Suwannamajo who assisted in photo-taking around Bangkok's insect outlets at night.

References

Asia Pacific Migrations Research Network (APMRN), 2003. Migration Issues in the Asia Pacific: Thailand. (URL: http://www.unesco.org/most/apmrnw14.htm). Retrieved August 5, 2003.

Bukkens S.G.F. 1997. The nutritional value of edible insects. Ecol. Food Nutr. 36: 287–319.

Costa-Neto E.M. 2002. The use of insects in folk medicine in the State of Bahia, Northeastern Brazil, with notes on insects reported elsewhere in Brazilian folk medicine. Human Ecol. vol. 30, no 2.

Dept. Agriculture Rural Development Bac Kan Province. 1997. Market opportunities, appropriate technologies and financial viability for demonstration farms. Tech. Rept. Vietnam-Finland Forestry Sector Co-operation Program, no. July. 6, URL: http://www.fao.org/DOCREP/005/AB598E/AB598E27.htm#TopofPage. Accessed April 2003.

Hirsch P. and Cheong G. 1996. Natural Reource Management in the Mekong River Basin: Perspectives for Australian Development Cooperation. Final overview report to AusAID, April 2, 1996. URL: http://www.usyd.edu.au/su/geography/hirsch. Retrieved August 24, 2003.

Institute of Nutrition, Mahidol University, Thailand, 2002. Food Composition Database for INMUCAL Program.

Jamornman S. 2003. Ten questions on edible insects. Leaflet distributed in the National Seminar on Edible Insects. Fac. Agric., Kasetsart Univ., Thailand, February 3, 2003.

Joint FAO/WHO Expert Committee. 1973. Energy and Protein Requirements. WHO Technical Report Series no. 522. World Health Organization, Geneva.

Jongleebhan S. 2003 Happy life at old ages. Population and Development Newsletter. Institute for Population and Social Research, Mahidol University, Thailand, no. 23.

Kaniknant L. 2002. Crickets: Economic Insect. Newsl. Entomol. Zool. Assoc., Thailand, vol. 10, no. 1.

Kaniknant L. 2003. Growing Bamboos: Breeding Worms as by-products. Educational Leaflet prepared and distributed in the 41st Annual Workshop on Agriculture, 3–7 February 2003. Kasetsart University, Thailand.

Klinhom U., Rasirattana C., and Jitjumnong S. 1984. Nutritive values, parasites and poisonous matters in edible insects in northeast Thailand. Srinakarinwiroj University, Mahasarakram, Thailand.

Lewvaniach A., Chunram S., Charoentesprasit N., and Hanboonsong Y. 2000. Edible Insects in Thailand. Twelfth Entomol. Zool. Div. Conf., Bangkok, Thailand, pp. 355–365.

Lindroth R.L. 1993. Food conversion efficiencies of insect herbivores. Food Insect Newsl. 1993; vol. 6, no. 1. URL: http://www.hollowtop.com/finl_html/conversion.htm. Accessed August 2003.

Meusch E. and Yhoung-aree J. 2003. Participatory Assessment of the Role and Nutrition Value of Aquatic Resources in the Livelihoods of Rural People in Attapeu Province, Laos PDR. World Conservative Union and Food and Agriculture Organization of the United Nations, Regional Office for Asia and the Pacific. Bangkok, Thailand.

Meyer-Rochow V.B. 1976. The use of insects as human food. Food Nutr., Bengkok, Thailand, 33: 151–152.

Meyer-Rochow V.B. 1978/79. The diverse uses of insects in traditional societies. Ethnomedicine 5: 287–300.

Meyer-Rochow V.B. 2004. Traditional food insects and spiders in several ethnic groups of Northeast India, Papua New Guinea, Australia, and New zealand (this volume).

National Statistics Office. 2003. Table 1: Number and percentage of population by migration status, current residence and area. Bangkok, Thailand, URL: http://www.nso.go.th/eng/stat/migrant/migtab1.htm. Retrieved August 2003.

Ngunboonsri P., Suphiphat V., and Puwastein P. 1989. Nutritive value of pupae in bamboo. J. Nutr. Assoc. Thailand, 23.

Pemberton Robert W. 2004. Contemporary use of insects and other arthropods in traditional Korean medicine (this volume).

Phillips J. and Burkholder W. 1995. Allergies related to food insect production and consumption. Food Insect Newsl. vol. 8, no. 2. URL: http://www.hollowtop.com/finl_html/allergies.htm. Accessed August 2003.

Sing P. 1981. Traditional Recipes of Laos: Royal Recipes Book of Luang Prabang. Alan and Jennifer Davidson (trans.). Prospect Books Publ., London, UK.

Sirichakwal P. and Sungpuag. 1982. Insect as source of protein. Mahidol University Annual Research Abstracts, Mahidol University, Thailand.

Subansenee W. 2003. Major non-wood products of Thailand. Beyond timber, social, economic and cultural dimensions of non-wood forest products in Asia and the Pacific. RAP Publ. 1995/15. FAO Regional Office for Asia and the Pacific. URL: http://www.fao.org/DOCREP/005/AB598E/AB598E25.htm. Accessed April 2003.

Subansenee W., Denrungruang P., Sroithongkam P., Nilkamhaeng N., and Kangkarm Y. 2003. Promotion of Tropical NWFPs in Thailand. URL: http//www.iufro.bobu.ac.at/iufronet/d5/wu51100/pub3–5.11.htm. Accessed April 16, 2003.

Subchareon P., Deewiset K., Termwiset P., Tangsukruetai P., Saksakulpornlerd W., and Chairuk D. 2001. Insects in Thai folk medicine. Inst. Thai Traditional Med., Ministry of Public Health, Thailand.

Tembo M.S. 1993. Delicious Insects: Seasonal delicacies in diet of rural Zambians. URL: http://www.worldanti.com/public/1993/october/cu02.cfm. Released date October 1993, New York, Washington, DC.

Viwatpanich K. 1999. Insects: Future human foods. Folk Thai Medicine Foundation and Institute of Language and Culture, Mahidol University, Thailand.

Viwatpanich K. 2003. Insects: Cultural heritage and traditional foods in Northeast Thailand: natural interactions between humans and the environment (unpubl. paper). Fac. Medicine, Thammasat University, Thailand.

Washirarungsi J. and Sitimanodhamma J. 1982. Laotian Folk Medicine. Council of Science and Technology. State Publ., Laos.

WREN media. 2003. Edible insects—a culinary curiosity? URL: http://www.new-agri.co.uk/99-3/focuson/focuson4.html. Accessed April 2003.

Yhoung-Aree J., Puwastein P., and Attig A.G. 1997. Edible insects in Thailand: An unconventional protein source? Ecol. Food. Nutr. 36: 133–149.

Zimian Ding, Yonghua Zhao, and Xiwu Gao. 2004. Medicinal terrestrial arthropods in China (this volume).

21

Lessons from Traditional Foraging Patterns in West Papua (Indonesia)

MILA TOMMASEO-PONZETTA[1] AND MAURIZIO G. PAOLETTI[2]

[1]*Dept. of Zoology, University of Bari, Via Orabona 4, 70125 Bari, Italy.*
E-mail: m.tommaseo@biologia.uniba.it
[2]*Dept of Biology, University of Padova, Via U. Bassi 58/b, 35121 Padova, Italy.*
E-mail: paoletti@bio.unipd.it

Abstract

Insects and terrestrial invertebrates are presumed to have consistently contributed to the diet of our ancestors and it is estimated that nearly 2,000 insects still contribute worldwide to human nutrition.

As elsewhere in the tropical world, insects are a part of the diet of West Papua (ex-Irian Jaya) populations. The distribution of insect food consumption on the western half of the island of New Guinea follows different patterns, depending on the environment and population subsistence economy.

In the forest biome of the peripheral montane areas, many different insect species are collected but in small numbers, mainly by women and children. Insect collection occurs by chance, but is frequent, and possibly helps those most vulnerable to food shortages to complement their diet with some extra protein and fat.

In the western lowlands, a complex cultural system is centered around the traditional staple: sago (*Metroxylon sagu* and *M. rumphii*). The pith of this palm, which grows spontaneously in the swampy forest plains, contains a high amount of starch and is easily gathered by the local people. The beetle *Rhynchophorus ferrugineus papuanus* is strictly associated with the sago palm and oviposits especially on damaged or felled trees. Villagers so manage the sago palms to increase palm-worm oviposition and to collect a consistent number of larvae about 6 weeks later. This ancient sago culture, favored by an abundant and reliable resource, has developed a rich ceremonial life connected with sago larvae consumption.

Ecological, economic, and cultural factors influencing the variety of insect consumption modalities are taken into account here.

The development of local, small-scale invertebrate breeding systems can contribute to the development of sustainable, renewable resources, and protect tropical forests from degradation.

Key Words: entomophagy, West Papua, New Guinea, *Rhynchophorus ferrugineus papuanus*, Asmat, Papua, sago (*Metroxylon rumphii, M. sagu*)

Introduction

Four-fifths of all living species on the planet—estimated to be between 7.2 and 82 million (Erwin, 1988; Stork, 1988; Wilson, 1988; May, 1990, 1992; Paoletti et al., 1992)—are terrestrial invertebrates, in particular insects. Small invertebrates are considered to dominate in tropical forests: it is possible that their number exceeds 80 times that of known species (Paoletti, 1995). Clearly, insects and terrestrial invertebrates deserve greater attention as a potential food reserve (Meyer-Rochow 1973, 1975; DeFoliart, 1989, 1992, 1997).

Insects and terrestrial invertebrates are supposed to have constituted an important part of the diet of early hominids and to have played a key role in human evolution (Kliks, 1978; Mann,1981; Hamilton, 1987; Harris and Ross, 1987; Stini, 1988; Skinner, 1991; McGrew, 2001; Tommaseo-Ponzetta, 2003 and this volume).

The cultural and socioeconomic revolution that followed the spread of agriculture in the Neolithic gradually reduced the importance of gathering in human nutrition, especially in the Northern Hemisphere. This may have led to a loss of knowledge of the comestibility of semidomesticated and wild plants and small animals, such as insects and other terrestrial invertebrates. Limitations related to small size, distribution, gathering or breeding of insects and terrestrial invertebrates, and also psychological reasons, promoted rejection of small creatures. Considering insects as noxious pests and therefore repulsive human parasites greatly reduced their consumption as food items (Harris and Ross, 1987).

Today, few insects in temperate developed countries are included in the alimentary spectrum: cheese maggots, which are considered edible in Southern Mediterranean regions, are amongst the few to pass the test (Paoletti and Dreon, this volume).

However, insect and terrestrial invertebrate consumption by those populations that still follow a hunter-gatherer—or, at least, a traditional—subsistence economy is widely reported (Bodenheimer, 1951; Elkin, 1964; Grottanelli, 1965; Clastres, 1972; Meyer-Rochow, 1973; Santos Oliveira et al., 1976; Posey, 1978, 1987; Hawkes et al., 1982; Dufour, 1987; DeFoliart, 1990; Meyer-Rochow and Changkija, 1997). Ramos-Elorduy (1992) estimates that the number of insect species consumed in the world nowadays is no less than 1,386, particularly in tropical areas, in both savanna and rainforest environments and rural areas. In addition, some snails, spiders, scorpions, and earthworms are consumed in regions such as subtropical southern China and Amazonia.

The idea that insects may be a valid food item even for technologically advanced societies is evoking increased interest in both scientific and public domains. Since 1988 a newsletter dedicated to "Food Insects" has been published and the number of scientific studies dealing with insect consumption is increasing. International symposia have recently been devoted to this subject, e.g., in 1992 in the Philippines (Hardouin and Stievenart, 1993), 1995 in Beijing (*Biodiversity in Agriculture for a Sustainable Future*, September 19–21, 1995; Paoletti and Bukkens, 1997), and in 2000 in Paris (*Les "insectes" dans la tradition orale*, October 3–6, 2000; Motte Florac and Thomas, 2003). The growing popularity of insects as food is proven by the existence of several recipe books (Taylor and Carter, 1976; Comby, 1990; Menzel and D'Aluisio, 1998) and an insect-eaters' club in Washington DC. Many different insects are also offered, as delicacies, in restaurants in the largest cities of Mexico, Thailand and China.

Insect consumption among New Guinea populations has already been mentioned. The island of New Guinea is situated a few degrees below the Equator: both its montane interior and swampy riverine plains are covered by tropical rainforest, but savanna-like grassland also occurs in the highlands: the variety of its ecological niches has favored biological, cultural, and linguistic human diversity, while population isolation in the interior of the island and paucity of communication with the world outside are mainly responsible for maintenance of traditional subsistence.

New Guinea's tropical environment, particularly favorable to invertebrate abundance and diversity, offers a unique opportunity for documenting invertebrate collection and consumption by human groups. Various anthropological works mention the inclusion of grubs, beetles, spiders, and other invertebrates in the diet of both lowland and montane groups (Rappaport, 1967; Heider, 1970; Clarke, 1971; Littlewood, 1972; Morren, 1977; Brown, 1978), although this information is seldom circumstantial and animal species are not always identified.

More informative are works devoted to the island ecology and human subsistence patterns (Dornstreich, 1977; Blum, 1979). Detailed studies on insect eating have been carried out by Meyer-Rochow (1973, 1976), Meyer-Rochow and Changkija (1997), and Mercer (1993, 1997) but deal mainly with the eastern part of the island (Papua New Guinea), while information on the western side, West Papua, is still scarce (Blum, 1979).

Caution must be exercised in generalizing information relating to one ethnic group to the whole island. For instance, earthworm collection and human consumption has been reported for the Bisorio, a nomadic people that live on the upper reaches of the Salumei River, a Sepik tributary, in Papua New Guinea (P. Agnoletto, pers. comm.). However, earthworm alimentary use has not yet been documented in the western half of the island.

This report constitutes the first pertaining to insect and invertebrate consumption among West Papua populations. Data were collected at different times (1984, 1990, 1995) in both montane and lowland areas, during anthropological surveys and field researches, whose results concern physical anthropological features (Tommaseo and Lucchetti, 1992), haematological findings (Re et al., 1989; Tommaseo et al., 1989, 1990, 1992), trace element analyses (Tommaseo-Ponzetta et al., 1998) osteology (Tommaseo-Ponzetta et al., 1997) and molecular genetics (Tommaseo-Ponzetta et al. 2000, 2002; Kayser et al., 2003). While this information is far from exhaustive, it is at least a first step toward better understanding of this land and its people by M.T.P.

Materials and Methods: Study Groups

Montane Area

Langda: This village is located 4°40' South, 139°58' East, at an altitude of about 2,000 m on the southwestern slopes of the Jayawijaya range, near Mts. Yarkon and Yamin (Dunerkon) (4,581 m). Mean temperatures range between 25°C (day) and 12°C (night). Fog is frequent and average rainfall reaches 6–7,000 mm y^{-1}. The inhabitants of the district number about 3,000 (village 185). They speak the Una language (Western Mek, Trans-New Guinea Phylum; Grimes, 1988) and call themselves Una or Uniang. Slash-and-burn agriculture is practiced, even along very steep mountain slopes. The main products are: sweet potato, taro, banana, sugarcane, tobacco, green bean, some fruits and other vegetables. Agriculture suffers from periodic scarcity (December-January are rainy months: the "hunger" period); a few pigs and poultry are raised, and the diet is largely supplemented by hunting-gathering in a territory ranging from the river valley floor (1,200 m) to the mid-montane environment (3,500 m).

Bime Valley: This valley follows a South-North orientation and extends from 4°25'–4°30' South to 140°12'–140°15' East, in the central Eastern Highlands near the Papua New Guinea border. Mean temperatures range between 27°C (day) and 17°C (night) and average rainfall is around 5,000 mm y^{-1}. Research was carried out in two proximate villages: Turwe (1,350 m; area inhabitants 1,000 (village 350)) and Calap (1,890 m; area inhabitants 895 (village 200)). The Bime language belongs to the Ketengban (Eastern Mek, Trans-New Guinea Phylum; Grimes, 1988). Horticulture is practiced on the river valley floor and along the mountain slopes by means of slash-and-burn. The main products are: sweet potato, taro, peanut, and banana; some pigs and poultry are raised.

Lowland Plain Area

Piramanak: Located along the basin of the Brazza River, a tributary of the Siretsj, at the time of the first visit (1981) this village consisted of a few family houses built high in trees, far in the forest interior. Every few months the village changed its location according to the exhaustion of sago reserves. In 1984, the group was compelled by the Indonesian government to settle alongside the Brazza River in common houses on low piles. It was located 5°0' South, 139°20' East; in 1984 its inhabitants were 78. Mean temperatures ranged between 28°C (day) and 17°C (night), average rainfall around 4,500 mm y^{-1}. The Brazza people are closely related to the Citak and speak Kaunok, an Asmat dialect (Asmat-Kamoro, Trans New Guinea Phylum; Grimes, 1988). Their economy is based on gathering (mainly sago, but also eggs, larvae, vegetables and fruits), river fishing, and hunting (wild boar, cassowary, arboreal marsupials, small rodents, and birds).

Coastal Asmat Villages: Following Amelsvoort (1964), the Asmat territory extends from 140° East—137° West and 4° to 7° below the Equator. Several villages located along the Flamingo Bay and Casuarina coast were visited during this survey. Mean temperatures vary between 35° and 21°C, average annual rainfall may reach 4,600 mm (van Arsdale, 1978). The villages share some common features: they are large (500 to 1,000 inhabitants); the people speak Asmat (Asmat-Kamoro, Trans-New Guinea Phylum; Voorhoeve, 1965; Grimes, 1988); and the economy is based on sea and river fishing, and gathering mainly of sago (as in Piramanak). Because of frequent flooding, horticulture is not a traditional practice and attempts are made only in a few elevated spots.

Data Collection

In the montane areas (Langda and Bime valley) an interpreter asked the population to collect and display those insects and other terrestrial invertebrates that were usually included in their diet.

The people were initially reluctant to admit that they ate insects, invertebrates and other small animals. This attitude may be ascribed to contact with western cultural agents (missionaries or administrative officers), who consider such practices backward or primitive. While they acknowledged the existence of the custom, they emphasized it was mainly limited to women and children. Two men from the village accompanied the researcher into the forest, to show the trees in which grubs could be found and how to collect them.

In Bime Valley, little boys were the major informants.

Given the difficulty of preserving and transporting live animals, and the country's restrictive nature protection law, only macrophotographs were taken of each sample and their vernacular names recorded.

In the lowland area, consumption by the Citak and Asmat populations of sago larvae has been extensively documented (Zachmann, 1966; Trenkenschuh, 1970; Schneebaum, 1980; Konrad et al., 1981; Sowada, 1986, 1995). So we limited our research mainly to observation and recording of their mode of consumption under various conditions, according to both everyday and ritual life.

Results and Discussion

Tables 21.1 and 21.2 list the crops and animals consistently representative of the villager subsistence, compiled from direct "in situ" observations.

Table 21.3 gives a preliminary list of insects demonstrated as eaten by local populations (Figs. 21.1 to 21.5). Most insect orders are represented, in particular wood-boring, root-boring and defoliating groups are the food targets. Ants and termites seem to be less important. In addition to insects, spiders and frogs are usually collected.

In Langda, some larvae were brought to the researcher together with the bark and rotting wood that housed them and a pair of cerambycid beetles, which were indicated as the adult form of these larvae. The larvae were usually collected under the bark of a particular tree called "*juwan*", their presence ascertainable by small holes on the outer bark surface.

In Bime, the largest number of insects and other small prey (not only insects, but also frogs and other small animals) were indicated as edible.

In the mountains, larvae and insects were said to be collected occasionally, mostly by women and children while going to their gardens and whenever the

Table 21.1 Crops most common in the study area

English name	Scientific name	Locality 1	Locality 2	Locality 3	Locality 4
Sweet potato	*Ipomoea batatas*	***	***		
Taro	*Colocasia esculenta*	***	***		*
Peanut	*Arachis hypogaea*	*	***		*
Pandanus	*Pandanus jiulianettii* and *P. brosimos*	*	***	***	*
Sugarcane	*Saccharum officinarum*	***	***	**	**
Gourd	*Cucurbita* sp.	**	**	*	*
Squash	*Cucurbita* sp.	**	**	*	*
Cassava	*Manihot esculenta*	*	**		
Yam	*Dioscorea* spp.	*	**		
Bean	*Phaseolus* sp.	**	**		*
Winged bean	*Psophocarpus tetragonolobus*	**	**		*
Kutzu	*Pueraria lobata*	*	**		
Vive	*Nicotiana tabacum*	***	***	**	**
Fern	*(?)*	***	***	*	*
Papaya	*Carica papaya* L.	**	**		***
Pineapple	*Ananas* sp.		*		**
Bread fruit	*Artocarpus* sp.		*	*	*
Banana	*Musa* spp.	*	**	*	***
Coconut	*Cocos nucifera*				***
Water-melon	*Eugenia* sp.	**	**		*
Sago palm	*Metroxylon sagu* and *Metroxylon rumphii*			***	***

Places: 1 = Langda; 2 = Bime valley; 3 = Piramanak; 4 = Asmat coast.
Presence: low*, medium**, high***
Source: Rehm S. (1994); Kays S. and Silava Dias J.C. (1995)

Table 21.2: Most common domestic and game animals in the study area

English name	Scientific name	Locality 1	Locality 2	Locality 3	Locality 4
Pig: d=domestic, f=feral	*Sus scrofa papuensis*	d*, f*	d*, f*	f**	f*
Dog	*Canis familiaris*	*	*	***	**
Fowl	*Gallus gallus*	**	**		*
Tree kangaroo	*Dendrolagus* spp.	***	**	**	**
Cuscus	*Phalanger* spp.	***	**	**	**
Giant rat	*Uromys* spp. and *Mallomys* spp.	*	*	**	**
Ring tail possum	*Pseudocheirus* spp.	*	**	**	**
Cassowary	*Casuarius* spp.	*	*	***	**
Brush Turkey	*Aepypodius* spp.		**	***	**
	and *Megapodius* spp.	*	*	***	**
Wallaby	*Dorcopsis*	*	*	*	*
Fruit bat	*Pteropus* spp.		*	**	**
Frog	*Hyalid* spp.	**	**	**	**
Spider	*Nephila maculata*	**	**	**	**
Lizard	(?)	**	**	**	*
Paradise bird	*Paradisaea* spp.	**	***	**	*
Black cockatoo	*Probosciger aterrimus*		**	***	***
White cockatoo	*Cacatua galerita*		**	***	***
Crowned pigeon	*Goura cristata*	*	*	***	***

Places: 1 = Langda; 2 = Bime valley; 3 = Piramanak; 4 = Asmat coast.
N.B. Fishes could not be identified.

Fig. 21.1: *Titia* (Coleoptera; Cerambycidae) indicated by Bime people as the adult form of the *mokah* larvae (photo M. Tommaseo-Ponzetta).

Table 21.3: Insects consumed in Irian Jaya

Place	Local name	Family	Species	Notes
1	mokah	Coleoptera Cerambycidae	larvae of *Batocera* sp. and *Dihamnus* sp.	At least 5 species have been indicated as edible
1	bolkaya	Blattodea Polyzosteriinae	*Polyzosteria* (?) sp (?).	Especially on decayed wood
2	dunkula	Lepidoptera, Sphingidae	yellow caterpillars of *Acherontia achesis*	Several caterpillar species are used as food
2	duh	Coleoptera Scarabaeidae	larvae of Melolonthinae	At least 6 species of Scarabeidae are consumed
2	dyk dyk	Coleoptera, Passalidae	adult and larvae	10 species have been described in New Guinea
2	bulutnamgme	Coleoptera Cerambycidae	larvae	At least 3 species have been shown to be edible
2	wisin	Orthoptera Tettigoniidae	adults	At least 5 species have been indicated as edible
2	pho	Orthoptera, Gryllidae	adults	At least 3 species have been indicated as edible
2	philipalala	Phasmatodea Podacanthinae	adults *Extatosoma* (?)	At least 3 species are commonly consumed
3, 4	tu / to	Coleoptera Curculionidae	larvae of *Rynchophorus ferrugineus*	Possibly 2 species of curculionid grubs are collected

Places: 1 = Langda; 2 = Bime valley; 3 = Piramanak; 4 = Asmat coast.
Source: C.S.I.R.O. (1970), DeFoliart (1996).

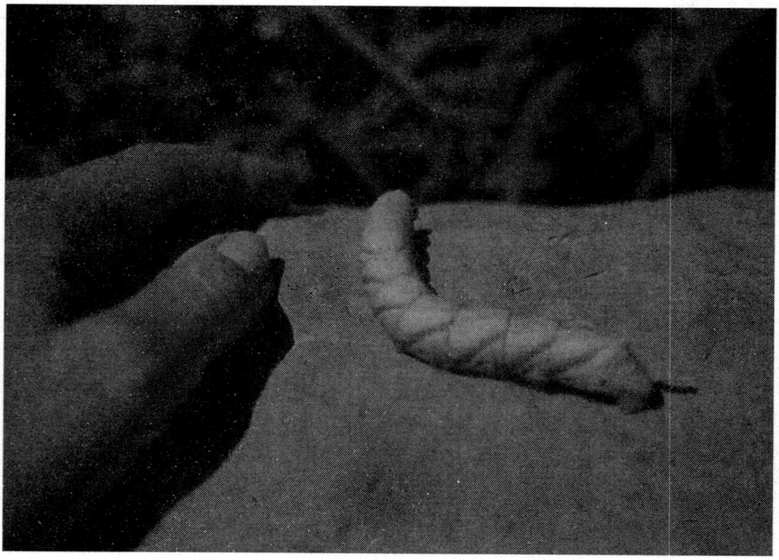

Fig. 21.2: Yellow caterpillar *dunkula* (Lepidoptera: Sphingidae) *Acherontia achesis*, consumed in Bime (photo M. Tommaseo-Ponzetta).

Fig. 21.3: *Duh* (Coleoptera: Scarabaeidae) Melolonthinae sp. (?) consumed in Bime (photo M. Tommaseo-Ponzetta).

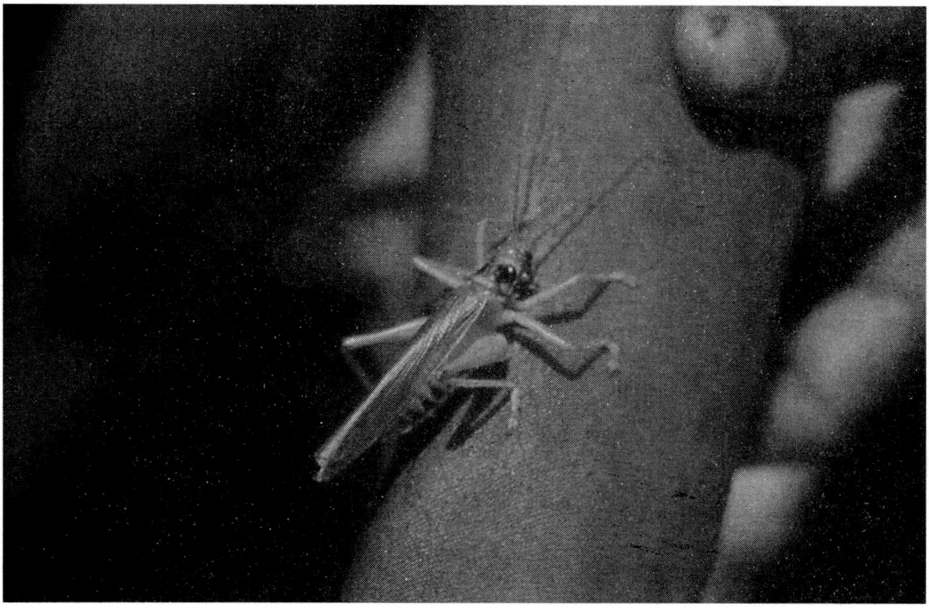

Fig. 21.4: Grasshopper named *wisin* (Orthoptera: Tettigoniidae) consumed in Bime (photo M. Tommaseo-Ponzetta).

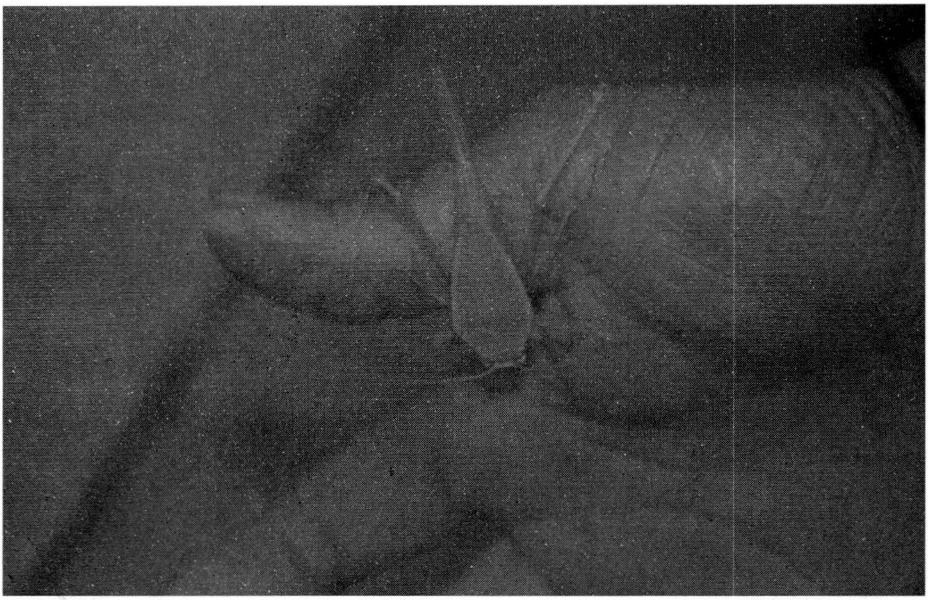

Fig. 21.5: Cricket named *pho* (Orthoptera: Gryllidae) consumed in Bime (photo M. Tommaseo-Ponzetta).

opportunity arose. In both Langda and Bime, larval and adult insects were said to be eaten raw, as snacks, when far from the village, or wrapped in small leaf bundles and roasted in the fire or cooked in a steamed pit, depending on the situation.

In the swampy lowlands, a shallow water table, periodic flooding, and low fertility in severely leached soils all favor sago palm culture (*Metroxylon sagu*, *M. rumphii*), which is the local staple, with its associated palm worm (*Rhynchophorus ferrugineus papuanus*). In this kind of environment, collecting other insects seems to be less important and they may also be less abundant.

The Asmats have evolved interesting palm management techniques for sago starch gathering and increased palm-worm production by clearing the palms and making holes in the stipe to entice *Rhynchophorus* egg laying and high larval development. (According to Schneebaum, 1980, at the time he wrote his work most Asmats were not aware of the beetles' role but believed they appeared through the magic performed when the holes were drilled.) Under the felled sago palm trunks scarabaeid grubs and hylid frogs are occasionally collected. According to van Arsdale (1978), other decaying trees, e.g. as giant anisoptera (*Anisoptera polyandra*), can harbor edible beetle larvae, similar to fallen sago palms.

In Piramanak and other small inland riverine villages along the Brazza, it was observed that the gathering of small animals, especially of sago larvae, is integrated into everyday meals. They are collected by everybody, especially the men, as prized delicacies, and kept in small leaf bundles. They are eaten raw, or strung on small spits, or mixed with sago flour and cooked in the ashes (see also Mayer-Rochow and Cerda et al. in this volume).

In Asmat coastal villages, large numbers of *Rhynchophorus* larvae are consumed on ritual occasions, when they assume strong symbolic importance (Plate X, 4). The ritual exchange of larvae ratifies a pact of friendship which cannot be broken and which has an essential role in reconciliation feasts. Larvae which have thus been made "sacred" during a feast are so imbued with power that they cannot be eaten by children, the sick or old, or pregnant women, who would not be able to withstand their strength and could even die. Even a man, whose wife has recently had a baby, will not eat these larvae for fear of harming the child.

According to Sowada (1986, 1995), three feasts are especially dedicated to these larvae. Every feast is preceded by the felling of hundreds of sago palms, which are then left to rot on the ground for the larvae to grow inside and be collected about six weeks later.

The feasts are called *imui* or *imbui*, *an*, and *firauwi*. The latter is also known as *pir-jimi* (Keenok dialect) or *basusuamkus (basu suangkus)* (Keendakap dialect).

During *imui*, a pact is established between two men or two women who, by exchanging the larvae, become partners or *diwap*, and are linked by a very special friendship expected to last their life time.

The *an* feast reestablishes peaceful, friendly relations previously broken off by headhunting between families or villages.

Pir-jimi is another feast celebrating friendship among men or among women and in the past preluded *papisji*, or ritual wife exchange. During the latter feast, two cylindrical containers (*tir*) of palm leaves are built inside the men's long house. One, the "men's tube", is small, but the other reaches the roof. For the occasion, the roof is opened so the tube can be filled with larvae by the women who have climbed up with vessels full of grubs, by means of an external scaffold. There is much drumming, singing and dancing. When the women's tube is full, a cut is made at its base so that the larvae tumble out on the floor of the feast house, are collected, shared out, and exchanged by the men. According to Sowada (1986, 1995), the larvae embody some sort of vital force which helps to establish the necessary harmony between the living and their ancestors—invoked during the ceremony—and among the living themselves.

A *basu suangkus* feast has been recorded in a few villages along the Siretsji river (Schneebaum, 1980, 1985; Konrad et al., 1981, Konrad G. and U., 1995). The name of the feast means "making visible the heads of men who have been killed in battle". It is reported that during the feast in Jaosokor in 1979, more than 3,000 sago palms were felled for the larvae to be gathered a few weeks later. In preparation for the feast, four logs are carved, representing the heads of those who must be avenged. In the final stage, the men bring bowls of larvae into the men's house and pour them into the square internal space formed by the four carved logs of the *basu suangkus*, which thus represents a ceremonial vessel. The larvae are distributed according to the importance of the family, the largest portion going to those who should avenge the deceased: nevertheless, they cannot eat them until they have taken their revenge. On that occasion, in Jaosokor, pigs were hunted instead of men.

Among the coastal Asmat, larvae are also represented as stylized decorative motifs on shields and various artistic objects are devoted to them, e.g., special bowls to be used on ritual occasions. In earlier times, such bowls carried human brains, presently substituted by sago grubs.

Insect consumption in West Papua therefore seems to be of various kinds, depending on ecological and cultural factors.

In the montane regions, Langda and Bime, a few traditional but mostly imported crops are grown. Nevertheless, especially in Langda, the people work hard for their subsistence; gardens are created on steep slopes subject to heavy erosion; harvests are not sufficient to allow more than a few pigs to be raised (even the price of a wife cannot be paid in pigs, as elsewhere in the highlands, but consists of iron axes and other items). The villages are obviously small, due to poverty and the uncertainty of agricultural resources. Gathering, together with hunting, is largely practiced throughout the year. During recurrent periods of shortage of agricultural products, hunting (mainly cuscus) and gathering become the most important—if not the only way—of obtaining food. In this kind of situation—given their unfavorable social status in Papuan society—women and also children are the most vulnerable to nutritional inadequacies: in particular, their access to protein and fat is limited. This may explain in part their recourse to insects and invertebrates as necessary complements to a poor nutritional pattern. In the highlands, great insect and invertebrate diversity therefore leads to high prey diversification. In this peripheral montane area, because of the scarcity of resources and the small village sizes, ritual life has no possibility of attaining the richness of that of the coastal Asmat region.

A staple generally derives from seeds or roots obtained from annual crops. Let us take the unique case of a tree (sago) which produces this key staple. In the lowland area, sago, which grows spontaneously in the swampy forest soil and is available throughout the year, is a reliable resource (Townsend, 1974), not even requiring hard work to harvest. A sago palm can be felled in a few minutes and one day's work by a man and his wife (necessary to wash the pith of the sago to extract the starch) suffices to provide food for a whole family for a substantial period (Ulijaszek and Poraituk, 1993).

The reliability and copiousness of this resource, together with the possibility of estuary and sea fishing, allows the existence of large coastal villages (up to 1,000 inhabitants). Staple availability, the consequent population density, and relative abundance of time (not occupied by subsistence activities), all contribute to the flourishing of a rich artistic and ceremonial life.

Sago larvae, the animal correspondents of the main staple, transform a carbohydrate into a nutritionally rich resource: in addition to proteins and especially fats, these larvae are an important source of iron, calcium, and vitamins (Table 21.4), which are most needed by the population of a tropical developing country. Therefore, the important symbolic role they assume in ritual life is more than justified by their importance to nutritional balance.

Table 21.4: Composition and nutritional value of some common foods

Food constituents and energy	**Sweet Potato (white)**	**Sweet Potato (brown yellow)**	**Taro**	**Sago (dry)**	**Sago larvae**	**Other insects (larvae and adults)**
Water %	68.0	68.0	73.0	11.0	70.5	71.5
Proteins *	1.6	1.6	1.8	0.5	6.1	15.8
Lipids *	0.2	0.2	0.1	0.3	13.1	5.0
Glucids tot. *	28.0	28.0	24.0	88.0	9.0	4.4
Cellulose *	1.0	1.0	1.0	0.2	—	—
Calcium **	33	33	51	32	461	30
Iron **	2.0	2.0	1.2	0.8	4.3	1.3
Vit. A tot. (rethynol)***	12	400	traces	0	—	—
Vit. B_1 **	0.09	0.09	0.10	0.01	0.08	0.22
Vit. B_2 **	0.04	0.04	0.03	—	0.43	0.50
Vit. PP tot. (niacin) **	1.2	1.2	1.4	0.1	3.7	0.4
Vit. C **	37	37	8	0	0	0
Edible matter/ 100 g %	87	83	81	100	100	100
kJ.	506	506	427	656	757	565
Kcal.	121	121	102	316	181	135

*g/100 g edible matter, **mg/100 g edible matter, ***µg/100 g edible matter
Source: Peters (1957); Barrau (1959) and Jardin and Crosnier (1975).

To improve the sustainability of villagers in the highlands, where no natural groves of sago thrive, it would be interesting to help the people develop—as we are trying to do in Amazonian Venezuela—small-scale units to grow palm worm larvae using local vegetal biomass as dietary items.

Use and development of local resources from the rich tropical biodiversity, following patterns traditionally adopted by the local people, constitutes a new developmental model that might possibly increase the sustainable use of resources in fragile forested environments.

Acknowledgments

The authors are grateful to Gene DeFoliart and Benno Meyer-Rochow for their invaluable help in identifying insect samples. We also thank Gunter and Ursula Konrad for their kind permission to publish their photo of Asmat men.

Research in West Papua of M.T.P. was generously supported by the Unit of Physical Anthropology of Padua University, Department of Biochemistry and

Molecular Anthropology of Bari University, and the Ligabue Study and Research Centre of Venice.

References

C.S.I.R.O. 1970. Insects of Australia. Melbourne Univ. Press, Sydney, Australia 1029 pp.

Barrau J. 1959. The sago palms and other food plants of marsh dwellers in the South Pacific islands. Econ. Bot. 13: 151–162.

Blum J.P. 1979. Untersuchungen zur Tierwelt im Leben der Eipo im Zentralen Bergland von Irian Jaya (West Neuguinea), Indonesien. Dietrich Reimer Verlag, Berlin, 19 pp.

Bodenheimer F.S. 1951. Insects as Human Food. Dr W. Junk, The Hague, The Netherlands, 352 pp.

Brown P. 1978. Highland People of New Guinea. Cambridge Univ. Press, Cambridge, MA, 258 pp.

Clarke W.C. 1971. Place and People. An Ecology of a New Guinean community. Univ. California Press, Berkeley, CA, 265 pp.

Clastres D. 1972. The Guayaki. *In:* Bicchieri M.G. et al. (eds). Hunters and Gatherers Today. Hold, Rinehart and Winston, New York, NY pp. 138–174.

Comby B. 1990. Délicieux insectes. Les protéines du futur.Editions Jouvence, Genève, 156 pp.

Cortivo P., Tommaseo M., Caenazzo L., Crestani C., Scorretti C., and Benciolini P. 1987. Genetic studies on the Asmat population (Irian Jaya, Indonesia). Anthropologischer Anzeiger 45 (4): 323–329.

DeFoliart G.R. 1989. The human use of insects as food and as animal feed. Bull. Entomol. Soc. Amer. 35: 22–35.

DeFoliart G.R. 1990. Insects as food in indigenous populations. *In:* Posey D.A., Overnal W.L., Clement C.R., Plotkin M.J. et al. (eds.). Ethnobiology: Implications and Applications. (Proc. First Int. Congr. Ethnobiology (Belém, 1988)). Museum E. Goeldi, Manaus, pp. 145–150.

DeFoliart G.R. 1992. Insects as human food. Crop Protection 11: 395–399.

DeFoliart G.R. 1997. An overview of the role of edible insects in preserving biodiversity. *In:* Paoletti M.G. and Bukkens S. (eds). Minilivestock. Ecology of Food and Nutrition. 36, 2–4: 109–132.

Dornstreich M.D. 1977. The ecological description and analysis of tropical subsistence patterns: an example from New Guinea. *In:* Bayliss-Smith T. (ed.). Subsistence and Survival. Rural Ecology in the Pacific. Academic Press, London, UK, pp. 245–271.

Dufour D.L. 1987. Insects as food: a case study from the Northwest Amazon. Amer. Anthropol. 89: 383–397.

Elkin A.P. 1964. The Australian Aborigines. Doubleday, New York, NY.

Erwin T.L. 1988. The tropical forest canopy—the heart of biotic diversity. *In:* Wilson E.O. (ed.). Biodiversity. National Acad. Press, Washington DC, pp. 123–129.

Grimes B.F. 1988. Ethnologue. Languages of the World. Summer Institute of Linguistics, Inc. Dallas, TX, 748 pp.

Grottanelli V.L. 1965. Cibi, bevande e narcotici. *In:* Grottanelli V.L. (ed.). Ethnologica. L'Uomo e la civiltà. Ed. Labor, Milano, vol. III, pp. 403–486.

Hamilton W.J. III 1987. Omnivorous primate diets and human overconsumption of Meat. *In:* Harris M. and Ross E.B. (eds.). Food and Evolution: Towards a Theory of Human Food Habits. Temple University Press, Philadelphia, PA, pp. 117–132.

Hardouin J. and Stievenart C. 1993. Invertebrates (Minilivestock) Farming. CTA (Center of Technical Cooperation), Wageningen, The Netherlands, 223 pp.

Harris M. and Ross E.B. (eds). 1987. Food and Evolution: Towards a Theory of Human Food Habits. Temple University Press, Philadelphia, PA, 633 pp.

Hawkes K., Hill K., and O'Connell J. 1982. Why hunters gather: optimal foraging and the Achè of Eastern Paraguay. Amer. Ethnol. 9: 379–98.

Heider K.G. 1970. The Dugum Dani. Aldine Publishing Co., Chicago, ILL, 334 pp.

Jardin C. and Crosnier J. 1975. Un taro, un poisson, une papaye. F.A.O., Commission du Pacifique du Sud, Noumea, Nouvelle Caledonie, 476 pp.

Kays S. and Silava Dias J.C. 1995. Common names of commercially cultivated vegetables of the world in 15 languages. Econ. Bot., 49 (2): 115–152.

Kayser M., Brauer S., Weiss G., Schiefenhövel W., Underhill P., Shen P., Oefner P., Tommaseo-Ponzetta M., Stoneking M., 2003. Reduced Y-Chromosome, but Not Mitochondrial DNA, Diversity in Human Populations from West New Guinea. Am. J. Hum. Genetics., 72: 281–302.

Kliks M. 1978. Palaeodietetics. A review of the role of dietary fiber in preagricultural human diets. *In:* Spiller G.A. and Amen R.J. (eds.). Topics in Dietary Fiber Research, Plenum Press, New York, NY, pp. 181–202.

Konrad G., Konrad U., and Schneebaum T. 1981. Asmat, Leben mit den Ahnen. Friedrich Brückner, Glashütten/Ts.,W-Germany, 193 pp.

Konrad G. and Konrad U. 1995. *Je Ti.* Gesellschaftliche Grundlage der Emari Ducur. *In:* Konrad G. and Konrad U. (eds). Asmat. Mythen und Rituale Inspiration der Kunst. Erizzo Ed. Venezia, pp. 105–214.

Littlewood R.A. 1972. Physical Anthropology of the Eastern Highlands of New Guinea. Univ. Washington Press, Seattle, WA, 224 pp.

May R. 1990. How many species? Phil. Trans. R. Soc. Lond. b. 330: 293–304.

May R. 1992. How many species inhabit the Earth? Sci. Amer. 267: 42–48.

McGrew W.C. 2001. The other faunivory: Primate insectivory and early human diet. *In:* Stanford C.B. and Bunn H.T. (eds.). Meat-eating and Human Evolution. Oxford Univ. Press, Oxford, UK, pp. 160–178.

Menzel P. and D'Aluisio F. 1998. Man Eating Bugs. Ten Speed Press, Berkeley, CA.

Mercer C.W.L. 1993. Insects as food in Papua New Guinea. *In:* Hardouin J. and Stievenart C. (eds.) Proc. Seminar "Invertebrates (Minilivestock) Farming" (EEC-DGXII/CTA/IFS/DMMMSU/ITM, Philippines, November 1992), pp. D15–D20.

Mercer C.W.L. 1997. Sustainable production of insects for food and income by New Guinea villagers. *In:* Paoletti M.G. and Bukkens S. (eds) Minilivestock. Ecology of Food and Nutrition (s.i.), 36, 2–4: 151–157.

Meyer-Rochow V.B. 1973. Edible insects in three different ethnic groups of Papua New Guinea. Amer. J. Clin. Nutr. 26: 673–677.

Meyer-Rochow V.B. 1975. Can insects help to ease the problem of world food shortage? SEARCH 6: 261–262.

Meyer-Rochow V.B. 1976. The use of insects as human food. Food Nutr. 33: 151–152.

Meyer-Rochow V.B and Changkija S. 1997. Uses of insects as human food in Papua New Guinea, Australia, and North-East India: Cross-cultural considerations and cautious conclusions. *In:* Paoletti M.G. and Bukkens S. (eds.). Minilivestock. Ecology of Food and Nutrition (s.i.), 36, 2–4: 159–185.

Morren G.E.B. 1977. From hunting to herding. Pigs and control of energy in montane New Guinea (West Sepik Province). *In:* Bayliss-Smith T. (ed.). Subsistence and Survival. Rural Ecology in the Pacific. Acad. Press, London, UK, pp. 273–315.

Motte-Florac E. and Thomas J.M.C. (eds.). (2003). Les insectes dans la tradition orale—Insects in oral literature and tradition. Peeters-SELAF (Ethnosciences), Paris-Louvain.

Ohtsuka R. 1970. The ecology of food consumption. Major nutrients in Foods. *In:* Ohtsuka R. and Suzuki T. (eds). Population Ecology of Human Survival. Bioecological Studies of the Gidra of Papua New Guinea. Univ. Tokyo Press, Tokyo, pp. 71–77.

Oliveira J.F.S., Passos de Carvalho J., Bruno de Sousa R.F.X., and Madalena Simao M. 1976. The nutritional value of four species of insects consumed in Angola. Ecol. Food and Nutr. 5: 91–97.

Paoletti M.G. 1995. Biodiversity. Traditional landscapes and agroecosystem management. Landscape and Urban Planning 31: 117–128.

Paoletti M.G. and Bukkens S. 1997. Minilivestock: Sustainable use of biodiversity for human food. Ecol. Food Nutr. (s.i.); 36: 2–4.

Paoletti M.G., Pimentel D., Stinner B.R., and Stinner D. (1992). Agroecosystem biodiversity: Matching production and conservation biology. Agric. Ecosys. Environ. 40: 3–26.

Peters F.E. 1957. Chemical composition of South Pacific foods: An annotated bibliography. Technical paper 100,. South Pacific Commission, Noumea, New Caledonia.
Posey D.A. 1978. Ethnoentomological survey of Amerind groups in lowland Latin America. Florida Entomologist, 61: 225–229.
Posey D.A. 1987. Ethnoentomological survey of Brazilian Indians. Entomol. Gener. 12: 191–202.
Ramos-Elorduy J. 1992. Proteinas de insectos comestibles como recurso actual y potencial. Proc. VII Seminario Nacional de Sanidad Vegetal. Riobamba, Ecuador.
Ramos-Elorduy J. and Pino J.M. 1990. Contenido calorico de algunos insectos comestibles de Mexico. Rev.ta Soc. Quim. Mexico 34: 56–68.
Rappaport R.A. 1967. Pigs for the Ancestors: Ritual in the Ecology of a New Guinea People. Yale Univ. Press, New Haven, CT.
Re, M.C., Tommaseo M., Furlini G., and La Placa M. 1989. High prevalence of serum antibody against human T Cell Leukemia Virus Type I (HTLV-I) among the Bisman Asmat population (Indonesian New Guinea). AIDS Res. Human Retroviruses, 5 (5): 551–554.
Rehm S. (ed.) 1994. Multilingual Dictionary of Agronomic Plants. Kluwer Acad. Publ., Dordrecht, The Netherlands, 286 pp.
Santos Oliveira J.F., Passos de Carvalho J., Bruno de Sousa R.F.X. and Madalena Simão M., 1976. The nutritional value of four species of insects consumed in Angola. Ecol. Food Nutr. 5: 91–97.
Schneebaum T. 1980. A Basu Suangkus feast. *In:* Trenkenschuh F. (ed.). An Asmat-Sketch Book. Asmat Musem of Culture and Progress, Agats (Irian Jaya) vol. 7. pp. 68–79.
Schneebaum T. 1985. Asmat Images, from the collection of the Asmat Museum of Culture and Progress. Asmat Museum of Culture and Progress, Agats (Irian Jaya), 198 pp.
Skinner M. 1991. Bee brood consumption: an alternative explanation for hypervitaminosis A in KNM-ER 1808 (*Homo erectus*) from Koobi Fora, Kenya. J. Human Evol. 20: 493–503.
Sowada A.A. 1986. Il culto del sago dei Papua asmat. *In:* AA.VV., Indonesia, la grande deriva etnica. Erizzo Editore, Venezia, pp. 135–139.
Sowada A.A. 1995. Sago und Sagolarven im Ritual. *In:* G. and Konrad U. (eds.). Asmat, Mythen und Rituale Inspiration der Kunst. Erizzo Editore, Venezia, pp. 91–104.
Stini W.A. 1988. Food, seasonality and human evolution. *In:* de Garine I. and Harrison G.A. (eds.). Coping with Uncertainty in Food Supply. Clarendon Press, Oxford, UK, pp. 32–51.
Stork N.E. 1988. Insect diversity: facts, fiction and speculation. Biol. J. Linn. Soc. 35: 321–337.
Sutton M.Q. 1988. Insects as food: aboriginal entomophagy in the Great Basin. Ballena Press Anthropol. Papers No. 33, Ballena Press, Menlo Park, CA, 115 pp.
Taylor R.L. and Carter B.J. 1976. Entertaining with insects, or: the original guide to insects cookery. Woodbridge Press, St. Barbara, CA, 160 pp.
Tommaseo M. and Lucchetti E. 1992. Anthropometrical study of the Asmat. Anthropologischer Anzeiger 50 (3): 217–234.
Tommaseo M., Doglioni L., and Alciati G. 1989. Haematochemical data of the Asmat population. Homo (1990) 3–4: 117–133.
Tommaseo M., Baroni A.G., and Altinier G. 1990. Asymptomatic hepatitis among the Asmat. Int. J. Anthropol. 5 (3): 35–44.
Tommaseo M., Hauser G., Vienna A., and Konrad G. and U. 1996. Fattori biologici e culturali nello studio di crani Asmat (Nuova Guinea Indonesiana). Museologia Scientifica XII: 133–151.
Tommaseo M., Lucchetti E., Alciati G., and Altinier G. 1992. Contribution to study of Asmat genetic variability. Int. J. Anthropol. 7 (1): 59–67.
Tommaseo-Ponzetta M. 2003. Rôle alimentaire des insectes dans l'evolution humaine. *In:* Motte-Florac E. and Thomas J.M.C. (eds.). Les insectes dans la tradition orale—Insects in oral literature and tradition. Peeters—Selaf (Ethnosciences 5), Paris, pp. 269–286.
Tommaseo-Ponzetta M., Hauser G., and Vienna A. 1997. Auditory hyperostosis and the environment: an update. Int. J. Anthropology, 12, 2: 29–40.
Tommaseo-Ponzetta M., De Robertis M., Tanzariello F., Attimonelli M., and Saccone C., 2000. Morphological and molecular variability of West New Guinea pygmoid groups. Rivista di Antropologia, suppl. vol.78: 253–260.

Tommaseo-Ponzetta M., De Robertis M., Tanzariello F., Attimonelli M., and Saccone C., 2002. Mitochondrial DNA variability in West Papua populations. Am. J. Phys. Anthrop., 117(1): 49–67.

Tommaseo-Ponzetta M., Hauser G., and Vienna A., 1997. Auditory hyperostosis and the environment: an update. International Journal of Anthropology, 12, 2: 29–42.

Tommaseo-Ponzetta M., Nardi S., Calliari I., and Carniel G., 1998. Trace elements in human hair and soil in Irian Jaya (Indonesia). Biological Trace Element Research, 62: 199–212.

Townsend P.K. 1974. Sago production in a New Guinea economy. Human Ecol. 2(3): 217–236.

Trenkenschuh F. 1970. Sago gathering practices. *In:* Trenkenschuh F. (ed.). An Asmat Sketch Book, The Asmat Museum of Culture and Progress, Agats (Irian Jaya) vol. 1, pp. 74–80.

Ulijaszek S.J. and Poraituk S.P. 1993. Making sago in Papua New Guinea: Is it worth the effort? *In:* Hladik C.M., Hladik A., Linares O.F., Pagezy H., Semple A., and Hadley M. (eds.) Tropical Forests, People and Food. Unesco, Paris and Parthenon Publ. Group, NewYork, NY, pp. 271–280.

van Amelsvoort V.F.P.M., 1964. Culture, stone age and modern medicine. Assen: Van Gorcum.

Van Arsdale P.W. 1978. Perspectives on Development in Asmat. *In:* Trenkenschuh F. (ed.). An Asmat Sketch Book,. Asmat Museum of Culture and Progress, Agats (Irian Jaya) vol. 5 (2): (1st ed. in 1975), 403 pp.

Voorhoeve C.L. 1965. The Flamingo Bay Dialect of the Asmat language. Martinus Nijhoff, The Hague, Netherlands.

Wilson E.O. 1988. The current state of biological diversity. *In:* Wilson E.O. (ed.). Biodiversity, Natl. Acad. Press, Washington DC, pp. 3–18.

Zachmann E. 1966. Journal (cited in F. Trenkenschuh, 1970).

22

Contemporary Use of Insects and Other Arthropods in Traditional Korean Medicine (*Hanbang*) in South Korea and Elsewhere

ROBERT W. PEMBERTON

Invasive Plant Research Laboratory, Agricultural Research Service, United States Department of Agriculture, 3205 College Ave., Ft. Lauderdale, FL 33314, USA.
E-mail: bobpem@saa.ars.usda.gov

Abstract

Insects and other arthropods have appeared in pharmacopeias of traditional Korean medicine but little is known about their use in modern South Korea. Interviews were conducted with 20 traditional medicine doctors at clinics in South Korea's *Kyeong Dong Shijang* in Seoul (one of the world's largest traditional drug markets) in 1993 to learn about the current patterns of usage. Seventeen products were prescribed and the use of arthropodal medicines either stable or increasing. Centipedes (*Scolopendra* spp.), used primarily to treat arthritis, and the silk moth fungus (*Beauveria bassiana* which infects silk moth larvae) used mostly to treat stroke, were the two most frequently prescribed and medically important arthropod drugs. Most of the arthropod drugs were traditionally collected or reared on the Korean Peninsula but are now imported, mainly from China. Folk logic appears to be the basis for some arthropod drugs use (i.e., centipedes, which have many legs, were used for leg problems). But many of the arthropods used have venoms and other defensive chemicals that are biologically and probably pharmaceutically active. The Korean use of arthropods as drugs (as well as for food and enjoyment) is due, in part, to more positive attitude towards these animals compared to many cultures. Use of traditional Korean medicine has expanded dramatically in the last 30 years and spread with Korean immigrants to the United States and other countries where Koreans have settled, such as Australia. Arthropod-based medicine is consumed by Koreans living overseas, a use facilitated by internet marketing, primarily by Chinese mainland companies. Terrestrial arthropods appear to be an unexplored and unexploited source of drugs for modern medicine.

Key Words: antiglobalization, arthropod, insect, drug, medicine, *Hanbang*, E-Commerce, Korea.

Introduction

In South Korea, people freely integrate modern scientific medicine with traditional Korean medicine, known as *hanbang*. *Hanbang* tends to be used to prevent illness through the use of medicinal tonics and to treat chronic health conditions, whereas modern medicine is more frequently used for traumatic injuries, illnesses requiring surgery, and infectious disease (Han, 1997). Both systems have a large obvious presence and prominence in the country. Seoul houses large modern university hospitals with high-tech medicine as well as one of the world's largest traditional medicine markets. This fascinating market, located in the Chegidong District of the city and called *Kyeong Dong Shijang*, comprises more than 900 businesses devoted to various aspects of traditional medicine (H. Park,* pers. comm., 1993). Among these are clinics with traditional medicine doctors who also sell prescribed drugs. Many other businesses are devoted to the processing of whole drug source materials into products that are also sold. Most of these are small shops with rows of processing equipment, and frequently with workers sitting in front of the shops chopping dried plants and other materials. The entire area is redolent of the smell of liquorice, sage-like *Artemisia* species, and other aromatics.

Traditional Korean medicine has undergone rapid growth in the last 30 years. Before the introduction of Western medicine by Christian missionaries in the 1880s, it was the medicine of Korea. In the first half of the 20th Century, the Japanese systematically suppressed traditional Korean medicine during the colonial period, commencing in 1907 with closure of the state-run *hanbang* medical school (Han, 1997). *Hanbang* doctors were still active during the colonial period but with no formal training their numbers declined. In 1951, the Korean government reopened the *hanbang* medical school. The number of schools offering *hanbang* medical training had increased to 2 by 1971, 11 by 1989 (Han, 1997), and 17 by 2000 (Kang, 2003). *Hanbang* hospitals have been built and *hanbang* medicine used more frequently as a curative medicine (Han, 1997). The increased popularity of Korean traditional medicine is due, in part, to patriotic feelings and increasingly to antiglobalization sentiments, which prompts some Koreans

*Mr. Hoon Park. Korea Pharmaceutical Traders Association, Room 1801 Korea World Trade Center, 159-1 Sam Sung-Dong, Kangnam-Ku, Seoul, South Korea.

to view modern-science-based medicine as invasive Western medicine (Kang, 2003). This parallels the antiglobalization sentiments directed towards imported foods, and enhanced valuation of foods gathered from the wild in Korea or produced on Korean farms (Pemberton, 2002). Traditional Korean medicines are both expensive and highly profitable, as evidenced by the legal wrangles in the 1990s over who has the right to sell them (Cho, 2000).

Traditional Korean medicine is similar to traditional Chinese medicine but with some unique theoretical and practical aspects. One difference is the Theory of Physical Constitution, which classifies people into one of four categories according to the size of their lungs, liver, spleen, and kidneys (Kang, 2003). Each category is specific in temperament, propensity for disease, and particular drugs needed for maintaining health and curing disease. In 1610 AD, the *Tong-Eui Pogam* (Treasures of Eastern Medicine) was published by the court physician Heo Jun (1610). This 25-volume work is still used today. Like Chinese traditional medicine, Korean traditional medicine uses a great array of mineral and biological substances as drugs. Plants are the most common source of drugs but many animals are also used. One of the marked differences between the drug materials in East Asian traditional and modern scientific medicines is the use of insects, other terrestrial arthropods, and their products as drugs (Read, 1935; Kim, 1984; Namba et al., 1988; Ding et al., 1997).

One of the standard pharmacopeias of South Korea (Kim, 1984) lists 13 terrestrial arthropod or arthropodal products. This encyclopedia, like many, is a fascinating and informative compendium of accumulated knowledge about drug materials and their uses, but provides little information about what is actually in use at the present time. My 1993 study sought to detect and describe the arthropod drug products currently in use in South Korea (Pemberton, 1999). This chapter is based largely on that study.

Methodology

Interviews with Korean traditional medicine doctors were conducted in Korean to determine what species of insects, other terrestrial arthropods, and their products they prescribed (if any). These doctors were practicing in small clinics that typically have an area for patient examination and diagnosis, and a drugstore that prepares and dispenses the drug mixtures ordered by the doctors. Traditional medicine doctors in South Korea attend six-year university programs that typically combine undergraduate-graduate degrees. Interviews were undertaken in 20 randomly selected clinics in the large *Kyeong Dong Shijiang* traditional medicine area of Seoul, South Korea, in November, 1993. The doctors interviewed ranged widely in age, from late 20s to 60s, and were, with one exception, all men.

The first part of the interview was based on the Korean language "Illustrated Natural Drugs Encyclopedia" (Kim, 1984). Photocopies of 13 pages, each illustrating and describing a different arthropod drug, were shown to the doctors, one page or product at a time, and he (she) asked whether it was currently prescribed. Next the doctors were asked whether they prescribed six other arthropod drugs not in the Kim Encyclopedia. Three of these were products I had encountered at various markets or seen offered by street peddlers and the other there discovered in preliminary sampling prior to the survey. The doctors were also asked if they knew of other arthropod drugs and which did they prescribe.

The second part of the interview concentrated on determining which of the arthropod drugs each doctor in his (her) own practice most commonly prescribed, the ones most medically valuable, and the most expensive. Additional questions were asked about whether arthropod drugs were mostly domestic or imported, and whether they were wild-gathered or cultured. Finally, they were asked whether the use of arthropod drugs was increasing, decreasing, or stable in their experience.

To facilitate a meaningful interpretation, the answers given in the above questionnaire and those extracted by Kim (1984) [translated from Korean *Hangul* and Chinese characters] are summarized in Appedix 22.1.

Results

The arthropod drugs used in Korean traditional medicine are listed in Table 22.1 in descending order by commonality (number of the 20 clinics visited in which they were found). The doctors prescribed 17 of the 19 arthropod drugs about which they were asked and indicated that they used no others. The degree of use of these products varied greatly. Three of the products, centipedes (Plate X, 5), cicada nymphal skins, and the silk moth fungus were prescribed by all the doctors. About half (9 of 19) of the products were sold by 10 of the 20 clinics. Two products, rice field grasshoppers and predatory water beetles, both listed in Kim's Encyclopedia, had yet to be prescribed by any of the doctors. Contrarily, 11 of 13 (84.6%) products in Kim's Encyclopedia are in current use, as well as 6 products not included in this reference work. Three of the products absent from the Encyclopedia, namely, praying mantis egg cases, scarab larvae, and *Cordyceps sinensis* (Berkeley) infected larvae, are common today, prescribed by 17 of 20 (85%), 14 of 20 (70%), and 9 of 20 (45%) of the doctors respectively.

The most important arthropod drugs in terms of "the most prescribed", "stated medical importance", and "cost" are shown in Table 22.2. Centipedes were the most frequently prescribed drug in 60% of the clinics, followed by the silk moth larvae infected with fungus (the most prescribed arthropod drug in 25% of the clinics), then honey produced from honeybees, the most prescribed in 15% of the clinics. Silk moth larvae infected with fungus were considered the medically most important by 65% of the practitioners, followed by centipedes, scorpions, cicada nymphal skins, and scarab larvae. Doctors at 50% of the clinics [only 16 responded to questions about the relative cost of the products] said

Table 22.1: Insects and other arthropods used in Korean traditional medicine

	Species or products (in order of commonness)	Korean name Korean name in *Hangul* (Romanized)	Clinics prescribing*
1.	Centipedes (*Scolopendra* species)	*wangjine*	100% (20/20)
2.	Silk moth larvae (*Bombyx mori*) infected with fungus (*Beauveria bassiana* (Bals.) Vuill.)	*baeggangjam*	100% (20/20)
3.	Cicada nymphal skins (*Cyrptotympana* species)	*seontoi*	100% (20/20)
4.	Scorpion (*Buthus martensis* Karsch)	*jeongal*	95% (19/20)
5.	Honey from honeybees (*Apis* sp.)	*bongmil*	90% (18/20)
[b]6.	Praying mantis egg cases (Mantidae)	*samagui aljib*	84% (16/19)
[b]7.	Scarab beetle larvae (Scarabaeidae)	*kumbangi*	70% (14/20)
8.	Silk moth larvae dung (*Bombyx mori* L.)	*nueddong*	55% (11/20)
9.	Blister beetles (*Mylabris* sp.)	*banmo*	50% (10/20)
[b]10.	Hepialid moth larvae (*Hepialus oblifurcus* Chu et. Wang) infected with fungus *Cordiceps sinensis*	*dongjungchacho*	45% (9/20)
11.	Mole cricket (*Gryllotalpa africana* Pal.)	*ddanggangaji*	40% (8/20)
12.	Paper wasp nests (*Vespa* and *Polistes* spp.)	*nobongbang*	30% (6/20)
13.	Silk moth adults (*Bombyx mori*)	*nuenabang*	20% (4/20)
14.	Horse fly adults (*Tabanus* species)	*soidunge*	10% (2/20)
[b]15.	Paper wasp larvae (Vespidae)	*malbeol*	10% (2/20)
[b]16.	Paper wasp adults (Vespidae)	*malbeol*	5% (1/20)
[b]17.	Bee glue (*Apis* spp.)	*milrab*	5% (1/20)
18.	Rice field grasshoppers (*Oxya* species)	*byomeddugi*	0% (0/20)
19.	Predatory diving beetle (*Cybister tripunctatus* (Oliv.))	*mulbanggae*	0% (0/20)

[a]Number of the 20 clinics interviewed prescribing the drug.
[b]Not in Kim's drug Encyclopedia (1984).

that scorpions are the most expensive arthropod drugs. Others stated that the "most expensive products" in some clinics included centipedes, scarab larvae, the *Cordyceps* fungus infecting *Hepialis* moth larvae, silk moth larvae infected with fungus, and horse flies.

Most doctors [17 of 20] indicated that arthropod drugs were usually mixed with other drug materials to produce the desired effects. A few doctors [3 of 20] always mixed arthropod drugs, though most doctors used some arthropod drugs without admixtures. Centipedes were most frequently used alone, but the silk moth larvae infected with fungus, scarab larvae, and *Cordyceps* fungus were also used singly.

Most doctors [13 of 15] believe that most arthropod drugs were collected from the wild, except for silk moth and honeybee products. They indicated that arthropod drugs were mostly [14 of 20 clinics] or entirely [5 of 20] imported. All 14 doctors who mentioned a source country for imports said China was the primary source. According to these three doctors, imports also came from Thailand, while one doctor indicated that Malaysia was a source.

Table 22.2: Most important arthropod drugs in traditional Korean medicine

Species or products	Most prescribed[a]	Medical importance[b]	Most expensive[c]
Centipedes	no. 1 in 12 of 20	no. 1 in 3 of 20	no. 1 in 2 of 16
Bombyx mori larvae infected with silk moth fungus *Beauveria bassiana*	no. 1 in 5 of 20	no. 1 in 13 of 20	no. 1 in 1 of 16
Cicada nymphal skins		no. 1 in 1 of 20	
Scorpions		no. 1 in 2 of 20	no. 1 in 8 of 16
Honey from honeybees (*Apis* species)	no. 1 in 3 of 20		
Scarab larvae		no. 1 in 1 of 20	no. 1 in 2 of 16
Hepialis larvae infected with *Cordyceps* fungus			no. 1 in 2 of 16
Horse flies			no. 1 in 1 of 16

[a]Number of the 20 clinics studied in which the product was prescribed most often.
[b]Number of the 20 clinics studied in which the product was considered the most important medically.
[c]Number of the 20 clinics studied in which the product was the most expensive.

With regard to the trend in the use of arthropod drugs in their practices, only one doctor [1 of 17] indicated that it was declining. Some doctors [4 of 17] indicated that they were prescribing some products more, others less than they used to, but most [12 of 17] indicated that their use of these products was either stable [6 of 17] or increasing [16 of 17]. An increase in use of arthropod based medicines is supported by importation data discussed below.

In Appendix 22.1, the products are divided into various types, including whole arthropods, arthropod parts and products, and disease-infected arthropods. The whole dried arthropods are the most common product type, comprising 11 of the 19 items listed, parts or products 6, and diseased insects, 2.

Discussion

Arthropods as Medicinal Preparations

All of these arthropod medicaments, except bee glue, are recorded in Read's compendium of Insect Drugs of China (Read, 1935). Today bee glue is one of the honeybee products dispensed by China's honeybee sting treatment centers (Pemberton, 2003). While arthropod medicaments are similar in China and Korea, medicinal uses often differ (Read, 1935; Namba et al., 1988; Ding et al., 1997).

Rice-field grasshoppers (*Oxya* spp., Acrididae), a product listed by Kim (1984) as a home remedy, were not used by the surveyed practitioners. These grasshoppers were the traditional food that has recently experienced a revival

(Pemberton, 1994). Grasshoppers were used in Amerindian and ancient Mexican medicine (Kevan, 1979; Ramos-Elorduy de Conconi and Moreno, 1988)

The predatory water beetle (*Cybister tripunctatus* Greshew), one of the products apparently not in current use, is a well-known insect in Korea, used in a traditional water beetle roulette-like game (Pemberton, 1990a).

Two products not in Kim's book but used today, are vespid wasp adults and larvae. The author encountered a living colony and nest of a vespid wasp (a hornet, *Vespa manchurica* Smith, in this case) on the street in downtown Seoul. It was offered for sale by a farmer who declared that if one larva was eaten every day, one would not feel the cold so much in winter. Previous observations by the author indicate that many kinds of vespid adults and larvae are commonly seen in various bottled spirits offered for sale in Taiwan, where they are also used as a tonic. Wasps have also been reported to be used in medicine by various South American cultures and in ancient Mexico (Posey, 1986; Ramos-Elorduy de Conconi and Moreno, 1988). Another product not listed by Kim is scarab larvae, which the author occasionally encountered for sale among sidewalk peddlers and at fairs. The Korean name *kombangi* is applied to most soil-dwelling scarab larvae, and probably many species are collected for medicine, as they are in China (Read, 1935).

A number of the doctors commented on the toxicity of blister beetles and mentioned the poisoning death of a pregnant woman due to an overdose. Some practitioners indicated that they no longer prescribed it, precisely because of toxicity problems. It is worth mentioning that blister beetles were also used in ancient times by Greeks and Mexicans (Beavis, 1988; Ramos-Elorduy de Conconi and Moreno, 1988).

Importation Aspects in South Korea

The belief of the doctors interviewed that most imported arthropod drugs are from China was confirmed by the Korean Pharmaceutical Traders Association, Trade Department manager, Mr. H. Park (per. comm.). Importation of products in South Korea is a sensitive issue, however. There is a strong preference for Korean drug and food products over imported ones (Pemberton, 1995a; 2002). A Korean belief called *shint'o puri*, which says "that things from Korean soil are best for Korean bodies" has been promoted and strengthened by the Korean government and business interests, channeled through consumer groups, to limit imports. However, while some clinics declared that they try to sell only Korean arthropod drugs, the reality is that most are now selling imported products. Among the reasons is that in recent years labor costs have risen very steeply in South Korea, which makes imported products from China much cheaper. Since the arthropod faunas of Korea and China at similar latitudes are comparable, the same or almost identical products that were formerly collected in Korea are now obtained from China.

Not long ago, in the 1980s, some products, such as centipedes, were not allowed to be imported, but around 1986, the regulations were changed to permit their import (H. Park, pers. comm.). In 1989, the Korean Pharmaceutical Traders Association began to publish annual reports detailing imports, which show a clear upward trend for the import of arthropodal drugs. From 1989 to 1992, the quantity of scorpions imported increased from 7,115 to 14,635 kg (Anonymous, 1989, 1990, 1991, 1992). Imports of centipedes, which are sold and tracked as individual units, increased from 15,804,400 units in 1990 to 50,209,220 units in 1991; it dipped to 34,804,060 units in 1992, when centipede imports were worth $1,850,137.

The Korean sensitivities concerning imports and trade relations made this research extremely difficult. During the period in which the interviews took place, the Uruguay Round of the world trade talks was being held. Many Koreans were angered by these talks and other actions to encourage or coerce a greater opening of their markets to imports. It is ironic that people in an industry now dependent on imports, were reluctant to talk with the author because of their discomfort that international talks might encourage Korea to open its markets. Some doctors declined to be interviewed because they thought the author was an investigator looking for internationally protected rare species, such as products derived from a tiger, which have long been medicinal material in East Asian traditional medicine. When the author explained he was only interested in learning the use of insects in their practice, some assumed his membership in a "protect the insects" organization. During field trials to test interview techniques, it was discovered that approaching clinics and doctors formally, with business card presentations, self-introductions, and stated purpose, evoked almost universal suspicion and rapid refusal. An informal approach of entering the clinics, showing the doctors photocopies of the medicinal arthropods, then, asking questions about the products, stimulated curiosity and a willingness to participate in the interviews. One-fourth of the interviews with doctors were obtained through referrals from friendly doctors, who participated in the interviews.

Folk Logic

Independent of the actual efficacy of these arthropodal drugs, an issue not addressed here, is an interesting folk logic that relates some characteristic of the product to its medical use. One aspect is similar to the European Doctrine of Signatures that sees an object's medicinal use in its shape or form. Thus, centipedes with their numerous legs, feet, and articulated body segments are used for leg, foot, and joint problems. Another related kind of reasoning sees the medical use in the negative interactions the arthropod has with people. For instance, blister beetles, which cause human skin to blister, are used to treat skin diseases. Scorpions, whose stings cause pain, are used to treat pain. Korean folk logic also relates reproductive parts or products to enhance sexual potency. Thus, praying mantis egg cases are used to stimulate male sexual stamina; cooked dog penis is eaten by some Korean men for the same purpose. Other apparently similar but unverified folk logic is the use of cicadas (whose loud calls are prominent parts of Korea's summer) to treat hearing problems and pharyngitis, and the use of blood-feeding horse flies (*Tabanus* species) to treat blood problems. Another interpretation of the similarity between arthropod characteristics and drug use is that the arthropod characteristics serve as mnemonic devices to mark and recall specific uses of these medicines (Nina Etkin, pers. comm.).

Biologically Active Chemistry

The biological or behavioral characteristics of these arthropods suggest the presence of biologically active chemicals that may be pharmacologically active. Wasps, scorpions, and centipedes have venoms used to defend themselves or to secure prey (Whitman et al., 1990). Cantharidin, used by blister beetles as a defensive chemical, has considerable activity in vertebrates including people, who can be fatally poisoned by as little as 0.5 mg kg^{-1} body weight (Eisner, 1970). Arthropods that live in close proximity, such as social wasps and bees, are subject to epidemic diseases. To limit disease activity, they incorporate antimicrobial compounds into their nests (Beattie et al., 1986). Likewise, soil dwelling insects, such as mole crickets, scarab larvae, cicada nymphs (which spend years in the soil), and centipedes are subject to microbial attack; therefore, they should produce chemicals to fight such attacks. Indeed, soil-dwelling ants have been shown to make and employ compounds that kill both fungi and bacteria in their underground nests (Holldobler and Wilson, 1990). Even the predatory water beetle (*Cybister tripunctatus*), listed in Kim's Encyclopedia, but not prescribed by the interviewed doctors, is known to use phenolic compounds to repel microbial attacks (Rodriquez and Levin, 1976). Fungi commonly make antibiotics and other toxins (Griffin, 1981), and have known pharmacological activity (Lewis and Elvin-Lewis, 1977). This suggests that *Beauveria bassiana* which infects silk moth larvae and *Cordyceps sinensis* fungus infecting *Hepialis* moth larvae may also contain pharmacologically useful chemicals. *Cordyceps sinensis*-infected *Hepialis* moth larvae do indeed contain an array of biologically active chemicals, which make it highly valued in Chinese medicine and of promise to Western medicine (Steinkraus and Whitefield, 1994).

The diverse defensive chemicals of arthropods differ in origin. Some are products manufactured by the arthropods themselves while others are defensive chemicals obtained from plants or prey, which then are sequestered, concentrated, or transformed for the arthropod's own defense (Eisner, 1970). Interestingly arthropods and plants frequently utilize the same chemical repellents for defense (Rodriguez and Levin, 1976; Harborne, 1988). Since plants or their chemicals constitute one of our largest sources of drug material (Balick and Cox, 1996), it is reasonable to expect pharmacological activity from arthropods, which feed on these drug-producing plants and incorporate phytochemicals or produce similar chemicals.

Cultural Values Related to Insects

The stable or upward trend in the use of arthropods in traditional Korean medicine is very interesting in light of South Korea's shift from a rural undeveloped nation to a modern industrial power. In this case, economic modernization and increasing standards of living have not meant loss of traditional medicines. A similar reality exists with regard to the continued use and value of wild-gathered food plants in South Korea (Pemberton, 1995a).

The author believes that the use of many insects and other arthropods as drugs is due to very different attitudes about these animals in Korea and other parts of eastern Asia, compared to the West. In general, East Asian cultures have a more balanced perspective regarding insects than do Western cultures. Entomophobia is uncommon in Asia, as is the idea held by many Westerners that most insects are related to filth or are dangerous. In Eastern Asia, some insect species are indeed disliked because they are pests, but others are considered aesthetically pleasing, good to eat, interesting pets, subjects of sport, enjoyable to listen to, and useful in medicine. This traditional perspective (Hearn, 1905; Laufer, 1927; Bodenheimer, 1951; Read, 1935) continues to persist in modern-day East Asia (Mitsuhashi, 1984; Pemberton, 1990a, 1990b, 1994, 1995b, 1995c, 2003).

Arthropod Drugs Go Global with Traditional Korean Medicine

The so-called Korean Diaspora has resulted in an estimated five million ethnic Koreans living outside Korea, a number equivalent to about 10% of South Korea's population (Kwon, 1997). China and the United States, with about two million each, are home to most Korean expatriates. Traditional Korean medicine has spread with Korean immigrants to the United States and other countries where Koreans have settled such as Australia (Pourat et al., 1999; Han, 2000). To determine whether or not the arthropod-based drug component of traditional Korean medicine is being used in the United States, *hanbang* pharmacy-clinics were queried by telephone by a native Korean speaker in April, 2003. Single, randomly selected *hanbang* pharmacy-clinics in New York and Los Angeles, the cities with the largest Korean communities, and in Philadelphia, suburban Atlanta, and Oakland, California were called. Each was asked if it sold the four arthropod drugs most frequently prescribed by *hanbang* pharmacy-clinics in Seoul during 1993 (centipedes, silk moth larvae infected with *Beauveria bassiana*, shed skins of cicada nymphs, and scorpions). All except the Oakland clinic indicated that they did. The Atlanta clinic said that they had centipedes for their own use but could order more to sell. Several clinics indicated that a prescription from a *hanbang* doctor was needed for the arthropod drugs. A *hanbang* wholesale importer in Los Angeles was located, which offers large quantities of these arthropods. The same telephone queries were made to a *hanbang* clinic in Sydney,

Australia and another in Auckland, New Zealand, cities with Korean immigrant communities. The Australian clinic indicated that centipedes, cicada nympal skins and the silk moth larvae with *B. bassiana* were available but scorpions were not. The New Zealand responder indicated that the products are illegal in New Zealand.

Internet searches were made during April, 2003 to locate suppliers of these Korean arthropod drugs. E-commerce suppliers of most important arthropod drugs were readily located, and sources of 14 of the 17 arthropod based drugs sold in Seoul in 1993 were found during a two-hour search. Most of the suppliers were in China, which is also the source for most arthropods drugs sold in South Korea. Both whole arthropods and extracts are available. Most of the suppliers' web pages are bilingual (English and Chinese) or trilingual (Chinese, English and Korean), reflecting the overlap in the use of these products by traditional Chinese medicine and traditional Korean medicine, as well as international marketing. Prices are listed in US dollars and credit cards are accepted.

Conclusions

Drugs from arthropods continue to be important in Traditional Korean and Chinese medicine, but are virtually absent in modern Western medicine. The types of arthropods used for medicine in Korea and China also occur in Europe, North America, and elsewhere. The species are different but the families and many of the genera are the same (Brues et al., 1954). Arthropod chemicals such as venoms, defensive secretions, and antimicrobials, that appear to be the basis of pharmacological activity, are also widespread (Eisner, 1970; Holldobler and Wilson, 1990). The absence of arthropod-based drugs in the West is probably related to negative cultural attitudes toward arthropods. The enormous richness and diversity of arthropods, and the use of many species as drugs against common and important diseases in traditional medicine in Korea and elsewhere, suggest that arthropods are a large, unexplored and unexploited source of potentially useful compounds for modern medicine.

Acknowledgments

The author thanks the many traditional medicine doctors (K.W. Choi, J.H. Chang, M.S. Han, C.Y. Hwang, P.I. Hwang, S.H. Joo, C.K. Kim, M.H. Kim, S.S. Lee, J.W Lee, P.H. Lee, S.W. Lee, S.Y. Lee, M.H. Song, an anonymous doctor, Hwal-Seng Dang Inc., Kum San Clinic, and Woo Nam Oriental Medicine Clinic) for granting interviews. Jeong-Sook Kim is acknowledged for her excellent efforts in persuading people to give interviews and for translation during the interviews. Hoon Park is thanked for information relating to the importation of arthropod drugs and the *Kyeong Dong Shijang* market, and Gil Soo Han for providing information on the use of traditional Korean medicine in Australia. Cho Yeon Hwang kindly assisted with translation of Korean language materials and Shu-Hui Wan made helpful translations of Chinese characters related to medical use of the arthropod drugs. Cho Yeon Hwang also assisted with Korean language telephone interviews. Thanks are also given to N. Etkin who provided rich discussion, and to D. Moerman for a helpful review of the Journal of Ethnopharmacology manuscript.

References

Anonymous. 1989, 1990, 1991, 1992. Pharmaceutical Exports and Imports. Korean Pharmaceutical Traders Association, Seoul, South Korea, 1989, pp. 617–629; 1990, pp. 743–757; 1991, pp. 642–655; 1992, pp. 577–591. (in Korean).

Balick M.J. and Cox P.A. 1996. Plants, People, and Culture, the Science of Ethnobotany. American Scientific Library, W.H. Freeman and Co., New York, NY, 228 pp.

Beattie A.J., Turnbull C.L., and Hough T. 1986. Antibiotic production: a possible function for the metapleural glands of ants (Hymenoptera: Formicidae). Ann. Entomol. Soc. Amer. 79: 448–450.

Beavis I.C. 1988. Insects and Other Invertebrates in Classical Antiquity. Univ. Exeter Press, Exeter, Devon, UK, 269 pp.

Bodenheimer F.S. 1951. Insects as Human Food. Dr. W. Junk Publ., The Hague, The Netherlands, 352 pp.

Brues C.T., Melander L.A., and Carpenter F.M. 1954. Classification of Insects. Bull. Mus. Comp. Zool. 108: 1–917.

Cho H-J. 2000. Traditional medicine, professional monopoly and structural interests: a Korean case study. Soc. Sci. Med. 50: 123–135.

Ding Z., Zhao Y., and Gao X. 1997. Medicinal insects in China. Ecol. Food Nutr. 36: 209–220.

Eisner T. 1970. Chemical defenses against predators in arthropods. *In:* Sondheimer E. and Simeone J.B. (eds.). Chemical Ecology. Acad. Press, New York, NY, pp. 157–217.

Griffin D.H. 1981. Fungal Physiology. John Wiley and Sons, New York, NY, 383 pp.

Han G.S. 1997. The rise of Western medicine and the review of traditional medicine in Korea: a brief history. Korean Studies 21: 96–121.

Han G.S. 2000. Traditional herbal medicine in the Korean community in Australia: a strategy to cope with health demands of immigrant life. Health: Interdisciplinary J. Soc. Study Health, Illness Med. 4: 426–454.

Harborne J.B. 1988. Introduction to Ecological Biochemistry. Acad. Press, London, UK, 356 pp.

Hearn L. 1905. Exotics and Retrospectives. Little, Brown, and Co., Boston, MA, USA, 299 pp.

Heo Jun 1610. Tong-Eui Pogam. 1966 Reprint. Pun Nyon Sa, Seoul, 2323 pp. (in Korean)

Holldobler B. and Wilson E.O. 1990. The Ants. Belknap Press Harvard Univ., Cambridge, MA, USA, 732 pp.

Kang G-L. 2003. A Korean skeptic's report: New ager-occupied territory. Committee Scientific Investigations of Claims for the Paranormal. www.csicop.org/sb/2000–03/korea.html (April 12).

Kevan K.M. 1979. The place of grasshoppers and crickets in Amerindian cultures. Proc. Second Triennial Meeting Pan-American Acridology Society, July 23–August 3, 1979, Bozeman, Montana. 74 pp.

Kim J.G. 1984. Illustrated Drugs Encyclopedia. Nam San Dang, Seoul, 503 pp. (in Korean)

Kwon T-H. 1997. Koreans in the World. Korean J. Population Development 26:1–18.

Laufer B. 1927. Insect musicians and cricket champions of China. Field Museum of Natural History Anthropological Leaflet 22: 1–27.

Lewis W.H. and Elvin-Lewis M.F. 1977. Medical Botany. John Wiley and Sons, New York, NY, 515 pp.

Mitsuhashi J. 1984. Edible Insects of the World. Kokinshoin, Tokyo, 270 pp. (in Japanese)

Namba T., Ma Y.H., and Inagaki K. 1988. Insect-derived crude drugs in Chinese Song Dynasty. J. Ethnopharmacology 24: 247–285.

Pemberton R.W. 1990a. The Korean water beetle game. Pan-Pacific Entomol. 66: 172–174.
Pemberton R.W. 1990b. The selling of singing *Gampsocleis gratisoa* Brunner (Orthoptera: Tettigoniidae) as singing pets in China. Pan-Pacific Entomol. 66: 93–95.
Pemberton R.W. 1994. The revival of the rice-field grasshoppers (*Oxya* spp.) as human food in South Korea. Pan-Pacific Entomol. 70: 323–327.
Pemberton R.W. 1995a. Wild food plants in South Korea: market presence, new crops, and exports to the United States. Econ. Bot. 50, 57–70.
Pemberton R.W. 1995b. Catching and eating dragonflies in Bali and elsewhere in Asia. Amer. Entomol. 41, 97–99.
Pemberton R.W. 1995c. Insects: old food in new Japan. Amer. Entomol. 41: 227–229.
Pemberton R.W. 1999. Insects and other arthropods used as drugs in Korean traditional medicine. J. Ethnopharmacology 65: 207–216.
Pemberton R.W. 2002. Wild-gathered foods as countercurrents to dietary globalization in South Korea. *In:* Cwiertka K. and Walraven B. (eds.), Asian Food, the Global and the Local. Curzon Press, Richmond, Surrey, UK, pp. 76–94.
Pemberton R.W. (2003). Persistence and change in traditional uses of insects in contemporary Asian cultures. *In:* Motte-Florac E. and Thomas J.M.C. (eds.). Les insects dans la tradition orale-insects in oral literature and tradition. (Ethnosciences 5) Peeters-Selaf, Paris.
Posey D.A. 1986. Topics and issues in ethnoentomology with some suggestions for the development of hypothesis-generation and testing in ethnobiology. J. Ethnobiol. 6: 99–120.
Pourat N., Lubben J., Wallace S.P., and Moon A. 1999. Predictors of use of traditional Korean healers among elderly Koreans in Los Angeles. Gerontologist 39: 711–719.
Ramos-Elorduy de Conconi J. and Moreno J.M. 1988. The utilization of insects in the empirical medicine of ancient Mexicans. J. Ethnobiol. 8: 195–202.
Read B.E. 1935. Insect drugs. Peking Nat. Hist. Bull. 94: 8–85.
Rodriguez E. and Levin D.A. 1976. Biological parallelisms of repellents and attractants in higher plants and arthropods. *In:* Wallace J.W. and Mansell R.L. (eds.). Biochemical Interaction between Plants and Insects. Recent Advances in Phytochemistry. Plenum Press, New York, NY, 10: 214–270.
Steinkraus D.C. and Whitefield J.B. 1994. Chinese caterpillar fungus and world record runners. Amer. Entomol. 41: 235–239.
Whitman D.W., Bloom M.S., and Alsop, D.W. 1990. Allomones: Chemicals for defense. *In:* Evans D.L. and Schmidt J.O. (eds.). Insect Defenses. State Univ. New York Press, Albany, New York, NY, pp. 289–351.

Appendix 22.1: Medical use of insects and other arthropod drugs in Korean traditional medicine[1]

Species or products	**Medical uses**
WHOLE ARTHROPODS	
Arachnids:	
Scorpion, Scorpionida, Buthidae (*Buthus martensis*)	Child fright disease, pain killer, convulsions, palsy, stroke, facial paralysis, migraine, lymph damaged by tuberculosis, tetanus, parotis, edema, carbuncle
Centipedes:	
Scolopendromorpha, Scolopendridae (*Scolopendra* species)	Joint problems (most important), stroke, convulsions, tetanus, lymphangitis, lumps or masses, poisonous tumor or carbuncle, neoplasm, alopecia areata (baldness with white specks), snake bite
Insects:	
Blister beetles, Coleoptera, Meloidae (*Mylabris* species)	External: boils (most important), fungal infection, facial paralysis due to stroke; internal: gonorrhea, lymphangitis, neoplasm, rabies, syphilis, edema
Horse fly adults, Diptera, Tabanidae (*Tabanus* species)	Amenorrhea, abdominal blood stasis, indigestion, sickness due to fighting
Mole cricket, Orthoptera, Gryllotalpidae (*Gryllotalpa africana*)	Urine rentention, urolithiasis (stones in bladder), edema, lymphangitis, furuncles
Paper wasp adults, Hymenoptera, Vespidae (*Vespa* and *Polistes* species)	Tonic: protection from winter cold
Paper wasp larvae, Hymenoptera, Vespidae (*Vespa* and *Polistes* species)	Tonic: protection from winter cold
Diving beetle, Coleoptera, Dytiscidae (*Cybister tripunctatus*)	Polyuria (in elderly), enuresis in children (nocturnal urination), all types of blood stasis
Rice field grasshoppers, Orthoptera, Acrididae (*Oxya* species)	Child fright disease, tetanus, whooping cough, cough, weakness
Scarab beetle larvae, Coleoptera, Scarabaeidae	Liver cirrhosis
Silk moth adult males, Lepidoptera, Bombycidae (*Bombyx mori*)	Impotence, premature ejaculation, pyuria (white turbid urine), ulceration
ARTHROPOD PARTS OR PRODUCTS	
Bee glue, Hymenoptera. Apidae (*Apis* species)	
Cicada nymphal skin, Homoptera, Cicadidae (*Cprytotympana* species)	Hearing problems (most important), cough that results in loss of voice; tonsillitis or pharyngitis, child fright disease, tetanus, itchy skin rash, eye diseases, carbuncle, rashes, and measles
Honey from honeybees, Hymenoptera, Apidae (*Apis* species)	Weakness, abdominal pain, angina pectoris, dry cough, rebellious *qi* (*chi*) resulting in cough, constipation, stomatitis or gingivitis, nasal polyp

Appendix 22.1: (Contd.)

Appendix 22.1: (Contd.)

Species or products	Medical uses
Paper wasp nests, Hymenoptera, Vespidae (*Vespa* and *Polistes* spp.)	Child fright disease, arthritis, lymphangitis, stroke, recurrent rash, tinea capitis (fungal rash of scalp), mastitis, toothache, intestinal worms
Praying mantis egg case, Orthoptera, Mantidae	Claimed to improve men's stamina
Dung of silk moth larvae, Lepidoptera, Bombycidae (*Bombyx mori*)	Diabetes (most important), arthritis, numbness in body, neuralgia, coldness and pain in feet and lumbar area, body paralysis, skin rash, conjunctivitis
DISEASED INSECTS	
Hepialid moth larvae, Lepidoptera, Hepialidae (*Hepialis oblifurcus*), infected with fungus *Cordyceps sinensis*	Lung disease, stamina booster
Silk moth larvae, Lepidoptera, Bombycidae (*Bombyx mori*), infected with fungus *Beauveria bassiana*	Stroke (most important), headache, convulsion, stroke-induced speech problems, tremor, tonsillitis or pharyngitis, urticaria, lymphedema or mastitis, rubella, mastitis, tubercular lymphaldenitis.

[1]Use information extracted primarily from Kim (1984) and secondarily from interviewed doctors cited in the Acknowledgments.

23

Insects as Traditional Food in China

Luo Zhi-Yi

Shanghai Institute of Entomology, Chinese Academy of Sciences, Chungkin Road (S) 225 Shanghai 200 025, P.R. China

Abstract

Consideration is given to insects as food in both ancient and present-day China. The most important food insects in China are the ant species *Polyrhachis vicina* Rogen, which is consumed in various forms such as dried bodies, ant powder, and ant wine, and the honeybee (*Apis cerana* Fabr. and *Apis mellifera* L.) and its products. It has been estimated that there are some 6.3 million hives in China with 150 thousand tons of bee products produced annually. Other insects such as the larvae of *Bombyx* and the bamboo weevil are also consumed but only in restricted areas of the country. Finally consideration is given to the future of insects as a food source in China. More research is urged so that they can be further identified for food or medical use and commercially exploited by controlled production but without disturbing the natural balance.

Key Words: edible insects, entomophagy, food insects, food in China

Introduction

In China, records of insects used as food dated as early as 1200 BC have been found. However, insect consumption was, and still is, a localized custom and only in recent years has the idea of using insects as a high-quality food source gained popularity. Interest in entomophagy now derives both from the need to preserve traditional customs and the necessity for developing new food sources. Whatever the motivation, a detailed and systematic investigation of insect consumption seems timely.

Early research on insects focused on evaluating whether insects were beneficial or harmful to humans. The earliest research on insects recorded in China concerned large-scale breeding of silkworms (*Bombyx mori*), which has a history of 5,000 years and presently millions of people are engaged in the industry. However, compared to the history of research on crop cultivation, animal husbandry, forestry and medicine, research on insects as human food has scarcely received attention. Even now, the selection of insects for use as human food is based on the experience of our ancestors in finding food for mere subsistence that was handed down through generations. Scientific investigations into whether these choices were nutritionally sound have only recently been started by *ad hoc* research.

Insect bodies have proven to be rich in proteins and to contain various essential amino acids. While in some developed countries, such as the United States, insects have gained popularity as a fancy delicacy, in China medicated diets containing insects have attracted many people. Both trends indicate that the breeding of insects for food in the form of small-scale animal husbandry may become a prosperous activity in the near future.

Insects as Food in Ancient China

In the books Zhou Li-Tian Guan-Bei Ren and Li Ji-Nei Ze, it is written that in 120 BC emperors used ants, cicadas, and wasps as foods and that this custom was kept alive and handed down through generations in various forms. Liu Xun and Duan Gonglu of the Tang Dynasty wrote that the tribal heads in GuangXi offered their guests a marmalade made of ants as a delicacy. Later, Lu You of the Song Dynasty and Kuang Lu of the Ming Dynasty expounded the eating of ants by people in GuangXi. Additional descriptions in the book Zhuang Zi-Da Shen show an old hunchback who captured cicadas and introduced eating and methods of collecting cicadas in the Chun-Qiu Dynasty. From the Chun-Qiu dynasty to the Tang Dynasty, people in Guang Xi took young wasps and pupae as food. Other food insects were added to the ants, cicada, and wasps in the Han and Wei dynasties, such as flies, larvae of Ephemeridae and larvae of chafer (Scarabaeidae). Silkworm and dragonfly came into use during the Yuan and Min Dynasties while the imago of Sphingidae, true water beetle, larvae of Tenebrionidae, and larvae of Cerambycidae were added during the Qin Dynasty rule of Qian Long.

We may thus conclude that from the very distant past to the early Qin Dynasty eating of selected insects was a habit. Table 23.1 summarizes the use of food insects in various parts of China through time and their methods of preparation and collection.

Table 23.1: Food insects in China through time

Insect	Region	Preparation	Collection
Locusts (Acrididae)	Most of China	Dried in sun and made into porridge or cake; fried with viscera, head, and limbs removed	Collected from the wild with graminale
Pupae of silkworm (*Bombyx mori)*	Jiangsu and Zhejiang	Deep-fried in oil, or the dry body stir-fried with chives	Obtained from silk-reeling factories
Wasps (*Vespa* spp.)	Most of China	Parched or canned	Collected from nests in trees, tree caves, and under mud
Sphinx larvae (*Sphinx* spp.)	Shangdong, Henan, Hebei, Anhwei, Jiangsun	Parched after soaking in salt water and cooked with noodles	Collected from soybean fields
Termites (Isoptera)	Yunnan, Guangxi, Guangdong, Fujian	Parched	Collected from timbers and from underground nests
Larvae of fly (Muscidae)	South Yangtse river area	Cleaned and made into 'Ba Zhen Cake' with rice powder	Culled from runnels near manure pits
Cordyceps (parasitic fungus on Hepialidae caterpillars)	Yunnan, Szechuan, Tibet	Stewed with chicken, used as tonic	Found under grass roots in high mountains
Litchi bug *(Tessaratoma papillosa)*	Guangxi, Guangdong, Fujian	Head, wings, legs, and viscera removed, body ?? and wrapped in cabbage leaves, then instantly cooked in hot ash	Found on litchi and longan trees
True water beetle (Dytiscidae)	Guangxi and Guangdong	Soaked in salty water and dried. Legs and wings removed.	Collected from flooded fields and pools
Mole cricket (Gryllotalpidae)	Guangxi and Guangdong	Limbs and viscera are removed	Collected from fertile arable lands
Belastoma (Belastomatidae)	Guangdong	Soaked in salty water and dried. Legs and wings removed.	Collected from flooded fields and pools
Bamboo weevil larvae (Curculionidae)	Guangxi and Szechuan	Imago: head, wings, legs and viscera removed, soaked into condiment and parched on ash. Larvae: stir-fried with condiment	Collected from bamboo forest
Larvae of chafer (Scarabaeidae)	All China	Head, legs and viscera are removed, then stir-fried with oil and salt	Collected from dry land and piles of poultry droppings

Table 23.1: (Contd.)

Table 23.1: (Contd.)

Insect	Region	Preparation	Collection
Larvae of Ephemeridae	All China	boiled	Collected from streams and pools
Cricket (Gryllidae)	China	Chained with steel thread, baked with sauce and sugar	Collected from caves near fields, from bean and vegetable fields, and under stones in grassfields
Larvae of Cerambycidae	Szechuan, Hunan, and Northeast China	Eaten raw or parched	Collected from tree stems
Larvae of bag worm (Psychidae)	Jiangsu, Shangdong	Parched or made into marmalade and sauce	Collected from forests and fruit gardens
Larvae of pink boll worm	Jiangsu, Shangdong	Pressed for oil; or eaten parched	Found in cotton storages
Chafer (Scarabaeidae)	Jiangsu	Parched and ground into powder	Collected from forests and fruit gardens
Blattaria	Guangdong	Cooked	Countryside

Insects as Food in Present-day China

Polyrhachis vicina Rogen

Probably *Polyrhachis vicina* Rogen is one of the most important food insects in present-day China. This ant species is widely distributed and is used for human consumption in subtropical, southeast China, especially in Yunnan, Guangdong, Guangxi, Fujian, and Zhejiang provinces. *Polyrhachis vicina* Rogen is available on the market in several forms, such as dried ant bodies, ant powder, and ant wine. Dishes made from ants are well known in Hangchou, such as "ants hugging eggs" and "ants climbing trees". The reason why ants have easily found their way into the human diet may be their valuable amino acid composition. Researchers from the Gurangxi College of Traditional Chinese Medicine have shown that *Polyrhachis vicina* Rogen contains 26 different free amino acids, among them the essential amino acids threonine, phenylalanine, valine, leucine, and lysine. The collection and production of this ant in well-known and on places such as Guangxi, Yunnan, and Zhejian are taking on increasing significance. Research is needed to study the ecology of this ant species to ensure that excessive collection will not disturb natural balances.

Apis cerana and Its Products

The honeybee and its products are also important insect foods in present-day China. In China the dominant species are *Apis cerana* Fabr. and *Apis mellifera* L.; the latter is mainly kept in northeast China, Inner Mongolia and Xinjiang, while the former is widely distributed in Szechuan, Yunnan, Guizhou, Guangdong, Guangxi, and Fujian. Presently, there are some 6.3 million hives in China with 150 thousand tons of honeybee products produced yearly. Indeed, China is one of the top two producers of honeybee products in the world. The products include honey, honey wax, propolis, honey jelly, and pollen. Honey and honey jelly are the most important products, but recently pollen has become popular as well.

Other Food Insects

According to surveys on the eating habits of the people of China, food insects cannot yet be considered a common food item. Some insect foods are highly valued but only in restricted areas of the country. For instance, larvae of Bombycidae are edible to people in Zhejiang and Jiangsu provinces where mulberry and *Bombyx mori* are produced. However, this food insect is not acceptable to people from many other areas in China. Dishes made of larvae of *Sphinx* are presented as a treat to guests in the North Huai River area. Raw Formicidae served with salt, vinegar and bamboo weevil are presented to guests by the Daizll people in Xishuang Banna. Drinking water with the true water beetle in it is preferred over plain water by people in Guangdong.

Insects as a Future Food Resource in China?

The development of insect foods in China should be carefully monitored and guided by experts in entomology, ecology, nutrition, and medicine. On the one hand, encouragement of insect foods may be desirable as many insect foods contain valuable nutrients and/or have medicinal value. On the other hand, given the large human population and rapid economic development in China, the collection, exploitation and breeding of insects for food should be carefully monitored to prevent problems of overcollection and disturbance of natural balances. In this context the following priorities can be set: 1) identify those insect species that have nutritional and/or medicinal value; 2) study the ecology of food insect species to preclude possible overcollection and to explore the possibility of controlled production; 3) augment research on the industrial development of insect food products, their commercialization and harmonization of the market to balance supply with demand of food insect products; and 4) advertise food insects to inform the people of their nutritional benefits and to improve their acceptability.

References (in Chinese)

Chu H.F. and Kao C.S. 1950. A commentary on the insects mentioned in the Pen TS'ao Kang Mu. Anuales Entomologici Sinici 1 (2): 234–262.

Liu Li Chun. 1995. The amusing report on insect foodstuff. Entomological Knowledge, 32 (2): 126-127.

Meng Zhao Lian. 1993. Chinese Insect Culture. TianJin People's Publ., Tianjin, pp. 99-108,

Zhang Chuan Xi. 1990. Resource Insects. Shanghai Sci. Tech. Publ., Shanghai, pp. 107-137.

Zhou Shu Wen. 1981. A History of Chinese Entomology. Science Press, Beijing, pp. 160–168.

Zhou Yao. 1988. A History of Chinese Entomology. Science Press. Tim Ze press, Xi-An, pp. 44–46.

24

Medicinal Terrestrial Arthropods in China

DING ZIMIAN[1], ZHAO YONGHUA[1], AND GAO XIWU[2]

[1]*Institute of Medicinal Plants, Chinese Academy of Medical Sciences, 100 094 Beijing, P.R. China*
[2]*Dept. of Entomology, Beijing Agricultural University, 100 094 Beijing, P.R. China*

Abstract

The medicinal insects in China are reviewed—the history of their use, reference sources currently available, and various forms in which they are consumed. The present situation of medicinal insect research is discussed and aspects for future investigation proposed.

Key Words: insects, arthropods, medicine, insect consumption, China, pharmacology.

Introduction

Traditional Chinese medicine is a unique aspect of the Chinese cultural heritage. Traditional Chinese medicine considers drug and food to be homologous in function for human health. As an important component of traditional medicine, insects have long been used in China along with other animals and plants. Indeed, over the past 5,000 years, it has been verified that many insects can be directly or indirectly used to cure common human diseases.

According to statistics of AnGuo city, the largest trading center in medicinal animals and herbs in China, fifteen kinds of medicinal insects are currently being used on a regular basis by the Chinese people (Chen Yongzeng, 1995). They belong to the following invertebrate groups: Hymenoptera, Hemiptera, Orthoptera, Lepidoptera, Mantodea, Coleoptera, Chilopoda, Scorpionida, Oligochaeta.

Today, many researchers in China are engaged in research on medicinal insects and remarkable progress has been made in their inventory, classification, and isolation of medicinal compounds. Here the use of medicinal insects in China, including its history and prospects is reviewed. It should be noted that the Chinese ideogram for medicinal insects includes some other kinds of arthropods such as scorpions and centipedes, as well as some worms, and earthworms.

History of Medicinal Insects in China

China has a long history of using insects for both food and medicine. According to ancient Chinese references, silkworms (*Bombyx mori* L.) were used as medicine as early as 3,000 years ago, bees (*Apis mellifera* L.) have been used for medicinal properties since the Xizhou Dynasty (about 1100–771 BC), and the lac-varnish insect (*Laccifer lacca* Kerr) since the 3rd century. "Shennong Bencaojing" is the first known monograph[1], dating back about 2,000 AD, on the pharmacology of Chinese medicine. In this book, 21 kinds of insects were recorded as having medicinal properties, and were divided into three grades according to the properties and functions perceived. Grade 1 insects include bee larvae, bee wax, bee honey, and mantis (*Hierodula saussurei* Kirby) egg-capsules, all of which are innocuous and can be used as food without harm to human health. Grade 2 insects include honeycomb, cicadas (*Scapsipedus aspersus* Walker), and silkworms (*Bombyx mori* L.), considered somewhat toxic and safely eaten in limited quantities. Grade 3 insects include northeast giant black chafers (*Holotrichia domphalia* Bates), oriental moths (*Cnidocampa flavescens* Walker), Chinese blistering cicada (*Lycorma delicatula* White), lesser blister beetles (*Mylabris cichorii* L.), and Asiatic cockroaches (*Blatta orientalis* L.), all of which are poisonous and can be consumed only occasionally in small amounts. The classification of medicinal insects into grades was based on the perceived toxicity of insects for humans. In ancient times, obviously people had no means of determining the degree of toxicity of insects in the laboratory nor the nature of the mechanism involved. Determination of toxicity was based entirely on practical experience. Therefore, while the classification of medicinal insects into grades has some practical value, it lacks scientific basis. "Mingyi Bielu", a book written by Tao Hongjing (Southern and Northern Dynasties, 420–589 AD), added nine kinds of medicinal insects to "Shennong Bencaojing". The "Compendium of Materia Medica", a book written by Li Shizhen (1587), recorded 73 kinds of medicinal insects. The supplement to the "Compendium..." written by Zhao Xuemin "Bencao Gangmu Shiyi" (Qing Dynasty, 1616–1911 AD) in 1983, recorded up to 105 medicinal insects.

In ancient China, the classification of insects lacked a scientific basis and insect names were hence often confused. For example, in "Shennong Bencaojing",

[1]"Shennong Bencaojing" is an ancient Chinese medical monograph. Neither the author(s) nor the exact year of completion is known. The book was handed down from one generation to another. Recently, a contemporary Chinese scholar, Cao Yuanyu, revised the book and renamed it "Bencaojing" (Cao Yuanyu, 1987).

bee larvae, bee honey, bee wax, and honeycomb are treated as different insects, but according to modern taxonomic concepts they are phases and products of the same insect species. After extensive revisions of the ancient Chinese monographs on medicinal insects, 143 medicinal insects, comprising 13 orders and 48 families, were recorded in the modern "Handbooks of Chinese Medicinal Animals" (1979, 1983).

Sources of Medicinal Insects in China Currently Available

As mentioned, according to the "Handbooks of Chinese Medicinal Animals" (1979, 1983), 143 medicinal insects have been recognized, the most important of which are listed in Table 24.1 along with their medicinal properties.

While earlier research on medicinal insects was predominantly concerned with common ailments, recent efforts have also focused on "modern", chronic diseases, notably cancer. In China, several insects and insect products have been found effective in killing cancer cells (Yang Jinxiang, 1984; Chang Minyi, 1987; Zhang Chuanxi, 1990). As many as 77 kinds of insects, belonging to 8 different orders and 14 families, are found to possess some form of antitumoral activity. The antitumoral activity of some of these insects has been clinically verified, that of others has been demonstrated in vitro. The most important insect species with antitumoral activity are listed in Table 24.2. Some of these are also included in Table 24.1 as they have multiple medicinal functions.

Some medicinal insects have relatives with similar curative effects that can be used as substitutes. This will augment the development of medicinal insects, especially since some insect species are rare and/or difficult to rear.

Methods of Using Insects in Chinese Medicine

With their long-standing practical experience in the use of medicinal insects, Chinese ancestors and their descendants carefully selected different parts, stages or products of the various medicinal insects to cure certain diseases. Even though this selection process lacked a solid pharmacological basis, it certainly had some rationale. In recent years, progress has been made in the isolation of active compounds and evaluation of medicinal parts of insects, notably with respect to ants (*Polyrhachis vicina* Roger) (Zhao Yi, 1983; Xu Qing, 1989). In general, however, the different forms in which medicinal insects were and are still consumed can be roughly divided into six categories:

(1) The main method in Chinese medicine of using insects to cure diseases is to consume the whole insect body. Chinese ancestors thought that food and medicine were homologous and used whole insects as food for survival and/or as medicine to cure disease. This has remained the basic approach among Chinese people to taking advantage of insects in medicine.

Table 24.1: Medicinal arthropods used in China

Order	Scientific name	Common English name	Function/indication
Thysanura	*Lepisma saccharina* L.	silverfish	rheumatism
Odonata	*Anax parthenope* Selys	giant dragonfly	detoxification,
Blattaria	*Periplaneta americana* L.	American cockroach	detoxification, detumescence
	Periplaneta australasiae L.	Australian cockroach	"
	Blatta orientalis L.	Oriental cockroach	"
	Blattella germanica L.	German cockroach	"
	Eupolyphaga sinensis Walker	Sand cockroach	pain relief
Mantodea	*Hierodula saussurei* Kirby	Saussurei's mantis	impotence
	Mantis religiosa L.	European mantid	"
	Statillia maculata Thunberg	maculated mantis	"
Orthoptera	*Locusta migratoria* L.	Asiatic migratory locust	pain and cough relief
	Oxya velox Thunberg	long-winged rice grasshopper	"
	Oxya chinensis Thunberg	Chinese rice grasshopper	"
	Oxya intricala Stal	small rice grasshopper	"
	Ceraeri kiangsu Tsai	bamboo locust	"
	Patanga japonica I.B.	Japanese ground grasshopper	"
	Scapsipedus aspersus Walker	cricket	swelling (inflammation) and fever relief, detoxification
	Mecopoda elongata L.	cane giant katydid	"
	Brachytrupes portentonus Lc.	larger brown cricket	"
	Gryllotalpa africana P. et B.	African mole cricket	swelling relief, detoxification
	Gryllolalpa formosana Sh.	Taiwan mole cricket	"
	Gryllus testaceus Walker	field cricket	"
Homoptera	*Huechys sanguinia* De Geer	red ladybug	detoxification, cough relief
	Huechys thoracica Dis.	short-winged ladybug	"
	Huechys philamata Fabr.	brown-winged ladybug	"
Hemiptera	*Platypleura kaempferi* F.	Kaempferi cicada	swelling relief
	Lycorma delicatula White	Chinese blistering cicada	"
	Laccifer lacca Kerr.	lac insect	"
	Nezara viridula L.	green stink bug	"
	Tessaratoma papillosa Drury	litchi stink bug	"
	Ericerus pe-la Chavannes	white wax insect	arrestation of bleeding, pain relief
Lepidoptera	*Cnidocampa flavescens* Walker	Oriental moth	detoxification
	Proceras venosatum Walker	striped stem borer	"
	Bombyx mori L.	silkworm	rheumatism
	Antheraea pernyi Geurin et Perny	silkworm	tranquilizer

Table 24.1: (Contd.)

Table 24.1: (Contd.)

Order	Scientific name	Common English name	Function/indication
	Papilio machaon L.	swallow-tailed butterfly	gastric problem
	Hepialus armoricanus Ober.	moth	cough relief
Diptera	*Chrysomyia megacephala* F.	Oriental latrine fly	detoxication, digestion
	Tabanus mandarinus Schiner	Chinese horse fly	swelling relief
	Tabanus budda Portsh	Buddha horse fly	"
	Tabanus kiangsesis Krober	Jiangsu horse fly	"
Coleoptera	*Cybister tripunctatus* L.	small diving beetle	blood circulation
	Cybister japonicus Sharp et Lewis	diving beetle	"
	Mylabris cichorii L.	lesser blister beetle	"
	Mylabris phalerata Pallas	large blister beetle	"
	Apriona germari Hope	mulberry longicorn	"
	Batocera horsfieldi Ch.	white-striped longicorn	"
	Holotrichia diomphilia Bates	northeast block chafer	pain relief
	Holotrichia morosa W.	brown mulberry chafer	"
	Holotrichia titanis Reiter	brown chafer	"
	Anomala corpulenta M.	metallic-green beetle	"
	Anomala exoleta F.	red-brown beetle	"
	Allomyrina dichotoma L.	horned beetle	"
	Liocola brevitarsis Lewis	white-spoiled flower chafer	"
	Martianus dermestoides Ch.	drug darkling beetle	tonic
Hymenoptera	*Apis cerana* Fabr.	honeybee	rheumatism, menstrual syndrome, arthritis, hepatitis
	Apis mellifera L.	honeybee	"
	Vespa ducalis Sm.	small yellowjacket	"
Scorpionidae	*Buthus martensi* Karsch	scorpion	swelling relief
Chilopoda	*Scolopendra subspinipes mutilans* L.	centipede	"

(2) Insect eggs, such as mantis eggs, have long been consumed as medicine in China.

(3) Insect nests, such as wasp and bee, are traditionally valued as medicine.

(4) Certain insect "secretions" are commonly consumed for their medicinal properties, such as honey from bees and excreta from silkworms.

(5) A combination of certain insects and bacteria is used as medicine. For instance, "jiangcan," which consists of the larvae of the silkworm (*Bombyx mori* L.) infected by *Beauveria bassiana* (Bals) Vuill, is a famous Chinese medicine.

(6) Several products have been developed in recent years that consist of compounds isolated from insects, such as melitin and cantharidin. Also, industrial products have become available, such as oral ant liquid (Hong Xingke and Mao Yanjun, 1994) and *Cordyceps sinensis* capsules.

Table 24.2: Insects with anticancer property in China

Order	Scientific name	Common English name	Function/indication
Blattaria	*Blatta orientalis* L.	Oriental cockroach	Renal cancer
	B. germanica L.	German cockroach	"
	Periplaneta americana L.	American cockroach	"
	P. australasiae Fabricius	Australian cockroach	"
	P. fuliginosa Serville	smoky brown cockroach	"
Orthoptera	*Gryllotalpa unispina* Sauss.	giant mole cricket	Liver cancer
	G. gryllotalpa L.	European mole cricket	"
Homoptera	*Cryptotympana japonensis* K.	blakish cicada [blackish?]	Thyroid cancer
	Hueckys sanguea De Geer	red ladybug	Skin cancer
	Nurudae sinica Tsai et Tang	Chinese gall aphid	Liver cancer
	N. shiraii Mat.	Sumac gall aphid	"
	N. rosea M.	rosy gall aphid	"
Hemiptera [Heteroptera?]	*Cyclopelta parva* Dallas	dadap bug	Esophagus cancer
Coleoptera	*Mylabris phalerata* Pallas	large blister beetle	Lung cancer
	M. cichorii L.	lesser blister beetle	"
Lepidoptera	*Hepialus armoricanus* Ober.*	swing moth	"
	H. yushuensis Chu et W.*	yushun swing moth	"
	H. sichuanus Chu et W.*	Sichuan swing moth	"
	H. yunlongensis Chu et W.*	Yunlong swing moth	"
	Bombyx mori L.**	silkworm	"
Diptera	*Tabanus mandarinus* Schiner	Chinese horse fly	Liver cancer
	T. kiansuensis Krober	Jiangsu horse fly	"
Hymenoptera	*Apis mellifera* L.	honeybee	Lung cancer
	Polistes chinensis Fabr.	long-legged wasp	"

*In Chinese medicine the larvae of Hepialidae (Lepidoptera) infected by *Cordyceps sinensis* (Berkeley) have medical value.

**In Chinese medicine the larvae of *Bombyx mori* L. (Lepidoptera) infected by *Beauveria bassiana* (Bals) have medical value.

Today in China, medicinal insects are still most commonly consumed as the whole insect body, although the other forms listed above are also conventional modes. Chinese people also gather even today some kinds of insects from the wild, such as blister beetle, mole cricket, and ants. Medicinal insect usage differs between rural and urban people. In rural areas people often collect insects from the wild while in urban areas people usually have no choice but to purchase industrial products. Some arthropods, such as the scorpion and silkworm, are reared artificially for urban dwellers and can be purchased in the form of ready-to-consume arthropod preparations in the market.

Current Situation of Medicinal Insect Research in China

China has a long history of research on medicinal insects. Many ancient books on pharmacology include medicinal insects. They also record the progress made during different dynasties and are mainly concerned with morphology, capturing, production, and the relation between collecting season and curative effects of medicinal insects. In general, the results are based on practical experience. With the development of modern science and technology, however, emphasis has shifted to the isolation of medicinal compounds of insects in pharmacological industries and to the artificial breeding and multiplication of medicinal insects (Handbooks of Chinese Medicinal Animals, 1979, 1983; Bai Qingyu, 1988; Lu Xiyou, 1983).

At present, the following insects are bred in China: sand cockroach (*Eupolyphaga sinensis*), scorpion (*Buthus martensi*), drug darkling beetle (*Martianus dermestoides*), grasshopper (*Oxya chinensis* Thunberg), ants (*Polyrhachis vicina* Roger), honeybee (*Apis cerana* Fabr.), and silkworm (*Bombyx mori* L.). Insect breeding is a labor-intensive industry in China, and is mostly done on a small scale.

Pharmacological research is mainly concentrated on the following insects (Handbooks of Chinese Medicinal Animals, 1979, 1983):

(1) *Malaphis chinensis* (Bell). The excreta of this insect, known as gallnut, have been verified to possess astringent activity, to arrest bleeding, and to kill many bacteria. It is currently used in China on a commercial basis, and is mainly distributed in Hunan, Hubei, Sichuan, and Shanxi provinces.

(2) Silkworm (*Bombyx mori*). The content of crude protein, fat, and carbohydrates in pupae are 51.5%, 29.3%, and 2.2% respectively (weight/weight). Silkworms products, pupae infected by *Beauveria bassiana* Vuill, and silkworm excreta, have the function of stimulating the adrenal gland, acting as a sedative, and decreasing the serum cholesterol level. The cholesterol lowering effect is due to the high content of unsaturated fatty acids in the silkworm pupae (Chen Xinqian and Jin Youyu, 1985).

(3) Winter larvae summer herb. This medicine consists of the larvae of *Hepialus armoricanus* Ober, infected by *Cordyceps sinensis* (Berkeley) Sac. The contents of protein, fat, carbohydrate, and cordyceptic acid in this product are 25%, 8.4%, 29%, and about 7% respectively. It has the function of calmative and is an astringent to womb and heart. This medicine is mainly distributed in Tibet, Qinghai, and Xinjiang.

(4) Blister beetle (*Mylabris cichorii* L.). The adults contain cantharidin, resin, and mineral elements. It stimulates animal skin and has been found to slow heart rate.

(5) Scorpion (*Buthus martensis*). Main active compounds of the scorpion include buthotoxin, lecithine, trimethylamine, betaine, and taurine. Scorpions have anticonvulsive activity and decrease the blood pressure.

In recent years, much analytical work on the active chemical components of medicinal insects has been done in China (Ding Zimian, 1994a, 1994b). Many active compounds, hormones, and toxins have been isolated from insects (Jiang Sanjun, 1993). Some active components have already been synthesized artificially and are used as clinical substitutes. For example, Chinese researchers have artificially synthesised sodium cantharidin, which has been used to cure lung and liver cancer in clinical studies (Handbook of Chinese Medicinal Animals, 1979).

Future Prospects of Medicinal Insect Research

We believe that the following aspects deserve more attention in future medicinal insect research.

(1) Traditionally, research on medicinal insects has been more of a practical nature than theoretical. As a consequence, more theoretical and elementary work needs to be done in the future, especially in the fields of toxicology, pharmacology, and chemical components of medicinal insects.

(2) Identification of insects and their taxonomic classification needs special attention. In ancient Chinese references, names of insects were often confused. Some insects with the same name are actually different insects, and vice versa. This can cause serious complications in clinical work.

(3) Medicinal insect resources should be carefully protected from overcollection for two main reasons. First, to guarantee their sustainable use and this especially applies to endangered insect species such as *Hepialus armoricanus* Ober (Lepidoptera) (winter larvae summer herb). Second, to avoid destruction of natural food chains, especially where natural enemies of pests are involved. We believe there are two main ways of protecting natural insect resources from overexploitation: one is to develop mass-rearing methods to enhance the output of insects to meet growing market need, and the second, finding natural substitutes and developing artificially synthesized substitutes through basic research work.

(4) Applications of pesticides can be reduced by developing more efficient methods for collecting insect pests that are traditional food and medicine, such as locusts and grasshoppers. This would decrease environmental pollution and other adverse impacts of pesticides while at the same time providing valuable insect resources for use as medicine and food. Efficient methods to collect insect pests should be developed as well as ways to make use of insect pests. For instance, cotton bollworm (*Helicoverpa armigera*), which is difficult to control by chemical insecticides, could be collected and the insect body used for extraction of proteins and chitin.

(5) There are presently 15 kinds of medicinal insect products on the Chinese market (Chen Yongzeng, 1995). There is an urgent need to develop a quality

control standard for these medicinal insects to guarantee the safety and effectiveness of the medicine for the people.

References (All references in Chinese, unless stated otherwise)

Bai Qingyu 1988. Medical Animals' Rearing. Chinese Forestry Press, Beijing.

Cao Yuanyu 1987. Bencaojing. Shanghai Sci. Tech. Press, Shanghai.

Chang Minyi. 1987. Anti-cancer Herbs. Hunan Sci. Tech. Press, Changsha, P.R. China, 118 pp.

Chen Xinqian and Jin Youyu 1985. The New Pharmacology. People's Hygiene Press, Beijing, 373 pp. (12 ed.).

Chen Yongzeng. 1995. Chinese Medicine Information (AnGuo city) 10: 5.

Ding Zimian. 1994a. The biology, nutrients, propagation and medical effects of spined ant *(Polyrrachis vicina). In:* Bull. 3rd Intl. Conf. Ethnopharmacology, D-63. Entomol. Soc. China, Beijing. (in English)

Ding Zimian. 1994b. Current situation of edible and medicinal ants in China. *In:* Bull. 50th Anniv. Chinese Entomol. Assoc. Entomol. Soc. China, Beijing, 598 pp.

Handbook of Chinese Medicinal Animals (Book one). 1979. Tianjin Sci. Tech. Press, Tianjin.

Handbook of Chinese Medicinal Animals (Book two). 1983. Tianjin Sci. Tech. Press, Tianjin.

Hong Xingke and Mao Yanjun. 1994. The Ants. China Agric. Sci. Tech. Press, Beijing, pp. 44–47.

Jiang Sanjun. 1993. Toxic insects and the application of insect toxins in China. J. Chinese Medicinal Materials 9: 38–40.

Li Shizhen (Ming Dynasty, 1368–1044 AD). (1587). Compendium of Materia Medica People's Hygiene Press, Beijing, Book 4, pp. 2217–2315. (revised ed.)

Lu Xiyou. 1983. Biology and rearing of scorpion. Nat. Res. Research 1: 66–74.

Polis G.A. 1990. The Biology of Scorpions. Standford Univ. Press, Palo Alto, CA, 588 pp.

Tao Hongjing (Southern and Northern Dynasties, 420–589 AD). 1986. Mingyi Bielu. People's Hygiene Press, Beijing.

Xu Qing. 1989. The effects of extracts of spined ants (*Potyrhachis vicina* Roger) on the sexual and immunological organs of white mouse. Guangxi J. Trad. Chinese Med. Pharm. 4: 46.

Yang Jinxiang. 1984. Reagent of Chinese Anti-cancer Medicine. People's Hygiene Press, Beijing, 192 pp.

Zhang Chuanxi. 1990. Resource insects. Shanghai Sci. and Tech. Press, Shanghai, 6 pp.

Zhao Xuemin (Qing Dynasty, 1616–1911 AD). 1983. Bencao Gangmu Shiyi. People's Hygiene Press, Beijing.

Zhao Yi. 1983. Study on the pharmacology and toxicity of spined ant (*Polyrhachis vicina* Roger). Guangxi J. Trad. Chinese Med. Pharm. 6: 39.

25

Nutritive Value of Earthworms*

Sun Zhenjun

College of Resources and Environmental Sciences, China Agricultural University, 100 094 Beijing, P.R. China. E-mail: sun108@cau.edu.cn

Abstract

The chemical compositions of the earthworm *Eisenia foetida*, its casts and body fluids were investigated and compared with those of a variety of common foods and animal feeds. Nutrient analyses showed that *Eisenia foetida* meal has a high protein content in the range of 54.6 to 71.0% dry matter. The protein content and amino acid composition were close to those of fish meal and eggs, and higher than cow milk powder and soybean meal. Casts of *E. foetida* had a protein content of 7.9% dry matter, similar to that of corn meal, and hence earthworm casts could be used not only as an excellent organic fertilizer, but also for partial replacement of corn meal or wheat bran in animal diets. Earthworm body fluids contained 9.4% protein and 78.79 free amino acids per liter and were rich in vitamins and minerals, in particular iron (Fe). Our nutrient analyses suggest that the earthworm (*Eisenia foetida*) could be an excellent protein supplement for animal feed and human food.

Key Words: earthworm, earthworm casts, nutritive value, protein supplement, animal diets, human food, amino acids, vitamins, minerals, *Eisenia foetida, Allolobophora caliginosa, Pheretima guillemi*

Introduction

In recent years, animal protein sources have become increasingly scarce and expensive in many countries and new protein sources need to be developed. Previous research has demonstrated that earthworms are rich in protein and

*This research was financially supported by State Natural Science Foundation of China (Grant No: 30270195).

various amino acids, and can be used as replacements for fish meal feed given to chicken, fish and pig (Harwood, 1976; Sugimura, 1984; Koh, 1985; Harwood and Sabine, 1978; Aklyaman, 1984; Stafford and Tacon, 1985; Edwards 1985; Velasquez, 1991; Sun, 1995). Earthworm casts reportedly contain 3.43% humus, 0.184% total nitrogen, 0.248% total phosphorus, and 29.93% organic matter; they have been used as organic fertilizer in crop, vegetable, and flower production with excellent results (Huang and Zhang, 1980; Curry, 1988). Use of earthworm casts as an ingredient of animal feed has not been reported in the literature to date.

In ancient times, people in Fujian and Guangdong provinces of China had the habit of eating earthworms. Even now, in Taiwan and Heinan and Guangdong provinces, some local people prepare special dishes featuring the earthworm as a basic ingredient (Zeng et al., 1982; Faduti, pers. comm.). Earthworms are also consumed by Ye'kuana Amerindians of the Alto Orinoco of Venezuela (Paoletti et al., 2003).

However, the most common uses of earthworms are treatment of certain diseases in traditional Chinese medical practices. "Earth dragon" is the Chinese name given to the earthworm, or "white neck earthworm", recorded early in an ancient medicinal book "Shen-nong-ben-cao-jing". Another famous ancient medicinal book, "Ben-cao-gang-mo", contains 42 entries relating to earthworms in the 42 volumes devoted to insects and describes their medicinal use in detail. Now the "earth dragons" used as medicine are mainly two species: *Pheretima aspergillum* (E. Perrier) and *Allolobophora caliginosa* (Duges). Besides these two species of earthworm, however, several are also used as Chinese animal materia medica: *Pheretima carnosa, P. medioca, P. hupeiensis, P. posthuma, P. praepinguis, P. tschiliensis, P. tschiliensis lanzhouensis, P. guillemi, P. vulgaris, P. peetinifera,* and *Eisenia foetida* (Gao, 1996). Aside from the medicinal use of whole earthworms, some pharmaceutical ingredients and special active proteins or peptides have been studied in recent years (Sun, 1997, 1998).

This paper assesses the potential of earthworms commonly-cultured in China as a protein source for use in animal and human nutrition, by comparing the nutrient composition of ground earthworm bodies, earthworm body fluids, and earthworm casts with those of common animal feeds and foodstuffs.

Comparison of Chemical Composition of *E. foetida* and Its Casts with Common Foods and Animal Feeds

Live earthworms, *E. foetida*, collected from the Muyu vermiculture farm in Laiyang, Shandong province, were kept in glass containers with moist toilet paper for 36 hours to ensure total evacuation of their gut contents. The earthworms and their casts were then freeze-dried, milled, and homogenized and stored in desiccators. Samples of wheat bran, two types of fish meal (from Peru and China), eggs and cow milk, were obtained from the Laiyang animal feed company.

The protein and amino acid contents of these samples were determined as standard units. Analyses were conducted in two specialist centers for chemical analysis with 4–6 replications of each sample. Crude protein (CP), dry matter (DM), crude fat, ash, calcium (Ca), phosphorus (P) and vitamins A, B_1, B_2, E and C were determined. Analytical procedures were those of standard methods for nutrient analysis (Xu, 1979, AOAC, 1975). A Hitachi 835-50 amino acid autoanalyzer was used for determination of the amino acid content (except tryptophan), A PH-360 atomic absorption spectrophotometer for mineral element determinations, and an H-850 fluorescence-photometer for vitamins A, B_1, B_2, E and C determination.

The data obtained were tested for variance and covariance according to Snedecor and Cochran (1980) and Mo (1984). T tests and Duncan's new multiple range test, cited from Steel and Torrie (1960), were applied to separate means when treatment effects were significant. Tables 25.1 to 25.5 present meant standard deviation (SD) data.

Table 25.1: Nutrient composition of earthworm (*E. foetida*) and some feed and foods. Data are presented as % (g/100 g on dry matter basis)

Feed or food	**DM**	**CP**	**Fat**	**Ash**	**Ca**	**P**	**ME (kcal g^{-1})**
Fresh earthworm (*E. foetida*)	15.7 (± 1.47)	11.02 (± 0.46)	1.89 (± 0.50)	1.4 (± 0.01)	0.22 (± 0.10)	0.65	
Earthworm meal (*E. foetida*)	90.6 (± 2.56)	54.6 (± 0.92)	7.34 (± 0.60)	21.2 (± 0.05)	1.55 (± 0.12)	2.75	2.99*
Earthworm casts (*E. foetida*)	82.2 (± 1.58)	7.9 (± 0.26)	1.1 (± 0.43)	34.2 (± 0.07)	1.42 (± 0.05)	0.28	0.95*
Peruvian fish meal	90.8 (± 0.72)	62.0 (± 0.28)	9.7 (± 0.48)	14.4 (± 0.13)	3.91 (± 0.04)	2.90	2.90
Chinese fish meal	88.5 (± 1.81)	53.9 (± 0.09)	9.3 (± 0.48)	18.9 (± 0.12)	4.59 (± 0.09)	2.15	2.35
Cow milk	12.7 (± 0.43)	3.5 (± 0.29)	3.5 (± 0.11)	0.7 (± 0.08)	0.12 (± 0.04)	0.09	0.65*
Egg	26.3 (± 0.34)	12.9 (± 0.17)	11.5 (± 0.08)	1.0 (± 0.01)	0.05 (± 0.08)	0.21	1.63**
Soybean meal	88.1 (± 1.28)	43 (± 0.70)	5.4 (± 0.13)	5.9 (± 0.08)	0.32 (± 0.11)	0.50	2.64
Corn meal	86.5 (± 0.49)	8.6 (± 0.31)	3.5 (± 0.09)	1.4 (± 0.02)	0.04 (± 0.08)	0.21	3.32
Wheat bran	82.2 (± 1.62)	14.2 (± 0.10)	2.0 (± 0.33)	4.4 (± 0.04)	0.14 (± 0.21)	1.06	1.78

*Calculated figure (Xiao, 1984). **Food energy.
DM = dry matter, CP = crude protein, Ca = calcium, P = phosphorus, ME = metabolic energy.

Table 25.1 summarizes the nutrient composition of the earthworms, earthworm casts, and some common animal food and feed ingredients. A comparison of the amino acid contents of earthworm meal and casts with that of fish meal, eggs, cow milk, wheat bran, and two kinds of insect is given in Tables 25.2 and 25.3.

It can be seen that the protein content of earthworm meal was 54.6%, which is close to that of Peru fish meal and not significantly different from that of Chinese fish meal, but higher than that of eggs and soybean meal. The crude fat content of earthworm meal was lower than that of fish meal but higher than that of soybean meal and corn meal. Except for corn meal, earthworm meal had the

Table 25.2: Contents of amino acids in earthworm and common feeds and foods (g/100 g on dry matter basis)

Amino acid	Earthworm meal	Earthworm casts	Peruvian fish meal	Chinese fish meal	Hen egg	Raw cow milk	Wheat bran
Thr*	2. 72 (± 0.09)	0.46 (± 0.01)	2.88 (± 0.28)	2.22 (± 0.24)	2.42 (± 0.06)	1.20 (± 0.08)	0.45 (± 0.038)
Ser	2.71 (± 0.08)	0.46 (± 0.09)	2.63 (± 0.14)	2.01 (± 0.24)	3.64 (± 0.14)	1.57 (± 0.09)	0.74 (± 0.036)
Gly	3.12 (± 0.24)	0.49 (± 0.03)	4.26 (± 0.09)	3.26 (± 0.25)	1.58 (± 0.32)	0.54 (± 0.11)	0.84 (± 0.033)
Cys	0.42 (± 0.10)	0.09 (± 0.03)	0.56 (± 0.18)	0.42 (± 0.22)	1.16 (± 0.20)	0.22 (± 0.09)	0.33 (± 0.028)
Val*	2.39 (± 0.27)	0.44 (± 0.01)	2.80 (± 0.25)	2.29 (± 0.37)	3.26 (± 0.18)	1.57 (± 0.14)	0.67 (± 0.014)
Met*	1.01 (± 0.42)	0.19 (± 0.03)	1.65 (± 0.57)	1.64 (± 0.33)	1.6 (± 0.09)	0.68 (± 0.16)	0.15 (± 0.003)
Ile*	2.40 (± 0.12)	0.38 (± 0.026)	2.42 (± 0.48)	2.23 (± 0.40)	2.99 (± 0.011)	1.28 (± 0.18)	0.37 (± 0.042)
Leu*	3.94 (± 0.15)	0.78 (± 0.034)	4.28 (± 0.22)	3.85 (± 0.19)	4.20 (± 0.10)	2.58 (± 0.15)	0.80 (± 0.042)
Tyr	1.73 (± 0.08)	0.24 (± 0.42)	2.12 (± 0.26)	1.63 (± 0.12)	1.98 (± 0.24)	1.28 (± 0.017)	0.52 (± 0.042)
Phe*	2.12 (± 0.81)	0.31 (0.038)	2.68 (± 0.28)	2.10 (± 0.45)	2.73 (± 0.25)	1.46 (± 0.19)	0.48 (± 0.034)
Lys*	4.26 (± 0.50)	0.68 (± 0.033)	4.35 (± 0.34)	3.64 (± 0.27)	3.32 (± 0.22)	2.11 (± 0.20)	0.47 (± 0.017)
HiS*	1.36 (± 0.24)	0.12 (± 0.014)	1.66 (± 0.21)	0.90 (± 0.32)	1.16 (± 0.23)	0.72 (± 0.25)	0.35 (± 0.002)
Arg	3.27 (± 0.33)	0.64 (± 0.17)	3.87 (± 0.37)	3.02 (± 0.51)	2.90 (± 0.13)	0.89 (± 0.12)	0.95 (± 0.027)

*Essential amino acids for humans.

Table 25.3: Amino acid content of *motto* and *kuru* (mg g^{-1} dry weight); values are the means of two determinations on two different samples

Amino acid	*Kuru* body (n = 2)	*Kuru* gut organs[a] (n = 2)	*Motto* body (n = 2)	*Motto* smoked (n = 2)
Asp	71.3	35.5	62.5	68.1
Thr	34.2	23.2	30.1	32.4
Ser	35.8	18.6	32.2	34.8
Glu	124	66.2	107	109
Pro	26.2	15.9	23.2	23.1
Gly	49.4	28.2	39.1	34.2
Ala	42.9	26.6	37.1	36.9
Val	33.9	21.4	31.5	32.5
Met	17.0	11.2	14.4	16.0
Ile	33.8	25.3	29.6	30.5
Leu	62.0	34.9	55.3	55.4
Tyr	21.4	9.52	20.2	19.9
Phe	28.9	18.3	26.7	27.3
His	18.7	11.1	15.8	14.7
Lys	54.2	32.1	49.7	48.5
Arg	60.5	32.9	55.7	53.5
Cys	7.17	5.02	5.81	5.86
Trp	8.23	5.54	8.51	9.64
Total protein content	729.6	421.5	644.5	652.3

[a] Parts not eaten
Source: Paoletti et al., 2003.

highest metabolic energy content (2.99 kcal g^{-1}) of the foods and feeds tested. Similar trends were found for other nutritional parameters, such as ash, Ca, and P content.

The contents of amino acids show that earthworm meal compared well with fish meal, egg, and milk for essential amino acids (Table 25.2). Notably, the lysine content of earthworm meal was higher than that of Chinese fish meal, eggs, and cow milk. Generally, the nutritional parameters of wheat bran were better than those of earthworm casts, except for ash content (Table 25.1) and some essential amino acids such as lysine and methionine (Table 25.2).

Since Lawrence and Miller first reported the protein content of earthworm in 1945, many nutritional evaluations of the earthworm have been published (French 1957; Needham, 1957; Mclnroy, 1971; Graff, 1981; Sabine, 1978; Blair, 1985; Boushy, 1986; Edwards, 1985; Edwards and Neuhauser, 1988). It can be concluded from the literature that earthworms contain high levels of protein and that this protein is rich in amino acids, considered essential for domesticated animals and humans. However, the protein content varies with the earthworm species and experimental food of earthworms, and the protein contents reported range from 48 to 71% (dry weight basis). *E. foetida* is purportedly a species with a relatively high protein content of 58–71% dry weight (Sabine, 1978) or about 9.7% of its live weight (Zeng et al., 1982).

Our results showed the protein content of *E. foetida* to be ca. 55% dry weight, similar to the results of Fosgate and Babb (1972) but lower than those reported in Sabine's review (Sabine, 1983).

Comparison of Protein Content of Four Earthworm Species

Four species of earthworms were collected for analysis of protein and amino acid content. Specimens of *Eisenia foetida*, fed on animal manure and crop straw, were obtained from the Muyu vermiculture farm. Wild samples of *Allolobophora caliginosa, Pheretima guillemi*, and *Eisenia foetida* were collected from fields in the Muyu region in Laiyang County. All 280 earthworm samples (70 samples each of the four species) were put under tap water for 30–40 minutes to remove impurities from the body surfaces, then dried with hygroscopic paper. Half the earthworm samples were freeze-dried with their gut contents intact, the other half freeze-dried after evacuation of their gut contents was ensured by keeping them in glass containers on wet soft toilet paper for 36 hours.

A comparison of the protein and EAA contents of meal made from *E. foetida* with that made from two other common earthworm species is shown in Table 25.4. The protein content of species *E. foetida* was significantly higher than that of the larger species *A. caliginosa* and *P. guillemi* ($p < 0.01$). *A. caliginosa* and *P. guillemi* (without gut contents) contained 49.7% and 50.1% crude protein respectively on a dry matter basis. These protein contents are lower than that of fish meal but higher than that of soybean meal. No significant differences were found in the protein content of wild and cultured species of *E. foetida* ($p > 0.05$). Inclusion of the intestinal contents lowered the protein content of the earthworm species significantly ($p < 0.01$), especially in *A. caliginosa* and *P. guillemi*. The

Table 25.4: Contents % (g/100 g on dry matter basis) of protein and amino acids in meal of different earthworm species

	***E. foetida* wild**	***E. foetida* cultured**	***A. caliginosa***	***P. guillemi***
Protein	59.11 (39.9) ± 2.96 (± 2.56)	59.41 (44.8) ± 2.23 (± 1.49)	49.7 (31.9) ± 1.71 (± 2.95)	50.11 (31.22) ± 1.36 (± 2.37)
Lys	4.17 (3.22)	4.25 (3.77)	3.04 (2.05)	2.86 (1.65)
Met	1.13 (0.81)	0.99 (0.95)	0.82 (0.50)	0.75 (0.40)
Ala	5.22 (2.94)	5.39 (2.75)	3.42 (1.94)	3.10 (1.83)
Arg	4.06 (2.95)	3.89 (3.26)	1.42 (1.02)	3.02 (1.74)
Cys	0.73 (0.47)	0.65 (0.35)	0.50 (0.30)	0.61 (0.44)
Val	2.89 (2.09)	3.29 (2.30)	2.27 (1.25)	2.08 (1.19)
Phe	2.38 (1.60)	2.46 (1.91)	1.70 (1.11)	1.52 (0.89)
Thr	3.40 (2.08)	2.94 (2.66)	2.18 (1.48)	2.01 (1.50)
His	1.56 (0.85)	1.74 (1.22)	0.62 (0.40)	1.04 (0.62)
Ile	2.36 (1.83)	2.60 (2.10)	2.03 (1.54)	1.91 (1.52)

Data in parentheses refer to earthworm meal with gut inclusions.

crude protein content of the latter species with gut inclusion amounted to only 62.3–64.2% of that found in the same species without intestinal contents.

The differences in protein content found for the same earthworm species is mainly due to variations in gut inclusions, aside from differences in experimental methods and the waste in which the earthworm is grown. Our analyses showed that gut inclusion lowered the protein content of *E. foetida* by at least 20–30% (Table 25.2). The lower protein content found in large earthworm species, such as *A. caliginosa* and *P. guillemi*, may have been the result of incomplete evacuation of gut contents, since larger species take more time to excrete their gut contents completely. So, actually speaking, there may be no significant differences in protein contents among the various species.

Contents of Free Amino Acids, Minerals, and Vitamins in Earthworm Body Fluids

Specimens of *E. foetida* were hydrolyzed with proteinase (Zhang, 1984) into a fluid. The amino acid composition of the fluid produced was analyzed and compared with that of fresh earthworm bodies. The mineral and vitamin contents of earthworm fluid were assessed in chemical analysis centers. Contents of free amino acids, minerals, trace elements, and vitamins in the body fluids of *E. foetida* are summarized in Table 25.5. The protein content of the body fluids of live *E. foetida*, dissolved by proteinase, ranged from 8.22 to 9.45%, and that of live earthworm bodies from 8.94 to 11.37%. The free amino acid (FAA) content in the earthworm body fluids reached 78.73 g L^{-1} (Table 25.5). This means that more than 90% of the earthworm protein was hydrolyzed into FAA.

The iron content of the earthworm body fluid was 0.33 g L^{-1}, which is 10 times higher than that of soybean and fish meal (Zeng et al., 1982). Most other elements in earthworm fluids, except for calcium (Ca), were 3 to 6 times higher than in fish meal or soybean meal (Zeng et al., 1982). Earthworm body fluids

Table 25.5: Contents of free amino acids (FAA) in body fluids and fresh body of *E. foetida*

FAA	Body fluids (g/l)	Fresh bodies (g/kg)	FAA	Body fluids (g/l)	Fresh bodies (g/kg)
Asp	5.9 ± 0.09	9.2 ± 0.28	Ile	4.7 ± 0.11	4.2 ± 0.38
Thr	4.3 ± 0.08	4.7 ± 0.14	Leu	9.4 ± 0.24	5.3 ± 0.36
Ser	3.1 ± 0.33	4.4 ± 0.08	Tyr	3.2 ± 0.22	4.1 ± 0.32
Glu	10.4 ± 0.15	3.3 ± 0.22	Phe	4.2 ± 0.08	4.9 ± 0.12
Gly	2.7 ± 0.42	4.2 ± 0.19	Lys	7.2 ± 0.17	7.2 ± 0.21
Ala	6.0 ± 0.12	9.0 ± 0.26	His	1.8 ± 0.23	2.9 ± 0.15
Cys	/	0.6 ± 0.09	Arg	5.5 ± 0.14	8.0 ± 0.09
Val	5.4 ± 0.10	6.1 ± 0.08	Prs	2.1 ± 0.24	3.3 ± 0.18
Met	3.0 ± 0.09	2.9 ± 0.17			

were rich in vitamin B_1 and B_2 (Table 25.6) and reached 54.65 and 83.06 mg per liter of raw earthworm body fluids.

Table 25.6: Contents of some elements and vitamins in body fluids of *E. foetida*

Element	mg/l	Element	g/l	Vitamin	mg/l
Mn	3.27 ± 1.01	Ca	0.33 ± 0.046	A	13.46 ± 0.37
Zn	6.90 ± 0.27	Na	0.31 ± 0.038	B_1	54.65 ± 0.80
Cu	1.08 ± 0.30	K	0.99 ± 0.042	B_2	83.06 ± 1.05
Pb	0.30 ± 0.15	Fe	0.33 ± 0.072	E	31.64 ± 0.64
Se	0.30 ± 0.12	Mg	0.11 ± 0.025	C	292.00 ± 2.35

Few reports are available on the protein content of earthworm fluids because earthworms are usually applied as livestock feed in the form of dry earthworm meal. Zou (1993) reported that earthworm body fluids contained a range of 7–9% protein.

Results from the literature show that the contents of essential elements such as copper (Cu), manganese (Mn) and zinc (Zn) are between one and six times higher in earthworm meal than in soybean meal and fish meal (Zeng et al., 1982; Beyer, 1987; Zou, 1993). A similar trend was found in the present study regarding the mineral element contents of earthworm body fluids, which contained 3 to 6 times more of most elements than did fish meal and soybean meal (Table 25.5). Earthworms seem able to tolerate large concentrations of some metals in their tissues, which may be related to specific metal-binding proteins in them (Suzuki et al., 1980). These metal-binding proteins may affect nutrient absorption when earthworms are introduced into feeds and foods; for example, a 90% phosphorus utilization rate from an earthworm-based diet was reported (Zeng et al., 1982).

Earthworm bodies were rich in vitamin A and vitamin B compounds; 0.25 mg vitamin B_1, and 2.3 mg vitamin B_2, per 100 g dry weight have been reported (Zeng et al., 1982). Our results demonstrated that the vitamin B_1 and vitamin B_2 contents of earthworm body fluids are 54.65 and 83.06 mg per liter fluids respectively. Since Zeng et al. (1982) did not supply information on the earthworm species and substrate his earthworms fed upon, his results cannot be compared in detail with those of our study. It may nonetheless be said that earthworms have potential for mineral and vitamin supplementation, as well as protein, in animal feed and human food production.

Actual Nutritional Value of Earthworm Protein

The nutritional value of protein depends on its specific amino acid composition. Comparing the amino acid contents of *E. foetida* reported in our study with those in other reports (Mclnroy, 1971, Sabine, 1983, Zeng et al., 1982, Zhang, 1984), and with those in *Eudrilus eugeniae* (Graff, 1981), *Lumbricus rubellus* (Taboga, 1980,

Nadazdin, 1988), *A. caliginosa*, and *P. guillemi* (Zeng et al., 1982; this volume), considerable variability can be seen among species and even within the same species. Nevertheless, Graff (1981) suggested that the contents of individual amino acids differ between species by no more than 17% and usually by considerably less.

Sabine (1983) reviewed the amino acid composition of earthworm protein reported by five authors, and compared these data with that of two common sources of protein supplements, meat meal and fish meal. His results showed that valine, leucine, and isoleucine were higher in earthworm meal than in fish meal, but lower than in meat meal. The methionine content of earthworm meal was close to that of meat meal, but 200% of that in fish meal. Arginine, histidine, and phenylalanine contents in earthworm meal were close to those in meat meal and slightly higher than in fish meal. Threonine, cystine, and tryptophan in earthworm meal were significantly higher than in fish meal and meat meal. Therefore, Sabine (1983) suggested that since earthworm protein was high in essential amino acids, including the sulphur-containing amino acids, it should be very suitable for animal feed.

In our studies, the contents of EAA (on a dry matter basis) in earthworm meal were close to or slightly lower than those in fish meal and eggs. The methionine content of our earthworm meal was significantly less than that of fish meal, soybean meal and hen egg ($p < 0.05$), which contrasts with data in Sabine's review (1983). The lysine content of our earthworm meal was significantly higher than that of eggs and cow milk ($p < 0.05$).

From the literature cited and the results of our present study, it can be concluded that *E. foetida* (especially when cultured) is relatively high in most essential amino acids, compared to other common foods and feeds. Biological value and net protein utilization are the two most important parameters used conventionally to evaluate protein quality of feed materials. Schulz and Graff (1977) reported a biological value of 84% and a net protein utilization of 79% in a rat-growth assay with *E. foetida* protein. These results were verified in fish and chicken tissues (Edwards, 1985; Koh, 1985; Velasquez, 1991; Sun, 1995).

It can be seen that earthworm protein is easily dissolved by enzymes into free amino acids. This suggests that earthworm protein is easily metabolized by animals. Thus, earthworms seem a promising source of protein for supplementation of animal feed and human food.

Nutritional Evaluation of Earthworm Casts

A nutritional evaluation of earthworm casts as a feed ingredient has not yet been published although they are considered an excellent organic fertilizer (Zeng et al., 1982; Curry, 1988). From the comparison of the protein content (7.9% dry matter) in dry earthworm casts with that of wheat bran and corn meal (Tables 25.1 and 25.2), it can be seen that the protein content is close to that of corn meal.

The amino acid contents of earthworm casts are generally lower than those of wheat bran, although some essential amino acids such as lysine, methionine, and isoleucine are richer in earthworm casts (Table 25.2). It may be concluded that earthworm casts could be partially substituted for wheat bran or corn meal in animal feed. However, there is one shortcoming, namely the lower total energy content of earthworm casts as a feed material.

Safety of Earthworm Used as Protein Source

Locally in China, people have been using earthworms as a food for several centuries. Records from the ancient book "On Guo Yi Gong" say that the people who lived in Fujian, considered different from other people, treated earthworms as a delicacy. They cut them into small pieces and mixed them with meat filling to make their food more tasty. Even now, earthworm soup, a traditional delicacy, is still offered in some restaurants of Guangdong province. In recent years, some countries in Western Europe and the Southern East have produced various earthworm products such as canned earthworms, mushroom-earthworms, and earthworm biscuits and bread. In California, a company composed of several earthworm farms in northern America, held an exhibition and competition on earthworm food in 1975. In some African and South American countries earthworms are commonly eaten. Given the high content and good quality of earthworm protein, the high content of vitamin B and other bioactive substances, it is very likely that earthworms could become an important source of animal protein in human nutrition should other sources become limited.

Due to the assimilation of metals from their environment, earthworms often contain high levels of metals, including some heavy metals that could be harmful to animals and human beings (Andersen, 1979, Zeng et al., 1982; Ireland, 1983; Fischer and Koszorus, 1992). However, in experiments with fish and chicken, no significant increments in heavy metals were found in the carcasses of the trial animals fed earthworms (Sun, 1995). Earthworm consumption may pose other possible hazards. For instance, it has been suggested that in natural food chains earthworms might carry some parasitic nematodes of chickens and pigs (Zeng et al., 1982). However, earthworms were not found to be significant in the natural distribution of parasitic nematodes in either chickens (Augustine and Lund, 1974) or pigs (Jakovijevic, 1975). Increasing proportions of earthworms and earthworm casts in fish and chicken diets produced no significant changes in the organoleptic qualities of the meat produced (Sun, 1995), although the earthworm meal itself reportedly had a garlic taste (Edwards and Lofty, 1977). Continuous monitoring of the organic substrates used in vermiculture, as well as of earthworm composition and contamination, will be necessary to safeguard human health, especially where large-scale commercial earthworm production is considered for human consumption.

Conclusions

(1) *Eisenia foetida* meal has a high protein content of 54.6–59.4% of its dry weight. Its protein content and amino acid composition are close to those of Peru fish meal and eggs, and higher than those of Chinese fish meal, cow milk, and soybean meal. The crude fat content of earthworm meal was 7.34%. The ash and energy content of earthworm meal are the highest of all the materials tested except for the higher energy content of corn meal.

(2) Casts of *E. foetida* contain 7.9% protein (dry weight), which is close to the protein content of corn meal but lower than that of wheat bran. However, earthworm casts are richer in lysine, methionine, and isoleucine. Earthworm casts could very well be used as a partial replacement of wheat bran or corn meal in animal feeds.

(3) Earthworm protein is readily hydrolyzed into free amino acids. Earthworm body fluids contain 9.34% protein and 78.73 mg of free amino acids per liter of raw fluid and are rich in vitamins and minerals, especially iron (Fe).

(4) The protein content of earthworms shows considerable variability among species and among experimental treatments in the same species, possibly due to variability in the extent of gut inclusion.

(5) Based on its nutrient content, *E. foetida* could be an excellent source of both protein and minerals in animal and human diets.

References

Aklyaman T. 1984. Supplementation of various meals to fish meal diet for chum salmon fry. Aquaculture 37 (3): 217–222.

Andersen C. 1979. Cadmium, lead and calcium content, number, and biomass, in earthworms (Lumbricidae) from sewage sludge treated soil. Pedobiologia, p. 309.

AOAC (Association of Official Analytical Chemists). 1975. *In:* Horwize W. (ed.). Official methods of Analysis, AOAC, Washington DC.

Augustine P.C. and Lund E.E. 1974. The fate of eggs and larvae of Ascaridia galli in earthworms. Avian Dis. 18: 394–398.

Beyer W.N. 1987. Relation of PH and other soil variables to concentration of Pb, Cu, Zn, Cd and Se in earthworms. Pedobiologia, p. 167.

Blair B. 1985. Unusual nutritions: a look at some feed alternatives. Milling 168 (6): 21–22.

Boushy E. 1986. Biological conversion of poultry and animal waste to a feedstuff for poultry. Poultry Adviser 19 (6): 51–65.

Curry J.P. 1988. The ecology of earthworms in reclaimed soils and their influence on soil fertility. *In:* Edwards C.A. and Neuhauser E.F. (eds.). Earthworms in Waste and Environmental Management. SPB Acad. Publ., The Hague, The Netherlands, pp. 21–261.

Edwards C.A. 1985. Production of feed protein from animal waste by earthworms. Phil. Trans. R. Soc. Lond. B 310: 299–307.

Edwards C.A. and Lofty J.R. 1977. Biology of Earthworms. Chapman and Hall, London, pp. 330–333. (2nd ed.)

Edwards C.A. and Neuhauser E.F. (eds.). 1988. Earthworms in Waste and Environmental Management. SPB Acad. Publ., The Hague, The Netherlands.

Fischer E. and Koszorus L. 1992. Sublethal effects, accumulation capacities and elimination rate of As, Hg and Se in the manure earthworm, *Eisenia foetida* (Oligochaeta, Lumbricidae). Pedobiologia, pp. 172–178.

Fosgate O.T. and Babb M.R. 1972. Biodegradation of animal waste by *Lumbricus terrestris*. J Dairy Sci. 55: 870–872.

French C.E. 1957. Nutrient composition of earthworms. J. Wildlife Mgmt. 21: 348.

Gao 1996. Records of Chinese materia medica. Jilin Sci. Jilin, Tech. Press, pp. 18–27. (in Chinese)

Graff O. 1981. Preliminary experiments of vermicomposting of different waste materials using Eudrilus eugeniae Kinberg. *In:* Appelhof M. (ed.). Workshop on the Role of Earthworms in the Stabilization of Organic Residues (Proc.), Beech Leaf Press, Kalamazoo, MI, vol. 1, pp. 179–191.

Harwood M. 1976. Recovery of protein from poultry waste by earthworms. In: Proc. First Australian Poul. Stockfeed Conv., Melbourne, Victoria, Australia, pp. 138–143.

Harwood M. and Sabine J.R. 1978. The nutritive value of earthworm meal. Proc. 2nd Austr. Poul. Stockfeed Conv., Sydney, NSW, Australia, pp. 164–171.

Huang F.-Z. and Zhang Y.-Z. 1980. Distribution of earthworms in China and their functions for substance cycles and nutrient accumulation in soil. Shanxi Sci. Tech. (Agric.), pp. 18–24. (in Chinese)

Ireland M.P. 1983. Heavy metal uptake and tissue distribution in earthworms. *In:* Satchell J.E. (ed.), Earthworm Ecology from Darwin to Vermiculture. Chapman and Hall, London, UK, pp. 247–256.

Jakovijevic D.D. 1975. Some aspects of the epizootology and economical significance of ascariasis in swine. Acta Vet. Beogr. 25: 315–325.

Koh T.S. 1985. Biological utilization of earthworm (*Eisenia foetida*) cake protein supplemented with methioine, histidine or tryptophan in chicks. Korean J. Animal Sci., 27 (5), 305–309.

Lawrence R.D. and Miller H.R. 1945. Protein content of earthworms. Nature, p. 155, 157.

Mclnroy D.M. 1971. Evaluation of the earthworm *Eisenia foetida* as food for man and domestic animals. Feedstuffs 43: 37–46.

Mo H.-D. 1984. Experimental Statistics of Agriculture. Shanghai Sci. Tech. Press, Shanghai, pp. 618–624.

Nadazdin M. 1988. Essential amino acid values in dried Californian earthworms with their possible use in poultry nutrition. Veterinarski Glasnik 42 (11/12): 745–751.

Needham A.E. 1957. Components of nitrogenous excreta in the earthworms *Eisenia foetida* and *Lumbricus terrestris*. J. Exp. Biol. 34: 425–446.

Paoletti M.G., Buscardo E., VanderJagt D.J., Pastuszyn A. et al. 2003, Nutrient content of earthworms consumed by Ye'kuana Amerindians of the Alto Orinoco of Venezuela, Proc. R. Soc. London, 270: 249–257.

Sabine J. 1978. The nutritive value of earthworm meal. Utilization of soil organisms in sludge management. Natl. Tech. Inform Serv., Springfield, VA, No. PS 286932.

Sabine J.R. 1983. Earthworms as a source of food and drugs. *In:* Satchell J.R. (ed.). Earthworm Ecology from Darwin to Vermiculture. Chapman and Hall, London, UK, pp. 285–296.

Schulz E. and Graff O. 1 977. Zur bewertung von Regenwurmmehl aus *Eisenia foetida* (Savigny, 1826) das Eiweissfuttermittel. Anndbau Forsch. Volkenrode 27: 216–218.

Snedecor G.W. and Cochran W.G. 1980. Statistical Methods. Iowa State Univ. Press, Ames, IA, 507 pp.

Stafford E.A. and Tacon A.G.J. 1985. The nutritive evaluation of dried earthworm meal (*Eisenia foetida* Savigny, 1826) included at low levels in production diets for rainbow trout, *Salmo gairdneri* Richardson. Aquac. Fish. Mgmt. 16: 213–222.

Steel R.G.D. and Torrie J.H. 1960. Principles and Procedure of Statistics. McGraw-Hill, New York, NY.

Sugimura K. 1984. Nutritional value of earthworms and grasshoppers as poultry feed. Jpn. Poul. Sci. 21 (1): 1–7.

Sun Z.-J. 1995. Production and Nutritive Value of Earthworm. "Domin" Press, Novi Sad, Yugoslavia.

Sun Zhenjun. 1997. Studies on earthworm active protein. Post-doctoral research report Agricultural Univ., Beijing, China.

Sun Zhenjun. 1998. Studies on earthworm active protein II. Extraction, properties and inducement of the antibacterial peptide from earthworms. 6th Int. Symp. Earthworm Ecology, Aug. 31–Sep. 4. 228 pp.

Suzuki K.T., Yamamura M., and Mori T. 1980. Cadmium-binding proteins induced in the earthworm. Arch. Environ. Contam. Toxicol., 9: 415–424.

Taboga L. 1980. The nutritional value of earthworms for chickens. Brit. Poul. Sci. 21: 405–410.

Velasquez L. 1991. A note on the nutritional evaluation of earthworm meal (*Eisenia foetida*) in diets for rainbow trout. Animal Prod. 53(1): 119–122.

Xiao Z.-D. 1984. A survey of earthworm metabolic energy. J. Jilin Agric. Univ., Jilin, 4: 28–32.

Xu Z.-Y. 1979. Experimental Handbook for Nutrition Analysis. Publ. House Agric., Beijing, P.R. China, 198 pp.

Zeng Z.-P., Zhang G.-Q., and Xu Q. 1982. Earthworm Cultures. Hubei People's Publ. House, Wuhan, China, 146 pp.

Zhang H.-Z. 1984. Nutritional composition and evaluation of earthworms. Chinese J. Zool., 19 (2): 18–19. (in Chinese)

Zou W.-C. 1993. Nutritive composition of earthworm's body fluids and its exploitation and utilization. Chinese J. Zool. 28 (3): 7–11.

26

Pharmaceutical Value and Use of Earthworms*

Sun Zhenjun and Cheng Wenling

College of Resources and Environmental Sciences, China Agricultural University, 100 094 Beijing, P.R. China. E-mail: sun108@cau.edu.cn; E-mail: chengwenling88@sina.com

Abstract

Earthworms have been a traditional medicine in China for at least 2,300 years. In ancient China, dilong (the Chinese medicinal name for earthworm) was used as an antipyretic and anaesthetic, for detoxification, treatment of hypertension and expediting childbirth, as well as in treatment of many common ailments, such as arthritis, itching, burns, carbuncles, erysipelas, and inflammation. With recent development of modern science, some active compounds, such as lumbritin and terrestrolumbrolysin, have been isolated from earthworm (Oligochaeta: Lumbricidea, *Lumbricus* spp.) bodies. In 1986, a Japanese scientist extracted an enzyme from earthworms (*Eisenia foetida*) that could dissolve thrombi in experimental conditions. This enzyme preparation has been made into an oral medicine by the pharmaceutical industry for use in the prevention of cardiovascular disease in Hong Kong, Japan, Korea, and China.

The authors review here the methods of collection and processing of medicinal earthworms, their pharmaceutical composition and pharmacological and clinical functions, including the effects of earthworms on the nervous system, blood system, cardiovascular system, respiratory system, uterus smooth muscle and their anticancer function. Clinical application of earthworms is recorded in detail, including treatments of tracheitis and bronchial asthma, epilepsy, high blood pressure, schizophrenia, leg ulcers, mumps, eczema, urticaria and anaphylactic diseases, chronic prostatitis, burns and scalds, fractures, erysipelas, sequelae of encephalitis B, chronic lumbago, skin crevices, blood-deficiency

*This research was financially supported by Doctoral Research Foundation (Grant No: 20020019029).

apoplexy, acute injury of soft tissues, vertigo, hematemesis and hematuria, digestive ulcer, vesical calculus, and cancer.

Key Words: medicinal earthworm, collection and processing, pharmaceutical composition, pharmacological and clinical functions, clinical application.

Introduction

For at least the last 2,300 years, earthworms have been recorded as a traditional medicine in China. Results from the studies of Zeng et al. (1982), Alumets et al. (1979), and Reynolds and Reynolds (1972) show that some nitrogenous substances extracted from the earthworm can make human lungs and bronchi dilate and can be used as an anti histamine to treat asthma. In ancient Chinese medicine books, such as *Qianjinfang, Danxinfang, Jixiaofan,* and *Bencaogangmu*, it is recorded that dilong (Chinese medicinal name for earthworm) was used as an antipyretic and anesthetic, for detoxification, treating hypertension and hastening parturition, as well as in the treatment of many common ailments, such as arthritis, itching, burns, carbuncles, erysipelas, and inflammation. With the development of modern science, some active compounds of the earthworm, such as lumbritin and terrestrolumbrolysin, have been isolated. Even more recently, an enzyme was extracted from earthworms that dissolved thrombi under experimental conditions. It has been made into an oral medicine by the pharmaceutical industry for use in the prevention of cardiovascular disease. This paper reviews the methods of collecting and processing of medicinal earthworms, the pharmaceutical composition and pharmacological and clinical functions of earthworms, as well as their clinical application in China.

Earthworm Collection

The collection of earthworms (*Pheretima* spp. and *Allolobophora* spp.) from fields occurs in spring, summer and autumn in China, for which a water solution of plants, *Polygonum hydropiper* and *Camellia sinensis* (fruit), is used. The solution is sprayed on the field to stimulate emergence of earthworms from the soil (similar to the formalin method). The worms collected are put in warm water, washed, then transferred to plant ash to choke and bored. The dead earthworms are cut, the guts flushed out of the body, and the processed worms sun-or heat-dried, then stored in a cool, dark, dry place to minimize mildew and insect pest attacks.

Processing of Medicinal Earthworms

Earthworms are processed by any one of the following methods:

(1) Cut into 1–2 cm sections, soaked in water for half a minute, rinsed for 1–2 minutes, and made into meal under strong sunshine. The dried earthworm tissues are kept in wooden boxes stored in a dry, ventilated place.

(2) Placed in bamboo baskets and sprayed with a little water, then immediately cut into 3–5 mm sections with a medicine-cutter. The sections are transferred to a bowl and the soil and sand rinsed off with water; they are then sun- or heat-dried. By this method the earthworms are quickly and completely sliced into distinct sections, the soil and sand easily and quickly rinsed off, and the earthworm fragments retain more medical ingredients.

(3) Moistened by sprinkling with Shaoxing wines and steeped for one hour. Meanwhile wheat bran is heated by frying, the earthworms added after completion of marination, and the mixture fried until the worms turn deep yellow. The mixture is removed from the pan, sieved to remove the wheat bran, and the fried earthworms spread out, cooled, and stored.

(4) Moistened with old rice vinegar in which they soak for one hour. When the vinegar has been absorbed, the earthworms are spread in an enamel dish to a depth of 3 cm and placed in an oven at 100°C for 2 hours. When the toasted earthworms have turned a deep yellow, they are cooled and stored in bottles.

Pharmaceutical Composition of Earthworms

Earthworms (*Pheretima* spp., *Allolobophora* spp., and *Eisenia* spp.) contain lumbrofebrine, terrestrolumbrolysin, lumbritin, hypoxanthine, and other purines, pyrimidines, choline, and guanidine. The fat of earthworms is composed of octade acids, palmitic acids, high-chain unsaturated fatty acids, linear and odd carbon fatty acids, branched fatty acids, phosphatide, cholesterin etc. The yellow chloragenous cells and organs of *Lumbricus terrestris* contain rich amounts of carbohydrates, lipids, protein, pigments, and some basic amino acids. The yellow pigment perhaps consists of riboflavin or its analogues (Jiangsu New Medical College, 1985).

The tissues of *Pheretima* spp. in southern China contain large amounts of microelements, Zn 59.1 μg/g, Cu 25.4 μg/g, Fe 1735.5 μg/g, Cr 10.93 μg/g, Mo 0.25 μg/g, Ca 1019.2 μg/g, and Mn 1143 μg/g (Zhao, 1988). *Allolobophora caliginosa* tissue contains crude protein 57.96%, crude fat 6.53%, crude ash 21.09%, crude fiber 0.36%, N extract 14.06%. *Eisenia foetida* contains crude protein 64.61%, crude fat 12.29%, crude ash10.16%, crude fiber 0.27%, N extract 12.67%. *E. rosea* contains crude protein 63.71%, crude fat 12.29%, crude ash 10.66%, crude fiber 0.21%, and N extract 12.67% (Zhang, 1984).

The blood and body fluids of *Lumbricus terrestris* contain small concentrations of glucose (0.01–0.05 μg/ml) (Prento, 1987), considerable lipids, including 35.14% neutral fat, 41.74% glucolipid, and 23.12 % phosphatide. The C-chain of the fatty acid is between 10° to 22°C. The neutral fat consists mainly of esters of lauric, oleic, myristic, and decanoic acids. The fatty acids of glucolipids are decanoic acid and some short chain fatty acids. The fatty acids of phosphatide are oleic, decanoic, linoleic and behenic acids. The proportion of unsaturated fatty acids is higher than that of saturated fatty acids and saccharides (Lee, 1986; Weber, 1985). A P-peptide substance exists in the gut wall of *L. terrestris* (Kaloustain, 1986).

Some active enzymes occur in the yellow chloragenous cells and organs of *L. terrestris* in high concentrations, including catalase, peroxidase, dismutase, β-D-glucosyl enzyme, alkaline phosphatase, esterase, S-amino-γ-ketoglutaric dehydrogenase, and porphyrin synthetase. The body fluids of *Eisenia spp.* contain at least 18 proteins with molecular weights between 1,000 and 95,000 Da (Valembois, 1982; Cheng, 1985).

The dormant species *A. caliginosa* contains a protein which can hydrolyze collagen (Kaloustain, 1986). Scientists from Japan, China, and Korea found and separated enzymes from the earthworm (*L. terrestris, E. foetida*) gut and body fluids which can dissolve fibrin. These enzymes have been developed as innovative medicines to treat cerebral thromboses and myocardial infarction (Cheng, 1989). Sun (1998) found and separated a kind of acid antibacterial peptide, tetradecapeptide, which has produced a disease-resistant, nutrient earthworm preparation which can be used in plant and animal production. There is also an enzyme in earthworm body tissue which can dissolve the earthworm under certain conditions (Sun, 1997).

Pharmacological and Clinical Functions of Earthworms

Effects on Nervous System

Zhang first reported in the 1950s that earthworms can reduce blood pressure (Zhang, 1959). Xu observed the phenomenon of a significant reduction in blood pressure of anesthetized dogs that were injected with a hot water and ethanol solution of macerated earthworm extracts. To understand the depressive mechanisms of earthworms, Xu conducted a medicine-pouring experiment with a rabbit heart separated from the body. The results showed that 0.001% of earthworm macerated extracts dissolved in hot water and ethanol solution can increase heartbeats, but when the dose was increased to 0.05–1%, heartbeats decreased. Up to a dose of 0.1 g/kg body weight, the heart rate of the rabbit decreased progressively. At a dose 15 to 17 times higher than a clinical enema to the dog, there were no changes in heart rate and electrical activity. Xu pointed out that the depressive mechanisms of earthworms cannot be explained solely by their effect on the heart (Xu, 1963). Wang (1963) separated an effective depressive component B_1 from earthworm tissues in an ion-exchange column. It significantly reduced a type of high blood pressure caused by excitation of the sciatic nerves from the central-area-forwards and confirmed that earthworms can cause antiexcitation similar to that produced by caffeine. Wang concluded that the main function of the depressant was located in the central nerve of the upper spinal cord and can directly influence the central nervous system by some inner receptors and cause dilatation of some internal blood vessels (Wang, 1963). Kuang (1984) reported that a 100% extract of earthworm (*Allolobophora* spp.) tissues can improve metabolism of dopamine (DA) and 5-hydroxy tryptamine (5-HT), monoamine neurotransmitter of the central nervous system; thus it can have a protective function against blood-deficiency brain death (Kuang, 1984).

Effects of Earthworms on Blood

Rao (1986) reported that the enzymes in earthworm body fluids can dissolve fibrin. Li (1988) studied extracts from earthworms to restrain the formation of thrombi, by comparing six indices of thrombosis including viscosity angle, development time of pro-thrombosis, formation time of a characteristic thrombus, dissolving time of the fibrin thrombus, length of the thrombus, and dry weight. Cheng studied the effects of different extracts of earthworms on the rates of decomposition of experimental thrombus of rabbits, with whole blood coagulum, blood plasma with platelets, and with pure fibrin coagulum of humans. The results showed that crude extracts of earthworms and protein extracts had significant effects in dissolving various kinds of thrombi of rabbit and humans. The thrombus-dissolving function was demonstrated by directly hydrolyzing fibrin (Cheng, 1989).

Effects of Earthworms on Cardiovascular System

Shen (1982) reported that injections of earthworm extract (0.5 g/ml) could act against arrhythmia and alleviated arrhythmia of various experimental models made with chloroform-adrenaline, ectisine, barium chloride, respectively. It also caused short conduction stops between the atrium and ventricle, not attributable to K+ in the agent.

Effects of Earthworms on Respiratory System

Early in the 1930s, Zhao (1930) separated an effective asthma-calming component from earthworms. This component was used in experiments with rabbit lungs and reportedly caused bronchiectasis. Hence it could be used to resist asthma caused by histamine and pilocarpine. When this component was mainlined into the body cavity of experimental animals, 50% of the animals could withstand the lethal dose of histamine (Zhang, 1990).

Effects on Uterus Smooth Muscle

Xu (1964) separated a kind of substance which can make the uterus smooth muscle contract. Experimental results showed that this substance significantly increased the tension of a pregnant or nonpregnant uterus. In 1984, he reported earthworm injections increased contractions of the mouse uterus more than the standard solution of pituitrin (0.01 mg/ml) (Xu, 1984).

Anticancer Function

Earthworm extracts were successfully used to cure cancer transplanted into S-180 cell of rats (Wang, 1986) and significantly reduced the cancer after treatment for 88 days with a 5 mg/ml enema with no adverse side effects (Wang, 1988). Han et al. (1991) separated some components by a dialytic method and observed their effects on MGc803 gastric cancers in participation of 3H-TdR. The results showed that some earthworm components could inhibit 3H-TdR participation of MGc803 gastric carcinoma ($p < 0.01$), and still had an inhibitory function even when the component was heated up to 56°C for half an hour ($p < 0.05$). This means that the dialytic components of earthworms have a strong heat resistance on a limited scale (Han, 1991). Further results showed that some other components could directly inhibit the growth of cancerous cells (Han, 1991). Sun et al. (1989) compared the cancer-killing ability of four treatments, including cancer cell suspension, earthworm extract-blood porphyrin derivative-laser, blood porphyrin derivative-laser, and earthworm extracts. The toxicity rate on cancer cells was highest in the treatment with an earthworm extract-blood porphyrin derivative-laser. With the chemical-luminous method, Sun (1991) concluded the mechanism by which the earthworm extract increased the cancer-killing capability of blood porphyrin, derivative-laser is by increasing active oxygen (Sun, 1991). A new kind of anticancer medicine, "Fu Nai Kang" has been developed in China (Wenlai and Zuyu, 1998).

Sperm-killing Function of Earthworms

Succinic and hyaluronic acids in earthworm tissue can agglutinate and kill sperms (Zhang, 1987, 1988).

Shi (1981) reported the toxicity of earthworms. Results showed that earthworms (*Allolobophora caliginosa*) can contain 200 ppm of arsenic. This arsenic toxicity can be decreased by irrigation and was comparatively low in experiments with rabbits, rats and dogs that had the enema or were intravenously injected with earthworm extracts. The toxic and side effects are below the level harmful to human health. The sperm-killing function of earthworms will be used for family planning in China (Zhang, 1998).

Clinical Applications of Earthworms

Clinical applications of earthworms have been used for treating the following diseases.

Treatment of Tracheitis and Bronchial Asthma

Fried hot earthworm powder was taken orally 3–4 times a day in doses of 3–4 g each for treatment of bronchial asthma (Cui, 1964). Mixtures of earthworm, inner bone of *Sepiella maindroni*, and coagulum of *Bambusa textilis* were ground to a fine powder for treatment of bronchial asthma using a single dose of 1.5 g once, together with some medicinal decoctions to nourish the lungs and dissolve phlegm. The therapeutic effects were very good (Zhu, 1964). An earthworm preparation, Chuan-shu-ning pill, was used for treatment of 44 cases of asthmatic patients and 84.09% responded favorably to the treatment. This method is characteristic of lasting and moderate chronic asthma (Ling, 1961). A single earthworm injection was used to treat 275 cases of bronchial asthma and asthmatic bronchitis and 78% of the patients, children in particular, recovered fully. The therapeutic effects were better for children than for adults (Shanghai Cooperative Group for Asthma Treatment, 1971) According to a report by Huang (1956), a 30% earthworm injection was used to treat child acute asthma and adult chronic asthma with a single dose of 0.1–2 ml for children and 2 ml for adults once a day when the asthma set in. After a 10- to 30-minute treatment, breathing became easier. Asthma wheezing diminished and phlegm was coughed. With two to four treatments, the symptoms of asthma disappeared entirely (Huang, 1956). Usually a 1 ml earthworm extract preparation (equal to 1 g earthworm) for adults was used as an intramuscular injection on the first day and 2 ml per day for a second dose if no side effects appeared on the first day. The course of treatment was ten days (Internal Medicine of Jiangsu Hospital, 1971). Some reports said earthworm mixtures from several species were better than single species earthworm preparations in curing 101 cases of acute asthma. A dose of 2 ml intramuscular injection per day, once every other day, resulted in 88.1% of the patients responding to the treatment in one to two weeks (Shanghai Cooperative Group for Asthma Treatment, 1971). A mixture of earthworm and toad powder was used to treat 107 cases of chronic tracheitis and the rate of effective treatment reached 92.5%. The results of germ cultures and bacterial checks showed that the earthworm tissue components were effective in controlling inflammation and repairing mucosa membranes (Chronic Tracheitis Treatment Group of PLA, 1974, 1986). Earthworm powder was used in the treatment of 100 cases of children with asthma and the therapeutic effects were very good, especially for acute asthma (Liang, 1984).

Treatment of Epilepsy

An earthworm pill, a secret recipe handed down from generation to generation, was composed mainly of earthworms (*A. caliginosa*) and vermillion. The pill looked like a green grain and was wrapped in gold leaf. Tests over 50 years confirmed its therapeutic effects against epilepsy (Gao, 1985). An earthworm mixture soup was effective in treating epilepsy. This consisted of earthworms 9 g, whole scorpions (*Buthus martensi*) 5 g, dry fruit of *Forsythia suspensa* 9 g, flowers of *Lonicera esquirolii* 12 g, stems of *Uncarria rhynchophylla* 15 g, and plaster stone 30 g. A 50% earthworm injection was used intramuscularly 3 to 5 times to control epileptic attacks and in 85.7% cases the disorder was effectively controlled (He, 1973). Zhu concocted a preparation of 3 to 6 g of earthworms, used once a day, to treat 20 cases of partial epilepsy; when coordinated with chemical treatments, 16 cases recovered fully, 3 improved, and only one showed no effect (Zhu, 1983). Zhang used another earthworm soup to treat 12 cases of epilepsy, dosed once a day for 10 to 20 days. In four cases there were no epileptic attacks for one year, in five no attacks for six months, and in the other patients there were obvious decreases in attack frequency (Zhang, 1984).

Treatment of High Blood Pressure

An earthworm tincture applied twice a day at 20 ml per dose was used to cure 34 cases of hypertension in patients who were treated with other medicines without effect. High blood pressure was usually reduced within 4 to 10 days by this earthworm treatment (He, 1973). An earthworm extract named Earth dragon B1 was used 3 times per day (2 ml dose) to treat 11 cases of hypertension; the results showed an effective ratio of suppression of 90.9% with no obvious side effect (Mao, 1959). An earthworm soup mixture was also used to treat 17 cases of hypertension with very good therapeutic effects (Huo, 1983). A K factor extracted from earthworms was injected intramuscularly to cure high blood pressure and an 86.6% improvement rate was found in 30 cases, in which the depressant function showed in various phases of hypertension and proved better than chemical treatments in controlling the disorder (Mao, 1966).

Treatment of Schizophrenia

A batch of 300 fresh earthworms was stewed with eggs and eaten as a dish to cure schizophrenia (Cai, 1966). Yan (1979) also reported that earthworms were used to treat 110 cases of schizophrenia. These 110 cases were divided into two groups, with 60 in the first group treated with earthworm powder and 50 in the second group with an earthworm injection. During a 60-day course of treatment, 18 patients were cured in the first group and 11 in the second, thereby indicating that fresh earthworms were more effective than dried ones (Yan, 1979).

Treatment of Leg Ulcers

An earthworm ointment for external use, "Xin-fu-shuang" in Chinese, was used to treat 50 cases of leg ulcers: 17 cases recovered fully and 37 improved. The ointment was effective because it arrested the pain, deposed rotten tissues, removed pus, and improved the growth of muscle buds (Shanghai Jinshan Hospital, 1982). A syrup made of earthworms and powdered sugar was spread on the ulcers with good results (Dermatological Department of the First Hospital Attached to Hubei Medical College, 1960). Earthworms were also used orally to cure leg ulcers. The earthworms were rinsed under cold water, wrapped in a skin of soybean milk, and consumed before dinner. A patient took 300 earthworms and recovered fully (Wan, 1963).

Treatment of Mumps

Earthworms were rinsed and put in a container, an equal quantity of powdered sugar added to the container, and the earthworms submerged. The earthworms gradually secreted a yellow-white mucus which was spread on the affected parts, then covered with gauze. The mucus was changed once every 2 to 3 hours. Observation of 20 cases showed this to be the best method of treating mumps because of its fast detumescence and antipyretic effects (Lu, 1961). Another paper reported that 170 cases of mumps were fully cured within 1 to 3 days of using this method (Xie, 1966).

Treatment of Eczema, Urticaria, and Anaphylactic Diseases

Earthworm tissue extracts were used to treat 35 cases of eczema with injections at acupuncture points; 14 patients recovered fully, 13 improved, 5 responded to treatment, while 3 showed no effect (Guangzhou 173 Hospital of PLA, 1974). A sample of 60 g earthworms was mixed with 30 g sugar and the patients recovered after application of the mixture to the affected parts 4 to 5 times daily. This method was used to treat itches and repeated eczema attacks (Hu, 1980).

Earthworm injections of 2 ml were used to treat 100 cases of urticaria once a day and resulted in an 84% cure rate (Li, 1976). In another 50 cases of treatment with this method, 15 patients recovered fully, 24 improved, 9 responded to treatment, while 2 showed no effect (Sichuan First Hospital, 1980). A liniment of earthworms and glycerol extract was applied externally to treat strip bled. Of the 9 cases, 3 patients were fully cured in three days and another recovered in four to ten days (Ye, 1982). Fresh earthworms were pounded with chive roots to a pulp which was stirred evenly with sesame oil. Liu (1983) reported that 26 cases of strip bled recovered fully using this method. Earthworms can also be used to treat regular lupus erythematosus. The method consists of an infusion of fresh earthworms and powdered sugar applied to the affected areas; 15 cases were cured with 2 to 4 treatments (Zhang, 1976). A formula consisting of whole bodies of *Pheretima aspergillum* or *Allolobophora caliginosa trapezoides*, roots of *Rehmannia glutinosa*, and root of *Paeonia suffruticosa* satisfactorily treated some anaphylactic diseases (Zhou, 1978).

Treatment of Chronic Prostatitis

Of 232 chronic prostatitis patients treated with earthworm mixtures, the therapeutic effects were such that 128 healed fully, 62 improved and 42 responded to treatment but relapsed when the treatment was stopped. This prescription comprised dry earthworms of *Pheretima aspergillum* or *Allolobophora caliginosa trapezoides* 20 g, roots of *Paeonia suffruticosa* 20 g, stem of *Aristolochia manshuriensis* 15 g, seeds of *Plantago aristata* 15 g, roots of *Astragalus membranaceus* 30 g and 10 g roots of *Glycyrrhiza uralensis* (Gao, 1996).

Treatment of Burns and Scalds

A sample of 15 earthworms was placed in a sugar solution and soaked for 10 hours to produce an infusion. Using this earthworm infusion on the wound surface, 50 cases of burns and scalds (10 to 110) recovered fully in one week (Zhang, 1974). According to another record, mixtures of dried earthworms 120 g, sugar 60 g, and a little borneol were stirred into an earthworm-sugar solution. The solution was put on the burned or scalded limb 4 to 6 times for one day and the wound usually recovered in 5 to 7 days with no scarring or other harmful effects (Zhuxi Hospital, 1973). Burn ointment 101, a Chinese traditional medicine, contains tissues of *Allolobophora caliginosa* and other medicinal herbs. Application to 5,011 cases of burns and scalds (first and second degree) showed full recovery of 98.7% wounds. In 32 cases of serious burns and scalds, 23 were completely cured (Li, 1980).

Treatment of Fractures

Using earthworms to treat 63 cases of femur fracture, the pain stopped within one hour of treatment, tumescence disappeared within 24 hours, and the bone mended over an average of 38.7 days. Li (1986) observed 264 cases of new fractures of femur stem (within 7 days) and found that fracture healing time was 3.6 days earlier after earthworm treatments than after other treatments ($p < 0.05$). He considered this effect might be related to lipid and nucleic acids in the

earthworm bodies (Li, 1986). Zheng (1988) and Xu (1984) reported that fresh earthworms can increase bone growth.

Treatment of Erysipelas

A mixture of fresh earthworm tissues and red sugar was applied to parts affected with erysipelas, to cure 11 patients in 3 to 5 days (Zaolin Hospital in Hongan County, 1973).

Treatment of Sequelae of Encephalitis B

Fresh earthworms were stewed in a soup. The soup was taken orally to treat 10 cases of sequelae of encephalitis B for 30 days and satisfactory therapeutic results were obtained (Luo, 1983).

Treatment of Chronic Lumbago

An earthworm powder consisting of worms and *Caesalpinia sappan,* was used to treat 50 cases of chronic lumbago. Of the 50 patients, 48 recovered fully (Pan, 1983).

Treatment of Skin Crevices

A skin-healing ointment made of earthworm powder and pearl powder was used to treat skin crevices twice a day. Of 42 cases treated, 41 recovered fully (Pan, 1983).

Treatment of Blood-deficiency Apoplexy

Guo (1988) reported that extracts from fresh earthworms were used to treat 381 cases of blood-deficiency apoplexy, with an average effective cure rate of 79%.

Treatment of Acute Injury of Soft Tissues

The prescription for treatment of acute injuries of soft tissues consists of 3 g dried bodies of *Pheretima aspergillum* or *Allolobophora caliginosa trapezoides,* 3 g central stem of *Casalpinia sappan,* 3 g dry fruit of *Daemonorops draco* spp., 3 g *Oynanchum chinense* plant, 3 g dry fruit of *Forsythia suspensa,* 3 g root-tuber of *Aconitum carmichaeli,* 3 g dry fruit of *Zanthoxylum bungeanum,* 9 g dry flower of *Carthamus tinctorius* and 9 g stem of *Cinnamomum camphora.* This medicinal earthworm/herb mixture was soaked in 500 ml of 75% Ftorl solution for 48 hours.

The infusion was applied directly to the affected part(s) 3 to 4 times daily. Of 122 cases of acute injury of soft tissues treated in this manner, 104 recovered fully and 18 improved. The cure took from 5 to 70 days (Li, 1987).

Treatment of Vertigo

An earthworm tissue mixture soup was used to treat vertigo. Of 32 cases treated, 20 were fully cured, 7 improved, 2 responded to treatment, and 3 reported no effect (Zheng, 1985).

Treatment of Hematemesis and Hematuria

A batch of 50 fresh earthworms was mixed with 250 g brown sugar and a yellow exudate emerged from the body pores of the worms. Patients took 20 ml of this secretion orally per treatment and stopped spitting blood within 2 hours. The disease was fully healed when 100 ml of the secretion had been taken (Wu, 1985; Liu, 1981).

Treatment of Digestive Ulcers

A dried earthworm powder was taken orally by 40 patients suffering from digestive ulcers, at 2 g per dose 3 to 4 times per day; 34 cases recovered fully and 6 cases improved (Mu, 1988).

Treatment of Vesical Calculcus

Chen reported that earthworms were applied to treat 5 cases of cystolith with significant therapeutic effects (Chen, 1987).

Treatment of Cancer

Earthworm soup mixtures termed I, II, and III were taken orally for 2 to 4 months as treatment of 44 cases of cancer of the alimentary tract. The average cure rate was 89.9%, suggesting this prescription was workable in both theory and clinical practice (Chen, 1983).

References

Alumets J. et al. 1979. Neuronal localisation of immunoreactive enkephalin and B-endorphin in the earthworm. Nature, 279: 805–806.

Asthma-treating Group of Shanghai Huashan Hospital. 1971. Pharmaceutical Materials of Nanjing Pharmaceutical College 1: 25.

Cai Zhongqi. 1966. Zhejiang J. Trad. Chinese Med. 3: 8.

Chen Jianming. 1983. Acta Chinese Med. Pharm. 2: 32.
Chen Quansheng. 1987. Zhejiang J. Trad. Chinese Med. 1: 42.
Cheng G.F. 1985. J. Biochem. 1 (5/6): 161–168.
Cheng Zhongliang. 1989. J. Shanxi Medical College 4: 203.
Chronic Tracheitis Treatment Group, PLA Third Hospital. 1986. Shanxi J. New Chinese Med. 3: 22.
Chu Yanliang. 1985. Zhejiang J. Trad. Chinese Med. 7: 328.
Cui Bichao. 1964. Shanghai J. Trad. Chinese Med. 11: 16.
Dermatological Dept., first hospital attached to Hubei Medical College. 1960. The Intermediate Medical, 6: 48.
Gao Yiyun. 1985. Jiangsu J. Trad. Chinese Med. 5: 631.
Gao Yiyun. 1996. Records of Chinese materia medica. Jilin Sci., Tech. Press, Jilin, pp. 18–27.
Guang Zhou 173 Army Hospital, PLA. 1974. Chinese Med. J. 4: 245.
Guo Fengtao. 1988. J. Integrated Chinese Western Med. 8(7): 400.
Han Wei 1991. J. Fourth Medical Univ. PLA 12(6): 409.
Han Wei, Wang Kewei, and Zhang Fake. 1991. J. Fourth Medical Univ. PLA, 12 (4): 302.
He Chunzheng. 1973. Wuhan New Medicine Pharm. 3: 66.
Hu Dongliang. 1980. Guizhou J. Med. Pharm. 3: 46.
Huang Wenda. 1956. J. Trad. Chinese Med. 4: 32.
Huo Junan. 1983. Sichuan J. Trad. Chinese Med. 2: 55.
Internal Medicine of Jiangsu Hospital. 1971. Pharmaceutical Materials 1: 20.
Jakovijevic D.D. 1975. Some aspects of the epizootology and economic significance of ascariasis in swine. Acta Vet. Beogr. 25: 315–325.
Jiangsu New Medical College. 1985. Dictonary of materia medica. Shanghai Sci. Tech. Press, Shanghai, 111 pp.
Kaloustain K.V. 1986. Comp. Pharm. Toxicol. 83 C (2): 329–333.
Kuang Peigeng. 1984. Shanxi J. Trad. Chinese Med. 13 (3): 133.
Li Junruo. 1986. J. Integrated Chinese Western Med. 5 (7): 402.
Li Lianrong. 1965. Zhejiang J. Trad. Chinese Med. 9: 278.
Li Shaojun. 1987. Hexin J. Trad. Chinese Med. 2: 13.
Li Tingshen. 1980. Research Chinese Trad. Patent Med. 1: 123.
Li Yongji. 1988. Acta Chinese Med. and Pharm. 2: 46.
Li Zhenji. 1976. New Trad. Chinese Med. 4: 178.
Liang Yuanli. 1984. Sichuan J. Trad. Chinese Med. 4 (7): 15.
Ling Xizhi. 1961. Jiangsu J. Trad. Chinese Med. 8: 78.
Liu Honghua. 1981. New J. Trad. Chinese Med. 12: 37.
Liu Xinfu. 1983. Henan J. Trad. Chinese Med. 6: 14.
Lu Hong. 1961. Yunnan J. Med. Pharm. 1: 10.
Luo Hanzhong. 1983. New J. Trad. Chinese Med. 4: 7.
Mao Wenhong. 1959. Shanghai J. Trad. Chinese Med. 4: 39.
Mao Wenhong. 1983. Chinese J. Cardiology, 11 (4): 290.
Mu Dejun. 1988. J. Trad. Chinese Med. 29 (7): 21.
Pan Gang. 1983. J. Trad. Chinese Med. 24 (11): 6.
Prento P. 1987. Comp. Biochem. Physiol.B: Comp. Biochem. 86 B (2): 333–341.
RaoYuxiong. 1986. Practical Medical J. 2 (6): 41.
Reynolds J.W. and Reynolds W.M. 1972. Earthworms in medicine. Amer. J. Nurs. 72: 1273.
Shanghai Cooperative Group for Asthma Treatment. 1971. New Chinese Med. 8: 62.
Shanghai Jinshan Hospital. 1982. Selections of Res. Materials III (11): 168.
Shen Liangshou. 1982. Hunan Medicine J. 6: 68.
Shi Jicun. 1981. Research Chinese Trad. Patent Med. 5: 17.
Sichuan First Hospital. 1980. Chongguang Med. and Pharm. 2: 36.
Sun Shufen. 1989. J. Fourth Medical Univ. PLA 19 (6): 376.
Sun Shufen. 1991. J. Fourth Medical Univ. PLA 21 (3): 141.
Sun Zhenjun. 1997. Studies on earthworm active protein. Post-doctoral research rept.

Sun Zhenjun. 1998. Studies on earthworm active protein II. Extraction, properties and inducement of the antibacterial epptide from earthworm. 6th Int. Symp. Earthworm Ecology, Aug. 31–Sep. 4, p. 228.
Valembois P. et al. 1982. J. Invert. Pat. 40 (1): 21–27.
Wang Jiguang. 1963. J. Lanzhou Medical College 1: 59.
Wang Kewei. 1986. J. Fourth Medical Univ. PLA 7(2): 85.
Wang Kewei. 1988. J. Fourth Medical Univ. PLA 9(1): 61.
Weber R.E. et al. 1985. Mol. Physiol. 7 (2): 93–106.
Wenlai L. and Zuyu W. 1998. Inhibitory effect of "FU NAI KANG" on advanced oesophageal cancer. 6th Int. Symp. Earthworm Ecology, Aug. 31–Sep. 4, p. 208.
Wu Qihai. 1985. Shanghai J. Trad. Chinese Med. 9: 34.
Xie Xin. 1966. J. Trad. Chinese Med. 2: 13.
Xu Shuyun. 1963. Acta Pharmaceutica Sinica 11 (1): 15.
Xu Shuyun. 1964. Acta Pharmaceutica Sinica 11 (11): 729.
Xu Zhenshan. 1984. Res. Chinese Trad. Patent Med. 3: 42.
Yan Shide. 1979. Zhejiang J. Trad. Chinese Med. 11: 440.
Ye Zhong. 1982. Selections of Sci. and Tech. Materials 11: 172.
Zaolin Hospital in Hongan County. 1973. Hubei Sci. Tech. 4: 20.
Zeng Z.-P., Zhang G.-Q., and Xu Q. 1982. Earthworm Cultures. Hubei People's Publishing House, Wuhan, China, 146 pp.
Zhang F. 1998. Earthworm powder for male infertility. 6th Int. Symp. Earthworm Ecology, Aug. 31–Sep. 4, p. 204.
Zhang Fufu. 1987. Shanxi J. Trad. Chinese Med. 10 (11): 45.
Zhang Fufu. 1988. J. Shanxi Trad. Chinese Med. College 11 (2): 30.
Zhang Hua. 1984. Sichuan J. Trad. Chinese Med. 2 (1): 57.
Zhang Liping. 1990. Fujian J. Trad. Chinese Med. 6: 52.
Zhang Peisheng. 1976. New Med. J. 3: 150.
Zhang Peitan. 1959. J. Lanzhou Medical College 4: 61.
Zhang Wangen 1974. J. Rural Doctors 4: 43.
Zhao Chungui. 1988. Bull. Chinese *materia medica* 13 (9): 38.
Zheng Jun. 1985. J. Hubei Trad. Chinese Medical College 5 (3): 4.
Zheng Peiyi. 1988. Fujian J. Trad. Chinese Med. Pharm. 19 (3): 617.
Zhou Xuehong. 1978. J. New Chinese Med. Pharm. 3: 9
Zhu Liangchun. 1964. J. Trad. Chinese Med. 11: 442.
Zhu Wenzheng. 1983. Hebei Med. Pharm. 3: 48.
Zhuxi Hospital. 1973. Hubei J. Hygiene 3: 53.

27

House Cricket Small-scale Farming

Alberto Collavo[1], Robert H. Glew[2], Yung-sheng Huang[3], Lu-Te Chuang[3], Rebecca Bosse[4], and Maurizio G. Paoletti[1]

[1]*Dept. of Biology, Padova University, Via U. Bassi 58/b, 35121 Padova, Italy*
[2]*Dept. of Biochemistry and Molecular Biology, University of New Mexico School of Medicine, Albuquerque, NM, USA*
[3]*Ross Products Division, Abbott Laboratories, Columbus, OH, USA*
[4]*NIOSH Laboratory, Cincinnati, OH, USA*

Abstract

The potential nutritional value of insects in general and the common house cricket, *Acheta domesticus*, in particular in human diets has long been recognized. In addition to providing a rich source of high quality proteins for human consumption, crickets and other related insects such as grasshoppers and locusts offer several other advantages as human food sources: they have a short life span, produce numerous offspring, are amenable to human cultivation, and can flourish under a wide range of environmental conditions. The main aims of this study were two: compare the yield of crickets raised on four different diets, and determine the amino acid, fatty acid, and mineral and trace element content of crickets grown under the best of these diets. The four diets were: aromatic-arboreal (AAD), dairy cow diet (DCD), DCD supplemented with yeast, and human refuse diet (HRD). The greatest yield (0.45 g per 10 g of feed) and highest survival (47.5%) of *A. domesticus* was achieved with HRD when grown for 9 weeks in 24 hours daylight. The protein content of crickets raised on all four diets ranged from 56.2 to 60.0% dry weight, and in all cases the essential amino acid score of the proteins approximated or exceeded the World Health Organization protein standard. The crickets contained 63–122 mg fatty acid per g dry weight, most of which was accounted for by palmitic acid, oleic acid, and the two fatty acids essential for humans, namely linoleic acid and α-linolenic acid. Crickets grown on any one of the diets contained significant quantities of the following minerals or trace elements: calcium (366–480 μg per g dry weight), copper (8.5–9.2 μg per g), iron (16.2–26.7 μg per g), and magnesium (255–306 μg per g). These data support the contention that crickets contain quantities of many nutrients that are essential to humans and show that the insect represents a commercially feasible source of food for human populations.

Key Words: *Acheta domesticus*, cricket, minilivestock, small-scale farming, food web optimization, edible insects, recycling wastes, natural resources, biodiversity, sustainable farming, food security, nutrients, fatty acids, minerals, amino acids

Introduction

The main purpose of this study was to test the hypothesis that breeding of the common cricket, *Acheta domesticus* (Plate X, 6), might offer human populations worldwide an economically feasible, sustainable, and nutritionally sound source of food. Several investigators have simulated large, insect-based production systems such as *Tribolium confusum* (Kok et al., 1990), using a variety of food sources (Larde, 1989). Several non-European human populations use different species of Orthoptera as a food source in Africa (Malaisse, 2004, this volume), Australia (Meyer-Rochow and Changkija, 1997) Latin America, (Onore, 1997; Ramos-Elorduy 1997; Ruddle, 1973), China (Luo Zhi-Yi, this volume), Thailand (Yhoung-Are and Viwatpanick, this volume), and North America (Menzel and D'Aluisio, 1998).

The high nutrient quality of the proteins of *Acheta domesticus* (L.) is widely recognized. For example, Nakagaki and coworkers (1987) demonstrated that crickets are an excellent food source for chicks and Finke and collaborators (1989) showed that weanling rats grow well on crickets. One of these same research groups (Finke, 2002) also demonstrated that invertebrates in general are an excellent food source for raising various insectivores. Furthermore, protein commonly accounts for as much as 60% dry weight of these invertebrates (DeFoliart, 1992).

This study sought to identify a food source that would be suitable for raising crickets in an economically feasible and environmentally sensitive and sustainable manner. Four different insect diets were compared: (1) an aromatic-arboreal diet (AAD), (2) a dairy cow diet (DCD), (3) the dairy cow diet supplemented with yeast (DCD+Y), and (4) a human refuse diet (HRD) (Table 27.1). The aromatic-arboreal diet was constituted from organic matter found in abundance in the woody/bushy territory of the Mediterranean, from which common plants with high nitrogen content and bacteriostatic/bactericidal activity were selected (Fenaroli, 1963). The DCD and DCD + Y diets were based largely on cereals fed to dairy cows (without the antiobiotics and integrators usually added). We used yeast as an additive in the DCD-based diet because preliminary studies had shown that it contains important growth factors (McFarlane et al., 1959; Patton, 1967) which would enhance cricket yield. The human refuse diet was selected because of the availability of large quantities of this material and because of agricultural politics. Other considerations that factored into our choice

Table 27.1: Diet composition (in grams)

Ingredients **Aromatic-Arboreal Diet (AAD)**		grams
False acacia (*Robinia pseudacacia*)		4.1
Yeast (*Saccharomyces cerevisiae*)		2.9
Basel (*Ocimum basilicum*)		1.3
Sage leaves (*Salvia officinalis*)		1.0
Hazel leaves (*Corylus avellana*)		0.5
Maple leaves (*Acer campestre*)		0.2
	Sum	10.0
Ingredients **Dairy Cow Diet with Yeast (DCD+Y)**		grams
Soybean flour (*Glycine max*)		2.07
Lucern (*Medicago sativa*)		1.78
Corn flour (*Zea mays*)		1.46
Wheat flour (*Triticum durum*)		1.31
Yeast (*Saccharomyces cerevisiae*)		1.15
Sugar beet (*Beta vulgaris* var. *esculenta*)		1.13
Silo		1.10
	Sum	10.00
Ingredients **Dairy Cow Diet (DCD)**		grams
Soybean flour (*Glycine max*)		2.26
Lucern (*Medicago sativa*)		1.97
Corn flour (*Zea mays*)		1.65
Wheat (*Triticum durum*)		1.50
Sugar beet (*Beta vulgaris* var. *esculenta*)		1.32
Silage corn		1.30
	Sum	10.00
Ingredients of **Human Refuse Diet (HRD)**		grams
Fruits and vegetables (peel and leftover)		3.4
Rice and pasta		2.7
Pork and beef meat		1.1
Bread		1.1
Cheese skins		1.1
Yolk		0.6
	Sum	10.0

of diets were cost, impact on the environment, and sustainability. In comparing the various diets used to raise the crickets, we assessed different cricket crowding conditions and the sustainability of cropping systems to produce the needed feed.

The second specific aim of the study was to compare the nutrient content of crickets raised on the four diets, with specific attention to essential amino acids, fatty acids, and mineral and trace elements. Our data identified a diet that supports a high yield of crickets and that can provide human diets with substantial quantities of essential nutrients.

Materials and Methods

Organisms

The crickets employed in this study belong to the species *Acheta domesticus* (L.). They were purchased from Livefood UK, The Acres, Gills Lane, Rooks Bridge, Somerset; www.livefood.co.uk. Two colonies of 70 individuals each were reared to obtain eggs and for a preliminary assessment of food preferences. At the time of egg hatching, individual organisms of the same day were collected in order to establish colonies and to measure food preferences, optimal population density, optimal rearing diets, and protein content and quality. Overall, 18 colonies comprised 50 to 250 individuals were established over a period of 18 months. Two of our aims were to compare the quality of the various feeds in terms of cricket production and to compare the energy efficiency conversion index (ECI) of our cricket populations with that reported for more conventional livestock.

Rearing Conditions

Colonies were reared in terrariums 0.50 m × 0.25 m × 0.35 m (total capacity 44 liters) and constructed of glass and wood. Each rearing box was covered with a 0.75 mm metal screen to prevent insects from escaping. The terrariums contained a nesting box (10 cm × 10 cm × 10 cm) filled with a mixture of sand, earth, and peat. Four egg cardboards to increase the walking surface were also settled inside the terrariums, together with test tubes filled with water to ensure humidity and a 25-watt lamp to ensure light and to maintain a suitable temperature (30.5°C). The crickets were exposed to 24 hours of light to optimize productivity, as suggested by Comby (1991) and Nakagaki and DeFoliart (1991). *A. domesticus* were fed *ad libidum* water and dry, finely ground feed. The period of observation was between 2 and 3 months for each colony.

During the rearing time, crickets were periodically counted and weighed to determine the extent of crowding and growth rates.

Assessment of Feed Preferences

Two colonies per diet were established to conduct a preliminary test of feed preference. Fifty newly hatched crickets of the same age were used to initiate each colony. We chose to use relatively few individuals so as to provide optimal access to food and water and to minimize crowding that might elicit abnormal behavior. Finely ground feed was given to prevent rotting and to optimize hygienic conditions. The feeds were oven dried at 90°C for 48 hours, then finely ground before placing in the terrariums. Feed consumption was monitored over a 60-day period, offering 40 different foods in groups of 16 substrates, and changing them randomly every 5 days; that is, each group of foods was left in the

1

4

2

5

6

3

Plate X

1—Klong Tuey Market, Bangkok, Thailand (photo K. Viwatpanich). 2—Raw male Giant Water Bugs (photographed in Khon Kaen Insect Shop; cost 5 Baht each) (photo B. Mukthabhan). 3—Street vendor of fried edible insects in Khoa San Road, Bangkok, Thailand (photo K. Viwatpanich). 4—Large numbers of *Rhynchophorus* larvae are consumed on ritual occasions by the Asmat. Here, Asmat men share the larvae during a feast, Papua New Guinea. (photo G. and U. Konrad). 5—*Acheta domesticus*, female. A good candidate for breeding (photo A. Collavo). 6—South Korean traditional medicine doctor holding centipedes (*Solopendra* sp.), one of the most important arthropod component used in South Korea to treat arthritis, stroke, snake bite, and many other maladies (photo R. Pemberton).

1

2

3

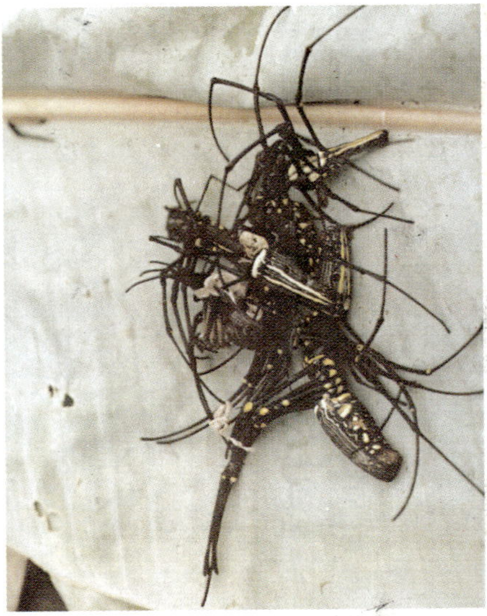

4

5

6

Plate IX

1—Woodborer larvae and Naga chilies for sale at a Kohima market, India (photo S. Changkija). 2—Red woodborer larvae from *Butea minor* for sale at a Kohima market, India (photo S. Changkija). 3—Live *Nephila* sp. spiders for the cooking pot on sale at local market in Kohima, India (photo S. Changkija). 4—Onabasulu of the Southern Highlands (Papua New Guinea) in front of the place for cooking sago-palm grubs and cooked sago-palm grubs (measuring ca. 3–4 cm each), wrapped in banana leaves (photo B. Meyer-Rochow). 5—Delousing among Kiriwina locals. The lice collected serve as indicators of good health and are consumed (photo B. Meyer-Rochow). 6—Central Australian honey-pot ants (*Camponotus inflatus*) on a slice of bread (photo B. Meyer-Rochow).

1

2

3

4

5

Plate VIII

1—*Platycoelia lutescens*. Larvae ("cuzos") from highlands in Ecuador (photo G. Onore). 2—*Panacea prola*. Last instar caterpillar larvae. Ecuador, Pichincha (photo G. Onore). 3—*Panacea prola*, dorsal view, exlarva, Ecuador (photo G. Onore). 4—*Panacea prola*, ventral view, exlarva, Ecuador (photo G. Onore). 5—Empty pupal case of a hepialid moth that emerged from the ground at the base of a River Red Gum tree, Australia (photo A.L. Yen).

3

4

5

Plate VII (Contd.)

3—*Ser servato* (*Tropidacris cristata*) roasted, Sabaneta Guaiabal, Amazonas, Venezuela (photo M.G. Paoletti). 4—Locusts (*Rhammatocerus* sp.) collected by Guajibo in Sabaneta Guaiabal, Amazonas, Venezuela (photo F. Torres). 5—Spider (*Theraphosa apophysis*) collecting and roasting inside a palm leaf, Achacoa, Alto Orinoco, Venezuela (photo M.G. Paoletti).

6

7

8

9

Plate VI (Contd.)

6—*Inago* (*Oxya yezoensis*), Japan (photo S. Mitshuashi). 7—Larvae of the caddisfly, *Stenopsyche griseipennis*, Japan (photo S. Mitshuashi). 8—*Magotaro-mushi*: larva of the dobsonfly (*Protohermes grandis*), Japan (photo S. Mitshuashi). 9—Skewered *Magotaro-mushi*: larvae of the dobsonfly (*Protohermes grandis*), Japan (photo S. Mitshuashi).

1

2

Plate VII

1—*Ser servato* wings (*Tropidacris cristata*), Alcabale da Guajibo, Amazonas, Venezuela (photo M.G. Paoletti). 2—*Ser servato* (*Tropidacris cristata*) Sabaneta Guaiabal, Amazonas, Venezuela (photo M.G. Paoletti).

1

2

3

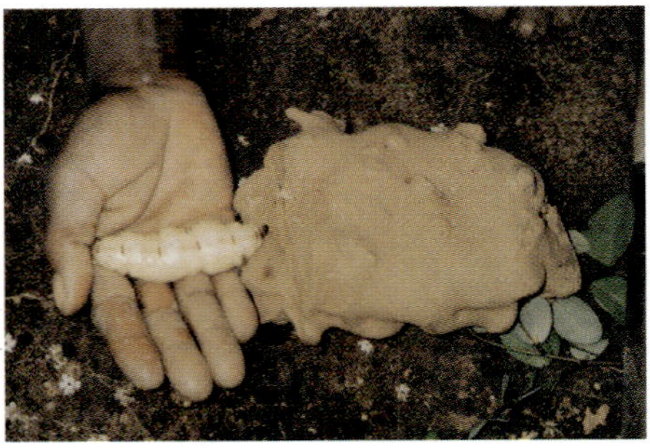

4

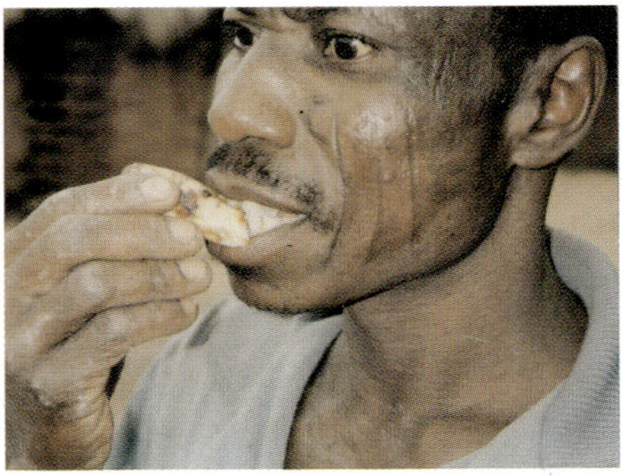

5

Plate VI

1—Collection of *Oryctes* spp. from refuse pile, Benin (photo S. Tchibozo). 2—*Oryctes* spp. collected from rolling trunk of palm trees, Benin (photo S. Tchibozo). 3—Looking for a queen in a termitarium of *Macrotermes*, Benin (photo S. Tchibozo). 4—Queen of *Macrotermes* from Mercell, Benin (photo S. Tchibozo). 5—Grilled queen of *Macrotermes*, Benin (photo S. Tchibozo).

1

2

3

4

5

Plate V

1—*Gonimbrasia zambesina* caterpillars at Luanshya market, Zambia (photo F. Malaisse). 2—*Macrotermes falciger* soldiers at Lubumbashi market, D.R. Congo (photo F. Malaisse). 3—*Phymateus viridipes*, at Lusaka market, Zambia (photo F. Malaisse). 4—*Ruspolia differens*, environs of Lubumbashi, D.R. Congo (photo F. Malaisse). 5—Woman selling larvae, Cameroon (photo A. van Huis).

1

2

3

4

5

Plate IV

1—*Achatina achatina*, dorsal view, Ghana. 2—*Achatina achatina*, ventral view, Ghana 3—*Archachatina marginata*, dorsal view, Nigeria. 4—*Archachatina marginata*, ventral view, Nigeria. 5—*Helix aspersa*, Lazio, Italy. The photos were kindly provided by the Associazione Nazionale Internazionale, Land Snail Museum, Sovana, Tuscany, Italy.

1

2

3

4

5

Plate III

1—A paca kept from very young age by local people in a Piaroa village, Alto Orinoco (Photo G. Govoni). 2—Pacas (*Agouti paca*) in a Piaroa village (photo G. Govoni). 3—An example of a traditional Piaroan community, Alto Orinoco (photo G. Govoni). 4—Internal view of paca enclosure in Caño Veneno, Alto Orinoco (photo G. Govoni). 5—*Yoyo* (*Bufo marinus*) hunted by the Yanomamo, Alto Orinoco. The same species introduced in Australia is considered poisoning pest.

Plate II

1—Cage bred grasscutter (*Thryonomys* sp.) in Gabonese farm (photo F. Jori). 2—Contrast between two cane rat youngsters: a natural brown-furred individual (top right) and a pale-furred individual issued from captive breeding in Gabon (photo P. Houben). 3—An adult brush-tailed porcupine (*Atherurus africanus*) being manipulated (photo F. Jori). 4—An adult brush-tailed porcupine (*Atherurus africanus*) (photo P. Houben). 5—Young Emin's rat (*Cricetomys emini)* (photo F. Berti).

6

7

8

9

10

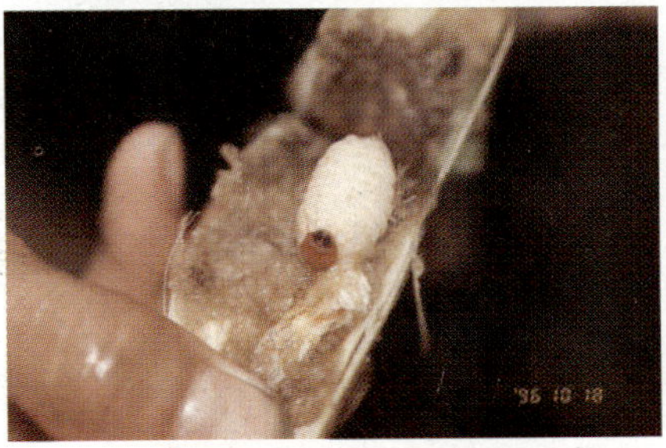

11

12

13

Plate I (Contd.)

6—*Geochelone denticulata*, tamed specimen inside an enclosure in a Piaroa village, Alto Orinoco (photo M.G. Paoletti). 7—Muscovy duck (*Cairina moschata*) trials at FUDECI in Puerto Ayacucho, Venezuela (photo M.G. Paoletti). 8—Black curassow (*Crax alector*), Alto Orinoco, Venezuela, (photo G. Govoni). 9—*Wokia* Spix's guan (*Penelope jacquacu*) in a Yanomamo village, Alto Orinoco (photo M.G. Paoletti). 10—Edible weevil larva inside a bamboo shoot in Vietnam (photo M.G. Paoletti). 11—Scorpions (*Buthus merthensi*) eaten currently in China are considered healing species (photo M.G. Paoletti). 12—Poisonous caterpillar *mamocorisina* Sphingidae (*Isognathus leachi*) at Mavaka, Motorema village, Alto Orinoco, Venezuela (Informant Ignacio) (photo M.G. Paoletti). 13—Poisonous caterpillar *mamocorisina* Sphingidae (*Isognathus leachi*) at Mavaka, Alto Orinoco, Venezuela (photo M.G. Paoletti).

Plate I

1—Robin (*Erithacus rubecola*) illigally trapped in Italy (photo L. Dreon). 2—Some "edible" colorful butterflies traditionally eaten in Friuli. a,b: *Syntomis phegea* male and female; c: *Zygaena ephialtes*; d. *Zygaena transalpina* (photo Ł. Dreon; det: Paolo Mazzei). 3—*Iguana iguana* meat much appreciated by the Guajibo and other Amerindians (photo M.G. Paoletti). 4—*Crocodylus intermedius* called *wobo* by the Guajibo, Alto Orinoco (photo M.G.Paoletti). 5—*Peltocephalus dumerilianus* hunted in a Piaroa village, Alto Orinoco, Venezuela. (photo M.G. Paoletti).

colony for 5 days. Accordingly, we were able to make 192 measurements, testing each food at least four times, most of them five times. Food consumption was expressed as the difference between the offered and unconsumed foods over a five-day interval. We noticed that approximately 10% of the offered food was dispersed in the box; this loss was measured when the terrariums were cleaned at the end of each week and taken into consideration when food consumption was calculated.

Feed Consumption, Efficiency Conversion Index (ECI), and Mortality

Thirteen colonies, each starting with 250 individuals, were grown to test these diets. Every 2–3 days cricket diet consumption was measured as just described to ascertain "feed preferences". Based on these measurements, efficiency conversion and mortality were determined.

The conversion efficiency of ingested feed was calculated according to Waldbauer (1968): ECI = 100 * [biomass gained/feed consumed]. At least 30 crickets where taken randomly and weighed. Mortality was recorded as decrease in population in each colony on the day of weighting.

Nutritional Value

The amino acid, fatty acid, and mineral and trace element composition of the adult insects, their diets, and their excreta were analyzed. The insects were collected when the population was 60 days old; however, one day before collection, they were starved for 24 hours to eliminate gut content. The specimens were finely dissected and lyophilized for three days. They were then finely pulverized using a ceramic mortar and liquid nitrogen, and again lyophilized until a constant weight was achieved. The powdered samples were stored at –80°C.

The cricket excreta deposited over a period of one month were collected from the cardboard containers in the breeding boxes. Most of this waste was made up of feed and spermatophores. The excrement samples were then dried at 60°C for 5 days, like the four diets offered during the period of breeding.

Protein Content and Amino Acid Composition

The amino acid pattern of the dietary protein source has a marked influence on the utilization of feed protein by organisms. The basis of this paradigm is that the need for protein is explained in terms of a need for individual amino acids, some of which are indispensable and must be supplied in the diet, the rest requiring only a dietary source of nitrogen and carbon skeletons from which cellular synthesis can occur. Furthermore, it is implicit in this paradigm that the magnitude of the need for the indispensable amino acids (IAA) is such that different proteins fed at isonitrogenous intakes may vary in ability to satisfy the

dietary need for protein. Thus, dietary proteins can be classified in terms of their quality, a function of their amino acid pattern. The consequence of this approach is that nutritional adequacy is likely to be met by feed in which the overall amino acid pattern is optimized by complementation, i.e., balancing inadequacies in one protein with abundance in others, to achieve an overall pattern which matches needs. In this respect, traditional feed intake practices that provide a satisfactory supply of most micronutrients also do the same for dietary protein (Milt-Ward et al., 1992).

The "chemical index" (CI) shall mean the lowest of the ratios between the quantity of each essential amino acid of the test protein and the quantity of each corresponding amino acid of the reference protein. To calculate CI it is necessary to analyze the feed protein for its nitrogen content. Protein content is calculated by multiplying the nitrogen content by 6.25 and the feed protein analyzed to determine its IAA content. The CI is calculated by dividing the milligrams of a particular IAA in one gram of the test protein by the milligrams of the indispensable amino acids in one gram of the reference protein which is the amino acid requirement pattern for a 2- to 5-year-old child. The quality of proteins refers to the CI, that is, the essential amino acid of the sample and the FAO pattern ratio (FAO/WHO, 1991).

To determine the chemical index, the protein was first hydrolyzed to yield individual amino acids and these amino acids then analyzed. Techniques for determination of amino acids were performed according to methods described in the literature (Bidlingmeyer et al., 1984; Cohen et al., 1988; Hariharan et al., 1993; Hirs, 1967; Hugli and Moore, 1972).

$$\text{Chemical Index} = \frac{\begin{array}{c}\text{Essential amino acid content of sample}\\ \text{(mg / g protein)}\end{array}}{\text{Pattern FAO amino acid in mg / g}} \times 100$$

An amino acid with a CI less than 100 is considered falling below the standard for that amino acid.

Fatty Acid Composition

Fatty acid content was determined according to methods described by Chamberlain et al. (1993) and Morrison and Smith (1964), and as specified by Paoletti et al. (2003).

Trace Element and Mineral Analysis

The insects, diets, excreta, and specimens were dried in a vacuum desiccator until a constant weight was obtained. Portions (approximately 0.2 g) were weighed into 125 ml beakers and wet-ashed using 10 ml concentrated nitric acid

and 1 ml perchloric acid. The samples were covered with watch glasses and allowed to reflux overnight at 150ºC. The next morning they were taken to near dryness at 150ºC. Samples that retained a darker color or showed remaining residue were treated with 1 mL of 4:1 nitric/perchloric acids and redigest. They were cooled and diluted to 10 mL in 4% nitric acid/1% perchloric acid. Sample solutions were analyzed for trace metal content by ICP-AES (optical emission spectrometry with inductively coupled plasma). This digestion technique does not dissolve any siliceous material in the samples.

Since the calcium content of some samples was out of the analytical range, some of the specimens had to be diluted and reanalyzed. Even with this second analysis, however, several samples remained outside the analytical range for calcium. In some cases the sample supply was depleted so a third dilution and analysis could not be completed; hence the calcium content was reported as > 2.5%.

Elements reported as n.d. were below the limit of detection. The limit of quantitation was 3.33 times the limit of detection.

Results

Rearing

Under the rearing conditions of 30.5°C and 24 h light, eggs hatched in 13 days and the adult stage was usually reached in 45 days; thus, the egg-adult cycle span was about 57 days. It is noteworthy that mortality was greatest during the first three molts (until day 9–10 at 30°C).

In comparing the average weight gained using the different diets (Fig. 27.1), it is apparent that the best growth rate was obtained with the human refuse diet. In one month crickets fed the HRD had doubled in weight and reached maximum weight (0.45 g per cricket) in 9 weeks. Other diets in order of performance vis-à-vis growth of the cricket population were DCD + Y (0.43 g per cricket), DCD (0.40 g per cricket) and AAD for which the mean weight at 10 weeks was 0.35 g per cricket.

In terms of survival, determined on day 61, the following data were obtained: for the colonies fed on ADD survival was 24%, DCD + Y 43.2%, DCD 27.1%, and HRD 47.5% (Fig. 27.2).

With the DCD + Y the optimum period for collecting crickets for consumption was between 9 and 10 weeks; for the HRD, the highest cricket yields were obtained at 8–9 weeks.

ECI Obtained with Human Refuse Diet

Rearing crickets using the HRD was more efficient with respect to time and weight gain than were the other diets. For this reason, we decided to

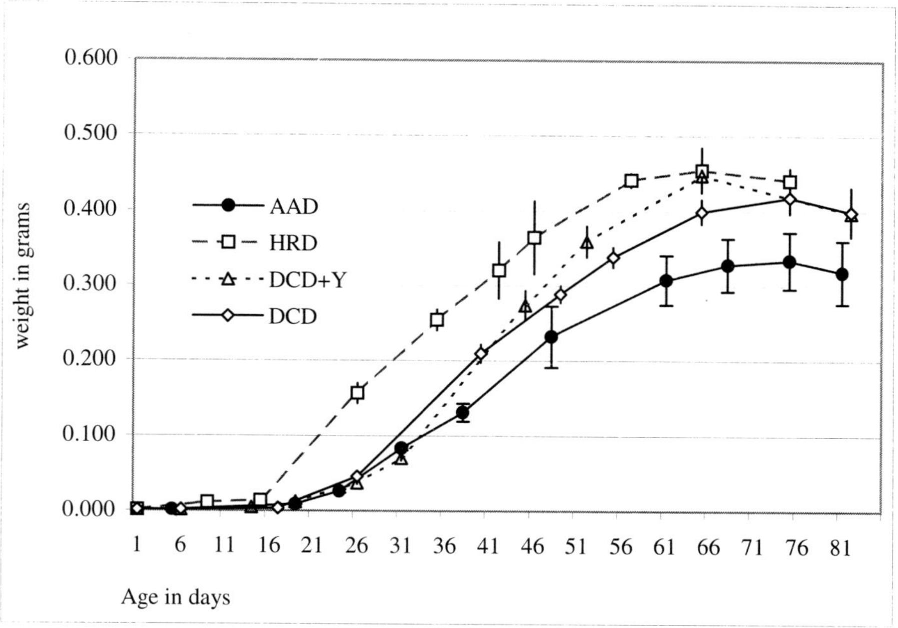

Fig. 27.1: *Acheta domesticus*, average weight gained using different diets (AAD—Aromatic-Arboreal Diet; DCD+Y—Dairy Cow Diet with Yeast; HRD—Human Refuse Diet; DCD—Dairy Cow Diet.

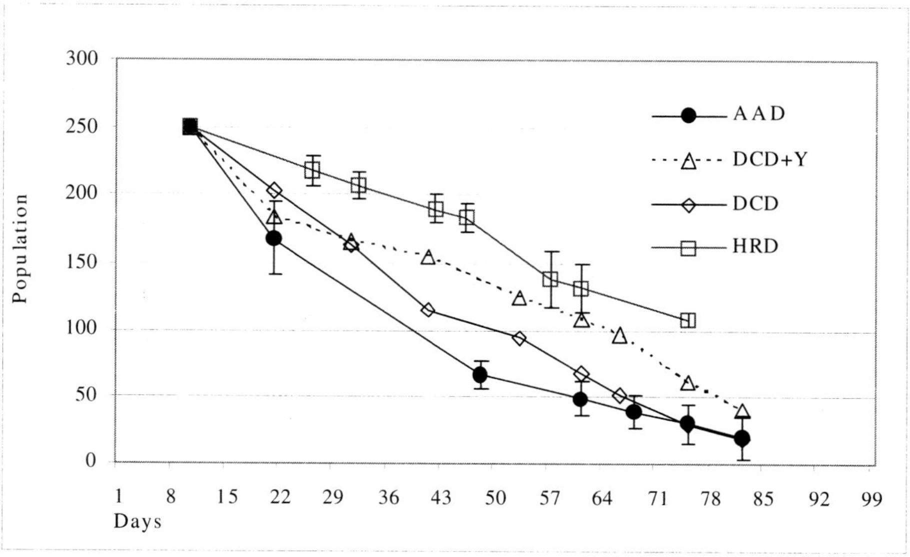

Fig. 27.2: *Acheta domesticus*, average mortalities in rearing colonies given different diets (AAD—Aromatic-Arboreal diet; HRD—Human Refuse Diet; DCD+Y—dairy Cow diet with Yeast; DCD—Dairy Cow Diet).

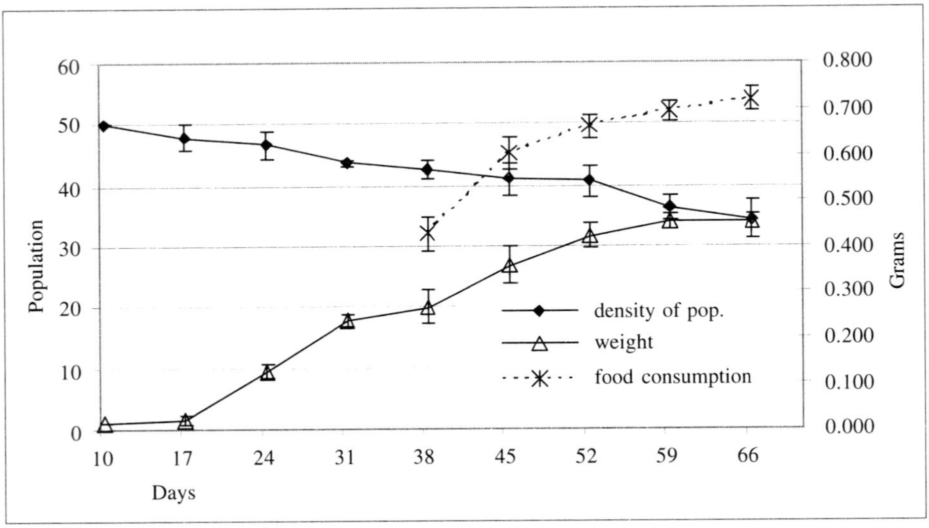

Fig. 27.3: Average density of population, weight, and feed consumption in colonies raised on Human Refuse Diet.

concentrate our efforts in carefully assessing feed input, mortality, and weight in three colonies raised using the human refuse diet. The three colonies were each composed of 50 individuals and data collected for 10 weeks, as shown in Fig. 27.3. From these data we obtained an ECI of 64-65 at week 8 (potentially best harvesting time) as reported in Table 27.11 (see "Discussion" for more details).

Protein and Amino Acid Content

Table 27.2 is a summary of the amino acid content of crickets raised under three different conditions; data for insects raised on the DCD cow diet (without yeast supplement) are not shown due to lack of exemplars at the time of analysis. Also included in this Table are the amino acid contents of the excreta and various diets. Overall, the differences in the contents of the various amino acids in the crickets were small among the three diets. The limiting amino acid (Table 27.3) was leucine for those raised on the AAD and leucine and lysine for the HRD. No essential amino acid was limiting in the case of crickets fed the DCD.

Fatty Acid Content

Table 27.4 summarizes the results of fatty acid analyses (expressed in $\mu g\ g^{-1}$ dry sample) of the total lipid fraction of the crickets raised on the three diets. The fatty acid fractions were rich in polyunsaturated fatty acid with a polyunsaturated/saturated ratio close to the recommended diet intake (stated as 1;

Table 27.2: Mean amino acid composition of *Acheta domesticus* (mg g^{-1} sample) and SD; AAD, aromatic-arboreal diet, DCD, dairy cow diet and HRD, human refuse diet

Amino acid	*A. domesticus* AAD (n=3)	*A. domesticus* DCD (n=3)	*A. domesticus* HDR (n=3)	Excrement AAD (n=3)	Excrement DCD (n=3)	Excrement HDR (n=3)	Diet AAD (n=3)	Diet DCD (n=3)	Diet HDR (n=3)
Asp	71.5 ± 7.3	67.3 ± 7.3	63.1 ± 5.8	5.5 ± 1.3	7.5 ± 3.5	6.2 ± 2.1	20.9 ± 0.4	27.5 ± 1.8	22.3 ± 0.4
Glu	86.5 ± 4.0	86.4 ± 7.1	83.8 ± 1.3	11.4 ± 2.3	12.3 ± 2.8	13.0 ± 2.3	17.3 ± 0.8	43.5 ± 0.6	60.2 ± 0.7
Ser	33.1 ± 2.6	32.0 ± 1.3	29.7 ± 1.0	5.1 ± 0.5	5.8 ± 1.5	6.2 ± 2.4	11.6 ± 0.0	10.6 ± 0.1	12.7 ± 0.1
Gly	27.9 ± 1.6	29.3 ± 1.5	28.4 ± 1.3	12.2 ± 1.6	12.5 ± 4.4	10.1 ± 1.8	8.1 ± 0.1	7.9 ± 0.1	6.7 ± 0.1
His	17.2 ± 0.6	17.4 ± 0.7	16.5 ± 0.5	3.1 ± 0.6	3.5 ± 0.2	3.5 ± 0.7	6.9 ± 0.2	6.4 ± 0.1	7.4 ± 0.1
Arg	41.3 ± 0.7	42.8 ± 1.6	39.6 ± 0.9	5.0 ± 0.6	4.7 ± 2.4	5.5 ± 2.2	13.6 ± 0.00	12.5 ± 0.3	10.5 ± 0.2
Thr	23.4 ± 0.3	24.9 ± 1.2	22.8 ± 2.0	5.2 ± 0.6	4.0 ± 1.2	3.4 ± 0.4	8.6 ± 0.0	8.9 ± 0.1	9.5 ± 0.1
Ala	46.6 ± 2.7	48.0 ± 2.7	45.8 ± 4.9	6.6 ± 0.4	10.3 ± 6.7	9.2 ± 4.0	10.6 ± 0.1	10.9 ± 0.1	9.1 ± 0.2
Pro	33.9 ± 1.2	34.2 ± 1.2	33.0 ± 1.5	6.1 ± 0.5	6.7 ± 2.2	6.9 ± 1.5	11.9 ± 1.0	13.3 ± 0.2	17.9 ± 0.4
Tyr	28.1 ± 3.3	24.9 ± 1.5	22.2 ± 2.3	3.2 ± 0.2	3.6 ± 1.1	3.5 ± 1.8	7.7 ± 0.1	6.9 ± 0.1	8.7 ± 0.6
Val	35.1 ± 0.7	35.6 ± 1.2	33.3 ± 1.8	6.3 ± 0.4	6.6 ± 1.9	6.3 ± 1.7	12.3 ± 0.1	11.6 ± 0.0	12.7 ± 0.2
Ile	24.1 ± 0.3	24.8 ± 1.2	22.3 ± 1.1	4.4 ± 0.3	4.5 ± 2.1	4.3 ± 1.5	8.4 ± 0.1	8.4 ± 0.0	9.3 ± 0.2
Leu	38.2 ± 0.5	39.7 ± 2.1	36.1 ± 1.2	6.5 ± 0.4	6.5 ± 2.7	6.2 ± 1.8	12.3 ± 0.0	15.3 ± 0.0	16.9 ± 0.1
Phe	20.4 ± 0.6	21.8 ± 0.9	20.0 ± 0.4	4.8 ± 0.3	3.8 ± 1.1	3.8 ± 0.7	11.3 ± 0.1	10.9 ± 0.1	11.2 ± 0.1
Lys	33.8 ± 0.9	34.6 ± 1.5	31.8 ± 0.2	6.5 ± 1.1	4.3 ± 1.4	4.8 ± 1.1	12.3 ± 0.1	10.9 ± 0.2	13.9 ± 0.1
Trp	7.1 ± 1.2	7.9 ± 0.8	7.5 ± 0.7	2.3 ± 0.3	1.8 ± 1.2	1.3 ± 0.2	4.3 ± 0.3	3.7 ± 0.4	3.0 ± 0.1
Cys	15.6 ± 2.4	14.0 ± 1.2	12.0 ± 0.9	3.7 ± 1.1	4.7 ± 1.0	3.3 ± 0.4	5.8 ± 0.4	7.2 ± 0.3	4.0 ± 0.1
Met	16.3 ± 1.5	14.0 ± 0.7	13.6 ± 0.4	2.6 ± 0.4	3.3 ± 1.1	2.0 ± 0.2	6.0 ± 0.2	4.5 ± 0.2	6.9 ± 0.0
Sum	600.1	599.0	561.7	100.2	105.7	99.9	189.6	221.0	242.6

Table 27.3: Chemical index of proteins in AAD, DCD, and HRD

	A. domesticus AAD	*A. domesticus* DCD	*A. domesticus* HRD
His	151	153	155
Ile	143	148	142
Leu	96	100	97
Lys	97	100	97
Cys+Met	213	187	182
Phe+Tyr	128	124	119
Thr	114	122	119
Trp	107	114	127
Val	167	170	169

FAO/WHO, 1991). Among the various colonies, the best fatty acid quality was provided by crickets fed the DCD (Table 27.5). The differences in quantities of saturated and unsaturated fatty acid present in *A. domesticus,* rearing diet, and excreta are reported in Fig. 27.4.

Mineral Content

The results of mineral analysis are shown in Table 27.6. Calcium content of crickets is higher than conventional food such as salmon, pig, or beef. The calcium:phosphorus ratio of the crickets fed the dairy cow diet was 1.0:4.4, the human refuse diet 1.0:6.4, and the aromatic-arboreal diet 1.0:1.54. The iron content of crickets was relatively high, especially in colonies reared on the DCD, with values similar to those for beef; Table 27.6 reports a higher iron content for frogs.

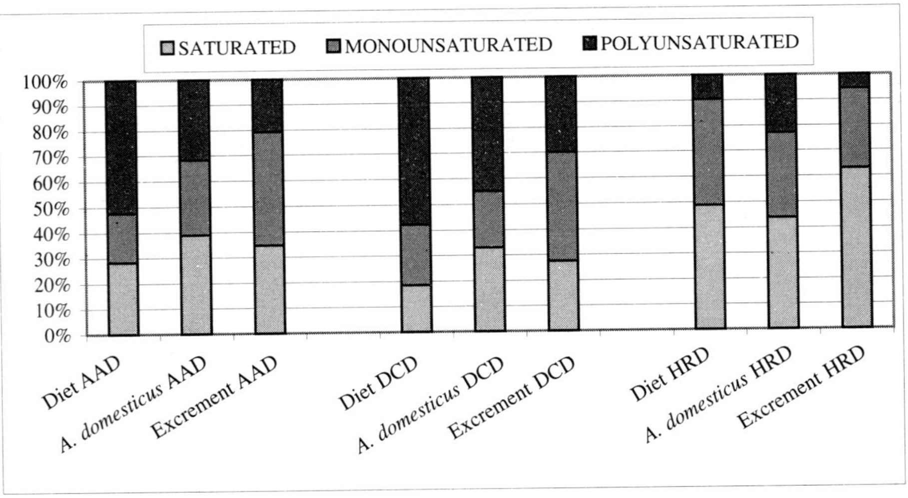

Fig. 27.4: Percent of saturated and unsaturated portions of fatty acid in samples. (Legends Fig. 27.1)

Table 27.4: Mean composition of fatty acid in *Acheta domesticus*, in collected excrement, and in diet offered—with SD. Values reported as μg g⁻¹ of dry weight

Fatty acid		*A. domesticus* AAD (n=3)	*A. domesticus* DCD (n=3)	*A. domesticus* HRD (n=3)	Excrement AAD (n=3)	Excrement DCD (n=3)	Excrement HRD (n=3)	Diet AAD (n=1)	Diet DCD (n=1)	Diet HRD (n=1)
Saturated										
C12:0	Lauric acid	0.26	0.08	0.16 ± 0.00	ND	ND	0.06 ± 0.04	ND	0.01	1.02
C14:0	Myristic acid	0.45 ± 0.07	0.58 ± 0.16	2.96 ± 1.45	0.17 ± 0.09	0.04 ± 0.02	1.12 ± 0.14	0.15	0.05	5.31
C15:0	Pentadecanoic acid	0.22 ± 0.06	0.15 ± 0.02	0.46 ± 0.21	0.04 ± 0.02	0.02 ± 0.01	0.15 ± 0.01	0.024	0.02	0.58
C16:0	Palmitic acid	21.18 ± 3.88	22.07 ± 3.92	31.82 ± 10.73	1.21 ± 0.35	0.68 ± 0.08	5.13 ± 0.54	1.84	6.02	23.26
C18:0	Stearic acid	6.05 ± 0.71	8.04 ± 1.43	8.87 ± 2.26	0.39 ± 0.13	0.25 ± 0.02	1.64 ± 0.17	0.30	1.59	6.20
C20:0	Eicosanoic acid	0.19 ± 0.04	0.27 ± 0.01	0.23 ± 0.06	0.06 ± 0.02	0.02 ± 0.00	0.06 ± 0.00	0.05	0.12	0.18
C22:0	Behenic acid	0.04	0.07 ± 0.04	0.06 ± 0.04	0.05 ± 0.01	0.02 ± 0.00	0.02 ± 0.01	0.03	0.10	0.07
C24:0	Lignoceric acid	ND	ND	ND	0.03 ± 0.01	0.01 ± 0.00	ND	ND	0.03	ND
Subtotal		28.39	31.27	44.56	1.95	1.042	8.18	2.40	7.94	36.62
Monounsaturated										
C14:1	Myristoleic acid	0.03 ± 0.01	0.05 ± 0.01	0.18 ± 0.12	0.02 ± 0.01	0.01 ± 0.00	0.05 ± 0.01	0.01	0.02	0.39
C16:1	Palmitoleic acid	3.93 ± 1.05	0.63 ± 0.14	2.22 ± 0.79	0.72 ± 0.28	0.11 ± 0.07	0.21 ± 0.06	0.24	0.18	1.48
C18:1ω9	Oleic acid	16.62 ± 2.80	18.96 ± 1.94	29.56 ± 10.16	1.46 ± 0.52	1.15 ± 0.16	3.00 ± 1.07	1.14	9.32	27.33
C18:1ω7	Vaccenoic acid	0.47 ± 0.12	1.01 ± 0.14	1.50 ± 0.60	0.15 ± 0.05	0.23 ± 0.05	0.37 ± 0.06	0.12	0.55	1.58
C20:1ω9	D11-Eicosenoic acid	0.09 ± 0.00	0.11 ± 0.02	0.18 ± 0.02	0.06 ± 0.04	0.03 ± 0.01	0.44 ± 0.06	0.02	0.07	0.18
C20:1ω7	D13-Eicosenoic acid	0.04 ± 0.01	0.05 ± 0.01	0.11 ± 0.05	0.02 ± 0.01	0.01 ± 0.00	0.04 ± 0.02	0.02	0.01	0.02
C22:1ω9	Euric acid	0.05 ± 0.03	0.11 ± 0.01	0.08 ± 0.01	0.07 ± 0.02	0.06 ± 0.04	0.05 ± 0.01	0.12	0.03	0.02
C24:1	Nervonic acid	0.17	0.12 ± 0.08	0.05 ± 0.02	0.04 ± 0.02	0.02 ± 0.01	0.02 ± 0.01	ND	ND	ND
Subtotal		21.39	21.03	33.87	2.54	1.62	4.18	1.65	10.19	31.00
Polyunsaturated										
C18:2ω6	Linoleic acid	18.97 ± 2.74	39.11 ± 3.09	20.15 ± 2.92	0.40 ± 0.13	0.98 ± 0.06	0.60 ± 0.26	0.93	21.50	5.87
C18:3ω6	γ-Linolenic acid	0.03 ± 0.00	ND	0.05 ± 0.01	ND	ND	ND	0.01	ND	0.03
C18:3ω3	α-Linolenic acid	3.71 ± 0.71	2.91 ± 0.31	1.39 ± 0.69	0.75 ± 0.28	0.13 ± 0.01	0.09 ± 0.02	3.46	3.13	0.88

Table 27.4: (Contd.)

Table 27.4: (Contd.)

Fatty acid		*A. domesticus* AAD (n=3)	*A. domesticus* DCD (n=3)	*A. domesticus* HRD (n=3)	Excrement AAD (n=3)	Excrement DCD (n=3)	Excrement HRD (n=3)	Diet AAD (n=1)	Diet DCD (n=1)	Diet HRD (n=1)
C20:2ω6	D11,14-eicosadienoic acid	0.05 ± 0.03	0.05 ± 0.01	0.10 ± 0.03	0.01 ± 0.01	0.01 ± 0.00	0.02 ± 0.01	0.01	0.02	0.08
C20:3ω6	Dihomo-g-linolenic acid	ND	ND	0.06 ± 0.02	ND	ND	0.01 ± 0.00	ND	ND	0.06
C20:4w6	Arachidonic acid	ND	0.09	0.33 ± 0.05	ND	0.01 ± 0.00	0.02 ± 0.01	ND	ND	0.16
C20:5ω3	Timnodonic acid	ND	ND	0.13 ± 0.06	0.02 ± 0.01	0.01	0.01	ND	ND	0.02
C22:4ω6	Adrenic acid	ND	ND	ND	ND	ND	ND	ND	ND	0.02
C22:5ω6	D4,7,10,13,16-docosapentaenoic acid	ND	ND	ND	ND	ND	ND	ND	ND	ND
C22:5ω3	D7,10,13,16,19-docosapentaenoic acid	ND	ND	ND	ND	ND	ND	ND	ND	0.03
C22:6ω3	Clupanodonic acid	ND	ND	ND	ND	ND	ND	ND	ND	ND
Subtotal		22.76	42.15	22.21 ± 3.77	1.17	1.13	0.75	4.42	24.65	7.13
Total fatty acid		**72.23**	**94.29**	**100.53**	**5.66**	**3.79**	**13.10**	**8.47**	**42.78**	**74.75**

ND: Absence of values means they were not detectable by analysis.

Table 27.5: Saturated and unsaturated fractions of fatty acid in *Acheta domesticus*, excrement, and diets (in %)

	Acheta domesticus			Excrement			Diets		
	AAD	DCD	HRD	AAD	DCD	HRD	AAD	DCD	HRD
Saturated	39.1	33.0	43.9	34.6	27.4	63.0	28.3	18.6	49.0
Monounsaturated	29.4	22.2	33.2	44.7	42.8	31.5	19.5	23.8	41.5
Polyunsaturated	31.5	44.8	22.9	20.7	29.8	5.5	52.2	57.6	9.5
Pol./Sat. ratio	**0.81**	**1.36**	**0.52**	**0.60**	**1.09**	**0.09**	**1.84**	**3.10**	**0.19**

Table 27.6: Mineral content in *A. domesticus* and other food; AAD, aromatic-arboreal diet, DCD, dairy cow diet and HRD, human refuse diet, results reported as µg/gram of element per sample

	A. domesticus AAD	*A. domesticus* DCD	*A. domesticus* HRD	*A. domesticus* (adult)[1]	*A. domesticus* (nymph)[1]	Salmon (*Salmo salar*)[2]	Mackerel (*Scomber scombrus*)[2]	Sardines (*Sardina pilchardus*)[2]	Mussel (*Mytilus edulis*)[2]	Frog (*Rana esculenta*)[2]	Chicken (*Gallus gallus*)[2]	Pig (*Sus scrofa domestica*)[2], beef	Horse (*Equus cabalus*)[2]	Adult bovine (*Bos taurus*)[2], filet	Corn (*Zea mays*)[2]	Wheat (*Triticum durum*)[2]	Rice (*Oryza sativa*)[2]	Tapioca (*Manihot esculenta*)[2]
Ag	n.d.	n.d.	n.d.	—	—	—	—	—	—	—	—	—	—	—	—	—	—	—
Al	9.86	12.58	10.72	—	—	—	—	—	—	—	—	—	—	—	—		—	—
As	0.01	0.08	0.07	—	—	—	—	—	—	—	—	—	—	—	—	—	—	—
Ba	0.37	0.56	0.27	—	—	—	—	—	—	—	—	—	—	—	—	—	—	—
Be	0.02	0.02	0.01	—	—	—	—	—	—	—	—	—	—	—	—	—	—	—
Ca	480	366	384	407	275	270	380	330	880	200	50	80	40	40	150	300	240	120
Cd	0.01	0.02	0.04	—	—	—	—	—	—	—	—	—	—	—	—	—	—	—
Co	0.02	0.02	0.01	—	—	—	—	—	—	—	—	—	—	—	—	—	—	—
Cr	0.68	1.02	0.68	—	—	—	—	—	—	—	—	—	—	—	—	—	—	—
Cu	9.2	8.7	8.5	6.2	5.1	—	7	2	12	—	0.9	1.3	2	0.8	3.1	4	1.8	—
Fe	19.5	26.7	16.2	19.3	21.2	7	12	18	58	60	6	8	39	19	24	36	8	10
K	2,812	2,589	2,642	3,470	3,520	3,100	3,600	6,300	3,200	—	3,000	2,900	3,310	3,300	2,870	4,940	920	200
La	0.06	0.04	n.d.	—	—	—	—	—	—	—	—	—	—	—	—	—	—	—
Li	0.02	0.04	0.01	—	—	—	—	—	—	—	—	—	—	—	—	—	—	—
Mg	306	267	255	337	226	—	210	100	440	—	240	170	290	200	1200	1600	200	—
Mn	6.1	8.0	4.6	11.5	8.9	—	—	—	—	—	—	—	—	—	—	—	—	—
Mo	0.21	0.40	0.17	—	—	—	—	—	—	—	—	—	—	—	—	—	—	—
Na	2,045	2,376	2,474	1,340	1,350	980	1,300	660	2,900	—	620	560	740	410	350	—	50	40
Ni	0.16	0.32	0.13	—	—	—	—	—	—	—	—	—	—	—	—	—	—	—
P	2,605	1,607	2,442	2,950	2,520	2,800	2,640	2,150	2,360	4,300	1,600	1,600	2,310	2,000	2,560	3,300	940	120
Pb	0.20	0.06	0.08	—	—	—	—	—	—	—	—	—	—	—	—	—	—	—
Sb	0.83	n.d.	n.d.	—	—	—	—	—	—	—	—	—	—	—	—	—	—	—

Table 27.6: (Contd.)

Table 27.6: (Contd.)

Se	n.d.	n.d.	n.d.	0.19	0.1	—	—	—	—	—	—	—	—	—	—	—	—	—
Sr	1.25	0.71	0.80	—	—	—	—	—	—	—	—	—	—	—	—	—	—	—
Te	n.d.	0.11	n.d.	—	—	—	—	—	—	—	—	—	—	—	—	—	—	—
Ti	0.10	0.14	0.09	—	—	—	—	—	—	—	—	—	—	—	—	—	—	—
Tl	0.16	n.d.	0.21	—	—	—	—	—	—	—	—	—	—	—	—	—	—	—
V	0.01	0.01	0.01	—	—	—	—	—	—	—	—	—	—	—	—	—	—	—
Y	0.01	0.01	0.01	0.21	0.28	—	—	—	—	—	—	—	—	—	—	—	—	—
Zn	61.4	52.7	65.8	67.1	68.0	—	—	20	8	22	10.8	16	37.2	28	22.1	29	13	—
Zr	0.06	0.04	0.06	—	—	—	—	—	—	—	—	—	—	—	—	—	—	—

[1]Finke, 2002. Values for commercially raised crickets.

[2]Carnovale E. and Marletta L. 2000. Composizione degli alimenti. Istituto Nazionale di Ricerca per gli alimenti e la nutrizione.

n.d. = non detectable

— = not reported

The high sodium content probably reflects NaCl supplementation of the diets. Silver (Ag), antimony (Sb), selenium (Se), and tellurium (Te) were not detectable.

Discussion

Rearing

From the standpoint of weight gain and survivorship, the best diets with which to rear crickets appeared to be dairy cow supplemented with yeast, followed closely by the human refuse diet. The poorest insect yield was obtained using the aromatic-arboreal diet, attributable largely to the fact that many of the crickets were cannibalized. Apparently the aromatic-arboreal diet was not well-balanced nutritionally and hence the stronger crickets supplemented their alimentation by consuming the weaker ones.

Since magnesium is known to reduce the activity of crustaceans (Morris and Spicer, 1993) and because it has a narcotic effect that reduces aggression in molted exemplars (Chouduri and Bagh, 1974), in an effort to reduce cannibalism we added magnesium chloride to the insect drinking water: surprisingly, magnesium had no effect either on the extent of cannibalism or the insect singing performances.

From the data gathered, it appears that the best time to collect crickets for consumption is before they become full adults, that is, at about week 4 or week 5 (Fig. 27.1). Adults, compared with nymphs, have a more coriaceous exoskeleton and develop wings. From the standpoint of weight gain, the optimum harvesting time was between week 8 and 9.

Chitin

Cricket exoskeleton is partiallly composed of chitin, a biopolymer that cannot be hydrolyzed by humans lacking the appropriate enzymes. Thus chitin acts like dietary fiber without offering any calories. So, whole insects as a source of protein are of somewhat lower quality than vertebrate animal products because of the indigestibility of chitin (Phelps et al., 1975; Dreyer and Wehmeyer, 1982).

Chitinase, an enzyme that digests chitin, is found in fish (Rehbein et al., 1986), reptiles, birds (Weiser et al., 1997) insectivorous mammals (Smith et al., 1998) and cetaceans (Olsen et al., 2000), animals that feed on insects, and crustaceans including krill. Chitinase is also present in association with other enzymes such as papain and bromelain in some tropical plants, e.g. pineapple (*Ananas comosus*), mango (*Mangifera indica*), and pawpaw (*Carica papaya*) (Azarkanb et al., 1987; Subroto et al., 1999) where it seems to perform antimycotic activities. Apparently in carnivorous *Bromeliaceae,* most of the enzymes used to release nutrients contained in animal and vegetal tissues are not secreted directly by the

plants but produced by the bacteria and saprophytic fungi hosted by them (Benzing, 1980). Papain and bromelain assist in protein degradation by releasing free amino acids that are ready for absorption and enabling optimization of fat assimilation.

It would be interesting to study whether consumption of insects combined with the tropical fruits cited above or other material would lead to chitin softening or even digestion. According to Goodman (1989), insect chitin has anticancer properties. Various studies (Muzzarelli, 1997) report on the successful use of chitin and derivates in ulcer and lipid absorption.

Diets

Results of the present study indicate that in terms of weight gain and mortality, the best diet for rearing crickets is the HRD (Fig. 27.2). This indicates that its nutrient content and balance are superior to those of the other diets.

The great difference in yield of crickets obtained for the two dairy cow diets led us to speculate that the yeast supplement probably provides the insects with critical growth factors. The nutritional value of yeast for the growth of insects has been documented by McFarlane and coworkers (1959) and Patton (1967).

Amino Acid Composition

Except for aspartic acid and tyrosine, both of which are nonessential amino acids, there were few differences in the amounts of the various amino acids contained in the crickets raised on the three diets (Table 27.2). With the exception of glutamic acid, the foods offered to the insects did not vary greatly in their content of the various amino acids. The Chemical Index (Table 27.3) of each of the three diets was close to 100 in all samples, thereby indicating these foods contained good-quality proteins.

Fatty Acid Composition

The amount of fatty acid in the three colonies of crickets was relatively high and ranged from 6 to 12% dry weight. Good quantities of the two essential fatty acids, linoleic acid (18:2ω-6) and α-linolenic (18:3ω-3) were present: linoleic acid at a percentage of 26.24 ± 0.71 in AAD, 41.50 ± 1.30% in DCD, and 20.84 ± 4.09% in HRD, α-linolenic acid at a percentage of 5.13 ± 0.71 in AAD; 3.08 ± 0.17% in DCD, and 1.31 ± 0.36% in HRD. These fatty acids represent the substrates for the synthesis of longer, more unsaturated fatty acids. Usually the major dietary sources of linoleic acid and α-linolenic acid are polyunsaturated fatty acid-rich vegetable oils. Recommended dietary intakes of linoleic acid and α-linolenic

acid are usually about 3–5% and 0.5% of dietary energy respectively (Ziegler and Filer, 1996).

Interestingly, arachidonic acid was either absent or present at very low levels in all the cricket specimens, regardless of the food upon which they were raised. Also absent or lacking were the long-chain polyunsaturated fatty acids, eicosapentaenoic acid and docosahexaenoic acid in particular. The excreta contained very little fatty acid, indicating that crickets do an efficient job of digesting and absorbing the lipids in their diets.

The polyunsaturated/saturated ratio in the HRD is not close to unit, probably because the quantities of fat varied in the meat leftovers used in preparing the diet. This value is confirmed from the high amount of lipid found in the excreta. Simply reducing the quantity of animal fat in the leftovers used as ingredients in preparing the diets should diminish this value.

Minerals and Trace Elements

As mentioned earlier, Table 27.6 compares the mineral and trace element composition of *A. domesticus* with that of some common foods. The quantities of calcium, copper, and magnesium in the crickets exceeded those of other animals except mussel. The potassium content of the crickets was slightly lower than in fish, mollusks, amphibians and other vertebrates. The phosphorus content varied greatly among the crickets grown on the three diets; in general, however, the crickets contained more phosphorus than some common livestock consumed by humans. The zinc content of crickets was high and comparable to that reported for other common animal and plant foods (Table 27.6). The relatively high sodium content of the crickets was likely the result of supplementation of the diets with sodium chloride, as suggested by other investigators (Nakagaki et al., 1987; Luckey and Stone, 1968).

ECI Obtained Using the Human Refuse Diet

Total body composition data (Table 27.7) show that crickets have a higher water content, similar protein content, and significantly less fat than most common livestock. The ECI for the diets in our trials was much higher than for any of the vertebrate species, i.e., poultry 1.5 times higher, pig and sheep 2 times higher and beef 4 times higher. The differences between our results and those of Nakagaki and DeFoliart (1991) are probably due to differences in insect harvesting times; they harvested at 21 days, we harvested at 45 days.

The favorable conversion efficiency showed by crickets relative to domesticated animals, and by insects in general, is probably due to their unusual physiology; insects do not have to spend energy maintaining body temperature. Another advantage of insect breeding is the short span of productive cycle. These

Table 27.7: Comparison of current breeding animals and *Acheta domesticus* (in vertebrates, entrails are not included) and ECI

Species and references	% Water	% Protein	% Fat	% ECI
A. domesticus				
Woodring et al. 1977	68.4	15.0	10.3	—
Nakagaki and DeFoliart, 1991	74.2	—	—	92
Finke, 2002 [1]	73.1	17.9	5.1	—
This work[2]	72.3	16.3	—	59
Chicken				
Ensminger, 1980	64.0	18.8	14.2	—
Meyer and Nelson, 1963	—	28.6	4.9	35
Lovell, 1979	—	—	—	48
Pig				
Ensminger, 1980	50.0	13.0	34.4	—
Meyer and Nelson, 1963	—	12.7	38.0	28
Whittemore & Elsey, 1976	52.6	14.7	29.5	—
Lovell, 1979	—	—	—	31
Sheep				
Ensminger, 1980	53.2	15	29.0	—
Meyer & Nelson, 1963	—	15.2	25.2	18
Beef				
Ensminger, 1980	53.5	17.0	26.0	—
Meyer and Nelson, 1963	—	18.2	21.1	16
Berg and Butterfield, 1976	52.0	17.1	26.9	—
Lovell, 1979	—	—	—	13

[1] Average values of adults and young.
[2] Colonies fed with human refuse diet.

points should to be taken into consideration in the context of commercial production.

Environmental Impacts and Diet Sustainability

In breeding crickets using the dairy cow diet, we were able to identify possible environmental factors related to: tillage (soil erosion, organic soil loss), chemical fertilizers, insecticides, herbicides and fungicides applications and their potential effects on nontarget organisms and the environment.

Table 27.8 reports such impacts under conditions of conventional cultivation. We tried to measure the possible environmental impacts of various crops under conventional agricultural practices. Activities such as tillage, use of chemical fertilizers, insecticide, herbicide, and fungicide applications were "weighed" on the basis of assessed agronomic practices: the "weight" of every activity, expressed on a scale of 1 to 4 (where 1 represent lowest impact) was summed to obtain an index of total impact. In this manner we were able to recognize that in

Table 27.8: Environmental impacts of different crops under conventional agricultural practices

	Tillage	Chemical fertilizers	Fungicides	Insecticides	Herbicides	Total impact
Corn flour (*Zea mays*)	xxx	xxxx		x	xxxx	12
Silage corn (*Zea mays*)	xxx	xxx		x	xxx	10
Soybean flour (*Glycine max*)	xx	x			xxx	6
Lucern (*Medicago sativa*)	x	x			x	3
Sugar beet (*Beta vulgaris* var. *esculenta*)	xxx	xxx	xx	xx	xxxx	14
Wheat (*Triticum durum*)	xx	xx	x		x	6

cultivating sugar beet or corn greater impacting activities are involved than in cultivating alfalfa. Wheat and soybean have mean inpacts. Alfalfa is the more sustainable crop.

Most of the ingredients used in the diets tested in this work have a low impact under conventional methods of production and might have an even lower impact were biological or integrated agriculture were the goal.

An estimate of the resources required to produce 1 ton of *A. domesticus* using conventional agriculture is reported in Table 27.9. As for of the yeast-based diet, we were not able to estimate the environmental cost of cricket production due to lack of information about yeast production and impacts.

Based on diet consumption, we estimated that 2.78 tons were required to produce one ton of crickets. The quantities of the various dietary components and cultivable surface needed to produce one ton of crickets are shown in Table 27.10.

Table 27.9: Annual production of different crops per hectare (ha); 1 q = 100 kg (pers. comm. G. Cozzi, 2002)

Lucern (*Medicago sativa*), hay	130 q/ha;
Corn flour (*Zea mays*)	120 q/ha;
Silage corn (*Zea mays*)	800 q/ha;
Soybean flour (*Glycine max*)	40 q/ha;
Wheat (*Triticum durum*)	70 q/ha;
Sugar beet (*Beta vulgaris* var. *esculenta*)	650 q/ha.

Table 27.10: Resources and cultivable surface needed to produce 1 ton of *Acheta domesticus* with dairy cow diet (DCD)

	Soybean flour	Lucern	Corn	Wheat	Sugar beet	Silage corn	Total
Resources needed	6.4 q	5.5 q	4.6 q	4.1 q	3.6 q	3.5 q	27.8 q
Surface needed	1596 m^2	425 m^2	382 m^2	593 m^2	56 m^2	44 m^2	3096 m^2

Hygienic Hazard and Potential Constraints

The vernacular and scientific names of *Acheta domesticus* suggest infestation of human dwellings and in the past they were common inhabitants of kitchens, bakeries, and places where grains and flour were stored; in recent years they have become rare if not altogether absent, depending largely on the standards of hygiene acquired by humans. Moreover, insects can be vectors of disease, spreading various forms of parasite and pathogen: virus, rickettsiae, bacteria, protozoa, and nematodes (Busvine, 1980).

In this study neither viral and bacteriological susceptibilities of bred *A. domesticus*, nor their potential allergenic properties were assessed. Insects, as arthropods, could present the same problems as shellfish (i.e., shrimp, lobster, crayfish), which are well known for their ability to induce mild to severe allergic reactions in susceptible individuals. Thus, this risk should not be taken lightly. Usually the popular image of insect allergies is that associated with the bites and stings of venomous species such as bees, ants, and wasps (injectant allergens). More common allergic reactions attributable to insects include those caused by contacting body parts or waste products (contact allergens) or inhaling microscopic dust particles composed of pulverized carcasses, cast skins and excreta (inhalant allergens). Allergies caused by contacting or inhaling insect material can have significant health consequences in the home or work environment with symptoms ranging from eczema and dermatitis, to rhinitis, congestion, and bronchial asthma. In severe cases, sensitivity to insect material might be heightened to the extent that the victim experiences anaphylactic shock, a potentially life-threatening condition often involving rapid swelling, acute respiratory distress, and collapse of circulation (Phillips and Burkholder, 1984). Since most insect allergies are of the contact and inhalant type, it would be reasonable to assume that the greatest health risk associated with food insects would be to workers involved in their production. Given the small and obscure nature of the food insects industry, virtually nothing is known of such problems. A case has been reported for Bulgaria in which employees working on nuts acquired eczema and dermatitis due to contact with substances secreted by larvae of *Plodia interpunctella* feeding on the nuts (Phillips and Burkholder, 1995).

Reactions to Orthoptera (grasshoppers, crickets, locusts, cockroaches, etc.) are also common. LeClercq (1969) reported that workers rearing locusts suffered rhinitis, itching skin, bronchitis, and ultimately asthma in general sequence. Wirtz (1980) recounted one study of migratory locusts where all of the workers became allergic to the insect. We know of a researcher who suffered dyspnea (labored breathing) during a prolonged session of grinding crickets into meal to supplement chicken feed. Three cases of anaphylactic shock involving orthopterans were reported by Wirtz (1980). Good ventilation, protective clothing, gloves, and masks are common sense preventive measures.

Our major interest focused on ingestant allergens, that is, eating or unintentionally swallowing allergenic insect material. Unfortunately direct evidence

for allergies to food insects is practically nonexistent. Some insight can be gained from controlled experiments on human subjects done with preparations of common food-infesting insects. A classic study by Bernton and Brown in 1967 utilized dialized extracts of seven such insects in skin sensitivity tests of subjects with and without known allergies. Test extracts included those of the rice weevil (*Sitophilus oryzae*), fruit fly (*Drosophila melanogaster*), Indian meal moth (*Plodia interpunctella*), saw-toothed grain beetle (*Oryzaephilus surinamensis*), red flour beetle larvae and adults (*Tribolium castaneum*), flour beetle (*Tribolium confusum*), and lesser grain borer (*Rhyzopertha dominica*). Of the 230 allergic patients, 68 (29.6%) reacted positively to one or more of the dialized insect extracts. Surprisingly, of the 194 nonallergic subjects, 50 (25.8%) showed sensitivity to at least one extract. A total of 333 positive reactions were observed. The degree of overall sensitivity was practically the same for both groups, with the Indian meal moth extract eliciting the most positive reactions followed by extracts of red flour beetle larvae, red flour beetle adults, rice weevils, fruit flies, confused flour beetles, saw-toothed grain beetles, and lesser grain borers.

The only reported case of *A. domesticus* as a vector of diseases in vertebrates is mentioned for *Geopetitia aspiculata*, a nematode found in 12 species of Passeriformes, one species of Charadriiformes and one species of Coraciiformes which died at the Assiniboine Park Zoo, Winnipeg, Canada. Fortunately, this nematode is harmless for humans (Bartlett et al., 1984).

There may be processes that can effectively diminish the potential threat of food allergies. One school of thought suggests that insect allergens in food are deactivated by cooking, yet when five of the aforementioned insect extracts were heated at 100°C for one hour, positive skin reactions were still observed, albeit deemed less vigorous than those of the unheated treatments. In a 1964 study, Bernton and Brown heat-treated extracts of cockroaches at 100°C for one hour and found that these allergens resisted deactivation. The idea that insect allergens are deactivated in the highly acidic environment of the stomach is also appealing until one considers the number of normally eaten foods that have been identified as potentially allergenic and whose allergens obviously survive digestion and cooking.

For most people, working with or eating food insects should pose little if any health risk, especially if they have no history of allergy to insects or other arthropods. Nonetheless, since sensitivity can be acquired with repeated exposure to an allergen, a measure of vigilance is in order. A person with known insect or arthropod allergies would be wise to exercise some caution. Cross-reactivity among related as well as taxonomically dispersed groups of insects has been established. There is also evidence for cross-reactivity among distantly related members of Arthropoda, suggesting the existence of common allergens within the phylum. So, if you are allergic to shellfish, you might want to reconsider the idea of eating insects.

Conclusion

Crickets are a good resource of animal protein, fat, and other nutrients.

Some considerations about costs should be made in cricket breeding. When a human refuse diet is adopted, the input starts at a level of domestic waste. If not used, this material represents a cost for waste processing and disposal and a loss of organic matter. Otherwise, by converting this organic refuse into animal proteins, the waste becomes a resource (Table 27.11).

Table 27.11: Mean density of population (and SD), fresh weight of *A. domesticus*, diet consumed, and Efficiency Conversion Index (ECI) in colonies fed human refuse diet (HRD)

Days of life	Density of population (no. of crickets)	Weight of *A. domesticus* (in g)	Amount of diet consumed (in g)	ECI
10	50 ± 0	0.010 ± 0.001		
17	48 ± 2	0.019 ± 0.009		
24	47 ± 2	0.125 ± 0.013		
31	44 ± 1	0.236 ± 0.011		
38	43 ± 2	0.266 ± 0.038	0.426 ± 0.039	63 ± 6.11
45	41 ± 3	0.358 ± 0.041	0.605 ± 0.036	59 ± 4.16
52	41 ± 3	0.422 ± 0.026	0.663 ± 0.027	64 ± 2.52
59	36 ± 2	0.451 ± 0.017	0.694 ± 0.022	65 ± 1.73
66	34 ± 3	0.452 ± 0.016	0.721 ± 0.027	63 ± 2.08

Where the dairy cow diet is used, one needs to consider in the production system costs for agricultural ingredients composing the diet and the relative costs of herbicides, fungicides, insecticides, and fertilizers causing high environmental impacts and potential hazards.

The aromatic arboreal diet does not seem favorable for the low cricket population density it can sustain; however it appears environmentally more friendly.

The protein quality and linoleic and a-linolenic acid content of crickets are high. The fatty acid profiles most likely reflect the fatty acid composition of the feed, which is commonly seen in monogastric animals (Finke, 2002).

The chitin in crickets presents a problem because of its nondigestibility in humans, as well other livestock, due to their lack of chitinase. Future studies should investigate whether this problem can be solved by consuming insects together with fruits such as Bromeliaceae and Caricaceae that contain enzymes with an activity similar to chitinase. To reduce the quantity of chitin, insect harvesting could be done before the last molt.

Excreta produced during the breeding cycle, still high in nitrogen, can be used as fertilizer or input for crops.

To purchase crickets, contact: Livefood UK: http://www.livefood.co.uk/

References

Azarkanb M., Amranib A., Nijsb M., Vandermeersa A. et al. 1997. Carica papaya latex is a rich source of a class II chitinase. Phytochem. 46: 1319–1325.

Bartlett C.M., Crawshaw G.J., and Appy R.G. 1984. Epizootiology, development, and pathology of *Geopetitia aspiculata* Webster, 1971 (Nematoda: Habronematoidea) in tropical birds at the Assiniboine Park Zoo, Winnipeg. Canada J. Wildl. Dis. 20: 289–99.

Benzing D.H. 1980. The Biology of Bromeliads. Mad River Press, Eureka, CA, USA, 305 pp.

Bernton H.S. and Brown H. 1964. Insect allergy—preliminary studies of the cockroach. J. Allergy 35: 506–513.

Bernton H.S. and Brown H. 1967. Insects as potential sources of ingestant allergens. Ann. Allergy 25: 381–387.

Bidlingmeyer B.A., Cohen S.A., and Tarvin, T.L. 1984. Rapid analysis of amino acids using precolumn derivitization. J. Chromatogr. 336: 93–104.

Busvine J.R. 1955. Simple methods for rearing the crickets (*Gryllus domesticus* L.) with some observations on speed of development at different temperatures. Proc. R. Ent. Soc. Lond. (A) 30: 15–18.

Busvine, J.R. 1980. *Insects and Hygiene*. Chapman and Hall, London, 568 pp.

Chamberlain J., Nelson G., and Milton K. 1993. Fatty acid profiles af major food sources of howler monkeys (*Aloutta palliata*) in the Neotropics. Experientia 49: 820–823.

Choudhuri D. and Bagh R. 1974. On the sub-social behaviour and cannibalism in *Schizodactylus monstrosus* (Orthoptera: Schizodactylidae). Rev. Ecol. Biol. Sol. 11: 569–573.

Cohen S.A. and Strydom D.J. 1988. Amino acid analysis utilizing phenylisotiocyanate derivatives. Ann. Biochem. 174: 1–16.

Comby B. 1991. Insetti che bontà. Piemme ed. Milano, 158 pp.

DeFoliart G.R. 1992. Insects as human food. Crop Protection, Elsevier, New York, NY, vol. 11, pp. 395–399.

DeFoliart G.R. and Nakagaki B.J. 1991. Comparison of diets for mass-rearing *A. domesticus* (Orthoptera: Gryllidae) as a novelty food, and comparison of food conversion efficiency with values reported for livestock. J. Econ. Entomol. 84: 892–896.

Dreyer J.J. and Wehmeyer A.S. 1982. On the nutritive value of Mopanie worms (Sur la valeur nutritive des vers de Mopanie). S. Afr. J. Sci. 78: 33–35.

FAO/WHO. 1991. Report on protein quality evaluation. Paper no. 51, FAO, UN, Rome.

Fenaroli G. 1963. Le sostanze aromatiche. Hoepli Ed., Milano, Italica, vol. 11, 1004 pp.

Finke M.D. 2002. Complete nutrient composition of commercially raised invertebrates used as food for insectivores. Zool. Biol. 21: 269–285.

Finke M.D., DeFoliart G.R., and Benvenga N.J. 1989. Use of a four-parameter logistic model to evaluate the quality of the protein from three insect species when fed to rats. J. Nutr. 119: 864–871.

Goodman W.G. 1989. Chitin, a magic bullet? Food Insects Newsl. 2(3): 1–6.

Hariharan M., Naga S., and VanNoord T. 1993. Systematic approach to the development of plasma amino acid analysis by high-performance liquid chromatography with ultraviolet detection with precolumn derivatization using phenyl isothiocyanate. J. Chromatography: Biomed. Appl. 621: 15–22.

Hirs C.W.H. 1967. Performic acid oxidation. Methods Enzymol. 11: 197–199.

Hugli T.E. and Moore S. 1972. Determination of tryptophan content of proteins by ion exchange chromatography of alkaline hydrolysates. J. Biol. Chem. 247: 2828–2834.

Kok R., Shivhare U.S., and Lomamliza K. 1990. Mass and component balances for insect production. Can. Agric. Eng., 185–192

Larde G. 1989. Investigations on some factors affecting larval growth in a coffee-pulp bed. Biol. Wastes 30: 11–19.

LeClercq M. 1969. Entomological Parasitology. Pergamon Press, New York, NY, 158 pp.

Luckey T.D. and Stone P.C. 1968. Cricket nutrition: supplementation of grass by known nutrients. J. Insect Physiol. 14: 1533–1538.

Malaisse F. 2004. Human consumption of Lepidoptera, termites, Orthoptera, and ants in Africa (this volume).

McFarlane J.E., Neilson B., and Ghouri A.S.K. 1959. Artificial diets for the house crickets. Can. J. Zool. 37: 913–916.

Menzel, P. and D'Aluisio F. 1998. Man Eating Bugs. Ten Speed Press, Berkeley, CA, 192 pp.

Meyer-Rochow V.B. and Changkija S. 1997. Uses of insects as human food in Papua New Guinea, Australia, and North-East India: cross-cultural considerations and cautious conclusions. Ecol. Food Nutr. 36: 159–185.

Milt-Ward D.J., Newsholme E.A., Pellett P.L., and Uauy R. 1992. Amino acid scoring in health and disease. *In:* Protein-Energy Interactions: Proc. Workshop Int. Dietary Energy Consultancy Group. United Nations Univ. Switzerland. Online. Internet. June 30, 2003. Available www.unu.edu/unupress/food/foodnutrition.html

Morris D. and Spicer J.I. 1993. A brief re-examination of the function and regulation of extracellular magnesium and its relationship to activity in crustacean arthropods. Comp. Biochem. Physiol. 106: 19–23.

Morrison W.R. and Smith L.M. 1964. Preparation of fatty acid methyl esters and dimethylacetals from lipids with boron trifluoride-methanol. *J. Lipid Res.*, 5: 600–608.

Muzzarelli R.A.A. 1997. Human enzymatic activities related to the therapeutic administration of chitin derivates. Cell. Mol. Life Sci. 53: 131–140.

Nakagaki B.J. and DeFoliart G.R. 1991. Comparison of diets for mass-rearing *Acheta domesticus* (Orthoptera: Gryllidae) as a novelty food, and comparison of food conversion efficiency with values reported for livestock. J. Econ. Entomol. 84 (3): 891–896.

Nakagaki B.J., Sunde M.L., and DeFoliart G.R. 1987. Protein quality of the house cricket, *Acheta domesticus,* when fed to broiler chicks. Poult. Sci. 66: 1367–1371.

Onore, G. 1997. A brief note on edible insects in Ecuador. *Ecology of Food and Nutrition,* 36 (2–4): 277–285.

Olsen M.A., Blix A.S., Utsi T.H.A., Sormo W., and Mathiesen S.D. 2000. Chitinolytic bacteria in the mink whale forestomach. Can. J. Microbiol. 46 (1): 85–94.

Paoletti M.G., Buscardo E., Vanderjagt D.J., Pastuszyn A. et al. 2003. Nutrient content of termites (*Syntermes* soldiers) consumed by Makiritare Amerindians of the Alto Orinoco of Venezuela. Ecol. Food Nutr. 42: 173–187.

Patton R.L. 1967. Olicidic diets for *Acheta domesticus* (Orthoptera: Gryllidae). Ann. Entomol. Soc. Amer. 60: 1238–1242.

Phelps R.J., Struthers J.K., and Mayo S.J.L. 1975. Investigations into the nutritive value of *Macrotermes falciger*, Isoptera, Termiditae Recherches sur la valeur nutritive de Macrotermes falciger, Isoptère Termiditae. Zool. Afr. 10: 23–132.

Phillips J.K. and Burkholder W.E. 1984. Health hazards of insects and mites in food. In: Baur F.J. (ed.). Insect Management for Food Storage and Processing. Amer. Assoc. Cereal Chemists, St. Paul, MN, pp. 279–292.

Phillips J.K. and Burkholder W.E. 1995. Allergies Related to Food Insect Production and Consumption. USDA-ARS Stored-Product Insects Research Laboratory, vol. 8.

Ramos-Elorduy J. 1997. Insects: a sustainable source of food? Ecol. Food Nutr. 36: 247–276.

Rehbein H., Danulat E., and Leineman M. 1986. Activities of chitinase and protease and concentration of fluoride in the digestive tract of Antarctic fishes feeding on krill (*Euphausia superba,* Dana). Comp. Biochem. Physiol. 85: 545–551.

Ruddle, K. 1973. The human use of insects: examples from the Yukpa. Biotropica, 5: 94–101.

Smith, S.A., Robbins L.W., and Steier J.G. 1998. Isolation and characterization of a chitinase from the nine-banded armadillo, *Dasypus novemcinctus*. J. Mammalogy 79: 486–491.

Subroto T., Sufiati S., and Beintema J.J. 1999. Papaya (*Carica papaya*) lysozyme is a member of the family 19 (basic, class II) chitinases. J. Molec. Evol. 49: 819–821.

Waldbauer G.P. 1968: The consumption, digestion and utilization of solanaceous and non-solanaceous plants by larvae of the tobacco hornworm, *Protoparcasxta* (Johan.) (Lepidoptera: Sphingidae). Entomol. Exp. Appl. 7: 253–269.

Weiser J.I., Porth A., Mertens D., and Karasov W.H. 1997. Digestion of chitin by Northern Bob-whites and American robins. Condor 99: 554–556.

Wirtz R.A. 1980. Occupational allergies to arthropods—documentation and prevention. Bull. Entomol. Soci. Amer. 26: 356–360.

Ziegler E.E. and Filer Jr. L.J. 1996. Present Knowledge in Nutrition. ILSI Press, Washington, DC, 684 pp.

28

Insects in the Human Diet: Nutritional Aspects

Sandra G.F. Bukkens

Istituto Nazionale di Ricerca per gli Alimenti e la Nutrizione (INRAN), Rome, Italy, INRAN, Via Ardeatina 546, 00178 Rome, Italy. E-mail: sbukkens@hotmail.com

Abstract

An overview of the nutritional quality of food insects, with special emphasis on their role as a source of animal protein for humans is presented. Data available on the content of protein, essential amino acids, fat, fatty acids, and micronutrients of the most common food insects are summarized. It is concluded that food insects are generally an excellent source of protein, fat, and micronutrients. The amino acid composition of most food insects compares favorably with the reference standard recommended by FAO/WHO/UNU.

Key Words: Amino acid content, antinutritional factors, chitin, cholesterol, energy content, essential amino acids, fat content, fatty acid composition, fiber content, mineral content, pesticide contamination, protein content, protein digestibility, seasonal ataxic syndrome, toxic factors, trace elements, vitamin content.

Introduction

In many parts of Africa, Asia, South America, and Australia a wide range of animal products are eaten that may not be common or known to researchers from Europe or North America. These animals include many different insects, such as locusts, grasshoppers, termites, ants, beetles, and caterpillars. The insects consumed generally have high protein content and may significantly contribute to the total protein intake of indigenous populations, at least during certain seasons of the year. However, thus far little attention has been paid to insect consumption in dietary surveys among indigenous populations. There may be several reasons for this. First, food insects are usually collected from the wild

and often eaten raw on the spot, and this casual eating may go unnoticed by outside observers (Robson and Yen, 1976; Posey, 1987). Second, insect consumption is usually seasonal, depending on the appearance of certain stages of the insect (Dufour, 1987). Third, the attitude of outside observers that entomophagy is either a curiosity or simply repulsive may cause the indigenous populations to conceal their consumption of insects (DeFoliart, 1989).

In cultures in which insect consumption is common, insects form a regular part of the diet, as a side dish, snack, or ingredient of composite dishes, whenever available during the year. Although generally insects are not merely eaten to avoid starvation, some studies show that they are most often collected and consumed when other animal foods are available in limited quantities, or not at all (Dufour, 1987; Moreno-Black and Somnasang, 2000). Nevertheless, depending on the culture, many insects are valued as delicacies in their own right, such as the *Rhynchophorus* larvae (Curculionidae) and alate ants (Formicidae) by the Tukanoan Indians in the northwest Amazon (Dufour, 1987).

Here the focus is on intentional insect consumption, especially among indigenous populations of tropical countries, and the role of insects in providing animal protein, essential amino acids, and micronutrients to these populations. This does not mean that insects are not consumed in other countries. On the contrary, in some developed countries, such as Japan, insect consumption is part of the traditional diet (Pemberton and Yamasaki, 1995; Mitsuhashi, 1997, and this volume). Insects are also consumed by certain immigrant groups in developed countries, for instance the Thai giant waterbug *(Lethocerus indicus;* Belastomatidae) is imported from Thailand and sold in California, USA (Pemberton, 1988). Furthermore, several insect species have gained some popularity as a health food or fancy delicacy in countries such as the United States, Mexico, and P.R. China. Last but not least, some insects and insect fragments are eaten unintentionally by every single person when, by chance, the insect material lies unperceived in foods of either plant or animal origin.

Insects as Part of the Human Diet

Selection of Insect Species for Consumption

Why are certain insect species and stages consumed and others not? According to the optimal foraging theory, it is the overall efficiency of foraging which will determine the popularity of insect species and stages in the diet of a society (Ardeshir, 1990; Dufour, 1987). In other words, the insect species and stages collected and consumed most commonly and in the largest quantities are the most predictable resources in space (living in nests, feeding selectively on specific plants) and time and with the highest nutritional value for human consumption. These species and stages are actively sought and collected in large quantities in any single attempt. Some insect species may even be managed resources,

such as the palm weevil *(Rhynchophorus;* Curculionidae) larvae that are literally 'harvested' from the pith of felled palms (e.g. Cerda et al., 2001, and this volume). Nevertheless, other species and stages less predictable in space and time or occurring in smaller aggregations are often collected opportunistically and in small quantities (Dufour, 1987).

In general, the stage of the life cycle collected is the soft-bodied (and often largest) form, i.e., the form that contains relatively little exoskeleton (Dufour, 1987). For instance, for Coleoptera and Lepidoptera this is the larval stage, for Formicidae the female alates (bearing eggs).

Holt (1988) mentioned that a common criterion as to whether an animal is/is not fit for human alimentation is the food it lives upon. The great majority of insects live entirely upon vegetable matter in one form or another (and often on our most-valued vegetable crops); they are thus relatively clean feeders compared to commonly consumed animals such as lobsters, eel, and pig. It has been observed throughout the Amazon that the consumption of invertebrates is greatest among those dependent on forest leaves and litter, such as caterpillars, *Atta* ants, and termites (Paoletti et al., 2000).

Major Insects Consumed

Insect consumption has been documented for various countries in the world. Most of the insects consumed in significant quantities belong to one of these six orders: Lepidoptera, butterflies and moths, usually consumed in the larval or nymphal stage (caterpillars); Coleoptera, beetles, also predominantly consumed in the larval stage ("grubs"); Orthoptera, locusts, crickets, and grasshoppers; Isoptera, termites; Hymenoptera, ants, bees, and wasps; and Hemiptera (bugs) (Table 28.1).

Of course, Table 28.1 is far from exhaustive and many more insect species from different families and orders are known to be consumed, albeit in less important quantities. For instance, Meyer-Rochow (1973) documented the consumption of such insects as the praying mantid *(Hierodula* sp.; Mantidae) and stick insects (Podacanthinae) in Papua New Guinea, and Ban (1978) described the consumption of a wide range of aquatic insect larvae in Thailand. The best-studied country with regard to entomophagy is undoubtedly Mexico, where Ramos-Elorduy and coworkers have identified over 100 edible insect species (Ramos-Elorduy de Conconi, 1991; Ramos-Elorduy and Pino Moreno, 2002). No less impressive are the detailed studies of Malaisse and coworkers (Malaisse 1995, 1997; Malaisse and Parent, 1980) on the consumption of caterpillars, termites, and other insects in the African woodlands (notably Zaire and Zambia). Note that not all of the data on the nutritional composition of insects provided by Malaisse and Parent (1980) and Malaisse (1995) are reported in this review. The interested reader is advised to consult the original publications.

Excellent monographs on insect species consumed through time by human populations worldwide have been provided by, among others, Bodenheimer

Table 28.1: Major orders and families of insects consumed by human populations and geographic areas of consumption

Order and Common name	Main families	Area of consumption (reference)
Order: Lepidoptera Butterflies and moths (caterpillars)	Saturnidae, Noctuidae, Notodontidae, Cossidae, Bombycidae. Also, less common, Limacodidae, Thaumetopoeidae, Castnidae, Brassolidae, Hepialidae, Pyralidae, Sphingidae	Angola (Santos Oliveira et al., 1976), Nigeria (Ashiru, 1988), South African Pedi Nation (Quin, 1959), Sudan (Dirar, 1994), Zaire (Kodondi et al., 1987a,b; Malaisse and Parent, 1980; Malaisse, 1995), Zimbabwe (Benhura and Chitsiku, 1992; Glew et al., 1999), sub-Saharan Africa (van Huis, 1996), Australian Northern Territory (Cherikoff et al., 1985), Central Australia (Naughton et al., 1986), Papua New Guinea (Meyer-Rochow, 1973; Tommaseo and Paoletti, 1997), P.R. China (Luo Zhi-Yi, 1997; Simoons, 1991), Thailand (Yhoung-aree et al., 1997), Ecuador (Onore, 1997), Mexico (Ramos-Elorduy, 1991; Ramos-Elorduy and Moreno, 2002), Colombian Vaupés region (Dufour, 1987), Venezuela (Marconi et al., 2002; Paoletti et al., 2000).
Order: Coleoptera Beetles (grubs)	Curculionidae, Cerambycidae, Scarabaeidae	Angola (Santos Oliveira et al., 1976), Congo (Kinkela and Bézard, 1993), Ghana (Dei, 1989, 1991), sub-Saharan Africa (van Huis, 1996), Papua New Guinea (Ferguson et al., 1989; Mercer, 1997; Meyer-Rochow, 1973; Ohtsuka et al., 1984; Tommaseo and Paoletti, 1997), Philippines (Colting., 1985; Robson and Yen, 1976), Thailand (Yhoung-aree et al., 1997), Colombian Vaupés region (Dufour, 1987), Colombian-Venezuelan border region (Ruddle, 1973), Venezuela (Cerda et al., 2001, Marconi et al., 2002), Ecuador (Onore, 1997), Mexico (Ramos-Elorduy, 1991; Ramos-Elorduy and Moreno, 2002)
Order: Orthoptera Locusts, crickets, grasshoppers	Acrididae, Gryllidae, Tettigoniidae	Malawi (Ferguson et al., 1989), Sudan (Dirar, 1994), Zimbabwe (Benhura and Chitsiku, 1991, 1992), sub-Saharan Africa (van Huis, 1996), Philippines (Colting, 1985), Thailand (Yhoung-aree et al., 1997), Papua New Guinea (Meyer-Rochow, 1973; Tommaseo and Paoletti, 1997), Colombian-Venezuelan border region (Ruddle, 1973), Ecuador (Onore, 1997), Mexico (Ramos-Elorduy, 1991; Ramos-Elorduy and Moreno, 2002)

Table 28.1: (Contd.)

Table 28.1: (Contd.)

Common name	Main families	Area of consumption (reference)
Order: Isoptera		
Termites	Termitidae	Angola (Santos Oliveira et al., 1976), Cameroon (Chevassus-Agnes and Pascaud, 1972), Kenya (Murphy et al., 1991), Nigeria (Ukhun and Osasona, 1985), Zimbabwe (Benhura and Chitsiku 1992; McGregor, 1995), sub-Saharan Africa (van Huis, 1996), Philip pines (Colting, 1985), Colombian Vaupés region (Dufour, 1987), Venezuela (Marconi et al., 2002, Paoletti et al., 2000; Paoletti et al., 2003)
Order: Hymenoptera		
Ants	Formicidae	Sub-Saharan Africa (van Huis, 1996), Philippines (Colting, 1985), PR China (Luo Zhi-Yi, 1997), Thailand (Yhoung-aree et al., 1997), Papua New Guinea (Meyer-Rochow, 1973; Ohtsuka et al., 1984), Colombian Vaupés region (Dufour, 1987), Ecuador (Onore, 1997), Mexico (Ramos-Elorduy, 1991; Ramos-Elorduy and Moreno, 2002), Venezuela (Paoletti et al., 2000)
Bees and wasps (eggs, pupae, larvae, and adults)	Apidae, Vespidae	Thailand (Yhoung-aree et al., 1997), Ecuador (Onore, 1997), Mexico (Ramos-Elorduy 1991; Ramos-Elorduy and Moreno, 2002)
Order: Hemiptera		
Bugs (eggs, nymphs, adults)	Belastomatidae, Coreidae, Corixidae, Pentatomidae	Sub-Saharan Africa (van Huis, 1996), P.R. China (Simoons, 1991), Thailand (Yhoung-aree et al., 1997), Papua New Guinea (Meyer-Rochow, 1973), Mexico (Ramos-Elorduy, 1991)

(1951), Taylor (1975), and Holt (1988). But undoubtedly the most comprehensive work on entomophagy worldwide is being undertaken by Gene DeFoliart (in litt).

Protein Content of Insects

Crude Protein Content of Common Food Insects

The crude protein content for major insect species consumed is given in Table 28.2. In all studies examined, protein content was determined as total nitrogen (N) using a conversion factor of 6.25. As can be seen from Table 28.2, caterpillars have a high protein content of about 50–60 g/100 g dry weight, except the witchetty grub that has a protein content of 22 g/100 g dry weight. The protein

Table 28.2: Energy, protein, fat, ash, and fiber content of some common food insects

Food insect	Country	Moisture (g/100 g edible portion)	Energy (kcal/ 100 g)	Crude protein[1] (% w/w)	Total fat (% w/w)	Ash (% w/w)	Crude fiber (% w/w)
(1)	(2)	(3)	(4)	(5)	(6)	(7)	(8)
Lepidoptera: Saturniidae							
Caterpillar of moth, *Imbrasia ertli*[2]	Angola	9.02	375	48.7	11.1	14.4	N.A.
Caterpillar of moth, *Usta terpsichore*[2]	Angola	9.24	371	44.1	8.6	11.8	N.A.
Caterpillar, *Nudaurelia oyemensis*[3]	Zaire	7.0	N.A	56.8	11.3	3.5	N.A.
Caterpillar, *Imbrasia truncata*[3]	Zaire	7.3	N.A.	60.0	15.2	3.7	N.A.
Caterpillar, *Imbrasia epimethea*[3]	Zaire	7.0	N.A.	58.1	12.4	3.7	N.A.
Caterpillars, mean (range) of 17 Saturniidae species[4]	Zaire	81.8 (73–91)	458 (417–504)*	66.6 (51.9–79.6)*	13.9 (8.1–21.5)*	5.9 (3.8–8.8)*	N.A.
Mopane worm, *Gonimbrasia belina*[5]	South Africa	N.A.[5]	N.A.	62*	16*	7.6*	11.4*
Mopane worm (dried), *Gonimbrasia belina*[6]	Africa	6.1	444	56.8	16.4	6.9	9.6
Mopane worm (dry-roasted), *Gonimbrasia belina*[27]	Zimbabwe	N.A.	N.A.	48.3*	20.6*	N.A.	N.A.
Lepidoptera: Notodontidae							
African silkworm, *Anaphe venata*[7]	Nigeria	6.61	610*	60.03*	23.22*	3.21*	N.A.
Caterpillars, mean (range) of 5 species[8]	Zaire	77.7 (72–82)	447 (397–485)*	55.3 (51.6–61.0)*	19.0 (10.1–26.0)*	5.6 (4.3–7.7)*	9.0 (6.5–11)*
Lepidoptera: Thaumetopoeidae							
Caterpillar, *Anaphe panda*[9]	Zaire	73.9	543*	45.6*	35.0*	3.7*	6.5*
Lepidoptera: Limacodidae							
Caterpillar, unidentified species[9]	Zaire	82	397*	69.6*	9.2*	8.5*	8.0*

Table 28.2: (Contd.)

Table 28.2: (Contd.)

(1)	(2)	(3)	(4)	(5)	(6)	(7)	(8)
Lepidoptera: Noctuidae							
Bogong moth, whole, *Agrotis infusa*[10]	Australia	49.2	301	26.8	19.8	2.7	N.A.
Bogong moth, abdomen, *Agrotis infusa*[10]	Australia	35.2	457	21.7	38.8	1.9	N.A.
Lepidoptera: Cossidae							
Witchetty grub[11]	Australia	38.8	417	13.2	36.2	1.2	N.A.
Witchetty grub, *Xyleutes* sp.[12]	Australia	N.A.	N.A.	N.A.	38	N.A.	N.A.
Lepidoptera: Bombycidae							
Silkworm, *Bombyx mori*[13]	East Asia	60.7	229	23.1	14.2	1.5	N.A.
Spent silkworm larvae[14]	India	18.9	N.A.	48.7	30.1	8.6	N.A.
Coleoptera: Curculionidae							
Palm grubs, *Rynchophorus* sp.[15]	Colombia	13.7	661	24.3	55.0	1.0	N.A.
Palm worm, *Rynchophorus palmarum*[28]	Venezuela	71.7	583*	25.8*	38.5*	2.1*	N.A.
Palm worm, *Rhynchophorus phoenicis* Fabr.[16]	Zaire	N.A.	N.A.	N.A.	42 2*	N.A.	N.A.
Sago grub larvae, *Rhynchophorus schach* Olive[17]	Papua New Guinea (PNG)	62.9[(a)] 79.2[(b)]	240[(a)] 126[(b)]	12.1[(a)] 7.4[(b)]	19.1[(a)] 9.1[(b)]	1.0[(a)] 0.9[(b)]	N.A.
Palm weevil larvae, *Rhynchophous phoenicis*[18]	Angola	10.75	562	20.34	41.73	2.39	N.A.
Coleoptera: Cerambycidae							
Grub ("aruk"), larvae[19]	PNG	55.9	266	20.2	19.6	2.2	N.A.
Beetles, *Polycleis* spp., *Sternocera* spp.[6]	Africa	56.2	192	27.1	3.7	1.8	6.4
Orthoptera: Acrididae							
Locust, *Locusta migratoria manillensis* Meyen[20]	Philippines	66.3	147	13.7	4.3	2.3	N.A.

Table 28.2: (Contd.)

Table 28.2: (Contd.)

(1)	(2)	(3)	(4)	(5)	(6)	(7)	(8)
Locust, *Locustana* spp.[6]	Africa						
- raw		57.1	N.A.	18.2	21.5	N.A.	N.A.
- fried		48.0	N.A.	30.0	10.0	N.A.	N.A.
- flour		7.1	436	47.5	22.9	N.A.	4.9
Grasshoppers, *Zonocerus* sp.[6]	Africa						
- raw		62.7	170	26.8	3.8	1.2	2.4
- grilled, ground		7.0	420	62.2	10.4	4.0	N.A.
Rice-hopper, dried, *Oxya verox*[21]	East Asia	29.8	296	64.2	2.4	3.4	N.A.
Orthoptera: Gryllidae							
Crickets, raw, *Brachytrypes membranaceus*[6]	Africa	76	117	13.7	5.3	2.1	2.9
Hymenoptera: Formicidae							
Ant, female sexuals, *Atta sexdens*[22]	Colombia	6.1	628	39.7	34.7	1.6	N.A.
Ant, female sexuals, *Atta cephalotes*[22]	Colombia	6.9	580	48.1	25.8	2.3	N.A.
Flying ants (raw), females, *Carebara* sp.[6]	Africa	60	N.A.	3.0	9.5	N.A.	N.A.
Flying ants (raw), males, *Carebara* sp.[6]	Africa	60	N.A.	10.1	1.3	N.A.	N.A.
Ant eggs, "Itlog langgam"[20]	Philippines	71.0	128	17.4	3.8	2.8	N.A.
Tree ants, "burgjog", *Oecophylla* spp.[19]	PNG	51.5	122	16.8	4.0	22.9(?)	N.A.
Tree ants, *Oecophylla virescens*[17]	PNG	78.3	111	8.9	5.8	1.3	N.A.
Hymenoptera: Vespidae							
Bee maggots, canned, *Vespa singulata*[21]	East Asia	42.6	234	20.3	7.9	9.5	0
Isoptera: Termitidae							
Termites, soldiers, *Syntermes* sp.[22]	Colombia	10.3	467	58.9	4.9	4.8	N.A.
Termites, soldiers, *Syntermes aculeosus*[29]	Venezuela						
- thorax plus abdomen		77.8	93	12.6	2.3	1.9	N.A.
- heads		73.1	114	20.3	2.3	1.2	N.A.
Termites, cooked[23]	Cameroon	5–8	N.A.	N.A.	61.1*	N.A.	N.A.

Table 28.2: (Contd.)

Table 28.2: (Contd.)

(1)	(2)	(3)	(4)	(5)	(6)	(7)	(8)
Termites, *Termes* spp.[24]	Kenya	40	414	28.8	32.3	0.9	N.A.
Termites, mature alates, *Macrotermes subhyalinus*[15]	Angola	0.94	612	38.42	46.1	6.56	N.A.
Termites, *Macrotermes bellicosus*[25]	Nigeria	6.0	N.A.	34.8*	46.1*	10.2*	N.A.
Termites, *Termes* spp.[6]	Africa						
- raw		44.5	356	20.4	28.0	2.9	2.7
- dried		1.7	656	35.7	54.3	4.8	N.A.
- smoked		7.8	579	36.5	44.4	5.4	3.4
- fried		14.7	542	31.8	42.6	5.1	5.2
Diptera: Culicidae							
Lake flies (cake), *Chaoborus edulis*[6]	Africa	15.7	382	48.6	10.3	4.2	N.A.
Lake flies, flour[26]	Uganda	9.8	454	67.0	4.2	11.6	6.7

N.A. = Not available.

*Results expressed on dry-matter basis.

[1]Crude protein content was determined in all studies as 6.25 × nitrogen content.

[2]Santos Oliveira et al. (1976). Traditional preparation of caterpillars: viscera removed, then caterpillars boiled, roasted or sun dried. Salt is added. Results per 100 g food as consumed.

[3]Kodondi et al. (1987a). Caterpillars purchased on local market, traditional preparation of smoking and drying. Results per 100 g edible portion.

[4]Malaisse and Parent (1980). Species: *Athletes semialba, Bunaea alcinoe, Bunaeopsis aurantiaca, Cinabra hyperbius, Cirina forda, Gonimbrasia hecate, Gonimbrasia richelmanii, Gonimbrasia zambesina, Gynanisa maja ata, Imbrasia epimethea, Imbrasia macrothyris, Lobobunaea saturnus, Melanocera parva, Imbrasia dione, Imbrasia rubra, Tagoropsis flavinata* and one unidentified species. Fresh caterpillars, intestinal contents and hairs removed according to tradition. Samples freeze dried prior to analysis. Results on dry-matter basis.

[5]Dreyer and Wehmeyer (1982). Gastrointestinal contents of Mopanie worms squeezed out and remains dried in sun. Pulverized prior to analysis. Results on dry-matter basis. According to Quin (1959), the moisture content of fresh, whole mopanie worms is 83.1%.

[6]Wu Leung et al. (1968). Results per 100 g edible portion.

[7]Ashiru (1988). Dried, pulverized larvae without hairs. Results on dry-matter basis.

[8]Malaisse and Parent (1980). Species: *Elaphrodes lactea, Drapedites uniformes, Antheua insignala,* and two unidentified species of the Notodontidae family. Fresh caterpillars, intestinal contents and hairs removed according to tradition. Samples freeze dried prior to analysis. Results on dry-matter basis.

[9]Malaisse and Parent (1980). Fresh caterpillars, intestinal contents and hairs removed according to tradition prior to analysis. Samples freeze dried. Results on dry-matter basis.

[10]Cherikoff et al. (1985). Roasted.

[11]Cherikoff et al. (1985). Cossid larvae found in witchetty bush (*A. kempeana*), Australian Northern Territory. Roasted.

[12]Naughton et al (1986). Raw or lightly cooked? Results on wet-matter basis.

[13]Wu Leung et al. (1972), Dignan et al. (1994). Raw caterpillars. Results per 100 g edible portion.

[14]Rao (1994). By-product of Indian silk industry. Raw, whole larvae.

Table 28.2: (Foot notes contd.)

[15]Dufour (1987). Smoke dried.

[16]Kinkela and Bezard (1993). Whole larvae, freeze dried for analysis. Results on dry-weight basis.

[17]Ohtsuka et al. (1984). Foods collected in villages in state in which they are usually eaten (inedible parts removed). For the sago grub larvae, two samples (a and b) were analyzed. Results per 100 g edible portion.

[18]Santos Oliveira et al. (1976). Palm weevil larvae bodies incised and fried whole in palm oil. Termites: wings removed and fried in palm oil. Results per 100 g food as consumed.

[19]Ohtsuka et al. (1984), Dignan et al. (1994). In state in which usually eaten (raw?). Results per 100 g portion consumed.

[20]Abdon et al. (1990). Results per 100 g portion consumed.

[21]Wu Leung et al. (1972). Results per 100 g portion consumed.

[22] Dufour (1987). Prepared by dry toasting. Energy values reported by the author show considerable discrepancy with energy values calculated from reported macronutrient values using the classic Atwater factors of 4, 9, 4 kcal/g for protein, fat, and carbohydrate.

[23]Chevassus-Agnes and Pascaud (1972). Sample freeze dried prior to analysis.

[24]Murphy et al. (1991). Results on wet-weight basis.

[25]Ukhun and Osasona (1985). Dewinged, raw. Results on dry-weight basis.

[26]Bergeron et al. (1988). Sun-dried, finely stone-ground flour of three different lake flies of genera *Chaoborus* and *Povilla* (Chironomidae).

[27]Glew et al. (1999). Larvae of the Emperor moth (*Gonimbrasia belina*), or mopane worms, purchased in the market and dry roasted by local market women. Samples dried prior to analysis. Results on dry-matter basis.

[28] Cerda et al. (2001). Raw adult larvae reared on Moriche palm (*Mauritia flexuosa*). Original results expressed on dry-matter basis.

[29] Marconi et al. (2002). Results in g per 100 g weight consumed. Fresh, raw samples collected in Venezuelan Amazonia.

content of palm weevil larvae, the most important food insect among the beetles, varies between 23 and 36 g/100 g dry weight. Within the order Orthoptera, protein content varies widely among species with a low of 41 g/100 g dry weight for the locust *Locusta migratoria manillensis* and a high of 91 g/100 g dry weight for the rice hopper *Oxya verox*. The protein content of ants also varies considerably, with a low of 7.5–25 g/100 g dry weight for the African species and a high of 42–52 g/100 g dry weight for the Colombian species. The protein content of the various termite species ranges between 35 and 65 g/100 g dry weight, except for a high of 75 g/100 g dry weight for termite soldier heads.

Ramos-Elorduy de Conconi (1991), analyzing the protein content of edible insect species of several different orders in Mexico, reported a protein content (g/100 g dry weight) of 34–72% for the order Lepidoptera, 52–77% for the order Orthoptera, and 20–69% for order Coleoptera. Among order Hemiptera, the typical Mexican insect dishes "ahuahutle," "axayacatl", and "jumiles de Taxco" have approximate protein contents (g/100 g dry weight) of 58–72%, 54–70%, and 51–70% respectively.

In general, the protein content of insects is comparable to that of conventional meats (e.g. beef and pork), which typically ranges from 40 to 75 g/100 g dry weight (INN, 1997).

Protein Digestibility

Few studies have attempted to evaluate insect protein. Dreyer and Wehmeyer (1982) reported for dried, traditionally prepared mopanie worms (caterpillars of the moth *Gonimbrasia belina*) a protein digestibility (D) of 85.8%, an assimilability (A) of 78.8%, and a net protein utilization (NPU = D × A/100) of 67.6%. The latter figure is definitely lower than the NPU of whole hen's egg protein (93.5%, the upper point of the scale), but similar to the NPUs of soybean (61%) and wheat germ (67%) (FAO, 1970). Bergeron et al. (1988) estimated the in-vitro protein digestibility of an aquatic insect flour at 91%. Ramos-Elorduy de Conconi et al. (1981) assessed the in-vitro protein digestibility for several edible insect species and dishes and found that values ranged from 77.9% to 98.9% [*Atizies taxcoensis* ("jumiles") 77.9%, *Sphenarium histrio* ("chapulines") 85.6%, *Atta mexicana* (ants) 87.6%, "ahuahutle" 89.3% *Cossus redtenbachi* ("gusano rojo de maguey") 92.4%, *Eucheria socialis* 93.5%, *Liometopum apiculatum* ("escamol") 93.9%, "axayacatl" 98.0%, and *Laniifera cyclades* 98.9%].

Essential Amino Acids

The 8 essential (indispensable) amino acids found in some common food insects are listed in Table 28.3: isoleucine (Ie), leucine (Leu), lysine (Lys), methionine (Met), phenylalanine (Phe), threonine (Thr), tryptophan (Trp), and valine (Val), as well as cystine (Cys), the total S-containing amino acids (SAA = Met + Cys), tyrosine (Tyr), the total aromatic amino acids (AAA = Phe + Tyr), and arginine (Arg) and histidine (His). Also shown is the first limiting amino acid (not considering Arg and His) and its corresponding amino acid score.

The amino acid score is a *chemical grading* of the quality of a (food) protein obtained by comparing its amino acid content with a reference pattern. The score is calculated for all the essential amino acids and the lowest score taken. The reference pattern utilized here is based on the pattern of requirement suggested in 1985 by FAO/WHO/UNU for preschool children (2–5 years) and recommended for use to evaluate dietary protein quality for all age groups, except infants (< 1 year) (FAO, 1991) (Table 28.3). Other reference patterns are in use (e.g., hen's egg protein) which may result in different amino acid scores for the same foods, notably with regard to the sulfur-containing amino acids. The amino acid score is not to be confused with biological value or net protein utilization, which require biological tests.

In the majority of the food insects, either tryptophan or lysine is the first limiting amino acid. However, in some cases lysine and tryptophan are well represented. For instance, several of the caterpillars from the Saturniidae family, the palm weevil larva, and aquatic insect flour have an amino acid score for lysine well over 100. In general, the (limiting) amino acid scores for food insects range from 0 to 102. There is no obvious similarity in amino acid pattern among

Table 28.3: Amino acid content (in mg amino acid/g crude protein) of some common food insects

Food insect	Ile	Leu	Lys	Met	Cys	SAA	Phe	Tyr	AAA	Thr	Trp	Val	Arg	His	Limiting EAA	Amino acid score
Reference pattern	**28**	**66**	**58**			**25**			**63**	**34**	**11**	**35**				
Lepidoptera																
Caterpillar, *Nudaurelia oyemensis* (Saturniidae)[a]	25.6	82.7	79.8	23.5	19.7	43.2	58.6	75.7	134	44.5	36.0	96	63.5	18.1	Ile	91
Caterpillar, *Imbrasia truncata* (Saturniidae)[a]	24.2	73.1	78.9	22.2	16.5	38.7	62.2	76.5	139	46.9	16.5	102	55.5	17.4	Ile	86
Caterpillar, *Imbrasia epimethea* (Saturniidae)[a]	28.6	81.0	74.2	22.4	18.7	41.1	65.0	75.0	140	48.0	16.0	102	66.2	19.7	Ile	>100
Caterpillar, *Imbrasia ertli* (Saturniidae)[b]	36.0	36.7	39.3	15.8	13.4	29.2	17.4	13.2	30.6	40.5	8.1	41.9	N.A.	N.A.	AAA	49
Caterpillar, *Usta terpsichore* (Saturniidae)[b]	108.7	91.3	91.0	11.3	12.9	24.2	55.9	33.0	88.9	50.8	6.6	75.8	N.A.	N.A.	Trp	60
Mopane worm, *Gonimbrasia belina* (Saturniidae)[j]	44.5	64.6	74.2	20.7	21.5	42.2	52.8	63.8	117	56.8	11.6	57.0	59.0	31.1	Leu	98
Maguey worm, *Aegiale hesperiaris* (Megathymidae)[c,d]	49	52	36	10	N.A.	N.A.	37	42	79	33	9	47	30	16	Lys	62
"Gusano rojo de maguey," *Cossus redtenbachi* (Cossidae)[c,d]	51	79	49	8†	13	21†	40†	53	93†	47(c) 53(d)	6	61(c) –(d)	60	16	Trp	55
Caterpillar meal, *Bombycomorpha* sp.[e]	46.1	62.1	64.5	17.9	29.9	47.8	59.8	81.4	141.3	42.9	13.0	60.5	41.8	31.0	Leu	94
Caterpillars cooked, *Bombycomorpha* sp.[e]	44.3	77.4	47.8	16.2	21.8	37.9	93.4	63.4	157	49.1	9.4	55.0	37.0	16.2	Lys	82

Table 28.3: (Contd.)

Table 28.3: (Contd.)

Food insect	Ile	Leu	Lys	Met	Cys	SAA	Phe	Tyr	AAA	Thr	Trp	Val	Arg	His	Limit-ing EAA	Amino acid score
African silkworm larvae, *Anaphe venata*[f]	21.44	13.12	8.80	0	0	0	21.44	24.96	46.4	3.84	0	17.6	3.20	7.84	Trp	0
Spent silkworm pupae[g]	57	83	75	46	14	60	51	54	105	54	9	56	68	25	Trp	82
Coleoptera: Curculionidae																
Palm weevil larvae, *Rhynchophorus phoenicis*[b]	77.5	58.9	63.9	12.0	10.6	22.6	32.8	13.6	46.4	28.6	5.1	54.9	N.A.	N.A.	Trp	46
Palm worm, *Rhynchophorus Palmarum*[k]	29.1	62.8	66.7	10.5	8.9	19.4	28.3	37.6	65.9	44.6	9.7	31.4	62.8	39.5	SAA	77
Larvae of *Sciphophorus acupunctatus*[d]	48.2	78.2	53.5	20.2	26.7	46.9	46.1	63.5	109.6	40.4	8.1	62.0	44.0	14.7	Trp	74
Orthoptera: Acrididae																
"Chapulin," *Sphenarium histrio*[c]	53	87	57	7	13	20	44	73	117	40	6	51	66	11	Trp	55
"Chapulin," *Sphenarium purpurascens*[d]	42	89	57	25	18	43	103	63	166	38	6.5	57	60	22	Trp	59
Mexican "Chapulines"[c]	46	64	52	8	N.A.	N.A.	36	32	68	49	10	54	42	21	Lys	90
Hymenoptera: Formicidae																
Ants, *Atta mexicana*[c,d]	53	80	49	19[†]	15	34[†]	41[†]	47	88[†]	43	6	64	47	25	Trp	55
"Escamol," *Liometopum apiculatum*[c,d]	49	76	58	18[†]	14	32[†]	39[†]	68	107[†]	42	8	60	50	29	Trp	73
Isoptera: Termitidae																
Termites, *Macrotermes bellicosus*[h]	51.1	78.3	54.2	7.5	18.7	26.2	43.8	30.2	74.0	27.5	14.3	73.3	69.4	51.4	Thr	81

Table 28.3: (Contd.)

Table 28.3: (Contd.)

Food insect	Ile	Leu	Lys	Met	Cys	SAA	Phe	Tyr	AAA	Thr	Trp	Val	Arg	His	Limiting EAA	Amino acid score
Termites, mature alates, *Macrotermes subhyalinus*[b]	37.1	79.7	35.4	12.9	9.0	21.9	43.1	36.8	79.9	41.9	7.7	51.4	N.A.	N.A.	Lys	61
Termites, *Syntermes aculeosus*[l], thorax plus	56.1	88.8	67.4	19.2	9.0	28.2	32.5	45.0	77.5	49.2	8.0	61.4		29.0	Trp	73
abdomen and heads	52.2	78.8	42.6	9.4	11.8	21.2	26.6	55.2	81.8	51.1	8.1	68.5	N.A.	26.3	Lys,Trp	73
Hemiptera																
"Ahuahutle" or Mexican caviar, eggs of water bugs (Corixidae)[c]	50	80	35	15	N.A.	N.A.	34	111	145	40	11	60	77	33	Lys	60
"Axayácatl", adults and nymphs of water bugs (Corixidae and Notonectidae)[c]	59	80	43	16	N.A.	N.A.	32	45	77	44	16	55	55	24	Lys,	74
"Jumiles de Taxco", nymphs of *Atizies taxcoensis* (Pentatomidae)[c]	41	77	31	17	10	27	36	66	102	42	1	73	51	18	Trp	9
"Jumiles", nymphs of several species of *Pentatomidae*[c]	45	62	38	15	N.A	N.A.	25	40	65	28	15	48	29	30	Lys,	66
Diptera																
Aquatic insect flour[i]	52	79	78	24	8.5	32.5	47	58	105	46	N.A. or 0?	47	69	32	Trp? Leu	0? >100

N.A. = data not available.
[a]Kodondi et al. (1987a). Original data reported in mg amino acid/g N, conversion factor used 1/6.25 = 0.16.
[b]Santos-Oliveira et al. (1976). Original data in g AA/16 g N, conversion factor used × 10.

[c]Ramos-Elorduy (1991). Original data in mg AA/16 mg N, conversion factor used 10.

[d]Ramos-Elorduy et al. (1987). Original data in mg AA/16 mg N, conversion factor used 10.

[e]FAO (1970). Protein content 15.2 g/100 g meal of caterpillar and 51.6 g /100 g cooked caterpillar. Original data in mg/g N. Conversion factor used 1/6.25 = 0.16.

[f]Ashiru (1988). Original data in g amino acid/100 g crude protein, conversion factor used 10. The author explicitly mentions that methionine and tryptophan are not present in *Anaphe* larvae, nor is cystine listed with the amino acids present.

[g]Rao (1994). By-product of Indian silk industry. Original data in g AA/16 g N, conversion factor used 10.

[h]Ukhun and Osasona (1985). Dewinged, fresh termites.

[i]Bergeron et al. (1988). Flour made by stone grinding a mixture of three aquatic insects (lake flies) from Chironomidae genera *Chaoborus* and *Povilla*. Original data reported on a % w/w (amino acid/crude protein) basis, conversion factor used 10.

[j]Glew et al. (1999). Original data reported as mg amino acid/g dry weight of sample. Total protein content 482.7 mg/g dry weight. Conversion factor used × (1/0.4827).

[k]Cerda et al. (2001). Data refer to raw adult larvae reared on Moriche palm. Original data reported as g amino acid/100 g dry weight of sample. Total protein content 25.8 g/100 g dry weight of sample. Conversion factor used (1000/25.8).

[l]Marconi et al. (2002). Termite soldier thorax plus abdomen (upper results) and heads (lower results). Collected in Venezuelan Amazonia. Samples collected fresh and then dry frozen in laboratory.

[†]Data not consistent in the two publications of the same author. What is listed as methionine in Ramos-Elorduy et al. (1987) corresponds to total S-containing AA in Ramos-Elorduy (1991). Similarly, what is listed as phenylalanine in Ramos-Elorduy et al. (1987) corresponds to total aromatic AA in Ramos-Elorduy (1991). Data listed correspond to those in Ramos-Elorduy (1991).

the food insects listed, not even among species from the same order. Also Yhoungaree et al. (1997) report that the limiting amino acids vary widely according to the type of insect.

In most developing countries, lysine is the first limiting amino acid due to its low content in cereal staple foods (Table 28.4) (Hoshiai Kazuo, 1995). In some countries such as Mexico, where maize (corn) is an important staple food, tryptophan is the first limiting amino acid. However, local consumption patterns within countries may vary and the limiting amino acids depend on the local staple food(s), especially where the diet is monotonous.

Kodondi et al. (1987a) reported consumption of caterpillars *(Nudaurelia oyemensis, Imbrasia truncata,* and *Imbrasia epimethea*) in a local community in Zaire where cereals are the staple food. The caterpillars analyzed proved to be a rich source of lysine (Table 28.3) and are likely to be of importance in complementing the lysine-poor staple protein. In Papua New Guinea, palm weevil larvae (*Rynchophorus* sp.) are reported to be an important part of the diet and are consumed in combination with staples such as sago, sweet potato, yam, and taro (Meyer-Rochow, 1973; Ohtsuka et al., 1984; Mercer, 1997). Even though the overall amino acid score of the palm weevil larva appears to be poor (46), its amino acid composition nicely complements that of tubers (Tables 28.3 and 28.4). While the protein of tubers is limiting in lysine and leucine, the palm worm is a good source of both lysine (amino acid score 110) and leucine. The limiting amino acids in palm weevil larvae are tryptophan and the aromatic amino acids, which are well represented in the tubers. Note, however, that the amino acid composition of the palm weevil larva listed in Table 28.3 refers to a species of *Rynchophorus* collected in Angola, which may not be similar to that of the species eaten in Papua New Guinea.

In many African communities (e.g. Kenya, Nigeria, and Zimbabwe) where maize is the staple food, consumption of termites is common (Ukhun and Osasona, 1985; Murphy et al., 1991; McGregor, 1995). Maize protein is a poor source of lysine and tryptophan (Table 28.4). The data in Table 28.3 suggest that some termite species, e.g. *Macrotermes bellicosus* collected in Nigeria, may be valuable in complementing maize protein (amino acid score 93 and 130 for lysine and tryptophan resp.), while others may not (e.g. *Macrotermes subhyalinus* collected in Angola with first and second limiting amino acids lysine and tryptophan resp.).

Given the variability in amino acid composition of similar species, it is recommended that available data in the literature be used with caution and that wherever possible in dietary surveys, the protein quality of food insects be analyzed in relation to the protein quality of the dietary staple.

Fiber Content

The exoskeleton of insects is partly composed of chitin, a structural, nitrogen-containing polysaccharide, which consists of *N*-acetylglucosamine in β-1,4

Table 28.4: Essential amino acids in some staple foods

Food item	Ile	Leu	Lys	Met	Cys	SAA	Phe	Tyr	AAA	Thr	Trp	Val	Arg	His	1st lim. EAA*	2nd lim. EAA*
Grains																
Maize (grain, whole meal)	37	125	27	19	16	35	49	38	87	36	7	48	42	27	Lys (47)	Trp (64)
Millet (grain)	41	96	34	25	24	48	48	32	81	39	20	55	53	24	Lys (59)	Thr (115)
Rice (milled, polished)	44	86	38	22	16	38	51	34	85	35	14†	61	79	25	Lys (66)	Thr (103)
Sorghum (grain)	39	133	20	14	15	29	49	27	76	30	12	50	31	21	Lys (34)	Thr (88)
Wheat (grain)	35	72	31	16	27	43	48	32	80	31	12†	47	49	25	Lys (53)	Thr (91)
Starchy roots, tubers																
Cassava (manioc) meal	28	40	41	13	14	27	25	16	41	26	12	33	109	21	Leu (61)	AAA (65)
Sweet potato	37	54	34	17	11	28	39	23	62	38	17†	45	49	13	Lys (59)	Leu (82)
Taro tuber	35	74	39	13	26	40	51	36	87	41	14	61	89	18	Lys (67)	Leu (112)
Yam tuber	37	65	41	16	12	28	48	32	80	36	13	47	76	19	Lys (71)	Leu (98)

Data from FAO (1970). Original data in mg/g N, column chromatographic method. Conversion factor used: × (l/N conversion factor), where the N conversion factor = 6.25, except for rice 5.95 and wheat 5.83.

†Trp not determined with column chromatographic method, value listed is based on chemical or microbiological method.

*Based on the 1989 FAO/WHO scoring pattern reported in Table 28.3 (FAO, 1991). Amino acid score in parentheses.

linkage. Biochemically speaking, chitin is like the plant fiber cellulose, an unbranched polymer of glucose residues joined by β-1,4 linkages. Chitin, like cellulose, is not hydrolyzed in the human intestinal tract because of the absence of the relevant enzyme, chitinase.

Poor digestibility of chitin may lead one to expect high values for fiber content of insects, especially those with a hard exoskeleton. Data on insect fiber content are scarce. Raw termites, raw crickets, and raw grasshoppers, insects with a hard exoskeleton, have fiber contents of 4.9, 12.1, and 6.4 g/100 g dry weight respectively, similar to the fiber content of caterpillars (with a soft exoskeleton!), which ranges between 6.5 and 11.4 g/100 g dry weight (Table 28.2). These figures suggest that the fiber content of insects is indeed higher than that of other animal products—"conventional" meat generally does not contain fiber—and similar to that of grains (containing cellulose) (e.g. whole wheat flour 9.6 g/100 g) (INN, 1997).

Fat Content and Fatty Acid Composition

Insects are generally rich in fat (Table 28.2). The total fat content for caterpillars ranges from 8.1 to 59 g/100 g dry weight, the highest value referring to the witchetty grub eaten by Australian Aborigines. Among the beetles, palm weevil larvae (grub or worm) have a very high fat content of about 50 g/100 g dry weight (range 42–64 g/100 g dry weight) (Table 28.2). Termites are also a good source of fat with a fat content of about 50 g/100 g dry weight. The fat content of grasshoppers and related species and that of ants is lower and data appear more variable.

The fatty acid composition of some food insects is reported in Table 28.5. Data are scarce and limited to caterpillars, palm weevil grub, and termites, insects known to be a significant source of fat. The data in Table 28.5 show that there is little similarity in the fatty acid composition of related insect species (from the same taxonomic family) collected in different locations. On the other hand, the fatty acid composition of related species collected in the same location is similar. This suggests that the fatty acid composition of food insects is largely influenced by the host plant on which they feed.

All the food insects analyzed are a significant source of the essential fatty acids linoleic acid (C18:2,n-6) and linolenic acid (C18:3,n-3) (Table 28.5). Insects do not appear to be a significant source of arachidonic acid (C20:4,n-6) and docosahexaenoic acid (C22:6,n-3), long-chain polyunsaturated fatty acids that have received much attention in recent years for their presumably essential role in human neural development (e.g. Broadhurst et al., 1998). For the insects listed in Table 28.5, arachidonic acid was detected (0.3%) only in the crude lipid fraction of *Nudaurelia oyemensis* (Lepidoptera: Saturniidae) (Kodondi et al., 1987).

Few studies report the cholesterol content. Cerda et al. (2001) measured 188 mg/100 g dry weight of sample for palm worm (*Rhynchophorus palmarum*) collected in Venezuela. Marconi et al. (2002) reported cholesterol levels for several

insect species collected in Venezuela: 51 and 61 mg/100 g edible weight for fresh termite thorax plus abdomen and termite head respectively (*Syntermes aculeosus*), 9.1 mg/100 g edible weight for "Makoya" grub (Scarabaeidae: *Pelidnota* sp.), and 18 mg/100 g edible weight for caterpillars (Sphingidae: *Erinnyis ello*).

Micronutrients

The few data available on the mineral and vitamin content of food insects are given in Tables 28.6 and 28.7. Food insects appear to be a good source of iron (Fe). For comparison, beef has an iron content of about 6 mg/100 g dry weight (2.1 mg/100 g fresh weight), while the iron content of most food insects lies well above this value (INN, 1997). No data are available on whether this iron is readily accessible. The calcium content of food insects is certainly higher than that of conventional meats, although lower than that of whole milk (920 mg/100 g dry weight or 120 mg/100 g fresh weight) (INN, 1997).

The B vitamins appear well represented in food insects. Thiamine levels range from 0.1 to 4 mg/100 g dry weight, and riboflavin levels from 1 to 8 mg/100 g dry weight (*Termes* spp. not considered) (Table 28.7). For comparison, the thiamine and riboflavin content of whole meal bread are 0.16 and 0.19 mg/100 g dry weight respectively, and for hen's egg 0.42 and 1.2 respectively (INN 1997). Vitamin B_{12} (cyanocobalamin) levels are reported only for caterpillars of the Saturniidae family and range from 0.015–0.027 μg/100 g dry weight (Kodondi et al., 1987) (see Table 28.7). Vitamin B_{12} only occurs in foods of animal origin.

Retinol and β-carotene levels are very variable for the insects analyzed. Retinol levels range from 6 to 356 μg/100 g dry weight and β-carotene levels from 6 to 1,800 μg/100 g dry weight. Cerda et al. (2001) reported a high vitamin E level for palm worm (*Rhynchophorus palmarum*): 34.7 and 9.2 mg/100 g dry weight of sample respectively for α-tocopherol and β+γ-tocopherol. Marconi et al. (2002) reported alpha tocopherol levels for the following insects collected in Venezuela: 0.48 and 0.15 mg/100 g edible weight for fresh termite thorax plus abdomen and termite head respectively (*Syntermes aculeosus*), 0.0039 mg/100 g edible weight for "Makoya" grub (Scarabaeidae: *Pelidnota* sp.,), and 0.60 mg/100 g edible weight for green "Opomoschi" caterpillar (Sphingidae: *Erinnyis ello*).

Antinutritional Factors

Little is known about the possible presence of toxic or antinutritional factors in food insects. It has been suggested that consumption of the larvae of the African silkworm *(Anaphe venata)* in southwest Nigeria may be implicated in the etiopathogenesis of a seasonal ataxic syndrome (Adamolekun, 1993; Adamolekun and Ibikunle, 1994). The presence of thiaminases in the silkworm may cause an acute exacerbation of marginal thiamine-deficiency in the period of availability

Table 28.5: Fatty acid composition (%) of some food insects

Food insect	C14:0	C16:0	C18:0	Other SFA	Total SFA	C16:l (n-7)	C18:l (n-9)	Other MUFA	Total MUFA	C18:2 (n-6)	C18:3 (n-3)	Other PUFA	Total PUFA
Lepidoptera													
Smoked caterpillar, *Nudaurelia oyemensis* (Saturniidae)[1]	0.2	21.8	23.1	0.2	45.3	0.6	5.6		6.2	5.7	35.6	2.1	43.4
Smoked caterpillar, *Imbrasia truncata* (Saturniidae)[1]	0.2	24.6	21.7	trace	46.5	0.2	7.4		7.6	7.6	36.8	-	44.4
Smoked caterpillar, *Imbrasia epimethea* (Saturniidae)[1]	0.6	23.2	22.1	0.2	46.1	0.6	8.4		9.0	7.0	35.1	0.4	42.5
Caterpillar, *Imbrasia ertli* (Saturniidae)[2]	1.0	22.0	0.4	38.0% C20:0, 1.5% other		22.0	2.0	0.8		20.0	11.0	0.2	
Caterpillar, *Usta terpsichore* (Saturniidae)[2]	2.3	274	0.1	29.7% C17:0, 7.5% C20:0		27.4	1.7	0.2		27.2	2.8	0.1	
Mopane worm, *Gonimbrasia belina* (Saturniidae)[8]	—	25.5	10.2	0.3% C12:0, 0.3% C20:0	36.3	0.9	13.3	—	14.2	11.5	38.0	—	49.5
Witchetty grub, *Xyleutes* sp. (Cossidae)[3]	—	29.4	3.1		32.5	tr	67.1		67.1	0.4	—	—	0.4
Spent silkworm pupae (Bombycidae)[4]		26.2	7.0		33.2		36.9		36.9	4.2	25.7		29.9
Coleoptera: Curculionidae													
Palm weevil larvae *Rhynchophorus phoenicis*[2]	2.5	36.0	0.3	2.1		36.0	30.0	0.6		26.0	2.0	trace	

Table 28.5: (Contd.)

Table 28.5: (Contd.)

Food insect	C14:0	C16:0	C18:0	Other SFA	Total SFA	C16:1 (n-7)	C18:1 (n-9)	Other MUFA	Total MUFA	C18:2 (n-6)	C18:3 (n-3)	Other PUFA	Total PUFA
Palm worm (entire larvae) *Rhynchophorus phoenicis* Fabr.[5]	1.8	38.0	4.5		44.3	2.5	46.2		48.7	5.0	1.5		6.5
Isoptera: Termitidae													
Termites, *Macrotermes bellicosus*[6]	0.18	46.54			46.7	2.09	12.84		14.9	34.42	3.85		38.3
Termites, mature alates, *Macrotermes subhyalinus*[2]	0.9	33.0	1.4	3.8		33.0	9.5	1.2		43.1	3.0	3.7	
Termites, boiled[7]	1.3	28.0	8.5	+	37.8	3.4	48.0		51.4	9.5	1.4		10.9

SFA: saturated fatty acids, MUFA: mono unsaturated fatty acids, PUFA: poly unsaturated fatty acids.

[1]Data from Kodondi et al. (1987a). Caterpillars purchased in market in Kinshasa, Zaire, traditional preparation consisting of smoking and drying. Other PUFA reported by the authors include C20:2, C20:4 (arachidonic acid), and C20:5.

[2]Data from Santos-Oliveira et al. (1976) referring to Angola. Foods analyzed as consumed (traditional preparation). Note that the traditional preparation of termites and palm weevil larvae involves frying in palm oil. Data as reported by authors in percentage of total fat. Note that the sums of the fatty acids listed in the original paper for the four insects analyzed do not add up to 100%. Other PUFA reported by the authors include only C20:2.

[3]Data from Naughton et al. (1986) referring to Australian Aborigines. Central Australia.

[4]Data from Rao (1994) referring to India. The spent silkworm is a by-product of the silk industry. Whole pupae were analyzed.

[5]Data from Kinkela and Bézard (1993) referring to Congo (Central Africa).

[6]Data from Ukhun and Osasona (1985) referring to Nigeria. Dewinged, fresh termites.

[7]Data from Chevassus-Agnes and Pascaud (1972) referring to northern Cameroon.

[8]Data from Glew et al. (1999) referring to Zimbabwe.

Table 28.6: Mineral and trace element contents (mg/100 g dry matter) of some common food insects[1]

Food insect	Na	K	Ca	P	Fe	Mg	Zn	Cu	Mn
Lepidoptera: Saturniidae									
Caterpillar *Imbrasia ertli*[2]	2,418*	1,204	55	600	2.1	254	N.A.	1.5	3.4
Caterpillar *Usta terpsichore*[2]	3,340*	3,259	391	766	39.1	59	25.3	2.6	6.7
Caterpillar *Nudaurelia oyemensis*[3]	140	1,107	149	871	9.7	266	10.2	1.2	5.5
Caterpillar *Imbrasia truncata*[3]	183	1,349	132	842	8.7	192	11.1	1.4	3.2
Caterpillar *Imbrasia epimethea*[3]	75	1,258	225	666	13.0	402	11.1	1.2	5.8
Caterpillars, mean ± sd (range) of 15 different species**[4]			141 ±118 (50–500)	975 ± 515 (500–2,300)	75 ± 61 (10–200)				
Mopanie worms, *Gonimbrasia belina*[5]	1,032	1,024	174	543	31	160	14	0.91	3.95
Mopane worm, *Gonimbrasia belina*[6]			488	613	76.9				
Mopane worm, *Gonimbrasia belina*[20]	?	1,580	273	634	30.4	185	14.2	0.71	4.0
Lepidoptera: Notodontidae									
Caterpillars, 4 species (*Elaphrodes lactea, Drapedites uniformes, Antheua insignata*, 1 unidentified species)[4]			85 (20–200)	774 (500–1,500)	50 (10–80)				
African silkworm larvae, *Anaphe venata B.*[7]	30	1,150	40	730	10	50	10	1	40
Lepidoptera: Thaumetopoeidae									
Caterpillar, *Anaphe panda*[4]			200	450	10				
Lepidoptera: Limacodidae									
Caterpillar, unidentified species[4]			1,600	900	20				

Table 28.6: (Contd.)

Table 28.6: (Contd.)

Food insect	Na	K	Ca	P	Fe	Mg	Zn	Cu	Mn
Lepidoptera: Noctuidae									
Bogong moth (whole), *Agrotis infusa*[8]	43	554	431	N.A.	23.6	260	14	1.6	
Bogong moth (abdomen), *Agrotis infusa*[9]	40	488	174	N.A.	11	123	16.9	1.1	
Lepidoptera: Bombycidae									
Silkworm, *Bombyx mori*[10,11]			15	641	3.1				
Lepidoptera: Cossidae									
Witchetty grubs[12]	5.7	293	29	N.A.	15.0	37	0.4	3.4	N.A.
Coleoptera: Curculionidae									
Sago grub[13]	13	304	18	96	N.A.	29	4.9	2.1	0.55
Sago grub larvae,	7.5[(a)]	633	47.2	314	8.4	145	5.23	2.09	10.5
Rhynchophorus schach[14]	87.5[(b)]	1,112	81.7	488	9.5	254	13.75	3.06	2.07
Palm weevil larvae, *Rhynchophorus phoenicis*[2]	44.8	2,209	208	352	14.7	33.6	26.5	1.6	0.8
Coleoptera: Cerambycidae									
Grub ("aruk")[11,14]	16.1	550	21	402	13.3	49.7	3.61	0.55	0.21
Orthoptera: Acrididae									
Locust, *Locusta migratoria manillensis*[15]			303		3.0				
Locust, roasted[16]	55	545	90	424	N.A.	62	8.4	3.0	1.46
Locust, fried, *Locustana* sp.[6]			288		9.6				
Rice hopper, dried, *Oxya verox*[10]			74	866	33.1				
Grasshoppers (raw/grilled), *Zonocerus sp.*[6]			107/190	449 (gr.)	29 (raw)				
Orthoptera: Gryllidae									
Crickets, raw, *Brachytrypes membranaceus*[6]			75		54				

Table 28.6: (Contd.)

Table 28.6: (Contd.)

Food insect	Na	K	Ca	P	Fe	Mg	Zn	Cu	Mn
Isoptera: Termitidae									
Termites, mature alates, *Macrotermes subhyalinus*[2]	1,988*	480	40	442	7.6	421	N.A.	13.7	64.4
Termites, *Macrotermes bellicosus*[17]		117	44.6			28.0			
Termites, dried, smoked, fried, *Termes* spp.[6]			144 dr 99 sm 94 fr	793 dr 71 sm 609 fr	53 dr 23 sm 20 fr				
Hymenoptera: Vespidae									
Bee maggots (canned), *Vespa singulata*[10]			14	366	19				
Hymenoptera: Formicidae									
Tree ants, *Oecophylla* sp.[11,14]	180	541	48	517	21.8	70	10.1	0.87	9.06
Tree ants, *Oecophylla virescens*[14]	270	957	79.7	936	109.0	122.1	16.9	2.17	6.30
Ant eggs[18]			252	749	6.9				
Diptera: Culicidae									
Lake flies (cake), *Chaoborus edulis*[6]			166		78				
Aquatic insect flour[19]	433	1,106	296	1,220	1,442	187	14.5	5.8	16.4

[1]Sodium (Na), potassium (K), calcium (Ca), total phosphorus (P), iron (Fe), magnesium (Mg), zinc (Zn), copper (Cu), and manganese (Mn).

[2]Santos-Oliveira et al. (1976). Traditional preparation, see Table 28.2. Original data in mg per 100 g food as consumed. Moisture content see Table 28.2, conversion factors used: 1.0095 *(Macrotermes subhyalinus),* 1.0991 *(Imbrasia ertli),* 1.1018 *(Usta terpsichore),* 1.1204 *(Rhynchophorus phoenicis).*

[3]Kodondi et al. (1987a). Caterpillars from local market of Kinshasa, Zaire, traditional preparation (smoking and drying). Original data reported in mg /100 g smoked caterpillar. Moisture contents of caterpillars: 7.0 *(N. oyemensis),* 7.3 *(I. truncata)* and 7.0 *(I. epimethea),* conversion factors used: 1.075, 3.079, and 1.075 respectively.

[4]Malaisse and Parent (1980). Fresh, removal of intestinal contents and hairs according to tradition. Freeze-dried. Original data on dry-matter basis.

[5]Dreyer and Wehmeyer (1982). Gastrointestinal contents of caterpillars squeezed out and remains dried in sun, then ground. Original results on moisture-free basis.

[6]Wu Leung et al. (1968). Original results reported in mg per 100 g edible portion. Conversion factors used: crickets 4.1667; lake flies 1.186; grasshoppers raw 2.681. grilled (flour) 1.0753, fried locust 1.923; mopane worm 1.0650; termites dried 1.017, smoked 1.085, fried 1.172.

[7]Ashiru (1988). Powdered larvae without hairs. Original data on dry matter basis.

[8]Cherikoff et al. (1985). Whole moth, water content 49.2%, conversion factor 1.97.

[9]Cherikoff et al. (1985). Abdomen only, water content 35.2%, conversion factor 1.54.

[10]Wu Leung et al. (1972). Original results in mg per 100 g edible portion. Conversion factors used: bee maggots 1.742, rice hoppers 1.425, and silkworm 2.545.

Table 28.6: (Foot notes contd.)

[11]See also Dignan el al. 1994.

[12]Cherikoff et al. (1985). Cossid larvae found in witchetty bush *(A. kempeana)*, roasted, water content 38.8% (conversion factor 1.63).

[13]Ferguson et al. (1989). Sago grubs from local market Papua New Guinea. East Sepik Province. Moisture content 64%. Original data given in mg/100 g edible portion on a dry-weight basis.

[14]Ohtsuka et al. (1984). Samples collected in villages in state in which usually eaten. Original results in mg/100 g edible portion. Conversion factors used: grub 2.268; sago grub 2.695 (a) and 4.808 (b), tree ants *Oecophylla* sp. 2.062, and *Oecophylla virescens* 4.608. Several reported values appear unusual.

[15]Abdon et al. (1990). Fresh locust, edible portion, moisture content 66.3%, conversion factor 2.97.

[16]Ferguson et al. (1989). Locust purchased from local market in Malawi, Zomba district. Moisture content of fresh (roasted?) locusts 50%. Original data given in mg/100 g edible portion on a dry-weight basis.

[17]Ukhun and Osasona (1985). Dewinged, fresh, moisture content 6%. Original data in mg/100 g sample, conversion factor 1.0638.

[18]Abdon et al. (1990). Fresh, moisture content 71.0%, conversion factor 3.45.

[19]Bergeron et al. (1988). Flour made by stone-grinding a mixture of three aquatic insects from Chironomidae genera *Chaoborus* and *Povilla*. Moisture content 9.8 g/100 g flour (conversion factor = 1.109)

[20]Glew et al. (1999). Original data reported in µg/g dry weight. Conversion factor used: × 0.1. Also reported arsenic, chromium, molybdenum, nickel, selenium, all of which were not detectable (sensitivity of method employed 5 µg/g dry weight).

*Probably salt was added in the preparation of the caterpillars.

***Athletes semialba, Bunaea alciroë, Bunaeopsis aurantiaca. Cinabra hyperbius, Cirina forda, Gonimbrasia hecate, Gonimbrasia richelmanii, Gonimbrasia zambesina, Imbrasia epimethea, Imbrasia macrothyris, Lobobunaea saturnus, Imbrasia dione, Imbrasia rubra, Tagoropsis flavinata*, and one unidentified species.

and consumption of the larvae (Nishimune et al., 2000). Thorough heating of the African silkworm is mandatory to make the worm safe for consumption. Much lower thiaminase activity has been measured in Japanese silkworms (*Bombyx mori*) (Nishimune et al., 2000).

Contamination of food insects also deserves some attention. For example, Saeed et al. (1993) have argued that pesticide contamination of locusts may carry health risks for the people regularly consuming these insects. Green et al. (2001) recently reported that annual migration of Bogong moths, *Agrotis infusa* (Noctuidae), has been shown to bring toxic levels of arsenic to the caves in the Australian Alps where the moths estivate in large numbers. Arsenic has historically been used in agricultural pesticides. Historically, adult Bogong moths formed an important part of the summer diet of Aborigines from around the Australian Alps (Flood, 1980), and mass harvesting of the Bogong moth in the mountains for communal feasting by Australian Aborigines has been described by several authors (e.g. Cherry, 1991; Yen this volume).

Brief reviews of the potential hazards of *indiscriminate* consumption of insects have been provided by Blum (1994) and Berenbaum (1993). Blum (1994) discussed the potential hazards of compounds synthesized within the body of the insect, while Berenbaum (1993) focused on compounds sequestered by the insect from plants.

Table 28.7: Vitamin content of some common food insects (in mg/100 g dry matter unless stated otherwise)

Food insect	**Vit A (retinol)**	**β-carotene**	**α-tocopherol**	**Thiamine (B_1)**	**Riboflavin (B_2)**	**Niacin**	**Pyridoxine (B_6)**	**Folic acid**	**Pantothenic acid**	**Biotin**	**Cyanocobalamin (B_{12})**
Lepidoptera: Saturniidae											
Mopane worm, *gonimbrasia belina*[1]	6.5 μg	1.71 i.u.		0.58	4.98	11.9					
Mopane worm, dried, *gonimbrasia belina*[2]	53 μg	634 μg		0.55	1.99	11.6					
Caterpillar, *Usta terpsichore*[3]				4.04	2.09	0.33					
Caterpillar, *Nudaurelia oyemensis*[4]	32 μg	6.8 μg	ND	0.21	3.4	10.1	54 μg	21.5 μg	9.5	32.0 μg	0.015 μg
Caterpillar, *Imbrasia truncata*[4]	33 μg	7.1 μg	ND	0.32	5.5	11.8	151 μg	40.0 μg	11.0	48.5 μg	0.027 μg
Caterpillar, *Imbrasia epimethea*[4]	47.3 μg	8.2 μg	ND	0.21	4.3	11.8	86 μg	6.8 μg	7.8	24.7 μg	0.016 μg
Lepidoptera: Cossidae											
Witchetty grubs[5]				1.5							
Lepidoptera: Sphingidae											
Caterpillar: *Erinnyis ello*[11]	ND	122 μg	0.6*								

Table 28.7: (Contd.)

Table 28.7: (Contd.)

Food insect	Vit A (retinol)	β-carotene	α-tocopherol	Thiamine (B_1)	Riboflavin (B_2)	Niacin	Pyridoxine (B_6)	Folic acid	Pantothenic acid	Biotin	Cyanocobalamin (B_{12})
Coleoptera: Curculionidae											
Palm weevil larvae *Rhynchophorus phoenicis*[3]				3.38	2.51	3.36					
Palm worm *Rynchophorus palmarum*[10]		1800 µg	34.7								
Coleoptera: Scarabaeidae											
Makoya grub, *Pelidnota* sp.[11]	7.2 µg	7.8 µg	0.004*								
Orthoptera: Acrididae											
Rice-hoppers, dried, *Oxya verox*[6]	356 µg	78 µg		0.34	7.84	10.0					
Isoptera: Termitidae											
Termites, mature alates, *Macrotermes subhyalinus*[3]				0.132	1.15	4.63					
Termites, *Termes* spp.[2]											
- dried	N.A.			0.03	6.07	5.9					
- smoked	0			0.11	0.07	1.95					
- fried	N.A.			0.14	3.79	9.81					

Table 28.7: (Contd.)

Table 28.7: (Contd.)

Food insect	**Vit A (retinol)**	**β-carotene**	**α-tocopherol**	**Thiamine (B_1)**	**Riboflavin (B_2)**	**Niacin**	**Pyridoxine (B_6)**	**Folic acid**	**Pantothenic acid**	**Biotin**	**Cyanocobalamin (B_{12})**
Syntermes aculeosus soldier[11]											
- thorax plus abdomen	N.D.	231 μg	2.2								
- termite head		15 μg	0.56								
Hymenoptera: Formicidae											
Ant eggs[7]				1.48	2.55						
Tree ants, whole, *Oecophylla* sp.[8]				0.44	0.98						
Hymenoptera: Vespidae											
Bee maggots, canned, *Vespa singulata*[6]				0.70	1.08	11.3					
Diptera: Culicidae											
Lake flies (cake) *Chaoborus edulis*[2]				1.5	4.1	21.7					
Aquatic insect flour[9]	—			1.8	8.9	28.8					

i.u. = international units, N.A. = data not available, N.D. = not detectable.

[1]Dreyer and Wehmeyer (1982). Gastrointestinal contents of caterpillars squeezed out and remains dried in sun, then ground. Original results on moisture-free basis and expressed in i.u.: 21.6 iu retinol and 1.71 iu β-carotene. Conversion factors used: 1 i.u. vitamin A = 0.3 μg retinol.

[2]Wu Leung et al. (1968). Original results in mg per 100 g edible portion; conversion factors used: lake flies 1.186, Mopanie worms 1.0650, termites dried 1.071, smoked 1.085, and fried 1.172.

[3]Santos-Oliveira et al. (1976). Foods analyzed as consumed (traditional preparation, see Table 28.2). Original data in mg per 100 g food as consumed. Moisture contents, see Table 28.2; conversion factors used: 1.0095 (*Macrotermes subhyalinus*), 1.1018 (*Usta terpsichore*), and 1.1204 (*Rhynchophorus phoenicis*).

[4]Kodondi et al. (1987a, b). Caterpillars purchased in local market of Kinshasa, Zaire. Traditional preparation of smoking and; drying. Original data in mg/100 g smoked caterpillar, moisture contents of caterpillars: 7.0 *(N. oyemensis)*, 7.3 (*I. truncata)* and 7.0 (*I. epimethea)*, conversion factors used: 1.075, 1.079, and 1.075, respectively. Authors also report vitamin values for 100 g fresh *Imbrasia truncata*: 23.1 μg retinol, 229 μg β-carotene, and 3 mg α-tocopherol; no value is given for the moisture content of the fresh caterpillar.

[5]Cherikoff et al. (1985). Cossid larvae found in witchetty bush *(A. kempeana)*. Roasted, water content 38.8%, conversion factor 1.63.

[6]Wu Leung et al. (1972). Original results in mg per 100 g edible portion, conversion factors: bee maggots 1.742, rice hoppers 1.425.

[7]Abdon et al. (1990). Fresh ant eggs, moisture content 71.0%, conversion factor 3.45.

[8]Dignan et al. (1994). Original values in mg/100 g edible portion, conversion factor 2.083.

[9]Bergeron et al. (1988). Flour made by stone-grinding a mixture of three aquatic insects from Chironomidae genera *Chaoborus* and *Povilla.* Moisture content 9.8 g/100 g flour, conversion factor 1.109.

[10]Cerda et al. (2001). Authors also report 9.2 mg β+γ -tocopherol per 100 g dry weight of palm worm.

[11]Marconi et al. (2002). Original results in μg per 100 g edible weight. For moisture contents see Table 28.2; conversion factors used: 4.5 for termite thorax plus abdomen, and 3.7 for termite head.

*Results for Makoya grub and Opomoshi caterpillar on a fresh weight basis.

Discussion

Data on the nutrient composition of edible insects are quite scarce. For instance, the widely used food composition table for use in Latin America (Wu Leung and Flores, 1961) lists no insects. Even the newly issued national food composition table for Mexico (Munoz de Chavez et al., 1996) lists only a few edible insects ('Ahuahutle', 'Gusanos de Maguey', 'Jumiles', and 'Sats orugas') despite the extensive work of Ramos-Elorduy on edible insects in this country.

The data available show high variability in nutrient composition for related insect species. There may be two reasons for this. First, the chemical composition of insects may depend on the (plant) host on which they feed and hence be location-specific. Second, variability in chemical composition may be due to analytical variability among studies. In general, the studies cited provide little, if any information on the precision of the analyses and in most cases single samples were analyzed. In using data on insect nutrient composition, special attention should be paid to the moisture content of the insects analyzed which appears highly variable and affects the chemical composition when expressed on a fresh-weight basis.

As a food group, insects appear to be nutritious. They are rich in protein and fat and provide ample quantities of minerals and vitamins. The amino acid composition of the insect protein is in most cases better than that of grains or legumes and in several cases the food insects may be of importance in complementing the protein of commonly consumed grain staples among indigenous populations.

Insects are unlikely to compete with conventional animal products such as beef and pork, but as a side dish, snack or delicacy—the way they are typically consumed by indigenous populations—they should be given due importance. Given their high nutritional value, efforts should be made to retain the tradition of entomophagy where it is still alive. Unfortunately, availability of several insects may be in decline due to destruction of insect host trees/plants (e.g. Ashiru, 1988; McGregor, 1995). Controlled production of selected food insects is a possibility (e.g., Cerda et al., 2001; Cerda et al., this volume; Yhoung-Aree and Viwatpanick, this volume), but requires extreme caution with regard to its possible ecological consequences.

References

Abdon I.C., Del Rosario I.E., Aguinaldo A.R., Lontoc A.V., and Alejo L.G. 1990. Food Composition Tables Recommended for Use in the Philippines. Food and Nutrition Research Institute, Manila, Philippines.

Adamolekun B. 1993. *Anaphe venata* entomophagy and seasonal ataxic syndrome in southwest Nigeria. Lancet, 341(8845): 629.

Adamolekun B. and Ibikunle F.R. 1994. Investigation of an epidemic of seasonal ataxia in Ikare, western Nigeria. Acta Neurol. Scand. 90 (5): 309–311.

Ardeshir H. 1990. Grub to eat. Nutr. Food Sci. 123: 10–11.

Ashiru M.O. 1988. The food value of the larvae *of Anaphe venata* Butler (Lepidoptera: Notodontidae). Ecol. Food Nutr. 22: 313—320.

Ban P. 1978. On some edible aquatic flora and fauna of Northern Thailand—their morphology, ecological distribution and their utilization. Food (Bangkok) 10 (1): 57–80.
Benhura M.A.N. and Chitsiku I.C. 1991. Food consumption patterns of the people of Dangamvura in Mutare. Central Afr. J. Med. 37 (11): 346–352.
Benhura M.A.N. and Chitsiku I.C. 1992. Seasonal variation in the food consumption patterns of the people of Mutambara district of Zimbabwe. Central Afr. J. Med. 38 (1): 8–13.
Berenbaum M.R. 1993. Sequestered plant toxins and insect palatability. Food Insects Newsl. VI (1): 6–9.
Bergeron D., Bushway R.J., Roberts F.L., Kornfield I., Okedi J., and Bushway A.A. 1988. The nutrient composition of an insect flour sample from Lake Victoria, Uganda. J. Food Comp. Analysis 1: 371–377.
Blum M.S. 1994. The limits of entomophagy: A discretionary gourmand in a world of toxic insects. Food Insects Newsl. VII (1): 1–8.
Bodenheimer, F.S. 1951. Insects as Human Food. Dr. W. Junk, The Hague, Netherlands.
Broadhurst C.L., Cunnane S.C., and Crawford M.A. 1998. Rift Valley lake fish and shellfish provided brain-specific nutrition for early Homo. Brit. J. Nutr. 79: 3–21.
Cerda H., Martinez R., Briceno N., Pizzoferrato L. et al. 2001. Palm worm (*Rhynchophorus palmarum*) traditional food in Amazonas, Venezuela—Nutritional composition, small scale production and tourist palatability. Ecol. Food Nutr. 40: 13–32.
Cherikoff V., Brand J.C., and Truswell A.S. 1985. The nutritional composition of Australian Aboriginal bushfoods. 2. Animal foods. Food Tech. in Australia 37 (5): 208–211.
Cherry R.H. 1991. Use of insects by Australian Aborigines. Amer. Entomol. 37: 9–13.
Chevassus-Agnes S. and Pascaud M. 1972. Composition en acides gras de quelques plats cuisines du Nord-Cameroun (Adamaoua). Ann. Nutr. Aliment. 26: 7–31.
Colting L.M. 1985. Preliminary result on the utilization of insects as human food in northern Luzon (Philippines). 16th Ann. Conv. Pest Control Council Philippines, La Trinidad, Benguet, Philippines, 2–4 May 1985. Mountain State Agri. College, La Trinidad, Benguet, Philippines.
DeFoliart G.N. 1989. The human use of insects as food and as animal feed. Bulletin of the ESA, Spring 1989: 22–35.
DeFoliart G.N. (in litt.). The Human Use of Insects as a Food Resource: A Bibliographic Account in Progress. See http://www.food-insects.com/.
Dei G.J.S. 1989. Hunting and gathering in a Ghanaian rain forest community. Ecol. Food Nutr. 22: 225–243.
Dei G.J.S. 1991. The dietary habits of a Ghanaian farming community. Ecol. Food Nutr. 25: 29–49.
Dignan C.A., Burlingame B.A., Arthur J.M., Quigley R.J., and Milligan G.C. 1994. The Pacific Islands Food Composition Tables. Swiftprint, Palmerston North.
Dirar H.A. 1994. Commentary: The fermented foods of the Sudan. Ecol. Food Nutr. 32(3/4), 207–218.
Dreyer J.J. and Wehmeyer A.S. 1982. On the nutritive value of Mopanie worms. S. Afr. J. Sci. 78: 33–35.
Dufour D.L. 1987. Insects as food: A case study from the Northwest Amazon. Amer. Anthropol. 89 (2): 383–397.
FAO. 1970. Amino-acid content of foods and biological data on proteins. FAO Nutritional Studies no. 24. FAO, United Nations, Rome, Italy.
FAO. 1991. Protein Quality Evaluation. FAO Food and Nutrition Paper no. 51. FAO, United Nations, Rome, Italy.
Ferguson E.L., Gibson R.S., Weaver S.D., Heywood P., Heywood A., and Yaman C. 1989. The mineral content of commonly consumed Malawian and Papua New Guinean foods. J. Food Comp. Analysis 2: 260–272.
Flood J.M. 1980. The Moth Hunters. Australian Institute of Aboriginal Studies, Canberra, NSW, Australia.

Glew R.H., Jackson D., Sena L., VanderJagt D.J., Pastuszyn A., and Millson M. 1999. *Gonimbrasia belina* (Lepidoptera: Saturniidae): A nutritional food source rich in protein, fatty acids, and minerals. Amer. Entomol. 45 (4): 250–253.

Green K., Broome L., Heinze D., and Johnston S. 2001. Long distance transport of arsenic by migrating Bogong moths from agricultural lowlands to mountain ecosystems. Victorian Naturalist 118 (4): 112–116.

Holt V.M. 1988. Why Not Eat Insects? British Museum (Natural History), London. (originally printed 1885)

Hoshiai Kazuo 1995. World balance of dietary essential amino acids relative to the 1989 FAO/WHO protein scoring pattern. Food Nutr. Bull. 16 (2): 166–177.

INN (Istituto Nazionale della Nutrizione). 1997. Tabelle di Composizione degli Alimenti. INN, Rome, Italy.

Kinkela T. and Bézard J. 1993. Les lipides de quelques produits alimentaires Congolais. Sciences des Aliments 13: 567–575. (in French.)

Kodondi K.K., Leclercq M., Bourgeay-Causse, Pascaud A., and Gaudin-Harding F. 1987a. Interet nutritionnel de chenilles d'attacides du Zaire: Composition et valeur nutritionnelle. Cahiers Nutr. Diet. XXII (6): 473–477.

Kodondi K.K., Leclercq M., and Gaudin-Harding F. 1987b. Vitamin estimations of three edible species of *Attacidae* caterpillars from Zaire. Int. J. Vit. Nutr. Res. 57 (3): 333–334.

Luo Zhi-Yi. 1997. Insects as food in China. Ecol. Food Nutr. 36 (2–4): 201–207.

Malaisse F. 1995. Diversite et importance des chenilles dans l'alimentation des populations du Zaire. Tropicultura, 13(2): 72–73.

Malaisse F. 1997. Se Nourrir en Foret Claire Africaine—Approche ecologique et nutritionnelle. Centre Tech. Coop. Agri. Rurale (CTA), Wageningen, Les Presses Agronomiques de Gembloux (A.S.B.L.), Gembloux.

Malaisse F. and Parent G. 1980. Les chenilles comestibles du Shaba meridional (Zaire). Les Naturalistes Belges, 61: 2–24.

Marconi S., Manzi P., Pizzoferrato L., Buscardo E. et al. 2002. Nutritional evaluation of terrestrial invertebrates as traditional food in Amazonia. Biotropica 34 (2): 273–280.

McGregor J. 1995. Gathered produce in Zimbabwe's communal areas: changing resource availability and use. Ecol. Food Nutr. 33: 163–193.

Mercer C.W.L. 1997. Sustainable production of insects for food and income by New Guinea villagers. Ecol. Food Nutr. 36 (2–4): 151–157.

Meyer-Rochow V.B. 1973. Edible insects in three different ethnic groups of Papua and New Guinea. Amer. J. Clin. Nutr. 26: 673–677.

Mitsuhashi J. 1997. Insects as traditional foods in Japan. Ecol. Food Nutr. 36 (2–4): 187–199.

Moreno-Black G. and Somnasang P. 2000. In times of plenty and times of scarcity: Nondomesticated food in Northeastern Thailand. Ecol. Food and Nutr., 38 (6), 563–586.

Munoz de Chavez M., Chavez Villasana A., Roldan Amaro J.A., Ledesma Solano J.A. et al. 1996. Tablas de Valor Nutritivo de los Alimentos de Mayor Consumo en Mexico. Editorial Pax Mexico, Mexico D.F.

Murphy S.P., Weinberg-Andersson S.W., Neumann C., Mulligan K., and Calloway D.H. 1991. Development of research nutrient data bases: An example using foods consumed in rural Kenya. J. Food Comp. Anal. 4: 2–17.

Naughton J.M., O'Dea K., and Sinclair A.J. 1986. Animal foods in traditional Australian Aboriginal diets: Polyunsaturated and low in fat. Lipids 21 (11): 684–690.

Nishimune T., Watanabe Y., Okazaki H., and Akai H. 2000. Thiamin is decomposed due to *Anaphe* spp. entomophagy in seasonal ataxia patients in Nigeria. J. Nutr. 130: 1625–1628.

Ohtsuka R., Kawabe T., Inaoka T., Suzuki T. et al. 1984. Composition of local and purchased foods consumed by the Gidra in lowland Papua. Ecol. Food Nutr. 15: 159–169.

Onore G. 1997. A brief note on edible insects in Ecuador. Ecol. Food Nutr. 36 (2–4): 277–285.

Paoletti M.G., Dufour D.L., Cerda H., Torres F., Pizzoferrato L., and Pimentel D. 2000. The importance of leaf- and litter-feeding invertebrates as sources of animal protein for the Amazonian Amerindians. Proc. R. Soc. London B 267: 2247–2252.

Paoletti M.G., Vanderjagt D.J., Pastuszyn A., Buscardo E. et al. 2003. Nutrient content of termites (*Syntermes* soldiers) consumed by Makiritare Amerindians of the Alto Orinoco of Venezuela. Ecol. Food Nutr. 42: 173–187.

Pemberton R.W. 1988. The use of the Thai giant waterbug, *Lethocerus indicus* (Hemiptera: Belostomatidae), as human food in California. Pan-Pacific Entomol. 64 (1): 81–82.

Pemberton R.W. and Yamasaki T. 1995. Insects: Old food in new Japan. Amer. Entomol. 41: 227–229.

Posey D.A. 1987. Ethnoentomological survey of Brazilian Indians. Entomol. Gener. 12: 191–202.

Quin P.J. 1959. Foods and Feeding Habits of the Pedi. Witwatersrand Univ. Press, Johannesburg, S. Africa.

Ramos-Elorduy J. and Pino Moreno J.M. 2002. Edible insects of Chiapas, Mexico. Ecol. Food Nutr. 41: 271–299.

Ramos-Elorduy de Conconi J. 1991. Los Insectos como Fuente de Proteinas en el Futuro. Editorial Limusa, Mexico, D.F.

Ramos-Elorduy de Conconi J., Pino Moreno J.M., and Gonzalez Meza O. 1981. Digestibilidad in *vitro* de algunos insectos comestibles en Mexico. Folia Entomol. Mexicana 49: 141–154.

Ramos-Elorduy de Conconi J., Pino Moreno J.M., and Romero Suarez L.A. 1987. Determinacion del valor nutritivo de algunas especies de insectos comestibles del Estado de Puebla. Anales Inst. Biol. UNAM 58, Ser. Zool. (1): 355–372.

Rao P.U. 1994. Chemical composition and nutritional evaluation of spent silk worm pupae. J. Agric. Food Chem. 42 (10): 2201–2203.

Robson J.R.K. and Yen D.E. 1976. Some nutritional aspects of the Philippine Tasaday diet. Ecol. Food Nutr. 5: 83–89.

Ruddle K. 1973. The human use of insects: Examples from the Yukpa. Biotropica 5 (2): 94–101.

Saeed T., Dagga F.A., and Saraf M. 1993. Analysis of residual pesticides present in edible locusts captured in Kuwait. ARAB Gulf J. Sci. Res. 11 (1): 1–5.

Santos Oliveira J.F., Passos de Carvalho J., Bruno de Sousa R.F.X., and Madalena Simao M. 1976. The nutritional value of four species of insects consumed in Angola. Ecol. Food Nutr. 5: 91–97.

Simoons F.J. 1991. Food in China: A Cultural and Historical Inquiry. CRC Press, Boca Raton, FL, pp. 321–326.

Taylor R.L. 1975. Butterflies in My Stomach. Or: Insects in Human Nutrition. Woodbridge Press Publ. Co., Santa Barbara, CA.

Tommaseo Ponzetta M. and Paoletti M.G. 1997. Insects as food of the Irian Jaya populations. Ecol. Food Nutr. 36 (2–4): 321–346.

Ukhun M.E. and Osasona M.A. 1985. Aspects of the nutritional chemistry of *Macrotermis bellicosus*. Nutr. Repts. Int. 32 (5): 1121–1130.

van Huis A. 1996. The traditional use of arthropods in Sub Saharan Africa. *In:* Sommeijer M.J. and Francke P.J. (eds.). Proc. Experimental and Applied Entomology of the Netherlands Entomological Society, Uitgeverij Nederlandse Entomologische Vereniging (NEV), Amsterdam, the Netherlands, vol. 7, pp. 3–20.

Wu Leung W.T. and Flores M. 1961. Food Composition Table for Use in Latin America. INCAP, Guatemala City/ICNND, Bethesda, MD.

Wu Leung W.T., Busson F., and Jardin C. 1968. Food Composition Table for Use in Africa. FAO, Rome/U.S. Dept. Health, Education and Welfare, Bethesda, MA.

Wu Leung W.T., Butrum R.R., Chang F.H., Rao M.N., and Polacchi, W. 1972. Food Composition Tables for Use in East Asia. FAO, Rome/U.S. Dept. Health, Education and Welfare, Bethesda, MD.

Yhoung-aree J., Puwastien P., and Attig G.A. 1997. Edible insects in Thailand: an unconventional protein source? Ecol. Food Nutr. 36 (2–4): 133–149.

29

Hygiene and Health Features of "Minilivestock"

VALERIO GIACCONE

Dept. of Public Health, Comparative Pathology and Veterinary Hygiene, Padova University, Padova, Italy. E-mail: valerio.giaccone@unipd.it

Abstract

Frogs, snails, rodents, and insects could be a very interesting source of food for humans but unfortunately at present constitute no or a minimal part of the daily diet in most developed countries. The advantages and, of course, the problems which could arise for humans from the hygienic point of view, by the consumption of this "uncommon" source of meat are elucidated in this paper. All these animals could produce meats with very high protein content and with a quite interesting content of monounsaturated and polyunsaturated fatty acids. On the other hand, they could present a variety of hygienic problems if consumed raw. New techniques of food handling and preservation which are able to reduce or eliminate these problems are proposed.

Key Words: frog meat, snails as food, insects as food, rodents as food, food hygiene

The Concept of "Food"

Humans have eaten many of the plant and animal species found on the planet (Harris, 1985). Through the ages, our eating habits have progressively changed and certain types of food have been replaced by others in our daily diet. In most cases this has occurred through a very slow process of evolution, thereby making it difficult to fully understand why these changes actually took place. However, for certain types of food the concept of history repeating itself seems to apply, i.e., foods which were very much in vogue in prehistoric or ancient times

were increasingly excluded from the human diet for various reasons, so that nowadays they are no longer considered "food", except for those cases which in recent decades have been reevaluated as such. For example, the ancient Romans regularly consumed birds such as swans and peacocks: the dormouse was considered a special delicacy and among aromatic plants silphium (possibly *Ferula* sp.), held in high regard, was imported at a costly price from the Middle East. The Romans valued this plant so much and consumed it in such vast quantities that it eventually became extinct (Flandrin and Montanari, 1997). In his book *De re coquinaria* (one of the first books written about cooking and diet) Apicius (Apicio, 1994) mentions a recipe for a dessert prepared with rose petals. The Romans were also known to use a sauce made of fermented fish as a dressing, something that would certainly not appeal to most Westerners today, although very salty and aromatic fish sauces are quite commonly eaten with rice and vegetables in the Southeast Asian cuisine.

It is quite indicative that in a continent such as Europe the various populations would have made a progressive selection among all the foods nature has made available, choosing some and excluding others. The Italians and French are very fond of snails, frogs, and garlic, which are not greatly appreciated by the majority of people of Anglo-Saxon descent. Vice versa, the Scots are known to eat haggis (boiled sheep stomach) which is not the most appealing of dishes for people in other parts of Europe.

Changes in the dietary habits of a population can certainly be attributed to important historical, religious, geographical, social, and economic events, which are not explored in depth here. However, it should be remembered that often different "human societies" have appealed to man's religious conscience to prohibit the consumption of certain foods and encourage others. It is common knowledge that followers of the Jewish and Islamic religions are prohibited from eating pork and horse meat, and that Catholic Christians observe the precept of eating fish on Fridays. Less well known perhaps is the fact that toward the end of the Medieval period, two popes prohibited the consumption of horse meat, stating that it would "corrupt the spirit and transmit leprosy" (Parisi and Giaccone, 1992). Consumption of horse meat was totally stopped in Europe (Harris, 1985). Only at the beginning of the 1800s did this prohibition (which by then had become outdated) disappear and slowly the custom of slaughtering horses and eating their flesh returned to continental Europe. However, the Anglo-Saxon populations never followed this custom, so that today the British and Americans do not consider horses edible (the same applies to rabbits). These cultures perceive horses and rabbits as animals for sport or as pets and the very idea of eating their flesh is abhorrent, as is the eating of dogs, cats, rats or insects for Europeans. Therefore, it appears evident that our diet starts primarily in our thinking, before reaching our mouth, and that dietary traditions are still today, as in the past, conditioned by mental attitudes arising from historical, geographical, and social factors as mentioned previously. In this paper I identify the primary hygiene and health aspects for humans that need to be considered if

rodents, amphibians, mollusks and insects are to be consumed. The methodological approach to this aspect differs markedly from the social and psychological one; in other words, we need to go right back to the definition of the word "food".

Generally speaking, food refers to "any substance eaten by man that enables him to maintain not only his basic metabolism, but also the many activites of everyday life". It should also be remembered that in nearly every country of the world there are specific regulations which govern the production, transformation, sale and use of food for human consumption.

Consequently, a precise legal definition of "food" does exist and it corresponds with the definition given by health authorities who control the hygiene-health conditions of food for humans. From this specific point of view we can include in the term "food" any substance that is edible, provided it has: (1) exemplary features of hygiene and (2) characteristics of taste, smell, color, and consistency which are typical for the product. In order to be permitted for human consumption therefore, a food (1) may not contain any microorganisms, their toxins or chemical compounds, which could cause a specific disease to humans (food diseases), and (2) must have its own sensory characteristics, which correspond to the standard ones—those generally considered the "norm" for each food.

One example is milk, which should be white in color, have its own particular faint smell, and a slightly sweet taste; meat, on the other hand, should be either a deep red color or a light red (pinkish), according to the type of animal it comes from, with a faint typical smell. Any deviation from these ideal characteristics would constitute a defect and mean that the food was no longer suitable for human consumption. However, it should be pointed out that these characteristics should always be seen in relation to the specific food in question. A classic example of this is cheeses such as Gorgonzola or Roquefort, in which the presence of selected mold is essential. The growth of these types of molds in other foods, such as bread, meat or fruit on the other hand, suffices to exclude that product from human consumption (as moldy food).

In defining "food" therefore, there is no particular restriction concerning the animal or plant species supplying the protein, carbohydrates, fats or essential microelements required for human metabolic needs. The only limitations from the legislative point of view are that these potentially edible substances need to have optimal hygiene-health and sensory characteristics and must not endanger human health. It is also evident that apart from this type of restriction, the food must be accepted by the people, i.e. it must be within the normal dietary customs of a population. Although applicable in Europe to frog meat, it is not for insects, spiders and rodents such as rats and mice. On the other hand, these groups may be considered a form of nutrition for other populations, notably in Central and South America, Africa, India, China, and several other countries in Southeast Asia.

This paper is restricted to considering the use of rodents, amphibians, reptiles, mollusks, and insects as human food, only in relation to their hygiene-health and nutritional aspects.

The Consumption of Rodent Meat

Various rodents are eaten in tropical and temperate countries and some are addressed as potential candidates for new domesticates (Govoni et al., this volume; Jori et al., this volume). Furthermore domestic mice and rats have been used as human resources. For instance, the name of a sauce that Italians are particularly fond of, the classic "ragu" or "ragout" (French) has an interesting origin. Etymological dictionaries show that the term ragout derives from the verb *ragouter* (to whet the appetite, to be appetizing) so that the participle *ragoutant*, used as an adjective, indicates the quality of being appetizing. However, there is another version of the origin of this word, perhaps more intriguing, which dates back to the siege of Paris (Sept 18, 1870–January 28, 1871) by the Prussians in the Franco-Prussian war. The duration and ferocity of the siege caused an incredible food shortage and the Parisians were literally starving as their normal food resources came to an end. So, they began to kill off the animals in the Zoo, including the elephants and giraffes. Once this limited supply of meat was consumed, legend has it that the Parisians trapped mice and rats. To overcome any feelings of disgust in eating such animals, they stewed them in an aromatic sauce that concealed the type of meat used. From this sauce "tasting of rat" (ratgout) cames the term ragout, which refers to a dish based on meat chopped into large or smaller pieces and aromatized with vegetables. To our knowledge, this situation remained unique in Europe and no similar situations of rats and mice being eaten have been found, not even during the Middle Ages. However, it is a well-known fact that mouse meat is consumed rather frequently in China and other countries of Southeast Asia (Flandrin and Montanari, 1997; Luo Zhi-Yi, this volume).

The data garnered from the literature regarding the hundredth composition of the muscular mass of mice is rather limited. Table 29.1 reports data from some Internet websites on the composition of mouse meat (*Mus domesticus*) and rat meat (*Rattus norvegicus*). It can be seen that rat meat is on average richer in protein

Table 29.1: Composition of mouse meat (*Mus domesticus*) and rat meat (*Rattus norvegicus*)

	Mus domesticus	*Rattus norvegicus*
Water content %	68.3	66.1
Dry residue %	32.7	33.9
Raw protein % ss	55.8	61.8
Raw fats % ss	23.6	32.6
Ash % ss	11.8	9.8
Energy value (kcal/g)	5.25	6.37

and fats than mouse meat (which, in turn, contains more water) and consequently can supply a greater number of calories.

From the point of view of food safety, mice and rats, like but perhaps more so than insects, have always been seen as dangerous sources in the spread of many human infectious diseases. Medical literature abounds with references that confirm this role, but interestingly in most cases the rodents merely play the role of "live carriers" of the disease's microorganism agent (the most well-known example is the association between hematophagous fleas and rats in the diffusion of *Yersinia pestis*, bubonic plague). To be considered nonedible, however, the flesh of the mice and rats would need to be infested with pathogenic agents of food diseases, such as *Salmonella, Shigella, Campylobacter*, etc., or to have accumulated residues of natural or xenobiotic chemical compounds which are potentially dangerous for human health (residues of pesticides, antibiotics and chemotherapy drugs, heavy metals).

The microbiological aspect of this type of meat is very much conditioned by the farming situation, but also by the type of slaughter practiced and the subsequent conservation of the carcasses, almost always done by refrigeration. The slaughter of rodents must necessarily observe the technical and especially the legislative regulations in force on the subject. A search for facilities already carrying out a similar type of processing led us to the possibility of using quail slaughterhouses.

An alternative solution could be the setting up of *ad hoc* processing plants; it should be pointed out, however, that these facilities have to be authorized by the relevant health authorities. From the legislative point of view, mice and rats reared for human consumption would be considered "farmed wild game". According to the Italian rules, in Italy the production of their meat would therefore come under the Decree of the President of the Republic 31.12.1992, no. 559, also with reference to any related factors, such as hygiene, structural and functional requirements of the plant, and laboratories for the dissection of the animal. It should also be remembered that it is already an established custom that the meat of mice and rats produced in other countries may be imported into the EU in the form of fresh meat (frozen or refrigerated), or cured, seasoned, or canned meat products. This possibility would mean an adjustment to current legislation regulating the importation of foodstuffs into the EU from the rest of the world. Once these problems have been overcome and a hygienically proper slaughtering for these animals assured (including regular health visits by veterinarians both ante- and post-mortem), in theory there should be nothing to oppose the addition of this meat into the daily diet of Europeans, provided consumers are able to modify their food habits.

Amphibians

Basically four species of amphibians are eaten in Europe: two are mainly aquatic (*Rana esculenta* or green frog, and *Rana ridibunda* or swamp frog) and two mainly

land frogs (*Rana temporaria* and *Rana graeca*). In America both the leopard frog (*Rana pipiens*) and bullfrog (*Rana catesbeiana*) are captured and reared. In Asia the largest frog, also exported in large quantities, is the *Rana tigrina* (up to 20 cm in length) but many other frog species are captured as well (see Negroni, this volume). The giant of all frogs lives in Africa, *Conraua goliath* (over 30 cm long and up to 3 kg in weight).

Frog consumption goes back to ancient times in Italy but the methods of capture and culinary preparation have changed radically in the few decades. Up until the middle of the lst century, the custom of eating frogs was limited to the Po River valley, in particular those areas where rice was cultivated. However, nowadays frog legs have started appearing on the menu in many restaurants in central and southern Italy as well. Once, frogs and rice paddies were an inseparable twosome: the capture of frogs was very prevalent among the poorer families, with some eaten at home and the others sold in the market, thereby entering the commercial restaurant circuit. Prior to the introduction of chemical herbicides and insecticides, the demand for frog meat in Italy was totally based on the capture and direct sale to the consumer, with all the hygiene—health concerns that could arise from procedure. However, up to the 1950s frogs were taken to the markets live in baskets or jute sacks, and killed directly in front of the buyer, thereby making refrigeration unnecessary. It also used to be common practice to break their feet immediately upon capture to preclude their escape. But this provoked extensive hemorrhaging of the legs, causing rapid deterioration of the meat, which is totally unacceptable today.

Intensive application of herbicides and insecticides in rice paddies (1960–1970) led to the almost total disappearance of frog colonies. To satisfy the increasing market demand, importation of frogs from overseas was initiated, either live (to be slaughtered in Italy) or as frog legs or fillets, frozen or deep frozen. This commercial tendency still exists today: Italy currently imports approximately 9,000 tons of live frogs from Mediterranean countries in particular Egypt and Turkey. The frogs are slaughtered on arrival in Italy. Another large quantity of meat (an estimated 7,000–8,000 tons per year) comes from east Asia; in this case the frogs have already been slaughtered and filleted, or the legs (which is their most highly valued part) frozen or deep frozen. Up to a few years ago, China was the main supplier of frozen frog legs to the Italian market. However, in 2002 Italian health authorities placed a ban on their importation, along with all fish products from China, due to health concerns after the discovery of traces of antibiotics in frozen shrimps from that country. In addition to these concerns, there is also an important issue from the hygiene-health point of view, namely the fact that the frogs consumed nowadays are nearly all from non-European countries could signify a risk to the consumer. Bearing in mind the conditions that these animals are caught in and the use of plant protection products and other chemical compounds in these countries, the consumer could be put at risk through an accumulation in the frog meat of residues of substances dangerous to human health.

Sensory and Nutritional Features of Frog Meat

Like all foods of animal origin, frogs must undergo strict health controls by public health veterinarians before they can be considered fit for human consumption. The health authorities firstly need to check the freshness of the product and its health; secondly they must verify that no fraudulent substitutions of toads have occurred (toads are generally not considered comestible in Italy) and ascertain any possible changes in the sensory characteristics of the meat. Fresh frog meat should be a pinkish color and have no unpleasant smell: it has been noted that a faint urine-like smell may emanate if the frogs are not skinned immediately following slaughter. It has also been observed that toad meat (*Bufo bufo*) will always exude a urine-like smell. In this respect it should be remembered that it is not difficult to distinguish between frogs and toads if the animals are still whole and the skin intact; contrarily, distinguishing parts, such as legs, in skinned animals can be problematical. In this case, X-ray of the back legs (the different skeletal structure of frogs and toads can be easily noted) and electrophoretic tests of the muscle proteins give results that are completely reliable.

Frogs have an abundant weight after slaughter, between 65% and 69% of the weight of the live animal, which compares favorably with the weight percentages of our most important edible animal species. With regard to the chemical composition of the meat, the Table supplied by the Istituto Superiore per la Nutrizione (Giaccone et al., 1988) shows that frog meat has an average water content of 81%, 15.5 to 16% protein and about 1–2% of nonproteinous nitrogen compounds. As claimed by traditional folklore several decades ago, this meat has a low fat content (0.2–0.3%) and low carbohydrate content (<0.1%) and is therefore suitable for people with digestive problems and those who are overweight. Table 29.2 shows hundredth chemical composition of frog flesh and compares it with the meat of other animals consumed. It can be seen that frog meat compares more than favorably with all other types of meat.

Table 29.2: Comparison of chemical composition and energy levels of some types of fresh meat and fish products (values shown in 100 g of edible portion)

Food	**Edible portion**	**water (g)**	**protein (g)**	**lipids (g)**	**energy (kcal)**	**energy (kJ)**
Adult bovine meat	100	73.8	21.4	3.7	119	497
Calf meat	100	76.9	20.7	1.0	92	384
Rabbit (leg)	83	72.0	21.0	5.9	137	574
Snail	24	82.8	12.9	1.7	67	280
Lean pork (leg)	87	75.2	20.2	3.2	110	459
Fatty pork (leg)	90	72.9	20.4	5.1	128	533
Frog	—	81.9	15.5	0.2	64	267
Fillet of bass	100	69.9	21.3	6.8	146	611
Sole	48	79.5	16.9	1.4	83	347
Seabream fillet	100	73.2	20.7	3.8	117	490

Hygiene–Health Aspects in the Processing and Consumption of Frog Meat

To our knowledge, northern Italy has at least three frog processing plants; the systems in these facilities follow the health criteria imposed by the European Union for all processing plants that produce meat for human consumption. Before slaughtering, as a precautionary measure frogs must be first of all hosed down with cold tap water, to induce a sort of loss of consciousness. They are then stunned by an electrical shock, and decapitated. Skinning and complete degutting are by further washing of the bodies in tap water at 0°C. They are then packed in polystyrene boxes and covered with crushed ice to keep the temperature constant at 0°C during their transportation to the fish market and from there to the place of sale to the public.

Interestingly, the skin of even edible frogs contains poisonous muciferous glands but the mucus is not a health risk for those who eat the meat. However, the danger could be for those who handle the animals in the slaughtering process: factory workers could develop a rather serious form of conjunctivitis or dermatitis due to continual contact with the slightly poisonous mucus of the amphibians (Giaccone et al., 1988).

By using the usual Good Manufacturing Practices (GMP) in the skinning and gutting stages, the frog meat undergoes no negative repercussions in this aspect, and also microbial contaminants are limited as much as possible. One of the main reasons a veterinary check is essential at the slaughter and sale, is to ensure that the meat is not already microbiologically contaminated at the slaughter, and consequently during transportation and sale, especially bearing in mind the normal intestinal microbial flora of these cold-blooded animals.

While alive, frogs can pick up many infections (bacterial, viral, protozoan, and mycotic). Among these infectious illnesses, the most dangerous arise from *Aeromonas hydrophila* (the so-called "red-leg" that can cause a large number of deaths within each colony), *Eberthella ranicida, Achromobacter desmolyticum, Pseudomonas* spp., and *Flavobacterium hydrophilum*. In the literature, it is also noted that some types of atypical mycobacteria, agents of minor zoonoses for humans, such as *Mycobacterium piscium* and *M. ranae*, can induce miliary lesions in the liver and other internal organs (Giaccone et al., 1988). Therefore, theoretically, frogs could be vectors to humans of mycobacteria, with the possible development of a specific granulomatous dermatitis.

Based on the results of specific experiments held in the 1940s, frogs are not considered a danger in spreading *Leptospira*, but no up-to-date confirmation of this is available (Giaccone et al., 1988). Live and slaughtered frogs can harbor enterobacteria in their gut, especially *Salmonella* spp., total and fecal coliforms, *Escherichia coli*, sulfite-reducing clostridia, and *Enterococcus*. At least 9 bibliographical references document the presence of *Salmonella* spp. in batches of fresh and frozen frogs (the percentage is between 5.1 to 35–40% of the individuals in the batch). Among the most extensive serotypes are *S. newport, S. arizonae,*

S. paratyphi, B. java. Apart from *Salmonella*, it is possible to isolate other enterobacteria in the intestinal contents of frogs, such as *Shigella* spp., *Yersinia enterocolitica, Edwarsiella, Proteus, Citrobacter*, and *E. coli*. Most of these bacteria are simple saprophytes, that by multiplying excessively could be responsible for bringing about a decomposition of the meat. On the other hand, some can be agents of food diseases in humans (*Shigella, Y. enterocolitica*, some specific serotypes of *E. coli*), with consequences that could be very serious, and in rare cases, even mortal.

Their presence in frog meat could therefore be the cause of foodborne infections and intoxications in humans. In particular, we currently lack any indicator of the possible presence of verocytotoxic strains of *E. coli* in frog meat (the so-called VTEC agents of a hemorrhagic colitis HC and an hemolytic uremic syndrome HUS).

Frog meat consumed raw or undercooked could be the vessel for carrying some of the Metazoa parasites to humans. As far as Platyhelminthes are concerned, it is known that frogs can harbor in their flesh plerocercoidal larvae of *Diphyllobotrium latum* and *D. ranarum*. Panebianco and Giuffrida (1993) observed the presence of whitish nodules of *Gnatostoma spinigerum* in frozen frog legs imported from overseas. Frog connective and adipose tissue can be infested by plerocercoidal larvae of *Spirometra* and other Pseudophyllidae. If eaten by humans, these larvae can cause a specific eosinophilic enteritis, the so-called "sparganosis", which however is generally rather benign.

Frogs, through their capture and breeding (usually extensive, but also intensive) can accumulate residues of natural or xenobiotic toxic compounds. This is perhaps the hygiene-health aspect that has currently assumed major importance as nowadays frogs consumed in Italy mostly come from non-European countries where problems of environmental pollution from heavy metals, plant protection products, agricultural pesticides, and other compounds toxic to humans are very common.

Between the end of the 1800s and 1940 at least two cases of poisoning from eating frog meat were recorded. These cases, due to their characteristics, can be considered "historical" and as such it is not very probable that they would occur today. In both cases, those who became ill were from a group of soldiers training in the field who had consumed frog meat from animals they themselves had caught. The amphibians had priorly eaten a large quantity of cantharis Coleoptera and accumulated a very irritating active ingredient, cantharidin, in their flesh. A large number of the soldiers became intoxicated, showing signs of asthenia, priapism, and hematuria (Giaccone et al., 1988).

Nowadays, frogs can accumulate in their flesh traces of heavy metals, (especially mercury, lead, and cadmium) as well as traces of plant protection products (organochlorinate pesticides such as DDT, DDD and DDE, aldrin and deildrin, still used in Southeast Asian countries). Ajmerito et al. (1998) found traces of mercury and lead in frogs from the fish market, in quantities above the limits indicated by current national and EU legislation.

This confirms the need for frog meat to undergo strict health checks in the processing stage, espcially if imported from non-European countries. Out of interest, it is useful to emphasize that even though amphibians may have more or less developed aquatic habits, frogs are not considered "a product of fishing" and as such do not come under the competence of Act 30.12.1992 no. 531 which regulates the production and selling of fish products in Italy.

Currently in Italy the slaughtering of frogs and their sale is controlled solely through the hygiene-health section of Act 30.4.1962 no. 283 according to its regulation D.P.R. no. 327/80.

Terrestrial Gastropoda (Snails)

Snails are one of the items currently undergoing an exciting gastronomic rediscovery. This delicacy is being sold not only at fairs and local festivals, prepared according to old traditional recipes, but also at an ever-increasing number of restaurants which now propose snails as part of "special" menus or as an alternative dish to be inserted in their standard menu. Figures relating to their wholesale consumption confirm that this demand is on the increase and is only being partly met by national production (Novelli et al., 2001; Elmslie, this volume).

Similar to frogs, and perhaps even more so, the overall configuration of snails and their biology have an important influence on the hygiene-health characteristics of the product destined for human consumption. This is especially so if we bear in mind that contrary to almost all other animals, snails are sold live and do not undergo proper slaughtering. Therefore, the health checks hygienist veterinarians carry out on batches in the process of being sold are often cursory and do not take into consideration the actual biology of these mollusks.

Edible snails in European countries are nearly all exclusively of order Styllomatophora, Helicidae family. Furthermore, these gastropods are now the subject of an active and profitable farming system (Elmslie, this book); however in some cases (when they are imported from non-European countries) they may have been merely collected, with all the related hygiene-health consequences (see the points above pertaining to frogs).

The most popular types of snails for sale are *Helix adspersa* (the common snail), *H. pomatia* (the original and most important edible snail), *Achatina fulica*, and *Archachatina marginata* (very large African snails valued for their size). In Italy the most sought after snails belong to genera *Helix*, *Eobania*, and *Euparipha*, but those farmed belong to family Helicidae which has three different genera: *Helix* (*H. pomatia*, *H. lucorum*, *H. cineta*, *H. ligata*), *Cryptomphalus* (*C. mazzullii* and *C. aspersa*), and *Cantareus* (*C. aperta*). The majority of snails farmed are herbivores (feeding on algae, leaves, mushrooms, stalks and roots of plants). Farmers sometimes feed them with real composite fodder, similar to that used for animals such as pigs or chickens. Snails show a preference for alkaline soils, rich in carbonate; it is also necessary on these farms to supply calcium carbonate to

maintain their shells in good health and to prevent acts of cannibalism. With reference to their most interesting anatomical and physiological features from the health point of view, we should first of all mention the epiphragm, a whitish membrane of parchment consistency that closes the opening of the shell during winter hibernation and summer estivation. Composed of a mucus produced by the coat and dried (sometimes it is rich in calcareous salts), the epiphragm allows the filtering of air, but not water, from the animal, thereby preventing body dehydration.

In fact, snails alternate periods of an active life style (moving snails) with periods of hibernation (epiphragmated or operculated snails); hence snails reaching the markets and our tables may be either moving snails or ones in hibernation. Hibernation is induced by excessive heat or cold, and even by an extremely dry spell. In both conditions, the animal stops eating and empties its intestine, retracts from its shell and forms an epiphragm. All this justifies the slight difference in the hygienic conditions of the two types of snails, especially from the microbiological point of view: moving snails are much more exposed to environmental contamination than the epiphragmated (both of the microbiological and chemical types) and have an abundant and composite microbial flora in their intestine.

Epiphragmated snails on the other hand, can nearly all be considered microbiologically sterile, provided they are still alive.

Like all other food products of animal origin in Italy, snails too must meet precise legislative standards to be edible, but contrary to all other animals, no specific regulation applies solely to them. To carry out sanitary checks on the batches sold, the state health authorities must therefore refer to the Act of 30.4.1962 no. 283 and its regulation (D.P.R. no. 327/1980) which sets out the more generic aspects of hygiene in the production and sale of food. The more detailed aspects of these checks are set out in the Newsletter of the Ministry of Health no. 30 of 3.7.1987. According to the standards indicated, it is essential that edible snails be kept under veterinary control during transportation and at the place of sale, whereas no actual health checks are required at the farming stage.

At these checks, an evaluation is made of the general appearance of the animals, their color, smell, and the presence of feces, mucus or foam, or any other anomalies in appearance and behavior. The most important point in the case of moving snails is that the animals all be alive and moving; hence their reflexes are tested with mechanical stimulation of the tentacles, foot or coat (the pin or stick test) or by dusting some with salt or vinegar (if the animals are alive and well, they will react by emitting a large quantity of mucus). Healthy snails have the open foot extended and erect tentacles that react to touch, the body surface is smooth, glossy and damp, with compact flesh and a fragrant odor. If ill, they will be withdrawn inside their shells or the foot externally visible, everted, but limp and without tone; they do not react to stimulation and emit abundant foamy mucus that has an unpleasant odor. Those that have been dead for a short while will be withdrawn inside their shells, will not react to stimulation,

and will emit a moldy odor. Those that have been dead for a longer time will give off a very bad odor.

For epiphragmated snails, on the other hand, when the animals are healthy the epiphragm is whole, unbroken, thick and closes the opening of the shell, which has a glossy smooth appearance. In animals suffering from malnutrition or sick, the epiphragm is fragile, concave, and depressed in the shell: if it has holes and/or is seeping, this signals that the animal is already dead. Needless to say, only those snails that show the aforesaid health characteristics should be destined for the market and human consumption. Batches containing dead, sick or dying animals, on the contrary, must be withdrawn and destroyed according to the regulations that apply in these cases. In the last few years, frozen epiphragmated snails have appeared on the market. In these cases, the health checks anticipate first a verification that the epiphragm is whole; if it is, this is a good sign. In case of doubt, it is appropriate to thaw the batch and evaluate the appearance, odor, and texture of the flesh of some of the animals in the batch, as done with fresh epiphragmated snails.

Like all other living creatures, snails are liable to various types of pathologies. The majority do not directly affect the hygiene-health aspect of the animals as they are not transmitted to humans who eat them; but it is appropriate to mention this fact since pathologies can sometimes cause serious damage in farms. The bibliographical references dealing with this topic are very few compared to the number found on other types of mollusks.

In other words, an abundant amount of data exist on the infectious and noninfectious diseases of Lamellibranchia mollusks, but information on the pathologies that may strike the pulmonate ones is scant (Cooper and Knowler, 1991). Most of the research carried out to date on the pathologies of gastropods concerns parasitic diseases, in particular those that could potentially be transmitted to humans and other animals. There is some information on the diseases of gasteopods sustained by Protozoa and acid-resistant bacteria but this research was carried out in the 1960s. Records of animal neoplasms held in Washington DC show some cases of tumors found in snails, but they mostly refer to unknown etiological forms. Table 29.3 summarizes some of the main causes of death among farmed snails.

The data available in the literature on aspects of public health and food microbiology that affect edible snails is likewise limited and not very recent. From these references and through personal experience, it is possible to note that healthy moving snails have a microbial flora that is reasonably restricted, formed mostly by saprophytic Gram-positive bacteria (*Micrococcus, Staphylococcus, Corynebacterium, Bacillus,* and *Lactobacillus*). If during the rearing and capturing stages the basic hygiene regulations have been observed, the Gram-negative spoiling bacteria (mainly made up of enterobacteria) are always quantitively rather low. With respect to the microorganisms that are pathogenic agents of the major food diseases, it is rare that a batch of fresh moving snails carries *Salmonella* and *Listeria monocytogenes*, but the risk cannot be completely excluded.

Table 29.3: Physiological factors associated with cases of mortality in farmed snails

Factor	Effect	Comments
Injuries	Physical injuries, at times mortal. Loss of body parts that sometimes grow again.	The shell can be fragile due to a lack of minerals in the diet. Handle the snails with care and not on rough surfaces.
Poisoning	Sensory exhaustion leading to torpor, death of animals.	Many substances toxic to snails: molluschicides, insecticides, detergents, disinfectants, copper salts. Danger of traces in the animal muscles.
Nutritional deficiencies	Weight loss and death of animals. Shell fragility. Cannibalism.	Preventable by correct distribution of food rations. Not a risk for human health.
Nematode infestation (*Alloionema, Rhabditis, Angiostoma, Nemhelix*)	Few species can cause real lesions of tissues or death of animals. Others live as saprophytes and attack the mollusk body only when dead.	Regular checks of farms; elimination of dying or dead animals can drastically reduce the level of infestation.
Trematodes	Usually behave as saprophytes, but in cases of infestation can cause serious damage to mollusk body.	Gasteropods are intermediary guests of many trematode parasites that have higher vertebrates as definitive guests.
Larval forms of Diptera other insects	Species of parasites damage and can kill snails. Others which are nonpathogenic lay eggs only on dead animals.	Carry out proper checks of diffusion of insects in the farms.
Flies (*Piophila*)	Lay eggs only on dead animals.	Can disseminate pathogenic bacteria to batches.
Fungi	Some molds are pathogenic and kill snails.	Check dampness in farms; remove dead animals in order to prevent spreading of saprophytic molds.
Bacteria	*Aeromonas* and *Pseudomonas* are pathogenic to snails; cause weight loss, torpor and yellowing of foot. Snails can carry many other types of bacteria, also pathogenic to humans.	Respect rules of hygiene in farms to avoid spreading.

From Cooper and Knowler (1991), modified.

In batches of dying snails or those already in the process of decay, there is an obvious increase in the spoiling Gram-negative microflora (coliforms and other enterobacteria, *Pseudomonas* spp.).

Lastly, in epiphragmated snails microbial bacteria are almost completely absent, as suggested above. Other bacteria which cause typical forms of

foodborne intoxication (*Staphylococcus aureus, Bacillus cereus, Clostridium perfringens*, and *C. botulinum*) are, in theory, able to contaminate a fresh live product, since the microorganisms occur everywhere in the environment: however it should be remembered that these bacteria are dangerous to human health only when they reach significant microbial levels (> 10^3–10^4 CFU/g food) and it is difficult for them to actively proliferate if a composite saprophyte microbial flora is present. It is assumed, therefore, that these microorganisms are of significance only when the product has already been transformed, i.e. when the snails have already been cooked and shelled. Contamination of the product after cooking by the toxic bacteria mentioned earlier could in fact bring about their considerable proliferation, possibly causing food poisoning, if the intrinsic conditions of the food are favorable to this increase (pH and a_w values, aerobiosis or anaerobiosis of the food).

Snails are sensitive to various parasitic diseases but only one is transmissible to humans; namely the one caused by the nematode *Angiostrongylus cantonensis* that can become a parasite of mollusks of the *Achatina* family and perhaps even cause serious forms of parasitic meningoencephalitis in humans. This serious illness has been recorded mostly in Far Eastern countries (China, Malaysia, Indochina); as far as we know it has never caused health problems in Western countries.

Nutritional Features of Snails

Research was recently carried out by Novelli et al. (2001) into the composition of the muscular tissue in operculated snails collected during the winter season in the foothill zones of northern Italy and in agricultural areas in central and southern Italy. The species of snails investigated were: *Helix lucorum, H. adspersa, H. pomatia, H. vermiculata, H. aperta*. The average weight of the snails was about 14 grams, apart from *H. aperta* specimens, which oscillated between 4–6 grams. The water content in the edible portion of the 5 species studied was between 80 and 82%. The raw protein varied from a little less than 15% in *H. lucorum* to a minimum of 10.2% in *H. vermiculata*. The lipid content was consistently below 1% while ash varied between 1.4 and 2.5%. Nitrogenous extracts varied from 1.4 to a little under 5%. The distribution of saturated, monounsaturated, and polyunsaturated fatty acids was respectively 21%, 14.3%, and 40.5%. Among the most abundant were palm oil and stearic acid that added up to 8.6% and 10.4% of the total; among the polyunsaturated ones, arachidonic acid represented a little under 12% while eicosopentaenoic acid came to 1.95%.

According to the aforesaid authors, it may be deduced that with regard to composition, the muscular tissue of land snails of genus *Helix* shows some unique features. While the protein content can be considered superior to that found in most marine mollusks, the ponderal quantity of fat is lower. Considering the low values found, it is reasonable to hypothesize that the phospholipidic structure

of cell membranes is nearly always exclusively composed of lipids with a structural valency.

The profile of the composition of fatty acids in which the polyunsatured ones are more common (especially linoleic and arachidonic acids) agrees with the above-given hypothesis. The ash content was high compared to the muscular tissue of other species of animals consumed by humans, but reasonably similar to that found in snails of the same species by Gomot (1998). The snails examined were rich in calcium, however, an element needed to metabolize the cells and to build the shell itself, and also rich in copper: the latter is not surprising if one recalls that among all mollusks, the snail has one of the highest copper contents in its hemocynin. The results obtained bring us to the conclusion that snail meat is extremely rich in proteins of high biological value and concomitantly is extremely low in fats. Furthermore, the fats are mostly of the mono- and polyunsatured fatty acid type, which are particularly useful for a balanced diet.

Edible Insects

In modern Western societies, a certain cultural evolution has brought us to consider insects as a primitive food and consequently develop a sort of repulsion for them. However, in the rest of the world another cultural concept includes insects as an integral part of the diet (Ramos-Elorduy, 1997, and this volume). And just recently, some exclusive Western restaurants have begun to add insect-based dishes to the menu, as a gastronomic novelty (Ramos-Elorduy de Conconi, 1996).

Developed nations, where there is no pressure to find food to satisfy people's hunger, are currently not able to view insects as a possible alternative form of nutrition. Nonetheless insects, with their high content of proteins, vitamins and minerals, in certain phases of their biological cycle could contribute significantly to a population's energy requirements (Ramos-Elorduy, 1997). Insects can be consumed in any period of their biological cycle (eggs, larvae, pupae, and adult individual) but the younger stages of growth are preferred in that they have a thinner and softer cuticle (Ramos-Elorduy, 1982). Hence it is necessary to move toward intensive farming of insects, with obvious advantages under the following aspects: efficiency of converting them into food, possibility of using organic substances not suitable for conventional agriculture, and reduction or elimination of the need for fertilizers, herbicides and insecticides, thereby making it a coherent practice within the principles of sustainable agriculture which has a lower environmental impact (Paoletti and Bukkens, 1997).

True, some insect rearing facilities have been established in the USA and Europe but for the moment these insects are essentially used to supply experimental prospects and in the biological control against insects which threaten agriculture. Concomitantly research is underway to use insects as an alternative source of protein in cattle rearing fodder and in a still rather distant future, may also became a food source in Western countries (Collavo, this volume).

Taking into account the relative scarcity of literary information on the nutritional values and hygiene and health features of edible insects, I carried out a series of experiments described below (Giaccone et al., 2000). The samples used for chemical-physical and microbiological analyses (25 in all) comprised about 200 g of larvae and 500 adult individuals. Five types of insects were analyzed, of which four were in the larval stage and only one in the adult: *Zophobas morio* and *Tenebrio molitor* (Coleoptera larvae), *Chilecomadia moorei* and *Galleria melonella* (Lepidoptera larvae), and *Acheta domesticus*, adult stage of the common cricket. These insects, except the larvae of *C. moorei* imported from Chile, all came from an Italian closed cycle rearing farm. Microbiological analyses quantitated the total mesophile microbial levels, the enterobacteria, the total and fecal coliform bacteria (indicators of environmental pollution and fecal contamination), sulfite-reducing clostridia, *Micrococcus* and *Staphylococcus*, *Salmonella* spp., *Listeria monocytogenes*, and *Bacillus cereus*.

As for the chemical-physical analyses, the amount of water, raw protein, raw fat, ashes, and fiber was determined: furthermore, the fatty acid profile was established for each sample. Results of the chemical analyses are reported in Tables 29.2, 29.4 and 29.5.

The microbial flora of the insects analyzed was quantitatively very high (on average levels over 10^5–10^6 CFU/g of the product) and furthermore made up of Gram-negative bacteria; in particular, high amounts of total and fecal coliform bacteria were found, a sign that the insects (both larvae and adults) carry a rich microbial flora on the body surface and in their intestinal contents.

In spite of this, neither *Salmonella* spp. nor *L. monocytogenes* were ever identified in the samples tested; this information, although encouraging, does not mean we can exclude a priori the fact that insects could at times also harbor these two species of bacterial agents of food-related illnesses which can affect humans. By comparison, it may be noted that *Salmonella* spp. was occasionally identified on the surfaces of flies and in the intestinal contents of African bugs. As for the Gram-positive bacterial flora, this was mainly composed of *Micrococcus* spp. and *Lactobacillus* spp. (in amounts that could exceed 10^5 CFU/g). However, the amounts of *Staphylococcus* ($< 10^3$ CFU/g) were much lower; nor were any cases of enterotoxic strains of *S. aureus* ever isolated.

Even with the limited number of samples analyzed, the microbiological conditions of the raw material is of relatively little concern, in particular in view of

Table 29.4: Analysis of per cent composition (average values of two batches)

	Zophobas morio	*Chilecomadia morei*	*Galleria melonella*	*Tenebrio molitor*	*Acheta domesticus*
Water content %	57.78	51.75	60.07	—	34.74
Raw protein % ss	43.80	33.45	40.34	56.76	69.19
Raw fats % ss	45.13	62.33	53.69	30.01	20.73
Ash % ss	2.58	1.93	2.23	4.08	4.98
Fiber % ss	4.72	3.86	3.42	5.90	8.03

Table 29.5: Fatty acids profile. Composition in percentages (average values of 2 batches)

Fatty acid	*Zophobas morio*	*Chilecomadia moorei*	*Galleria melonella*	*Tenebrio molitor*	*Acheta domesticus*
C12:0	0.05	0.23	—	0.14	0.04
C14:0	1.02	0.32	0.12	2.08	0.72
C14:1	0.01	0.04	—	0.04	0.03
C15:0	0.23	0.01	0.05	0.14	0.10
C16:0	29.72	26.20	28.61	20.79	23.99
C16:1	1.28	5.30	1.09	1.36	1.06
C17:0	0.44	0.04	0.13	0.22	0.23
C17:1	0.15	0.01	0.24	0.02	0.07
C18:0	7.76	1.03	1.23	5.06	8.01
C18:1-n9	38.21	57.07	42.94	38.16	24.49
C18:2-n6	15.67	2.96	17.61	27.37	33.23
C18:3-n6	0.11	0.15	0.01	0.23	0.07
C18:3-n3	0.70	0.17	1.91	1.17	1.39
C20:0	0.22	0.14	0.04	0.32	0.65
C20:1-n9	0.03	0.04	4.36	0.13	0.18
C20:2-n6	0.03	—	0.04	0.06	0.05
C20:3-n6	—	—	—	0.08	
C20:4-n6	0.03	0.01	—	—	0.10
C20:3-n3	—	—			0.27
C20:5-n3	—	0.04	—	—	0.04
C22:0	0.06	0.06	—	0.09	0.11
Saturated	39.5	28.03	30.18	28.84	33.85
Monounsaturated	39.68	62.46	48.63	39.71	25.83
Polyunsaturated	16.54	3.33	19.57	28.83	35.23

its transformation in the food industry (e.g. by pasteurization or sterilization of the fresh product). On the other hand, the results confirm that by adopting sufficient environmental health measures in closed cycle rearing it is possible to produce insects that attract a microbial flora that contains no (or a very limited amount of) microbial forms pathogenic to humans.

The hundredth chemical composition showed a reduced water content in the larval forms and a significantly low water content in crickets, the only imaginal forms considered. The ponderal increase of the cuticular matrix brought about a reduction in the water content, in accordance with a natural and progressive water loss in the progression from larval form to pupa to mature adult. The amount of dry substances found in *A. domesticus* is approximately double that found by Nakagaki and DeFoliart (1991) in the same species but at the nymph stage.

The total nitrogen, converted into raw protein with a coefficient of 6.25 in accordance with that reported by Koo et al. (1980) demonstrates values in the larval forms that are very close to those of vertebrate meat but more than double in the case of *A. domesticus*. Little information is available about the digestibility

as well as the biological value of these proteins. The exocuticula is made up of a mosaic of chitin micelles composed of a proteinic substance called sclerotin. The quality of the proteins found in insects in terms of digestibility, use, and amino acid values is lower than these values for vertebrates. Elimination of the chitinous component raised these parameters to values close to those of casein (Nakagaki and DeFoliart, 1991).

The lipid content is high in all the insects studied but the larval forms had the highest amount. As can be seen from the Tables, it is not immediately possible to extrapolate specific correlations between the lipid content and the water content, at least not between different genera. From the qualitative point of view, a reduced content of saturated fatty acids (made up of palm oil acid) was observed, and in the larval forms the abundance of palmitic acid depends on certain metabolic functions (Table 29.5). Compounds of the structural function (polyunsaturated with 18 to 20 atoms of carbonium) in the imaginal forms would tend to partially replace those with an energetic function. By comparison from this specific point of view it should be emphasized that insects have a fatty acid profile very similar to that of fish.

In brief, the raw fiber can be identified by the chitin: this molecule is chemically correlated to cellulose, in certain aspects roughly emulating its structural functions, and constitutes up to 40% of the cuticle in weight and 10% of the dry weight of the insects in toto. Ash shows somewhat of a correlation with the raw fiber content; in other words, the relative abundance and/or specialization of the cuticle seems to accompany an increase in the mineral content.

In considering the subject in more generic terms, from the personal data and literary information available, it is possible to assume that insects actually constitute a good source of nutrition for humans. From the hygiene-health point of view, at the moment there is no record of pathologies that insects may contract being directly transmissible to humans through their consumption.

Much has been written in medical literature about the role that many insects can play as carriers (active and passive) of various microorganisms, which are agents of infectious illnesses, whether they be viruses, rickettsia, bacteria or fungi. However, this aspect is in fact directly associated with the prevention of human infectious diseases and only marginally concerns food hygiene. On the other hand, my research has shown, as anticipated, that insects can contain an abundant and composite microbial flora, which is also composed of entirely saprophytic bacterial and fungal species that they pick up from the external environment.

Nevertheless, it cannot be excluded a priori that these animals could occasionally become carriers of microorganisms which are pathogenic for humans (*Salmonella, Shigella*, enteropathogenic *Escherichia coli*, enteropathogenic *Vibrio, Campylobacter, Clostridium perfringens* and *C. botulinum, Staphylococcus aureus*, and *Listeria monocytogenes*) should the opportunity arise.

Contamination of edible species by plant protection products has not been widely monitored and could create some problems, especially in areas of intensive agriculture.

All this leads to the conclusion, for the moment at least, that insects could enter into the human food chain, provided they undergo a transformation which will inactivate or reduce the microbial bacteria they contain. Treatments such as pasteurization or cooking should ensure this end result, provided they are continued for sufficiently long periods of time at sufficiently high temperatures to inactivate both the living forms and the spores of the above-mentioned microorganisms.

The efficiency of other systems of conservation such as salting, smoking and the use of ionizing radiation still needs to be evaluated. With a view to an eventual utilization of insects as a part of human food resources, it would be opportune to monitor the location, collection or rearing of these animals, in order to avoid any possible problems of an accumulation of chemical compounds that could be potentially toxic to humans.

References

Ajmerito P., Sigliuzzo P. and Julini M. 1988. Presensa di residui di metalli pesanti in cosce di rana commercielizzate. Jl Progresso veterinario, 31: 232–233.

Apicio. 1994. "La cucina dell'antica Roma". A cura di C. Vesco. Newton Ed., Roma.

Cooper J.E. and Knowler C. 1991. Snails and snail farming: an introduction for the vetrinary profession. Veter. Rec. 21/28: 541–549.

Elmslie L.J. 2004. Snail collection and small-scale production in Africa and Europe. (this volume)

Flandrin J.L. and Montanari M. 1997. Storia dell'alimentazione, Laterza Eds., Roma.

Giaccone V., Adriano D., and Parisi E. 1988. La commercializzazione delle rane in un agglomerato urbano industrializzato. Considerazioni igienico-sanitarie e ispettive. Industrie Alimentari 27: 980–989.

Giaccone V., Novelli E., Catellani P., and Ruggeri L. 2000. Insects as a possibile source of food for humans. Sis. Vet. Ann. Meeting Selected Abstracts Grafiche Scuderi, Messina 2: 157.

Gomot A. 1998. J. Moll. Stud., cited by Novelli et al., 2001.

Govoni Guido, Fielding Denis and Paoletti M.G. 2004. Rodent farming in the Amazon: Experience with Amerindians in Venezuela. (this volume)

Harris M. 1985. Good to Eat. Simon and Schuster, New York, NY.

Jori Ferran, Edderai D., and Nouben P. 2004. Potential of rodents for minilivestock in Arica (this volume)

Koo et al. (1980), cited by Giaccone et al. (2000).

Nakagaki B.J. and DeFoliart G.R. 1991. Comparison of diets for mass-rearing *Acheta domesticus* (Orthoptera: Gryllidae) as a novelty food, and comparison of food conversion effinciency with values reported for livestock. J. Econ. Entomol. 84(3): 891–896.

Negroni Gianluigi, 2004. Frogs as food. (this volume)

Novelli E., Bracchi P.G., Balzan S., and Grigoletto L. 2001. Investigation on muscle tissue composition of terrestrial snails belonging to the genus *Helix*. Atti Soc. Ital. Sci. Veter. 55: 333–334.

Panebianco A. and Giuffrida A. 1993. Larve di *Gnatostoma spinigerum* in cosce di rana congelate importate. Atti Convegno nazionale SISVET.

Paoletti M.G. and Bukkens S.G.F., 1997. Minilivestock, Ecology of Food and Nutrition, 368(2–4): 95–346.

Parisi E. and Giaccone V. 1992. Lezioni di Ispezione e Controllo delle derrate alimentari di origine animale I. Cooperativa Libraria Universitaria di Torino, Torino, Italy.

Ramos-Elorduy J. 1982. Los insectos como Fuente de Proteinas en el Futuro. Ed. Limusa (2nd ed.).
Ramos-Elorduy J. 1997. Insects: a sustainable source of food? Ecol. Food Nutr. 36: 247–276.
Ramos-Elorduy J. 2004. Insects: A hopeful food source. (this volume)
Ramos-Elorduy de Conconi J. 1996. Insectos Comestibles. Platillo de Gourmets Ed. Fondo de Cultura Economica, Mexico City, Mexico.

Minilivestock and BEDIM Association

The Bureau for Exchange and Distribution of Information on Minilivestock (BEDIM; same initials in French) was created circa 1990 as a special Unit of the Tropical Animal Production Service at the Institute of Tropical Medicine, Antwerp-Belgium. Later, BEDIM became an international nonprofit association under Belgian legislation, status approved by Royal Decree and issued in the Belgian Official Gazette on 12 December 1996. Since 1990, BEDIM collected properly authored documents dealing with any animal living on land (at least in part, such as frogs or birds), used locally for years either for consumption or otherwise (local medicaments, sales, etc.), usually collected in the wild (which frequently means poached) but apparently susceptible to true breeding under human control.

The easiest and cheapest way to distribute information is evidently a good mailing list for printed information. Hence BEDIM decided to produce a semestral bulletin divided into three parts: news on the association and on minilivestock (French and English), very brief notes (either French or English with résumé or summary), and survey of the literature identified since the previous bulletin (English only). Some years ago, with more information coming in from Latin America, a few pages in Spanish were added for matters dealing specially with Central and South America. The Animal Production and Health Division of the Food and Agriculture Organization has recognized and given some support to BEDIM since its inception.

The semestral Bulletin published its 13th volume in 2004 and is sent to about 50 African countries, 30 in the Americas, 15 in the Asia and Pacific region, and a dozen European countries. It is worth noticing that, thanks to FOA/AGAP assistance, the Bulletins are sent free of charge to any institutional (research, education, etc.) library and to large groups, NGOs, etc. where one Bulletin can be used by many people. Personal subscriptions for individual mailings are also possible. Semestral Bulletins are available on BEDIM website as well.

Bulletin coverage has changed over the years and nowadays is focused on events, seminars, facts, etc. and subjects such as Human Nutrition, Edible Rodents, Guinea Pigs, Birds, Reptiles, Frogs, Edible Snails, Manure Worms and earthworms, Insects, Cultural Anthropology, and Minilivestock (in general).

The BEDIM association is also involved in the development of minilivestock activities in tropical rural areas where extensionists are too often absent or lack competence. Many incoming letters ask for advice before starting some project or when the writers are faced with problems. The role of BEDIM is then to provide suggestions, documentation, references or names of local people competent in the matter and already known as such by the Secretariat through previous contacts.

Repeated requests for advice on how to set up a breeding unit for a particular minilivestock led the association to issue very cheap leaflets on construction and operation of small backyard farms. Today our Series Information and Documentation under the standard title of Guide-booklet for Breeding Techniques contains 8 titles (1—Grasscutter; 2—Giant snails; 3—Frogs; 4—Guinea pigs; 5—Brush-tailed Porcupine; 6—Manure worms *Perionyx excavatus*; 7—Maggots; 8—Cricetomas).

At present these booklets are available only in French (aulacodes, escargots géants, grenouilles, cobayes, athérures, vers de compost *Perionyx excavatus*, asticots, cricétomes). These "Guides techniques d'élevage" are available free of charge by post or e-mail on simple written request.

It was recently decided to transfer from paper form to a new computerised database the most important old documents as well as the recent ones dealing with minilivestock in general and for given species. This compendium called BEDIM DATABASE is run by the Central Library of the Faculté universitaire des Sciences agronomiques (FUSA) of Gembloux, Belgium and can be called free on its website. The BEDIM association is managed by a Board of nine (maximum) Trustees elected for a period of four years, renewable by the General Assembly of all members held every second year. Meanwhile the international association BEDIM is very pleased with the publication of this book on minilivestock containing up-to-date information and appropriate author contributions.

Administrative information

Registered Office
6 Sur-les-Waides, B-4800 Petit-Rechain (Verviers). Belgium
Secretary of the Association
Dr E. Thys, Veterinary Department, Institute of Tropical Medicine.
Nationalestraat 155. B-2000 Antwerp. Belgium.
ethys@itg.be
Secretariat
Mrs Goorickx Muriel. FUSA, Bibliothèque centrale. Passage des Déportés, 2. B-5030 Gembloux, Belgium.
Goorickx.m@fsagx.ac.be
Website: http://www.bib.fsagx.ac.be/bedim

BEDIM database: http://www.bib.fsagx.ac.be/cgi-bin/db/search?bedim
Membership: yearly
Effective member (with right to vote): 15 Euros
Adherent member (no right to vote): 3 Euros
From developing country: 2 Euros
Bulletin subscription: yearly
Ordinary: 6.25 Euros
Members: 5 Euros
Payable in Euros only, all extra costs covered by the sender, in the name of BEDIM Intern. Assoc.

- Banque FORTIS, Brussels n. 001-2949595-95

Code IBAN: BE 76 0012 9495 9595
Code BIC/SWIFT: GE BA BE BB

- Banque de la Poste, Brussels n. 000-0574065-19

Code IBAN: BE 39 0000 5740 6519
Code BIC/SWIFT: BP OT BE B1
NB. Payments can be made for several years

Video cassette

- On sale at the S1ecretariat (FUSA Gembloux)
- Basic price: 8 Euros
- Packing and mailing costs: + 7 Euros
- In PAL. Available on request by pack of 20 units in SECAM or NTSC

Subject Index

B

L

Q

R

T

Species Index (Latin names: *Genus_species*)

A

D

H

N

O

P

U

V

W

Colour Plates Section

I-X